IET MATERIALS, CIRCUITS AND DEVICES SERIES 45

Characterisation and Control of Defects in Semiconductors

Other volumes in this series:

Volume 2	**Analogue IC Design: The current-mode approach** C. Toumazou, F.J. Lidgey and D.G. Haigh (Editors)
Volume 3	**Analogue–Digital ASICs: Circuit techniques, design tools and applications** R.S. Soin, F. Maloberti and J. France (Editors)
Volume 4	**Algorithmic and Knowledge-Based CAD for VLSI** G.E. Taylor and G. Russell (Editors)
Volume 5	**Switched Currents: An analogue technique for digital technology** C. Toumazou, J.B.C. Hughes and N.C. Battersby (Editors)
Volume 6	**High-Frequency Circuit Engineering** F. Nibler *et al.*
Volume 8	**Low-Power High-Frequency Microelectronics: A unified approach** G. Machado (Editor)
Volume 9	**VLSI Testing: Digital and mixed analogue/digital techniques** S.L. Hurst
Volume 10	**Distributed Feedback Semiconductor Lasers** J.E. Carroll, J.E.A. Whiteaway and R.G.S. Plumb
Volume 11	**Selected Topics in Advanced Solid State and Fibre Optic Sensors** S.M. Vaezi-Nejad (Editor)
Volume 12	**Strained Silicon Heterostructures: Materials and devices** C.K. Maiti, N.B. Chakrabarti and S.K. Ray
Volume 13	**RFIC and MMIC Design and Technology** I.D. Robertson and S. Lucyzyn (Editors)
Volume 14	**Design of High Frequency Integrated Analogue Filters** Y. Sun (Editor)
Volume 15	**Foundations of Digital Signal Processing: Theory, algorithms and hardware design** P. Gaydecki
Volume 16	**Wireless Communications Circuits and Systems** Y. Sun (Editor)
Volume 17	**The Switching Function: Analysis of power electronic circuits** C. Marouchos
Volume 18	**System on Chip: Next generation electronics** B. Al-Hashimi (Editor)
Volume 19	**Test and Diagnosis of Analogue, Mixed-Signal and RF Integrated Circuits: The system on chip approach** Y. Sun (Editor)
Volume 20	**Low Power and Low Voltage Circuit Design With the FGMOS Transistor** E. Rodriguez-Villegas
Volume 21	**Technology Computer Aided Design for Si, SiGe and GaAs Integrated Circuits** C.K. Maiti and G.A. Armstrong
Volume 22	**Nanotechnologies** M. Wautelet *et al.*
Volume 23	**Understandable Electric Circuits** M. Wang
Volume 24	**Fundamentals of Electromagnetic Levitation: Engineering sustainability through efficiency** A.J. Sangster
Volume 25	**Optical MEMS for Chemical Analysis and Biomedicine** H. Jiang (Editor)
Volume 26	**High Speed Data Converters** A.M.A. Ali
Volume 27	**Nano-Scaled Semiconductor Devices** E.A. Gutiérrez-D (Editor)
Volume 28	**Security and Privacy for Big Data, Cloud Computing and Applications** L. Wang, W. Ren, K.R. Choo and F. Xhafa (Editors)
Volume 29	**Nano-CMOS and Post-CMOS Electronics: Devices and modelling** Saraju P. Mohanty and Ashok Srivastava
Volume 30	**Nano-CMOS and Post-CMOS Electronics: Circuits and design** Saraju P. Mohanty and Ashok Srivastava
Volume 32	**Oscillator Circuits: Frontiers in design, analysis and applications** Y. Nishio (Editor)
Volume 33	**High Frequency MOSFET Gate Drivers** Z. Zhang and Y. Liu
Volume 34	**RF and Microwave Module Level Design and Integration** M. Almalkawi
Volume 35	**Design of Terahertz CMOS Integrated Circuits for High-Speed Wireless Communication** M. Fujishima and S. Amakawa
Volume 38	**System Design With Memristor Technologies** L. Guckert and E.E. Swartzlander Jr.
Volume 39	**Functionality-Enhanced Devices: An alternative to Moore's law** P.-E. Gaillardon (Editor)
Volume 40	**Digitally Enhanced Mixed Signal Systems** C. Jabbour, P. Desgreys and D. Dallett (Editors)

Volume 43 **Negative Group Delay Devices: From concepts to applications** B. Ravelo (Editor)
Volume 47 **Understandable Electric Circuits: Key concepts, 2nd Edition** M. Wang
Volume 53 **VLSI Architectures for Future Video Coding** M. Martina (Editor)
Volume 58 **Magnetorheological Materials and Their Applications** S. Choi and W. Li (Editors)
Volume 60 **IP Core Protection and Hardware-Assisted Security for Consumer Electronics** A. Sengupta and S. Mohanty
Volume 68 **High Quality Liquid Crystal Displays and Smart Devices: Vol. 1 and vol. 2** S. Ishihara, S. Kobayashi and Y. Ukai (Editors)
Volume 69 **Fibre Bragg Gratings in Harsh and Space Environments: Principles and applications** B. Aïssa, E.I. Haddad, R.V. Kruzelecky, W.R. Jamroz
Volume 70 **Self-Healing Materials: From fundamental concepts to advanced space and electronics applications, 2nd Edition** B. Aïssa, E.I. Haddad, R.V. Kruzelecky, W.R. Jamroz
Volume 71 **Radio Frequency and Microwave Power Amplifiers: Vol. 1 and vol. 2** A. Grebennikov (Editor)
Volume 73 **VLSI and Post-CMOS Electronics Volume 1: Design, modelling and simulation and volume 2: Devices, circuits and interconnects** R. Dhiman and R. Chandel (Editors)

Characterisation and Control of Defects in Semiconductors

Edited by
Filip Tuomisto

The Institution of Engineering and Technology

Published by The Institution of Engineering and Technology, London, United Kingdom

The Institution of Engineering and Technology is registered as a Charity in England & Wales (no. 211014) and Scotland (no. SC038698).

First published 2019

The Institution of Engineering and Technology
Michael Faraday House
Six Hills Way, Stevenage
Herts, SG1 2AY, United Kingdom

www.theiet.org

British Library Cataloguing in Publication Data
A catalogue record for this product is available from the British Library

ISBN 978-1-78561-655-6 (hardback)
ISBN 978-1-78561-656-3 (PDF)

Typeset in India by MPS Limited
Printed in the UK by CPI Group (UK) Ltd, Croydon

Contents

Preface xv

1 Characterizing electrically active defects by transient capacitance spectroscopy **1**
Lasse Vines

1.1 Introduction 1
1.2 Characteristics of electrically active defect levels 3
1.3 Junction spectroscopic techniques for characterizing deep-level defects 7
1.3.1 Deep-level transient spectroscopy 7
1.3.2 Variations of capacitance spectroscopic techniques 13
1.3.3 Deep-level optical spectroscopy 16
1.4 Examples: characterizing electrically active defects in semiconductors 19
1.4.1 Irradiation-induced defect complexes and vacancy migration in silicon 20
1.4.2 Electrically active defects in zinc oxide and the involvement of hydrogen 25
1.5 Extending junction spectroscopic techniques beyond traditional methodology 36
1.6 Conclusions 38
Acknowledgments 38
References 38

2 Luminescence from point defects in wide-bandgap, direct-gap semiconductors **45**
Michael A. Reshchikov

2.1 Introduction 45
2.2 Types of transitions related to photoluminescence 46
2.3 Phenomenological models 48
2.3.1 Rate equations model 49
2.3.2 Configuration-coordinate model 63
2.3.3 Diagonal transitions between localized states 67
2.3.4 Summary 70

2.4 Experiment: photoluminescence from defects 71
2.4.1 Radiative defects in GaN 71
2.4.2 Defect-related photoluminescence in other wide bandgaps 83
2.4.3 Summary 88
2.5 Conclusions and outlook 88
References 89

3 Vibrational spectroscopy **97**
Matthew D. McCluskey

3.1 Theory 97
3.1.1 Units 97
3.1.2 Linear chain 98
3.1.3 Defect vibrational modes 102
3.1.4 Anharmonic potential 103
3.1.5 IR activity 103
3.1.6 Number of atoms 104
3.1.7 Infrared absorption 105
3.1.8 Linewidth and temperature dependence 106
3.1.9 Wave functions and symmetry 108
3.1.10 Anharmonic coupling 110
3.1.11 Raman scattering 111
3.2 Experiment 112
3.2.1 Raman spectroscopy 112
3.2.2 Fourier transform infrared spectroscopy 113
3.2.3 IR pump–probe 114
3.2.4 Applied stress 115
3.3 Examples 116
3.3.1 Interstitial oxygen 116
3.3.2 Impurities in GaAs 121
3.3.3 Resonant interaction in AlSb 124
3.3.4 Impurities in CdTe 125
3.3.5 Hydrogen in silicon 125
3.3.6 Impurity-hydrogen pairs: trends 126
3.3.7 MgH complex in GaN 127
3.3.8 NH complex in ZnO 128
3.3.9 Hydrogen donors in oxide semiconductors 130
3.3.10 Vacancy-hydrogen complexes 133
3.3.11 Hydrogen in strontium titanate 134
3.3.12 Hydrogen molecules 136
3.3.13 From LVM to phonon 136
3.4 Summary and outlook 138
Acknowledgements 139
References 139

4 **Magnetic resonance methods** **151**
Jean-Louis Cantin

4.1 Electron spin resonance spectroscopies 151
4.1.1 What is EPR used for? 152
4.1.2 Spin Hamiltonian formalism 153
4.1.3 Hyperfine interactions 158
4.1.4 Resonance 159
4.1.5 Transition probability: relaxation phenomena 161
4.1.6 Experimental setup 165
4.1.7 Electron nuclear double resonance 170
4.1.8 Pulsed spectroscopies 171
4.1.9 Optically detected magnetic resonance, electrically detected magnetic resonance 172
4.2 Illustrative examples: structural and chemical control 173
4.2.1 The SiC/oxide interface defects 173
4.2.2 The N dumbbell in GaN 183
4.3 Examples: electrical and optical activities 186
4.3.1 P-Type doping of GaN 186
4.3.2 Origin of the residual conductivity of Ga_2O_3 188
4.3.3 SiC defects as quantum bits 190
4.4 Summary and outlook 192
References 193

5 **The role of muons in semiconductor research** **199**
Patrick. W. Mengyan

5.1 Introduction 199
5.2 Muon spin research (μSR): the techniques 201
5.2.1 General principles 203
5.2.2 Polarisation functions 210
5.2.3 Transverse field μSR 223
5.2.4 Longitudinal field μSR 226
5.2.5 Zero-field μSR 227
5.2.6 Resonance-based μSR techniques 228
5.2.7 Low-energy muons (LEM or LE-μSR) 231
5.2.8 Illumination and muon spectroscopy of excited states 231
5.3 Putting muons to use in semiconductors 232
5.3.1 Muonium states 233
5.3.2 Identifying muonium centres 234
5.3.3 Processes and dynamics involving muonium centres 237
5.3.4 Measuring donor and acceptor levels 245
5.3.5 Effects of carrier concentrations 247
5.4 Select examples of μSR contributions to the field 247
5.4.1 Group-IV 249
5.4.2 Group III–V 250

5.4.3 Group II–VI 252
5.4.4 Group II–IV–V_2 chalcopyrites 252
5.4.5 Oxides 253
5.5 Final thoughts 254
Acknowledgements 255
References 255

6 Positron annihilation spectroscopy, experimental and theoretical aspects 263
Jonatan Slotte, Ilja Makkonen and Filip Tuomisto

6.1 Introduction 263
6.2 Positrons in crystalline semiconductors 264
6.2.1 Positrons obtained from radioactive sources 264
6.2.2 Slow monoenergetic positron beams for thin-layer studies 264
6.2.3 Positron thermalization and diffusion 265
6.2.4 Positron trapping 266
6.2.5 The kinetic trapping model and positron lifetime spectroscopy 267
6.3 Doppler broadening techniques 269
6.3.1 Doppler broadening spectroscopy 269
6.3.2 Coincidence Doppler broadening spectroscopy 270
6.3.3 Supporting theory for defect identification 271
6.4 Temperature-dependent measurements in narrow bandgap semiconductors 273
6.4.1 Vacancy annealing in Ge 273
6.4.2 Positron traps in low temperature irradiated GaSb 274
6.5 Acceptor-like defects in wide bandgap semiconductors 277
6.5.1 Cation vacancies and acceptor impurities in GaN and ZnO 277
6.5.2 Cation and oxygen vacancies in metal oxides 279
6.6 Summary and outlook 284
References 284

7 First principles methods for defects: state-of-the-art and emerging approaches 289
Elif Ertekin and Hannes Raebiger

7.1 Defect levels 289
7.2 Defect thermochemistry 292
7.2.1 Defect formation enthalpy 292
7.2.2 Phase stabilities and chemical potential space 293
7.2.3 Defect formation enthalpy diagrams 294
7.2.4 Configuration coordinate diagrams and optical excitation 295
7.2.5 Defect equilibria and carrier concentrations 296

7.3 Sources of uncertainty in defect formation energy calculations 297
7.4 Total energies using first principles electronic structure calculations 298
7.4.1 Overview and many-particle Schrödinger equation 298
7.4.2 Hartree–Fock 299
7.4.3 Density functional theory 301
7.4.4 Emerging approaches—quantum Monte Carlo 303
7.5 Achieving higher accuracy with density functional theory and other approaches 303
7.5.1 Observations 303
7.5.2 Generalized Koopman's theorem 305
7.5.3 LDA+U and PBE+U 307
7.5.4 Green's function approaches: the GW method 308
7.5.5 Hybrid functionals 311
7.5.6 Fitted chemical potentials 313
7.6 Finite size effects 314
7.6.1 Charged defects 315
7.6.2 Band filling for shallow defects 316
7.7 Recent examples 317
7.7.1 Mg and other possible acceptors in GaN 317
7.7.2 Oxygen vacancies and other defects in metal oxides 321
7.7.3 Transition metal impurities 326
7.7.4 Defect clustering and aggregation 329
7.8 Summary and outlook 333
References 333

8 Microscopy of defects in semiconductors **345**
Fabien C.-P. Massabuau, Jochen Bruckbauer, Carol Trager-Cowan and Rachel A. Oliver

8.1 Introduction 345
8.2 Basic principles of the microscopy techniques 346
8.2.1 Scanning probe microscopy 346
8.2.2 Scanning electron microscopy 353
8.2.3 Transmission electron microscopy 369
8.3 Examples of the application of microscopy to semiconductor materials 382
8.3.1 Dislocation densities and Burgers vectors in GaN 382
8.3.2 Imaging of the impact of dislocations on materials properties 390
8.3.3 A multi-microscopy example: structure, properties and interactions of dislocations in InGaN 396
8.3.4 Imaging defects in 2D materials 400
8.4 Conclusions and outlook 406
Acknowledgements 407
References 407

9 Three-dimensional atomic-scale investigation of defects in semiconductors by atom probe tomography **417**
Lorenzo Rigutti and Didier Blavette

9.1 Introduction 417
9.2 Atom probe tomography 419
9.2.1 Generalities and history 419
9.2.2 Principles and performances 419
9.2.3 APT versus SIMS 424
9.3 Three-dimensional defects 427
9.3.1 Nanocrystals, quantum dots 427
9.3.2 Impurity clusters 431
9.4 Two-dimensional defects 433
9.4.1 Segregation to stacking faults 433
9.4.2 Stacking faults in heterostructures 435
9.4.3 Intergranular segregation of dopants in silicon 438
9.4.4 Segregation of dopants to gate interfaces in MOSFET transistors 439
9.5 One-dimensional defects 440
9.6 Point defects 442
9.6.1 Three-dimensional field ion microscopy 442
9.6.2 In situ photoluminescence 442
9.7 Conclusion 444
Acknowledgements 445
References 445

10 Ion-beam modification of semiconductors **451**
Kai Nordlund and Sergei. O. Kucheyev

10.1 Introduction 451
10.2 Theoretical and experimental background 452
10.2.1 Theory of ion stopping and energy deposition 452
10.2.2 Basics of ion implantation experiments 457
10.3 Damage production and annealing 458
10.3.1 Low-fluence primary damage 459
10.3.2 Defect migration and interaction 461
10.3.3 High-fluence damage overlap and amorphization 462
10.3.4 Recrystallization of amorphous state 465
10.3.5 Defect evolution during recrystallization and dopant activation 465
10.4 Ripple formation on semiconductors 466
10.4.1 Experimental background 467
10.4.2 Theoretical models 469
10.5 Time-resolved experiments to probe defect evolution 473
10.5.1 Dynamic annealing 473
10.5.2 Traditional dose-rate effect experiments 473

10.5.3 Pulsed-ion-beam method 478
10.5.4 Comparison of radiation defect dynamics in different semiconductors 486
10.6 Conclusions 488
Acknowledgments 489
References 489

11 Characterizing defects with ion beam analysis and channeling techniques **501**
Lino Pereira, Andre Vantomme and Ulrich Wahl

11.1 Tutorial 501
11.1.1 Ion beam analysis 501
11.1.2 Channeling techniques 507
11.2 Examples 523
11.2.1 Damage in semiconductors 524
11.2.2 Lattice location of dopants 532
11.3 Outlook 550
11.3.1 Technical developments 550
11.3.2 Emerging applications 551
References 553

Index **563**

Preface

The 30th edition of the International Conference on Defects in Semiconductors (ICDS-30) took place in Seattle, Washington in July 2019, 60 years after ICDS-1 in Gatlinburg, Tennessee: Radiation Effects in Semiconductors. With the previous books focusing on identification and characterization of defects in semiconductors dating from the end of the twentieth century, the idea of composing a new edited volume arose among those of us who benefited so much from having an up-to-date reference book at hand when we began our scientific careers. The contributors to this volume wish to honor the past 60 years of progress and development and, in particular, the most recent advances. We hope that this volume will prove as useful and enlightening to the next generations of semiconductor scientists.

The functional properties of a semiconductor material are defined by the defects in it, and ultimately by the nature of the electronic states they create in the bandgap of the semiconductor. Hence, the control of semiconductor properties requires that the formation and the introduction mechanisms of these defects are understood. Their other physical properties, such as how they interact with other defects in the lattice and whether they can be made to move with the hope of them getting trapped at a neutralizing location or driven out of the region of interest, are also important. Many different experimental and theoretical methods need to be employed to gather even a fraction of the required knowledge. Finally, one needs to be able to manufacture and process the semiconductor material in such a way that desired defects are introduced but the harmful ones are not. This can be very challenging.

Often, it is rather straightforward to control the introduction of the desired impurities in the semiconductor matrix. Dopants can be added to the growth environment in various ways, or they can be diffused in or implanted after the growth process. The most important limitations are solubility in the case of in situ or diffusion doping, while implantation is mostly applicable to close-to-the-surface doping profiles. However, while the introduction of dopants is controllable and requires active measures, other kinds of point defects are formed either thermally, due to kinetic or chemical effects, or as radiation damage in the case of implantation processing. In addition, the growth environment may contain some unwanted impurities that are difficult to control: typical omnipresent elements are oxygen and hydrogen. Further, in the case of wide-bandgap semiconductors such as the III-nitrides, where native substrates are not easily available, the lattice mismatch between the thin film and substrate causes initial stresses and strain that are most often relaxed through the generation of dislocations and other extended defects. There are many ways to try to avoid the formation of the unwanted defects or to try to remove them by post-processing measures, such as thermal treatments. Even though many defect-related problems have been identified and solved over the past 60 years of semiconductor research, the constant quest for faster, cheaper, less-power consuming, and new kinds of electronics generates the need for new materials' properties and creates new defect-related challenges.

The wide variety of methods used to identify and characterize defects in semiconductors can be roughly divided into electrical measurements, optical spectroscopy, particle beam methods, and electron microscopy. Theoretical calculations provide significant insights into the physical properties of the defects. The scientific literature contains a wide selection of detailed reviews on the various experimental and theoretical approaches to studying defects in semiconductors. In this book, the focus is on point defects since they are easily thermally (and kinetically) generated during manufacturing and abundantly introduced during post-manufacturing processing steps. All chapters follow a similar structure: the first part provides a tutorial-like presentation of a method or of a collection of methods, while the second part demonstrates the possibilities of the method by discussing various examples from the recent research on defects in semiconductors. Each chapter is concluded with a discussion on the potential future developments and new application areas of the reviewed method.

This book is composed in such a way that each of the 11 chapters, written by experts in the respective fields, can be read as a separate review on the technique it focuses on. For the benefit of a reader who wishes to get a more thorough overview of the field, the examples of defect studies discussed in the chapters focus around two themes of recent interest: acceptor-type defects in nitride semiconductors and oxygen vacancies in oxide semiconductors. Also, other semiconductor systems and defects are discussed, including those of interest for quantum computing in SiC, those of interest for thermoelectric and long-wavelength optoelectronics in narrow gap semiconductors, and many others. Chapters 1 and 2 focus on the characterization of the electrical and optical properties of the defects. Methods reviewed in these chapters include deep-level transient spectroscopy and other capacitance spectroscopic techniques as well as photoluminescence spectroscopy. Identifying the microscopic (atomic) structures of the defects is at the focus in Chapters 3–6. Vibrational spectroscopy and magnetic resonance techniques reviewed in Chapters 4 and 5 are based on the interactions of electromagnetic radiation (microwaves, infrared, visible light) with the ions and electrons in the semiconductor lattice. Chapters 5 and 6 review the use of "exotic" particles, namely, muons and positrons, in the identification and quantification of point defects in semiconductors. Chapter 7 concludes the discussion on the basic physical properties of defects by reviewing the state-of-the-art theoretical approaches used in the field. Chapter 8 reviews the various electron-microscopy techniques and their uses for imaging individual point defects and interactions between extended and point defects. Chapter 9 reviews a recent addition to the family of defect-characterization techniques: 3D atom probe tomography that augments the capabilities of secondary ion mass spectrometry. Chapters 10 and 11 conclude the book by reviewing ion-beam-based modification and characterization methods of semiconductors and their defects.

Chapter 1

Characterizing electrically active defects by transient capacitance spectroscopy

Lasse Vines[1]

1.1 Introduction

Defects define to a large extent the functionalities of a semiconductor—they are the catalyst for electrical conduction and optical interaction but can at the same time be detrimental to device performance. These defects, or perturbations to a semiconductor lattice, can be of intrinsic nature, for example, a vacancy or self-interstitial, or extrinsic, such as an impurity atom. Some of these defects are electrically inactive and will not be considered in this chapter. The properties of electrically active defects have been intensely studied for decades, but the topic remains a crucial part of semiconductor science. Electrically active defects form states in the forbidden band gap, and these states can donate or accept electrons to or from the conduction band (the lowest range of available electron states) or the valence band (the highest range of electron energies in which electrons are normally present). Electronic states formed close to the bottom of the conduction band or top of the valence band are labeled shallow, typically formed by substitutional impurities and used as dopants. Deep defects are found further away from the band edges and are present in all semiconductors. Examples of such defects are impurity atoms, intrinsic defects (e.g., vacancies, interstitials, and antisite defects), or complexes between them. The present chapter will discuss properties and techniques to electrically characterize deep-level defects in semiconductors. In particular, spectroscopic characterization techniques based on measuring capacitance transients will be a main focus, with deep-level transient spectroscopy (DLTS) as a prime example.

In traditional and well-developed semiconductors such as silicon (Si) and gallium arsenide (GaAs), many of the deep levels have been identified and thoroughly studied. This identification and characterization of the deep levels by DLTS and related techniques have given us a unique insight into point defects and their behavior, but even deep-level defects in Si remain an active field of research where important defect-related topics remain unresolved. Further, as new semiconducting materials are being explored, a crucial part of understanding their properties is the

[1]Department of Physics, University of Oslo, Oslo, Norway

formation of electrically active deep defects—since these may ultimately determine the electrical performance of the material.

In early studies of deep-level defects, techniques such as Glow curve [1] and thermally stimulated current and capacitance were utilized [8,9]. However, the usefulness of these techniques was often limited due to long time constants or available temperature range, and mainly suitable for the study of dopants. With the introduction of DLTS by Lang in 1974 [10], many of the early limitations were overcome, and since then, a wide range of related techniques has been developed, including the optical counterpart deep-level optical spectroscopy (DLOS) proposed by Chantre *et al.* [11]. Today, these techniques are some of the most sensitive methods to study defects and widely used for both traditional and novel semiconductors. A critical factor for this success is that the sensitivity is related to the charge carrier concentration in the material. Hence, by deliberately choosing materials with a low doping density, an impressive sensitivity is achieved, where studies of defects in a genuinely dilute regime is possible. For example, conventional DLTS typically has a sensitivity limit 4–5 orders of magnitude below that of the charge carrier concentration, thus defect densities as low as 10^8 cm^{-3} are detectable. However, the techniques typically rely on a p–n junction or Schottky barrier contact of sufficient quality. Further, the observable region within the band gap may be limited due to the available temperature and type of conductivity, restricting the usefulness of the conventional DLTS method. As a response to this, several related techniques have been developed. For example, many semiconductors show an asymmetric suitability of doping leaving only one type of conduction available, limiting conventional DLTS to study only half of the band gap. However, utilizing optical capture or emission, the probing region may be extended. The techniques normally rely on transient capacitance measurements, although measurements based on current measurements are also possible. With the many variations, the terms are sometimes used differently in the literature. Here, we will refer to them in accordance with the framework described by Blood and Orton [12].

In this chapter, characterization of electrically active defects is discussed and transient capacitance measurement techniques are addressed. The chapter starts by introducing the main properties describing the electrical activity of a defect, before techniques to measure these properties using thermal and optical emission are discussed. Key techniques based on measuring capacitance transients are reviewed, with particular emphasis on DLTS, including how DLTS can be utilized for direct measurements of capture cross sections and defect profiles. The latter part of the chapter shows examples of use, where defect studies in silicon is chosen as a well-established material where most of the defect levels are identified, but where fundamental knowledge of the defects can still be gained. Defects in zinc oxide (ZnO) is also discussed and chosen as a less studied material where direct identification of the origin of the defect levels remains an important challenge.

1.2 Characteristics of electrically active defect levels

To fully understand the influence of a defect on the electrical properties of a material or a device, a range of parameters needs to be obtained. Consider a point defect having a state in the band gap with an energy E_t below the conduction band edge (E_c). According to Shockley, Read [13], and Hall [14], this state can capture or emit an electron or hole from the conduction or valence band, respectively, as illustrated in Figure 1.1. Depending on the capture and emission rates, some of the processes shown in Figure 1.1 will be dominating. If the dominating process is trapping of electrons from the conduction band with the subsequent reemission to the conduction band, the center acts as a trap, and if the trapping of the electron is followed by a trapping of a hole from the valence band, i.e., a recombination of an electron hole pair, the level acts as a recombination center, whereas in a generation process, trapping of an electron from the valence band is followed by emission of the electron to the conduction band. The dominating process depends on the position of the level in the band gap and the capture cross sections for electrons and holes. Defect levels located in the upper half of the band gap often have larger rates for the electron emission (e_n) compared to that of the hole emission (e_p) because $(E_c - E_t) < (E_t - E_v)$, where E_v is the valence band edge. Therefore, they mostly interact with the conduction band and capture and emit electrons. It is then labeled an electron trap and often with E as a prefix. The converse holds for hole traps. It is also useful to distinguish between majority and minority carriers, relating electron and hole traps to the prevailing conductivity in the material. In n-type materials, electron traps act as majority carrier traps, and hole traps are labeled minority carrier traps, while the opposite holds for p-type material.

Deep levels are also described as having donor or acceptor character according to their charge state before or after capturing an appropriate carrier. The convention

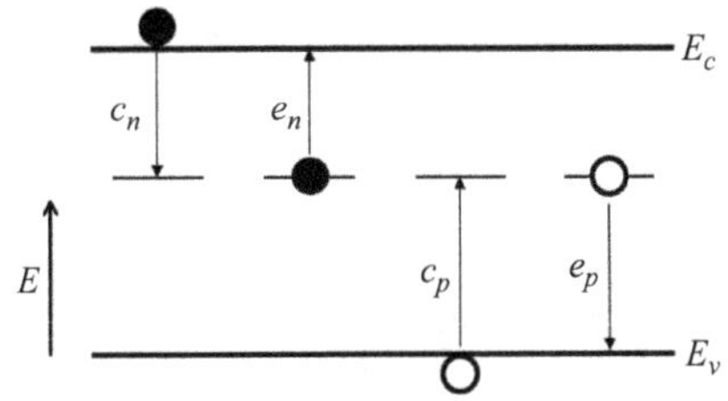

Figure 1.1 Schematics of the charge carrier processes to and from a defect level, showing the capture (c_n) and emission (e_n) of electrons from the conduction band (E_c), and capture (c_p) and emission (e_p) of holes from the valence band (E_v)

Table 1.1 Charge states of donors and acceptors when occupied or not occupied by an electron

	Full	**Empty**
Donor	Neutral	Positive
Acceptor	Negative	Neutral

used is found in Table 1.1, where a donor is neutral when filled with an electron, while an acceptor is negative.

Thus, important electrical parameters include the energy position and concentration of the defect level, the thermal and optical capture rates for electrons and holes, and the thermal and optical emission rates for electrons and holes.

Trapping of electrons from the conduction band and holes from the valence band occur at rates of c_n and c_p, respectively. The capture rates can be defined as

$$c_n = n\sigma_n v_{th,n}; \quad c_p = p\sigma_p v_{th,p}$$

where n and p are the concentrations of electrons in the conduction band and holes in the valence band, respectively, $\sigma_{n,p}$ is the capture cross section of an electron or a hole, $v_{th,n(p)} = \sqrt{3k_bT/m^*_{n(p)}}$ is the electron (hole) thermal velocity, k_b is the Boltzmann constant, T is the temperature, and m^* is the effective mass of electrons or holes. The capture cross sections are dependent on the charge state of the defect center; a Coulomb attractive defect center normally has a significantly larger cross section than an initially neutral center which, in turn, has a larger cross section than a Coulomb-repulsive center. As a rule of thumb, an attractive, neutral, and repulsive center has corresponding capture cross sections on the order of 10^{-14} or higher, 10^{-15}–10^{-17} and 10^{-19} cm^2 or lower. If there are N_t traps in which n_t are occupied by electrons, the change in n_t can be expressed as

$$\frac{dn_t}{dt} = (c_n + e_p)(N_t - n_t) - (e_n + c_p)n_t$$

At thermal equilibrium, $dn_t/dt = 0$ must be satisfied and detailed balance demands that the emission and capture rate of the "individual" charge carriers must be the same. Thus, one can separate the equations for electrons and holes. In the case of an electron trap, the capture and emission of electrons from a trap can be described as $c_n(N_t - n_t) = e_n n_t$. Further, the thermal occupancy of the trap can be expressed by the Fermi–Dirac function, and the emission rate becomes

$$e_n = c_n N_c e^{-(E_c - E_t)/k_bT} = \sigma_n v_{th,n} N_c e^{-(E_c - E_t)/k_bT}$$

where N_c is the density of states in the conduction band. The corresponding emission rate for holes is $e_p = \sigma_p v_{th,p} N_v e^{-(E_t-E_v)/k_bT}$. Interestingly, the capture rate is dependent on the position of the Fermi level, E_f, while the emission rate is not. The emission rate for electrons (holes) depends on the energy difference $E_c - E_t$ (or $E_t - E_v$) and the capture cross sections σ_n and σ_p and are characteristic properties of the trap.

The energy required for an electron to be excited from the trap into the conduction band can be related to Gibbs free energy, $G(T)$, as $\Delta G(T) = \Delta H - T\Delta S = E_c - E_t$, where H is the enthalpy and S is the entropy. The emission rate can then be expressed as

$$e_n = \sigma_n v_{th,n} N_C e^{-\Delta G/k_bT} = e^{-\Delta S/k} \sigma_n v_{th,n} N_C e^{-\Delta H/k_bT} = \sigma_{na} \beta T^2 e^{-\Delta H/k_bT}$$

where β contains all the temperature-independent constants, and σ_{na} is called the apparent capture cross section. Note that the ratios between the degeneracy factors (before and after emission) of the center have been left out in the above equations, but may be significant.

The parameters can be extracted by measuring the emission rate at different temperatures. Dividing the emission rate by T^2 and rearranging the equation by taking the natural logarithm, one obtains

$$\ln\left(\frac{e_n}{T^2}\right) = -\frac{\Delta H}{k_bT} + \ln(\beta\sigma_{na})$$

where

$$\beta T^2 = v_{th,n} N_C = \sqrt{\frac{3k_bT}{m_n^*}} \left(\frac{2\pi m_n^* k_bT}{h^2}\right)^{3/2} = \sqrt{\frac{3k_b}{m_n^*}} 2\left(\frac{2\pi m_n^* k_b}{h^2}\right)^{3/2} T^2$$

Thus, a plot of $\ln(e_n/T^2)$ versus $1/T$, the so-called Arrhenius plot, can be drawn and should yield a straight line. ΔH can now be found from the slope of the line, and σ_{na} is found where the extrapolated line and the e_n/T^2-axis intersect. The thermal measurements of ΔH are often interpreted as $(E_c - \Delta H)$, that is, setting $\Delta H = E_t$, ignoring the entropy factor (i.e., $\Delta H \cong \Delta G$). This is not necessarily valid in all cases, as ΔS might give a significant contribution. The correct approach to find ΔG in thermal experiments would be to measure e_n and σ_n at a specific temperature and calculate ΔG. However, it has been shown that for shallow hydrogenic states in Si, there is no change in entropy and $\Delta H \cong \Delta G$, while for deep states, this is generally not true [12].

So far, we have only considered thermal excitation and emission. Techniques based on thermal emission are often insufficient for wide band gap semiconductors where thermal emission from a state to one of the bands may not be possible within the experimentally available temperature range. Another important restriction applies to semiconducting materials where only one type of charge carrier is available for conduction, i.e., where only n- or p-type material exists. For example, many of the semiconducting oxides have shown such a doping asymmetry. However, excitation and emission may also be induced optically, where the optical emission rate (e_n^o) of an electron from a deep level to the conduction band is given by a photoionization cross section (often called optical cross sections), σ_n^o, and the photon flux, Φ, according to

$$e_n^o = \sigma_n^o \Phi$$

The hole emission process is equivalent to an electron capture process in which a photon excites an electron from the valence band to the defect level. The optical cross section depends on the energy and a rough estimate for the cross section has been proposed by, for example, Lucovsky [15]:

$$\sigma_n^o \propto \frac{1}{E_i}\left[\left(\frac{E_\mathrm{i}}{h\nu}\right) \cdot \left(1 - \frac{E_\mathrm{i}}{h\nu}\right)\right]^{3/2}$$

However, more refined models are developed as well, including the so-called Bois–Vincent–Chantre [11] and Pässler [16] models. If the temperature is sufficiently low so that thermal release of trapped carriers can be neglected, the evolution of the occupancy of the level under illumination is given by a similar equation to that found for thermal capture and emission.

$$\frac{dn_t}{dt} = -\sigma_n^o \Phi n_t + \sigma_p^o \Phi (N_t - n_t)$$

Note here that an important difference compared to the thermal emission is that optical emission is also dependent on the intensity of the incident light, Φ, making quantitative assessment more challenging. The time constant for the capacitance transients, τ^o, will in this case correspond to

$$\tau^o = \left(\sigma_n^o \Phi + \sigma_p^o \Phi\right)^{-1}$$

Hence, the time constant is related to the sum of the two optical cross sections. Generally, both these optical cross sections are involved at a given energy, in contrast to the situation for the thermal emission. There are two interesting limiting cases, however, which are utilized in defect characterization; first, in a steady-state limit where $t \to \infty$, as pursued in photocapacitance measurements, the electrons on a trapped level are given by

$$(n_t)_\infty = N_t \frac{e_p^o}{e_p^o + e_n^o}$$

Second, in the opposite limiting case when $t \to 0$, and assuming that at time $t = 0$, all the centers are filled with electrons, i.e., $n_t(0) = N_t$ and $p_t(0) = 0$, the equation can be simplified to

$$\left(\frac{dn_t}{dt}\right)_0 = -\sigma_n^o \Phi N_t$$

This indicates that the σ_n^o can be obtained by measuring the first derivative of the capacitance transient. Similar arguments also hold for holes.

1.3 Junction spectroscopic techniques for characterizing deep-level defects

Point defects can have a profound impact on material and device performance, but this performance modification can be utilized to study and extract fundamental information about the defects and their electrical characteristics. For example, by measuring the current or capacitance characteristics of a p–n diode or a Schottky contact, the charge capture and emission properties can be revealed and thereby also characteristic features of electrically active defects as outlined in the section above. Measurements conducted on semiconductor junctions are called junction spectroscopy and cover a wide range of techniques. This chapter will mainly focus on capacitance-related methods; thus, a more correct term is capacitance spectroscopy. Note, however, that similar variations also exist for current measurements. Measurements can be carried out in steady-state, like thermally stimulated capacitance, or measuring capacitance transients, like in DLTS. Common for all of these techniques is the manipulation of the depletion region of a junction $W = \sqrt{(2\epsilon(V_0 - V)/q)((N_a + N_d)/N_a N_d)}$, where V_0 and V are the build in and applied voltage, respectively, ϵ is the permittivity, q is the elementary charge, and N_a and N_d are the acceptor and donor on the p- and n-sides, respectively. As discussed in the introduction, DLTS has been a crucial ingredient in the success of this class of analysis techniques for defect studies and will be particularly highlighted in the present section.

1.3.1 Deep-level transient spectroscopy

DLTS is an experimental technique utilizing the modification of the depletion region capacitance in temperature and time domains when altering the applied voltage and thereby providing information about the emission rate and concentration of a defect level.

In DLTS, the applied bias voltage is alternately fixed between a smaller (filling pulse) and a larger reverse bias voltage, as schematically shown in Figure 1.2(a) and (c). To illustrate the method, consider a deep acceptor type electron trap of concentration N_t, where $N_t \ll n$, close to a Schottky barrier in an n-type semiconductor. During the filling pulse, Figure 1.2(a), the steady-state depletion width, $W(0)$ is small, and outside $W(0)$ the level is below the Fermi level and will, therefore, trap charge. After the filling pulse, a sudden increase in reverse bias is applied and the depletion region increases to $W^*(V)$, Figure 1.2(b). The traps between $W(0)$ and $W^*(V)$ are now above the Fermi level, and at this moment, the occupied traps start to release the electrons to the conduction band at a rate given by the thermal emission rate for the given defect level. Importantly, the electric field in the depletion region sweeps the electrons away with a negligible amount of re-trapping, and the charge redistribution results in a decrease in depletion region toward the steady-state condition, $W(V)$. Thus, the reemission of carriers can be detected by the influence on the junction capacitance, ΔC, through a transient detection technique.

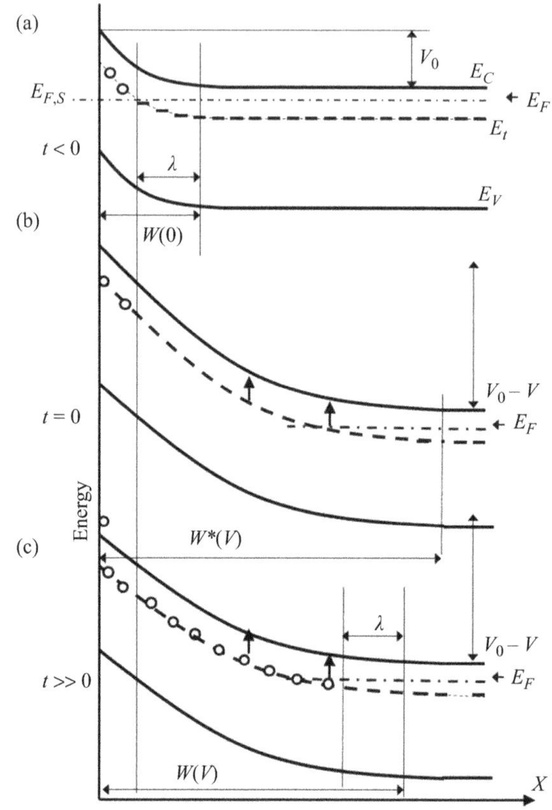

Figure 1.2 An illustration of the band bending in n-type Si versus the distance from a Schottky barrier metallurgical contact (a) during the filling pulse ($t < 0$), (b) immediately after the increase in reverse bias ($t = 0$), and (c) after emission of the trapped carriers, in the steady-state reverse bias regime ($t \gg 0$). $W(0)$, $W^*(V)$, and $W(V)$ represent the depletion region at $t < 0$, $t = 0$, and $t \gg 0$, respectively, while the built-in voltage and reverse bias are labeled as V_0 and V, respectively, λ is a transition region partially occupied by electrons, E_F, and $E_{F,S}$ is the Fermi levels of the bulk and Schottky contact, respectively

The conventional DLTS measurements as proposed by Lang are carried out while scanning the appropriate temperature range, changing the emission rate of the defect levels, and recording the capacitance transients. A DLTS signal, S, is then constructed by weighting the change in capacitance, ΔC, with a given time window, so that peaks in the DLTS spectrum occur when the emission rate of the trapped electron corresponds to the time window. Thus, one has a mean to separate the energy levels from each other in the form of characteristic peaks or signatures in the DLTS spectrum, as illustrated in Figure 1.3.

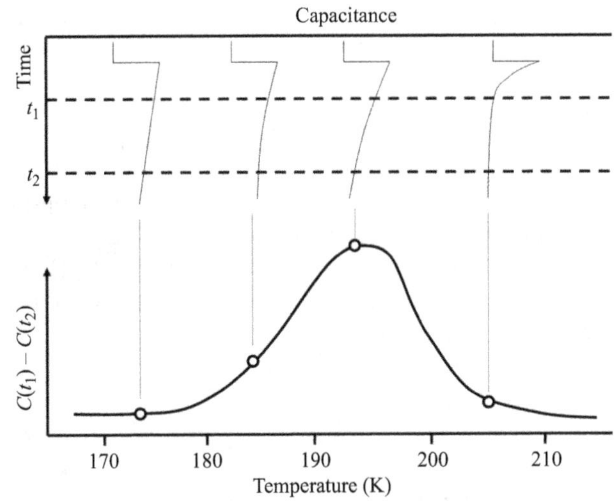

Figure 1.3 Schematics of the construction of a DLTS spectrum from the difference in junction capacitance transients between two fixed times t_1 and t_2 at different temperatures

The change in capacitance caused by the emission of the trapped electrons is proportional to the trap density, and for the temperature where the DLTS signal peaks, the change in capacitance can be shown to be

$$\left[\frac{\Delta C}{C}\right]_x = -\frac{n(x)}{N_d W^2} x \Delta x$$

If the amount of point defects are small so that $\Delta C \ll C$, that is, $W^*(V) - W(V) \ll W(V)$ (Figure 1.2), the capacitance signal becomes a simple exponential decay, and the amplitude is directly proportional to the trap density. However, as seen in Figure 1.2, there will be a region with partial filling of the trap level during the filling state of the measurement, λ, and this must be taken into account. The width of this region is given by

$$\lambda = \sqrt{\frac{2\epsilon(E_F - E_t)}{qN_d}}$$

If the trap density is uniformly distributed in the measurement region, an integration of the equation above from $W(0) - \lambda$ to $W(V) - \lambda$ results in a relation between the change in capacitance and the trap concentration. An often appropriate assumption is neglecting the edge regions, i.e., setting $\lambda = 0$ and $C(0) \gg C(V)$, where the relation between the change in capacitance and trap density can be expressed as $N_t = (2\Delta C/C)N_d$ [10]. However, in some cases, the bracketed terms can become substantial and must be taken into account:

$$\frac{\Delta C}{C} = -\frac{N}{2N_d}\left[1 - 2\frac{\lambda}{W(V)}\left(1 - \frac{C(V)}{C(0)}\right) - \left(\frac{C(V)}{C(0)}\right)^2\right]$$

Figure 1.3 shows how a series of transients taken at different temperatures giving rise to a peak in the difference between two measuring points on the transient. The change in capacitance between, for example, the two points, $C(t_1)$ and $C(t_2)$ for a specific time window, was originally found by applying a boxcar function, while a lock-in and more high-resolution weighting functions were later introduced. For example, the lock-in weighting function is defined as

$$w(t) = \begin{cases} 1 & t_d + 2^{t-1} < t < t_d + 2^t\tau_s \\ -1 & t_d < t < t_d + 2^{t-1}\tau_s \end{cases}$$

where t_d is the "delay time" before the measurement starts, and τ_s is the transient time interval. By varying the length of the time window, the peak is shifted in temperature giving a set of DLTS spectra with different time windows. Applying the weighting function to the measured capacitance transient results in a DLTS signal, S_i for a window length i, and given by

$$S_i(T) = \frac{1}{n_i}\sum_{t=t_d}^{t_d+t_i} \Delta C(T,t)w(t)$$

where the values of $S_i(T)$ are a sum of positive and negative values and taken of 2^i points. Thus, it can be seen as a difference in capacitance of the first and last measured values in the time-window. The DLTS spectrum can then be constructed by plotting the values of S versus temperature for the different time windows, where S relates to ΔC by $\Delta C = S_{i,peak}(T_{peak})/F_i$, where F_i is a numerical factor which can be calculated for the given window length i. The general idea is that at the peak in the DLTS signal $dS/dT = 0$, and the emission rate e_n can be found.

The expression for the emission rate of charge carriers from a trap can be found by measuring the change in concentration of filled traps. The apparent capture cross section, σ_{na}, of the traps and the activation enthalpy ΔH can be found from an Arrhenius plot, see Figure 1.4:

$$\ln\frac{e_n}{T^2} = \frac{\Delta H}{k_b T} + \ln(\beta\sigma_{na})$$

Hence, one can readily obtain the activation energy (ΔH) from the slope in the Arrhenius plot and the apparent capture cross section (σ_{na}) from the extrapolated offset of this line at $1/T = 0$.

A few words of caution are needed when using the above equations, however. The energy level position E_t is modified slightly by the inclusion of the degeneracy factors g_0 and g_1 for the level before and after electron capture,

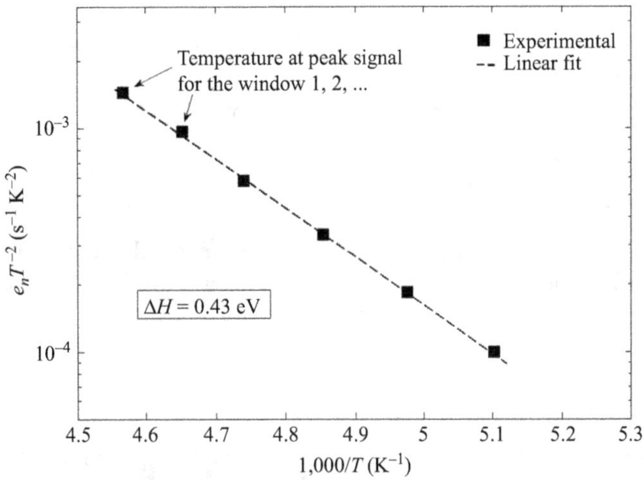

Figure 1.4 Arrhenius plot of ion implanted n-type Si made from DLTS spectra with time windows in the range 20–640 ms

i.e., $E_t(\text{measured}) = E_t(\text{true}) - kT\ln(g_0/g_1)$. Hence, unless the degeneracy factors are known, there is an uncertainty of $\pm kT$ in the trap energy. Further, several of the parameters may be temperature dependent, including E_t. If E_t can be written as $E_t^0(1+\alpha T)$, where α determines the temperature dependence, the slope of the typical Arrhenius plot gives the activation energy extrapolated to absolute zero, E_t^0. Similarly, the Arrhenius plot is based on the assumption that σ_n (σ_n) is independent of temperature but may in some cases vary strongly [17,18], i.e., may be of the order of 0.1 eV [19] or more [18]. This emphasizes the importance of measuring capture cross sections directly rather than deriving them from emission rates.

Note also that there may be a significant Franck–Condon shift of deep energy levels due to lattice relaxation. This implies a discrepancy between the position of the level before and after electron capture with modification to the occupation statistics and should also be modified accordingly, i.e., $E_t = E_c - (1/2)(E_{t,b} + E_{t,a})$, where $E_{t,b}$ and $E_{t,a}$ are the energies before and after relaxation, respectively.

The use of an Arrhenius plot to extract the activation energy, see Figure 1.4, assumes that the emission rate is independent of the measurement conditions, including the electric field. However, for certain traps and conditions, this may no longer hold and include extra complexity in the interpretation. However, this can also be used to extract valuable information about a defect level. In particular, field-enhanced emission can be caused by the so-called Poole–Frenkel effect, and direct and phonon-assisted tunneling. For the Poole–Frenkel effect, typically relevant for electric fields above 10^4 V cm^{-1}, the electric field distorts the electrostatic potential binding the charge carrier to the defect and thereby reducing the barrier height for thermal emission. In direct tunneling, as the name indicates, the charge carrier tunnel through the barrier and the effect becomes significant for electric fields above 10^7 V cm^{-1}. In phonon-assisted tunneling, the phonons effectively reduce the barrier and the probability to tunnel through (electric field $>10^4$ V cm^{-1}).

The emission rate from a trap with a barrier lowering due to the Poole–Frenkel effect can be written as

$$e_n(\varepsilon) = e_{n0}e^{(\Delta E_t(\varepsilon))/(kT)}$$

where ε is the electric field strength, ΔE_t is the energy lowering of the barrier height, and e_{n0} is the emission rate at zero fields. For a Coulomb potential in three dimensions, the barrier lowering can be described by

$$\Delta E_t(\varepsilon) = \frac{1}{2}\left(\frac{q^3\varepsilon}{\pi\epsilon}\right)$$

Thus, a logarithmic plot of the emission rate versus the square root of the electric field is often utilized to demonstrate the Pool–Frenkel effect of the defect level.

1.3.1.1 DLTS profiling

For nonuniform defect profiles, for instance, in samples subjected to ion implantation, the trap density relation $N_t = (2\Delta C/C)N_d$ does not hold. To quantitatively obtain the concentration of defects in the measurement region, one must integrate over the defect profile, that is, one must obtain a depth profile of the trap levels. Several schemes for obtaining trap profiles have been developed, where one approach is to systematically increase the probing volume by reducing the effective voltage applied during the filling sequence, while keeping the steady-state reverse bias voltage constant. A specific trap level is then studied by maintaining the temperature to that of the maximum response for the given time window. Assuming an asymmetric junction or a Schottky barrier, and that $\Delta C \ll C_{rb}$, an incremental change in the relative capacitance due to a small change in the pulse voltage, a depth profile is obtained and given by [10]

$$\frac{d\left(\frac{\Delta C}{C}\right)}{dV} = \left(\frac{\epsilon}{qW^2N_d}\right)\left(\frac{x}{x-\lambda}\right)\frac{N_t(x)}{N_d(x)}$$

The application of depth profiles to gain insight into the behavior of a specific defect will be given in a later section.

1.3.1.2 Measuring capture cross section

In the Arrhenius plot of the emission rate versus $1/T$, the apparent capture cross section (σ_∞) is derived from the extrapolated offset to $1/T = 0$ of the linear dependence of the emission rate. However, a more accurate and direct method for measuring the capture cross section, or in general studying the filling of a defect level, is to measure the time to fill a trap. The filling time can be measured by increasing the duration of the filling pulse while monitoring the DLTS signal keeping the temperature constant at the DLTS peak maximum for the given defect level. For example, during the filling of an electron trap (in n-type), the amount of trapped charge, n_t, will as a first approximation be $n_t = N_T(1 - e^{c_n \tau_p})$, where the capture cross section can be extracted from the capture rate. More refined theories take into account deviations from the exponential behavior shown above, in which injected free carrier concentration is not uniform. In this way, a very accurate measurement of the capture rate of a deep level is available.

These investigations are essential in many applications which make use of deep levels as recombination centers and may also reveal overlapping of levels with comparably similar emission rates as well as reveal energy barriers for electron capture, which will be exemplified in a later section.

1.3.2 Variations of capacitance spectroscopic techniques

After the proposed DLTS scheme by Lang in 1974, a long list of related techniques or variations of the original method have been developed. Common for many of these related techniques is that they try to overcome some of the drawbacks with the original (conventional) DLTS. Important drawbacks or limitations include (i) the energy resolution is limited/poor for differentiating between levels with similar emission rates, (ii) only a part of the band gap is probed for a given temperature range, and (iii) the upper limit for quantitative defect concentration assessment is limited to ~10% of the free charge carrier concentration. To meet these challenges, different techniques have been proposed, including Laplace-, constant-capacitance-, junction-, optical-, and minority carrier-DLTS, and some of these techniques will briefly be described in the following.

1.3.2.1 Laplace-DLTS

The DLTS concept using a weighting function results in a high sensitivity and a DLTS spectrum that is easy to communicate, but the possibility to distinguish levels of similar emission rates is limited due to the relatively broad peaks inherent in the technique. Different weighting functions can be utilized to improve the energy resolution, for example, GS4 and GS6 [20], but overlapping levels may still prove to be difficult to differentiate. A significant improvement to the energy resolution, however, can be found using Laplace-DLTS (LDLTS), developed by Dobaczewski, Peaker, and Bonde Nielsen [21]. LDLTS is an isothermal technique where the emission rates are found by numerical methods similar to the inverse Laplace transform. Realizing that the obtained capacitance transient contains a spectrum of emission rates, the spectral function from the measured transient is the integral

$$f(t) = \int_0^\infty F(s)e^{-st}\, ds$$

where *f*(*t*) is the measured transient, and *F*(*s*) is the spectral density function. To find the spectral function in the equation above, numerical methods are employed, typically based on the Tikhonov regularization method. That is, a mathematical algorithm is employed that effectively performs an inverse Laplace transform for the function $f(t)$, where the result is a spectrum of delta-like peaks for multi-, mono-exponential transients or a broad spectrum with no fine structure for continuous distribution, see Figure 1.5. In this method, it is not necessary to make any a priori assumptions about the functional shape of the spectrum, except that all decays are exponential in the same direction.

LDLTS provides an increase in energy resolution over conventional DLTS. For relatively shallow states that emit at low temperature, the reduction in line-width can give more than two orders of magnitude increase in resolution over conventional DLTS. This makes it possible to observe effects and processes which are impossible to see with the usual methods.

1.3.2.2 Junction DLTS

In conventional DLTS, the filling pulse does not change the occupancy of minority carrier traps, hence limiting the investigated energy region to no more than one-half of the band gap. In junction DLTS, minority carriers are injected and subsequently captured at a trap by a forward-biased filling pulse over a p–n junction. Thus, the main difference between junction DLTS and conventional DLTS is the method of priming the traps prior to the thermal relaxation, i.e., whether the occupancy of minority carrier traps is altered or not.

The depletion capacitance is altered when the free minority carriers from the forward biased p–n junction is captured at a deep state. During the subsequent reverse bias, the capture process is suppressed and the trapped minority carriers are released by thermal emission, similar to that of the majority carriers. This produces a transient in the depletion capacitance which can be detected by the conventional DLTS method, but with an opposite sign of the DLTS signal. Since majority carriers also enter the depletion region under forward bias, majority carrier emission

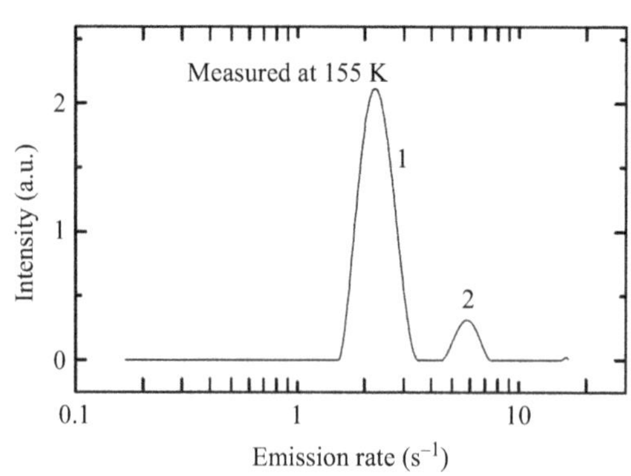

Figure 1.5 Laplace DLTS spectrum recorded at 155 K on a ZnO sample, displayed as signal intensity versus emission rate (s^{-1}). The figure illustrates the superior resolution obtained by LDLTS, where the peak normally labeled E3 in conventional DLTS spectra are shown to consist of two levels with similar emission rates

peaks also appear in a junction DLTS scan, and the resulting DLTS spectrum will be a superposition of signatures from minority and majority carrier traps. Hence, interpretation of a DLTS spectrum from a junction DLTS experiment can be more complex.

The observation of minority carrier traps in a junction DLTS experiment depends upon the injection efficiency of the junction and the relative magnitudes of the electron and hole capture cross sections. Since values of σ_n and σ_p are not easily obtained, a typical approach is to measure the DLTS peak as a function of the minority carrier density by varying the forward current, and ensuring that the filling time is sufficiently long to reach a steady state. Under low injection conditions, only majority carrier traps are observed. As the forward current is increased, the minority carrier injection increases while the majority carrier density is less affected, such that only the minority capture increases with increasing current. Consequently, if the peak height saturates at a constant value, it implies that the capture cross section of the minority carrier is substantially larger ($\sigma_{\text{min}} \gg \sigma_{\text{maj}}$) than that of the majority carrier, and the trap concentration can be deduced.

1.3.2.3 Optical DLTS (ODLTS)

In optical DLTS (ODLTS), illumination with sub-band gap light removes majority carriers from the trap leaving the trap occupied by minority carriers which are subsequently lost by thermal emission. While junction DLTS depends on the ratio of the thermal capture cross sections as discussed above, ODLTS depends on the ratio between optical capture cross sections.

For junction DLTS and ODLTS, there are at least two competing processes during the filling period; consequently, the trap occupancy at the start of the emission transient depends upon detailed properties of the trap (i.e., σ_n and σ_p) and the priming conditions. This means that it is difficult to accurately obtain N_t from the peak heights.

ODLTS was introduced by Mitonneau *et al.* in 1977 and typically performed by illuminating the sample with photons with energy below the band gap $\left((1/2)E_g < h\nu < E_g\right)$ while keeping the sample under reverse bias. This can be achieved using a thin Schottky contact or a p–n junction. In the case of an n-type sample, the illumination removes electrons from some of the minority carrier traps, and these are swept out of the depletion region before they are recaptured. When the optical filling pulse is removed, the traps can relax by thermal emission of holes and observed by a capacitance transient.

The below band gap light is only weakly absorbed; hence, one can assume that the photon flux $\Phi(\lambda)$ is close to constant through the sample, and e_n^o and e_p^o are similarly uniform. Thus, under illumination, the steady-state occupancy of the traps becomes

$$n_t^o(\infty) = \frac{e_p^o + e_n^{th}}{e_p^o + e_p^{th} + e_n^o} \cdot N_t$$

Therefore, the optical emission rates should be such that the trap occupancy is perturbed by removing electrons faster optically than holes are emitted optically and thermally. This means that $e_n^o > e_p^{th} + e_p^o$, and n_t will be reduced below N_t such that some of the traps can be regarded as being occupied by holes.

If the photon flux can be made sufficiently large, the optical rates will dominate and the equation above becomes

$$\frac{n_t^o(\infty)}{N_t} = \frac{e_p^o}{e_p^o + e_n^o} = \left(1 + \frac{\sigma_n^o}{\sigma_p^o}\right)^{-1}$$

or

$$p_t^o = N_t - n_t^o(\infty) = \left(1 + \frac{\sigma_p^o}{\sigma_n^o}\right)^{-1} N_t$$

Therefore, the ability to observe a minority carrier trap in ODLTS is determined by the ratio of its optical emission cross sections for minority and majority carriers, i.e., the ratio between σ_{maj} or σ_{min}. Thus, the absence of minority carrier signature cannot be used as evidence for the absence of minority carrier traps since the filling of the traps depends on the photoionization properties.

1.3.2.4 Minority carrier transient spectroscopy (MCTS)

Junction DLTS and ODLTS affect both majority and minority carriers and it can therefore be difficult to assess the trap and its concentration. If the charge carriers are generated outside the depletion region, however, and minority carriers are to be found one diffusion length from the depletion region, they may enter the probing region, while majority carriers will not. This leaves us with only minority carrier injection within the depletion region. For example, in materials with a sufficiently long diffusion length, the minority carriers can be generated by illumination at the back side and then diffuse over to the front side. In this case, the illumination is only used to generate carriers, and the trap filling follows thermal capture processes. This variation of DLTS is called minority carrier transient spectroscopy (MCTS) and only probe minority carrier traps. As a result, the capacitance transient will have the opposite sign and hence also the signatures in the DLTS spectrum, as shown in Figure 1.6. Hence, junction DLTS, ODLTS, and MCTS can all provide different spectra for minority carrier traps, Figure 1.6, reflecting both the sensitivity of each trap to a particular filling process and the competing temperature dependent processes.

1.3.3 Deep-level optical spectroscopy

Probing deep states using depletion capacitance measurements have been closely associated with thermal emission methods but apply also to studies of optical emission. To utilize the optically induced emission process, however, it requires that the optical emission rates exceed that of the thermal ones. To achieve this, the sample temperature is typically reduced. However, one still needs to be careful in

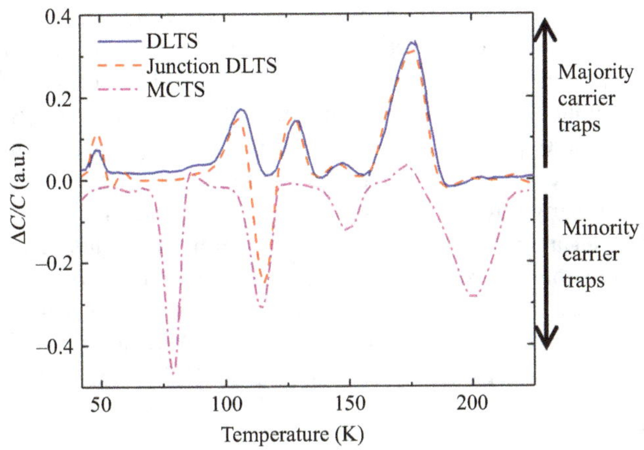

Figure 1.6 *The figure shows a DLTS, a junction DLTS and an MTCS spectrum, where the DLTS only exhibits majority carrier traps, MCTS only minority carrier traps, while the junction DLTS spectrum comprises both majority and minority carrier traps*

assuming that the optical emission dominates over thermal emission rates. In general, the steady-state occupancy within the depletion region will be given by

$$n_t^o(\infty) = \frac{e_p^o + e_p^{th}}{e_p^o + e_p^{th} + e_n^o + e_n^{th}} \cdot N_t$$

If the temperature is sufficiently low so that thermal emission rates can be neglected, the capture rate at the depletion edge can be defined as $c_n(x^o) = e_n^o + e_p^o$.

The capture rate will fall exponentially with distance moving into the depletion region from the depletion edge. Thus, the change in occupancy will be sufficiently abrupt, so that it can be represented as a step in the charge carrier concentration, with a distance λ^o from the depletion edge. The transition region can then be described by $\lambda^o = L_D\left[2\ \ln\left(n\langle v_n\rangle\sigma_n/e_n^o + e_p^o\right)\right]^{1/2}$, where $\langle v_n\rangle$ is the average drift velocity of electrons and L_D is the Debye length. When the photon energy is less than half the band gap, carrier emission to the nearest band edge is likely to dominate and only one optical emission rate applies to the equation above.

These considerations can be utilized in different techniques and applications. A steady-state approach is to measure the capacitance as a function of photon energy at a given temperature. This method is labeled steady-state photocapacitance (SSPC) and can be considered as an optical equivalent to thermal admittance spectroscopy. Although we are here concerned with transient techniques, it is instructive to discuss SSPC briefly in order to better understand the transient version of deep-level optical spectroscopy (DLOS).

SSPC measurements use optical excitation to probe deep levels in a semiconductor. The sample temperature is reduced to a temperature where optical emission rates dominate over thermal emission, while the traps are filled with carriers, i.e., by reducing the depletion region or using band to band illumination. Before the measurements, a reverse bias is applied, and the capacitance is monitored as the sample is illuminated by photons with increasing energy. The result can be shown as a plot of capacitance versus photon energy. The technique has the ability to reveal optically active traps in the sample over the entire band gap, which is a challenge in many novel and wide band-gap semiconductors.

If we consider a single majority carrier trap in n-type material, and a photon energy above the ionization energy of the trap, the removal of electrons is the only possible emission process causing an increase in net space charge density and therefore an increase in the measured capacitance. The (high frequency) capacitance is determined by the depletion depth, and the change in capacitance is given by

$$\frac{\Delta C}{C} = \frac{1}{2}\left[\frac{(x_d)^2 - (x_0)^2}{x_1^2}\right]\frac{N_t}{N_d}$$

where x_0, x_1, and x_d are the depletion widths during the filling pulse, before and after optical emission, respectively. Given that $N_t \ll N_d$. As the photon energy is increased above $(1/2)E_g$, one can no longer neglect transitions of holes to the valence band, and this must be taken into account when considering the capacitance response

$$\frac{\Delta C}{C} = \frac{1}{2}\left(\frac{(x_d)^2 - (x_0)^2}{x_1^2}\right)\left(\frac{e_n^o}{e_n^o + e_p^o}\right)\frac{N_t}{N_d}$$

Unfortunately, the transient approach utilized in DLTS cannot be used for optical emission rates since the emission is a function of the photon flux. However, the optical emission rates for electrons and holes can be separated by measuring the initial change of $n_t(t)$ and applying suitable starting conditions. Particularly interesting are starting conditions where the traps are either completely filled or empty. If $n_t(0) = N_t$, the change in trap concentration becomes $dn_t/dt = -e_n^o N_t$, and when $n_t(0) = 0$, it becomes $dn_t/dt = e_p^o N_t$. These rates can be measured from the capacitance transient so that the variation of dC/dt close to $t = 0$ with photon energy $h\nu$ follows the spectral variation of e_n^o and e_p^o. If $\Phi(h\nu)$ and N_t are known, the photoionization spectrum of $\sigma_n^o(h\nu)$ and $\sigma_p^o(h\nu)$ can be obtained.

There are several routes to perform such an experiment, which are labeled as electrical, thermal, and optical DLOS in accordance with Chantre [11]. In the case of electrical DLOS, as short circuit pulse fills the traps with majority carriers, while a forward bias injection pulse over a p–n junction can be used to fill the traps with minority carriers. In thermal DLOS, a temperature is chosen where the traps are emptied by thermal emission. In optical DLOS, an optical excitation is used with a suitable photon energy and sample temperature. Figure 1.7 shows an illustration of the optical cross-section spectrum of a defect level with an ionization level of 1 eV. An experimental example of a DLOS spectrum from defect levels observed in ZnO is shown in Figure 1.11.

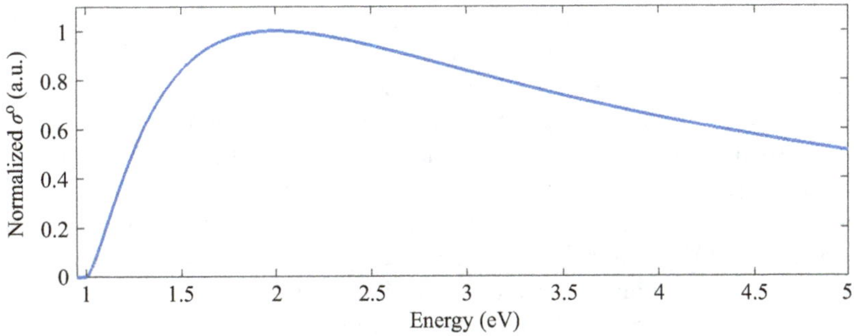

Figure 1.7 Illustration of the optical capture cross-section spectrum for a defect level with energy of 1 eV, and where the maximum is found around double this energy before it falls off as $\sim(h\nu)^{-3/2}$ *at higher photon energies*

1.4 Examples: characterizing electrically active defects in semiconductors

Transient capacitance spectroscopy is a family of powerful techniques for defect characterization that allow for the measurement of key parameters of defect states with an unparalleled sensitivity. However, they usually give no direct information about the chemical constituents or atomic arrangements giving rise to these deep levels. Therefore, careful preparation of experiments is of utmost importance. For example, preparing samples with known chemical impurities provides one way to learn about the constituents of deep-level defects. Alternatively, the introduction of controlled concentrations of deep levels by irradiation with high-energy electrons or other energetic particles may reveal the identity and behavior of important intrinsic defects. Heat treatment under various ambient conditions has also been extensively utilized, not only to identify precursors or constituents of a defect but also to determine the thermal stability and to monitor the evolution kinetics. In any case, correlation with other types of defect spectroscopies such as electron paramagnetic resonance (EPR), infrared spectroscopy (FTIR), and positron annihilation spectroscopy, from which the atomic arrangement and charge state of a defect can often be inferred, enables the DLTS signatures of many defects to be identified and cataloged. Finally, comparing experimental results with theoretical simulations is invaluable in identifying and understanding the defect behavior.

A prime example of combining transient capacitance spectroscopy with selective processing conditions, related experimental techniques and theoretical simulations, to reveal defect physics in semiconductors is silicon (Si). For Si, a vast amount of research have been put down and resulted in identification of a wide range of defects and a deep understanding of their behavior. This has further spurred our understanding of point defects in semiconductors in general. It is therefore appropriate in a chapter on transient capacitance spectroscopy to elaborate on defects in silicon. Indeed, it is one of the few materials where a large number of DLTS signatures have been unambiguously identified. However, even though the field is mature with several decades of development, important challenges remain. For example, point defects are of utmost importance to the efficiency of solar cells, where both light-induced degradation (LID) [22] and light and elevated temperature-induced degradation [23] are expected to be related to point defects. In particular, LID is expected to originate from a boron–oxygen complex, but the exact configuration is yet to be revealed. In the present chapter, a brief review of prominent point defects in Si will be given, before we demonstrate how DLTS can be used to reveal the migration energy of the mono-vacancy. Although the mono-vacancy is one of the most fundamental defects in semiconductor physics and utilized in semiconductor processing for decades, it is only recently that the migration energy of the −2 charge state confirmed in the dilute regime by DLTS [2], previously only observed by the pioneering work of Watkins using EPR.

Zinc oxide (ZnO) serves well for demonstrating transient capacitance spectroscopy in a less developed material. The renewed interest in oxide semiconductors has resulted in a considerable progress in the material development and in understanding the nature of intrinsic and impurity-related defects and dopants in oxides over the last decade. This holds particularly true for ZnO, a wide band gap semiconductor presently used in many diverse products, notably piezoelectric transducers, varistors, and transparent conductive films. However, the technological advances of ZnO have been hindered by the difficulty in controlling and understanding the electrical behavior of intrinsic and impurity-related defects. Several electrically active defects have been observed in ZnO, but several more have probably eluded direct observation due to the wide band gap and lack of stable p-type material. Here, the present understanding of prominent electrically active defects in ZnO will be reviewed, with examples of how processing and dedicated experiments may shed light on the origin and behavior of signatures observed by transient capacitance measurements [18–23]. Ultimately, the examples demonstrate how DLTS and related techniques can efficiently be utilized in the study of defects in semiconductors.

1.4.1 Irradiation-induced defect complexes and vacancy migration in silicon

Irradiation-induced point defects in crystalline silicon have been experimentally studied for more than half a century and have brought to life a deep understanding of the behavior of defects in Si and semiconductors in general. In this respect, DLTS has been decisive. Historically, irradiation-induced point defects have mainly been introduced through bombardment with MeV electrons but are also introduced using ions and neutrons. In high-energy electron bombardment, a uniform distribution of isolated point defects is generated over depths of several millimeters. For ion bombardment, on the other hand, nonuniform depth distributions are formed with localized regions of high defect concentration close to the projected range. Both cases, however, give us the opportunity to study fundamental defect complexes, their kinetics, and how they interact with impurities (Table 1.2).

Table 1.2 Characteristics and identity of typical irradiation-induced deep defect levels in Si

Energy level position (eV)	Capture cross section (cm^2)	Identity	Stability	Reference
E_c −0.17	~10^{-14}	VO(0/−)	≥300°C	[79]
E_c −0.23	10^{-15}–10^{-16}	V_2(−−/−)	200–300°C	[80,81]
E_c −0.42	~10^{-15}	V_2(0/−)	dependent	
E_v +0.20	~10^{-15}	V_2(+/0)	on purity	
E_c −0.44	10^{-15}–10^{-16}	E-center VP(0/−), VAs(0/−), VSb(0/−)	~150°C	[79]
E_c −0.10		C_i(0/−)	≤RT	[82,83]
E_v +0.27	~10^{-17}	C_i(+/0)		
E_v +0.35	~10^{-16}	C_iO_i(+/0)	≥300°C	[32]
E_c −0.38		C_iP_s(0/−)	~125°C	[82]
E_v +0.48		C_iP_s(+/0)		
E_c −0.13		B_i(0/+)	<RT	
E_c −0.45		B_i(0/−)		
E_c −0.26	10^{-13}–10^{-14}	B_iO_i	~175°C	[84]
E_v +0.30		B_iB_s	≥400°C	
E_v +0.29		B_iC_s	~400°C	

During irradiation, mono-vacancies (V) and the self-interstitials (I) are formed, two of the most fundamental building blocks for higher order defects of zero, one, two, and three dimensions in Si. After irradiation, a reaction sequence involving vacancies and interstitials will occur, and the resulting defect formation will depend on temperature and available defects and impurities. Many of these are electrically active and have been identified in the DLTS spectra. Vs in Si are highly mobile and not observed unless the sample is kept at temperatures well below 100 K during and after irradiation, which provide a challenge for conventional DLTS setups. Spectroscopic identification of Vs has therefore primarily been made via the pioneering work by Watkins *et al.* in the 1960s to 1980s using EPR as the main characterization technique [24–26]. The self-interstitial remains even more elusive, since it displays high mobility even at liquid helium temperature (~4 K) [26]. Thus, no unambiguous spectroscopic signal of I has been detected experimentally, and its existence is mainly verified indirectly by trapping at impurities like aluminum, boron, carbon, and oxygen [26,27].

Vacancies mainly form divacancies (V_2) and the vacancy–oxygen pair (VO) after irradiation and room temperature (RT) annealing, while interstitial defects are not confined to host lattice symmetry and multiple atomic configurations, which often exhibit similar total energies. The search for these configurations has led to the detection of a complex hierarchy of defect reactions and an insight into the relationship between reaction kinetics and structure. A Si self-interstitial generated by the irradiation can kick out a substitutional impurity. In p-type Si, the main competitors for trapping migrating Is are substitutional boron (B_s) and substitutional carbon (C_s), forming their interstitial counterparts B_i and C_i, respectively, via a Watkins replacement mechanism. B_i can exist in three different charge states (+, 0, −); however, it is expected to show negative-U properties so that the neutral charge state can only be seen after photoexcitation [28]. In DLTS, only the peak at ~0.45 eV is observed since the ionization process is limited by the first event (0.45 eV), and the second one (0.13 eV) follows immediately. B_i is mobile at RT and displays a long-range migration with an activation energy of ~0.6 eV [27], with subsequent trapping by impurities, forming complexes with oxygen, carbon, and substitutional boron. The energy levels E_c −0.26 eV, E_v +0.29 eV, E_v +0.30 eV appearing in the DLTS spectrum have been associated with B_iO_i, B_iC_s, and B_iB_s, respectively [29–31].

C_i is also mobile at RT [29], where one of the most prominent and abundant traps for mobile C_is in Si is oxygen interstitials (O_i), forming the electrically active C_iO_i complex. The C_iO_i complex has been identified by several different optical and electrical characterization techniques: infrared spectroscopy (IR) [32], photoluminescence spectroscopy [33], and DLTS (with an energy level about E_v +0.36 eV) [30]. C_iO_i is known to be stable at temperatures up to 300°C [31,32].

1.4.1.1 Vacancy-related defects and the migration of the mono-vacancy

DLTS has been decisive in the characterization and understanding of electrically active defects in Si. However, the mono-vacancy has escaped attention by DLTS, partly due to its high mobility. Hence, information about the mono-vacancy has relied on other characterization techniques, and in particular EPR and the pioneering work of Watkins *et al.* The mono-vacancy (V) can appear in at least five different charge states (2+, +, 0, −, 2−) where V^+ and V^- are paramagnetic and directly detectable by EPR [24–26]. From the EPR studies utilizing p-type and n-type materials of different resistivity, stress-induced alignment, and photo-excitation with different wavelengths, also indirect evidence for the existence of V^{2+}, V^0, and V^{2-} has been obtained. However, V exhibits a negative-U character, first predicted theoretically by Baraff *et al.* [34] and later confirmed experimentally [28]. The single donor level ($V^{+/0}$) is situated ~0.03 eV above E_v and the double donor ($V^{2+/+}$) ~0.1 eV deeper at ~E_v +0.13 eV [25]. Much less is known about the positions of the single and double acceptor states ($V^{0/-}$ and $V^{-/2-}$); however, they appear to be deeper than 0.17 eV below E_c since no levels shallower than that can be associated with V according to a DLTS study by Troxell and Watkins [35].

The evasive nature of V can partly be attributed to the activation energy for thermal migration of V, which is reported to ~0.45 and ~0.32 eV for V^0 and V^{2+}, respectively, as revealed by EPR and DLTS measurements [36]. For V^{2-}, which prevails in n-type material, EPR data find a migration barrier of ~0.18 eV [24], i.e., substantially lower than that for V^0 and V^{2+}. Hence, the energy level positions combined with the low migration barriers make direct observation of V by conventional DLTS challenging. However, V can be studied indirectly by monitoring one or more of the prominent trapping centers for V. For example, vacancy diffusion can be studied by low-temperature irradiation and monitoring the trapping of migrating V at O_i, forming substitutional oxygen, commonly known as the VO center (or A center) [37,38]. Indeed, oxygen is the dominant residual impurity in Czochralski grown (Cz) Si, and also a major impurity in float-zone (Fz) silicon with $[O_i]$ typically in the 10^{17} and 10^{15}–10^{16} cm^{-3} range, respectively. Another vacancy-related center is the divacancy (V_2); however, it has been shown that a substantial part of the irradiation-induced divacancies is formed directly during irradiation and not through migration of mono-vacancies [39]. The VO center is electrically active and gives rise to a characteristic single acceptor level at $\sim E_c - 0.17$ eV [29,35,40], readily observed by DLTS in n-type samples. Hence, by performing in situ DLTS analysis after irradiation at cryogenic temperatures, where V^{2-} remains immobile, the growth in [VO] during subsequent thermal treatment (annealing) can be monitored using the $E_c - 0.17$ eV level. Importantly, DLTS gives access to a low net doping concentration and a dilute [V], in contrast to that of EPR [24].

Figure 1.8 shows DLTS spectra using a GS4 weighting function of a sample irradiated with $\sim 1 \times 10^{12}$ H^+ cm^{-2} at ~30 K and which has undergone sequential heating and cooling cycles with increasing maximum temperature, where the first DLTS cycle ended at 62 K, and the temperature range increased with 10 K between each cycle. An offset to the DLTS signal has been added to increase readability. A broad shoulder with the indication of a peak at ~60 K appears already during the first heating scan but remains constant in the subsequent scans to higher

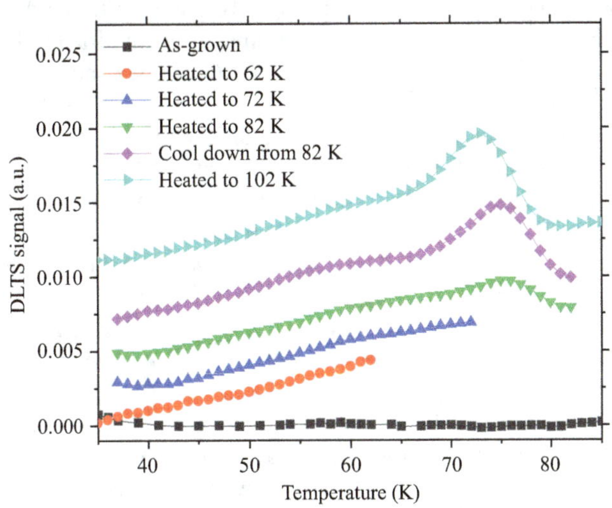

Figure 1.8 DLTS spectra of an unirradiated and irradiated Cz sample with $\sim 1 \times 10^{12}$ H^+ cm^{-2} at ~30 K. Cycling of the DLTS measurements has been performed with increasing peak temperature, and an offset has been added to the graphs for clarity

temperatures. The shoulder has tentatively been assigned to the trivacancy center (V_3) since it appears to be formed directly during the irradiation. V_3 gives rise to a shallow acceptor level at $\sim E_c - 0.075$ eV [3,41], i.e., consistent with the position of the shoulder peak. On the other hand, a distinct signal with a maximum at 75 K is revealed during the scan to 82 K, and the signal grows in amplitude with increasing cycling temperature. The DLTS peak is positioned at $\sim E_c - 0.17$ eV and assigned to the VO center [38]. The substantial increase in [VO] during temperature cycling confirms the presence of mobile Vs and a diffusion length on the order of 20 nm during the 14 min of scanning from 75 K to 82 K and back to 75 K.

Further, no decreasing DLTS peak with the growth of [VO] was found, at least for temperatures up 85 K using the rate window of 5.12 s^{-1}. This indicates that the $V^{-/2-}$ level is located deeper than $\sim$0.19 eV below E_c, assuming an electron capture cross section of $\sim 10^{-15}$ cm^2, which is a typical value for acceptor states due to vacancy-type defects in silicon. Similar results were obtained previously by Troxell and Watkins [35] and imply that the $V^{-/2-}$ level remains filled in the space charge region during the quiescent reverse bias stage of the DLTS measurements. With no reverse bias applied and temperatures below 90 K, the Fermi level is positioned at $\sim$0.07–0.08 eV below E_c in the probed volume of the studied samples.

First-order kinetics is anticipated for the reaction $V + O_i \rightarrow VO$ since the irradiation-induced concentration of V is in the 10^{12}–10^{13} cm^{-3} range while $[O_i]$ is in the 10^{17} cm^{-3} range for the Cz samples and expected to be in the 10^{15} cm^{-3} range for the Fz samples. Therefore, $[O_i] \gg [V]$ in both cases, and the temperature dependence of the growth rate of [VO] is given by the activation energy for migration of V^{2-} ($E_{migr}^{V^{2-}}$). Moreover, if $[O_i]$ is known, the pre-exponential factor of the V^{2-} diffusivity ($D_0^{V^{2-}}$) can be deduced. By carefully monitoring the DLTS peak value of VO for a given time window versus time, as exemplified in Figure 1.9(a), an isochronal annealing study for the growth of VO can be conducted. The growth of [VO] in Fz samples after isothermal annealing at temperatures between 70.5 K and 87.2 K are displayed in Figures 1.9(b). According to first-order kinetics, one could expect the concentration of VO to be described by $[VO] = [VO]_{sat}\left(1 - e^{-c(T)t}\right)$, where $[VO]_{sat}$ is the saturated concentration approached as the annealing duration becomes long, $c(T)$ is the temperature dependent rate constant, and T is the absolute temperature. Thus, a plot of $\ln(1 - [VO]/[VO]_{sat})$ versus t should exhibit a linear dependence where the slope is given by $c(T)$, Figure 1.9(b). Further, applying the theory for diffusion-limited reactions [42,43], E_{migr} was extracted and found to be $\sim$0.18 eV with $D_0^{V^{2-}} \approx 4 \times 10^{-3}$ cm^2 s^{-1} [3]. The $D_0^{V^{2-}}$ value is identical to that predicted for a simple jump process in a cubic diamond lattice structure with the lattice parameter of silicon. That is, no evidence of entropy effects is found for the migration of V^{2-}. Interestingly, the growth rate of [VO] is about a factor of 50 or higher in the Cz samples compared to the Fz ones, reflecting $[O_i]_{Cz} \gg [O_i]_{Fz}$. Importantly, both sets of results exhibit an activation energy, E_a, of $\sim$0.18–0.19 eV which is attributed to the migration of V^{2-}, and identical with that found by Watkins [24] in an EPR study of counter-doped n-type Cz samples with a net carrier concentration of $\sim 3 \times 10^{16}$ cm^{-3} and subjected to MeV electron fluences of $\geq 10^{17}$ cm^{-2}, finally confirming the pioneering work on the mono-vacancy in silicon by Watkins *et al.*

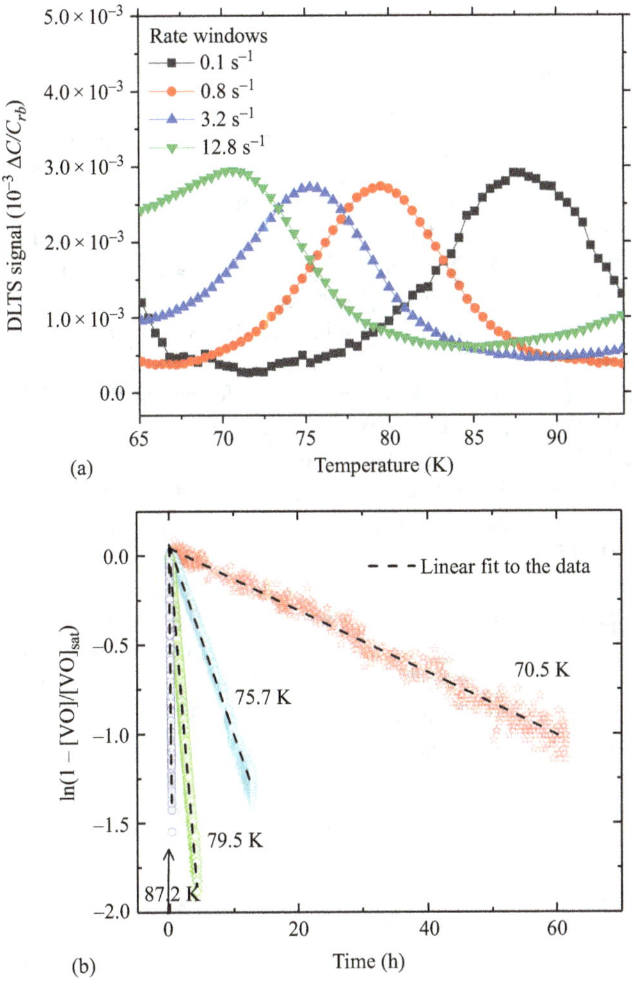

Figure 1.9 (a) *A DLTS spectrum around VO as observed using different time windows, forming the background for the chosen annealing temperatures. (b) ln(1 – [VO]/[VO]$_{sat}$) versus annealing time at 70.5 K, 75.7 K, 79.5 K, and 87.2 K, corresponding to peak value of VO level using four different rate windows (shown in inset)*

1.4.2 Electrically active defects in zinc oxide and the involvement of hydrogen

Zinc oxide (ZnO) is a wide band gap semiconductor (E_g ~3.4 eV) that has received considerable attention since the early 2000s and is presently used in many diverse products, notably piezoelectric transducers, varistors, and transparent conductive films. However, residual impurities and intrinsic defects drastically affect the electrical characteristics of the material and have limited the development toward applications like light emitting diodes and other bipolar components. Therefore, understanding intrinsic and impurity-related defects and their role for the electrical conductivity of semiconducting oxides in general, and ZnO specifically, has been a paramount challenge for materials development targeting efficient optoelectronic devices. Similar to that of silicon, some of the isolated elementary point defects, and in particular the zinc self-interstitial, are highly mobile and challenging to characterize by, e.g., DLTS. The identification of defect levels is further complicated by the lack of stable p-type material, making conventional DLTS only applicable to the upper part of the band gap. Hence, ZnO is a prime example to demonstrate the use of DLTS when studying a less mature material.

Electrically active defect levels have been observed in the 0.02–1.2 eV range below E_c, the accessible range for conventional DLTS in the material, and reported in the literature [4,44–48]. The labeling has been adopted to that proposed by Auret *et al.* as used in some of the early reports [44] following the traditional naming using E for electron trap and increasing numbers starting from the most shallow level. Figure 1.10 shows a typical DLTS spectrum of a bulk hydrothermally grown ZnO sample. In the 80 K–400 K temperature range, at least five defect levels are observed, labeled as stated above E1–E5, with energy positions corresponding to 0.1, 0.18, 0.30, 0.54, and 1.0 eV below E_c, respectively. The wide band gap and lack of p-type material also exemplifies the importance of optical excitations in capacitance spectroscopy. Figure 1.11 shows a DLOS spectrum of an MBE grown ZnO where defect levels in the lower part of the band gap are revealed [1], with energy positions of 2.1, 2.8, and 3.1 below E_c. The energy position was determined by Bois–Chantre fits (i.e., Chantre–Vincent–Bois) [11]. However, the microscopic origin of many of the levels has been challenging to reveal. Approaches such as

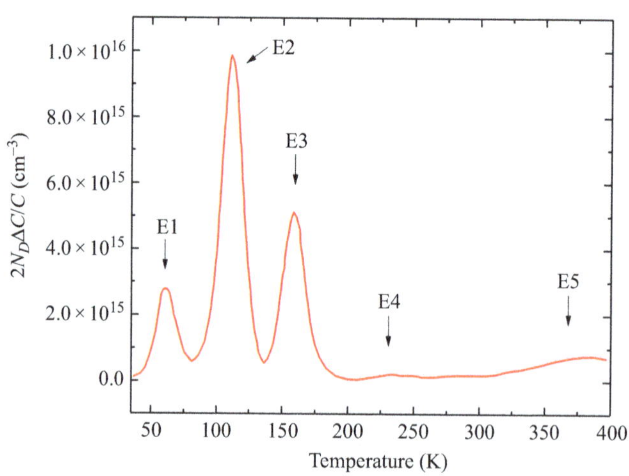

Figure 1.10 DLTS spectra from a typical ZnO sample using a rate window of 640 m s^{-1}. C_b represents the reverse bias capacitance, N_D is the carrier concentration, and ΔC is the amplitude of the capacitance transient

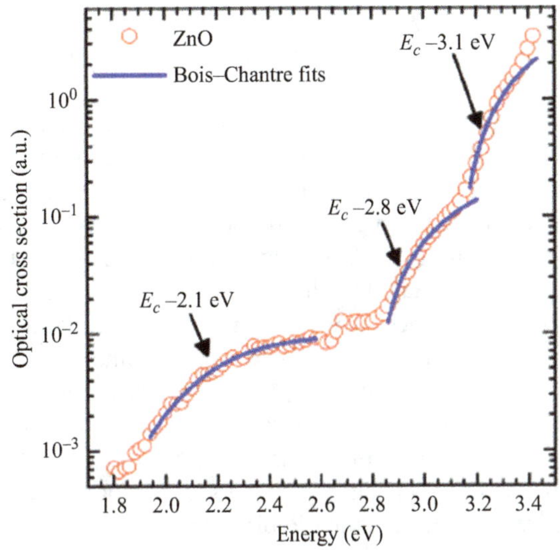

Figure 1.11 *DLOS transient analysis showing optical cross section versus photon energy obtained from MBE grown ZnO. © 2012. Reprinted, with permission, from [1]*

Table 1.3 *Overview of typical defect levels observed in ZnO by DLTS*

Defect	Energy position (eV)	Apparent capture cross section (cm^{-2})	Tentative assignment	Reference
E1	E_c −0.1	5×10^{-13}		
E2	E_c −0.19	$\sim 5 \times 10^{-17}$	Iron	[4]
T2	E_c −0.17	10^{-17}–10^{-16}		[55,85]
E3	E_c −0.30	1×10^{-15}	Hydrogen related [2]	[2,6,44]
E3′	E_c −0.30			
E4	E_c −0.54	5×10^{-17}	V_O^{++}/V_O^{+} [3]	[5,44,45]
T3	E_c −0.63	1×10^{-14}		[86]
E5	E_c −1.0	5×10^{-17}	Intrinsic	[47,54]
E6	E_c −1.2	5×10^{-17}	Intrinsic	[47,54]

varying the growth method, heat treatments including annealing ambience, electron, proton, and heavy ion irradiation have been pursued, but further investigations are still of utmost importance. Here, some of the investigations related to the most prominent defect levels are discussed to illustrate the different approaches to identify defect levels observed by DLTS (Table 1.3).

1.4.2.1 E2 (E_c −0.20 eV)—an iron-related defect level

A level labeled E2, with an energy level position around E_c −0.20 eV, can be found in both bulk single crystalline ZnO wafers and thin films. However, it is not consistently reported in the literature, and its presence depends strongly on the ZnO growth conditions [45,49,50]. This may indicate that E2 is due to an impurity-related defect, rather than one of the primary intrinsic defects. It was further noted that the annealing ambience had a profound impact on the presence of the E2 level, in particular, when annealing in an O-rich ambience gave rise to E2 in hydrothermally grown bulk material. On the other hand, after annealing in Zn-rich ambient, E2 disappeared. Hence, it was concluded by correlating annealing conditions with DLTS spectra that E2 is related to (or promoted by) an O-rich (Zn-deficient) defect like V_{Zn}, O_{Zn}, and O_i [48]. It was also shown that the rate of electron emission was enhanced by the external field due to distortion (lowering) of the defect potential well, as shown in Figure 1.12. A close proportionality between the shift in the apparent energy position and the square root of the electric field was observed, as predicted by the so-called Poole–Frenkel effect discussed previously (Figure 1.12) [51]. In n-type material, only donor-like states display field-enhanced emission and not (single) acceptor-like ones that become neutral after electron emission; hence, it was concluded that E2 arose from a donor-like defect.

Further, it was noted that a large difference in the E2 concentration occurred between samples annealed under similar conditions. Therefore, assuming that steady-state condition was reached during the heat treatment in O-rich ambient, where E2 is observed as previously noted, it was argued that an intrinsic nature of E2 could be excluded. An impurity-related origin of the defect level was therefore sought by correlating the concentration of the level with results from other techniques that are sensitive to impurities. In this case, the defect level was compared

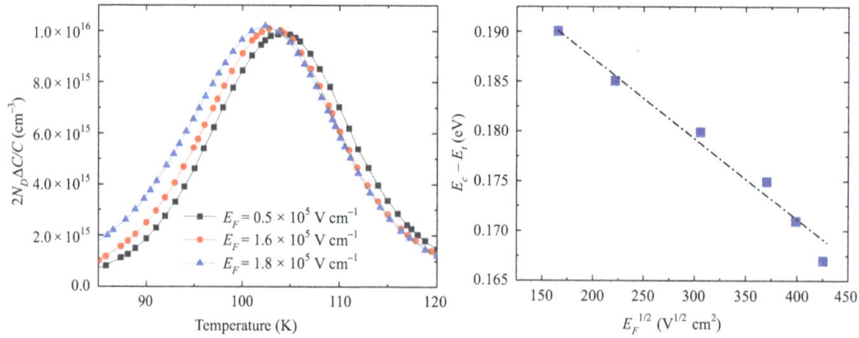

Figure 1.12 Left panel shows DLTS spectra of E2 taken for different reverse biases of −1, −3, and −4 V corresponding to peak electrical fields of 0.5×10^5, 1.6×10^5, and 1.8×10^5 V cm^{-1}, respectively, demonstrating the Poole–Frenkel effect, while the right panel shows the correlation between the E2 and the Fe concentration. © 2013. Reprinted, with permission, from [4]

with the results from secondary ion mass spectrometry (SIMS) to identify potential impurities. Indeed, the concentration of E2 followed closely that of iron in different samples, see Figure 1.13 (right panel), where E2 represented about 30% of the total Fe concentration in each sample. This proportionality held over a wide dynamic range is strong evidence that E2 involves Fe in a configuration enhanced by O-rich conditions, such as Fe on Zn site. Indeed, EPR measurements reported by Jiang *et al.* [52] estimated the ionization energy for the substitutional Fe donor (Fe_{Zn}^{2+}) to be $\sim$0.24 eV with an uncertainty of $\pm$0.05 eV. It should also be mentioned that the Fe concentration remained constant in all the samples after the different annealing treatments, as determined by SIMS. This may indicate a high migration energy of Fe and a lower binding energy of E2 which would explain a change in the defect configuration with the annealing conditions (activation under O-rich conditions and deactivation under Zn-rich conditions), while the Fe concentration is maintained. However, it was concluded that E2 originate from Fe_{Zn}^{2+} based on the correlation with iron, the energy position of E2, its donor-like behavior, and the response on annealing ambient.

1.4.2.2 E3 (E_c −0.30 eV)—a hydrogen-related defect level

The most commonly observed defect level in ZnO is found around E_c −0.3 eV, normally labeled E3. The E3 signature is observed in most ZnO materials irrespective of the growth method used, which has made it one of the most studied defect levels in ZnO. It was first reported by Simpson and Cordaro [53] and later characterized in more detail and labeled E3 by Auret *et al.* [44,46]. The strong presence in materials originating from a wide variety of growth techniques has been considered an argument for the defect level to be one of the fundamental primary intrinsic defects, for example, the oxygen vacancy. However, it was shown by Auret *et al.* that E3 is not influenced by MeV electron or proton irradiation, and

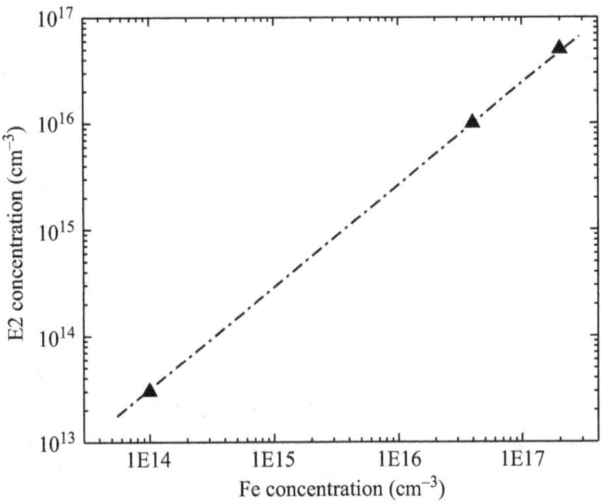

Figure 1.13 The correlation between the E2 and the Fe concentration. © 2013 Reprinted, with permission, from [4]

later it has also been demonstrated a similar behavior when implanting self-ions, i.e., O and Zn [54]. Different results are reported on the dependence on the annealing ambience, where Mtangi *et al.* [55] have studied the evolution of deep-level defects in melt grown samples during post-growth annealing at temperatures between 600°C and 900°C in Ar and O_2 ambient, where a variation in the E3 peak intensity by a factor of ~3 was found between samples. In addition, von Wenckstern *et al.* have showed that nearby defect levels may strongly overlap with E3 and thereby influence the evaluation of E3 [56]. They have also shown that E3 exhibit a negative-U behavior, where a second optical transition occurred for photon energies higher than 1 eV, which accounts for the lowering of the E3 DLTS signal at low temperatures under illumination [57]. The various observations have triggered both tentative assignments of intrinsic and extrinsic origin, including a relation to oxygen vacancy, interstitial zinc [58], and transition metals [59].

However, when implanting hydrogen in bulk ZnO samples, and monitoring the region around the projected range of the H ions, a strong and consistent increase in the E3 concentration is observed [6] as discussed in Figure 1.14(a). A similar implantation using He ions did not alter the E3 concentration, demonstrating that the presence of H is a necessary condition, and that only ion-induced defects are not sufficient to enhance the formation of E3. Interestingly, the H implanted samples show an increase of N_d in the implanted region that amounts to more than 95% of the implanted concentration, implying that almost all the implanted H atoms act as shallow donors with H_i as the prevailing configuration. Thus, both N_d and E3 concentrations are highly nonuniform and it must be emphasized that a quantitative conversion of the DLTS peak amplitude to concentration is strictly valid for uniform defect profiles only. Thus, depth profiling measurements are highly appropriate for such defects studies.

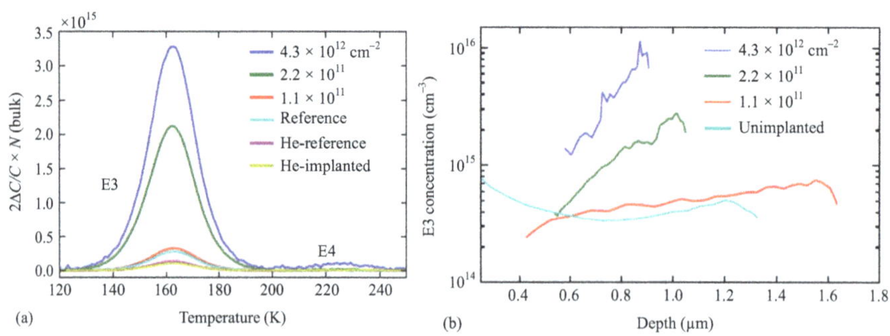

Figure 1.14 Part (a), left panel, shows DLTS spectra using a rate window of 640 m s^{-1} from a box-profile sample with different implantation doses, an unimplanted reference sample, and a sample before and after He implantation. Part (b), right panel, illustrates the influence of the hydrogen fluence on the charge carrier profile for single profile and box-profile implantations. © 2014. Reprinted, with permission, from [6]

Figure 1.14 (left panel) displays the DLTS spectra from samples implanted with a box-profile of H using three implantation energies, and with different overall implantation doses together with an unimplanted reference sample and a sample before and after He implantation [6]. Figure 1.14 (right panel) displays the E3 versus depth profiles for different implantation fluences. Before implantation, the E3 distribution is almost uniform while after implantation it increases with depth. The absolute concentration of E3 grows strongly with the H-dose, with an increase of more than a factor 20 at a depth of ~0.9 μm in a box-implanted sample to a fluence of 4.3×10^{12} cm^{-2} relative to an unimplanted control sample. However, the probing depth for E3 decreases with the H-dose because of an increase in N_d and limited applicable maximum reverse bias voltage.

Similar observations have been revealed for deuterium implantations, where the rate of increase for both N_d and E3 are a factor of ~5–10 higher in the H-implanted samples relative to the D-implanted ones [2]. It was further shown that the probability for trapping migrating D_is by implantation-induced intrinsic defects leading to loss of the shallow D_i donors is higher than that for H_i. For implantations at 158 K, D_i is almost immobile and a large fraction remains in the interstitial configuration without reacting with the ion-induced intrinsic defects. Furthermore, Zn_i, acting as a shallow double donor, is expected to become mobile around RT and interact with other defects, giving rise to a loss in the net carrier concentration [60]. That is, the initial balance between the generated Frenkel pairs of Zn_i and V_{Zn} is no longer valid, and hence, the actual fraction of D_i may be larger than that estimated, because of compensation by the deep V_{Zn} acceptors.

It is important to note that H implanted under the dilute (low dose) conditions gives rise to a defect concentration of the same order of magnitude as that of the implanted H ions. As previously discussed, H_i acts as a shallow donor in ZnO [61] and may influence the n-type conductivity. For instance, an estimation based on TRIM simulations using protons with an energy of 325 keV indicates that on average 7–8 Frenkel pairs are formed in total (Zn interstitials (Zn_i) and Zn vacancies (V_{Zn}) plus the corresponding pairs for oxygen (O_i, V_O)) [2,62]. Taking into account that only a few percent of the Frenkel pairs survive dynamic annealing [39], the resulting defect formation becomes less than 1 Frenkel pair per proton. Hence, a major fraction of the H atoms is anticipated to reside on interstitial sites directly after implantation, consistent with the large increase in N_d prior to the subsequent diffusion/redistribution of H_i. Moreover, a close resemblance between the evolution of E3 and N_d has been demonstrated as a function of ion dose, depth, and post-implant time, implying strongly that H_i is also decisive for the formation of E3. Based on the above considerations, it was suggested that E3 is associated with a defect configuration invoking H_i and a primary intrinsic defect like V_O, V_{Zn}, or O_i. However, it was noted that a center involving Zn_i was regarded as less likely since both Zn_i and H_i are expected to be positively charged and Coulomb repulsion will suppress the formation of such a center.

The implantation using in situ DLTS measurements after hydrogen (H) and deuterium (D) implantations at cryogenic temperatures also revealed that E3 starts to form at temperatures below 200 K. Hence, in contrast to that of the free charge carriers (electrons) induced by the implanted H_i, the formation of E3 involves a thermally activated process like migration and trapping of interstitial H or D (H_i/D_i). Thus, the kinetics of the E3 formation and evolution could be explored. Figure 1.15 shows the evolution of the charge carrier and E3 concentration after implantation at 158 K and subsequent annealing at RT (300 K) extracted from their depth-distributions.

In the initial stages after H and D implantations at cryogenic temperatures (<160 K), an increase in E3 is observed, with a corresponding decrease in the carrier concentration. After this initial increase, both the carrier concentration and E3 decrease in concentration, where N_d shows a larger rate constant compared to that of E3. The decay rate of N_d is about a factor of 1.4 higher in the H-samples compared to the D-samples and reflects presumably the difference in diffusivity between H_i and D_i given by the square root of their mass ratio. For E3, the rate constants differ by a factor of ~5.5 in the two samples and may indicate an influence on the annealing process by the ratio between the concentrations of implanted species and generated defects. From the isothermal annealing experiments, an Arrhenius plot can be constructed for the loss of E3, resulting in an activation energy of ~0.87 (±0.05) eV with a pre-exponential factor of $\sim 2 \times 10^{10}$ s^{-1}. A first-order kinetics process (exponential decay) for the loss of E3 can occur via either dissociation of the defect center or reaction with another defect/impurity having about one order of magnitude (or more) higher concentration than E3. In the first case, a pre-exponential factor between 10^{12} and 10^{13} s^{-1} is expected. Thus, despite a considerable uncertainty by up to one order of magnitude in the value extracted for the pre-factor, dissociation is excluded as a likely process for the loss of E3.

Instead, a reaction with a species X of sufficient concentration and diffusivity is favored and applying the theory for diffusion-limited reactions, a pre-factor of $\sim 10^{10}$ s^{-1} would result in $D_{X0}[X] \sim 10^{16}$ cm^{-1} s^{-1}. From SRIM simulation, the concentration of H in the implanted peak region was estimated to $\sim 4 \times 10^{16}$ cm^{-3}

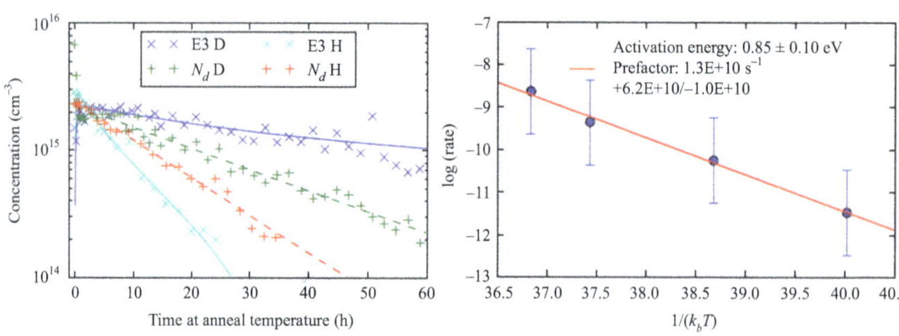

Figure 1.15 Time dependence of the peak concentrations of excess E3 centers and excess charge carriers during isothermal annealing at 300 K after deuterium and hydrogen implantation (1×10^{12} cm^{-2}, 158 K). © 2017. Reprinted, with permission, from [2]

in the reported samples, where H_i emerged as a viable candidate for X. Indeed, H_i is quite mobile around RT where the migration energy for interstitial hydrogen is reported to be in the range of ~0.5–0.9 eV [63,64]. Assuming E_m ~0.7 eV with a pre-exponential factor of ~3×10^{-2} $cm^2\ s^{-1}$ for the diffusivity, a diffusion length of ~0.5 μm is obtained after 12 h at 295 K. Here, the activation energy would represent the migration energy (E_m) of H_i plus a contribution from a possible reaction barrier between H_i and E3. It should also be underlined that the migration process must be of short range (<100 nm) since SIMS measurements did not reveal any broadening of corresponding D concentration-versus-depth profiles after the E3 annealing. Hence, by combining cryogenic H implantations and online DLTS some insight into the origin and behavior of the E3 center was finally revealed.

1.4.2.3 E4 (E_c −0.57 eV)—an oxygen vacancy-related defect level

In addition to the E3 center, a level at E_c −0.57 is commonly observed and extensively studied. In contrast to the E3 center, E4 is found in substantially lower concentrations in as-grown material but responds to an irradiation or implantation fluence of electrons, protons, and heavier ions [44,54]. Moreover a negative-U behavior of the E4 center has been reported by Frank *et al.* [45] and Ellguth *et al.* [65], although they find different energy positions for the E4 level. Based on these findings, several authors have tentatively assigned the level to the oxygen vacancy (V_O) [45,48].

Similar to that for E2, E4 depends strongly on the annealing conditions in a consistent manner, annealing in Zn atmosphere increases the DLTS signal from E4, while annealing in O_2 atmosphere removes it. Interestingly, it was shown that E4 displays almost identical concentrations in samples annealed with similar conditions, indicating that thermal equilibrium is indeed reached during the treatment in Zn-rich conditions and supporting an intrinsic origin of E4, like V_O. A formation energy was also extracted and found to be ~1.9 eV.

As discussed in [66], the generation rate of the E4 center extracted from the results for the MeV proton and helium irradiated samples by Auret *et al.* [44] are low and only on the order of ~0.2%–0.3% of the total vacancy (or interstitial) generation. These irradiations were performed at RT, and it was found that E4 starts to disappear already at postimplant annealing temperatures below 400 K [54]. This implies a low thermal stability of the irradiation-induced E4 centers and/or a high probability to react with other defects mobile at RT (or just above).

In deuterium implanted samples, Figure 1.16, the E4 peak amplitude grows almost linearly with the D fluence, as expected for a primary intrinsic defect [5]. DLTS spectra constructed by applying the high-energy resolution weighting functions of GS4 and GS6-type [20] reveal only one level contribution to the E4 peak, i.e., no evidence of multiple levels is found. However, as discussed for the implantation experiments related to the E3 center above, a quantitative conversion of the DLTS peak amplitude to concentration is strictly valid for uniform defect profiles only, and depth profiling measurements are required. Interestingly, after deuterium implantation, the concentration of E4 increases with depth up to the

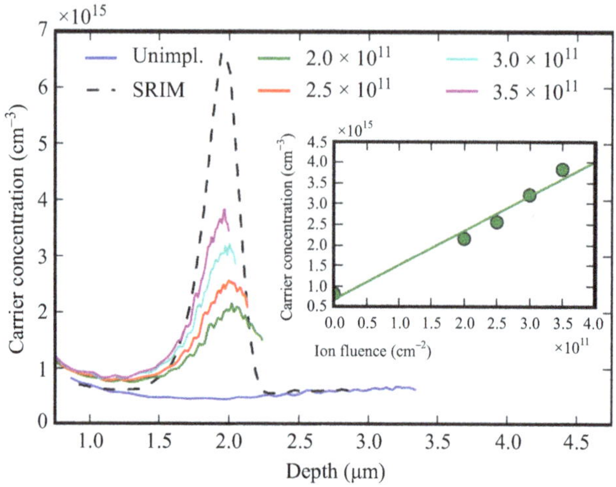

Figure 1.16 Influence of deuterium fluence (cm^{-2}) on the E4 center concentration versus depth profile. The calculated profile of the oxygen vacancy (V_O) concentration corresponds to the highest fluence implantation and was obtained using the SRIM-code by assuming a fraction of 1% the V_Os surviving dynamic annealing. The inset shows the fluence dependency of the maximum of the E4 profile after implantation. The highest fluence used for hydrogen (8 × 10^{11} cm^{-1}) is outside the range shown but falls on the solid line drawn for the H$^+$ ions. © 2016. Reprinted, with permission, from [5]

profile peak at ~2 μm and then decreases rapidly. In addition, the absolute concentration of E4 grows strongly with the D-fluence. Due to the implantation of D resulting in an increase in N_d, the probing depth for E4 decreases with the D-fluence because of the concurrent increase in N_d and limited applicable maximum reverse bias voltage.

The dashed profile in Figure 1.16 shows V_O according to SRIM simulations assuming a displacement energy threshold of 68 and 43 eV for substitutional O and Zn atoms, respectively. Assuming a fraction of ~1% of surviving V_Os, an excellent quantitative agreement was obtained between the measured E4 profiles and the calculated V_O ones, as illustrated for the highest fluence implant in Figure 1.16. The value of ~1% appears quite reasonable for a primary defect in comparison with that for silicon, and the close resemblance in shape between the experimental and calculated profiles corroborates the assumption of E4 arising from a low-order (primary) defect. This is also supported by the linear dependence of the E4 profile peak concentration on the ion fluence, shown in the inset of Figure 1.16. However, similar arguments can be used for any of the three other primary defects (V_{Zn}, Zn_i, and O_i), albeit with slightly different absolute values for the fraction surviving dynamic annealing. On the other hand, both V_{Zn} and O_i are deep double-acceptor centers with their charge state transitions occurring in the lower part of the energy band while Zn_i is a shallow donor with the level position located very close to E_c [67,68]. Therefore, it was concluded that V_O appears as the most plausible origin of the E4 level [5].

Ellguth *et al.* [65] utilized a variant of ODLTS where electrical pulses were used for trap filling and monochromatic light to excite carriers from the traps. Here, E4 was revealed to be a double donor center where an electron from the deeper-lying state, labeled E42, could be excited to the conduction band irrespective of whether the shallow state, labeled E41, was occupied or not. The results are consistent with a negative-U center. Also, theoretically, there are strong evidence that V_O is a negative-U double donor center [69,70], i.e., the energy position of the (+/0) transition lies below that of the (2+/+) transition. Most of the recent theoretical results based on density functional theory propose a position of E_c −1.0 eV for the thermodynamic (2+/0) V_O transition with a negative-U value of ~1.0 eV [71]. This places the (+/0) and (2+/+) transitions at ~1.5 and ~0.5 eV below E_c, respectively.

Interestingly, filling pulse measurements of the E4 amplitude, as extracted by the DLTS formalism, at sample temperatures between 210 K and 245 K reveal an increase in capture rate with increase in temperature, shifting the filling curves to shorter pulse lengths. In the inset of Figure 1.17, the filling pulse durations at half of the full peak amplitude are given versus the reciprocal thermal energy. The values exhibit an Arrhenius behavior and the capture is thermally activated with an energy of ~0.15 eV. Assuming that E4 is due to V_O, the measured barrier of ~0.15 eV for the electron capture by E4 is assigned to the thermal energy barrier for the $V_O^{2+} \rightarrow V_O^{+}$ transformation.

The evolution of E4 after implantation can be further studied during annealing around RT. Annealing experiments between 290 K and 315 K all showed a single

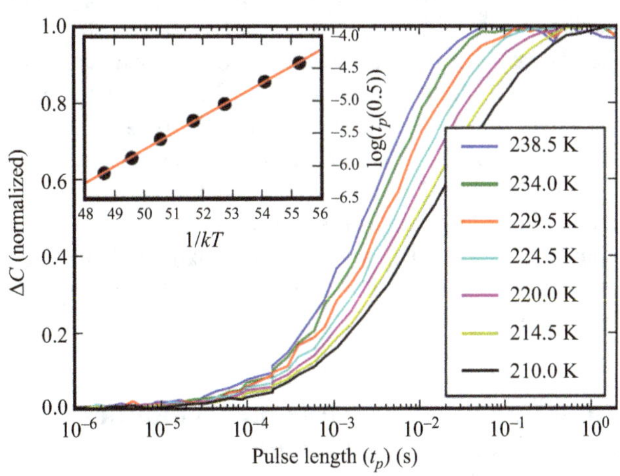

Figure 1.17 Normalized DC as a function of bias pulse length t_p. The logarithm of the 50% value of t_p shows an Arrhenius behavior with an activation energy of 0.15 eV. © 2016. Reprinted, with permission, from [5]

exponential decay and a first-order kinetics process, where the deduced rate constants obey an Arrhenius behavior. The value extracted for the activation energy is E_a ~0.6 eV with a pre-exponential factor of ~2×10^5 s^{-1}. Assuming that first-order annealing kinetics apply to E4, a defect center can arise either from a dissociation of the center or a diffusion limited reaction with another defect/impurity species having a concentration about one order of magnitude (or more) higher than that of the defect center itself. However, for a dissociative process, the pre-exponential factor is expected to be on the order of ~10^{13} s^{-1} and is therefore ruled out as a mechanism responsible for the loss of E4. Thus, diffusion limited reaction appeared as a plausible candidate.

It was also observed that the annealing rate of E4 increases strongly with the H^+ fluence. This implies an involvement of hydrogen in the annealing process with migration of interstitial hydrogen, H_i, as a likely ingredient. H_i is known to be rather mobile already at RT [63,64] and assuming the reaction $H_i + E4 \rightarrow HE4$, a rate equation can be derived for the loss of E4 by applying the theory for diffusion-limited reactions [42,43]:

$$\frac{d[\mathrm{E4}]}{dt} = -4\pi R D_{\mathrm{H_i}}[\mathrm{H_i}][\mathrm{E4}]$$

where square brackets denote concentration, t is the annealing time, R is the capture radius for the reaction, and $D_{\mathrm{H_i}}$ is the diffusivity of H_i. Since $[\mathrm{H_i}] \gg [\mathrm{E4}]_{t=0}$, the solution of the rate equation reduces to $[\mathrm{E4}] = [\mathrm{E4}]_{t=0}e^{-ct}$ with the annealing rate constant c being $c = 4\pi R D_{\mathrm{H_i}}[\mathrm{H_i}] = 4\pi R D_{\mathrm{H_{i0}}}[\mathrm{H_i}]e^{-E_m(\mathrm{H_i})/kT}$, where $E_m(\mathrm{H_i})$ represents the migration energy of H_i, and $D_{\mathrm{H_{i0}}}$ is the pre-exponential factor of $D_{\mathrm{H_i}}$. Thus, the experimentally determined E_a value of ~0.6 eV can be associated with $E_m(\mathrm{H_i})$, consistent with previous reports in the literature yielding values of $E_m(\mathrm{H_i})$ in the range of ~0.5–0.9 eV [63,64]. Further, putting $[\mathrm{H_i}]$ equal the carrier concentration immediately after the implantation, $N_d(t = 0)$, i.e., prior to when a deep acceptor-like center starts to evolve and reduce the net carrier concentration one obtains $RD_{\mathrm{H}} = 1.4 \times 10^{-11}$ cm^3 s^{-1} by equaling the pre-exponential factor for the annealing rate constant to the measured value of ~2×10^5 s^{-1}. Assuming $R = 5$ Å, based on geometrical considerations and omitting possible electrostatic interactions, $D_{\mathrm{H_{i0}}}$ becomes equal to ~3×10^{-4} cm^2 s^{-1}, which is close to that anticipated for a typical interstitial migration process. In particular, it was observed that the experimental annealing rate constant scaled approximately linearly with the H fluence. In summary, V_O is a probable candidate for the identity of E4, which is also corroborated by the E4 annealing results.

1.5 Extending junction spectroscopic techniques beyond traditional methodology

DLTS is a powerful technique with an unparalleled sensitivity. However, the microscopic configuration remains hidden making defect identification of a specific energy level challenging. Therefore, combining the DLTS methodology with a site selective characterization technique, or a site selective defect formation, is an intriguing idea for further expansion of this family of transient capacitance-based techniques. An example of the former is combining DLTS with site-selective excitation techniques such as those based on synchrotron radiation.

Several attempts have been undertaken to utilize capacitance spectroscopy in synchrotron radiation experiments to study the origin of defects and impurities in semiconductors. For example, Ishii *et al.* [72] monitored the capacitance of a barrier structure to obtain X-ray absorption fine structure (XAFS) spectra of diluted defects, where it was proposed that the capacitance-XAFS signal arose due to the direct transitions between the inner core level and the local defect level in the semiconductor band gap and carried the information of the local atomic environment of the defect. This interpretation was put in question by Weber *et al.* [73], where the deep-level defect labeled EL2 in GaAs was studied using the time constants of the photocapacitance transients under X-ray excitation. In their measurements, the local atomic arrangement of EL2-defect differed significantly from the one for ideal GaAs lattice, while the capacitance transient time constant spectrum coincided well with the X-ray absorption spectrum of the GaAs matrix. In another study, Fujioka *et al.* [74] successfully used monochromatic synchrotron excitation for site-selective DLTS measurements. Fujioka *et al.* extracted the ratio of the magnitude between two trap states in the gap change with X-ray excitation energy. When the excitation energy which corresponded to the binding energy of core electrons of a specific atom present in the sample, in this case, Si in an Al/AlN/Si heterostructure, the DLTS response of a trap changed significantly. The authors concluded from this observation that one of the levels, labeled HT1, consists of continues levels just above the valence band and is related to Si atoms at the AlN/Si interface, while a discrete level labeled HT2 is related to Al and N in the AlN film. While the interpretation may still be developed further, the results are promising for combining the extreme sensitivity of DLTS-related techniques with site selectivity of synchrotron-based techniques.

As demonstrated in some of the examples in the previous section, combining irradiation or ion implantation-induced defect formation with in situ (online) DLTS characterization enables fundamental and highly controlled experiments where diffusion is suppressed. This can be further explored when combining with a variable energy electron irradiation, where the onset of vacancy generation for, e.g., different sublattices in a compound semiconductor may be explored.

The conventional DLTS developed by Lang is a macroscopic technique, typically measured over a diode structure with a diameter in the mm range, where the response from an ensemble of point defects with the same emission rate of charge carriers are observed as a signature in the DLTS spectrum. However, nanostructures require a local inspection of electrical properties. A technique that has been successfully applied to measure both topographical and functional properties of nanostructures is scanning probe microscopy (SPM). Hence, several attempts have been made in developing SPM-based techniques that introduce DLTS or similar concepts for local, laterally resolved, defect investigations [75,76]. However, practical implementation has proven challenging, and the full potential has yet to be revealed. One challenge with local DLTS measurements is the low capacitance due to a probe tip radius of a few tens of nanometers or less. Hence, capacitance of a moderately doped semiconductor can typically be in the attofarad range, and challenging to utilize in a DLTS type of scheme. Another practical issue is the temperature scan used in conventional DLTS, which can be time consuming and pose high requirements on repeatability. Hence, isothermal techniques have mainly been reported so far. For example, Dózsa *et al.* applied SPM with a preamplifier to increase the sensitivity sufficiently to measure a self-assembled iron-silicide nanostructure [77]. Lányi and Nádaždy proposed isothermal charge-transient spectroscopy where the charge transient following the voltage pulse periodically applied to the sample, and where a few selected temperatures were recorded [76]. Chinone *et al.* have developed a technique they called super-higher-order scanning nonlinear dielectric microscopy [78], where they could measure SiO_2/SiC structures and obtain a local-DLTS signal. However, so far the techniques have not been widely adopted, partially because of the challenge in obtaining reliable results which is an inherent challenged in the SPM techniques.

1.6 Conclusions

The present chapter has reviewed the fundamental properties of electrically active defects in semiconductors and described how transient capacitance techniques can be utilized to characterize deep levels. A few examples have been given from a very well-studied material where most of the levels observed by DLTS have been identified, namely Si, and a material where defect identification is still controversial, but no less important, i.e., ZnO. The examples are intended to present strengths and weaknesses with this class of techniques and approaches to identify, characterize, and understand electrically active defects.

Acknowledgments

The author would like to acknowledge the essential role of many collaborators and students at the University of Oslo. I would also like to acknowledge the authors of our references for permission to reprint their figures.

References

[1] Bube, R.H., *Photoconductivity of solids*. 1960, New York: Wiley.

[2] Sah, C.T., L. Forbes, L.L. Rosier, and A.F. Tasch, *Thermal and optical emission and capture rates and cross sections of electrons and holes at imperfection centers in semiconductors from photo and dark junction current and capacitance experiments*. Solid-State Electronics, 1970. 13(6): pp. 759–788.

[3] Sah, C.T. and J.W. Walker, *Thermally stimulated capacitance for shallow majority-carrier traps in edge region of semiconductor junctions*. Applied Physics Letters, 1973. 22(8): pp. 384–385.

[4] Lang, D.V., *Deep-level transient spectroscopy – New method to characterize traps in semiconductors*. Journal of Applied Physics, 1974. 45(7): pp. 3023–3032.
[5] Chantre, A., G. Vincent, and D. Bois, *Deep-level optical spectroscopy in GaAs*. Physical Review B, 1981. 23(10): pp. 5335–5359.
[6] Blood, P. and J.W. Orton, *The electrical characterization of semiconductors: Majority carriers and electron states*. Techniques of Physics. Vol. 14. 1992, London: Academic Press.
[7] Shockley, W. and W.T. Read, *Statistics of the recombinations of holes and electrons*. Physical Review, 1952. 87(5): pp. 835–842.
[8] Hall, R.N., *Electron-hole recombination in germanium*. Physical Review, 1952. 87(2): pp. 387–387.
[9] Lucovsky, G., *On the photoionization of deep impurity centers in semiconductors*. Solid State Communications, 1965. 3(9): p. 299–302.
[10] Pässler, R., *Photoionization cross-section analysis for a deep trap contributing to current collapse in GaN field-effect transistors*. Journal of Applied Physics, 2004. 96(1): pp. 715–722.
[11] Pals, J.A., *Properties of Au, Pt, Pd and Rh levels in silicon measured with a constant capacitance technique*. Solid-State Electronics, 1974. 17(11): pp. 1139–1145.
[12] Henry, C.H. and D.V. Lang, *Nonradiative capture and recombination by multiphonon emission in GaAs and GaP*. Physical Review B, 1977. 15(2): pp. 989–1016.
[13] Mircea, A. and A. Mitonneau, *Study of electron traps in vapor-phase epitaxial GaAs*. Applied Physics, 1975. 8(1): pp. 15–21.
[14] Istratov, A.A., *New correlation procedure for the improvement of resolution of deep level transient spectroscopy of semiconductors*. Journal of Applied Physics, 1997. 82(6): pp. 2965–2968.
[15] Dobaczewski, L., P. Kaczor, I.D. Hawkins, and A.R. Peaker, *Laplace transform deep-level transient spectroscopic studies of defects in semiconductors*. Journal of Applied Physics, 1994. 76(1): pp. 194–198.
[16] Murin, L.I., E.A. Tolkacheva, V.P. Markevich, *et al.*, *The oxygen dimer in Si: Its relationship to the light-induced degradation of Si solar cells?* Applied Physics Letters, 2011. 98(18): p. 3.
[17] Bredemeier, D., D. Walter, S. Herlufsen, and J. Schmidt, *Lifetime degradation and regeneration in multicrystalline silicon under illumination at elevated temperature*. AIP Advances, 2016. 6(3): p. 8.
[18] Gur, E., G. Tabares, A. Arehart, *et al.*, *Deep levels in a-plane, high Mg-content MgxZn1-xO epitaxial layers grown by molecular beam epitaxy*. Journal of Applied Physics, 2012. 112(12): p. 7.
[19] Hupfer, A., C. Bhoodoo, L. Vines, and B.G. Svensson, *Formation and evolution of E3 centers in hydrothermally grown zinc oxide*. Materials Science in Semiconductor Processing, 2017. 69: pp. 13–18.
[20] Bhoodoo, C., L. Vines, E. Monakhov, and B.G. Svensson, *Diffusivity of the double negatively charged mono-vacancy in silicon*. Journal of Physics-Condensed Matter, 2017. 29(20): p. 6.

[21] Quemener, V., L. Vines, E.V. Monakhov, and B.G. Svensson, *Iron related donor-like defect in zinc oxide*. Applied Physics Letters, 2013. 102(23): p. 3.

[22] Hupfer, A., C. Bhoodoo, L. Vines, and B.G. Svensson, *Formation and annihilation of E4 centers in ZnO: Influence of hydrogen*. Journal of Applied Physics, 2016. 119(18): p. 6.

[23] Hupfer, A., C. Bhoodoo, L. Vines, and B.G. Svensson, *The E3 center in zinc oxide: Evidence for involvement of hydrogen*. Applied Physics Letters, 2014. 104(9): p. 4.

[24] Watkins, G.D., *Lattice defects in semiconductors*. Institute of Physics Conference Series, 1974. 23: p. 1.

[25] Watkins, G.D., *Intrinsic defects in silicon*. Materials Science in Semiconductor Processing, 2000. 3(4): pp. 227–235.

[26] Watkins, G.D., *A review of EPR studies in irradiated silicon, in radiation damage in semiconductors*. 1965, Paris: Dunod. pp. 97–111.

[27] Watkins, G.D., *Defects in irradiated silicon – EPR and electron-nuclear double-resonance of interstitial boron*. Physical Review B, 1975. 12(12): pp. 5824–5839.

[28] Watkins, G.D. and J.R. Troxell, *Negative-U properties for point-defects in silicon*. Physical Review Letters, 1980. 44(9): pp. 593–596.

[29] Watkins, G.D. and J.W. Corbett, *Defects in irradiated silicon. 1. Electron Spin resonance of Si-A center*. Physical Review, 1961. 121(4): p. 1001–1014.

[30] Asom, M.T., J.L. Benton, R. Sauer, and L.C. Kimerling, *Interstitial defect reactions in silicon*. Applied Physics Letters, 1987. 51(4): pp. 256–258.

[31] Kimerling, L.C., M.T. Asom, J.L. Benton, and P.J. Drevinsky, *Interstitial defect reactions in silicon*. Materials Science Forum, 1991. 38–41: pp. 141–150.

[32] Svensson, B.G. and J.L. Lindstrom, *Annealing studies of the 862 cm^{-1} infrared band in silicon*. Physica Status Solidi A-Applications and Materials Science, 1986. 95(2): pp. 537–542.

[33] Davies, G., A.S. Oates, R.C. Newman, *et al.*, *Carbon-related radiation-damage centers in Czochralski silicon*. Journal of Physics C-Solid State Physics, 1986. 19(6): pp. 841–855.

[34] Baraff, G.A., E.O. Kane, and M. Schluter, *Silicon vacancy – Possible Anderson negative-U system*. Physical Review Letters, 1979. 43(13): pp. 956–959.

[35] Troxell, J.R. and G.D. Watkins, *Interstitial boron in silicon – A negative-U system*. Physical Review B, 1980. 22(2): pp. 921–931.

[36] Watkins, G.D., J.R. Troxell, and A.P. Chatterjee, *Vacancies and interstitials in silicon*. Institute of Physics Conference Series, 1979. 46: pp. 16–30.

[37] Bemski, G., *Paramagnetic resonance in electron irradiated silicon*. Journal of Applied Physics, 1959. 30(8): pp. 1195–1198.

[38] Watkins, G.D., J.W. Corbett, and R.M. Walker, *Spin resonance in electron irradiated silicon*. Journal of Applied Physics, 1959. 30(8): pp. 1198–1203.

[39] Svensson, B.G., C. Jagadish, A. Hallen, and J. Lalita, *Generation of vacancy-type point defects in single collision cascades during swift-ion bombardment of silicon*. Physical Review B, 1997. 55(16): pp. 10498–10507.

[40] Corbett, J.W., G.D. Watkins, R.M. Chrenko, and R.S. McDonald, *Defects in irradiated silicon. 2. Infrared Absorption of Si-A center.* Physical Review, 1961. 121(4): p. 1015–1022.

[41] Markevich, V.P., A.R. Peaker, S.B. Lastovskii, *et al.*, *Trivacancy and trivacancy-oxygen complexes in silicon: Experiments and ab initio modeling.* Physical Review B, 2009. 80(23): p. 7.

[42] Waite, T.R., *Theoretical treatment of the kinetics of diffusion-limited reactions.* Physical Review, 1957. 107(2): pp. 463–470.

[43] Waite, T.R., *General theory of bimolecular reaction rates in solids and liquids.* Journal of Chemical Physics, 1958. 28(1): pp. 103–106.

[44] Auret, F.D., S.A. Goodman, M. Hayes, *et al.*, *Electrical characterization of 1.8 MeV proton-bombarded ZnO.* Applied Physics Letters, 2001. 79(19): pp. 3074–3076.

[45] Frank, T., G. Pensl, R. Tena-Zaera, *et al.*, *Energetically deep defect centers in vapor-phase grown zinc oxide.* Applied Physics A-Materials Science & Processing, 2007. 88(1): pp. 141–145.

[46] Auret, F.D., S.A. Goodman, M.J. Legodi, W.E. Meyer, and D.C. Look, *Electrical characterization of vapor-phase-grown single-crystal ZnO.* Applied Physics Letters, 2002. 80(8): pp. 1340–1342.

[47] Quemener, V., L. Vines, E.V. Monakhov, B.G. Svensson, *Electronic properties of vacancy related defects in ZnO induced by mechanical polishing.* Applied Physics Letters, 2011. 99(11): p. 3.

[48] Quemener, V., L. Vines, E.V. Monakhov, and B.G. Svensson, *Evolution of deep electronic states in ZnO during heat treatment in oxygen- and zinc-rich ambients.* Applied Physics Letters, 2012. 100(11): p. 4.

[49] Hayes, M., F.D. Auret, P.J.J. van Rensburg, *et al.*, *Electrical characterization of H+ ion irradiated n-ZnO.* Nuclear Instruments & Methods in Physics Research Section B-Beam Interactions with Materials and Atoms, 2007. 257: pp. 311–314.

[50] Scheffler, L., V. Kolkovsky, E.V. Lavrov, and J. Weber, *Deep level transient spectroscopy studies of n-type ZnO single crystals grown by different techniques.* Journal of Physics-Condensed Matter, 2011. 23(33): p. 4.

[51] Frenkel, J., *On pre-breakdown phenomena in insulators and electronic semiconductors.* Physical Review, 1938. 54(8): pp. 647–648.

[52] Jiang, Y.Q., N.C. Giles, and L.E. Halliburton, *Persistent photoinduced changes in charge states of transition-metal donors in hydrothermally grown ZnO crystals.* Journal of Applied Physics, 2007. 101(9): p. 8.

[53] Simpson, J.C. and J.F. Cordaro, *Characterization of deep levels in zinc-oxide.* Journal of Applied Physics, 1988. 63(5): pp. 1781–1783.

[54] Vines, L., J. Wong-Leung, C. Jagadish, *et al.*, *Acceptor-like deep level defects in ion-implanted ZnO.* Applied Physics Letters, 2012. 100(21): p. 4.

[55] Mtangi, W., M. Schmidt, F.D. Auret, *et al.*, *A study of the T2 defect and the emission properties of the E3 deep level in annealed melt grown ZnO single crystals.* Journal of Applied Physics, 2013. 113(12): p. 8.

[56] von Wenckstern, H., H. Schmidt, M. Grundmann, *et al.*, *Defects in hydrothermally grown bulk ZnO*. Applied Physics Letters, 2007. 91(2): p. 3.

[57] Pickenhain, R., M. Schmidt, H. von Wenckstern, *et al.*, *Negative-U properties of the deep level E3 in ZnO*. Physica Status Solidi B-Basic Solid State Physics, 2018. 255(7): p. 16.

[58] Frenzel, H., H. von Wenckstern, A. Weber, *et al.*, *Photocurrent spectroscopy of deep levels in ZnO thin films*. Physical Review B, 2007. 76(3): p. 6.

[59] von Wenckstern, H., G. Biehne, M. Lorenz, *et al.*, *Dependence of trap concentrations in ZnO thin films on annealing conditions*. Journal of the Korean Physical Society, 2008. 53(5): pp. 2861–2863.

[60] Bhoodoo, C., A. Hupfer, L. Vines, E.V. Monakhov, B.G. Svensson, *et al.*, *Evolution kinetics of elementary point defects in ZnO implanted with low fluences of helium at cryogenic temperature*. Physical Review B, 2016. 94(20): p. 11.

[61] Van de Walle, C.G., *Hydrogen as a cause of doping in zinc oxide*. Physical Review Letters, 2000. 85(5): pp. 1012–1015.

[62] Knutsen, K.E., A. Galeckas, A. Zubiaga, *et al.*, *Zinc vacancy and oxygen interstitial in ZnO revealed by sequential annealing and electron irradiation*. Physical Review B, 2012. 86(12): p. 5.

[63] Johansen, K.M., J.S. Christensen, E.V. Monakhov, A.Y. Kuznetsov, B.G. Svensson, *Deuterium diffusion and trapping in hydrothermally grown single crystalline ZnO*. Applied Physics Letters, 2008. 93(15): p. 3.

[64] Wardle, M.G., J.P. Goss, and P.R. Briddon, *First-principles study of the diffusion of hydrogen in ZnO*. Physical Review Letters, 2006. 96(20): p. 4.

[65] Ellguth, M., M. Schmidt, R. Pickenhain, H.V. Wenckstern, M. Grundmann, *Characterization of point defects in ZnO thin films by optical deep level transient spectroscopy*. Physica Status Solidi B-Basic Solid State Physics, 2011. 248(4): pp. 941–949.

[66] Monakhov, E.V., A.Y. Kuznetsov, and B.G. Svensson, *Zinc oxide: Bulk growth, role of hydrogen and Schottky diodes*. Journal of Physics D-Applied Physics, 2009. 42(15): p. 17.

[67] Janotti, A. and C.G. Van de Walle, *Native point defects in ZnO*. Physical Review B, 2007. 76(16): p. 22.

[68] Oba, F., A. Togo, I. Tanaka, J. Paier, G. Kresse, *Defect energetics in ZnO: A hybrid Hartree–Fock density functional study*. Physical Review B, 2008. 77(24): p. 6.

[69] Zhang, S.B., S.H. Wei, and A. Zunger, *Intrinsic n-type versus p-type doping asymmetry and the defect physics of ZnO*. Physical Review B, 2001. 63(7): p. 7.

[70] Janotti, A. and C.G. Van de Walle, *Oxygen vacancies in ZnO*. Applied Physics Letters, 2005. 87(12): p. 3.

[71] Alkauskas, A., *Advanced calculations for defects in materials electronic structure methods*. 2011, Weinheim, Germany: John Wiley and Sons.

[72] Ishii, M., Y. Tanaka, T. Ishikawa, *et al.*, *Site-selective x-ray absorption fine structure analysis of an optically active center in Er-doped semiconductor thin film using x-ray-excited optical luminescence*. Applied Physics Letters, 2001. 78(2): pp. 183–185.

[73] Weber, J., J. Bollmann, S. Knack, *et al.*, Capacitance-transient detection of X-ray absorption fine structure: A possible tool to analyze the structure of deep-level centres?, in Gettering and Defect Engineering in Semiconductor Technology, H. Richter and M. Kittler, Editors. 2004, Zurich-Uetikon: Trans Tech Publications Ltd. pp. 483–488.

[74] Fujioka, H., T. Sekiya, Y. Kuzuoka, *et al.*, *Synchrotron-radiation deep level transient spectroscopy for defect characterization of semiconductors.* Applied Physics Letters, 2004. 85(3): pp. 413–415.

[75] Toth, A.L., L. Dozsa, J. Gyulai, F. Giannazzo, V. Raineri, *SCTS: Scanning capacitance transient spectroscopy.* Materials Science in Semiconductor Processing, 2001. 4(1–3): pp. 89–91.

[76] Lányi, S., *On the lateral resolution of scanning capacitance microscope based C-V measurements.* Ultramicroscopy, 2010. 110(10): pp. 1343–1348.

[77] Dózsa, L., G. Molnar, V. Raineri, *et al.*, *Scanning tip measurement for identification of point defects.* Nanoscale Research Letters, 2011. 6: p. 5.

[78] Chinone, N., R. Kosugi, Y. Tanaka, *et al.*, *Local deep level transient spectroscopy using super-higher-order scanning nonlinear dielectric microscopy.* Microelectronics Reliability, 2016. 64: pp. 566–569.

[79] Brotherton, S.D. and P. Bradley, *Defect production and lifetime control in electron and gamma-irradiated silicon.* Journal of Applied Physics, 1982. 53(8): pp. 5720–5732.

[80] Evwaraye, A.O. and E. Sun, *Electron-irradiation-induced divacancy in lightly doped silicon.* Journal of Applied Physics, 1976. 47(9): pp. 3776–3780.

[81] Svensson, B.G. and M. Willander, *Generation of divacancies in silicon irradiated by 2-MeV electrons – Depth and dose dependence.* Journal of Applied Physics, 1987. 62(7): pp. 2758–2762.

[82] Chantre, A. and L.C. Kimerling, *Configurationally multistable defect in silicon.* Applied Physics Letters, 1986. 48(15): pp. 1000–1002.

[83] Song, L.W. and G.D. Watkins, *EPR identification of the single-acceptor state of interstitial carbon in silicon.* Physical Review B, 1990. 42(9): pp. 5759–5764.

[84] Mooney, P.M., L.J. Cheng, M. Suli, J.D. Gerson, J.W. Corbett, *Defect energy-levels in boron-doped silicon irradiated with 1-MeV electrons.* Physical Review B, 1977. 15(8): pp. 3836–3843.

[85] Schmidt, M., M. Ellguth, C. Czekalla, *et al.*, *Defects in zinc-implanted ZnO thin films.* Journal of Vacuum Science & Technology B, 2009. 27(3): pp. 1597–1600.

[86] Schmidt, M., M. Ellguth, F. Schmidt, *et al.*, *Defects in a nitrogen-implanted ZnO thin film.* Physica Status Solidi B-Basic Solid State Physics, 2010. 247(5): pp. 1220–1226.

Chapter 2

Luminescence from point defects in wide-bandgap, direct-gap semiconductors

Michael A. Reshchikov[1]

2.1 Introduction

Photoluminescence (PL) is a powerful tool for studying point defects (both native defects and impurities) in semiconductors. Between ~1930 and ~1960 the optical studies of color centers, such as *F* centers in alkali halide crystals, were the mainstream of research in this field. Investigation of luminescent properties of phosphors, mostly II–VI semiconductors such as ZnS and CdS, was the second wave of research in the 1940s–1980s, fueled by technological applications in television screens and radars. With the development of III–V semiconductors for solid-state lasers, defects in GaAs and GaP have attracted considerable interest. Finally, wide-bandgap semiconductors have attracted major attention in the last two decades. Among these, GaN is of paramount importance due to its remarkable applications for light emitters. It appears that point defects in GaN are understood better than in any other semiconductor or insulator. This knowledge can serve as a valuable tool for verification and improvement of modern first-principles theories.

In nonpolar semiconductors or semiconductors with indirect bandgaps, such as Si, SiC, GaP, and diamond, the low-temperature PL spectrum contains numerous sharp lines. The lines are often attributed to excitons bound to different defects. Many works, including several review papers, book chapters, and textbooks are devoted to detailed analyses of these PL lines [1–4]. However, in wide-bandgap, direct-gap semiconductors, such as GaN, ZnO, and AlN, defect-related luminescence bands are broad, often having a Gaussian shape. This is a result of strong electron–phonon coupling in materials with pronounced polar character. GaN, with the bandgap of 3.50 eV at low temperature, is an ideal material for analysis of point defects with PL, thanks to high radiative efficiency, high crystalline quality, availability of UV lasers for excitation of PL, and sensitive detectors for its analysis in the range of photon energies of practical interest [5].

Identification and understanding of point defects in *wide-bandgap* semiconductors are hampered by several factors. In particular, several broad PL bands

[1]Department of Physics, Virginia Commonwealth University, Richmond, VA, USA

related to different types of defects contribute to a PL spectrum, and it may be difficult to resolve them. Moreover, potential fluctuations due to nonuniform distribution of charged defects are common in high-resistivity semiconductors, and they strongly affect positions and shapes of PL bands. Finally, efficient radiative defects, even with low concentrations, can cause very strong PL, which may obscure weaker PL signals from defects with higher concentrations but lower carrier-capture cross sections. Numerous examples show that attributions of PL bands based on expectations from growth conditions or from predictions of first-principles calculations are often unreliable and lead to wrong interpretations. All these factors slow down the development of state-of-the-art first-principles calculation approaches that use multiple adjustment parameters and need reliable experimental data.

In this chapter, phenomenological theories of PL are presented and compared with experimental results on PL from wide-bandgap semiconductors, primarily GaN. Types of electron transitions leading to PL are defined in Section 2.2. The rate equations model and the configuration-coordinate (CC) model are presented in Section 2.3. In particular, we will show how to estimate the concentrations of defects and reveal their important characteristics such as the energy levels, carrier capture coefficients, electron–phonon coupling strength, and the presence of excited states. Properties of the yellow luminescence (YL) band in GaN, one of the most debated PL bands, will be explained based on these theories in Section 2.4.1.2. Other examples include analysis of defects in Mg-doped GaN (Section 2.4.1.3) and defects in unintentionally doped GaN (Section 2.4.1.1). A brief review of defect-related PL bands in ZnO and AlN is given in Section 2.4.2. PL from excitons, transition metals, and rare earths is beyond the scope of this chapter.

2.2 Types of transitions related to photoluminescence

The following electron transitions in semiconductors are traditionally considered (Figure 2.1). At low temperatures, numerous excitonic transitions, which cause sharp lines close to the bandgap, are observed in wide-bandgap semiconductors instead of band-to-band transitions. We will refer to them in general as the near-band-edge (NBE) emission (Figure 2.1(a)).

The discrete energy states in the gap, through which recombination takes place, were called activators (radiative defects) and poisoners (nonradiative defects) in early papers on luminescence from phosphors. They are also called recombination centers or traps. Transitions via these states occur in two or three steps. First, a charge carrier (free electron or hole) is captured very quickly by a defect (a nonradiative transition where the energy is released in the form of multiple phonons). In the final step, a charge carrier with the opposite sign is captured by the same defect with the emission of a photon (radiative recombination) or multiple phonons (nonradiative recombination). Sometimes, an additional, intermediate step occurs: between the first capture and the final radiative recombination,

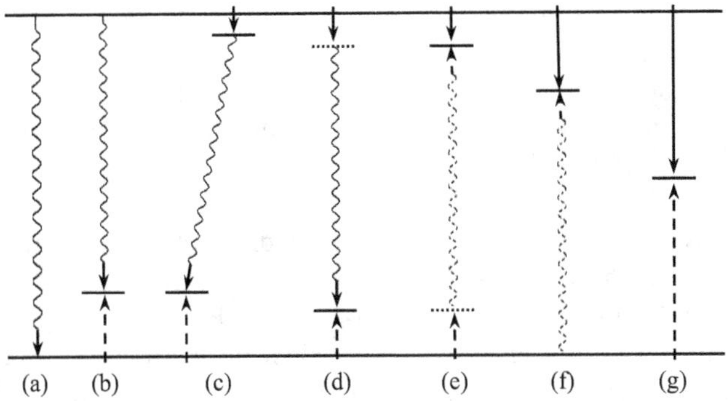

Figure 2.1 *Band diagram with electron and hole transitions: (a) NBE (band-to-band and excitonic), (b) e-A-type, (c) DAP-type, (d) internal transition for a deep donor, (e) internal transition for a deep acceptor, (f) h-D-type, and (g) nonradiative. The wavy and straight lines represent radiative and nonradiative transitions, respectively. The solid and dashed lines represent transitions of electrons and holes, respectively*

a charge carrier with the opposite sign is captured at an excited state of the charged defect. In this case, an electron and hole are bound to the defect, which can be considered a bound exciton. Most common types of transitions involving defect states (Figure 2.1) are explained below.

In n-type semiconductors, transitions from the conduction band to defects (usually acceptors), the so-called e-A transitions (Figure 2.1(b)), are responsible for the majority of defect-related PL bands. In time-resolved PL experiments, the PL signal from such transitions decays exponentially after a laser pulse, and the characteristic PL lifetime is inversely proportional to the concentration of free electrons. At low temperatures, when the concentration of free electrons is very low, these transitions are replaced by transitions from shallow donors to the same defects (acceptors), the so-called donor–acceptor-pair (DAP) transitions (Figure 2.1(c)). These are "diagonal" (tunnel) transitions that occur efficiently because the overlap of the delocalized wave function at shallow donors with that at acceptors is significant. The PL decay is *nonexponential* for the DAP transitions, because transition times for different pairs vary in a wide range due to the random distribution of DAP separations. In order to observe DAP-type transitions involving *deep* donors, very high concentrations of either donors or acceptors are needed, so that the average distance between the donor and acceptor becomes very small (such as for the blue luminescence band in GaN:Mg) [6].

In wide-bandgap semiconductors, especially in semi-insulating ones, transitions from an excited state to the ground state of the same defect are common (similar to what was observed from color centers in alkali halide crystals). We will call these transitions "internal", hereafter. For deep donors, an excited state is close to the conduction band (Figure 2.1(d)), and for deep acceptors, an excited state is close to the valence band (Figure 2.1(e)). The PL signal from such transitions decays *exponentially* after a laser pulse, even at very low temperatures. Thus, time-resolved PL measurements at low temperature allow us to distinguish the DAP and e-A transitions from internal transitions. Note that internal transitions considered in this chapter produce broad PL bands due to strong electron–phonon interaction. This makes them different from intra-center transitions in transition metals where PL lines are narrow [7].

From the symmetry point, one may expect radiative transitions of free holes to donor levels, the h-D-type transitions (Figure 2.1(f)). However, in reality, these transitions are not observed or rarely observed in PL spectra of wide-bandgap semiconductors. In n-type semiconductors, when the Fermi level is close to the conduction band, such transitions cannot compete with the much faster non-radiative capture of holes by acceptors. Thus, the related PL signal is expected to be very weak. In high-resistivity semiconductors, internal transitions (Figure 2.1(d) and (f)) dominate, because the nonradiative capture of free carriers at excited states is very fast. One would expect h-D-type transitions in p-type GaN. However, in order to achieve p-type conductivity, the high concentration of Mg is needed, because its energy level is relatively deep in GaN (0.22 eV). In such samples, almost all transitions occur via the Mg level, as will be shown in Section 2.4.1.3. Even more difficult to achieve p-type conductivity in other wide-bandgap semiconductors such as ZnO or AlN.

Nonradiative recombination is schematically shown with transitions via a deep defect level (Figure 2.1(g)), where the energy is released in the form of multiple phonons. The most efficient nonradiative recombination occurs via defects with levels close to the middle of the band gap or via defects with excited states [8]. Certain point defects, as well as structural defects and the surface states, can contribute to nonradiative recombination. Nonradiative transitions are not detected in PL experiments; however, they reveal themselves when they reduce PL quantum efficiency or cause nonlinear effects, such as population inversion and abrupt thermal quenching of PL in high-resistivity semiconductors [9].

2.3 Phenomenological models

A powerful phenomenological model, which uses rate equations to describe flows of charge carriers in semiconductors, is based on the Shockley–Reed–Hall approach [10,11]. This model allows calculating intensities and characteristic times of PL related to different defects in various experimental conditions, including wide ranges of temperature and excitation intensity. Another useful tool for understanding defects in wide-bandgap, direct-gap semiconductors is the CC model, which accounts for atomic relaxations in the vicinity of a defect. The CC model explains the shapes and positions of the defect-related PL bands. Some phenomena, such as thermal quenching of PL, can be explained by both models (yet by different mechanisms) and lead to different results [12]. Then, it may be difficult to find out which of the mechanisms is adequate in a particular case. Positions and shapes of the PL bands are also affected by potential fluctuations due to nonuniform distribution of impurities and defects in a semiconductor. The role of these effects is especially important in wide-bandgap semiconductors because of low electrical conductivity.

2.3.1 Rate equations model

Rate equations are commonly used to quantify flows of electrons and holes in semiconductors and to explain the behavior of PL. Often, to obtain analytical solutions, simplifying assumptions are made, such as a low concentration of free carriers as compared to the concentrations of trapped carriers. If such assumptions are based on intuition, rather than on experimental facts or exact numeric solutions, this may result in misleading, unjustified conclusions. On the other hand, too complex systems with many fitting parameters lead to ambiguous solutions and untrustworthy conclusions. The simplest and most versatile model will be considered below [9,13].

Three major types of defects should be included to describe adequately the kinetics of PL in GaN and other wide bandgaps: a shallow donor *D*, a radiative acceptor *A*, and the dominant nonradiative recombination center *S* (Figure 2.2). Shallow donors are always present and are responsible for n-type conductivity in undoped GaN and ZnO. Nonradiative defects must be accounted for to explain the low quantum efficiency of PL. Different channels of nonradiative recombination (including point and extended defects) can be usually replaced with an effective center, which is the most efficient in nonradiative recombination. These non-radiative centers quickly capture both electrons and holes, and they usually have very deep energy levels so that thermal emission from these levels to free bands can be ignored. For generality, we consider cases when the *S* center is a donor and when it is an acceptor. At least one radiative defect *A* must be included to explain a particular defect-related PL band. It can be a donor or an acceptor, yet acceptors

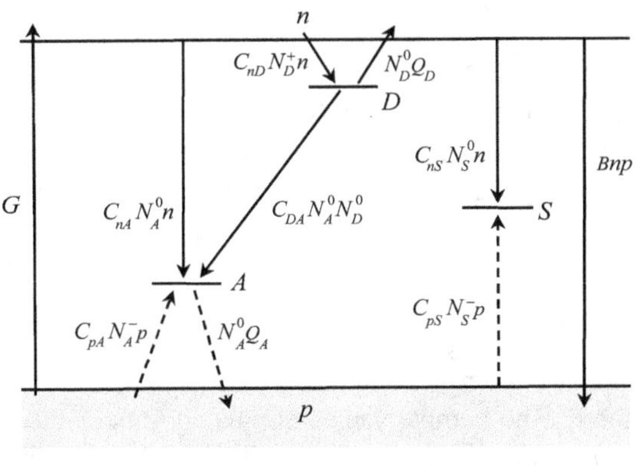

Figure 2.2 Band diagram with three types of defects: shallow donors D, radiative acceptors A, and nonradiative acceptors S. The solid arrows show transitions of electrons, and the dashed arrows show transitions of holes. Model parameters are defined in the text

usually play a more important role in PL from wide-bandgap semiconductors. The model can be easily expanded to include more defects [14]. The exciton annihilation and band-to-band transitions cause the NBE emission, and they are shown as the far-right arrow in Figure 2.2. These transitions can be ignored if their quantum efficiency is low in comparison with the nonradiative recombination.

The rate equations describing flows of electrons and holes can be written down from transitions shown in Figure 2.2 (where A and S centers are both acceptors) as

$$\frac{\partial n}{\partial t} = G - C_{nA}N_A^0 n - C_{nS}N_S^0 n - C_{nD}N_D^+ n + N_D^0 Q_D - Bnp, \tag{2.1}$$

$$\frac{\partial p}{\partial t} = G - C_{pA}N_A^- p - C_{pS}N_S^- p + N_A^0 Q_A - Bnp, \tag{2.2}$$

$$\frac{\partial N_D^0}{\partial t} = C_{nD}N_D^+ n - C_{DA}N_D^0 N_A^0 - N_D^0 Q_D, \tag{2.3}$$

$$\frac{\partial N_S^0}{\partial t} = C_{pS}N_S^- p - C_{nS}N_S^0 n, \tag{2.4}$$

$$\frac{\partial N_A^0}{\partial t} = C_{pA}N_A^- p - C_{nA}N_A^0 n - C_{DA}N_A^0 N_D^0 - N_A^0 Q_A. \tag{2.5}$$

In the case of steady-state PL, all derivatives are equal to zero, and in the case of time-resolved PL, differential equations must be solved. When S center is a deep donor, N_S^0 and N_S^- should be replaced with N_S^+ and N_S^0, respectively. Similarly, N_A^0 and N_A^- should be replaced with N_A^+ and N_A^0, respectively, if the radiative defect A is a donor. The charge neutrality requires that

$$p + N_D^+ = n + N_A^- + N_S^-, \tag{2.6}$$

if S center is a deep acceptor, and

$$p + N_D^+ + N_S^+ = n + N_A^-, \tag{2.7}$$

if S center is a deep donor. In (2.1)–(2.7), G is the electron–hole generation rate (in cm^{-3}/s); C_{nA}, C_{nD}, and C_{nS} are electron-capture coefficients for A, D, and S centers, respectively; C_{pA} and C_{pS} are hole-capture coefficients for A and S centers; C_{DA} is the effective coefficient of DAP recombination; B is the radiative recombination coefficient; n and p are the concentrations of free electrons and holes, respectively; N_A^0, N_A^-, N_A^+, N_D^0, N_D^+, N_S^0, N_S^-, and N_S^+ are the concentrations of A, D, and S centers in different charge states.

In the first approximation, the temperature dependence of the capture coefficients can be ignored. The comparison of simulated temperature dependences of the PL intensity and the PL lifetime with experimental data may help to find which coefficients have strong temperature dependence and which ones are temperature independent. The capture coefficients C_{ij} can be interpreted in terms of the capture cross sections σ_{ij} for defects. For example, $C_{pA} = \sigma_{pA}\langle v_p \rangle$, where σ_{pA} is the hole-capture cross section for defect A, and $\langle v_p \rangle$ is the mean thermal velocity of holes in the valence band. However, coefficients C_{ij} are more fundamental quantities than σ_{ij}, and they can be compared with the values calculated from first principles [15]. The parameters, C_{nA} and C_{DA}, describing radiative recombination, are small, typically in the 10^{-14}–10^{-12} cm^3/s range. We do not expect any significant temperature dependence for the radiative capture coefficients [1].

Vast experimental data suggest that nonradiative capture/recombination coefficients have values in a wide range for different defects: from ~10^{-5}–10^{-4} cm^3/s for attractive centers at very low temperatures (the so-called giant traps), to about 10^{-9}–10^{-8} cm^3/s for neutral centers, and to 10^{-15} cm^3/s for repulsive centers [16–18]. These coefficients correspond to capture cross sections that are close to the atomic cross section for neutral centers, much larger than that for attractive centers, and much smaller than that for repulsive centers. The capture cross section for giant traps decreases with increasing temperature as $\sigma \propto T^{-3}$, because charge carriers are likely to be emitted back to a free band from excited states of the attractive center [16,18]. For repulsive centers, the capture cross section increases with temperature, because free carriers have to overcome the repulsive potential barrier [16,19]. In the case of neutral centers, no significant temperature dependence is expected for σ. However, for defects with strong electron–phonon coupling, independent of their charge, σ may increase exponentially above some characteristic temperature. This temperature behavior is caused by the existence of a potential barrier between adiabatic potentials of the ground and excited states of the center. This effect will be discussed in more detail in Section 2.3.2. Finally, the radiative capture coefficient for NBE transitions, B, can be calculated from the first principles or found from the experiments for a particular semiconductor [20]. Note that at low temperature, the NBE emission is mostly caused by exciton annihilation, and it may not follow the bimolecular law (Bnp), because excitons may form very quickly from photogenerated electron–hole pairs before they separate.

We should bear in mind that many types of defects are always present in semiconductors due to contamination with impurities during growth or due to growth conditions favoring formation of certain native defects. Then, defects that capture photogenerated *minority* carriers with the highest efficiency will contribute more to the PL spectrum. For example, if defect A and defect B have the same concentration in an n-type semiconductor, but the hole-capture coefficient for defect A is orders of magnitude larger than that for defect B, we may not be able to see the PL signal from defect B because it may be obscured by much stronger PL signals from other types of defects or by a low-energy tail from NBE emission.

When the transition energy level for a defect is not too far from a free band, bound carriers may be thermally emitted to the free band (bound holes to the valence band and bound electrons to the conduction band). The probability of this process is described by the thermal emission coefficients having a dimension of inverse time. The thermal emission coefficients for electrons at donors D and for holes at acceptors A can be found from detailed balance as

$$Q_D = C_{nD} N_c g^{-1} \exp(-E_D/kT) \tag{2.8}$$

and

$$Q_A = C_{pA} N_v g^{-1} \exp(-E_A/kT), \quad (2.9)$$

respectively. Here, E_D and E_A are the ionization energies of the donor D and acceptor A, respectively, g is the degeneracy of the donor and acceptor states, N_c and N_v are the effective densities of states in the conduction and valence bands, respectively. Thermal emission of electrons or holes from nonradiative centers can also be included in the model. However, in the examples given below, the thermal emission of bound carriers from S centers will be ignored, in the assumption that the nonradiative defect level is very deep.

The PL intensity for transitions from the conduction band to an acceptor (the e-A transitions) is $I^{e-A} = C_{nA} N_A^0 n$, and that for DAP recombination involving the same acceptor is $I^{DAP} = C_{DA} N_D^0 N_A^0$ (Figure 2.2). For DAP transitions involving shallow donors and deep acceptors, the e-A and DAP transitions are often difficult to resolve because the PL bands from deep acceptors are broad. In this case, the defect-related PL is the sum of e-A and DAP components: $I_A = C_{nA} N_A^0 n + C_{DA} N_D^0 N_A^0$. The PL quantum efficiencies for radiative defect A and for NBE emission are defined as $\eta_A = I_A/G$ and $\eta_{NBE} = Bnp/G$, respectively.

The photon flux, P_0, is defined as the number of photons passing through the unit area of the sample surface per unit time: $P_0 = P_{exc}/\hbar\omega_{exc}$, where P_{exc} is the excitation power density, and $\hbar\omega_{exc}$ is the photon energy of the excitation light. We assume for simplicity that each absorbed photon with energy above the bandgap creates an electron–hole pair, and these nonequilibrium carriers are generated uniformly in a cylinder with a diameter equal to the diameter of the laser beam incident on the sample surface and the height equal to the PL effective depth d_{eff}. For semiconductors with relatively short diffusion length, the latter is customarily defined as $d_{eff} = 1/\alpha$, where α is the absorption coefficient at the laser wavelength. Then, $G = \alpha P_{exc}/\hbar\omega_{exc}$ in this "simple model." For example, for GaN illuminated by a continuous-wave HeCd laser with power density of 1 W/cm^2, $G = 1.6 \times 10^{23}$ cm^{-3}/s in the ~100 nm-thick near-surface region ($\alpha \approx 10^5$ cm^{-1} at $\hbar\omega_{exc} = 3.81$ eV), and $G = 0$ outside of this region. In a more realistic model (advanced model), the intensity of the laser light decreases as $\exp(-\alpha x)$, where x is the distance from the semiconductor surface. Then, $G(x) = \alpha P_{exc} \exp(-\alpha x)/\hbar\omega_{exc}$. This advanced model can be used for numerical solutions of (2.1)–(2.5), while the simple model with constant G is useful for obtaining analytical expressions.

2.3.1.1 Temperature dependence of PL intensity

Examples of dependences of the PL intensity (or PL quantum efficiency) on temperature for n- and p-type semiconductors are shown in Figure 2.3. The model parameters for these cases are given in Table 2.1. More examples can be found in [9,12,14,21].

Usually, at low temperatures, the PL intensity is independent of temperature, and it decreases with temperature above some critical value of T, the process called

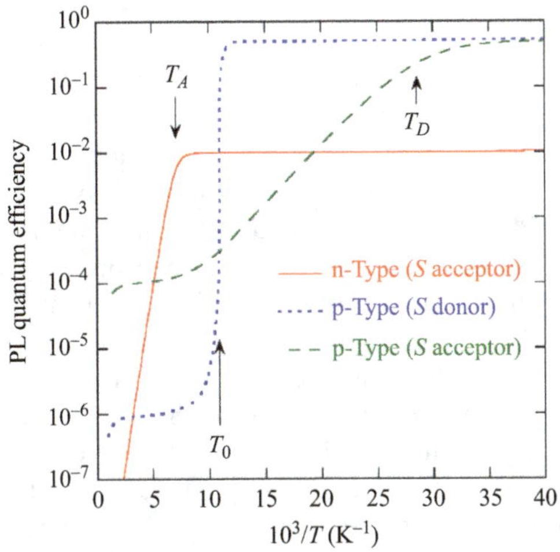

Figure 2.3 Temperature dependences of the defect-related PL quantum efficiency calculated using (2.1)–(2.9) with parameters from Table 2.1 and $G = 10^{20}$ *cm*$^{-3}$*/s*

Table 2.1 Model parameters

Parameter	n-Type (*S* acceptor)	p-Type (*S* donor)	p-Type (*S* acceptor)
C_{nA} (cm^3/s)	10^{-12}	10^{-12}	10^{-12}
C_{DA} (cm^3/s)	10^{-13}	10^{-13}	10^{-13}
C_{nD} (cm^3/s)	10^{-6}	10^{-6}	10^{-6}
C_{nS} (cm^3/s)	10^{-7}	10^{-5}	10^{-7}
C_{pA} (cm^3/s)	10^{-6}	10^{-6}	10^{-6}
C_{pS} (cm^3/s)	10^{-5}	10^{-7}	10^{-5}
B (cm^3/s)	10^{-10}	10^{-10}	10^{-10}
N_A (cm^3/s)	10^{16}	10^{19}	10^{19}
N_D (cm^3/s)	10^{18}	10^{17}	10^{17}
N_S (cm^3/s)	10^{17}	10^{18}	10^{18}
E_D (meV)	30	30	30
E_A (meV)	200	200	200

the thermal quenching of PL. In most cases, the following expression can be used to fit the temperature dependence of the PL quantum efficiency [12]:

$$\eta(T) = \frac{\eta_0}{1 + C\exp(-(E_A/kT))}, \tag{2.10}$$

where η_0 is the PL quantum efficiency in the limit of low temperatures, E_A is an activation energy, and C is a constant. There are different mechanisms of the thermal quenching of PL and therefore different meaning and expressions for the constants, E_A and C [12]. According to the Seitz–Mott mechanism, a defect may convert from a radiative center to a center of nonradiative recombination with increasing temperature (one-center mechanism). This mechanism is typical for color centers in alkali halides, and it can be better explained by using the CC model (Section 2.3.2). More common mechanism of the thermal quenching of PL in semiconductors is the Schön–Klasens mechanism [12]. According to the Schön–Klasens mechanism, PL quenching is caused by thermal emission of charge carriers from defects to the nearest free band and their recapture by different types of defects or recombination channels (multicenter mechanism). Let us consider a conductive n-type GaN as an example.

The Fermi level in n-type GaN is close to the conduction band, because the shallow donor energy is low ($E_D \approx 30$ meV for Si_{Ga} and O_N). At $T > 50$ K, the thermal (equilibrium) concentration of free electrons greatly exceeds the concentration of optically generated electrons ($n_0 > 10^{16}$ cm^{-3} and $n_{opt} \approx 10^{10}$ cm^{-3} at $G = 10^{20}$ cm^{-3}/s for parameters in Table 2.1). Photogenerated holes are quickly captured by radiative acceptors, with the characteristic time $\tau_{pA} = (C_{pA}N_A)^{-1} \approx 10^{-10}$ s, and by nonradiative defects, with the characteristic time $\tau_{pS} = (C_{pS}N_S)^{-1} \approx 10^{-12}$ s. At very low temperatures (below 50 K for GaN), DAP recombination dominates over e-A recombination, and the PL decay is nonexponential. At higher temperatures (but below the thermal quenching of PL), e-A transitions dominate, and the PL intensity decays exponentially after a laser pulse, with a characteristic time constant $\tau_0 = \tau_{nA} = (C_{nA}n)^{-1}$. The quantum efficiency of PL from center A in a conductive n-type semiconductor at low temperature, $\eta_0 = I^{PL}/G = p/(\tau_{pA}G)$, is independent of temperature and is governed by the capture rate of holes:

$$\eta_0 = \frac{C_{pA}N_Ap}{C_{pA}N_Ap + C_{pS}N_Sp} = \frac{\tau_{pS}}{\tau_{pA} + \tau_{pS}}. \qquad (2.11)$$

For parameters from Table 2.1 in the case of an n-type semiconductor, $\eta_0 \approx 0.01$ at $T < 100$ K, and it is independent of temperature. The thermal quenching of PL related to center A begins when the characteristic time of the thermal emission of holes from the A center to the valence band (τ_{therm}) becomes smaller than the PL lifetime (τ_0). It can be obtained from (2.1) to (2.9) that

$$C = (1 - \eta_0)C_{pA}N_v/g \qquad (2.12)$$

and

$$\tau_{therm} = \frac{g\exp(E_A/kT)}{(1 - \eta_0)C_{pA}N_v}, \qquad (2.13)$$

where E_A is the ionization energy of the defect [13,21]. The critical temperature T_A, at which the thermal quenching of defect-related PL begins, can be found from the condition $\tau_0 = \tau_{therm}(T = T_A)$. In a wide range of temperatures, the PL quantum efficiency and its lifetime have the following temperature dependences [22]:

$$\eta(T) = \frac{\eta_0}{1 + \tau_0/\tau_{therm}} \qquad (2.14)$$

and

$$\tau(T) = \frac{\tau_0}{1 + \tau_0/\tau_{therm}}. \tag{2.15}$$

In p-type semiconductors, the Fermi level is close to the shallow acceptor level, not far from the valence band. For GaN with $E_A \approx 0.2$ eV, the thermal (i.e., in dark) concentration of free holes is higher than the concentration of optically generated holes at $T > 100$ K for $G = 10^{20}$ cm^{-3}/s. Under these conditions, recombination channels compete for electrons as minority charge carriers. For some model parameters, thermal quenching can be observed with the activation energy equal to the shallow donor ionization energy (E_D) above a critical temperature, T_D (Figure 2.3). For other parameters, due to nonlinear effects in a semi-insulating semiconductor, unusual behavior of PL such as abrupt quenching at a characteristic temperature, T_0, may be observed (Figure 2.3). The latter will be discussed in more detail in Section 2.3.1.2.

An important feature of PL in a p-type semiconductor is that thermal emission of holes to the valence band leads to an increase of the NBE emission intensity. It is equal to Bnp and increases with temperature with the activation energy close to E_A, because n is temperature independent, while p increases with temperature as $p \propto \exp(-E_A/kT)$. Observation of rising NBE intensity and the independence of temperature for PL from shallow acceptors serve as a confirmation of p-type conductivity when electrical methods are not accessible [21].

2.3.1.2 Abrupt and tunable quenching of PL

Abrupt and tunable quenching of PL was reported in the early literature on PL from phosphors [23] but was fully understood only recently [9]. This type of thermal quenching is expected to be common in wide-bandgap semiconductors, where nonlinear effects and population inversion are possible due to the high resistivity of the material. The abrupt quenching of PL occurs at a critical temperature, T_0 (Figure 2.3). In contrast to the PL quenching described with (2.10), the "activation energy" of the quenching (the slope) has no physical meaning, and the critical temperature can be tuned by the excitation intensity (Figure 2.4).

This phenomenon is explained by saturation of nonradiative centers with photogenerated electrons and the ensuing blockade of this recombination channel at low temperatures. The blockade also results in increased concentration of electrons in the conduction band and inversion of the conductivity from p-type to n-type under illumination. The abrupt transition from high to low PL intensity occurs when the concentration of thermally emitted holes p_0 rises to the level when it exceeds the concentration of optically generated holes p_{opt}. The p_{opt} is proportional to G, and this explains why T_0 shifts to higher temperatures with increasing excitation intensity.

The critical temperature of this drop is

$$T_0 \approx \frac{E_A}{k \ln(B/G)} \tag{2.16}$$

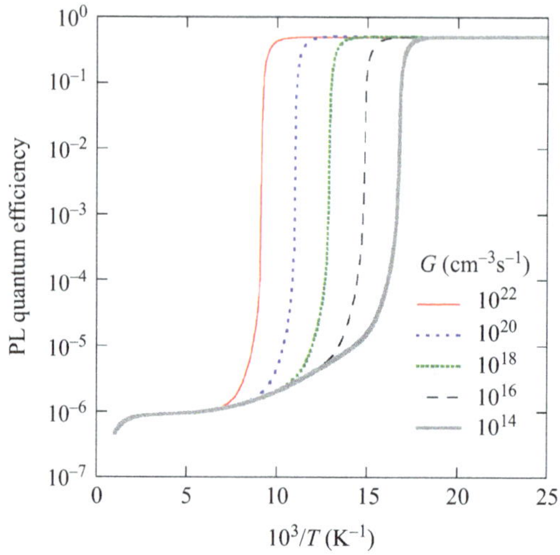

Figure 2.4 Temperature dependences of the defect-related PL quantum efficiency calculated using (2.1)–(2.9) for selected values of G (cm^{-3}s^{-1}). Parameters are from Table 2.1 for p-type with S center being a deep donor

with

$$B = C_{pS}N_S\left(\frac{N_A}{N_D} - 1\right)\frac{N_v}{g} \tag{2.17}$$

when $N_A > N_D$ [9]. Thus, the ionization energy of the acceptor, the thermal emission of holes from which leads to the sudden drop of PL intensity, can be determined from the $T_0(G)$ dependence and its fit with (2.16). The blockade of the dominant nonradiative channel at low temperature occurs due to specific parameters of the system, namely when $(C_{pS}C_{nD})/(C_{pA}C_{nS}) \ll 1$. When S center is acceptor, the abrupt and tunable thermal quenching is also possible; however, the drop of PL intensity is much smaller and may not be noticed [21].

2.3.1.3 PL lifetime

The experimentally measured PL lifetime, τ, strongly depends on the type of transition (Figure 2.1). It is important to measure PL at different time delays after a laser pulse and at different wavelengths (time-resolved PL spectra) to make sure that the PL lifetime corresponds to a particular PL band. Otherwise, the measured signal could be a tail from fast NBE emission, PL from another defect contributing in a wide range of photon energies, or even reflected laser light passing through a monochromator and detected as a parasitic signal. Further, the decay of the PL signal after a laser pulse may not be exponential. In this case, the definition of PL lifetime requires justification, and sometimes it cannot be defined at all. Several examples will be considered below.

At low temperatures, DAP-type transitions (transition (c) in Figure 2.1) are common in semiconductors, especially in n-type, where the transition rate depends on overlap of the delocalized wave function of an electron bound to a shallow donor D with a wave function of a hole bound to an acceptor A. Due to random distribution of DAP separations, slow and fast decays from distant and closer pairs contribute to the PL signal after the excitation has ceased, and the experimentally observed decay of PL is nonexponential, often a power dependence such as $I(t) \propto t^{-n}$ with $n = 1$–2 [24]. Note that DAP-type transitions involving a deep donor will also cause nonexponential decay of PL after an excitation pulse; however, such transitions can be observed only in heavily doped semiconductors where the distance between the components of DAP becomes very small.

With increasing temperature, electrons are thermally emitted from shallow donors to the conduction band. The concentration of free electrons increases, and e-A-type transitions (transition (b) in Figure 2.1) become dominant in an n-type semiconductor with the free-electron concentration n_0. Solutions of differential equations (2.1)–(2.5) show that the decay of PL in this case is exponential: $I(t) = I(0)\exp(-t/\tau)$, where

$$\tau = \tau_0 = \frac{1}{C_{nA} n_0}. \tag{2.18}$$

According to (2.18), PL lifetime decreases with increasing temperature in a nondegenerate semiconductor due to an increase in n_0 if C_{nA} is constant. At high enough temperatures ($T > T_A$), thermal emission of holes from acceptors A to the valence band results in PL quenching, which is described with (2.14). The measured PL lifetime has a similar temperature dependence (2.15), because the holes escape from acceptors A with the characteristic time τ_{therm} faster than the actual PL lifetime τ_0 [22]. Measurement of PL lifetimes as a function of n_0 for PL bands caused by e-A transitions is a good method to find electron-capture coefficients, C_{nA}, and determine their temperature dependences [25]. On the other hand, when the electron-capture coefficient is known for at least one of the observed PL bands, by measuring τ for this PL and using (2.18), we can find n_0. Note that in a wide range of temperatures, both the DAP and e-A transitions may contribute to PL decay, and it may be difficult or impossible to resolve the components. The PL decay may be *almost* exponential, but with a long tail in which instantaneous τ increases with time. In this case, an introduction of the effective lifetime τ^* may be useful. The τ^* is defined as a characteristic time for which the product of PL intensity $I^{PL}(t)$ and time t has a maximum at $t = \tau^*$ [22,26]. With increasing contribution of e-A transitions, τ^* approaches τ_0. The introduction of τ^* is more meaningful than an introduction of several components of τ in an arbitrary way, or an introduction of a stretched exponent with arbitrary parameters.

In a p-type semiconductor doped with a shallow acceptor A, the dominant PL band may be caused by e-A transitions due to the abundance of these acceptors. The electrons are minority carriers in p-type, and the PL decay is governed by the decay of the concentration of photogenerated electrons n, because PL intensity is $I_A = C_{nA} N_A^0 n$, and N_A^0 does not change much. The n decreases with time due to capture by neutral acceptors A with the characteristic time $\tau_{e-A} = \left(C_{nA}N_A^0\right)^{-1}$ and by the dominant nonradiative centers S with the characteristic time $\tau_{e\text{-}S} = (C_{nS}N_S)^{-1}$. Since nonradiative capture is usually much faster than radiative one ($C_{nS} \gg C_{nA}$), the PL decays very fast, with $\tau \approx \tau_{e\text{-}S}$ [21], and the *observed* "PL lifetime" is much shorter than $\tau_{e\text{-}A}$.

In the case of internal transitions (transitions (d) and (e) in Figure 2.1), PL decays exponentially with time after a laser pulse, and the experimentally determined PL lifetime is the longest of nonradiative capture of a free carrier by an excited state and internal transition from the excited state to the ground state [27]. If the excited state behaves as a giant trap due to a large wave function, the electron-capture coefficient for the nonradiative capture decreases with increasing temperature as T^{-3}. Then we may observe a power dependence for the measured PL lifetime: $\tau \propto T^3$ (provided that the internal transition is faster than the electron capture) [27].

We conclude that when the decay of PL after a laser pulse is exponential in a wide range of temperatures, including very low temperatures, the transition is most probably internal. Alternatively, DAP-type transitions may dominate for different radiative acceptors in an n-type semiconductor and exhibit nonexponential (often power) decay of PL at very low temperatures. These transitions are replaced at elevated temperatures (at $T > 50$ K for GaN) with e-A-type transitions involving the same acceptors. It is often impossible to spectrally resolve the DAP and e-A transitions since the ionization energy of the shallow donor is much smaller than the PL band width. However, a conversion of the nonexponential decay to an exponential one with increasing temperature is an indication of the transition from DAP to e-A recombination. In this case, the PL decay becomes nearly exponential at elevated temperatures (DAP transitions may still contribute as a long tail in the decay dependence), and the characteristic PL lifetime is inversely proportional to the concentration of free electrons, see (2.18). This can be verified by either changing temperature in a nondegenerate semiconductor or by measuring PL decay in samples with different n_0, which can be determined from the Hall effect measurements.

2.3.1.4 Determination of the absolute internal quantum efficiency of PL

The absolute internal quantum efficiency of PL, η, is an important parameter, which, in particular, is needed for the determination of the concentration of defects from PL. We define η as the number of photons emitted from a semiconductor volume divided by the number of electron–hole pairs generated in the same volume per unit time, i.e., $\eta = I^{PL}/G$. For example, the absolute internal quantum efficiency of e-A transitions shown in Figure 2.2 is $\eta_{e\text{-}A} = C_{nA}N_A^0 n/G$. Sometimes, η is estimated from external quantum efficiency obtained from the comparison of emitted power and incident power for a sample placed in the integrating sphere. This method may lead to errors, since it is difficult to account for the losses of PL before it exits the sample. Another method, the so-called ABC method [28], may lead to even larger mistakes (sometimes by orders of magnitude) because of unjustified simplifications of rate equations [29].

The most reliable method of finding η is based on the analysis of temperature dependences or excitation intensity dependences of PL intensities. In high-quality crystals, a PL band with a very high η saturates, and efficiencies of other recombination channels simultaneously increase [9,12,13,29–31]. Two examples are shown in Figures 2.5 and 2.6. The Zn-related BL1 band with a maximum at 2.9 eV is very strong in n-type GaN co-doped with Si and Zn due to the high concentration of Zn_{Ga} defects (~10^{18} cm^{-3}) and their large hole-capture coefficient [30]. The BL1 band is quenched at the temperatures above 250 K due to thermal emission of holes from the −/0 transition level of Zn_{Ga} to the valence band. Concurrently with this quenching, intensities of two other PL bands (YL1 and NBE) rise stepwise by a factor *R* (Figure 2.5). It can be shown by using rate equations or just by simple reasoning that the increase of PL intensities is caused by recapture of the released holes by different recombination channels (radiative and nonradiative) in the proportion of their capture efficiencies, because the relative efficiency of a particular recombination channel in an n-type semiconductor is determined by the product of its hole-capture coefficient and the concentration of related defect ($C_{pA}N_A$, $C_{pS}N_S$, and Bn_0 for the example shown in Figure 2.2). The quenching of the BL1 channel

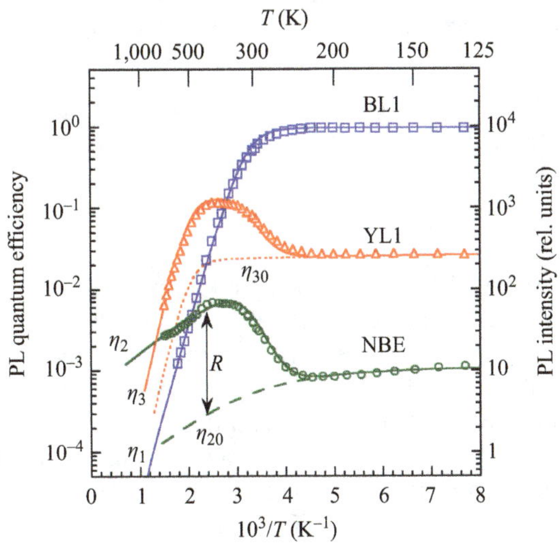

Figure 2.5 Temperature dependence of the absolute internal quantum efficiency of PL in GaN:Si,Zn. Points are experimental data and lines are calculated with the rate equations model. The rise of η_2 (NBE) and η_3 (YL1) by a factor $R \approx 10$ simultaneously with the quenching of the BL1 band is fitted with (2.20) with $\eta_{10} = 0.83$–0.95 for the BL1 band. © 2012 by the American Institute of Physics. Reprinted, with permission, from [30]

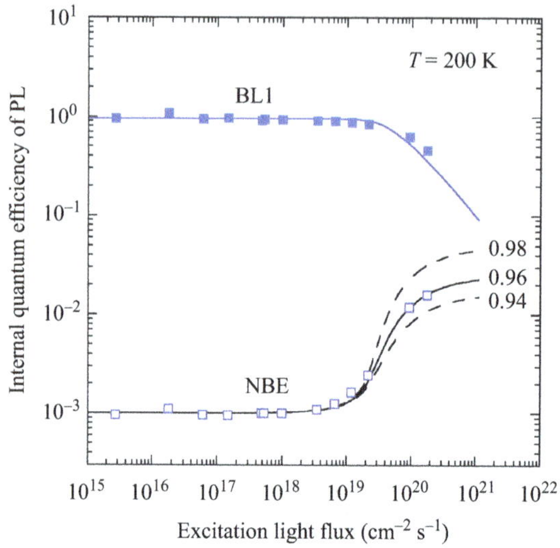

Figure 2.6 Dependence of the absolute internal quantum efficiency of PL on excitation light flux in GaN:Si,Zn. Points are experimental data and lines are calculated using (2.22) (for BL1) and (2.20) (for NBE) with $\eta_{10} = 0.94$*–*0.98 *for the BL1 band. © 2012 by the American Institute of Physics. Reprinted, with permission, from [30]*

with very high quantum efficiency (η_{10} before the quenching) causes rises in efficiencies of all other channels (YL1, NBE, and nonradiative) by a factor

$$R = \frac{1}{1 - \eta_{10}}. \tag{2.19}$$

The value of η_{10} for the BL1 band can be found from (2.19) after R is determined.

Internal quantum efficiencies of other PL bands (YL1 and NBE in the above example) are scaled to that of the BL1 (before the quenching) according to their relative integrated PL intensities. Remarkably, the complex temperature behavior of PL quantum efficiencies ($\eta_i(T)$ with $i = 2$ for the NBE and $i = 3$ for the YL1) can be reproduced by using only one fitting parameter η_{10} if the "quenching-free" temperature dependences of these efficiencies, $\eta_{i0}(T)$, are extrapolated from their low-temperature parts. In Figure 2.5, $\eta_i(T)$ for the NBE and YL1 bands are plotted by using the following expression [30]:

$$\frac{\eta_i(T)}{\eta_{i0}(T)} = \frac{1 - \eta_1(T)}{1 - \eta_{10}}. \tag{2.20}$$

Note that only relative PL intensities, available from experiment, are needed for this fit, because $\eta_1(T) = \eta_{10} I_1(T)/I_{10}$, where $I_1(T)$ and I_{10} are the relative PL intensities of the BL1 band in a wide range of temperatures and in the limit of low temperatures, respectively.

Similarly, a PL band with a very high quantum efficiency can be "quenched" by high excitation intensity. This also results in the redistribution of holes (as minority carriers in n-type) and a stepwise increase of efficiencies of all other recombination channels (radiative and nonradiative) in the proportion of their hole-capture efficiency. An example for the GaN:Si,Zn sample is shown in Figure 2.6. Equations (2.19) and (2.20) are also valid in this case, with T replaced by P_{exc}, and the limit of low temperatures replaced with the limit of low excitation intensities.

After internal quantum efficiency of a very strong PL is found by using the above approach, the quantum efficiencies of all other PL bands can also be found from a comparison of their integrated PL intensities in relative units. Moreover, the samples with the calibrated quantum efficiency can be used as standards to find η in the samples where the steps R cannot be observed due to low quantum efficiency. For this, the samples and the calibrated samples must be measured in identical conditions, and the integrated PL intensities for various PL bands compared. For GaN, different PL bands with a very high η were used as standards, including the YL1 band in GaN:Si,C, the BL1 band in GaN:Si,Zn, and the ultraviolet luminescence (UVL) band in GaN weakly doped with Mg [32]. Consistent results from different samples demonstrate the reliability of the proposed method.

2.3.1.5 Determination of the concentration of radiative defects

The concentration of defects can be determined from dependences of PL intensity on the excitation intensity [13]. At low excitation intensity, the concentration of holes bound to acceptors in an n-type semiconductor is proportional to the excitation light flux P_0. When defects of type i with the concentration N_i are saturated with holes, the related PL intensity saturates. The transition from the linear dependence, $I_i(P_0)$, to the saturation occurs at the characteristic generation rate

$$G_{i0} = \frac{N_i}{\eta_{i0}\tau_i}. \tag{2.21}$$

This equation is obtained by using the simple model, according to which the excitation light is absorbed in the near-surface layer, uniformly up to the depth $d_{eff} = \alpha^{-1}$, where α is the absorption coefficient for the semiconductor at the wavelength of the incident light. In a more realistic model, the light intensity decreases exponentially from the surface, as $P_0 \exp(-\alpha x)$. In this case, the saturation of defects at different depths with holes occurs at different G [33]. The linear dependence $I_i(P_0)$ converts to the logarithmic dependence at $G \approx G_{i0}$ instead of a complete saturation. In a wide range of the excitation intensity, the PL quantum efficiency has the following dependence [29,33]:

$$\eta_i(P_0) = \frac{N_i}{\alpha\tau_i P_0} \ln\left(1 + \frac{\alpha\tau_i\eta_{i0}P_0}{N_i}\right), \tag{2.22}$$

where η_{i0} is the absolute internal quantum efficiency of the ith PL band in the limit of low excitation intensities. PL lifetime τ_i can be found from time-resolved PL experiments, and the value of η_{i0} can be found as explained in Section 2.3.1.4.

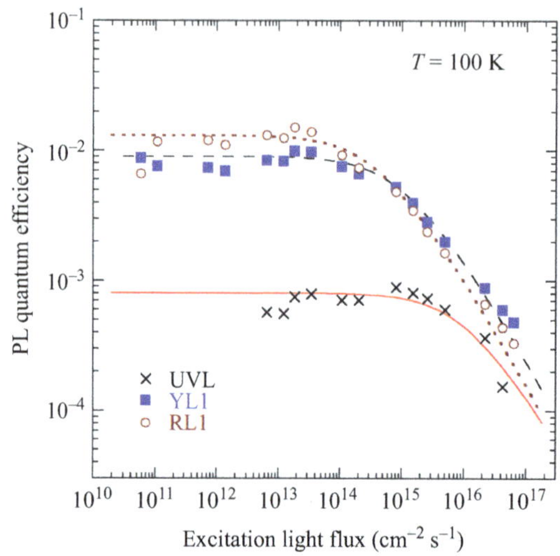

Figure 2.7 Dependence of PL quantum efficiency on excitation light flux for undoped GaN with a low concentration of defects. The lines are calculated using (2.22). $N_i = 7 \times 10^{12}$ cm^{-3} for the UVL band, and $N_i = 2.4 \times 10^{14}$ cm^{-3} for the YL1 and RL1 bands. Reprinted, with permission, from [34]

Figure 2.7 compares the experimental data for three PL bands in undoped GaN with the calculated dependences using (2.22) [34]. The only fitting parameter is N_i.

Relative concentrations of radiative point defects can also be found from a comparison of integrated PL intensities for different defects when the hole-capture coefficients for the defects are known. Indeed, it follows from (2.2) for an n-type semiconductor at low temperature ($Q_A = 0$) that [13,29,34]

$$\frac{N_i}{N_j} = \frac{C_{pj}\, I_i^{PL}}{C_{pi}\, I_j^{PL}}, \tag{2.23}$$

where C_{pi} and C_{pj} are the hole-capture coefficients for defects i and j. These coefficients can be found from the analysis of the thermal quenching of PL [13].

The concentration of radiative defects can also be estimated from time-resolved PL experiments [22]. At low excitation intensity, the PL intensity after a laser pulse is proportional to the excitation light flux P_0. At high excitation intensity, the PL intensity from the defects of type i flattens out when the defects are saturated with photogenerated holes. The critical generation rate for this transition is [22]

$$G_{i0} = \frac{N_i}{\eta_{i0} t_L}, \tag{2.24}$$

where t_L is the laser pulse length.

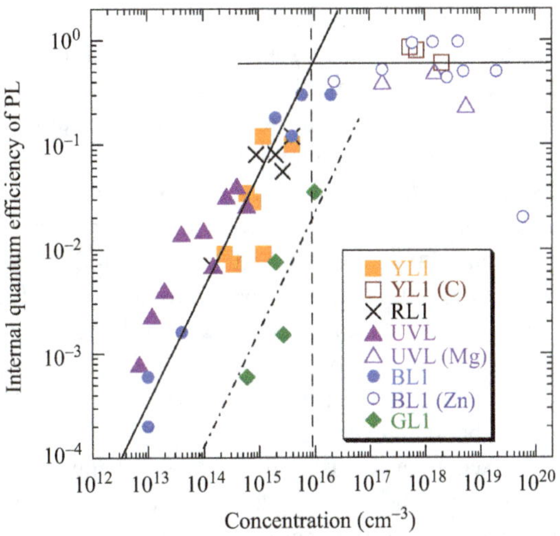

Figure 2.8 Dependence of PL quantum efficiency for defect-related PL bands in GaN on the concentration of related defects. The data are from undoped and doped (the impurity in parentheses) GaN. The solid lines are guides for eyes, which show a linear dependence and saturation. The dashed vertical line indicates the concentration of defects above which the PL intensity from dominant defects has no linear relationship with the defect concentration. The data for the GL1 band follow linear dependence (dash-dotted line) but lie well below the data for the dominant PL bands (RL1, YL1, BL1, and UVL), because the related defect has much lower hole-capture cross section. Reprinted, with permission, from [34]

The comparison of the internal quantum efficiency of PL and the concentration of related defects (Figure 2.8) shows that, for the majority of defect-related PL bands in a large set of GaN samples, η_i increases linearly with N_i up to $N_i \approx 10^{16}$ cm^{-3}, where it approaches unity. This is because the hole-capture coefficient is very high for defects contributing to the PL spectrum in n-type GaN: $C_{pA} = 10^{-7}$–10^{-6} cm^3/s for Mg_{Ga} (UVL band), Zn_{Ga} (BL1 band), C_N (YL1 band), and unknown defect (RL1 band). All these defects are acceptors and act as attractive centers for holes [34]. On the other hand, the green luminescence (GL1) band is attributed to the donor state of unknown defect [34]. It captures holes less efficiently, and a higher concentration is needed to observe the GL1 band.

2.3.2 *Configuration-coordinate model*

In early works on PL from luminescent solids (called phosphors), one-dimensional CC diagrams were used for qualitative interpretation of the experimentally observed absorption and emission spectra from alkali halides (such as KCl, NaCl,

KBr, and NaI) and II–VI semiconductors (such as ZnS, CdS, and ZnO) [35,36]. The quantum mechanical theory of the shapes and positions of broad PL and absorption bands in crystals originates with the works of Huang and Rhys [37] and Pekar [38]. In spite of its simplicity, the theory was successful in quantitative interpretation of many experimental results [17,39]. The empirical fitting parameters obtained from the quantitative description of experimental PL bands with the CC model can be compared with the parameters calculated from the first principles [40].

An example of a one-dimensional CC diagram is shown in Figure 2.9. The parabolas are adiabatic potentials that describe the total energy of the defect system containing a defect and surrounding atoms in a crystal lattice (the energy of atoms plus the energy of electrons). The ground state (potential 1) corresponds to a negatively-charged deep acceptor A in an n-type semiconductor with the Fermi level close to the conduction band. The energy of the system varies in one-dimensional space as $U_g = 0.5K_gQ^2$, where the coefficient K_g is the "spring" constant, and Q is the generalized coordinate [39]. The coefficient, K_g, can be presented as $K_g = m_g\Omega_g^2$, where m_g is the effective mass of the oscillating system involving the defect in the ground state, and Ω_g is the effective phonon frequency in this state.

After a photon with energy above the bandgap is absorbed, a hole and an electron appear in the valence and conduction bands, respectively, and the energy of the system increases by E_g (potential 2). Then a hole is captured by the A^- defect,

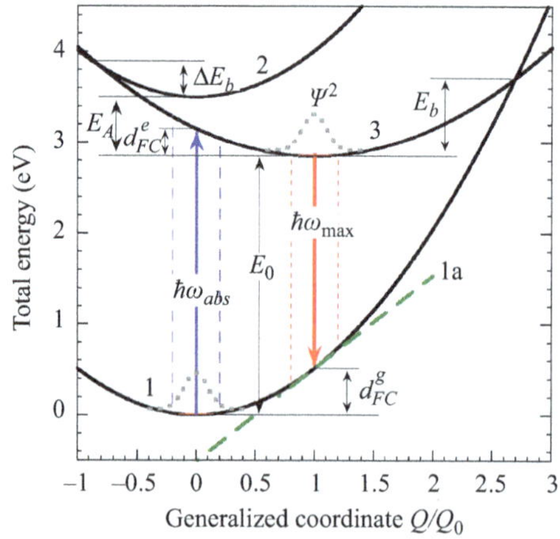

Figure 2.9 One-dimensional configuration-coordinate diagram for a deep acceptor A. The potential 1 represents the total energy of the system in the ground state A^-. The potential 2 is the same system plus a free electron in the conduction band and a free hole in the valence band. The potential 3 is the system energy after the acceptor captures a free hole and becomes A^0. The shift of the excited state minimum relative to the ground state minimum by $Q = Q_0$ describes the lattice relaxation. After the emission of a photon, the system relaxes to the minimum of the ground state by emitting multiple phonons

and the acceptor becomes neutral. The bound hole causes the distortion of the atomic arrangement, and the system relaxes to its equilibrium minimum at $Q = Q_0$ in potential 3 by emitting several phonons. The relaxation is fast, on a timescale of picoseconds [17,35]. The capture of a hole by the acceptor is thermally activated if a barrier ΔE_b exists between parabolas 2 and 3 as shown in Figure 2.9. E_A is the energy distance between the −/0 transition level of the acceptor and the valence band. The energy of the system in the excited state (potential 3) is $K_e = E_0 + 0.5m_e\Omega_e^2(Q - Q_0)^2$, where E_0 is the zero-phonon line (ZPL) energy ($E_0 = E_g - E_A$), Ω_e is the effective phonon frequency in the excited state, and m_e is not necessarily equal to m_g.

At low temperatures, the system remains at the zero vibrational level of the excited state during the characteristic time, which is called the PL lifetime (see Section 2.3.1.3), often on a timescale of microseconds. The optical transition is shown with a vertical downward arrow, because massive ions do not change their positions in a short time of transition (the Franck–Condon approximation). The PL band has a maximum at $\hbar\omega_{\max}$. The width of the band is determined by the size of the wave function Ψ in the excited state and the slope of the ground state potential near Q_0. The PL band shape is asymmetric if the slope changes significantly within the length of Ψ, and it is Gaussian if the slope in this region is constant (dashed line 1a). After the optical transition, the system relaxes to the minimum of the ground state by emitting several phonons. The Franck–Condon shifts in the ground (d_{FC}^g) and excited (d_{FC}^e) states (Figure 2.9) are defined as: $d_{FC}^g = 0.5K_gQ_0^2$ and $d_{FC}^e = 0.5K_eQ_0^2$. The dimensionless factors $S_g = d_{FC}^g/\hbar\Omega_g$ and $S_e = d_{FC}^e/\hbar\Omega_e$ are called the Huang–Rhys factors. In semiclassical approximation, the shape of a defect-related PL band is [41]

$$I^{PL}(\hbar\omega) = I_{\max}^{PL}\exp\left[-2S_e\left(\sqrt{\frac{E_0 + 0.5\hbar\Omega_e - \hbar\omega}{E_0 + 0.5\hbar\Omega_e - \hbar\omega_{\max}}} - 1\right)^2\right]. \tag{2.25}$$

In the case of very strong electron–phonon coupling (large S_e, S_g), the curvature of the parabola decreases and approaches a straight line, and the potential 1 near the point $Q = Q_0$ can be replaced with a tangent (line 1a in Figure 2.9). Then, the spectral dependence of the PL intensity becomes Gaussian:

$$I^{PL}(\hbar\omega) = I_{\max}^{PL}\exp\left(-\frac{(\hbar\omega - \hbar\omega_{\max})^2}{2\sigma^2}\right), \tag{2.26}$$

with $\sigma = S_g\hbar\Omega_g S_e^{-1/2}$. The full width at half maximum (FWHM) of the PL band at zero temperature is $W(0) = 2\sigma(2\ \ln 2)^{1/2}$. With increasing temperature, the band width increases as [17,42]

$$W(T) = W(0)\sqrt{\coth\left(\frac{\hbar\Omega_e}{2kT}\right)}. \tag{2.27}$$

Equation (2.25) reproduces the asymmetric shape of the PL bands with intermediate electron–phonon coupling strength (S_e = 3–7) and nearly Gaussian shape for $S_e > 10$. Since $0.5\hbar\Omega_e$ can be found from the $W(T)$ dependence (it can even be ignored due to its relatively small value), and the parameters of the PL band maximum ($\hbar\omega_{\max}$ and $I_{\max}$) are determined directly from the experiment, the only fitting parameters in (2.25), which are needed to describe the PL band shape, are S_e and E_0. Note that, due to simplifications used in the derivation of (2.25), S_e and $\hbar\Omega_e$ should be considered the effective Huang–Rhys factor and effective phonon mode. In the case of several phonon modes ($i = 1, 2, \ldots$), several sets of S_{ei} and $\hbar\Omega_{ei}$ can be found, as will be shown below.

The above semiclassical approach explains the shape of a defect-related PL band. However, it ignores the fine structure, which is sometimes observed in experiment. At low temperatures, the intensities of emission lines related to optical transitions from the zero vibrational level of the excited state to the n vibrational level ($n = 0, 1, 2, \ldots$) of the ground state are determined by the Poisson distribution,

$$I_{0\to n} = I_0 \frac{S^n}{n!}, \tag{2.28}$$

where I_0 is the intensity of the ZPL line. If several vibrational modes with phonon energies $\hbar\Omega_1, \hbar\Omega_2, \hbar\Omega_3, \ldots$ are coupled to the electron transition with strengths $S_1, S_2, S_3, \ldots$, respectively, intensities of transitions with the emission of $k + l + m \ldots$ phonons of different modes can be found as [39]

$$I_{0\to klm\ldots} = I_0 \frac{S_1^k}{k!} \frac{S_2^l}{l!} \frac{S_3^m}{m!} \cdots. \tag{2.29}$$

The phonon replicas corresponding to the emission of $k + l + m \ldots$ phonons create peaks in the PL spectrum that are shifted to lower energies from the ZPL by distances $k\hbar\Omega_1 + l\hbar\Omega_2 + m\hbar\Omega_3 \ldots$. The observation of the ZPL in the low-temperature PL spectrum allows finding the value of E_A with high accuracy. Note that both the shape of a broad PL band and phonon-related fine structure can be explained with the CC model irrespective of the transition type (i.e., e-A, DAP, or internal transition). The effect of the transition type can be noticed in photo-excitation spectra as will be shown below.

In PL excitation (PLE) experiments, a photon with an energy below the bandgap excites an electron from the acceptor to the conduction band (vertical upward arrow in Figure 2.9), after which the system relaxes to the minimum of potential 3. Some defects in wide-bandgap semiconductors and insulators have excited states below the conduction band, such as F center in alkali halides [17]. Resonant excitation of the bound electron from the ground state to the excited state (what we call the internal transition) produces an excitation band in the PLE (or absorption) spectrum with a maximum at $\hbar\omega_{abs}$ and Gaussian or asymmetric shape, often mirroring the PL band shape and sometimes showing the ZPL [39].

When an electron is excited to the conduction band (the external transition), the maximum of the PLE band shifts to higher energies from $\hbar\omega_{abs}$ due to the continuous spectrum of the conduction band. The shape of the PLE spectrum, or defect-related absorption spectrum, in this case is given by [43]

$$\sigma(\hbar\omega) \propto \int_0^{\infty} \frac{\exp\left[-(E+\hbar\omega_{abs}-\hbar\omega)^2/\Gamma^2\right]\sqrt{E}}{\hbar\omega(E+\hbar\omega_{abs})^2}\,dE, \tag{2.30}$$

where in the limit of low temperatures,

$$\Gamma = \frac{\Omega_g}{\Omega_e}\sqrt{2d_{FC}^e\hbar\Omega_g}. \tag{2.31}$$

It is often difficult or impossible to measure the PLE spectrum of a defect, because it may be located close or even above the bandgap. In old literature, due to insufficient experimental data and by analogy with F centers in alkali halides [17], it was often *assumed* that parabola 3 in Figure 2.9 corresponds to a center in which an electron is in an excited state located below the conduction band (and not in the conduction band) and that optical transition is internal, within the localized center [44–47]. The PLE spectra were simulated with a Gaussian curve or with a curve symmetric to the PL band. Such a view persisted long after and led to errors in parameters of the CC model that can be compared with the first-principles calculations (e.g., the energy of the PLE band maximum). Another widespread mistake is an assumption that the thermal quenching of a defect-related PL band is always caused by a conversion of a center from radiative to nonradiative by a transition of the system over barrier E_b in the excited state (Figure 2.9). These issues will be discussed in Section 2.4.1.2 with the YL1 band as an example.

The one-dimensional CC diagram is very useful for explaining shapes of PL bands related to deep-level defects. Equation (2.25) can be used to fit symmetric and asymmetric shapes of broad PL bands and extract the characteristic parameters. It is important to remember that the theoretical lineshape presents the number of emitted photons as a function of photon energy. However, PL spectra obtained by using grating monochromators, after correction for the spectral response of the measurement system, represent emitted power as a function of wavelength, λ. To plot the PL spectra in units proportional to the number of emitted photons as a function of photon energy, the original PL spectrum (in mW/nm) must be multiplied by λ^3. This correction changes the shape of a broad PL band and shifts its maximum. An example is given in Section 2.4.1.2 (Figure 2.14).

2.3.3 Diagonal transitions between localized states

Shifts of PL bands with excitation intensity are often explained with the DAP model or with a model accounting for potential fluctuations in a semiconductor with a random distribution of charged impurities [48–51]. Similar effects may be observed when PL originates from the near-surface depletion region or from regions where structural defects cause local electric fields. It may be difficult to distinguish these cases, because they have a similar origin (spatial localization of charge carriers and tunnel transitions) and similar observed effects: a PL band shifts to lower energies with decreasing excitation intensity or with a time delay after a laser pulse.

2.3.3.1 Donor–acceptor pairs

For distant DAPs with separation r, the energy of emitted photons is [49]

$$\hbar\omega(r) = E_g - E_A - E_D + E_C(r), \tag{2.32}$$

where E_g is the bandgap energy, E_A and E_D are the ionization energies of the acceptor and the donor, respectively. $E_C(r) = K/r = e^2(4\pi\varepsilon_0\varepsilon r)^{-1}$, where e is the electron charge, ε_0 is the permittivity of vacuum, and ε is the low-frequency dielectric constant, which describes the Coulomb interaction between the donor and acceptor ions with separation r. The DAP emission maximum, $\hbar\omega_{\max}$, corresponds to the effective Coulomb interaction for a DAP with effective separation $r = r_{eff}$. The emission peak shifts to higher energies with increasing excitation intensity, because emission from more distant pairs saturates, and the contribution from closer pairs with stronger Coulomb interaction increases [48,49]. In the limit of infinite separations between the donor and acceptor, $E_C(\infty) = 0$ and $\hbar\omega_\infty = E_g - E_D - E_A$. Zacks and Halperin [52] proposed a phenomenological model, in which $\hbar\omega_{\max}$, as a function of the excitation intensity P_{exc}, can be found from the following expression:

$$P_{exc} = D\frac{E_C{}^3}{E_B - E_C}\exp\left(-\frac{4E_B}{E_C}\right), \tag{2.33}$$

where D is a constant, $E_C = \hbar\omega_{\max} - \hbar\omega_\infty$, and $E_B = K/R_B$ is the Coulomb interaction energy for a pair with a separation equal to the Bohr radius, R_B, of a shallower component in the effective-mass approximation. An example of the dependence calculated with (2.33) is shown in Figure 2.10. The total shift of the DAP-related PL band in the very wide range of P_{exc} is not expected to exceed $0.5E_B$ [52]. For example, the UVL, BL1, and YL1 bands in GaN (with $E_B \approx 60$ meV) blue-shift by only 6–8 meV when P_{exc} increases from 10^{-5} to 0.3 W/cm^2 [53,54].

In time-resolved PL measurements, a DAP-related PL band red-shifts with time delay. This happens because electrons and holes at closer pairs with stronger Coulomb interaction recombine first, while those at more distant pairs recombine at longer time delays after a laser pulse [24,55]. The magnitude of the shift is similar to that due to excitation intensity.

2.3.3.2 Effect of potential fluctuations

Large shifts of PL bands can be explained in a model accounting for potential fluctuations. The potential fluctuations in semi-insulating or heavily doped semiconductors originate from nonuniform distribution of charged donors and acceptors [50,51]. For transitions from the conduction band (or shallow donors) to a defect A, the PL band maximum depends on the potential fluctuation amplitude γ as [5,50,51]

$$\hbar\omega_m = E_g - E_A - 2\gamma, \tag{2.34}$$

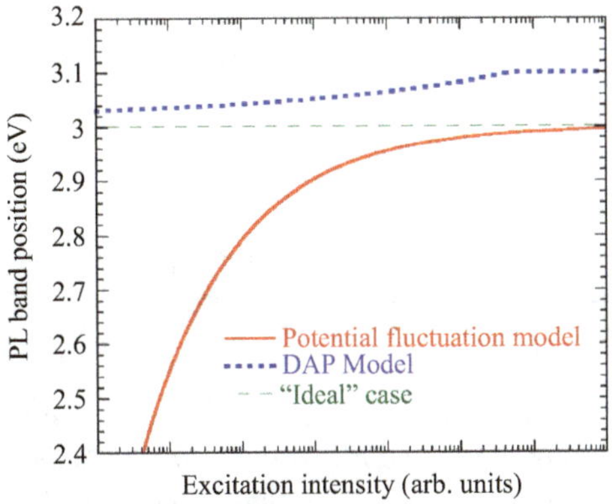

Figure 2.10 *Effect of excitation intensity on the position of a PL band caused by transitions from a donor with $E_D = 0.3$ eV to an acceptor with $E_A = 0.2$ eV in a semiconductor with $E_g = 3.5$ eV. The shift due to Coulomb interaction in DAP is shown with dotted line. The shift due to potential fluctuations is shown with the solid line. The dashed line shows the PL band position ignoring the Coulomb interaction and potential fluctuations*

where $\gamma = KN_A^{2/3}n^{-1/3}$, E_A is the defect ionization energy, and n is the concentration of free electrons or holes (whichever is higher). At very low excitation intensity and low temperature, the potential fluctuations may be very large in a wide-bandgap semiconductor ($\gamma < 0.5E_g$). With increasing excitation intensity and/or temperature, n increases, and the magnitude of γ decreases due to screening.

Although PL bands red-shift with decreasing excitation intensity and with a time delay after a laser pulse both in the case of DAP transitions and in the case of potential fluctuations, there are important differences between these two cases. In the first case, a PL band maximum is at a higher energy as compared to an ideal crystal with an infinitely low concentration of defects, and the shift is relatively small. In the second case, PL bands red-shift as compared to an ideal crystal with a flat potential (Figure 2.10). The shift due to potential fluctuations may be very large in wide-bandgap semiconductors, because they have a low concentration of free carriers, especially at low temperature.

Interestingly, in the samples exhibiting abrupt and tunable quenching of PL (Section 2.3.1.2), the shift may abruptly increase with increasing temperature above critical temperature T_0 (Figure 2.11) [21]. In this case, accumulation of electrons is expected in the conduction band at low temperatures ($T < T_0$) [9]. These free electrons partially screen potential fluctuations. Additionally, deep donors become saturated with electrons, and acceptors become partially saturated with holes, even at very low excitation intensity, the so-called population inversion [9]. As a result, significantly fewer defects remain charged, which removes the origin of potential fluctuations. Above the critical temperature ($T > T_0$), the concentration of electrons

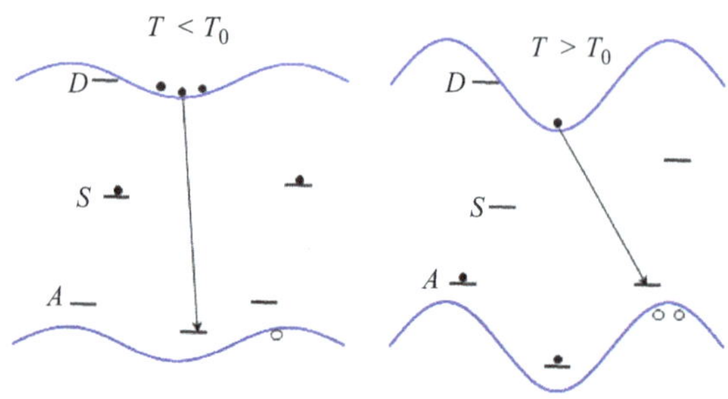

Figure 2.11 Schematics of a semiconductor bandgap in the presence of potential fluctuations. Radiative transitions causing defect-related PL are shown with arrows. © 2018 by the American Physical Society. Reprinted, with permission, from [21]

in the conduction band abruptly drops, nonradiative donors lose bound electrons, and shallow acceptors lose bound holes, both becoming charged. These processes cause potential fluctuations to rise in a narrow temperature range near T_0 [21]. The rise (and the associated shift of the PL bands) is especially large at very low excitation intensities when the material becomes nearly insulating at $T > T_0$.

Note that other sources of electric fields in a semiconductor may case the shifts similar to these due to potential fluctuations. For example, in a p-type semiconductor, there is a downward band bending near the surface (about 1–2 eV for GaN). Then, recombination of photogenerated electrons (free or bound to shallow donors at the surface) with holes bound to acceptors in the depletion region, or even beyond it, would cause diagonal transitions with reduced photon energy. With increasing excitation intensity, the band bending decreases and the defect-related band blue-shifts [21].

2.3.4 Summary

Phenomenological models are very useful for explanation of the behavior of PL from defects in wide-bandgap semiconductors. Several important parameters of defects or semiconductors can be obtained, such as electron- and hole-capture coefficients, concentrations of defects that are often below the secondary-ion mass-spectrometry (SIMS) detection limit, accurate positions of optical and thermodynamic transition levels, charge states, strength of the electron–phonon coupling, and types of transitions. Some of the parameters are useful for material growers and device engineers, others help to improve approaches and adjustments used in the first-principles calculations. By comparing the PL intensity with the concentration of the related defects (Figure 2.8), we find that unintentionally introduced impurities (such as Mg, Zn, and C in GaN with the concentrations below the SIMS detection limit) can cause very strong PL bands that dominate in a PL spectrum. This finding may cast doubt on certain previously made attributions of PL signals to intentionally introduced impurities or to native defects in semiconductors.

2.4 Experiment: photoluminescence from defects

In this section, the experimental data on PL from defects in GaN will be analyzed in more detail, because they are understood much better than defects in other wide-bandgap, direct-gap semiconductors. A few examples of defect-related PL in ZnO and AlN will be given in Section 2.4.2.

2.4.1 Radiative defects in GaN

The main PL bands observed in GaN and the parameters of related defects are listed in Table 2.2. Some of the PL bands appear only in GaN doped with corresponding impurities. They are labeled according to the luminescence band color with an index indicating the impurity (e.g., the blue luminescence due to Cd_{Ga} is labeled BL_{Cd}). Other PL bands appear in undoped GaN, and they may be related either to native defects or to unintentionally introduced impurities. They are labeled with the color and number if more than one PL band from the different types of defects appears in the same spectral region. The ZPL is shown for PL bands where the ZPL and phonon-related fine structure were observed. The parameters $E_0^* = E_0 + 0.5\hbar\Omega_e$ and S_e are found from the fitting of PL band shapes and positions with (2.25). The $\hbar\Omega_e$ is found from the fit of the temperature dependence of the PL band width with (2.27). The PL bands in undoped GaN will be reviewed in Section 2.4.1.1. The two most important examples of defects in n- and p-type GaN are given in Sections 2.4.1.2 and 2.4.1.3.

2.4.1.1 Defect-related PL bands in undoped GaN

In undoped GaN, several PL bands can be observed (Figure 2.12). The UVL band with the first peak at about 3.28 eV, followed by several longitudinal optical (LO) phonon replicas, is commonly observed in undoped GaN grown by different techniques. It is caused by electron transitions from shallow donors (at low temperatures) or from the conduction band (at $T > 40$ K) to the shallow Mg_{Ga} acceptor, which has the $-/0$ transition level at 223 ± 2 meV above the valence band [21]. The UVL band is relatively strong in undoped GaN grown by all major techniques, because the Mg_{Ga} acceptor has a very high hole-capture coefficient, so that even minute contamination of the Ga source with Mg or the "memory effect" of a growth system results in the appearance of this PL band. GaN samples grown by metal-organic chemical vapor deposition (MOCVD) or by hydride vapor phase epitaxy (HVPE) are often contaminated with Zn. In these samples, the BL1 band with a maximum at 2.85 eV is strong, because the Zn_{Ga} acceptor has a high hole-capture coefficient.

In conductive n-type GaN grown by MOCVD, the YL1 band with a maximum at ~2.2 eV is the major PL band due to the contamination of these samples with carbon. In n-type GaN grown by HVPE, the YL1 band is less common (which agrees with low concentration of carbon), and two other PL bands are observed: the RL1 band with a maximum at 1.7 eV is the dominant defect-related PL band at low excitation intensity, and the GL1 band with a maximum at 2.35 eV emerges at high excitation intensity. The RL1 band is caused by an unidentified acceptor (with the

Table 2.2 Parameters of radiative defects in GaN

PL band	Defect	$\hbar\omega_{max}$ (eV)	E_A (eV)	ZPL (eV)	E_0^* (eV)	S_e	$\hbar\Omega_e$ (meV)	d_{FC}^g (eV)	C_{nA} (cm^3/s)	C_{pA} (cm^3/s)	Reference
RL1		1.70	1.2		~2.3	9.5		~0.6	4×10^{-13}	3×10^{-7}	
RL2		1.76			~2.4	~20					
RL3		1.77	~0.2								
YL1	C_N	2.17	0.916	2.59	2.66	7.8	56	0.44	1×10^{-13}	4×10^{-7}	
YL3		2.07	1.130	2.38	2.41	6.2		0.31			
GL1		2.35	~0.5		2.97	10.3	42	0.6		5×10^{-8}	
GL2	V_N	2.33	~0.4		2.85	26.5	23	0.5			
BL1	Zn_{Ga}	2.86	0.400	3.10	3.14	3.2		0.24	7×10^{-13}	5×10^{-7}	
BL2	C_NH_i	2.96	0.15	3.33	3.35	3.7		0.37			
UVL	Mg_{Ga}	3.28	0.223	3.29		0.4	91	~0.1	3×10^{-12}	1×10^{-6}	
UVL_{Be}	Be_{Ga}	3.38	0.11								[56]
GL_{Ca}	Ca_{Ga}	2.5									[57]
BL_{Cd}	Cd_{Ga}	2.72	~0.53	2.937							[58]
BL_P	P_N	2.98	~0.29	3.20							[59]
BL_{As}	As_N	2.68	~0.54	2.95							[59]

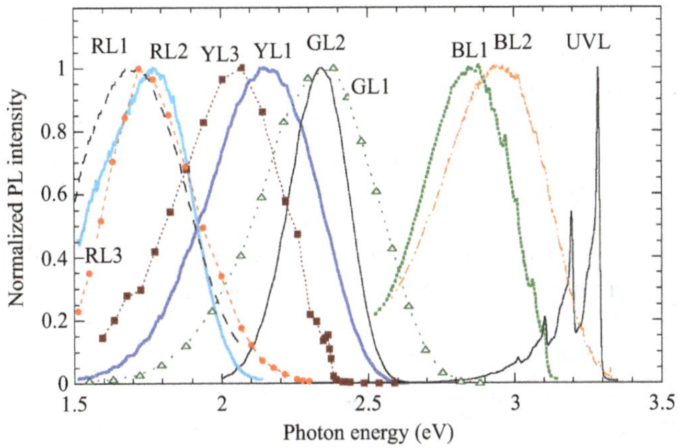

Figure 2.12 Normalized PL spectra from main radiative defects in undoped GaN. The RL3, YL3, and GL1 spectra (shown with symbols) are obtained from time-resolved PL measurements, while other spectra are from steady-state PL measurements. The intensity is calculated as the number of photons per eV. T = 18 K

capture coefficients similar to those of other acceptors). The GL1 band is related to a donor state of another unidentified defect. When this defect is positively charged, an excited Coulomb state appears near the conduction band [27]. This state behaves as a giant trap for electrons. The trapped electrons transition to the ground state after about 1 μs. The GL1 band behaves as a "secondary" PL band of some defect (i.e., requires two holes to be captured); however, we were not able to find the associated "primary" PL band (which should be observed when one hole is captured). We preliminarily attributed the GL1 band to the C_N^+ defect [27]; however, further studies invalidated this assumption.

In semi-insulating GaN samples (undoped or doped with acceptors), the RL2, GL2, and BL2 bands are often observed. The BL2 band with a maximum at 3.0 eV is attributed to the C_NH_i complex [60]. It is caused by an internal transition from an excited state near the conduction band to the 0/+ level of the C_NH_i defect. Under continuous UV illumination, the C_NH_i complex disintegrates due to light-induced defect reaction. As a result, the BL2 band bleaches, and the C_N-related YL1 band emerges. The BL2 band is observed in semi-insulating GaN samples grown by HVPE or MOCVD, where hydrogen is used in the growth process.

The GL2 band with a maximum at 2.33 eV is attributed to the isolated nitrogen vacancy (V_N) [41]. The RL2 band with a maximum at 1.7–1.8 eV is also likely related to the V_N (probably a complex), because it appears together with the GL2 in undoped GaN samples grown in severe Ga-rich conditions. The BL2, RL2, and GL2 bands are all caused by internal transitions of electrons from excited states close to the conduction band to the ground states of deep donors.

The RL3 (at ~1.8 eV) and YL3 (at 2.06 eV) bands appear in undoped, HVPE-grown GaN. It is likely that these bands originate from one defect, the RY3 center [61,62]. The unusual properties of the RY3 band, consisting of a fast (about 10 ns) RL3 component and a much slower (~0.2 ms) YL3 component, were explained in a model of an acceptor with two excited states [62]. An internal transition of a hole from a delocalized excited state near the valence band to the ground state located at 0.5–1.0 eV below the conduction band causes the RL3 component. Recombination of free electrons or electrons at shallow donors with a hole at a localized excited state at 1.13 eV above the valence band is responsible for the YL3 component.

2.4.1.2 Yellow luminescence

The YL band with a maximum at about 2.2 eV is the most common and important defect in undoped GaN and GaN doped with different impurities [5]. One of the first and one of the best papers about the YL band in GaN is the work by Ogino and Aoki [63]. By changing growth and doping conditions, these authors have found a correlation between the YL band and carbon, and they attributed this band to transitions from shallow donors (at low temperature) or from the conduction band (at elevated temperatures) to a carbon-related acceptor with the ionization energy $E_A = 0.86 \pm 0.04$ eV. By using the CC model described in Section 2.3.2, they also determined the following parameters of the YL band and related acceptor: $\hbar\omega_{\max} = 2.14$ eV, $\hbar\omega_{abs} = 3.19$ eV, $E_0 = 2.64 \pm 0.05$ eV, $W(0) = 0.43$ eV, $\hbar\Omega_g \approx \hbar\Omega_e = 40 \pm 5$ meV, $S_g = 12.8$, and $S_e = 13.4$. In the following four decades, the YL band has been the topic of hundreds of publications, in which many controversial results and assignments could be found. For a long time, there was a consensus in the nitride community that the YL band is caused by the V_{Ga}-related acceptor (such as the $V_{Ga}O_N$ complex), yet its relation to carbon was also noted by several authors [5]. A popular (and apparently false) viewpoint is that there are several YL bands, related to different defects, which cannot be resolved in the PL spectrum [64,65]. Below, the latest results on the YL band in GaN are summarized.

Shape and fine structure of the YL1 band

In spite of apparently different shapes, positions, and temperature behaviors of the YL band in different samples, significant evidence exists that it is related to the same defect, namely to the C_N [32,54,66,67]. To distinguish this defect from possible new findings, we label this PL band the YL1. The YL1 band is found in different GaN samples: undoped; doped with C, Si, or Fe; and grown by molecular-beam epitaxy (MBE), MOCVD, and HVPE techniques [54,62]. One of its fingerprints is the ZPL observed at 2.57 eV for $T < 20$ K and at 2.59 eV for $T = 40$–50 K and the characteristic phonon-related fine structure at the high-energy side of the band (Figure 2.13).

The 2.57 eV peak is attributed to electron transitions from shallow donors to the C_N acceptor, and the 2.59 eV peak is caused by transitions from the conduction band to the same $-/0$ level of the C_N [54]. The intensity of the ZPL is only 0.01 of the intensity of the YL1 maximum. The shape and fine structure of the YL1 band can be described using (2.29) with $S_1 = 4$, $\hbar\Omega_1 = 39.5$ meV, $S_2 = 2$, $\hbar\Omega_2 = 91.5$ meV for two phonon modes. The ZPL and the phonon structure can be

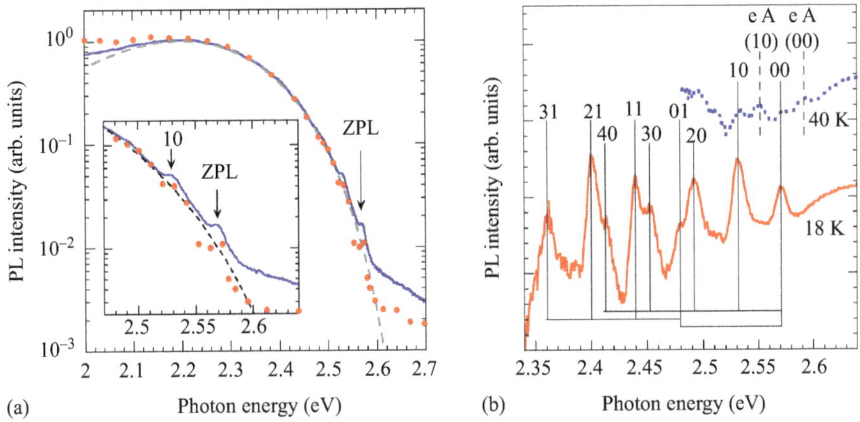

Figure 2.13 The YL1 band in undoped HVPE GaN (sample H2057). (a) The steady-state (solid line) and time-resolved (circles) PL spectra at $T = 18$ K. The dashed line is calculated using (2.25). (b) Fine structure of the YL1 band from DAP transitions at $T = 18$ K (solid line) and from e-A transitions at 40 K (dotted line) is obtained after subtraction of the smooth component which is calculated with (2.25). © 2018 by the American Physical Society. Reprinted, with permission, from [54]

observed only in high-quality samples and under certain experimental conditions (low temperature and low excitation intensity). Other fingerprints of the YL1 band, such as its shape and carrier capture coefficients ($C_{nA} = (1.1 \pm 0.3) \times 10^{-13}$ cm^3/s and $C_{pA} = (3.7 \pm 1.6) \times 10^{-7}$ cm^3/s [25,34]), help to recognize this band in other samples and conditions.

Figure 2.14 shows the shape of the YL1 band before and after corrections. We suspect that the appropriate corrections are made not in all publications, which may cause a discrepancy in the reported results. In our setup, the YL1 band has a maximum at 2.25 eV in as-measured spectrum, 2.20 eV after correcting for spectral response of the measurement system, and 2.15 eV after additional correction (multiplication by λ^3) to present the PL intensity as the number of photons per eV instead of mW/nm, see Section 2.3.2. The FWHM of the YL1 band is 370, 412, and 440 meV, respectively, in these cases. Note that the bandgap of GaN may increase by up to 0.02 eV in GaN layers grown on sapphire due to in-plane biaxial strain. As a result, the YL1 band blue-shifts by roughly the same value [54]. Other reasons that may affect the YL1 band position and width will be discussed later in this section.

The shape of the YL1 band can be fitted using (2.25) with the following parameters: $S_e = 7.8$, $E_0^* = 2.66$ eV, and $\hbar\omega_{\max} = 2.15$ eV. The value of $E_0 = E_0^* - 0.5\hbar\Omega_e$ in the fit slightly differs from the ZPL value (2.59 eV) due to the simplicity of the model. The effective phonon energy was determined from the fit of the temperature dependence of the FWHM with (2.27): $\hbar\Omega_e = 56 \pm 5$ meV (Figure 2.15). This is an effective energy of pseudo-local phonons obtained in the

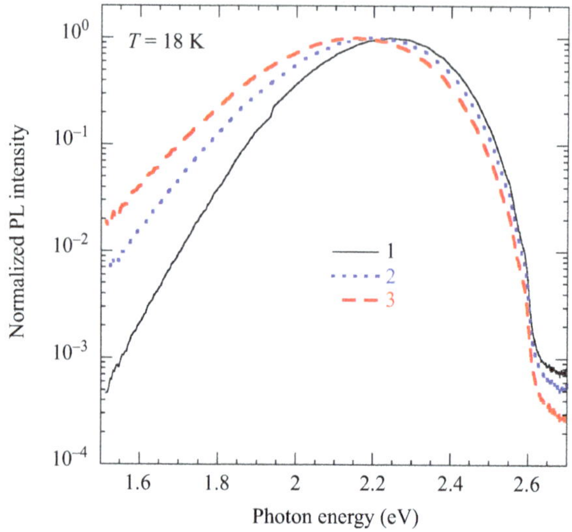

Figure 2.14 The shape of the YL1 band in GaN:Si (sample cvd3540) at $T = 18$ K. 1—as measured spectrum, 2—spectrum corrected for spectral response of the measurement system, 3—the corrected spectrum multiplied by λ^3 to present the PL intensity as the number of photons per eV

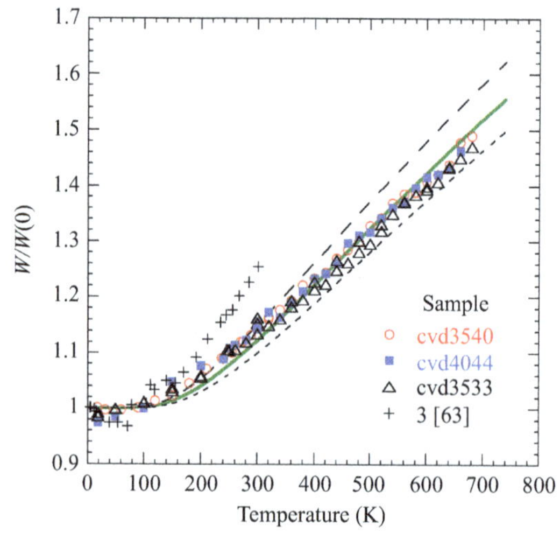

Figure 2.15 Temperature dependence of the YL1 band width for three MOCVD GaN:Si samples. The lines are fit using (2.27) with $\hbar\Omega_e = 56$ meV (the solid line), 51 meV (the upper dashed line), and 61 meV (the lower dashed line). $W(0) = 440$ meV. The data from [63] are shown with crosses

one-dimensional CC model, and it roughly agrees with $\hbar\Omega_1 = 39.5$ meV and $\hbar\Omega_2 = 91.5$ meV obtained from the fine structure of the YL1 band. Further, we obtain $d_{FC}^{e} = S_e\hbar\Omega_e \approx 0.44$ eV from the parameters determined above. The value of $d_{FC}^{g} = S_g\hbar\Omega_g$ can be found from the positions of the YL1 band maximum and its ZPL: $d_{FC}^{g} \approx E_0 - \hbar\omega_{\max} + 0.5\hbar\Omega_e \approx 0.44$ eV. Finally, assuming that $S_e/S_g = \hbar\Omega_e/\hbar\Omega_g$ [5], we conclude that $S_g \approx S_e$ and $\hbar\Omega_g \approx \hbar\Omega_e$. Note that the difference of these parameters from those found in [63] originates mainly from the difference in the parameter $\hbar\Omega_e$ (56 ± 5 meV versus 40 ± 5 meV in [63]). It is possible that additional broadening of the YL1 band in [63] with temperature (Figure 2.15) is caused by increased contribution from transitions in the near-surface depletion region of GaN:C powder studied in that work.

Ogino and Aoki have also examined the characteristic excitation band in the PLE spectrum of C-doped GaN [63]. By *assuming* that the characteristic excitation band has the Gaussian shape, they determined the value of $\hbar\omega_{abs} = 3.19$ eV. However, it is known that for excitation from a defect level to the conduction band (in contrast with the excitation to an excited level as in the case of *F* centers), the PLE band has an asymmetric shape described with (2.30) and (2.31). By using the parameters obtained above, we fit the available experimental data [63,68] and estimate $\hbar\omega_{abs} = 3.00 \pm 0.05$ eV (Figure 2.16).

With the use of the obtained parameters, a one-dimensional CC diagram can be constructed (Figure 2.17). Transitions in this system are explained in the

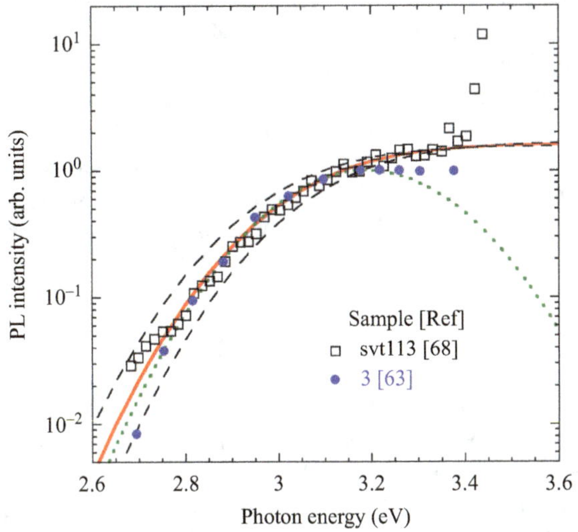

Figure 2.16 The PLE spectrum of GaN at $T = 4$–15 K [63,68]. The solid and dashed lines are calculated using (2.30) with $\Gamma = 0.22$ meV and $\hbar\omega_{abs} = 3.00$ eV (the solid line) and $\hbar\omega_{abs} = 2.95$ and 3.05 eV (the dashed lines). The dotted line is a Gaussian with $\hbar\omega_{abs} = 3.19$ eV [63]

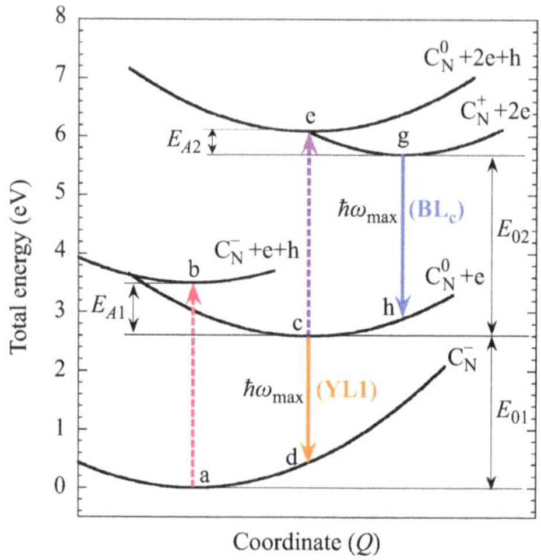

Figure 2.17 Configuration-coordinate diagram for the C_N defect in GaN. Parabola with a minimum at zero (point a) is the ground state of the system (C_N defect is filled with electrons: C_N^-). The transition a-b is the excitation of an electron from the valence band to the conduction band, with C_N remaining negatively charged. The transition b-c is the capture of a hole by the defect (the C_N^- becomes C_N^0). The transition c-d corresponds to the YL1 band maximum. The transition from d to a is the lattice relaxation after a photon is emitted. The transition from c to a causes the ZPL of the YL1 band with energy E_{01}. The parabola with the minimum at point e is identical to the parabola at c but shifted upward by $E_g = 3.503$ eV. The transition e-g corresponds to a capture of the second hole by the defect, which converts C_N^0 to C_N^+. The transition g-h corresponds to the maximum of the BL_C band. The transition from g to c causes the ZPL of the BL_C band with energy E_{02}. © 2018 by the American Physical Society. Reprinted, with permission, from [21]

figure caption. In addition to parabolas related to the – and 0 charge states of the C_N, a parabola for the positive charge state of this defect is included, as predicted by theory [69,70]. Until recently, optical transitions related to the 0/+ level of the C_N could not be found. The fact that the "secondary" PL band did not appear above background signal in this spectral region when the YL1 band saturated at high excitation intensity indicated that the hole-capture coefficient of this band is abnormally small or that this transition is nonradiative [71]. However, by increasing excitation intensity and after finding samples with a very low concentration of uncontrolled radiative defects, we have discovered a new PL band, the BL_C band, which is likely the sought "secondary" band of the C_N defect [32].

The BL_C band has a maximum at 2.85 ± 0.1 eV, and it is characterized with very short lifetime (between 0.27 and 2.5 ns in different samples) [32]. In conductive GaN samples co-doped with C and Si, this band appears only at high excitation intensity, which agrees with expected properties of the "secondary" band. In semi-insulating GaN doped with only C, the BL_C band can be observed also at low excitation intensities; however, the BL2 band obstructs the observation of this band at low temperatures. A large amount of experimental evidence indicates that the BL_C band is caused by electron transitions from the conduction band (or an excited state close to it) to the 0/+ level of the C_N [32]. The hole-capture coefficient for this donor state is extremely small ($C_{pA} = 10^{-10}$ cm^3/s). The BL_C band was observed in many samples where the YL1 band is strong, including undoped, Si-doped, and C-doped GaN samples grown by MOCVD [32]. Note that a blue band observed by Seager *et al.* in C-doped GaN [72,73] and proposed to be the "secondary" band for the C_N defect [70], in fact, shows the properties of the BL2 band assigned to the C_NH complex [32].

Thermal quenching of the YL1 band

With increasing temperature, the YL1 band is quenched at temperatures above ~500 K due to thermal emission of holes from the C_N acceptor to the valence band [5,63,74]. The temperature dependence of the YL intensity can be fitted with (2.10), and the ionization energy E_A can be found. The variation of E_A in different samples served in the literature as a proof that *different* defects contribute to the YL band; i.e., for defects with different thermodynamic transition levels, the optical transition levels are close and indistinguishable. In particular, Armitage *et al.* [64], based on the analysis of the quenching behavior of the YL band in two samples, concluded that the YL band with $E_A = 1040 \pm 60$ meV in C-doped GaN originates from a carbon defect, whereas a very similar YL band with $E_A = 650 \pm 40$ meV in undoped GaN is caused by the V_{Ga} defect.

However, as was shown recently [74], the thermal quenching behavior of the YL1 band (identified by its ZPL and other fingerprints) may be very different in different samples: the *apparent* E_A varies between 0.06 and 1.8 eV (Figure 2.18). One reason for this is that in the samples with high concentration of free electrons and high quantum efficiency of the YL1 band, the beginning of thermal quenching is shifted to higher temperatures (see (2.13) and (2.14)) and may not begin below 700 K. On the other hand, in semi-insulating GaN samples, the abrupt and tunable quenching of the YL1 band is observed (Section 2.3.1.2), and the slope of the quenching is not related to E_A. In such samples, the ionization energy can be determined from the shift of the abrupt region in the temperature dependence (2.16). For the majority of GaN samples, $E_A = 0.85 \pm 0.1$ eV, which is very close to the original data of Ogino and Aoki [63] (0.86 ± 0.04 eV). This value agrees with the depth of the hole trap H1 (or HT1) at ~0.9 eV above the valence band, which is often observed with deep-level transient spectroscopy (DLTS) in GaN grown by MOCVD [75]. Note that the most accurate value of E_A is obtained from the ZPL at low temperatures: $E_A = 916 \pm 3$ meV [54]. The accuracy of E_A obtained from thermal quenching of PL or from DLTS is much

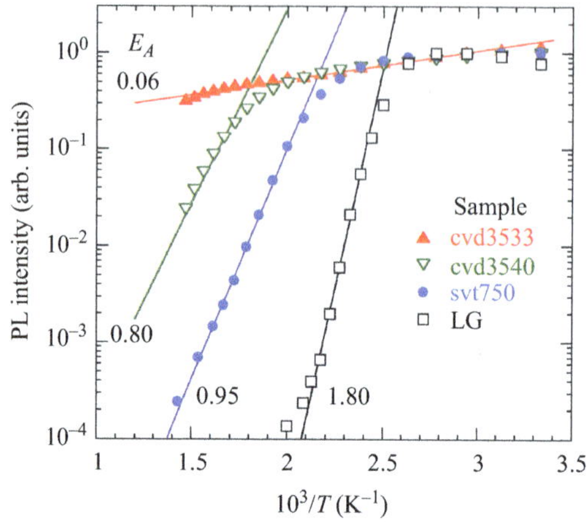

Figure 2.18 The temperature dependence of the PL intensity (normalized at T = 340 K) for the YL1 band in GaN. $P_{exc} \approx 0.001$ W/cm^2. The lines are calculated using (2.10) with the following parameters: E_A: 0.06 eV (sample cvd3533), 0.80 eV (cvd3540), 0.95 eV (svt750), and 1.80 eV (LG). © 2018 by the American Institute of Physics. Reprinted, with permission, from [74]

lower (±0.1 eV). Besides, E_A may decrease with increasing temperature, as the bandgap decreases.

The YL1 band in GaN heavily doped with Si

In the samples with a high concentration of Si (close to 10^{19} cm^{-3}), the YL band is quenched with much smaller activation energy: 0.4–0.6 eV [30,76]. This can be explained by potential fluctuations in heavily doped semiconductors [50,51]. In such semiconductors, diagonal transitions between localized states (Section 2.3.3) lead to a wide distribution of photon energies and PL lifetimes. Transitions with larger separation of carriers have longer τ_0, and their quenching begins at lower temperatures as compared to pairs with smaller separations, see (2.13) and (2.14). As a result, the PL quenching region stretches, and the slope of the quenching decreases. A similar effect can be observed in lightly doped but highly compensated semiconductors, where potential fluctuations are also expected [50,51]. Potential fluctuations may affect not only the thermal quenching dependence but they may also distort the shape of a broad PL band, including its width and position.

Figure 2.19 shows the effect of excitation intensity on the YL1 band in GaN heavily doped with Si. The YL1 band gradually red-shifts with decreasing excitation intensity, totally by about 0.1 eV. The band width increases with decreasing excitation intensity; however, there is no indication that more than one band contribute to its shape. In contrast, there is a very small shift (by only ~7 meV) for n-type GaN with low concentration of impurities (sample EM1256). In the former case, the large shift can be attributed to potential fluctuations, while in the latter

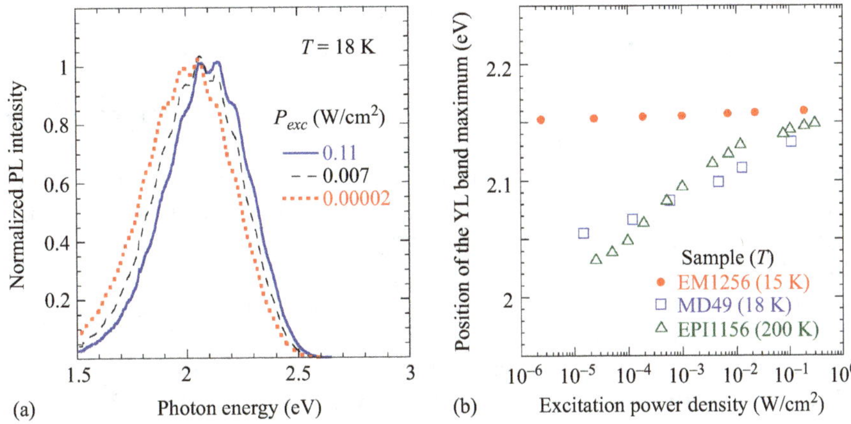

Figure 2.19 Evolution of the YL1 band with excitation intensity. (a) The YL1 band spectrum at $T = 18$ K for GaN:Si,C sample MD49 ([C] = 5×10^{18} cm^{-3}, [Si] = 6×10^{18} cm^{-3}, $n_0 \approx 2 \times 10^{16}$ cm^{-3}). (b) Position of the YL1 band maximum as a function of the excitation power density in the samples with low concentration of impurities ([C] ≈ [Si] ≈ 3×10^{16} cm^{-3}, $n_0 \approx 2 \times 10^{16}$ cm^{-3} in sample EM1256), high (sample MD49) and very high concentrations of Si ([Si] ≈ 1×10^{19} cm^{-3}, $n_0 \approx 1 \times 10^{19}$ cm^{-3} in sample EPI1156)

case, the small shift is caused by the DAP nature of transitions at low temperature (Section 2.3.3).

We demonstrate with above examples that different positions, shapes, and behavior of the YL band with temperature and excitation intensity cannot serve as the evidence that it is related to different defects in different samples. Thorough investigations are needed to conclude whether one or more defects contribute to this broad PL band. It appears that in a variety of GaN samples grown by different techniques, the same C_N defect is responsible for the YL band (the YL1). Of course, the YL1 band should not be confused with other defect-related bands that have slightly different positions but distinctive properties (such as the YL3, GL1, and GL2 bands, see Table 2.2).

2.4.1.3 Photoluminescence in Mg-doped GaN

Magnesium in GaN is the only dopant available to produce reliable p-type GaN, which is necessary for electronic and optical devices. The thermodynamic −/0 transition level of the Mg_{Ga} acceptor is located at 223 ± 3 meV above the valence band maximum [21,77]. Transitions of electrons from shallow donors (at $T < 30$ K) or from the conduction band (at $T > 50$ K) to the Mg_{Ga} acceptor are responsible for the UVL band with the main peak at about 3.28 eV, followed by several LO phonon replicas in undoped and Mg-doped GaN [21]. In n-type GaN, the UVL band intensity increases linearly with the concentration of the Mg_{Ga} defects, and it saturates at [Mg] > 10^{16} cm^{-3}, when its quantum efficiency approaches unity [21,34]. In GaN grown by MBE, the UVL band remains the dominant PL band when the concentration of Mg is much higher than 10^{18} cm^{-3}, and GaN becomes conductive p-type [21,41]. In some samples, the GL2 band is stronger than the UVL band. The GL2 band is assigned to the V_N defect and usually observed in semi-insulating GaN, often in Mg-doped GaN [41]. By analyzing the behavior of the UVL and NBE bands in Mg-doped GaN, it is possible to determine whether the material is conductive n-type, conductive p-type, or semi-insulating, as explained below [21].

In conductive n-type GaN:Mg, the UVL band is strong at low temperatures due to the very high hole-capture coefficient of the Mg_{Ga}. At $T > 100$ K, the UVL band is quenched, and its temperature dependence can be explained using (2.10)–(2.12) with $E_A = 150$–190 meV and $C_{pA} = 10^{-6}$ cm^3/s. The PL lifetime that can be measured directly with time-resolved PL follows the dependence described with (2.18), in which $C_{nA} = 3 \times 10^{-12}$ cm^3/s.

In conductive p-type GaN:Mg, the UVL band is not so strong, and its quenching begins at much lower temperatures than in n-type, with $E_A \approx 20$ meV corresponding to the ionization energy of shallow donors [21]. In contrast to n-type GaN:Mg samples, the UVL intensity in conductive p-type GaN:Mg has very weak temperature dependence up to room temperature, because the concentration of neutral Mg_{Ga} acceptors is high, much higher than the concentration of free holes up to 300 K. A distinctive feature of conductive p-type GaN:Mg samples is rising NBE intensity with increasing temperature. The rise occurs because the concentration of free holes increases proportionally to $\exp(-E_A/kT)$. This finding may serve as a proof of p-type conductivity when the Hall effect measurements are inconclusive or cannot be conducted. Another distinctive feature is that after a laser pulse, the UVL intensity decays very fast in p-type GaN:Mg. The *measured* "PL lifetime" in this case is, in fact, the characteristic time of electron capture by nonradiative defects [21].

In p-type GaN:Mg grown by MOCVD and containing a high concentration of Mg (above ~10^{19} cm^{-3}), the blue luminescence (BL_{Mg}) band with a maximum at 2.7–2.9 eV is usually observed, in addition to or instead of the UVL band. This PL band is attributed to DAP-type transitions from an unknown deep donor to the shallow Mg_{Ga} acceptor [6,21]. The deep donor is likely to be a complex involving hydrogen [78–80]. The BL_{Mg} band blue-shifts by about 0.2 eV with increasing excitation intensity. Such a large shift is attributed to DAP-type transitions involving a deep donor with a localized wave function [6,81]. The BL_{Mg} band appears only in GaN:Mg with a very high concentration of Mg, when the overlap between wave functions of the donor and acceptor becomes significant. The BL_{Mg} band in heavily doped GaN:Mg should not be confused with the Zn-related BL1 band and the C_NH-related BL2 band observed in other samples at about the same photon energy.

In semi-insulating GaN:Mg or in p-type GaN at low temperature, the abrupt and tunable quenching of PL is often observed [21,41]. Sometimes, in such

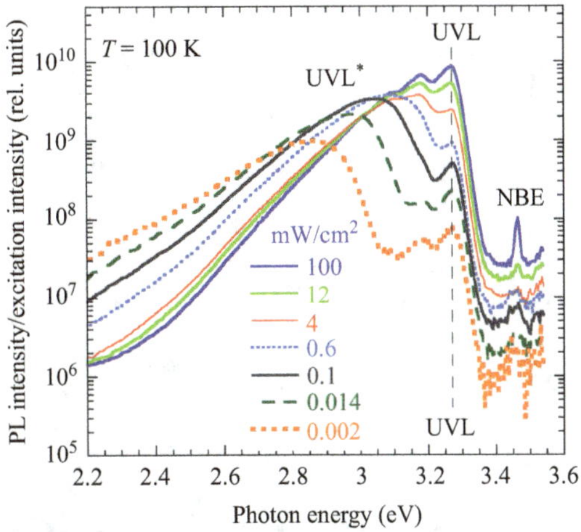

Figure 2.20 PL spectra from HVPE-grown GaN:Mg at different excitation power densities. The PL intensity is divided by the excitation intensity for the convenience of comparison. © 2018 by the American Physical Society. Reprinted, with permission, from [21]

samples, the UVL band may look similar to the BL_{Mg} band if strong potential fluctuations are present or the strong electric field is caused by the near-surface region [21]. In this case, the UVL band (called the UVL^* band to avoid ambiguity) does not show its characteristic fine structure and may red-shift by more than 0.5 eV in certain conditions. Figure 2.20 shows an example where the UVL and UVL^* bands coexist in one sample. It is likely that in this particular case, the "moving" UVL^* band is caused by diagonal transitions in the depletion region near the sample surface, while the "normal" UVL band originates from the region beyond this depletion region. The electric fields causing diagonal transitions may also be caused by potential fluctuations, as discussed in Section 2.3.3.

It is easy to confuse PL bands in Mg-doped GaN. For example, the shape and position of the UVL^* band at low excitation intensity (Figure 2.20) are very similar to the parameters of the BL_{Mg} band. In this situation, careful analysis is needed. Figure 2.21 shows the dependence of the PL band maximum on the excitation intensity for the UVL^* and BL_{Mg} bands in two GaN:Mg samples. The shifts of the bands can be explained with the DAP model (BL_{Mg}) and with the potential fluctuation model (UVL^*), see Section 2.3.3.

2.4.2 Defect-related photoluminescence in other wide bandgaps

Results on PL from defects in wide-bandgap semiconductors, other than GaN, are controversial and poorly understood. Below, a few examples of deep-level defects in ZnO and AlN will be analyzed.

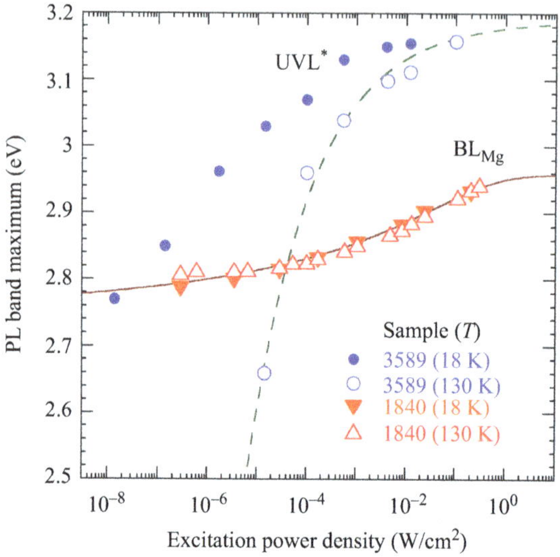

Figure 2.21 *Shift of PL band maxima with excitation power density at $T = 18$ K and 130 K. The solid line is calculated using (2.33) with $\hbar\omega_\infty = 2.72$ eV, and the dashed curve is calculated using (2.34) with $\hbar\omega_{max} = 3.19$ eV. © 2018 by the American Physical Society. Reprinted, with permission, from [21]*

2.4.2.1 PL bands in ZnO

Zinc oxide has many advantages among other wide-bandgap semiconductors (including large exciton binding energy, inexpensive growth and abundance of its components) [82]. Main difficulties with using ZnO in optoelectronic applications arise from the lack of reliable p-type conduction. Analysis of defects in ZnO can be found in the reviews by McCluskey [83,84]. An enormous number of publications are devoted to PL in ZnO, including ZnO nanostructures. In spite of this, systematic studies similar to those for defects in GaN are rare. Attempts to tabulate defect-related PL bands can be found [85], yet most of the PL bands remain unidentified. It appears that several PL bands in ZnO are caused by internal transitions, because PL decay is often exponential at low temperatures [85,86]. Below the most interesting cases are reviewed and existing problems revealed.

The Cu-related green band

It appears that several types of defects contribute to luminescence at 2.4–2.5 eV, the so-called GL band [87]. Among these bands, only one (GL_{Cu}) can be clearly recognized due to its fine structure. The GL_{Cu} band, with a maximum at 2.40–2.45 eV (510 nm), has a narrow ZPL at 2.859 eV, followed by phonon replicas [88]. The fine structure of this band is formed by a superposition of peaks from two phonon modes: a pseudo-local mode with the energy of about 32 meV and the LO phonon mode with the energy 72 meV [87,88]. The GL_{Cu} band is caused by the internal transition of a hole in the Cu_{Zn} acceptor from the excited state at E_V +0.4 eV to the ground state at E_C −0.2 eV [82,84,88,89] (transition e in Figure 2.1). The exponential decay of the GL_{Cu} band (PL lifetime is 440 ± 10 ns at 1.6 K and 10 K [87,88]) is a strong argument for such attribution. The thermal quenching of the GL_{Cu} band at $T > 250$ K with the activation energy of 380 meV is explained by the emission of holes from the excited state of the Cu_{Zn} acceptor to the valence band [89]. Lyons *et al.* [90] have found recently from the first-principles calculations that the 0/+ and −/0 transition levels of the Cu_{Zn} are located at E_V +0.46 eV and E_C −0.08 eV, respectively. These authors attributed the GL_{Cu} band to electron transitions from the conduction band to the 0/+ level of Cu_{Zn}. However, in this case, PL decay would be nonexponential at low temperatures (DAP transitions involving shallow donors are expected to dominate), and at higher temperatures, for e-A transitions, the PL lifetime would depend on the concentration of free electrons, which contradicts to experimental data [87,88]. Thus, the 50-year-old model of Dingle [88] continues withstanding the test of time.

The Li-related orange band

Substitutional lithium (Li_{Zn}) is a deep acceptor in ZnO, with the −/0 transition level at about 0.8 eV above the valence band [84,91]. Transitions from shallow donors to the Li_{Zn} acceptor are responsible for the orange luminescence (OL) band with a maximum at 1.95 eV and nonexponential decay of PL at 10 K in undoped and Li-doped ZnO [86,87,92]. Note that the more accurate value of $\hbar\omega_{max}$ is 1.88 eV for this band, because the spectra reported in [87,92] were not multiplied by λ^3 as discussed in Section 2.3.2. The band maximum blue-shifts monotonically with increasing temperature, totally by 0.15 eV with increasing temperature up to 320 K [92]. The values of $\hbar\omega_{max}$ for the Li-related band scatter in the literature, and sometimes this band is called the yellow band with a maximum at 2.25 eV [84]. This discrepancy can be partly explained by a contribution of other PL bands in this part of the spectrum that are difficult to resolve [93]. Doping with Li acceptor makes ZnO semi-insulating.

The frequently cited value of 0.8 eV for the binding energy of the Li_{Zn} acceptor originates from a pioneer work by Schirmer and Zwingel [91], where it was roughly estimated from the position and shape of the thermoluminescence band associated with Li. The value of $E_A \approx 0.5$ eV was obtained from the slope of the thermal quenching of the OL band [92]. However, it was established later that the quenching of the OL band occurs by the abrupt and tunable quenching mechanism (Section 2.3.1.2), and therefore, the slope of the quenching is not related to E_A [93]. The value of E_A was estimated as 0.65 ± 0.10 eV from the analysis of the PL quenching at different excitation intensities and by using (2.16) [93].

The N-related red band

Over a long period of time, substitutional nitrogen (N_O) was assumed to be a shallow acceptor in ZnO [83], as by theorists so by experimentalists. However, in 2009, Lyons *et al.* [94] calculated by using hybrid functionals that the −/0 transition level of the N_O acceptor is located at 1.3 eV above the valence band. They predicted a PL band with a maximum at 1.7 eV caused by electron transitions from the conduction band to the −/0 level of the N_O. The prediction was soon confirmed experimentally [95]. In as-grown N-doped ZnO, nitrogen forms a complex with hydrogen. However, after annealing at $T = 675–775°C$, the N–H complexes dissociate, and the RL band with a maximum at 1.70 eV emerges. In the PLE spectrum, absorption at photon energies above 2.2 eV was attributed to excitation of electrons from the −/0 level of the N_O to the conduction band [95]. A similar onset (2.1 ± 0.05 eV) for the excitation of electrons from the N_O acceptor to the conduction band has been found from photo-induced electron paramagnetic resonance experiments [96]. These results agree with early theoretical predictions [43,94].

Ironically, Lyons *et al.* [97] repeated the first-principles calculations of the N_O by using hybrid functionals and accounting more accurately for distortions of the defect in the neutral-state and charge-state corrections. This resulted in the −/0 level at 2.04 eV above the valence band and $\hbar\omega_{max} = 0.64$ eV for radiative transitions of electrons from the conduction band to this level. It was proposed that the RL band in N-doped ZnO is caused not by the transitions of electrons from the conduction band but by radiative capture of holes from the valence band to the −/0 level of the N_O acceptor [97]. However, the samples studied in [95] are n-type, and it seems unlikely that the radiative capture of holes (minority carriers in n-type) can compete with the nonradiative capture of holes by other acceptors. Time-resolved PL experiments with the red band in these samples may be useful to find the PL lifetime and resolve the controversy.

PL bands related to zinc vacancy

The zinc vacancy (V_{Zn}) is a double acceptor in ZnO, and it may form complexes with shallow donors, such as H, Ga, Si, and Al [84,98,99]. The RL band with a maximum at about 700 nm (1.7–1.8 eV) has been attributed to the V_{Zn} [98,100,101]. However, the exact structure of the defect responsible for the RL band and even the type of transitions remain unknown. The first-principles calculations predict numerous transitions involving isolated V_{Zn} in different charge states (from 2+ to 2−); however, none of them is likely to explain the RL band or other PL bands observed in the visible range [102]. Remarkably, the addition of shallow donors to the V_{Zn} may explain not only the RL band but also other unidentified PL bands in ZnO [99]. The main problem of analyzing PL from defects in ZnO is that several broad bands are present in undoped crystals and can hardly be resolved. Hydrogen is always present in ZnO, and it forms complexes with many defects [84]. Defects with up to three attached hydrogen atoms may form in ZnO [103].

2.4.2.2 PL bands in AlN

Compared to GaN and ZnO, there is very limited useful information about PL from defects in AlN. Among the reasons are the insufficient quality of AlN samples and lack of continuous-wave lasers with photon energy above 6 eV. In addition, the concentrations of intentionally and unintentionally introduced defects are typically high in the studied AlN samples. In these conditions, it is unlikely that PL intensity is proportional to the concentration of the related defect (Figure 2.8), and thus attributions of PL bands may not be reliable. Further, AlN is always semi-insulating (at least at low temperatures), and potential fluctuations can distort shapes and positions of PL bands. To date, the most comprehensive review on PL from AlN is the work of Koppe, Hofsäss, and Vetter [104]. Below, a few examples of defect-related PL in AlN are reviewed.

It is expected that substitutional acceptors in AlN (such as C_N, Mg_{Al}, and Zn_{Al}) have properties analogous to those of the acceptors in GaN (C_N, Mg_{Ga}, and Zn_{Ga}) but have larger ionization energies [105]. The group of Jiang [106,107] has demonstrated that the YL band gradually blue-shifts with increasing Al content in MOCVD-grown AlGaN samples and eventually converts into a broad band with a maximum at 3.9 eV in AlN. These authors assign the YL band in GaN and the 3.9 eV band in AlN to the $V_{Ga}O_N$ and $V_{Al}O_N$ complexes, respectively [106–108]. However, the attributions were largely based on early theoretical predictions [109] that later changed significantly [65].

In agreement with the modern view that the YL band is caused by the C_N acceptor (Section 2.4.1.1), the 3.9 eV band in AlN has also been recently attributed to the isolated carbon with the −/0 level at 1.9 eV above the valence band, in agreement with the first-principles calculations [110,111]. The samples in these works were grown by HVPE and contained a high concentration of carbon (between 7×10^{17} and 2×10^{19} cm^{-3}). The intensities of the 2.7 and 3.9 eV bands increased with increasing concentration of carbon. The 3.9 eV band could be excited resonantly with photon energies above 4.2 eV, and a flat maximum was observed at about 4.5–4.8 eV in the absorption and PLE spectra [111]. The PLE spectra for the 2.7 and 3.9 eV bands are very similar to each other, indicating the possible involvement of the C_N in both. The 3.9 eV band has been attributed to electron transitions from the conduction band to the −/0 level of the C_N, whereas the PL band with a maximum at 2.7–2.8 eV is assigned to DAP-type transitions from the deep V_N-related level to the C_N acceptor [111,112]. Note that for DAP transitions with deep and highly localized donors and acceptors, we expect significant blue-shift of the PL band with increasing excitation intensity, which apparently was not observed in [111].

In AlN grown by MOCVD and doped with Mg ([Mg] $\approx 10^{19}$–10^{20} cm^{-3}), two new PL bands appear: at 4.70 eV and at 5.54 eV [113,114]. The 5.54 eV band at $T = 10$ K is attributed to DAP transitions involving a shallow donor and the Mg_{Al} acceptor, the ionization energy of which was estimated to be 0.5 eV. The 4.7 eV band has been assigned to DAP transitions involving a deep donor (presumably V_N) and the same Mg acceptor [113,114]. In Zn-doped AlN ([Zn] $\approx 10^{20}$ cm^{-3}), this research group reported on the observation of the 4.50 and 5.40 eV bands and attributed them to transitions from the deep V_N donor (the 4.50 eV band) or from the conduction band (the 5.40 eV band) to the Zn_{Al} acceptor with the ionization energy of 0.74 eV [115].

Many broad PL bands were observed in the photon range between 1.4 and 4.7 eV from undoped or Si-doped AlN and attributed to various defects [104,116]. Unfortunately, the PL bands are very broad and often overlap. Deconvolution of overlapped bands with Gaussian curves [116] may not be justified, because the shapes of defect-related bands are often not Gaussian. Identifications based on the comparison with theoretical predictions can hardly be trusted, because the accuracy of the first-principles calculations for defects in wide-bandgap semiconductors is still insufficient and overstated.

2.4.3 Summary

Defect-related PL bands in GaN are classified, thoroughly studied, and can be reliably recognized by their distinctive features and behavior. Some of these bands (the YL1, GL2, BL1, BL2, and UVL in undoped GaN; and the UVL_{Be}, GL_{Ca}, BL_{Cd}, BL_{As}, BL_P, and BL_{Mg} in doped GaN) are identified with high confidence. For others (the RL1, RL2, RL3, YL3, and GL1), the defects are not identified, and only types of transitions and certain parameters are established. Some defects (such as V_{Ga}, $V_{Ga}O_N$, V_{Ga}-3H, $V_{Ga}O_N$-2H, C_NSi_{Ga}, C_NO_N, and many other complexes) are predicted by theorists [65,117] but not observed in the PL spectrum of GaN, possibly because they capture minority carriers inefficiently or behave as non-radiative centers. In less pure samples (semi-insulating or heavily doped), a PL band from a given defect may shift significantly, change its shape, and be confused with other defect-related bands. It is important to note that phenomenological models described in Section 2.3 work well for defects in GaN; the results are reproducible and predictable [9,13,14,21,22,25,27,30,34,41,54,62,74].

Defects in other wide-bandgap semiconductors are still poorly understood. Researchers have reached a general consensus about defect identity and electronic transitions for only a few defects in ZnO and AlN. High concentrations of defects and impurities, numerous broad and overlapped PL bands, semi-insulating behavior, and the lack of continuous-wave deep UV lasers (for above-bandgap excitation of AlN) hamper the more productive investigation of defects in wide-bandgap semiconductors with PL technique.

2.5 Conclusions and outlook

Point defects (native defects, impurities, and complexes) are responsible for broad PL bands in wide-bandgap and direct-gap semiconductors. Phenomenological models, including rate equations for flows of charge carriers, as well as one-dimensional CC diagrams, help to find important parameters of point defects, including energy levels (sometimes with very high accuracy), charge states, carrier capture coefficients, the strength of electron–phonon coupling, and types of electron transitions. These parameters can be compared with the predictions of the first-principles calculations and with experimental data obtained by complementary techniques. As a result, defects can be reliably identified and their properties and behavior understood. The concentration of defects can also be determined from PL measurements. Some defects are very efficient in capturing minority charge carriers, so that related PL bands can be detected and quantified when the concentration of the defects is as low as 10^{12} cm^{-3}.

Point defects in GaN are meticulously studied and well understood. In particular, the long-standing problem of the YL in GaN is solved. It is attributed to the isolated C_N defect, while numerous other attributions and a popular assumption that several defects cause very similar yellow bands are not confirmed. There is a good agreement between the properties of PL bands caused by the C_N, Zn_{Ga}, and Mg_{Ga} acceptors in GaN and the latest results of the first-principles calculations. These defects can serve as standards for the further development of theoretical calculation approaches and tuning phenomenological models that are used to analyze experimental data. A few defect-related PL bands in GaN remain unidentified. Joint forces of theorists and experimentalists using different techniques should resolve the last intriguing puzzles. Only after defects in high quality, bulk crystals are reliably recognized and thoroughly studied, can more complex materials (nanostructures, alloys, and heavily doped or compensated semiconductors) be successfully explored in terms of identification of point defects.

Identification of point defects in other wide-bandgap semiconductors, including ZnO and AlN, is less successful and will require more efforts. Shapes and positions of numerous defect-related PL bands in these materials are not always reproducible and are difficult to resolve and tabulate. In comparison to GaN, which was studied extensively for two to three decades, comprehensive investigations of point defects in ZnO and AlN are rare. Further progress in this field also calls for improving material quality and theoretical calculations accuracy.

References

[1] Pankove J. I. *Optical Processes in Semiconductors*, Dover Publ., New York, NY, USA, 1975.

[2] Dean P. J. 'Photoluminescence as a diagnostic of semiconductors'. *Prog. Cryst. Growth Charact.* 1982, 5, 89–174.

[3] Davies G. 'Optical Measurements of Point Defects', in Stavola M. (ed). *Identification of Defects in Semiconductors*, Semiconductors and Semimetals Series, Academic Press, San Diego, CA, USA, 1998, Vol. 51B, pp. 1–92.

[4] Pensl G., Choyke W. J. 'Electrical and optical characterization of SiC'. *Physica B.* 1993, 185, 264–283.

[5] Reshchikov M. A., Morkoç H. 'Luminescence properties of defects in GaN'. *J. Appl. Phys.* 2005, 97, 061301.

[6] Reshchikov M. A., Yi G.-C., Wessels B. W. 'Behavior of 2.8-and 3.2-eV photoluminescence bands in Mg-doped GaN at different temperatures and excitation densities'. *Phys. Rev. B.* 1999, 59, 13176–13183.

[7] Schulz H.-J., Thiede M. 'Optical spectroscopy of $3d^7$ and $3d^8$ impurity configurations in a wide-gap semiconductors (ZnO:Co,Ni,Cu)'. *Phys. Rev. B.* 1987, 35, 18–34.

[8] Alkauskas A., Dreyer C. E., Lyons J. L., Van de Walle C. G. 'Role of excited states in Shockley-Read-Hall recombination in wide-band-gap semiconductors'. *Phys. Rev. B.* 2016, 93, 201304.

[9] Reshchikov M. A., Kvasov A. A., Bishop M. F., *et al.* 'Tunable and abrupt thermal quenching of photoluminescence in high-resistivity Zn-doped GaN'. *Phys. Rev. B.* 2011, 84, 075212.

[10] Shockley W., Read J. W. T. 'Statistics of the recombination of holes and electrons'. *Phys. Rev.* 1952, 87, 835–842.
[11] Hall R. N. 'Electron-hole recombination in germanium'. *Phys. Rev.* 1952, 87, 387.
[12] Reshchikov M. A. 'Temperature dependence of defect-related photoluminescence in III-V and II-VI semiconductors'. *J. Appl. Phys.* 2014, 115, 012010.
[13] Reshchikov M. A., Korotkov R. Y. 'Analysis of the temperature and excitation intensity dependencies of photoluminescence in undoped GaN films'. *Phys. Rev. B.* 2001, 64, 115205.
[14] Reshchikov M. A. 'Two-step thermal quenching of photoluminescence in Zn-doped GaN'. *Phys. Rev. B.* 2012, 85, 245203.
[15] Alkauskas A., Yan Q., Van de Walle C. G. 'First-principles theory of nonradiative carrier capture via multiphonon emission'. *Phys. Rev. B.* 2014, 90, 075202.
[16] Abakumov V. N., Perel V. I., Yassievich I. N. *Nonradiative Recombination in Semiconductors*, Elsevier, Amsterdam, 1991, pp. 88–97.
[17] Stoneham A. M. *Theory of Defects in Solids*, Clarendon Press, Oxford, 1975, pp. 520–539.
[18] Lax M. 'Giant traps'. *J. Phys. Chem. Solids.* 1959, 8, 66–73.
[19] Bonch-Bruevich V. L., Landsberg E. G. 'Recombination mechanisms'. *Phys. Status Solidi.* 1968, 29, 9–43.
[20] Dmitriev A., Oruzheinikov A. 'The rate of radiative recombination in the nitride semiconductors and alloys'. *J. Appl. Phys.* 1999, 86, 3241–3246.
[21] Reshchikov M. A., Ghimire P., Demchenko D. O. 'Magnesium acceptor in gallium nitride: I. Photoluminescence from Mg-doped GaN'. *Phys. Rev. B.* 2018, 97, 205204.
[22] Reshchikov M. A. 'Time-resolved photoluminescence from defects in GaN'. *J. Appl. Phys.* 2014, 115, 103503.
[23] Klasens H. A. 'The temperature dependence of the fluorescence of photoconductors'. *J. Phys. Chem. Solids.* 1959, 9, 185–197, and references therein.
[24] Thomas D. G., Hopfield J. J., Augustyniak W. M. 'Kinetics of radiative recombination at randomly distributed donors and acceptors'. *Phys. Rev.* 1965, 140, A202–A220.
[25] Reshchikov M. A., McNamara J. D., Toporkov M., *et al.* 'Determination of the electron-capture coefficients and the concentration of free electrons in GaN from photoluminescence'. *Sci. Rep.* 2016, 6, 37511.
[26] Korotkov R. Y., Reshchikov M. A., Wessels B. W. 'Acceptors in undoped GaN studied by transient photoluminescence'. *Physica B.* 2003, 325, 1–7.
[27] Reshchikov M. A., McNamara J. D., Usikov A., Helava H., Makarov Y. 'Optically generated giant traps in high-purity GaN'. *Phys. Rev. B.* 2016, 93, 081202.

[28] Schroder D. K. 'The concept of generation and recombination lifetimes in semiconductors'. *IEEE Transactions on Electron Devices*. 1982, ED-29, 1336–1338.

[29] Reshchikov M. A. 'Internal Quantum Efficiency of Photoluminescence in Wide-Bandgap Semiconductors', in Case M. A., Stout B. C. (ed). *Photoluminescence: Applications, Types and Efficacy*, Nova Science Publishers, Inc., New York, NY, USA, 2012, pp. 53–120.

[30] Reshchikov M. A., Foussekis M. A., McNamara J. D., Behrends A., Bakin A., Waag A. 'Determination of the absolute internal quantum efficiency of photoluminescence in GaN co-doped with Zn and Si'. *J. Appl. Phys.* 2012, 111, 073106.

[31] Glinchuk K. D., Prokhorovich A. V. 'Effect of the 0.94, 1.0, 1.2, and 1.3 eV radiative centers on the intrinsic luminescence intensity in n-GaAs'. *Phys. Status Solidi A*. 1977, 44, 777–785.

[32] Reshchikov M. A., Vorobiov M., Demchenko D. O., *et al.* 'Two charge states of the C_N acceptor in GaN: Evidence from photoluminescence'. *Phys. Rev. B*. 2018, 98, 125207.

[33] Reshchikov M. A. 'Determination of acceptor concentration in GaN from photoluminescence'. *Appl. Phys. Lett.* 2006, 88, 202104.

[34] Reshchikov M. A., Usikov A., Helava H., *et al.* 'Evaluation of the concentration of point defects in GaN'. *Sci. Rep*. 2017, 7, 9297.

[35] Seitz F. 'Interpretation of the properties of alkali halide-Thallium phosphors'. *J. Chem. Phys.* 1938, 6, 150–162.

[36] Williams F. E., Eyring H. 'The mechanism of the luminescence in solids'. *J. Chem. Phys.* 1947, 15, 289–304.

[37] Huang K., Rhys A. 'Theory of light absorption and non-radiative transitions in *F*-centres'. *Proc. R. Soc.* 1950, A204, 406–423.

[38] Pekar S. I. *Untersuchungen über die Elektronentheorie der Kristalle*, Akademie-Verlag, Berlin, 1954, Ch. 5, 6.

[39] Rebane K. K. *Impurity Spectra of Solids*, Plenum Press, New York, NY, USA, 1970.

[40] Alkauskas A., Lyons J. L., Steiauf D., Van de Walle C. G. 'First-principles calculations of luminescence spectrum line shapes for defects in semiconductors: The example of GaN and ZnO'. *Phys. Rev. Lett.* 2012, 109, 267401.

[41] Reshchikov M. A., Demchenko D. O., McNamara J. D., Fernández-Garrido S., Calarco R. 'Green luminescence in Mg-doped GaN'. *Phys. Rev. B*. 2014, 90, 035207.

[42] Shionoya S., Koda T., Era K., Fujiwara H. 'Nature of luminescence in ZnO crystals'. *J. Phys. Soc. Jpn*. 1964, 19, 1157–1167.

[43] Alkauskas A., McCluskey M. D., Van de Walle C. G. 'Tutorial: Defects in semiconductors – Combining experiment and theory'. *J. Appl. Phys.* 2016, 119, 181101.

[44] Koda T., Shionoya S., Ichikawa M., Minomura S. 'Effect of pressure on the luminescence of Zinc Sulphide phosphors'. *J. Phys. Chem. Solids*. 1966, 27, 1577–1586.

[45] Bube R. H. 'Electron transitions in the luminescence of zinc sulphide phosphors'. *Phys. Rev.* 1952, 90, 70–80.
[46] Hwang C. J. 'Evidence for luminescence involving arsenic vacancy-acceptor centers in p-type GaAs'. *Phys. Rev.* 1969, 180, 827–832.
[47] Williams E. W. 'Evidence for self-activated luminescence in GaAs: The Gallium Vacancy-Donor Center'. *Phys. Rev.* 1968, 168, 922–928.
[48] Williams F. E. 'Theory of the energy levels of donor-acceptor pairs'. *J. Phys. Chem. Solids.* 1960, 12, 265–275.
[49] Thomas D. G., Gershenzon M., Trumbore F. A. 'Pair spectra and "Edge" emission in gallium phosphide'. *Phys. Rev.* 1964, 133, A269–A280.
[50] Levanyuk A. P., Osipov V. V. 'Edge luminescence of direct-gap semiconductors'. *Sov. Phys. Usp.* 1981, 24, 187–215.
[51] Shklovskii B. I., Efros A. L. *Electronic Properties of Doped Semiconductors*, Springer, Berlin, 1984, pp. 53–73 and 253–313.
[52] Zacks E., Halperin A. 'Dependence of the peak energy of the pair-photoluminescence band on excitation intensity'. *Phys. Rev. B.* 1972, 6, 3072–3075.
[53] Reshchikov M. A., Shahedipour F., Korotkov R. Y., Ulmer M. P., Wessels B. W. 'Photoluminescence band near 2.9 eV in undoped GaN epitaxial layers'. *J. Appl. Phys.* 2000, 87, 3351–3354.
[54] Reshchikov M. A., McNamara J. D., Zhang F., *et al.* 'Zero-phonon line and fine structure of the yellow luminescence band in GaN'. *Phys. Rev. B.* 2016, 94, 035201.
[55] Reshchikov M. A., Zafar Iqbal, M., Park, S.S., *et al.* 'Persistent photoluminescence in high-purity GaN'. *Physica B.* 2003, 340–342, 444–447.
[56] Dewsnip D. J., Andrianov A.V., Harrison I., *et al.* 'Photoluminescence of MBE grown wurtzite Be-doped GaN'. *Semicond. Sci. Technol.* 1998, 13, 500–504.
[57] Monteiro T., Boemare C., Soares M. J., Alves E., Liu C. 'Green and red emission in Ca implanted GaN samples'. *Physica B.* 2001, 308–310, 42–46.
[58] Lagerstedt O., Monemar B. 'Luminescence in epitaxial GaN:Cd'. *J. Appl. Phys.* 1974, 45, 2266–2271.
[59] Chen L., Skromme B. J. 'Spectroscopic characterization of ion implanted GaN'. *Mater. Res. Soc. Symp. Proc.* 2003, 743, L11.35.
[60] Demchenko D. O., Diallo I. C., Reshchikov M. A. 'Hydrogen-carbon complexes and the blue luminescence band in GaN'. *J. Appl. Phys.* 2016, 119, 035702.
[61] Reshchikov M. A., McNamara J. D., Usikov A., Helava H., Makarov Yu. 'Two yellow luminescence bands in GaN'. *Sci. Rep.* 2018, 8, 04853.
[62] Reshchikov M. A., Sayeed R. M., Özgür Ü., *et al.* 'Unusual properties of the RY3 center in GaN', *Phys. Rev. B.* 2019, 100, 045204.
[63] Ogino T., Aoki M. 'Mechanism of yellow luminescence in GaN'. *Jpn. J. Appl. Phys.* 1980, 19, 2395–2405.
[64] Armitage R., Hong W., Yang Q., *et al.* 'Contributions from gallium vacancies and carbon-related defects to the "yellow luminescence" GaN'. *Appl. Phys. Lett.* 2003, 82, 3457–3459.

[65] Lyons J., Alkauskas A., Janotti A., Van de Walle C. G. 'First-principles theory of acceptors in nitride semiconductors'. *Phys. Status Solidi B.* 2015, 252, 900–908.

[66] Lyons J. L., Janotti A., Van de Walle C. G. 'Carbon impurities and the yellow luminescence in GaN'. *Appl. Phys. Lett.* 2010, 97, 152108.

[67] Christensen S. G., Xie W. X., Sun Y. Y., Zhang S. B. 'Carbon as a source for yellow luminescence in GaN: Isolated CN defect or its complexes'. *J. Appl. Phys.* 2015, 118, 135708.

[68] Reshchikov M. A., Morkoç H., Park S. S., Lee K. Y. 'Photoluminescence study of deep-level defects in GaN', in Northrup J. E., Neugebauer J., Look D. C., Chichibu S. F., Riechert H. (ed). *GaN and Related Alloys-2001, Mater. Res. Soc. Symp. Proc.* 2002, 693, 365–370.

[69] Demchenko D. O., Diallo I. C., Reshchikov M. A. 'Yellow luminescence of gallium nitride generated by carbon defect complexes'. *Phys. Rev. Lett.* 2013, 110, 087404.

[70] Lyons J. L., Janotti A., Van de Walle C. G. 'Effects of carbon on the electrical and optical properties of InN, GaN, and AlN'. *Phys. Rev. B.* 2014, 89, 035204.

[71] Reshchikov M. A., Demchenko D. O., Usikov A., Helava H., Makarov Y. 'Carbon defects as sources of the yellow and green luminescence bands in undoped GaN'. *Phys. Rev. B.* 2014, 90, 235203.

[72] Seager C. H., Wright A. F., Yu J., Götz W. 'Role of carbon in GaN'. *J. Appl. Phys.* 2002, 92, 6553–6560.

[73] Seager C. H., Tallant D. R., Yu J., Götz W. 'Luminescence in GaN co-doped with carbon and silicon'. *J. Lumin.* 2004, 106, 115–124.

[74] Reshchikov M. A., Albarakati N. M., Monavarian M., Avrutin V., Morkoc H. 'Thermal quenching of the yellow luminescence in GaN'. *J. Appl. Phys.* 2018, 123, 161520.

[75] Polyakov A. Y., Smirnov N. B., Yakimov E. B., *et al.* 'Deep traps determining the non-radiative lifetime and defect band yellow luminescence in GaN'. *J. Alloys Compd.* 2016, 686, 1044–1052.

[76] Reshchikov M. A. 'Point Defects in GaN', in Jagadish C., Privitera V., Romano L. (ed). *Defects in Semiconductors*, Semiconductors and Semimetals Series, Academic Press, Burlington, UK, 2015, Vol. 91, pp. 315–367.

[77] Skromme B. J., Martinez G. L. 'Optical activation behavior of ion implanted acceptor species in GaN'. *Internet J. Nitride Semicond. Res.* 2000, 5S1, W9.8.

[78] Monemar B., Paskov P. P., Pozina G., *et al.* 'Properties of the main Mg-related acceptors in GaN from optical and structural studies'. *J. Appl. Phys.* 2014, 115, 053507.

[79] Shahedipour F., Wessels B. W. 'Investigation of the formation of the 2.8 eV luminescence band in p-type GaN:Mg'. *Appl. Phys. Lett.* 2000, 76, 3011–3013.

[80] Kamiura Y., Kaneshiro M., Tamura J., *et al.* 'Enhancement of blue emission from Mg-doped GaN using remote plasma containing atomic hydrogen'. *Jpn. J. Appl. Phys.* 2005, 44, L926–L928.

[81] Kaufmann U., Kunzer M., Maier M., Obloh H., Ramakrishnan A., Santic B. 'Nature of the 2.8 eV photoluminescence band in Mg doped GaN'. *Appl. Phys. Lett.* 1998, 72, 1326–1328.

[82] Özgür Ü., Alivov Ya. I., Liu C., *et al.* 'A comprehensive review of ZnO materials and devices'. *J. Appl. Phys.* 2005, 98, 041301.

[83] McCluskey M. D., Jokela S. J. 'Defects in ZnO'. *J. Appl. Phys.* 2009, 106, 071101.

[84] McCluskey M. D. 'Point Defects in ZnO', in Jagadish C., Privitera V., Romano L. (ed). *Defects in Semiconductors*, Semiconductors and Semimetals Series, Academic Press, Burlington, UK, 2015, Vol. 91, pp. 279–313.

[85] Reshchikov M. A., Nemeth B., Nause J., *et al.* 'Visible luminescence related to defects in ZnO'. *Mater. Res. Soc. Symp.* 2008, 1035, L03–L05.

[86] Reshchikov M. A. 'Defects in bulk ZnO studied by steady-state and time-resolved photoluminescence'. *Mater. Res. Soc. Symp.* 2008, 1035, L11–L15.

[87] Reshchikov M. A., Morkoç H., Nemeth B., *et al.* 'Luminescence properties of defects in ZnO'. *Physica B.* 2007, 401–402, 358–361.

[88] Dingle R. 'Luminescent transitions associated with divalent copper impurities and the green emission from semiconducting ZnO'. *Phys. Rev. Lett.* 1969, 23, 579–581.

[89] Reshchikov M. A., Avrutin V., Izyumskaya N., Shimada R., Morkoç H., Novak S. W. 'About the Cu-related green luminescence band in ZnO'. *J. Vac. Sci. Technol., B.* 2009, 27, 1749–1754.

[90] Lyons J. L., Alkauskas A., Janotti A., Van de Walle C. G. 'Deep donor state of the copper acceptor as a source of green luminescence'. *Appl. Phys. Lett.* 2017, 111, 042101.

[91] Schirmer O. F., Zwingel D. 'The yellow luminescence of zinc oxide'. *Solid State Commun.* 1970, 8, 1559–1563.

[92] Reshchikov M. A., Garbus J., Lopez G., Ruchala M., Nemeth B., Nause J. 'Acceptors in ZnO studied by photoluminescence'. *Mater. Res. Soc. Symp.* 2007, 957, K07–K19.

[93] McNamara J. D., Albarakati N. M., Reshchikov M. A. 'Abrupt and tunable quenching of photoluminescence in ZnO'. *J. Lumin.* 2016, 178, 301–306.

[94] Lyons J. L., Janotti A., Van de Walle C. G. 'Why nitrogen cannot lead to p-type conductivity in ZnO'. *Appl. Phys. Lett.* 2009, 95, 252105.

[95] Tarun M. C., Zafar Iqbal M., McCluskey M. D. 'Nitrogen is a deep acceptor in ZnO'. *AIP Adv.* 2011, 1, 022105.

[96] Stehr J. E., Hofmann D. M., Meyer B. K. 'Electron paramagnetic resonance and photo-electron paramagnetic resonance investigation on the recharging of the substitutional nitrogen acceptor in ZnO'. *J. Appl. Phys.* 2012, 112, 103511.

[97] Lyons J. L., Varley J. B., Steiauf D., Janotti A., Van de Walle C. G. 'First-principles characterization of native-defect-related optical transitions in ZnO'. *J. Appl. Phys.* 2017, 122, 035704.

[98] Knutsen K. E., Galeckas A., Zubiaga A., *et al.* 'Zinc vacancy and oxygen interstitial in ZnO revealed by sequential annealing and electron irradiation'. *Phys. Rev. B.* 2012, 86, 1211203.

[99] Frodason Y. K., Johansen K. M., Bjørheim T. S., Svensson B. G., Alkauskas A. 'Zn vacancy-donor impurity complexes in ZnO'. *Phys. Rev. B.* 2018, 97, 104109.

[100] Kappers L. A., Gilliam O. R., Evans S. M., Halliburton L. E., Giles N. C. 'EPR and optical study of oxygen and zinc vacancies in electron-irradiated ZnO'. *Nucl. Instrum. Methods Phys. Res., B.* 2008, 266, 2953–2957.

[101] Vlasenko L. S., Watkins G. D. 'Optical detection of electron paramagnetic resonance for intrinsic defects produced in ZnO by 2.5-MeV electron irradiation in situ at 4.2 K'. *Phys. Rev. B.* 2005, 72, 035203.

[102] Frodason Y. K., Johansen K. M., Bjørheim T. S., Svensson B. G., Alkauskas A. 'Zn vacancy as a polaronic hole trap in ZnO'. *Phys. Rev. B.* 2017, 95, 094105.

[103] Herklotz F., Hupfer A., Johansen K. M., Svensson B. G., Koch S. G., Lavrov E. V. 'Infrared absorption on a complex comprising three equivalent hydrogen atoms in ZnO'. *Phys. Rev. B.* 2015, 92, 155203.

[104] Koppe T., Hofsäss H., Vetter U. 'Overview of band-edge and defect related luminescence in aluminum nitride'. *J. Lumin.* 2016, 178, 267–281.

[105] Lyons J. L., Janotti A., Van de Walle C. G. 'Effects of hole localization on limiting p-type conductivity in oxide and nitride semiconductors'. *J. Appl. Phys.* 2014, 115, 012014.

[106] Nam K. B., Nakarmi M. L., Lin J. Y., Jiang H. X. 'Deep impurity transitions involving cation vacancies and complexes in AlGaN alloys'. *Appl. Phys. Lett.* 2005, 86, 222108.

[107] Fan Z. Y., Lin J. Y., Jiang H. X. 'Achieving Conductive High Al-Content AlGaN Alloys for Deep UV Photonics', in Razheghi M., Brown G. J. (ed). *Quantum Sensing and Nanophotonic Devices IV*, *Proc. SPIE.* 2007, 6479, 64791I.

[108] Sedhain A, Lin J. Y., Jiang H. X. 'Nature of optical transitions involving cation vacancies and complexes in AlN and AlGaN'. *Appl. Phys. Lett.* 2012, 100, 221107.

[109] Mattila T., Nieminen R. M. 'Point-defect complexes and broadband luminescence in GaN and AlN'. *Phys. Rev. B.* 1997, 55, 9571–9576.

[110] Collazo R., Xie J., Gaddy B. E., *et al.* 'On the origin of the 265 nm absorption band in AlN bulk crystals'. *Appl. Phys. Lett.* 2012, 100, 191914.

[111] Alden D., Harris J. S., Bryan Z., *et al.* 'Point-defect nature of the ultraviolet absorption band in AlN'. *Phys. Rev. Appl.* 2018, 9, 054036.

[112] Gaddy B. E., Bryan Z., Bryan I., *et al.* 'Vacancy compensation and related donor-acceptor pair recombination in bulk AlN'. *Appl. Phys. Lett.* 2013, 103, 161901.

[113] Nam K. B., Nakarmi M. L., Li J., Lin J. Y., Jiang H. X. 'Mg acceptor level in AlN probed by deep ultraviolet photoluminescence'. *Appl. Phys. Lett.* 2003, 83, 878–880.

[114] Nakarmi M. L., Nepal N., Ugolini C., Altahtamouni T. M., Lin J. Y., Jiang H. X. 'Correlation between optical and electrical properties of Mg-doped AlN epilayers'. *Appl. Phys. Lett.* 2006, 89, 152120.

[115] Nepal N., Nakarmi M. L., Jang H. U., Lin J. Y., Jiang H. X. 'Growth and photoluminescence of Zn-doped AlN epilayers'. *Appl. Phys. Lett.* 2006, 89, 192111.

[116] Lamprecht M., Jmerik V. N., Collazo R., Sitar Z., Ivanov S. V., Thonke K. 'Model for the deep defect-related emission bands between 1.4 and 2.4 eV in AlN'. *Phys. Status Solidi B.* 2017, 254 (8), 1600714.

[117] Matsubara M., Bellotti E. 'A first-principles study of carbon-related energy levels in GaN. II. Complexes formed by carbon and hydrogen, silicon or oxygen'. *J. Appl. Phys.* 2017, 121, 195702.

Chapter 3
Vibrational spectroscopy

Matthew D. McCluskey[1]

Vibrational spectra obtained with Raman or infrared (IR) spectroscopy provide information about the bonding properties of a solid. Vibrational spectroscopy is an especially important technique for detecting and studying impurities in semiconductors. Defect vibrational modes interact with light and the surrounding lattice, giving rise to well-defined peaks in the IR or Raman spectrum. In addition to their fundamental interest, vibrational properties provide insight into the microscopic structure and symmetry of defects. They also act as 'fingerprints' for identification. IR spectroscopy of oxygen in silicon, for example, is a standard characterization technique.

This chapter provides an overview of the theory, experimental methods, and examples of vibrational spectroscopy applied to defects in semiconductors. Additional information about IR spectroscopy can be found in Newman [1], Ramdas and Rodriguez [2], and Stavola [3], while Raman scattering is discussed by Cardona [4] and Yu and Cardona [5]. Reviews of local vibrational modes (LVMs) in semiconductors include Barker and Sievers [6], Murray and Newman [7], McCluskey [8], Pajot and Clerjaud [9], and Stavola and Fowler [10]. There are also reviews for specific materials systems such as defects in silicon [11], III–V semiconductors [12], and ZnO [13], and hydrogen in compound semiconductors [14–16].

3.1 Theory

3.1.1 Units

The frequency of an oscillator can be expressed as s^{-1} or Hz (ν), or rad/s (ω), where $\omega = 2\pi\nu$. As discussed further in Section 3.1.3, a photon of energy E_{photon} can excite a vibrational mode:

$$E_{\text{photon}} = h\nu = \hbar\omega \tag{3.1}$$

where h is Planck's constant ($h = 6.6 \times 10^{-34}$ Js, $\hbar = h/2\pi$). The photon wavelength is inversely proportional to the photon energy. In the literature, vibrational frequencies are often expressed in units of wavenumbers (cm^{-1}), defined as

$$\frac{1}{\lambda} = \frac{\nu}{c} = \frac{\omega}{2\pi c} \tag{3.2}$$

[1]Department of Physics and Astronomy, Washington State University, Pullman, WA, USA

where $c = 3.0 \times 10^{10}$ cm/s. The wavenumber is the number of waves per cm for the photon. Even though its unit is inverse length, it is nonetheless called a 'frequency'. I use the terms 'energy' and 'frequency' interchangeably, since they are proportional to each other. In this chapter, quoted frequencies are generally the values at cryogenic temperatures (2 K–15 K).

The absorption coefficient α is defined by the exponential decay of light in the medium. Neglecting reflections, the transmitted light intensity is

$$I = I_0 \exp(-\alpha d) \tag{3.3}$$

where I_0 is the incident intensity and d is the sample thickness. Absorbance is expressed in a base-10 system,

$$\text{Absorbance} = \log_{10}\left(\frac{I_0}{I}\right) = 0.434\alpha d \tag{3.4}$$

An absorbance of 2, for example, means that 1 per cent of the light is transmitted. In practice, an absorbance spectrum can be obtained by measuring transmission spectra of the sample (I) and a reference sample that does not contain the defect being studied (I_0).

3.1.2 Linear chain

To first order, a crystal can be approximated by masses connected by springs. A surprising amount of insight can be gleaned from a simple one-dimensional (1D) linear chain approximation [17]. The diatomic linear chain has alternating masses M_1 and M_2 connected by springs with a force constant C. This model yields a band of lattice wave, or phonon, frequencies. The lowest frequency modes are acoustic phonons, which travel at the speed of sound in the material. The high-frequency modes are optical phonons, which can absorb or reflect electromagnetic radiation. If the masses are different ($M_1 > M_2$), a forbidden gap opens up between the acoustic and optical bands. Semiconductors such as AlSb, GaP, and GaN have forbidden gaps. Semiconductors with equal or similar masses, such as Si, Ge, and GaAs, lack forbidden gaps.

Because a pure crystal has translational symmetry, phonons have well-defined momenta, labelled K. Photons also carry momentum but it is very small, practically zero on the scale of the Brillouin zone. Conservation of momentum therefore requires that only $K = 0$ phonons can be excited by light. Zone-centre phonon frequencies for several semiconductors are listed in Table 3.1. In a defective crystal, however, the translational symmetry is broken. There is no well-defined K value for an impurity vibrational mode.

When an impurity (or impurity complex) is introduced into a crystal, new vibrational modes may appear [6]. There are approximate guidelines for what happens with a substitutional impurity, assuming the spring constant does not change significantly. If the mass defect is lighter than the atom it replaces, its vibrational frequency will be higher than the maximum phonon frequency. This impurity vibration, an LVM, is highly localized and may give rise to a sharp peak in the IR or Raman spectrum. Examples of LVMs are listed in Table 3.2. If the

Table 3.1 Zone-centre transverse optical (TO) and longitudinal optical (LO) phonons at room temperature. After Barker and Sievers [6]

Structure	Material	TO (cm^{-1})		LO (cm^{-1})	
Diamond	C	1,332		1,332	
	Si	520		520	
	Ge	301		301	
Zincblende	AlSb	318		345	
	CdTe	140		167	
	GaAs	269		292	
	GaP	366		402	
	GaSb	225		236	
	InAs	219		243	
	InP	307		351	
	InSb	174		183	
	SiC	794		962	
	ZnS	271		352	
	ZnSe	209		250	
	ZnTe	190		210	
Wurtzite		$E//c$	$E\perp c$	$E//c$	$E\perp c$
	CdS	228	235	305	305
	CdSe	166	172	211	210
	GaN	533	560	744	746
	SiC	790	794	962	962
	ZnO	380	407	574	583

Table 3.2 Examples of LVM frequencies (cm^{-1}) due to substitutional impurities. Temperatures were 80 K or below except where noted (RT = room temperature). Data are primarily from Barker and Sievers [6] and Pajot and Clerjaud [9]; the GaSb:C frequency is from [153]

Host	Impurity						
	^{10}B	^{11}B	^{12}C	^{13}C	^{14}N	^{27}Al	^{28}Si
Si	646	622	608	589	653		
Ge	571	547	531	512	577	365 (RT)	
AlAs			632	609			
AlSb			592	573			
GaP	593 (B_{Ga})	570 (B_{Ga})	606		496	444	465 (Si_{Ga})
GaAs	540 (B_{Ga})	517 (B_{Ga})	582	561	473	362	384 (Si_{Ga})
	628 (B_{As})	601 (B_{As})					399 (Si_{As})
GaSb			540			317	
InP	544 (B_{In})	523 (B_{In})	547	527			
InAs			527	508			
InSb						298	

impurity mass is heavier than M_2, then it may exhibit a gap mode, the frequency of which lies in the forbidden gap. A gap mode is localized but not as much as an LVM. Finally, if the impurity is very heavy or if the spring constant is weak, it has a *resonant mode* that lies within the acoustic phonon band. Such a mode is not localized and, because its band of frequencies is broad, difficult to detect with spectroscopic methods.

To model impurity vibrational modes, we consider the linear chain model. The forces on M_1 and M_2 in unit cell s are given by

$$\begin{aligned} F_1 &= M_1\ddot{u}_s = C(v_{s-1} + v_s - 2u_s) \\ F_2 &= M_2\ddot{v}_s = C(u_s + u_{s+1} - 2v_s) \end{aligned} \tag{3.5}$$

These equations of motion are solved by complex exponential functions,

$$\begin{aligned} u_s &= u_{s0}\exp(i\omega t) \\ v_s &= v_{s0}\exp(i\omega t) \end{aligned} \tag{3.6}$$

where ω is the normal-mode frequency; u_{s0} and v_{s0} are the normal-mode amplitudes. Substituting these trial solutions into (3.5) yields

$$\begin{aligned} (2C/M_1 - \omega^2)u_s - C/M_1(v_{s-1} + v_s) &= 0 \\ (2C/M_2 - \omega^2)v_s - C/M_2(u_s + u_{s+1}) &= 0 \end{aligned} \tag{3.7}$$

The eigenvalues ω^2 are obtained by setting the determinant of the matrix of coefficients to zero,

$$\begin{vmatrix} 2C/M_1 - \omega^2 & -C/M_1 & 0 & 0 & \cdots & 0 & -C/M_1 \\ -C/M_2 & 2C/M_2 - \omega^2 & -C/M_2 & 0 & \cdots & 0 & 0 \\ \vdots & \vdots & \vdots & \vdots & \ddots & \vdots & \vdots \\ -C/M_2 & 0 & 0 & 0 & \cdots & -C/M_2 & 2C/M_2 - \omega^2 \end{vmatrix} = 0 \tag{3.8}$$

where periodic boundary conditions have been assumed. The number of rows (and columns) equals the number of atoms.

In practice, a computer program first defines the matrix in (3.8), but without the ω^2 terms. Then, the eigenvalues (ω^2) are obtained using an automatic routine. For each eigenvalue, there is a corresponding eigenvector that describes the relative motion of the atoms. A mass defect is created by replacing the M_1 or M_2 values in a given row by a defect mass m.

To illustrate with an old example [8], GaP was modelled as a linear chain of 256 atoms with masses $M_1 = 70$ (Ga) and $M_2 = 31$ (P). The spring constant C was adjusted so that the maximum phonon frequency matched the zone-centre transverse optical (TO) phonon mode (366 cm^{-1}). A list of frequencies was obtained numerically and plotted as a histogram. Then, a mass defect was introduced by replacing a phosphorus atom with carbon ($m = 12$). As shown in Figure 3.1, this

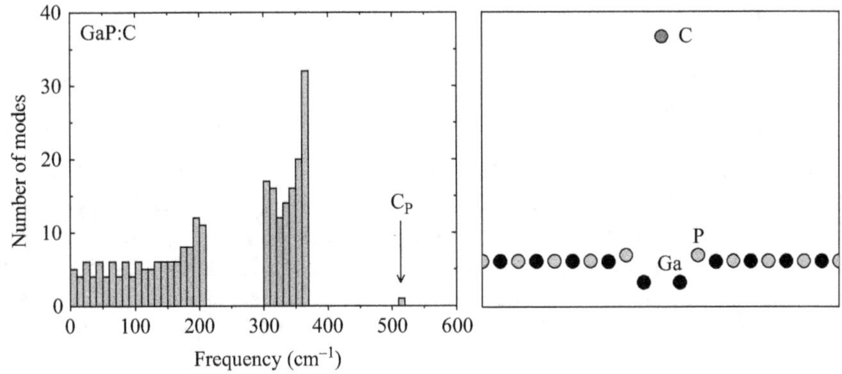

Figure 3.1 *Histogram of vibrational states for GaP:C from the linear-chain model. The normal-mode amplitudes for the C_P LVM are shown on the right*

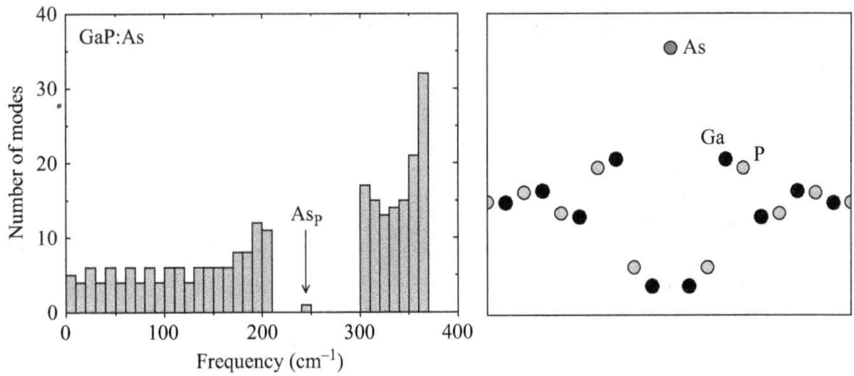

Figure 3.2 *Histogram of vibrational states for GaP:As from the linear-chain model. The normal-mode amplitudes for the As_P gap mode are shown on the right*

results in an LVM at 513 cm^{-1}. The corresponding normal mode displacements are shown. One can see that the vibrational amplitude is large for the carbon atom and much smaller for the neighbouring ones. The experimentally determined frequency for the $^{12}C_P$ LVM is 606 cm^{-1} [18]. The higher experimental frequency implies that the Ga–C spring constant is stronger than that of Ga–P.

Next, we consider the isoelectronic As_P impurity ($m = 75$), where arsenic is heavier than the phosphorus atom that it replaces. Here, a gap mode appears at 240 cm^{-1} (Figure 3.2). As shown in the figure, this vibrational mode is localized but not to the same degree as an LVM. The experimental As_P gap mode frequency is 269 cm^{-1} [19], in good agreement with the linear mass defect model.

The linear chain model can be extended to three dimensions, with 'bending' force constants [20]. The three-dimensional (3D) model has been applied, for example, to study the spatial extent of hydrogen vibrational modes [21]. Overall, the more sophisticated models replicate the qualitative results of the linear chain model.

What happens to an LVM when the impurity concentration gets large? As more impurity atoms are added randomly, pairs and larger clusters form, causing the vibrational peak to broaden. As the system enters the alloy regime (1 per cent impurity concentration or more), we can no longer consider the modes to be truly localized. For alloys, the zone-centre vibrational frequencies are given by the modified random element isodisplacement (MREI) model [22]. An example of this is given in Section 3.3.13.

3.1.3 Defect vibrational modes

The *diatomic model* has been used to describe LVM frequencies and how they depend on the masses of different isotopes [12]. In the diatomic model, an impurity of mass m is considered to be attached to a host atom M by a spring C (Figure 3.3). To account for the fact that the system is really not just two atoms, the mass M is multiplied by an empirical constant χ. The bond-stretching frequency of this quasi-molecule is

$$\omega = \sqrt{C\left(\frac{1}{\chi M} + \frac{1}{m}\right)} \equiv \sqrt{\frac{C}{\mu}} \tag{3.9}$$

where μ is the *reduced mass*.

To verify the identity of an LVM experimentally, the impurity may be replaced with a different isotope, causing a well-defined shift in the vibrational frequency.

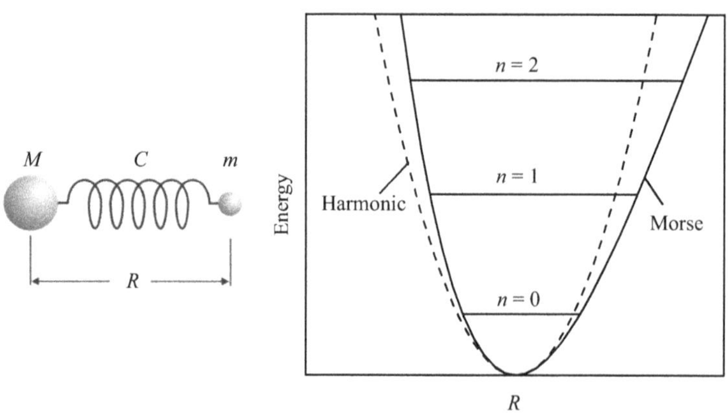

Figure 3.3 Diatomic model for an impurity m attached to a host atom M by a spring C. The anharmonic Morse potential is shown with the first three vibrational levels

The most dramatic isotope shift occurs when hydrogen ($m = 1$) is replaced by deuterium ($m = 2$). In that case, the isotopic frequency ratio is given by

$$r = \frac{\omega_H}{\omega_D} = \sqrt{2\frac{\chi M + 1}{\chi M + 2}} \tag{3.10}$$

where ω_H and ω_D are the hydrogen and deuterium frequencies, respectively. Because M is finite, the ratio r is slightly less than $\sqrt{2}$.

The energy of a vibrational state is

$$E_n \equiv \hbar\omega_n = \left(n + \frac{1}{2}\right)\hbar\omega \tag{3.11}$$

where n is a whole number. Light can excite the system from the ground state to an excited state. The transition from $n = 0$ to $n = 1$ is the 'fundamental' mode or 'first harmonic'. The transition from $n = 0$ to $n = 2$ is the 'overtone' or 'second harmonic'.

3.1.4 Anharmonic potential

The preceding discussion assumed a perfectly parabolic, or harmonic, potential. In reality, the potential energy begins to flatten out as the interatomic separation increases. The *Morse potential* is a popular anharmonic potential that yields good results for many vibrational systems [23]. As the amplitude of vibration gets larger, the deviation from the harmonic potential increases (Figure 3.3). Specifically, the energy levels are given by

$$\omega_n = \omega_e\left(n + \frac{1}{2}\right)\left[1 - x_e\left(n + \frac{1}{2}\right)\right] \tag{3.12}$$

The $n = 0 \rightarrow n = 1$ transition (fundamental) is given by

$$\omega_1 - \omega_0 = \omega_e - 2\omega_e x_e \tag{3.13}$$

The $n = 0 \rightarrow n = 2$ transition (overtone) is

$$\omega_2 - \omega_0 = 2\omega_e - 6\omega_e x_e \tag{3.14}$$

A light isotope has a larger vibrational amplitude than a heavy isotope. Therefore, its frequency is affected more by anharmonicity (i.e. its $\omega_e x_e$ value is larger).

3.1.5 IR activity

If a vibrational mode can be excited by electromagnetic radiation, it is said to be IR-active, since the relevant wavelengths lie in the IR region of the spectrum. Such a mode will absorb light at the vibrational frequency, resulting in a sharp dip in the transmitted intensity spectrum. To be IR-active, there must be a change in the dipole moment. Carbon monoxide (CO), for example, has a positively charged carbon and negatively charged oxygen. Its stretch mode causes the dipole to oscillate, so it is IR active. Nitrogen (N_2), on the other hand, is symmetric and therefore IR-inactive.

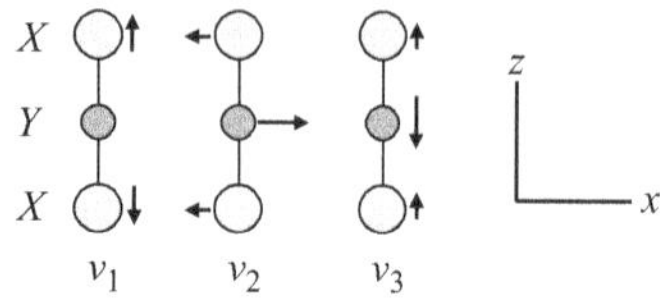

Figure 3.4 Vibrational modes for a linear XYX molecule: symmetric (ν_1), transverse (ν_2), and asymmetric (ν_3). Arrows indicate the relative displacements during the vibration

Consider a triatomic, linear *XYX* molecule like carbon dioxide. The *X* atoms have a different charge than the *Y* atom. This molecule has three vibrational modes: symmetric stretch (ν_1), transverse (ν_2), and asymmetric stretch (ν_3). In Figure 3.4, the relative displacements of the atoms are shown by arrows. For the ν_1 mode, when the top *X* atom is displaced upward, the bottom *X* atom is displaced downward by the same amount, and the *Y* atom does not move. Because the motion is symmetric, there is no dipole moment, so the mode is IR-inactive.

For the ν_2 mode, *X* and *Y* atoms move in opposite directions, inducing a dipole moment along the x direction. This mode is IR-active. Specifically, light polarized along the x-axis can excite this vibration. Similarly, the ν_3 mode involves the *X* and *Y* atoms moving in opposite directions and is IR-active. The induced dipole is along the z direction and can be excited by z-polarized IR light.

3.1.6 Number of atoms

Vibrational spectroscopy and isotope mixtures can be used to determine the number of atoms in a complex. This is especially true for hydrogen and deuterium, which have such different masses. A single hydrogen atom attached to an impurity or host atom has one stretch mode. When hydrogen is replaced by deuterium, the stretch mode shifts to lower frequency (Equation (3.9)). A sample treated in a hydrogen–deuterium mixture will show both peaks.

A multi-hydrogen complex, on the other hand, will exhibit new peaks corresponding to different isotope combinations. Consider a mass-and-spring model where two H atoms are attached to the host by strong springs and to each other by a weak spring (Figure 3.5). This complex has two stretch modes, asymmetric (the masses oscillate in phase) and symmetric (out of phase). The symmetric mode has a slightly higher frequency thanks to the stretching and compressing of the weak spring. A complex with two D atoms has the same modes but at lower frequencies. A complex with one H and one D, however, does not have symmetric/asymmetric modes. Instead, the modes are ‘H like’ or ‘D like’ depending on which atom has greater displacement. This results in two HD peaks that are different than the HH or DD peaks.

In the simple model shown in Figure 3.5, the two hydrogen atoms reside in environments that are equivalent by symmetry. The asymmetric HH and DD modes are IR-active while the symmetric modes are IR-inactive. An IR absorption

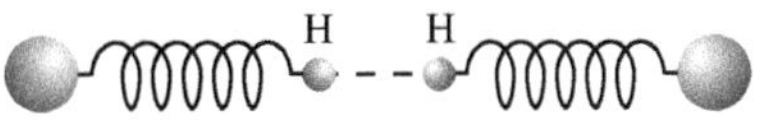

Figure 3.5 Model for a two-hydrogen centre where both hydrogen atoms reside in identical environments. Relative displacements for the H atoms are indicated for the asymmetric and symmetric stretch modes

spectrum would show one peak for the HH complex, one for the DD complex, and two for HD. If the hydrogen atoms were in inequivalent locations, then the IR spectrum would show two peaks for HH, two for DD, and two for HD. An example of this latter case is the H_2^* defect (Section 3.3.5).

3.1.7 *Infrared absorption*

IR absorption spectroscopy is a preferred technique for studying vibrational properties. An IR-active mode yields a peak in the IR absorption spectrum. Classically, the absorption of light comes about because the electric field pushes and pulls the impurity atom. The electromagnetic energy goes into driving mechanical motion. From the diatomic model, an impurity consists of two masses M and m attached to each other by a spring C. The masses are a distance $R + x$ apart, where R is the equilibrium separation, and have electric charges $\pm q$.

The equation of motion is given by

$$\ddot{x} + \gamma\dot{x} + \omega_0^2 x = \frac{qE}{\mu} \tag{3.15}$$

where γ is a damping constant, ω_0 is the vibrational frequency, and E is the electric field. At the dipole, let the electric field be

$$E = E_0 \cos(\omega t) \tag{3.16}$$

It is convenient to write this as a complex exponential with the understanding that we will take the real (Re) part of it later:

$$E = E_0 \exp(i\omega t) \tag{3.17}$$

The displacement x oscillates with the same frequency as the electric field:

$$x = x_0 \exp(i\omega t) \tag{3.18}$$

Plugging this into (3.15) yields the amplitude

$$x_0 = \frac{qE_0/\mu}{\omega_0^2 - \omega^2 + i\omega\gamma} \tag{3.19}$$

The velocity is the derivative of x (Equation (3.18)):

$$v \equiv \dot{x} = x_0 i\omega \exp(i\omega t) \tag{3.20}$$

The power absorbed by the dipole is given by force times velocity,

$$\begin{aligned} P &= q\mathrm{Re}(E)\mathrm{Re}(v) \\ &= \frac{q^2 E_0^2 \omega}{\mu} \cos \omega t \frac{\gamma\omega \cos \omega t - \left(\omega_0^2 - \omega^2\right) \sin \omega t}{\left(\omega_0^2 - \omega^2\right)^2 + \gamma^2\omega^2} \end{aligned} \tag{3.21}$$

Using the fact that the average of $\cos^2$ is 1/2 and the average of (cos $\times$ sin) is 0, the time-averaged power absorption is given by

$$\langle P \rangle = \frac{q^2 E_0^2}{2\mu\gamma} \frac{\gamma^2 \omega^2}{\left(\omega_0^2 - \omega^2\right)^2 + \gamma^2 \omega^2} \tag{3.22}$$

Equation (3.22) describes a peak in the absorption spectrum with a maximum near the resonant frequency ω_0. The damping coefficient γ is proportional to the width of the peak. Most LVMs give rise to a sharp peak with $\gamma \ll \omega_0$. For such a 'high-Q' oscillator, the power absorption is given by a Lorentzian function:

$$\langle P \rangle \approx \frac{q^2 E_0^2}{2\mu\gamma} \frac{\gamma^2}{4(\omega - \omega_0)^2 + \gamma^2} \tag{3.23}$$

How strong is the absorption peak? The peak strength measured by the integrated absorption,

$$A_I = \int \alpha(\omega)\, d\omega \tag{3.24}$$

where α is the absorption coefficient (cm^{-1}) [24]. As an order-of-magnitude estimate [25],

$$A_I\left(\mathrm{cm}^{-2}\right) \approx 10^{-16} N\left(\mathrm{cm}^{-3}\right) \tag{3.25}$$

where N is the impurity density. The calibration factor depends on the particular defect. For example, the C–H stretch mode in GaAs has a coefficient of 1.2×10^{-16} [26], while O–H in ZnO has 6×10^{-17} [27].

3.1.8 Linewidth and temperature dependence

At low temperatures, an LVM peak is usually sharp, with a width of ~1 cm^{-1} or less. As the sample is warmed up, the peak broadens due to a reduction in the vibrational lifetime. This broadening may be large enough to make detection impractical at room temperature. For this reason, many vibrational studies are performed at liquid helium temperatures (4 K–12 K).

There are two types of vibrational lifetimes that contribute to peak broadening. The first is the ground state recovery lifetime, given by $T_1 = 1/\gamma$. This is the time that it takes for the energy of a freely oscillating mass to decay by a factor of $1/e$ (37 per cent of its starting energy). Quantum mechanically, it is the mean time for the oscillator to decay from the $n = 1$ excited state to the $n = 0$ ground state. From (3.23), a short lifetime (large γ) results in a broad peak.

When an oscillator decays to the ground state, energy is conserved by producing a combination of lower frequency vibrational modes, which may be highly localized LVMs or less localized resonant or gap modes. The lifetime depends on the vibrational density of states (how many modes there are to decay into) and the coupling strength to these modes. Hydrogen in silicon, for example, has T_1 values from 4 to 295 ps [28,29]. Hydrogen that resides in an open volume has a long lifetime, while bond-centred hydrogen, which is squeezed between two silicon atoms, has a short lifetime.

At absolute zero temperature, the homogeneous vibrational linewidth is determined by T_1. Inhomogeneous broadening, which adds to the linewidth, can result from strains or random distributions of isotopes. As temperature increases, phonons become thermally populated. The jiggling of atoms near the impurity results in a dephasing lifetime T_2^*. This is the average time between 'collisions' with a neighbouring atom [30]. Most of the temperature-dependent broadening is due to dephasing.

In addition to broadening, the centre of the vibrational peak also shifts with temperature. This is because the jiggling atoms perturb the potential seen by the impurity, which thereby affects the vibrational energy levels. Most of the time, the frequency decreases as temperature increases, but the opposite is also observed. There is no simple way to predict the sign (positive or negative) of the shift. In general, impurities that are strongly coupled to the host atoms (e.g. bond-centred hydrogen) show more pronounced shifts than impurities that are not crowded (e.g. hydrogen in a vacancy). Although the lattice does expand upon warming, this only leads to a small shift that is typically neglected [31].

The frequency shift can be calculated by considering the diatomic model, but with another host atom nearby. Let x denote the random displacement of the host atom from equilibrium. This displacement perturbs the impurity frequency by $\delta\omega$:

$$\delta\omega(t) = ax + bx^2 + \cdots \tag{3.26}$$

Averaged over time, the linear term goes to zero. To lowest order, the shift is given by

$$\delta\omega = b\langle x^2 \rangle \tag{3.27}$$

where the brackets indicate a time-averaged value. For the host atom, $\langle x^2 \rangle$ is proportional to its energy E, so

$$\delta\omega \sim E = \left(n + \frac{1}{2}\right)\hbar\omega_0 \tag{3.28}$$

where ω_0 is the host atom's frequency. Given the Planck distribution,

$$\langle n \rangle = \frac{1}{e^{\hbar\omega_0/k_B T} - 1} \tag{3.29}$$

the temperature-dependent shift is

$$\delta\omega = \frac{\delta\omega_0}{e^{\hbar\omega_0/k_BT} - 1} \tag{3.30}$$

The temperature-dependent broadening, according to Persson and Ryberg [32], is given by

$$\delta\Gamma = \frac{2(\delta\omega_0)^2}{\eta}\frac{e^{\hbar\omega_0/k_BT}}{\left(e^{\hbar\omega_0/k_BT} - 1\right)^2} \tag{3.31}$$

where $\delta\omega_0$, ω_0, and η are the empirical constants.

This model (Equations (3.30) and (3.31)) does a good job describing the temperature dependence of hydrogen LVMs in a wide range of semiconductors [15]. Figure 3.6 shows hydrogen LVMs in ZnO at liquid-helium and room temperature. For the O–H mode, a good fit to (3.30) was obtained with $\omega_0 \approx 100\ \text{cm}^{-1}$ [33].

3.1.9 Wave functions and symmetry

An impurity in a crystal can be treated as a diatomic molecule with a frequency ω and reduced mass μ. Quantum mechanically, the wave function of a 1D oscillator is given by

$$\psi_n(x) \sim H_n(x)e^{-x^2/2} \tag{3.32}$$

where x displacement in units of $\sqrt{\hbar/\mu\omega}$ and $H_n(x)$ is a function called a Hermite polynomial. The first few Hermite polynomials are listed here:

$$\begin{aligned} H_0(x) &= 1 \\ H_1(x) &= 2x \\ H_2(x) &= 4x^2 - 2 \end{aligned} \tag{3.33}$$

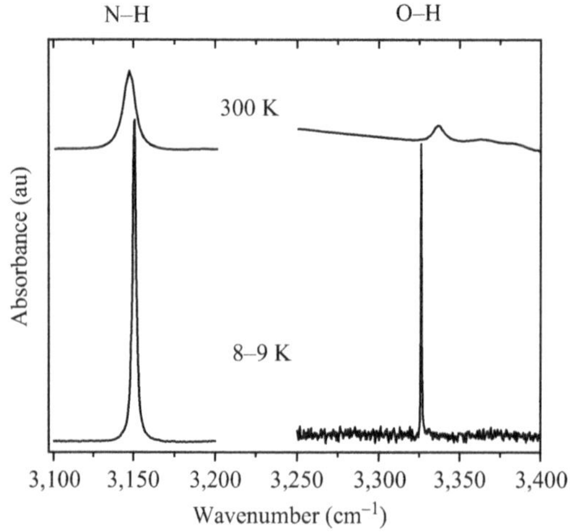

Figure 3.6 IR spectra of N–H and O–H complexes in ZnO (two different samples). The stretch modes shift and broaden with increasing temperature

To extend this model to three dimensions, we multiply the 1D wave functions for x, y, and z:

$$\psi(x,y,z) \sim H_{n_x}(x)H_{n_y}(y)H_{n_z}(z)e^{-(x^2+y^2+z^2)/2} \tag{3.34}$$

where n_x, n_y, and n_z are whole numbers. The energy levels of the 3D system are given by

$$\omega = \left(n_x + \frac{1}{2}\right)\omega_x + \left(n_y + \frac{1}{2}\right)\omega_y + \left(n_z + \frac{1}{2}\right)\omega_z \tag{3.35}$$

The product of the three Hermite polynomials in (3.34) tell us about the symmetry of the wave function. For the ground state ($n_x = n_y = n_z = 0$), the product of the Hermite polynomials is just 1. This is the most symmetric function, since it does not change upon reflection, rotation, or any other operation. (The exponential term in (3.34) also does not change.) This 'fully symmetric' state is designated A_1. The first excited state is proportional to x, y, or z. For example, consider the state $n_x = 0$, $n_y = 1$, $n_z = 0$. The product of Hermite polynomials is $(1)(y)(1) = y$. Physically, this corresponds to an impurity oscillating along the y direction. The second excited states involve functions such as x^2 or xy.

A substitutional impurity in a diamond or zincblende crystal has tetrahedral (T_d) symmetry [34]. The first excited state wave functions are proportional to x, y, and z. These states are degenerate (same frequency) and the symmetry is labelled T_2. The second excited states are labelled A_1, E, and T_2. A substitutional impurity in a wurtzite crystal has trigonal (C_{3v}) symmetry. Trigonal symmetry is lower than tetrahedral symmetry because the z-axis is distinct from the x- and y-axes. The first excited state is labelled A_1 if it is proportional to z; E if it is proportional to x or y. The A_1 and E modes have different frequencies, since the spring constant along the z-axis is different than that along the x or y directions.

Impurity-hydrogen pairs in diamond or zincblende crystals usually have C_{3v} symmetry. An example is shown in Figure 3.7 for the phosphorus–hydrogen complex in silicon. Hydrogen attaches to a silicon atom in an *antibonding*, or back-bonded,

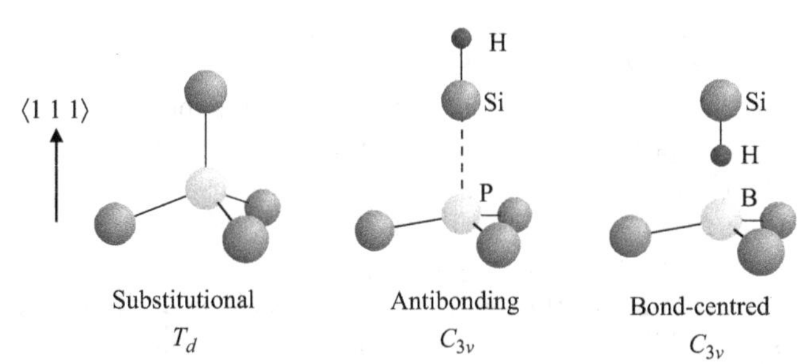

Figure 3.7 Ball-and-stick models for a substitutional impurity in a diamond or zincblende crystal, PH complex in silicon, and BH complex in silicon. Point-group symmetries are indicated

location. The first excited vibrational state has E or A_1 symmetry. The A_1 state is referred to as the *stretch mode* because the impurity-hydrogen bond is stretched and squeezed. The degenerate E modes are referred to *wag modes* (or bending modes) and involve motion in the x–y plane. Wag modes are softer (lower frequency) than stretch modes. The second excited states of wag modes are labelled A_1 and E. The boron–hydrogen complex in silicon also has C_{3v} symmetry. In that case, hydrogen is in a *bond-centred* position between silicon and boron (Figure 3.7).

3.1.10 Anharmonic coupling

The previous section considered stretch and wag modes to be independent of one another. The stretch mode involved motion along a longitudinal direction (z) while the wag mode was transverse (x–y plane). This is true only for a purely harmonic potential; i.e., a potential that does not contain cubic or higher order terms. In reality, the potential always has such anharmonic terms. Vibrational modes will interact with each other if they have the same symmetry. The stretch and wag modes acquire some degree of transverse and longitudinal motion, respectively.

As an example, consider the stretch mode (A_1) and the wag-mode second harmonic (A_1) of a hydrogen complex. In the absence of anharmonicity, the vibrational frequencies are ω_1 and ω_2. Now, consider an anharmonic interaction strength A. The frequencies are obtained by solving

$$\begin{vmatrix} \omega_1 - \omega & A \\ A & \omega_2 - \omega \end{vmatrix} = 0 \tag{3.36}$$

which leads to

$$\omega_\pm = \frac{1}{2}\left[\omega_1 + \omega_2 \pm \sqrt{(\omega_1 - \omega_2)^2 + 4A^2}\right] \tag{3.37}$$

The interaction prevents the frequencies from being degenerate, a phenomenon called *anticrossing* or *avoided crossing*. From (3.37), when $\omega_1 = \omega_2$, the anharmonic interaction splits the levels by $2A$. The modes hybridize so that each wave function is a linear combination of a stretch and wag mode.

This anharmonic interaction is called a *Fermi resonance*. The wag-mode second harmonic absorbs IR light weakly because it is an $n = 0$ to $n = 2$ transition. (Were it not for anharmonic terms, such a transition would be forbidden.) When the stretch and wag modes are nearly degenerate, the two states hybridize so that they both contain significant stretch-mode components. Then they can both absorb IR light, resulting in two IR peaks in the absorption spectrum. Note that the stretch mode does not interact with the E wag-mode because the symmetries are different.

A classic case of an anharmonic interaction was observed for donor–hydrogen complexes in silicon [35]. The stretch mode frequencies of ~1,600 cm^{-1} are quite close to twice the wag frequency (2 × 800 cm^{-1}). A Fermi resonance was also reported for the boron–deuterium complex in silicon [36]. In that case, the second harmonic of the B vibration is nearly degenerate with the B–D stretch mode. Additional examples of this interesting phenomenon are given in Sections 3.3.1.3 and 3.3.3.

3.1.11 Raman scattering

IR absorption spectroscopy is the most popular method for investigating LVMs in semiconductors. Another technique, especially important for IR-inactive modes or thin films, is Raman scattering. This process, discovered by Raman [37], involves inelastic scattering of light from a sample. In the *Stokes* process, the incoming photon loses energy, which goes into the creation of a vibrational mode. In the *anti-Stokes* process, a vibrational mode is destroyed, resulting in an increase in the photon energy.

The mechanism behind Raman scattering involves the polarizability α. When an electric field E is applied to an atom, the positive nucleus and electron cloud are pulled in opposite directions. This results in a polarization given by

$$p = \alpha E \tag{3.38}$$

Let the electric field of the incoming light be

$$E = E_L \cos(\omega_L t) \tag{3.39}$$

where E_0 is the amplitude and ω_L is the frequency. Inserting (3.39) into (3.38) yields

$$p = \alpha E_L \cos(\omega_L t) \tag{3.40}$$

Consider a diatomic molecule that vibrates with a frequency ω_0. The bond length is stretched or compressed by an amount Q,

$$Q = Q_0 \cos(\omega_0 t) \tag{3.41}$$

This change in bond length alters the polarizability. To first order,

$$\alpha = \alpha_0 + \alpha_1 Q_0 \cos(\omega_0 t) \tag{3.42}$$

where α_0 and α_1 depend on the particular atoms. Plugging (3.42) into (3.40) yields

$$p = \alpha_0 E_L \cos(\omega_L t) + \alpha_1 E_L \cos(\omega_L t)\, Q_0 \cos(\omega_0 t) \tag{3.43}$$

The first term on the right-hand side is elastic (Rayleigh) scattering. The second term is the Raman scattering term. By a trigonometric identity, it can be written as

$$p_{\text{Raman}} = \left(\frac{\alpha_1 E_L Q_0}{2}\right)\{\cos[(\omega_L + \omega_0)t] + \cos[(\omega_L - \omega_0)t]\} \tag{3.44}$$

This oscillating dipole emits radiation, like an antenna, of frequencies $\omega_L - \omega_0$ (Stokes) and $\omega_L + \omega_0$ (anti-Stokes).

This classical derivation shows that the emitted radiation frequency is up-shifted or down-shifted due to interactions between a vibrational mode and the incoming radiation. It assumes there is already a vibrational mode that has been excited. In the correct quantum mechanical picture, the Stokes photon is due to the *creation* of a vibrational mode, while an anti-Stokes photon results from the *destruction* of a vibrational quantum. The intensities of these lines will depend on the sample temperature. At the extreme cold end, there are no thermally populated vibrational modes, so the anti-Stokes intensity goes to zero.

3.2 Experiment

3.2.1 Raman spectroscopy

In a Stokes process, an incoming photon gives some of its energy to a vibrational excitation. The energy of that excitation is inferred from the difference between the incident and scattered photon energy.

A Raman experiment is conceptually simple. A monochromatic light source, usually a laser, impinges on the sample. Scattered light is collected by a monochromator, which provides the spectrum. The spectrum is displayed as intensity versus wavenumber (cm^{-1}), where wavenumber is the frequency difference between the laser photon and the scattered photon.

An experimental difficulty arises because the Raman peaks are close to the excitation wavelength. The Raman peak is typically weaker than the elastically scattered light by 4–6 orders of magnitude [5]. Raman peaks from trace impurities can be very weak indeed. A notch filter is employed to suppress the laser wavelength while transmitting other wavelengths. Double or triple monochromators can be used to further reject light from the excitation source. Sensitive detectors such as image intensifiers and cooled charge-coupled devices (CCDs) are used for single-photon detection.

By using polarizers, it is possible to determine the symmetry of a Raman-active vibrational mode. This is discussed by Cardona [4] and McCluskey and Haller [25]. Raman scattering spectra of the N–H complex in ZnSe are shown in Figure 3.8 [38]. In this experiment, an argon-ion laser beam (514.5 nm) was incident nearly normal to the (1 0 0) surface. The scattered light entered a single

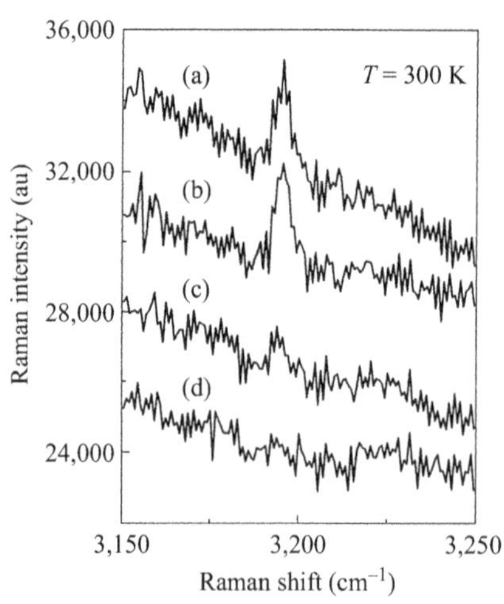

Figure 3.8 Raman spectra of ZnSe:N,H. Polarization conditions are (a) laser and scattered light [0 0 1], (b) laser and scattered light [0 1 1], (c) laser [0 0 1], scattered light [0 1 0], and (d) laser [0 1 1], scattered [0 1 $\bar{1}$]. © 1993. Reprinted, with permission, from [38]

monochromator and was detected by a photomultiplier tube. The figure shows the Raman intensities for various polarizations of incident and scattered light. From the strengths of the Raman line, the authors confirmed that the symmetry of the N–H complex is C_{3v}.

3.2.2 *Fourier transform infrared spectroscopy*

Fourier transform IR (FTIR) spectroscopy has high spectral resolution and the ability to measure a broad spectral region [39,40]. An FTIR spectrometer is based on the Michelson interferometer (Figure 3.9). A beam of collimated light from a light source (such as a heated ceramic) impinges on a beamsplitter. The beam is split into two beams that take different paths. One beam travels to a movable mirror while the other beam goes to a fixed mirror. The beams reflect off their respective mirrors and recombine. The recombined beam travels through the sample and is collected by a detector. The detector produces a signal that is proportional to the intensity of this beam. As one mirror is moved, the intensity versus optical path difference (δ) is plotted in an *interferogram*. Many interferograms can be taken and averaged in order to maximize the signal to noise ratio.

The interferogram $I(\delta)$ is peaked at zero path difference ($\delta = 0$), where all the wavelengths constructively interfere. As the moving mirror goes to nonzero δ values, some wavelengths interfere constructively while others interfere destructively. This results in oscillations in the interferogram. The spectrum (intensity versus frequency) is obtained by performing a fast Fourier transform on the interferogram,

$$I_0(k) = \int_0^\infty I(\delta)\cos(k\delta)\, d\delta \tag{3.45}$$

where $k = 2\pi/\lambda$. Note that spectra are conventionally plotted as intensity versus wavenumber ($1/\lambda$, in units of cm^{-1}).

Equation (3.45) describes a hypothetical mirror that adjusts the path difference from zero to infinity. In reality, of course, the mirror travel is finite. This limitation

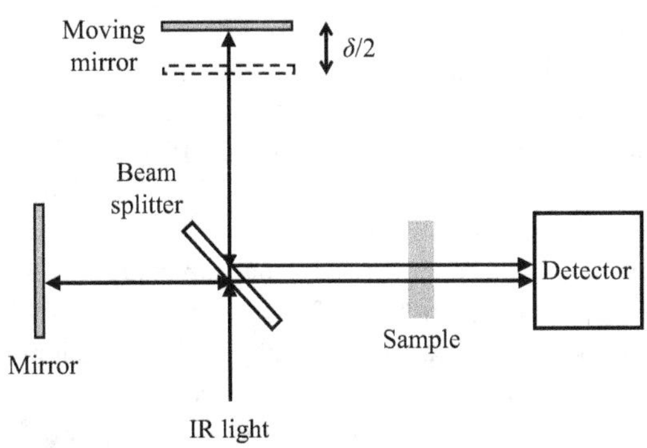

Figure 3.9 Schematic diagram of an FTIR spectrometer

introduces instrumental broadening. To see this, consider a monochromatic source (e.g. a laser). The interferogram is given by

$$I(\delta) = I_0 \cos(k_0\delta) \tag{3.46}$$

Given a maximum optical path difference L, the calculated spectrum is

$$I_0(k) = \int_0^L I_0 \cos(k_0\delta)\cos(k\delta)\, d\delta \sim \frac{\sin[(k-k_0)L]}{(k-k_0)L} + \frac{\sin[(k+k_0)L]}{(k+k_0)L} \tag{3.47}$$

Consider a mirror travel of $L \sim 1$ cm and mid-IR source with $k_0 \sim 1{,}000$ rad/cm. For these typical values, $k_0L \gg 1$, and we can safely drop the second term. The spectrum is then given by a sinc function,

$$I_0(k) \sim \mathrm{sinc}[(k-k_0)L] \tag{3.48}$$

where sinc $z = \sin(z)/z$. This line shape has a maximum at k_0, sidelobes, and a full width of $\Delta k = 2\pi/L$. The resolution in units of wavenumbers (cm^{-1}) is

$$\Delta\left(\frac{1}{\lambda}\right) \sim \frac{1}{L} \tag{3.49}$$

Thus, the resolution improves as the scanning length increases. Commonly, the interferogram is multiplied by an *apodization* function to reduce the sidelobes. A common apodization function is a straight line that is positive at $\delta = 0$ and zero at $\delta = L$.

3.2.3 IR pump–probe

Another experimental approach useful for studying vibrations is the *pump–probe* technique. In a pump–probe experiment, a short laser pulse is split into a pump and probe pulse, typically with an intensity ratio of 10:1. The probe pulse is delayed by a set amount of time by sending it on a longer path. The pump and probe beams impinge on the same area on the sample, and the intensity of the transmitted probe beam is measured. An optical chopper and a lock-in amplifier are used to maximize the signal-to-noise ratio. The lock-in signal is proportional to the change in transmission due to the pump pulse.

Short IR pulses can be produced by a free-electron laser or optical parametric oscillator. Pump–probe experiments provide a uniquely well-defined way to determine vibrational lifetimes of LVMs in semiconductors [29]. In these studies, the laser frequency is tuned to match a specific LVM. The pump pulse excites the impurity (typically hydrogen) into an excited vibrational state ($n = 1$). When it is in the $n = 1$ state, the impurity cannot absorb the pump photon because the $1 \rightarrow 2$ transition is off-resonance (Equation (3.12)). This results in an increase in transmitted probe intensity known as transient bleaching. The transient bleaching decays exponentially as the impurities return to their ground state. The time constant for this exponential decay gives the ground-state recovery lifetime.

Bond-centred hydrogen in silicon resides between two silicon atoms and has an LVM frequency of 1,998 cm^{-1}. Transient bleaching pump–probe experiments determined a lifetime of 8 ps [41]. Other hydrogen defects have much different lifetimes. The divacancy–hydrogen complex in silicon has an LVM at 2,072 cm^{-1} with a vibrational lifetime of 296 ps. This complex, which has two neighbouring silicon vacancies and two hydrogen atoms, has a weak interaction between the hydrogen atoms and the lattice. In essence, the hydrogen impurities are in a region of empty space, so they transfer vibrational energy to the host lattice slowly. Bond-centred hydrogen, on the other hand, is squeezed between two host atoms and transfers vibrational energy quickly.

3.2.4 *Applied stress*

By probing a semiconductor with mechanical stress, one can learn about the point-group symmetry of the defect and how strongly an impurity interacts with the lattice. The most common types of applied stress are *uniaxial stress* and *hydrostatic pressure*. To generate uniaxial stress, a gas-driven push rod compresses a crystal along a particular direction [2]. Paper is sometimes placed between the push rod and sample to prevent the sample from breaking. Owing to the low yield strength of semiconductor crystals, uniaxial stress is limited to 1–2 GPa.

When stress is applied, symmetry is lowered and hydrogen complexes preferentially orient along a specific direction. A famous example of this effect is the reorientation of boron–hydrogen complexes in silicon [42]. In the absence of stress, the B–H pairs are oriented randomly along equivalent ⟨1 1 1⟩ directions. As shown in Figure 3.10, stress that is applied along the [1 1 0] direction makes two ⟨1 1 1⟩

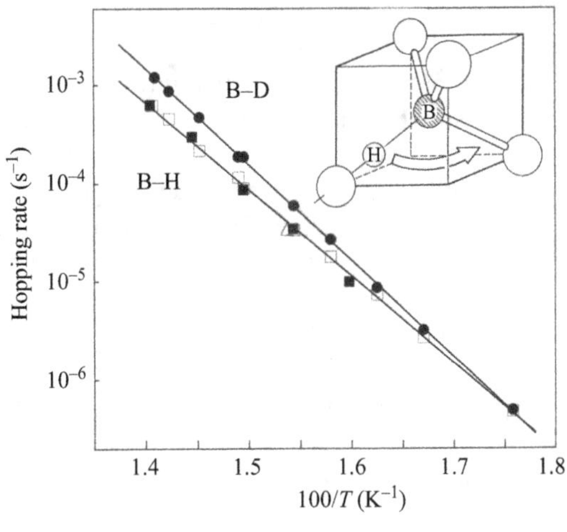

Figure 3.10 Reorientation rate for hydrogen in Si:B under uniaxial stress. Upon stress release, the dichroism (Equation (3.50)) decreases as hydrogen atoms return to their randomly oriented configurations. The decay rate for this process was measured for different temperatures and plotted. © 1988. Reprinted, with permission, from [42]

orientations inequivalent to the other two ⟨1 1 1⟩ orientations. At temperatures above 60 K, the hydrogen atom can quickly hop between the four sites and will preferentially occupy the lower energy sites. The resulting anisotropy is described by the *dichroism*

$$D = \frac{\alpha_{\perp} - \alpha_{//}}{\alpha_{\perp} + \alpha_{//}} \tag{3.50}$$

where $\alpha_{\perp}$ and $\alpha_{//}$ are the LVM absorption coefficients for IR light polarized perpendicular or parallel to the applied stress. Upon releasing the stress, the sites become equivalent and D decays with time. The temperature dependence of this decay for boron–hydrogen complexes indicates interesting reorientation kinetics that do not follow a simple Arrhenius model [43].

Large hydrostatic pressures are generated by a diamond-anvil cell [44]. Diamonds are mostly transparent in the visible and IR, enabling IR or Raman studies of materials up to megabar (100 GPa) pressures. A sample is placed in a hole drilled in a metal gasket, along with an ambient such as liquid argon, liquid nitrogen, mineral oil, or methanol–ethanol mixture. The flat parallel faces of two diamond anvils press on the metal gasket, transmitting pressure to the sample. Unlike uniaxial stress, hydrostatic pressure does not change the symmetry of the crystal.

3.3 Examples

3.3.1 Interstitial oxygen

3.3.1.1 Silicon

Oxygen is a prevalent impurity in Czochralski-grown silicon. Most of these impurities are in the interstitial form (O_i), in which oxygen binds to two silicon atoms. The minimum-energy location for the oxygen is off-centre. However, the ground-state vibrational wave function is centred on the ⟨1 1 1⟩ axis, so the defect really behaves like a linear *XYX* molecule [45]. Such a molecule has an IR-inactive symmetric stretch mode (ν_1), an IR-active transverse mode (ν_2), and asymmetric stretch mode (ν_3) (Section 3.1.5).

Due to silicon's enormous technological importance, the *asymmetric stretch* mode of O_i has been the subject of intense research [46]. The natural isotope abundances are 99.76 per cent oxygen-16, 0.04 per cent oxygen-17, and 0.20 per cent oxygen-18. The $^{16}O_i$, $^{17}O_i$, and $^{18}O_i$ asymmetric modes have frequencies of 1,136, 1,107, and 1,084 cm^{-1}, respectively (Table 3.3). Different silicon isotopes (92.2 per cent silicon-28, 4.7 per cent silicon-29, and 3.1 per cent silicon-30) result in frequency shifts of a few cm^{-1} (Figure 3.11).

Using a spring-and-mass approach for a bent *XYX* molecule, the isotopic shift of the ν_3 mode is given by [47]

$$\frac{\omega'}{\omega} = \frac{M_X M_Y \left(M'_Y + 2M'_X \sin^2 \alpha\right)}{M'_X M'_Y \left(M_Y + 2M_X \sin^2 \alpha\right)} \tag{3.51}$$

Table 3.3 Vibrational modes of $^{16}O_i$ in silicon. After Pajot [48] and McCluskey and Haller [25]

Frequency (cm^{-1})	Mode
29	Transverse (ν_2)
518	Resonant
613	Symmetric stretch (ν_1)
1,136	Asymmetric stretch (ν_3)
1,749	Combination ($\nu_1 + \nu_3$)

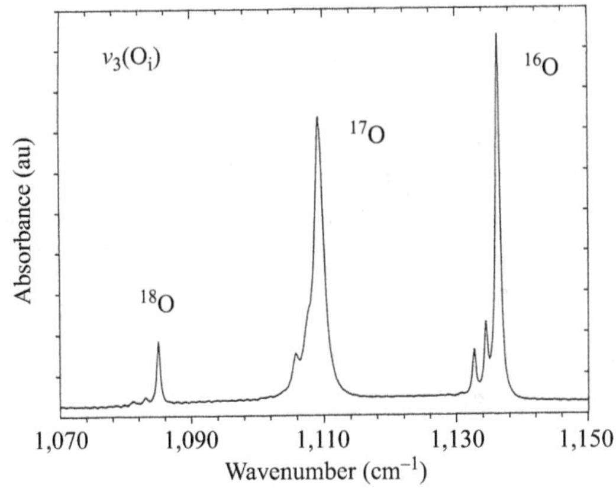

Figure 3.11 IR spectrum of interstitial oxygen in silicon deliberately doped with ^{17}O and ^{18}O. The small sideband peaks correspond to Si–O–Si where one silicon atom is ^{29}Si or ^{30}Si. © 1994. Reprinted, with permission, from [48]

where M_X and M_Y are the atomic masses of silicon and oxygen, primes refer to values after changing isotope masses, and 2α is the bond angle. The mass M_X is given by an average of the individual silicon masses M_1 and M_2,

$$M_X = \frac{1}{2}\chi(M_1 + M_2) \tag{3.52}$$

where χ is an empirical factor that accounts for the fact that the molecule is not free but rather embedded in a lattice. Using (3.50) to fit the experimental frequency shifts, a bond angle of $2\alpha = 164°$ was obtained [48].

While the asymmetric stretch mode is an oscillation in the $\langle 1\ 1\ 1 \rangle$ direction, *transverse* modes are in the $\{1\ 1\ 1\}$ plane. The transverse fundamental mode is very soft, with a frequency of 29 cm^{-1} [49]. The oxygen potential energy is similar to a two-dimensional (2D) harmonic potential, proportional to $x^2 + y^2$, but with a small bump in the centre. Although the minimum of this potential is off-centre, the wavefunctions are spread out and centred at the origin.

Interstitial oxygen in the ground state can be excited to an asymmetric stretch mode. This transition is labelled I (Figure 3.12). As temperature is increased, transverse-mode excited states become thermally populated. Transitions from these excited states are labelled II, III, and IV. Because there is anharmonic coupling between the asymmetric stretch mode and transverse modes, the frequencies of the various transitions are different. Peak II, for example, is 8 cm^{-1} below peak I. Yamada-Kaneta *et al.* [50] introduced a quartic potential that accurately modelled the experimental vibrational frequencies.

The *symmetric stretch* mode is IR-inactive, so its frequency must be inferred from those of other modes. Specifically, there is a weak absorption peak at 1,749 cm^{-1} that is a replica of the asymmetric-stretch 1,136 cm^{-1} peak. The 1,749 cm^{-1} peak is assigned to a combination mode: asymmetric (ν_3) plus symmetric (ν_1) stretch mode. This assignment implies $\nu_1 = 613$ cm^{-1} ($1{,}136 + 613 = 1{,}749$ cm^{-1}).

Finally, interstitial oxygen has a *resonant mode* at 518 cm^{-1}. It is resonant because the frequency lies in the phonon band. The vibrational mode is therefore delocalized, involving oscillations of many silicon neighbours. Unlike the 1,136 cm^{-1} peak, which shows a strong oxygen-isotope dependence, the resonant mode does not shift with oxygen mass. The insensitivity occurs because most of the vibrational motion involves silicon atoms rather than the oxygen impurity.

3.3.1.2 Germanium

Like silicon, interstitial oxygen in germanium attaches to two host atoms. Unlike silicon, where the Si–O–Si defect is essentially linear, Ge–O–Ge bond angle is $2\alpha = 111°$ [51]. This means that the oxygen atom is displaced from the $\langle 1\ 1\ 1 \rangle$ axis.

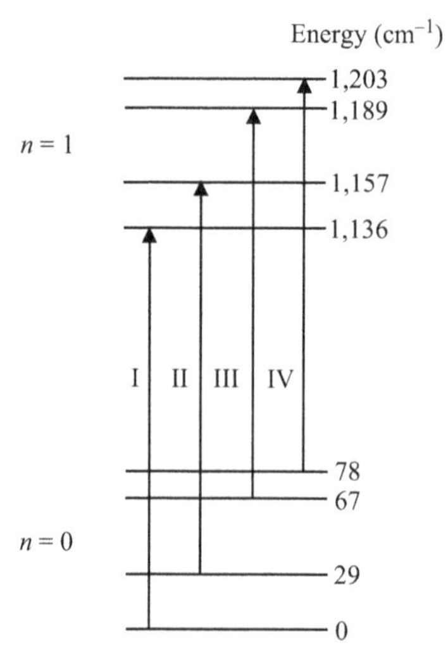

Figure 3.12 Energy levels for Si:O$_i$, where n refers to the asymmetric mode. The levels within each n level are transverse modes. Vertical arrows indicate IR-allowed transitions. After [49]

Instead of acting like a perturbed 2D oscillator, the transverse motion is rotational. The oxygen atom rotates about the ⟨1 1 1⟩ axis with quantum-mechanical levels given by

$$\omega = \frac{\hbar l^2}{2MR^2} \tag{3.53}$$

where l is the angular momentum quantum number, M is the oxygen mass, and R is the radius of rotation.

At room temperature, the asymmetric stretch mode of interstitial oxygen in germanium gives rise to a broad IR absorption band at 855 cm^{-1} at room temperature [52]. At liquid-helium temperatures, sharp peaks can be resolved (Figure 3.13). These peaks are due to the five stable germanium isotopes, which result in 11 average masses (Equation (3.52)). In addition, the transverse modes are thermally populated, resulting in transitions labelled I, II, III, and IV (Figure 3.12). The IR spectrum becomes much simpler when isotopically pure Ge:O samples are used. By clearing the spectral congestion, it is possible to measure the growth of II, III, and IV as temperature is raised. The temperature dependence is in excellent agreement with the rotor model (Equation (3.53)).

3.3.1.3 Silicon under pressure

Interstitial oxygen in silicon has served as a model system to study the vibrational properties of defects in semiconductors. The Si–O–Si defect is linear, aligned along a ⟨1 1 1⟩ direction. The oxygen wavefunction is centred at the ⟨1 1 1⟩ axis. Under hydrostatic pressure, however, the bonds get squeezed and the Si–O–Si defect

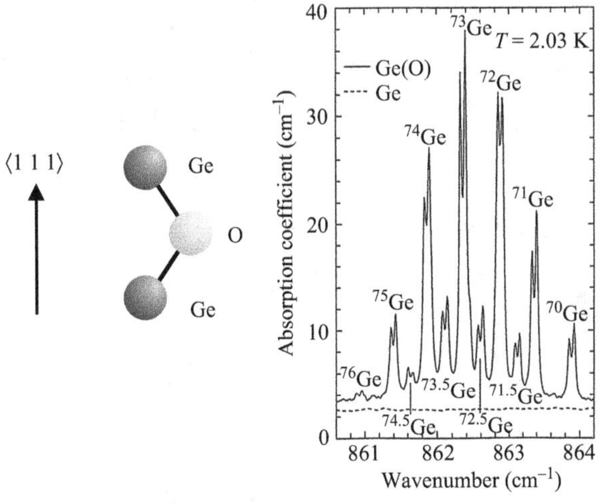

Figure 3.13 IR spectrum of Ge:O_i. The average mass of germanium in Ge–O–Ge is indicated. The splitting of each peak is due to I and II transitions (Figure 3.12). © 1994. Reprinted, with permission, from [51]

begins to buckle [53]. The O atom is pushed away from the ⟨1 1 1⟩ axis, similar to the case in germanium. With increasing pressure, the motion of the O atom becomes more rotational than vibrational and its excited-state energies follow (3.53). To my knowledge, this transition from vibrational to rotational motion has not been seen in any other material.

As the O buckles outward, the asymmetric mode (1,136 cm^{-1}) softens slightly. Meanwhile, the silicon phonon energies increase with pressure. This allowed us to study the effect of phonons on the vibrational lifetime [54]. Because the asymmetric-mode frequency is more than twice that of the maximum phonon frequency, it must decay into three or more phonons. Since more available states mean a shorter lifetime, the linewidth is proportional to the three-phonon density of states [55]. For pressures above 4 GPa, the $^{18}O_i$ asymmetric mode enters the two-phonon continuum. At that point, it can decay into only *two* phonons, resulting in a significant drop in lifetime and linewidth broadening (Figure 3.14).

In addition to this abrupt broadening due to phonons, Hsu, McCluskey, and Lindstrom [54] also observed a Fermi resonance between the asymmetric mode and the second harmonic of the resonant mode. This anharmonic interaction led to an avoided crossing of the two vibrational excitations. Because of the interaction, the states become hybrids between a localized and extended mode. This unusual phenomenon is similar to that seen in AlSb:Se,H, discussed in Section 3.3.3.

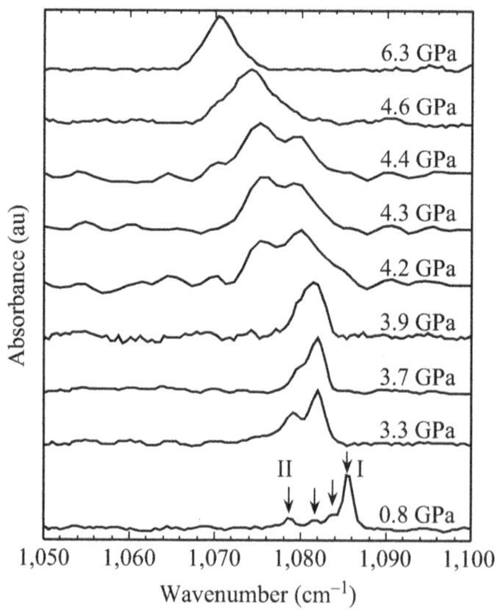

Figure 3.14 IR spectra of Si:^{18}O under pressure. An avoided crossing at 4.2 GPa is due to an interaction between the asymmetric mode and the second harmonic of a resonant mode. For pressures >4 GPa, the peak broadens as it enters the two-phonon band. © 2003. Reprinted, with permission, from [54]

3.3.2 Impurities in GaAs

3.3.2.1 Carbon

Carbon in GaAs is an acceptor that sits on the arsenic site, with LVM frequencies of 582 and 561 cm^{-1} for $^{12}C_{As}$ and $^{13}C_{As}$, respectively [56]. IR spectroscopy [57] shows fine structure that arises from the various isotope combinations of ^{69}Ga and ^{71}Ga neighbours (Figure 3.15). Because arsenic has only one stable isotope (^{75}As), the isotopic fine structure provides unambiguous evidence that carbon's nearest neighbours are gallium atoms; i.e., carbon occupies an arsenic site.

Like the fundamental mode, the overtone ($n = 0 \rightarrow 2$) at 1,162 cm^{-1} also shows fine structure due to gallium isotopes [58]. As temperature is raised, the excited $n = 1$ vibrational level becomes thermally populated [59]. This results in the appearance of an $n = 1 \rightarrow 2$ transition at 576 cm^{-1} (room temperature value). The $1 \rightarrow 2$ transition has a slightly lower frequency than the $0 \rightarrow 1$ transition (580 cm^{-1} at room temperature) because of the anharmonic potential (Section 3.1.4).

GaAs:C thin films grown by metalorganic chemical vapour phase deposition (MOCVD) contain carbon–hydrogen complexes. Hydrogen attaches to the carbon acceptor in a bond-centred site, forming a neutral C_{As}–H pair. From IR spectroscopy, the stretch modes of ^{12}C–H and ^{12}C–D complexes have LVM frequencies of 2,635 and 1,969 cm^{-1}, which gives an isotopic frequency ratio $r = 1.3386$. Lines that come from ^{13}C are also observed with the appropriate intensities (98.93 per cent carbon-12 and 1.07 per cent carbon-13) and frequency shifts.

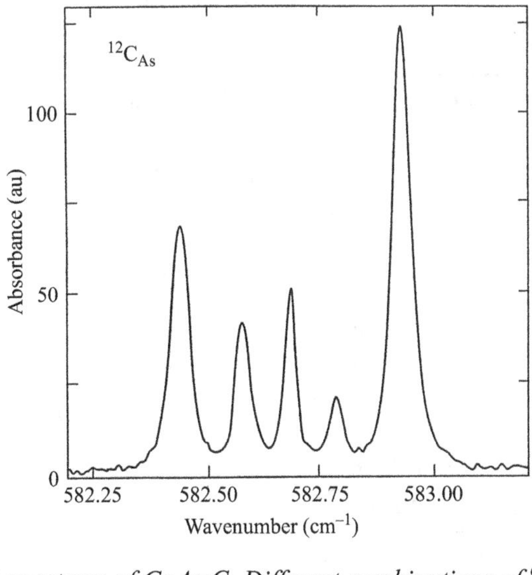

Figure 3.15 IR spectrum of GaAs:C. Different combinations of ^{69}Ga and ^{71}Ga give rise to the fine structure. © 1993. Reprinted, with permission, from [12]

The C–H complex in GaAs is a model system with well-understood vibrational modes [60]. As is typical, the *stretch mode* has A_1 symmetry and is characterized by the carbon and hydrogen atoms oscillating out of phase along the ⟨1 1 1⟩ direction. The carbon and hydrogen can also oscillate in phase along the ⟨1 1 1⟩ direction in the *longitudinal mode*. For the longitudinal mode, the C–H pair acts like a single unit, so the frequency depends on the C–H mass. The in-phase motion results in a much lower frequency than the stretch mode (Table 3.4).

Along with the two vibrational modes along the ⟨1 1 1⟩ direction, there are two modes in the {1 1 1} plane. Both have E symmetry. The *wag mode*, or bending mode, is where the carbon and hydrogen oscillate out of phase. This is like a small dog wagging its tail – in the case of deuterium, the tail wags the dog almost as much as the dog wags the tail! The *transverse mode* involves in-phase oscillation of the carbon and hydrogen. A fairly simple spring-and-mass model [60] explains the frequencies of all these modes (Table 3.4).

In addition to pairing with hydrogen, carbon can also interact with nitrogen to form a CN_2 complex with an IR absorption peak at 2,059 cm^{-1} [61]. By implanting GaAs with different nitrogen isotopes (^{14}N and ^{15}N), Alt *et al.* [62] observed peaks arising from $^{12}C^{15}N_2$ (2040 cm^{-1}) and $^{12}C^{14}N^{15}N$ (2050 cm^{-1}). Peaks at 1,972 and 1,964 cm^{-1} were tentatively ascribed to a CN pair ($^{12}C^{14}N$ and $^{12}C^{15}N$).

Table 3.4 Local vibrational modes of carbon–hydrogen complexes in III–V semiconductors. Unit vectors indicate the relative motion of the carbon and hydrogen atoms, where an 'up' arrow denotes a ⟨1 1 1⟩ direction

Mode	**Isotopes**	**Frequency (cm^{-1})**					
		AlAs [154]	**AlSb [155]**	**GaP [156]**	**GaAs [60]**	**InP [157]**	**InAs [158]**
Stretch (A_1)	^{12}C–H	2,558	2,567	2,660	2,635	2,703	2,687
↑	^{13}C–H	2,550		2,653	2,629		2,679
↓	^{12}C–D	1,903		1,981	1,969		
	^{13}C–D	1,894			1,958		
Longitudinal (A_1)	^{12}C–H	487		477	453	414	393
↑	^{13}C–H	477			438		379
↑	^{12}C–D	480			440		
	^{13}C–D	471			427		
Wag (E)	^{12}C–H	671			739		
→	^{13}C–H	653			730		
←	^{12}C–D	657			637		
	^{13}C–D	635			617		
Transverse (E)	^{12}C–H				563	521	518
→	^{13}C–H				548		503
→	^{12}C–D				466		
	^{13}C–D				464		

3.3.2.2 Silicon

Silicon in GaAs is an amphoteric impurity. It prefers to sit on the gallium site and act as a donor (Si_{Ga}) but occasionally occupies the arsenic site and acts as an acceptor (Si_{As}). The substitutional $^{28}Si_{Ga}$ donor has a sharp IR peak with a frequency of 384 cm^{-1} [63]. The narrow linewidth (0.4 cm^{-1}) does not show isotopic fine structure because the nearest neighbours are all ^{75}As. $^{28}Si_{As}$ acceptors show a peak at 399 cm^{-1} with fine structure resulting from isotopic combinations of ^{69}Ga and ^{71}Ga neighbours, similar to the case of carbon (Section 3.3.2.1).

In heavily doped GaAs:Si, silicon forms Si_{Ga}–Si_{As} nearest neighbour pairs [64]. The Si_{Ga}–Si_{As} pair is aligned along a $\langle 1\ 1\ 1 \rangle$ direction and has C_{3v} symmetry. Transverse and longitudinal vibrational modes have been observed at 464 and 393 cm^{-1}, respectively, where the two silicon atoms oscillate out of phase. The corresponding in-phase modes have not been observed, perhaps because they have low frequencies that are in the *reststrahlen* band. (The *reststrahlen* band between the TO and LO phonon frequencies has high reflectivity that effectively precludes transmission measurements.)

In GaAs under large hydrostatic pressures, shallow Si_{Ga} donors transform into deep-level defects called *DX* centres. In the *DX* configuration, silicon breaks a bond with arsenic and displaces along a $\langle 1\ 1\ 1 \rangle$ direction (Figure 3.16), reducing the symmetry from T_d to C_{3v} [65]. The Si impurity has vibrational modes along the $\langle 1\ 1\ 1 \rangle$ direction (A_1) and in the $\{1\ 1\ 1\}$ plane (E). The estimated ambient-pressure frequency of the E mode is 376 cm^{-1} [66]. This value agreed with first-principles calculations [67]. The A_1 mode was not observed experimentally, presumably because its low frequency lies in the phonon band.

Silicon donors can be passivated by hydrogen, forming Si_{Ga}–H pairs where the hydrogen attaches to silicon in an antibonding configuration (Figure 3.16). The $^{28}Si_{Ga}$–H complex has wag and stretch modes at 896 and 1,717 cm^{-1}, respectively [68]. Deuterated samples show $^{28}Si_{Ga}$–D frequencies at 647 and 1,248 cm^{-1}. Sideband peaks, corresponding to the rarer isotopes $^{29}Si_{Ga}$–H and $^{30}Si_{Ga}$–H, show frequency shifts that are consistent with the diatomic model. These isotopic frequency shifts proved that hydrogen forms a bond with silicon as opposed to a host arsenic atom.

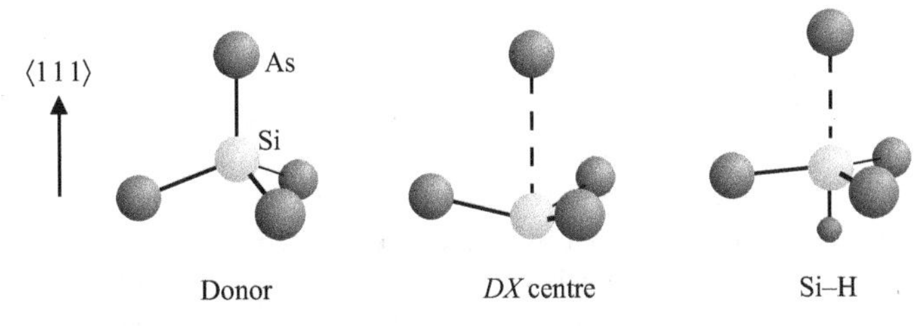

Figure 3.16 Ball-and-stick models for Si_{Ga} in GaAs: substitutional donor, DX centre, and Si–H complex

3.3.2.3 Oxygen

Interstitial oxygen in GaAs binds to Ga and As. Similar to oxygen in germanium (Section 3.3.1.2), it occupies an off-axis position and has an asymmetric mode at 845 cm^{-1} [69]. The mode splits into a doublet with each peak due to a ^{69}Ga or ^{71}Ga isotope. (^{75}As is the only stable arsenic isotope.) The corresponding $^{18}O_i$ mode has a frequency of 802 cm^{-1}.

Oxygen can also occupy an As vacancy, where it binds to two Ga atoms. This configuration is similar to the *A* centre in silicon [70]. The 0, −1, and −2 charge states give rise to respective vibrational frequencies of 731, 714, and 715 cm^{-1}, respectively [71]. Each of these peaks shows fine structure due to the three gallium isotope combinations. The 715 cm^{-1} mode, which is the strongest peak in semi-insulating GaAs, has a corresponding ^{18}O mode at 679 cm^{-1}.

Oxygen and hydrogen impurities come from wet boric oxide during liquid-encapsulated Czochralski growth of GaAs, forming OH defects with stretch modes near 3,000 cm^{-1} [72]. Heavy water (D_2O) was introduced to observe the OD lines [73]. The authors reported O–H (O–D) stretch modes at 3,079 (2,288) cm^{-1} and wag modes at 983 (729) cm^{-1}. The peaks were assigned to a neutral OH deep donor. Illumination caused the peaks to decrease in intensity and a new set of peaks to appear. The new peaks were attributed to the positive charge state.

3.3.3 *Resonant interaction in AlSb*

When AlSb:Se or AlSb:Te are annealed in hydrogen, the group-VI donors are passivated [74]. Hydrogen attaches to a host Al in an antibonding configuration. The hydrogen wag mode fundamental and second harmonic were observed in IR spectra. In deuterated samples, the Al–D stretch mode was observed. AlSb:Te also showed a single Al–H stretch-mode peak. However, AlSb:Se,H showed *two* peaks (Figure 3.17).

The presence of two peaks can be explained by a Fermi resonance between the stretch mode and a 'combination mode' that is the sum of several low-frequency modes [75]. Under pressure, these two modes show avoided crossing behaviour as they go into and out of resonance. The existence of the Fermi resonance is purely accidental. The stretch mode (1,613 cm^{-1}) just happens to be nearly degenerate with the combination mode (1,611 cm^{-1}). The other complexes (AlSb:Te,H; AlSb:Te,D; AlSb:Se,D) have different vibrational frequencies that do not result in resonant behaviour.

First-principles calculations were performed to determine the identity of the combination mode [21]. The results suggested that the combination is a second-harmonic wag mode (1,333 cm^{-1}) plus an Al–H transverse mode (278 cm^{-1}). The transverse mode involves in-phase motion of the Al and H atoms in the $\langle 1\ 1\ 1 \rangle$ plane, similar to C–H in GaAs (Section 3.3.2.1). The transverse mode is a gap mode and is much less localized than the wag or stretch modes.

When the stretch mode is in resonance with the combination mode, the two modes hybridize, and each mode has 'stretch-like' and 'combination-like' character. Unlike a conventional Fermi resonance, the combination mode is much more delocalized than the stretch mode. The vibrational wavefunctions are hybrids of localized and delocalized modes. I refer to these novel excitations as a *localons*.

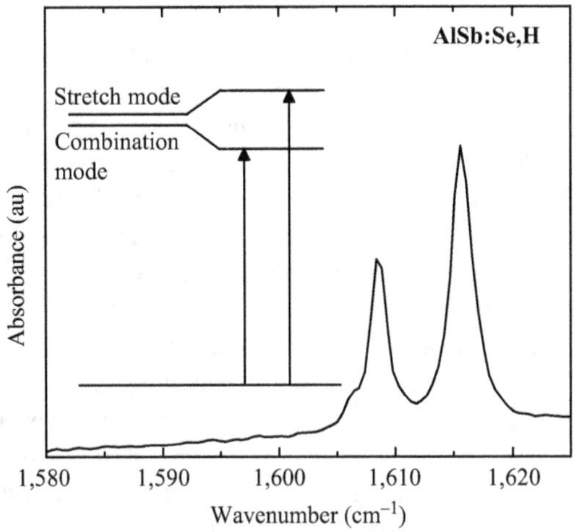

Figure 3.17 IR spectrum of SeH in AlSb. A resonant interaction between the stretch mode and a combination of lower frequency modes results in two IR peaks. After [21]

3.3.4 Impurities in CdTe

CdTe currently enjoys the distinction as the second most popular solar cell material (after silicon, of course). There are only a few studies of LVMs in CdTe, however. High-resolution IR absorption experiments found sharp lines corresponding to Mg_{Cd}, Ca_{Cd}, and S_{Te} impurities [76]. The LVM peak of Mg in a $Cd_{1-x}Zn_xTe$ alloy shows broadening that is maximized at $x = 0.5$. Ti_{Cd} in CdTe shows rich host-isotope structure due to the various Ti and Te isotope combinations [77].

Hydrogen passivates As acceptors in CdTe, leading to a stretch mode at 2,022 cm^{-1} [78]. In MOCVD-grown CdTe:As, this hydrogen passivation results in neutral As–H complexes and semi-insulating material. The As acceptors can be 'activated' by annealing the sample, dissociating the As–H pairs and producing p-type conductivity.

CdTe annealed in $CdSO_4$ shows IR absorption peaks at 1,097 and 1,108 cm^{-1} [79], attributed to an SO_2 molecule [80]. First-principles calculations [81] indicate that the molecule sits on a Te site, $(SO_2)_{Te}$, adjacent to a cadmium vacancy, V_{Cd}. Two different orientations of the molecule lead to the two IR peaks. As temperature increases, the molecule is free to rotate to the lowest energy configuration; at room temperature, the two peaks have merged into one.

3.3.5 Hydrogen in silicon

In silicon, bond-centred interstitial hydrogen sits between two silicon atoms and has positive charge. Its stretch mode has a frequency of 1,998 cm^{-1}. As temperature is raised, the hydrogen becomes increasingly mobile. Above 180 K, the 1,998 cm^{-1} IR peak disappears altogether as hydrogen atoms diffuse and become trapped by defects. One relatively stable form of hydrogen in proton-implanted silicon is labelled H_2^* [82], which consists of hydrogen atoms in the bond-centred and antibonding locations. These two hydrogen atoms result in Si–H stretch modes at 1,839 and 2,062 cm^{-1}.

The self-interstitial *I* consists of two Si atoms that share a single substitutional site. In proton-implanted silicon, hydrogen atoms attach to each Si, forming IH_2. The two Si–H bonds result in stretch modes at 1,987 and 1,989 cm^{-1}; wag modes at 743 and 748 cm^{-1} [83]. Vacancy-hydrogen complexes are discussed in Section 3.3.10.

Vibrational spectroscopy was important for determining the structure of dopant-hydrogen complexes in silicon. Pankove *et al.* [84] observed a room-temperature IR absorption peak in hydrogenated Si:B at 1,870 cm^{-1} arising from the Si–H stretch mode. Johnson [85] confirmed this assignment by exposing a sample to a deuterium plasma, resulting in a room-temperature Si–D stretch mode at 1,360 cm^{-1}. The isotopic frequency ratio r = 1,870/1,360 = 1.375 is characteristic for a Si–H bond (Section 3.1.3). The stretch-mode frequency agrees with calculations for the bond-centred configuration [86].

Other acceptor–hydrogen pairs have similar properties. BH, AlH, and GaH complexes show stretch modes at 1,907, 2,201, and 2,171 cm^{-1}, consistent with a Si–H bond that is perturbed by the neighbouring acceptor [87]. Further support for the bond-centred model was provided by the boron isotope shift. When ^{10}B is replaced by ^{11}B, the Si–H stretch mode decreases by 0.8 cm^{-1} [88], revealing a weak but noticeable B–H bond [89].

Donor–hydrogen complexes have the antibonding configuration (Figure 3.7, [90]). Because hydrogen is isolated from the donor, its LVM is weakly perturbed by the donor species. PH, AsH, and SbH have stretch modes near 1,600 cm^{-1} and wag modes near 809 cm^{-1} [91]. The wag modes are very insensitive to the donor species, shifting by <1 cm^{-1}. As discussed in Section 3.1.10, the stretch modes exhibit a Fermi resonance.

3.3.6 Impurity-hydrogen pairs: trends

The structures of hydrogen complexes, determined by LVM spectroscopy and first-principles calculations, follow certain trends [14]. We first consider conventional (non-nitride) III–V semiconductors. Acceptor–hydrogen complexes in these materials have hydrogen in the bond-centred site. For the carbon–hydrogen complex (Section 3.3.2.1), hydrogen attaches directly to the carbon.

For group-II acceptors (Be, Zn, Cd), hydrogen forms a bond primarily with the host anion (P, As) and resides in the bond-centred site, between the acceptor and the anion. This is similar to acceptor–hydrogen complexes in silicon (Figure 3.7). The LVM frequencies are characteristic of P–H and As–H bonds, but they are perturbed by the neighbouring acceptor. McCluskey *et al.* [92] reported systematic trends for these acceptors in GaAs, InP, and GaP. As the size of an atom increases, its equilibrium bond length with hydrogen increases. In a free molecule, a Cd–H bond is longer than a Be–H bond, and an As–H bond is longer than a P–H bond. By comparing these bond lengths to the lattice constant, one can get a measure of how compressed the acceptor–hydrogen bond is. Hydrogen in GaP:Cd is very compressed, since Cd is large and the GaP lattice constant is small.

As the compression of the acceptor–hydrogen bond increases, the LVM frequency increases due to the stiffening of the bond. Therefore, the stretch-mode frequency increases as one goes from Be to Cd. A high degree of compression also increases the coupling with the lattice, leading to a more pronounced temperature dependence of the LVM. The stretch mode of the ZnH complex in GaP, for example, shows a very large frequency shift and broadening as the sample is warmed from liquid-helium temperatures. Above ~80 K, the broadening is large such that the IR peak is no longer discernible [92].

These trends can be extrapolated to the ZnSe:As,H complex, in which a hydrogen attaches to an As acceptor in a bond-centred location [93]. Hydrogen's environment in this complex is quite similar to GaAs:Zn,H, where hydrogen sits between a host As and a Zn acceptor. This results in a comparable stretch-mode frequency (2,166 cm^{-1}) and temperature dependence.

Donor–hydrogen complexes have a different structure. Owing to Coulomb repulsion, the bond-centred site is not energetically favourable. Instead, hydrogen sits in an antibonding site. For silicon–hydrogen pairs in GaAs (Section 3.3.2.2), hydrogen attaches directly to the silicon donor. For group-VI donors (S, Se, Te) in GaAs, hydrogen bonds to a host cation (Ga) atom [31]. This latter case is similar to donor–hydrogen pairs in silicon (Figure 3.7).

These trends – bond-centred site for acceptors, antibonding site for donors – hold for most conventional (non-nitride, non-oxide) semiconductors. However, the rule is broken for Mg acceptors in GaN. In that case, hydrogen attaches to a host nitrogen in an antibonding location (Section 3.3.7). In ZnO:N, hydrogen also attaches to the acceptor in an antibonding location (Section 3.3.8). The higher ionicity of GaN and ZnO, as compared to the more covalent semiconductors, may be responsible for their different behaviour.

In general, it is challenging to distinguish between a bond-centred and antibonding site. The application of hydrostatic pressure is one way to probe the structure (Figure 3.18). For acceptor–hydrogen complexes in GaAs [94] and InP [95], the stretch mode showed a nonlinear (quadratic) increase versus pressure. This nonlinear shift was attributed to squeezing of the bond-centred acceptor–hydrogen bond. Donor–hydrogen complexes, in contrast, have the antibonding structure. Since they have more space, they are affected differently by pressure and show linear or sublinear pressure dependencies. Beyond these qualitative explanations, predictions by first-principles calculations have provided a quantitative means for testing different models (Sections 3.3.7 and 3.3.9.1).

3.3.7 MgH complex in GaN

The achievement of p-type GaN and subsequent light-emitting devices were made possible by Mg doping [96]. In MOCVD-grown GaN, Mg acceptors are passivated by hydrogen. In order to activate the Mg acceptors, thermal annealing is required to break up the MgH pairs [97]. While hydrogen was suspected to passivate Mg, vibrational spectroscopy was required to establish this unambiguously.

MOCVD-grown GaN:Mg thin films on sapphire substrates showed an IR absorption peak at 3,125 cm^{-1} [98], corresponding to an N–H stretch mode

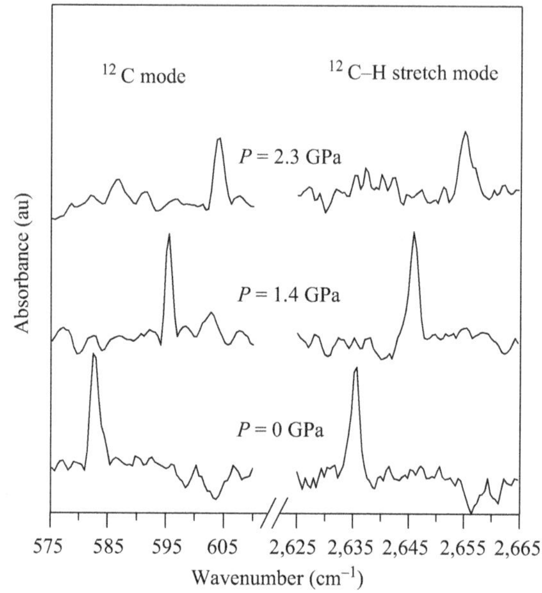

Figure 3.18 Carbon and carbon–hydrogen LVMs in GaAs under hydrostatic pressure. © 2001. Reprinted, with permission, from [94]

(Figure 3.19). The frequency is in good agreement with first-principles calculations [99], which predicted that hydrogen would attach to an N atom in an anti-bonding orientation. When the sample was annealed, the peak diminished and the conductivity dropped by several orders of magnitude, due to the activation of Mg acceptors. When the sample was exposed to a deuterium plasma, an N–D peak appeared with the appropriate isotope shift.

As noted in Section 3.3.6, the GaN:Mg,H complex is different than acceptor–hydrogen complexes in other III–V semiconductors in which hydrogen does not occupy the bond-centred location. This was confirmed with hydrostatic pressure measurements, which showed a small frequency shift versus pressure (~1 cm^{-1}/GPa) [100]. First-principles calculations indicated that, if H were bond-centred, the shift would be 15 cm^{-1}/GPa. Therefore, the bond-centred model could be excluded.

3.3.8 NH complex in ZnO

Nitrogen is an important impurity in oxide semiconductors. In ZnO, it acts as a deep acceptor [101,102]. While substitutional N_O LVMs have not been positively identified, N–H complexes have been characterized through IR absorption experiments. The identification of the stretch mode was important because it provides a way to track the dissociation of N–H pairs upon annealing. When the N–H pairs break up, substitutional N_O acceptors are left behind.

Jokela and McCluskey [103,104] introduced N–H complexes during chemical vapour transport (CVT) growth of single-crystal ZnO. In their CVT process, ZnO

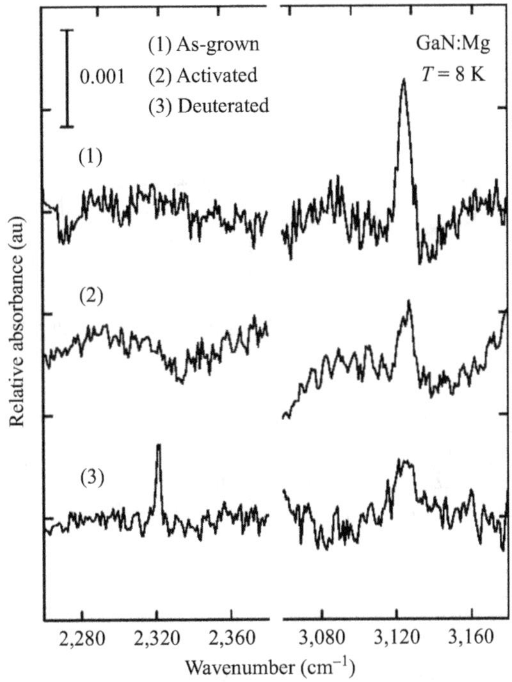

Figure 3.19 IR spectra of MgH complexes in GaN. (1) As-grown sample shows a peak at 3,125 cm^{-1}, corresponding to a N–H stretch mode. (2) Annealing dissociates the MgH pairs, activating the Mg acceptors. (3) Exposure to a deuterium plasma results in an N–D stretch-mode peak. Reprinted, with permission, from [98]

and graphite powder were placed at the hot end of a silica ampoule, while a single-crystal ZnO seed was placed at the cold end [105,106]. Ammonia (NH_3) was used as a carrier gas. This led to the growth of single-crystal ZnO doped with N–H pairs.

IR spectroscopy [104] showed a ^{14}N–H stretch mode at 3,151 cm^{-1} at liquid-helium temperatures (Figure 3.20). In addition to this fundamental mode, an overtone at 6,133 cm^{-1} was also observed. By performing a growth with isotopically enriched ^{15}N–H, the ^{15}N–H fundamental and overtone modes were observed at 3,144 and 6,121 cm^{-1}, respectively. From the Morse potential model (Section 3.1.4), these frequencies yielded an anharmonic parameter $x_e = 0.0253$ and frequencies ω_e (^{14}N–H) = 3,318 and ω_e (^{15}N–H) = 3,312 cm^{-1}. Finally, a sample was grown in a mixture of $^{14}ND_3$ and $^{15}NH_3$, yielding four isotope combinations (Figure 3.20). Taken together, the IR peaks provide unambiguous evidence that a single hydrogen atom attaches to the nitrogen acceptor.

Polarized IR experiments showed that the N–H dipole is oriented at ~114° to the *c*-axis. This corresponds roughly to a tetrahedral bonding direction that is not parallel to the *c*-axis (non-axial). The configuration is referred to as perpendicular

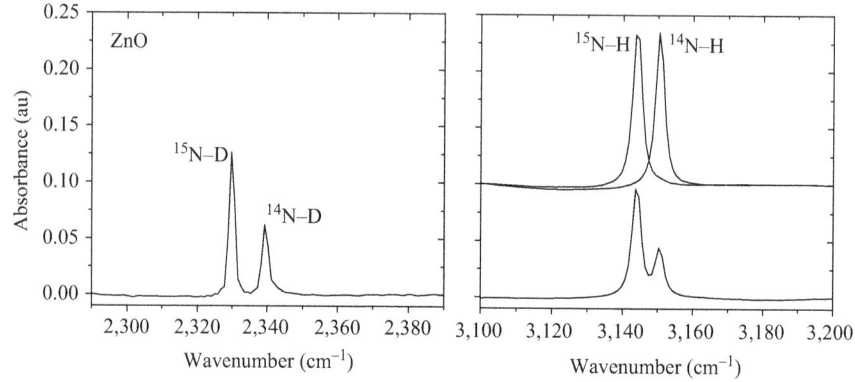

Figure 3.20 IR spectra of ZnO grown in ammonia of different isotopic composition. The top spectra were from samples grown in $^{15}NH_3$ and $^{14}NH_3$, respectively. The bottom spectrum was from a sample grown in a mixture of $^{15}NH_3$ and $^{14}ND_3$. After [104]

($\perp$), even though the angle is not really 90°. First-principles calculations indicated that hydrogen attaches to nitrogen in an antibonding ($AB_{\perp}$) orientation [107]. These calculations are in agreement with the experimental results presented here. It should be noted that the experimental spectra of [107] do not show strong evidence for ZnO:N–H complexes.

3.3.9 Hydrogen donors in oxide semiconductors

3.3.9.1 ZnO

Hydrogen is a common contaminant in ZnO and can act as a shallow donor. The substitutional hydrogen donor (H_O) is an oxygen vacancy with a proton in the centre [108]. It is stable up to ~500°C [109] and, along with group-III impurities, is an important source of n-type conductivity [110]. H_O is difficult to detect because its vibrational frequency lies in the two-phonon range where the crystal is strongly absorbing. Koch *et al.* [111] circumvented this difficulty by measuring the IR photoconductivity of ZnO:H. They observed dips in the photoconductive signal at 742 and 792 cm^{-1} corresponding to the A_1 and E vibrational modes of H_O (C_{3v} symmetry).

As predicted by Van de Walle [112], interstitial hydrogen (H_i) is a shallow donor in ZnO. Experiments found two different hydrogen donors with O–H stretch modes at 3,326 cm^{-1} [113] and 3,611 cm^{-1} [114]. The 3,326 cm^{-1} donor was tentatively assigned to hydrogen in an antibonding location near an isoelectronic calcium impurity [110,115,116]. This complex is unstable at room temperature, decaying with a time constant of a few weeks (Figure 3.21). The 3,611 cm^{-1} line, which is also unstable, was attributed to an isolated hydrogen donor that resides at a bond-centred site (H_{BC}) between zinc and oxygen, parallel to the *c*-axis. Other hydrogen-related defects are listed in Table 3.5.

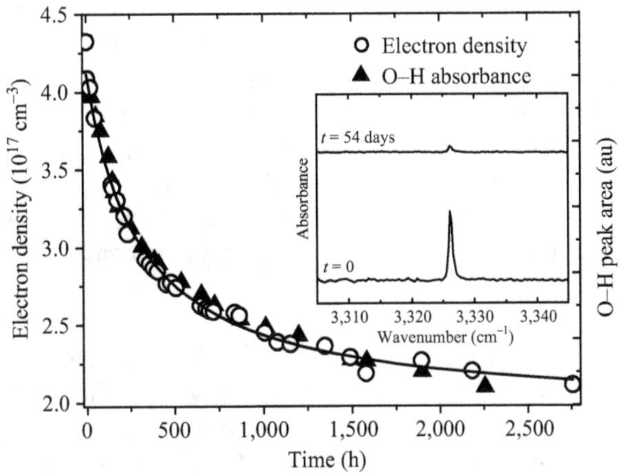

Figure 3.21 *Decay of the 3,126 cm^{-1} interstitial hydrogen donor in ZnO at room temperature. The time dependence of the electron density and peak area are similar, consistent with hydrogen acting as a shallow donor. Inset: Low-temperature IR spectra. After [27]*

Table 3.5 *Vibrational frequencies of impurities in ZnO. Data are from low-temperature IR spectroscopy unless noted otherwise (RT = room temperature). After McCluskey [13]*

Frequency (cm^{-1})	**Assignment**	**References**
275 (RT, Raman)	Zn cluster	Gluba *et al.* [117]
742, 792	H_O	Koch *et al.* [111]
2,269, 2,282 (RT, Raman)	$V_{Zn}N_2$	Khachadorian *et al.* [159]
2,306 (RT, Raman)	N_2	Soudi *et al.* [160]
2,783	Unknown	Li *et al.* [161], Lavrov and Weber [162]
3,151	NH	Jokela and McCluskey [103]
3,192	CuH	Gärtner and Mollwo [163]
3,226	KH	Parmar *et al.* [164]
3,303, 3,321	$V_{Zn}H_3$	Parmar *et al.* [134], Herklotz *et al.* [135]
3,312, 3,350	$V_{Zn}H_2$	Lavrov *et al.* [114]
3,329, 3,348	$V_{Zn}H_2^*$	Bastin *et al.* [133]
3,326	CaH	McCluskey and Jokela [110]
3,327	$V_{Zn}H$	Corolewski *et al.* [136]
3,347, 3,374	CuH_2	Lavrov *et al.* [165]
3,577	LiH	Lavrov [166], Halliburton *et al.* [167]
3,611	H_{BC}	Lavrov *et al.* [114]
3,677	Unknown	Corolewski *et al.* [136]
4,145 (Raman)	H_2	Lavrov *et al.* [148]

Hydrostatic pressure was useful in determining the structure of the 3,326 cm^{-1} hydrogen donor. First-principles calculations [115] predicted that if hydrogen resided in a bond-centred location, its LVM frequency would increase with pressure due to stiffening of the bond. If it was in an antibonding location, the O–H bond would weaken slightly, leading to a negative frequency shift. Experimentally, a negative shift with pressure was observed, leading to the conclusion that the complex was antibonding [27]. This finding concurred with the weak temperature dependence of the IR peak, which also suggested an antibonding location.

3.3.9.2 SnO_2

Similar to the case of ZnO, first-principles calculations predicted that interstitial hydrogen in tin oxide (SnO_2) acts as a shallow donor [118]. The prediction was verified by annealing single-crystal SnO_2 in hydrogen [119]. An IR absorption peak was observed at 3,156 cm^{-1} and assigned to the stretch mode of interstitial hydrogen. The O–H dipole was aligned along the *a*-axis, which is perpendicular to the *c*-axis in the rutile crystal structure. Along with the appearance of the O–H peak, the sample exhibited an increase in electrical conductivity and free-carrier absorption. These observations supported the theoretical model for interstitial hydrogen as a shallow donor. As in ZnO, the O–H peak is unstable, with a similar time constant of a few weeks.

Experiments by the Lehigh group also observed the interstitial hydrogen peak and confirmed that it is a shallow donor [120]. By correlating the integrated absorption of the O–H peak with free-carrier absorption, they revealed that a second donor, probably substitutional hydrogen (H_O), was stable to 600°C. Overall, the properties of interstitial and substitutional hydrogen donors are quite similar to those in ZnO and in good agreement with theory.

3.3.9.3 In_2O_3

Indium oxide is a widely used transparent conductor with tin as the preferred donor dopant. Interstitial hydrogen also acts as a shallow donor, giving rise to an O–H stretch mode at 3,306 cm^{-1} [121]. This peak is correlated with free-carrier absorption, as expected for a shallow donor. The intensity of the IR peak was used to study in- and out-diffusion of hydrogen in In_2O_3 near 700 K [122]. The authors also used stress-induced alignment (Section 3.2.4) to study diffusivity at a much lower temperature of 160 K.

3.3.9.4 TiO_2

Interstitial hydrogen in rutile TiO_2 has different vibrational modes depending on its charge state. The positive charge state dominates for samples annealed in air or water, which have a Fermi level well below the conduction band. H_i^+ has a single O–H stretch mode at 3,287 cm^{-1}, while the deuterium interstitial D_i^+ has a frequency of 2,445 cm^{-1}. The O–H (O–D) dipole is aligned perpendicular to the *c*-axis [123]. The neutral charge state H_i^0 shows several O–H peaks. Specifically, Herklotz *et al.* [124] attributed two O–H absorption peaks, at 3,288 and 3,292 cm^{-1}, to the neutral charge state. The corresponding O–D peaks are at 2,446 and 2,448 cm^{-1}.

The IR peak intensities of the neutral centre (H_i^0 or D_i^0) show a strong temperature dependence. A combined experimental and theoretical study attributed the modes and their temperature-dependent behaviour to small polarons [125]. Specifically, in the neutral charge state, the electron is localized on a neighbouring Ti atom. The electron is self-trapped, forming a small polaron. The O–H or O–D vibrational frequency depends on the particular Ti atom on which the electron resides. As temperature increases, electrons tend to occupy Ti atoms that are further away, leading to an increase in the intensity of the corresponding peak.

The vibrational lifetime of H_i^+ was studied in order to gain insight into hydrogen diffusion. IR pump–probe experiments determined the temperature dependence of the ground-state recovery lifetime [126]. By analysing the lifetime versus temperature, the authors concluded that the O–H stretch mode strongly couples to a wag mode. This wagging motion caused the hydrogen to move to a neighbouring oxygen site. In principle, IR light tuned to the O–H frequency could enhance hydrogen migration in TiO_2.

3.3.10 Vacancy-hydrogen complexes

Hydrogen passivates dangling bonds in vacancies. The best characterized defects are vacancy–hydrogen complexes in silicon, which were studied extensively by IR spectroscopy. The silicon vacancy can be decorated with 1, 2, 3, or 4 hydrogen atoms. To identify these different complexes, silicon samples were implanted with protons and deuterons in varying ratios. A vacancy with one hydrogen, VH (or VD), has a stretch mode at 2,039 (or 1,495 cm^{-1}) [127]. At the opposite extreme, a vacancy with four hydrogens has five possible isotope ratios that occur randomly: VH_4, VH_3D, VH_2D_2, VHD_3, and VD_4. Each of these isotopic combinations has a unique set of Si–H and Si–D stretch modes, near 2,200 and 1,600 cm^{-1}, respectively [128]. The stretch-mode frequencies increase with the number of hydrogen atoms in the vacancy. This is due to the shortening of Si–H bonds caused by the presence of many atoms in a small space.

The silicon divacancy–hydrogen (V_2H_2 or V_2D_2) complex has an IR peak at 2,072 or 1,510 cm^{-1}. In mixed proton/deuteron implantations, the V_2HD complex gives rise to two new peaks, verifying that it is a two-hydrogen centre (Section 3.1.6). A divacancy with a single hydrogen, V_2H, has a peak at 2,068 cm^{-1} [127].

In III–V semiconductors, the cation vacancy has three dangling bonds. InP is an example where the cation vacancy (V_{In}) is decorated by four hydrogens, resulting in an extra electron. This $V_{In}H_4$ complex acts as a donor and compensates acceptors in semi-insulating InP [129]. IR spectroscopy on InP [130], with varying hydrogen/deuterium ratios, was fully consistent with the $V_{In}H_4$ model.

Proton-implanted GaN shows N–H stretch modes near 3,100 cm^{-1} [131]. These modes were attributed to gallium vacancies with 1–4 hydrogen atoms, in good agreement with first-principles calculations [132]. As in the case of silicon, the vibrational frequencies increase with the number of hydrogen atoms. GaN implanted with deuterons showed N–D peaks in the 2,200–2,300 cm^{-1} range.

For II–VI semiconductors, cation vacancies have two dangling bonds. In ZnO, the zinc vacancy is passivated by two hydrogens [114]. In this $V_{Zn}H_2$ complex, one

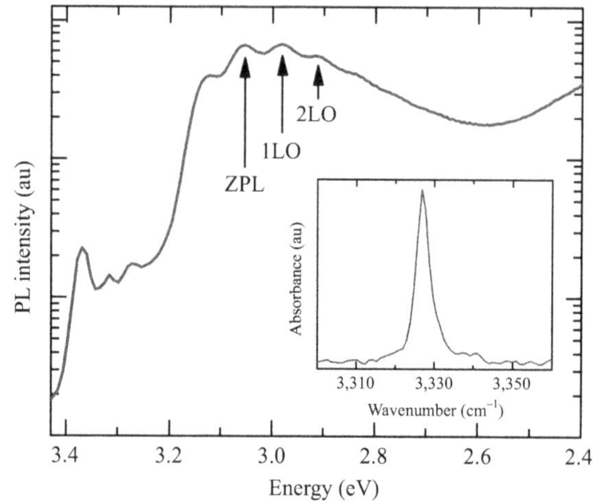

Figure 3.22 PL spectrum of ZnO. The 'blue band' has a zero phonon line (ZPL) of 3.05 eV and LO phonon replicas on the low-energy side. Inset: IR absorption spectrum, showing an O–H peak attributed to a zinc vacancy–hydrogen complex. After [136]

O–H dipole is parallel to the *c*-axis and the second O–H dipole is roughly perpendicular to the *c*-axis. The O–H bonds undergo stretch-mode vibrations, resulting in two IR absorption peaks. When ZnO was exposed to a plasma containing hydrogen and deuterium, new peaks appear, which correspond to O–H or O–D stretch modes in a V_{Zn}HD complex. A less stable complex, denoted $V_{Zn}H_2^*$, has both hydrogens in the perpendicular configurations [133]. Vibrational modes corresponding to a *three* hydrogen complex, tentatively attributed to a $V_{Zn}H_3$ donor, have also been observed [134,135].

Experiments performed by Corolewski *et al.* [136] provided evidence for a V_{Zn}H complex in ZnO, with an O–H stretch mode at 3,127 cm^{-1} (Figure 3.22). The mode was correlated with a structured blue PL band and the presence of Zn vacancies as revealed by positron annihilation spectroscopy. The O–H peak frequency showed a very weak temperature dependence, consistent with an open volume where the hydrogen has plenty of space and therefore couples weakly to the crystal. While these observations suggest a V_{Zn}H complex, the model is quite tentative.

3.3.11 Hydrogen in strontium titanate

Strontium titanate ($SrTiO_3$ or STO) is a complex oxide that has an indirect band gap at 3.2 eV [137]. It has the perovskite crystal structure at room temperature. Below 105 K, the crystal transforms to a tetragonal structure. While its dielectric constant rises as temperature goes down, the crystal does not become ferroelectric.

Most single-crystal samples contain the 'H_I' line near 3,500 cm^{-1}, due to a defect with one hydrogen atom [138]. The assumption in the literature was that H_I

corresponded to interstitial hydrogen. More recent first-principles calculations [139], however, suggest that it is actually a strontium vacancy decorated by one hydrogen ($V_{Sr}H^-$). STO annealed in hydrogen at 800°C show 'H_{II}' lines [140]. By annealing in a mixture of hydrogen and deuterium, it was shown that this complex contains two hydrogens. A theoretical study by Varley *et al.* [141] attributed the H_{II} centre to a titanium vacancy decorated with two hydrogen atoms ($V_{Ti}H_2^{-2}$).

Like other oxides [16], STO contains 'hidden hydrogen' that cannot be observed with IR spectroscopy. The identity of hidden hydrogen may be substitutional hydrogen (H_O), which is an oxygen vacancy occupied by a proton. Calculations indicate that H_O has a low vibrational frequency, in the range of two-phonon absorption where oxide crystals absorb strongly. Because of this, H_O is practically invisible. A second possibility is the H_2 molecule, which is IR-inactive.

An interesting example of hydrogen coming out from hiding is shown in Figure 3.23. First, a sample was annealed in deuterium, resulting in H_{II} centres (labelled DD) and H_I centres. The sample was then annealed in air at increasing temperatures. Around 400°C–500°C, an HD peak appeared, indicating that H was emerging from its hidden reservoir and replacing D in some of the H_{II} centres. At 600°C and above, hydrogen completely replaced deuterium. The displaced D atoms presumably formed hidden species inside the sample.

Vibrational spectroscopy played a key role in identifying the defects responsible for persistent photoconductivity (PPC) in STO. When STO is annealed at 1,200°C, with optimal partial pressures of hydrogen and oxygen, it exhibits large PPC at room temperature [142,143]. After exposure to >2.9 eV photons, the free-electron

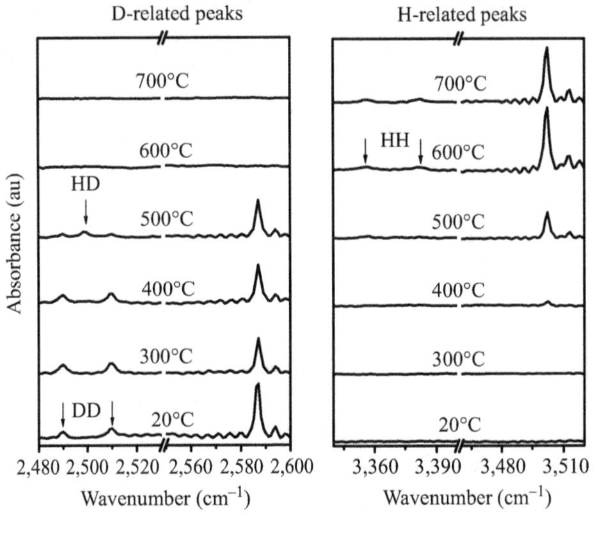

Figure 3.23 IR spectra (200 K) of deuterated $SrTiO_3$ after sequentially annealing in air. Peaks corresponding to HH, HD, and DD complexes are indicated. © 2014. Reprinted, with permission, from [140]

concentration increases by over two orders of magnitude, an increase that persists 'forever' at room temperature.

IR spectroscopy performed before and after exposure to a 405 nm light source showed that the H_I peak dropped in intensity, while peaks on the high-energy side (sidebands) increased [144]. Based on calculations by Varley *et al.* [141], the sidebands were assigned to neutral $V_{Sr}H_2$ complexes. The following model was proposed to explain the observations. Light causes H^+ to leave its substitutional site, resulting in an oxygen vacancy. Because V_O is a shallow double donor in STO, two electrons are donated to the conduction band. The diffusing H^+ then combines with $V_{Sr}H^-$, forming $V_{Sr}H_2$. The defect reaction is summarized as

$$H_O^+ + V_{Sr}H^- \xrightarrow{h\nu} V_O^{+2} + 2e^- + V_{Sr}H_2^0 \tag{3.54}$$

The authors of [144] noted that any oxide for which V_O is a shallow donor could, in principle, undergo this process. If it is indeed a phenomenon that exists in many materials, large room-temperature PPC could find its way into a variety of applications.

3.3.12 *Hydrogen molecules*

Hydrogen molecules had been speculated to exist inside semiconductors but, because they are non-polar, they are difficult to detect with IR spectroscopy. Raman spectroscopy, on the other hand, has different selection rules, which enabled researchers to observe hydrogen molecules in silicon [145], germanium [146], GaAs [147], and ZnO [148].

The isolated H_2 molecule is not IR active because its stretch mode is symmetric and does not induce a change in the dipole moment. In a crystal host, this vibration induces a small dipole moment in the surrounding atoms, resulting in a weak IR absorption peak [149]. In silicon, H_2 (D_2) resides in the interstitial T site (symmetry T_d) and has an IR absorption peak at 3,618 cm^{-1} (2,643 cm^{-1}). The molecule is free to rotate with a quantum number J. Because of selection rules, H_2 or D_2 vibrations with $J = 0$ are IR inactive [150]. The observed IR peaks therefore correspond to vibrational transitions with $J = 1$. HD molecules, in contrast, follow different selection rules. The IR absorption peak for HD at 3,265 cm^{-1} results from a vibrational transition with $J = 0$ in the ground state and $J = 1$ in the excited state. This unusual isotope effect caused confusion for many years.

3.3.13 *From LVM to phonon*

Localization is a theme that runs through vibrational phenomena. An LVM is highly localized, a gap mode is less localized, and a resonant mode is less localized still. A similar progression occurs as doping concentration increases. At dilute concentrations, a light-mass impurity has an LVM. As concentrations approach the alloy regime, 1 per cent or more, vibrational modes from neighbouring defects begin to overlap, causing the line to broaden. At sufficiently high concentrations, the vibrational modes are not localized at all, but are essentially phonons in a random alloy.

An example of this impurity-to-alloy transition was observed in zinc oxysulfide, $ZnS_{1-x}O_x$ [151]. At the dilute limit ($x \ll 1$), substitutional O impurities have

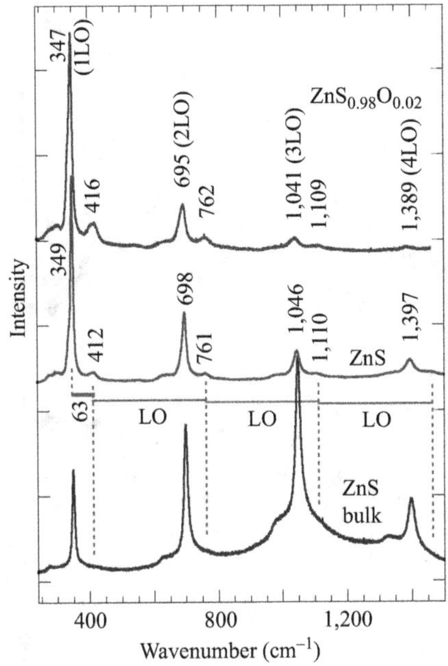

Figure 3.24 *Resonant Raman spectra of ZnS (O content less than 1 per cent) and $ZnS_{0.98}O_{0.02}$ films, and ZnS bulk crystal. The films show the characteristic ZnS LO mode at ~347 cm^{-1} and additional peaks, attributed to an O mode (~412 cm^{-1}) plus n LO modes. Reprinted, with permission, from [151]*

an LVM. As x gets larger, the vibration transitions into a phonon in what is called 'two mode' behaviour. The dependence of the vibrational frequencies on composition (x) is described by the MREI model [22].

Room-temperature Raman spectra are shown in Figure 3.24 for single-crystal ZnS (negligible O content) and sputtered $ZnS_{1-x}O_x$ thin films with $x < 0.01$ and $x = 0.02$. All samples showed fundamental and high-order LO modes. For the sputtered films, an O-related peak is observed near 412 cm^{-1}. Additional modes are observed, which correspond to n LO phonon + O-mode combinations. A linear chain model (Section 3.1.2) yielded an O_S LVM frequency of 433 cm^{-1}. A 3D spring-and-mass model gave a similar frequency, 436 cm^{-1}. The slight difference between experiment and theory was not surprising, since the models neglected changes in bond strength and anharmonicity.

The mass-defect models do not account for the difference in electronegativity between O and S. ZnO, in particular, has a large TO–LO splitting due to the polar nature of the crystal. The MREI model simulates this effect. As shown in Figure 3.25, with increasing O content, the O mode shifts in frequency until it reaches the LO mode of ZnO. In addition, the O mode splits into LO and TO modes, which shift towards their ZnO values as x approaches 1.

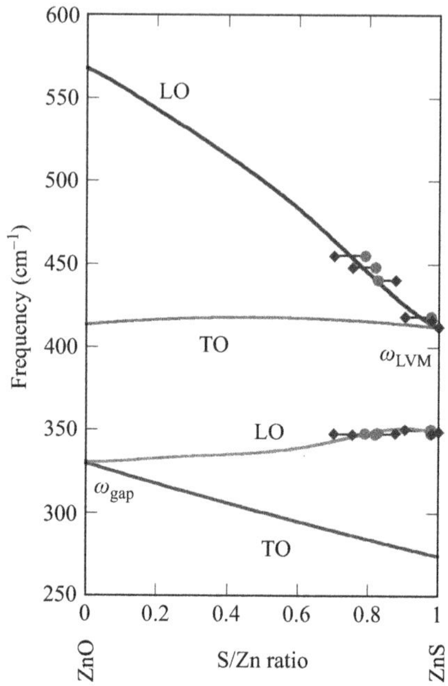

Figure 3.25 MREI model of phonon frequencies in zincblende $ZnS_{1-x}O_x$ (solid lines). Diamonds and circles indicate composition measured by two different techniques. © 2017. Reprinted, with permission, from [151]

3.4 Summary and outlook

Vibrational properties of defects in semiconductors are fundamentally interesting and provide a non-destructive means for accurately identifying impurities. The linear chain model suffices to explain the presence of phonons, LVMs, and gap modes in a crystal. Anharmonic effects result in small corrections to vibrational frequencies and, in the case of a Fermi resonance, can cause line splitting. IR spectroscopy is the predominant method for studying LVMs but Raman spectroscopy is also used, especially for dimers such as N_2 and H_2. Isotopic shifts, applied stress, and group theory enable researchers to deduce the microscopic structure of defects with a high degree of certainty.

A vast number of defects have been analysed with vibrational spectroscopy, a few of which were summarized in this chapter. Interstitial oxygen in silicon has enormous technological importance; the defect also appears in germanium, GaAs, and GaP. Substitutional impurities such as silicon and carbon have been characterized extensively in III–V semiconductors. Hydrogen can passivate dopants and defects or form electrically inactive H_2 molecules. In many oxide semiconductors, hydrogen acts as a shallow donor.

Vibrational spectroscopy will continue to be an essential characterization technique for studying defects in semiconductors. Wide-gap oxides such as Ga_2O_3 exhibit hydrogen O–H modes due to interstitial donors, vacancy–hydrogen complexes, and passivated acceptor dopants. The transition from an isoelectronic impurity to an alloy (e.g. Al in Ga_2O_3) is an interesting area with practical applications. Defects in nanocrystals and 2D materials play important roles but will require advanced characterization methods. As materials go to smaller scales, near-field Raman and IR probes could become increasingly important. Future improvements in techniques and theory will enable researchers to investigate defects in a broad range of emerging materials.

Acknowledgements

There are too many scientists to thank, but some of the most helpful in the field of vibrational spectroscopy included Joel Ager III, Leah Bergman, Bernard Clerjaud, Eugene Haller, Noble Johnson, Ron Newman, Bernard Pajot, Anant Ramdas, Michael Stavola, Chris Van de Walle, and Wladek Walukiewicz. Research on ZnO and related alloys was supported by the US Department of Energy, Office of Basic Energy Sciences, Division of Materials Science and Engineering under Award DE-FG02-07ER46386. Research on $SrTiO_3$ was supported by the National Science Foundation under Grant No. DMR-1561419.

References

[1] Newman R.C. *Infrared Studies of Crystal Defects* (London, Taylor and Francis, 1973)

[2] Ramdas A.K., Rodriguez S. 'Piezospectroscopy of semiconductors' in Seiler D.G., Littler C.L. (eds.). *The Spectroscopy of Semiconductors*, Semiconductors and Semimetals vol. 36 (San Diego, Academic Press, 1992), pp. 137–220

[3] Stavola M. (ed.). *Identification of Defects in Semiconductors*, Semiconductors and Semimetals vol. 51B (San Diego, Academic Press, 1999)

[4] Cardona M. 'Resonance phenomena' in Cardona M., Guntherodt G. (eds.). *Light Scattering in Solids II* (Berlin, Springer, 1982), pp. 19–98

[5] Yu P.Y., Cardona M. *Fundamentals of Semiconductors* (Berlin, Springer-Verlag, 1996)

[6] Barker A.S., Sievers A.J. 'Optical studies of the vibrational properties of disordered solids'. *Rev. Mod. Phys.* 1975, vol. 47, suppl. 2, pp. S1–S179

[7] Murray R., Newman R.C. in Madelung O., Schulz M. (eds.). *Landolt-Börnstein*, New Series vol. 22b (Berlin, Springer-Verlag, 1989)

[8] McCluskey M.D. 'Local vibrational modes of impurities in semiconductors'. *J. Appl. Phys.* 2000, vol. 87, pp. 3593–3617

[9] Pajot B., Clerjaud B. *Optical Absorption of Impurities and Defects in Semiconducting Crystals II: Electronic Absorption of Deep Centres and*

Vibrational Spectra, Springer Series in Solid-State Sciences Book 169 (Berlin, Springer-Verlag, 2013)

[10] Stavola M., Fowler W.B. 'Novel properties of defects in semiconductors revealed by their vibrational spectra'. *J. Appl. Phys.* 2018, vol. 123, pp. 1–16, 161561

[11] Newman R.C. 'Defects in silicon'. *Rep. Prog. Phys.* 1982, vol. 45, pp. 1163–1210

[12] Newman R.C. 'Local vibrational mode spectroscopy in III/V compounds' in Weber E.R. (ed.). *Imperfections in III/V Materials*, Semiconductors and Semimetals vol. 38 (San Diego, Academic Press, 1993), pp. 117–187

[13] McCluskey M.D. 'Defects in ZnO' in Buyanova I., Chen W., Stehr J. (eds.). *Advanced Electronic Materials and Novel Low Dimensional Structures* (Elsevier, 2018). ISBN 978-0081020548

[14] Chevallier J., Clerjaud B., Pajot B. 'Neutralization of defects and dopants in III-V semiconductors' in Pankove J.I., Johnson N.M. (eds.). Semiconductors and Semimetals vol. 34 (San Diego, Academic Press, 1991), pp. 447–510

[15] McCluskey M.D., Haller E.E. 'Hydrogen in III-V and II-V semiconductors' in Nickel N.H. (ed.). *Hydrogen in Semiconductors II*, Semiconductors and Semimetals vol. 61 (San Diego, Academic Press, 1999), pp. 373–440

[16] McCluskey M.D., Tarun M.C., Teklemichael S.T. 'Hydrogen in oxide semiconductors'. *J. Mater. Res.* 2012, vol. 27, pp. 2190–2198

[17] Kittel C. *Introduction to Solid State Physics*, 8th edition (New York, John Wiley & Sons, 2005)

[18] Hayes W., Wiltshire M.C.K., Dean P.J. 'Local vibrational modes of carbon in GaP'. *J. Phys. C* 1970, vol. 3, pp. 1762–1766

[19] Grosche E.G., Ashwin M.J., Newman R.C., *et al.* 'Nearest-neighbor isotopic fine structure of the As_P gap mode in GaP'. *Phys. Rev. B* 1995, vol. 51, pp. 14758–14761

[20] Keating P.N. 'Effect of invariance requirements on the elastic strain energy of crystals with application to the diamond structure'. *Phys. Rev.* 1966, vol. 145, pp. 637–645

[21] McCluskey M.D. 'Resonant interaction between hydrogen vibrational modes in AlSb:Se'. *Phys. Rev. Lett.* 2009, vol. 102, pp. 1–4, 135502

[22] Chang I.F., Mitra S.S. 'Application of a modified random-element-isodisplacement model to long-wavelength optic phonons of mixed crystals'. *Phys. Rev.* 1968, vol. 172, pp. 924–933

[23] Morse P.M. 'Diatomic molecules according to wave mechanics. II. Vibrational levels'. *Phys. Rev.* 1929, vol. 34, pp. 57–64

[24] Wilson E.B., Decius J.C., Cross P.C. *Molecular Vibrations* (New York, Dover, 1980)

[25] McCluskey M.D., Haller E.E. *Dopants and Defects in Semiconductors*, 2nd edition (Boca Raton, CRC Press, 2017)

[26] Davidson B.R., Newman R.C., Joyce T.B., Bullough T.J. 'A calibration of the H-C_{As} stretch mode in GaAs'. *Semicond. Sci. Technol.* 1996, vol. 11, pp. 455–457

[27] Jokela S.J., McCluskey M.D. 'Structure and stability of O-H donors in ZnO from high-pressure and infrared spectroscopy'. *Phys. Rev. B* 2005, vol. 72, pp. 1–4, 113201

[28] Lüpke G., Zhang X., Sun B., Fraser A., Tolk N.H., Feldman L.C. 'Structure-dependent vibrational lifetimes of hydrogen in silicon'. *Phys. Rev. Lett.* 2002, vol. 88, pp. 1–4, 135501

[29] Lüpke G., Tolk N.H., Feldman L.C. 'Vibrational lifetimes of hydrogen in silicon'. *J. Appl. Phys.* 2003, vol. 93, pp. 2317–2336

[30] Elliot R.J., Hayes W., Jones G.D., MacDonald H.F., Sennet C.T. 'Localized vibrations of H^- and D^- ions in the alkaline earth fluorides'. *Proc. R. Soc. A Math. Phys. Eng. Sci.* 1965, vol. 289, pp. 1–33, series A

[31] Vetterhöffer J., Weber J. 'Hydrogen passivation of shallow donors S, Se, and Te in GaAs'. *Phys. Rev. B* 1996, vol. 53, pp. 12835–12844

[32] Persson B.N.J., Ryberg R. 'Brownian motion and vibrational phase relaxation at surfaces: CO on Ni(1 1 1)'. *Phys. Rev. B* 1985, vol. 32, pp. 3586–3596

[33] McCluskey M.D., Jokela S.J., Hlaing Oo W.M. 'Hydrogen in bulk and nanoscale ZnO'. *Physica B* 2006, vol. 376–377, pp. 690–693

[34] Cotton F.A. *Chemical Applications of Group Theory*, 3rd edition (New York, John Wiley & Sons, 1990)

[35] Zheng J.-F., Stavola M. 'Correct assignment of the hydrogen vibrations of the donor-hydrogen complexes in Si: A new example of Fermi resonance'. *Phys. Rev. Lett.* 1996, vol. 76, pp. 1154–1157

[36] Watkins G.D., Fowler W.B., Stavola M., *et al.* 'Identification of a Fermi resonance for a defect in silicon: Deuterium-boron pair'. *Phys. Rev. Lett.* 1990, vol. 64, pp. 467–470

[37] Raman C.V. 'A change of wavelength in light scattering'. *Nature* 1928, vol. 121, pp. 619–620

[38] Wolk J.A., Ager J.W., Duxstad K.J., *et al.* 'Local vibrational mode spectroscopy of nitrogen-hydrogen complex in ZnSe'. *Appl. Phys. Lett.* 1993, vol. 63, pp. 2756–2758

[39] Bell R.J. *Introductory Fourier Transform Spectroscopy* (New York, Academic Press, 1972)

[40] Giffiths P.R., de Haseth J.A. *Fourier Transform Infrared Spectrometry* (New York, John Wiley & Sons, 1986)

[41] Budde M., Lüpke G., Parks Cheney C., Tolk N.H., Feldman L.C. 'Vibrational lifetime of bond-center hydrogen in crystalline silicon'. *Phys. Rev. Lett.* 2000, vol. 85, pp. 1452–1455

[42] Stavola M., Bergman K., Pearton S.J., Lopata J. 'Hydrogen motion in defect complexes: Reorientation kinetics of the B-H complex in silicon'. *Phys. Rev. Lett.* 1988, vol. 61, pp. 2786–2789

[43] Cheng Y.M., Stavola M. 'Non-Arrhenius reorientation kinetics for the B-H complex in Si: Evidence for thermally assisted tunneling'. *Phys. Rev. Lett.* 1994, vol. 73, pp. 3419–3422

[44] Jayaraman A. 'Diamond-anvil cell and high-pressure physical investigations'. *Rev. Mod. Phys.* 1983, vol. 55, pp. 65–107

[45] Artacho E., Lizón-Nordström A., Ynduráin F. 'Geometry and quantum delocalization of interstitial oxygen in silicon'. *Phys. Rev. B* 1995, vol. 51, pp. 7862–7865
[46] Hrostowski H.J., Kaiser R.H. 'Infrared absorption of oxygen in silicon'. *Phys. Rev.* 1957, vol. 107, pp. 966–972
[47] Herzberg G. *Molecular Spectra and Molecular Structure Vol. II—IR and Raman Spectra of Polyatomic Molecules* (New York, Van Nostrand, 1945)
[48] Pajot B. 'Some atomic configurations of oxygen' in Shimura F. (ed.). *Oxygen in Silicon*, Semiconductors and Semimetals vol. 42 (San Diego, Academic Press, 1994), pp. 191–249
[49] Bosomworth D.R., Hayes W., Spray A.R.L., Watkins G.D. 'Absorption of oxygen in silicon in the near and the far infrared'. *Proc. R. Soc. A Math. Phys. Eng. Sci.* 1970, vol. 317, pp. 133–152, series A
[50] Yamada-Kaneta H., Kaneta C., Ogawa T. 'Theory of local-phonon-coupled low-energy anharmonic excitation of the interstitial oxygen in silicon'. *Phys. Rev. B* 1990, vol. 42, pp. 9650–9656
[51] Mayur A.J., Sciacca M.D., Udo M.K., *et al.* 'Fine structure of the asymmetric stretching vibration of dispersed oxygen in monoisotopic germanium'. *Phys. Rev. B* 1994, vol. 49, pp. 16293–16299
[52] Kaiser W., Keck P.H., Lange C.F. 'Infrared absorption and oxygen content in silicon and germanium'. *Phys. Rev.* 1956, vol. 101, pp. 1264–1268
[53] McCluskey M.D., Haller E.E. 'Interstitial oxygen in silicon under hydrostatic pressure'. *Phys. Rev. B* 1997, vol. 56, pp. 9520–9523
[54] Hsu L., McCluskey M.D., Lindstrom J.L. 'Resonant interaction between localized and extended vibrational modes in Si:^{18}O under pressure'. *Phys. Rev. Lett.* 2003, vol. 90, pp. 1–4, 095505
[55] Sun B., Yang Q., Newman R.C., *et al.* 'Vibrational lifetimes and isotope effects of interstitial oxygen in silicon and germanium'. *Phys. Rev. Lett.* 2004, vol. 92, pp. 1–4, 185503
[56] Newman R.C., Thompson F., Hyliands M., Peart R.F. 'Boron and carbon impurities in gallium arsenide'. *Solid State Commun.* 1972, vol. 10, pp. 505–507
[57] Leigh R.S., Newman R.C., Sangster M.J.L., Davidson B.R., Ashwin M.J., Robbie D.A. 'Host and impurity isotope effects on local vibrational modes of GaAs:C_{As} and GaAs:B_{As}'. *Semicond. Sci. Technol.* 1994, vol. 9, pp. 1054–1061
[58] Leigh R.S., Sangster M.J.L. 'Host-isotope fine structure of the fundamental and second-harmonic transitions for the C_{As} local mode in GaAs'. *Phys. Rev. B* 1996, vol. 54, pp. 5562–5567
[59] Alt H.C. 'Anharmonicity of the C_{As} local oscillator in gallium arsenide'. *Mater. Sci. Forum* 1995, vol. 196–201, pp. 1577–1582
[60] Davidson B.R., Newman R.C., Bullough T.J., Joyce T.B. 'Dynamics of the H-C_{As} complex in GaAs'. *Phys. Rev. B* 1993, vol. 48, pp. 17106–17113
[61] Ulrici W., Jurisch M. 'Thermally induced carbon–oxygen complexes in GaAs'. *Phys. Status Solidi B* 2005, vol. 242, pp. 2433–2439

[62] Alt H.C., Wagner H.E., Glacki A., Frank-Rotsch C., Häublein V. 'Isotopic study of mid-infrared vibrational modes in GaAs related to carbon and nitrogen impurities'. *Phys. Status Solidi B* 2015, vol. 252, pp. 1827–1831

[63] Newman R.C. 'The lattice locations of silicon impurities in GaAs: Effects due to stoichiometry, the Fermi energy, the solubility limit and DX behaviour'. *Semicond. Sci. Technol.* 1994, vol. 9, pp. 1749–1762

[64] Theis W.M., Spitzer W.G. 'High-resolution measurements of localized vibrational mode infrared absorption of Si-doped GaAs'. *J. Appl. Phys.* 1984, vol. 56, pp. 890–898

[65] Chadi D.J., Chang K.J. 'Theory of the atomic and electronic structure of DX centers in GaAs and $Al_xGa_{1-x}As$ alloys'. *Phys. Rev. Lett.* 1988, vol. 61, pp. 873–876

[66] Wolk J.A., Kruger M.B., Heyman J.N., Walukiewicz W., Jeanloz R., Haller E.E. 'Local-vibrational-mode spectroscopy of DX centers in Si-doped GaAs under hydrostatic pressure'. *Phys. Rev. Lett.* 1991, vol. 66, pp. 774–777

[67] Saito M., Oshiyama A., Sugino O. 'Validity of the broken-bond model for the DX center in GaAs'. *Phys. Rev. B* 1992, vol. 45, pp. 13745–13748

[68] Pajot B., Newman R.C., Murray R., Jalil A., Chevallier J., Azoulay R. 'High-resolution infrared study of the neutralization of silicon donors in gallium arsenide'. *Phys. Rev. B* 1988, vol. 37, pp. 4188–4195

[69] Schneider J., Dischler B., Seelewind H., *et al.* 'Assessment of oxygen in gallium arsenide by infrared local vibrational mode spectroscopy'. *Appl. Phys. Lett.* 1989, vol. 54, pp. 1442–1444

[70] Watkins G.D., Corbett J.W. 'Defects in irradiated silicon. I. Electron spin resonance of the Si-*A* center'. *Phys. Rev.* 1961, vol. 121, pp. 1001–1014

[71] Alt H.C. 'Experimental evidence for a negative-*U* center in gallium arsenide related to oxygen'. *Phys. Rev. Lett.* 1990, vol. 65, pp. 3421–3424

[72] Pajot B., Song C.Y. 'OH bonds in gallium arsenide grown by the liquid-encapsulated Czochralski crystal-growth method'. *Phys. Rev. B* 1992, vol. 45, pp. 6484–6491

[73] Ulrici W., Jurisch M. 'Vibrational modes of a hydrogen-impurity centre in LEC-GaAs'. *Physica B* 2001, vol. 308–310, pp. 835–838

[74] McCluskey M.D., Haller E.E., Walukiewicz W., Becla P. 'Hydrogen passivation of Se and Te in AlSb'. *Phys. Rev. B* 1996, vol. 53, pp. 16297–16301

[75] McCluskey M.D., Haller E.E., Walukiewicz W., Becla P. 'Anti-crossing behavior of local vibrational modes in AlSb'. *Solid State Commun.* 1998, vol. 106, pp. 587–590

[76] Sciacca M.D., Mayur A.J., Shin N., Miotkowski I., Ramdas A.K., Rodriguez S. 'Local vibrational modes of substitutional Mg^{2+}, Ca^{2+}, and S^{2-} in zinc-blende and wurtzite II-VI semiconductors'. *Phys. Rev. B* 1995, vol. 51, pp. 6971–6978

[77] Sciacca M.D., Mayur A.J., Kim H., Miotkowski I., Ramdas A.K., Rodriguez S. 'Host-isotope fine structure of local and gap modes of substitutional impurities in zinc-blende and wurtzite II-VI semiconductors'. *Phys. Rev. B* 1996, vol. 53, pp. 12878–12883

[78] Clerjaud B., Côte D., Svob L., Marfaing Y., Druilhe R. 'Hydrogen-acceptor pairing in CdTe epitaxial layers grown by OMVPE'. *Solid State Commun.* 1993, vol. 85, pp. 167–170

[79] Chen G., Miotkowski I., Rodriguez S., Ramdas A.K. 'Stoichiometry driven impurity configurations in compound semiconductors'. *Phys. Rev. Lett.* 2006, vol. 96, pp. 1–4, 035508

[80] Lavrov E.V., Bastin D., Weber J., Schneider J., Fauler A., Fiederle M. 'Reassignment of the O_{Te}-V_{Cd} complex in CdTe'. *Phys. Rev. B* 2011, vol. 84, pp. 1–4, 233201

[81] T-Thienprasert J., Watcharatharapong T., Fongkaew I., Du M.H., Singh D.J., Limpijumnong S. 'Identification of oxygen defects in CdTe revisited: First-principles study'. *J. Appl. Phys.* 2014, vol. 115, pp. 1–6, 203511

[82] Holbech J.D., Nielsen B.B., Jones R., Sitch P., Öberg S. 'H_2* defect in crystalline silicon'. *Phys. Rev. Lett.* 1993, vol. 71, pp. 875–878

[83] Budde M., Nielsen B.B., Leary P., *et al.* 'Identification of the hydrogen-saturated self-interstitials in silicon and germanium'. *Phys. Rev. B* 1998, vol. 57, pp. 4397–4412

[84] Pankove J.I., Zanzucchi P.J., Magee C.W., Lucovsky G. 'Hydrogen localization near boron in silicon'. *Appl. Phys. Lett.* 1985, vol. 46, pp. 421–423

[85] Johnson N.M. 'Mechanism for hydrogen compensation of shallow-acceptor impurities in single-crystal silicon'. *Phys. Rev. B* 1985, vol. 31, pp. 5525–5528

[86] DeLeo G.G., Fowler W.B. 'Hydrogen-acceptor pairs in silicon: Pairing effect on the hydrogen vibrational frequency'. *Phys. Rev. B* 1985, vol. 31, 6861–6864

[87] Stavola M., Pearton S.J., Lopata J., Dautremont-Smith W.C. 'Vibrational characteristics of acceptor-hydrogen complexes in silicon'. *Appl. Phys. Lett.* 1987, vol. 50, pp. 1086–1088

[88] Pajot B., Chari A., Aucouturier M., Astier M., Chantre A. 'Experimental evidence for boron-hydrogen interaction in boron-doped silicon passivated with hydrogen'. *Solid State Commun.* 1988, vol. 67, pp. 855–858

[89] Estreicher S.K., Throckmorton L., Marynick D.S. 'Hydrogen passivation of shallow acceptors and donors in c-Si: Comparisons and trends'. *Phys. Rev. B* 1989, vol. 39, pp. 13241–13251

[90] Johnson N.M., Herring C., Chadi D.J. 'Interstitial hydrogen and neutralization of shallow-donor impurities in single-crystal silicon'. *Phys. Rev. Lett.* 1986, vol. 56, pp. 769–772

[91] Bergman K., Stavola M., Pearton S.J., Lopata J. 'Donor-hydrogen complexes in passivated silicon'. *Phys. Rev. B* 1988, vol. 37, pp. 2770–2773

[92] McCluskey M.D., Haller E.E., Walker J., Johnson N.M. 'Vibrational spectroscopy of group II acceptor – hydrogen complexes in GaP'. *Phys. Rev. B* 1995, vol. 52, pp. 11859–11864

[93] McCluskey M.D., Haller E.E., Zach F.X., Bourret-Courchesne E.D. 'Vibrational spectroscopy of arsenic-hydrogen complexes in ZnSe'. *Appl. Phys. Lett.* 1996, vol. 68, pp. 3476–3478

[94] McCluskey M.D., Haller E.E., Walker J., *et al.* 'Local vibrational modes in GaAs under hydrostatic pressure'. *Phys. Rev. B* 1997, vol. 56, pp. 6404–6407
[95] McCluskey M.D., Zhuravlev K.K., Davidson B.R., Newman R.C. 'Pressure dependence of local vibrational modes in InP'. *Phys. Rev. B* 2001, vol. 63, pp. 1–4, 125202
[96] Amano H., Kito M., Hiramatsu K., Akasaki I. 'P-type conduction in Mg-doped GaN treated with low-energy electron beam irradiation (LEEBI)'. *Jpn. J. Appl. Phys.* 1989, vol. 28, pp. L2112–L2114
[97] Nakamura S., Iwasa N., Senoh M., Mukai T. 'Hole compensation mechanism of p-type GaN films'. *Jpn. J. Appl. Phys.* 1992, vol. 31, pp. 1258–1266
[98] Götz W., Johnson N.M., Bour D.P., McCluskey M.D., Haller E.E. 'Local vibrational modes of the Mg-H acceptor complex in GaN'. *Appl. Phys. Lett.* 1996, vol. 69, pp. 3725–3727
[99] Neugebauer J., Van de Walle C.G. 'Hydrogen in GaN: Novel aspects of a common impurity'. *Phys. Rev. Lett.* 1995, vol. 75, pp. 4452–4455
[100] McCluskey M.D., Zhuravlev K.K., Kneissl M., *et al.* 'Vibrational spectroscopy of GaN:Mg under pressure'. *MRS Proc.* 2001, vol. 693, I2.4
[101] Lyons J.L., Janotti A., Van de Walle C.G. 'Why nitrogen cannot lead to p-type conductivity in ZnO'. *Appl. Phys. Lett.* 2009, vol. 95, pp. 1–3, 252105
[102] Tarun M.C., Iqbal M.Z., McCluskey M.D. 'Nitrogen is a deep acceptor in ZnO'. *AIP Adv.* 2011, vol. 1, pp. 1–7, 022105
[103] Jokela S.J., McCluskey M.D. 'Unambiguous identification of nitrogen-hydrogen complexes in zinc oxide'. *Phys. Rev. B* 2007, vol. 76, pp. 1–4, 193201
[104] Jokela S.J., McCluskey M.D. 'Structure and stability of N-H complexes in single-crystal ZnO'. *J. Appl. Phys.* 2010, vol. 107, pp. 1–5, 113536
[105] Jokela S.J., McCluskey M.D. 'Hydrogen complexes in ZnO grown by chemical vapor transport'. *Physica B* 2007, vol. 401–402, pp. 395–398
[106] Jokela S.J., Tarun M.C., McCluskey M.D. 'Nitrogen and hydrogen in bulk single-crystal ZnO'. *Physica B* 2009, vol. 404, pp. 4810–4812
[107] Li X., Keyes B., Asher S., *et al.* 'Hydrogen passivation effect in nitrogen-doped ZnO thin films'. *Appl. Phys. Lett.* 2005, vol. 86, pp. 1–3, 122107
[108] Janotti A., Van de Walle C.G. 'Hydrogen multicentre bonds'. *Nat. Mater.* 2007, vol. 6, pp. 44–47
[109] Shi G.A., Saboktakin M., Stavola M., Pearton S.J. '"Hidden hydrogen" in as-grown ZnO'. *Appl. Phys. Lett.* 2004, vol. 85, pp. 5601–5603
[110] McCluskey M.D., Jokela S.J. 'Sources of n-type conductivity in ZnO'. *Physica B* 2007, vol. 401–402, pp. 355–357
[111] Koch S.G., Lavrov E.V., Weber J. 'Photoconductive detection of tetrahedrally coordinated hydrogen in ZnO'. *Phys. Rev. Lett.* 2012, vol. 108, pp. 1–4, 165501
[112] Van de Walle C.G. 'Hydrogen as a cause of doping in zinc oxide'. *Phys. Rev. Lett.* 2000, vol. 85, pp. 1012–1015
[113] McCluskey M.D., Jokela S.J., Zhuravlev K.K., Simpson P.J., Lynn K.G. 'Infrared spectroscopy of hydrogen in ZnO'. *Appl. Phys. Lett.* 2002, vol. 81, pp. 3807–3809

[114] Lavrov E.V., Weber J., Börrnert F., Van de Walle C.G., Helbig R. 'Hydrogen-related defects in ZnO studied by infrared absorption spectroscopy'. *Phys. Rev. B* 2002, vol. 66, pp. 1–7, 165205

[115] Li X.B., Limpijumnong S., Tian W.Q., Sun H.B., Zhang S.B. 'Hydrogen in ZnO revisited: Bond center versus antibonding site'. *Phys. Rev. B* 2008, vol. 78, pp. 1–4, 113203

[116] Bang J., Choi E.-A., Chang K.J. 'The effect of impurities on hydrogen bonding site and local vibrational frequency in ZnO'. *J. Appl. Phys.* 2009, vol. 106, pp. 1–5, 053522

[117] Gluba M.A., Nickel N.H., Karpensky N. 'Interstitial zinc clusters in zinc oxide'. *Phys. Rev. B* 2013, vol. 88, pp. 1–8, 245201

[118] Singh A.K., Janotti A., Scheffler M., Van de Walle C.G. 'Sources of electrical conductivity in SnO_2'. *Phys. Rev. Lett.* 2008, vol. 101, pp. 1–4, 055502

[119] Hlaing Oo W.M., Tabatabaei S., McCluskey M.D., Varley J.B., Janotti A., Van de Walle C.G. 'Hydrogen donors in SnO_2 studied by infrared spectroscopy and first-principles calculations'. *Phys. Rev. B* 2010, vol. 82, pp. 1–4, 193201

[120] Bekisli F., Stavola M., Fowler W.B., Boatner L., Spahr E., Lüpke G. 'Hydrogen impurities and shallow donors in SnO_2 studied by infrared spectroscopy'. *Phys. Rev. B* 2011, vol. 84, pp. 1–8, 035213

[121] Yin W., Smithe K., Weiser P., *et al.* 'Hydrogen centers and the conductivity of In_2O_3 single crystals'. *Phys. Rev. B* 2015, vol. 91, pp. 1–7, 075208

[122] Qin Y., Weiser P., Villalta K., *et al.* 'Diffusivity of the interstitial hydrogen shallow donor in In_2O_3'. *J. Appl. Phys.* 2018, vol. 123, pp. 1–6, 161506

[123] Soffer B. 'Studies of the optical and infrared absorption spectra of rutile single crystals'. *J. Chem. Phys.* 1961, vol. 35, pp. 940–945

[124] Herklotz F., Lavrov E.V., Weber J. 'Infrared absorption of the hydrogen donor in rutile TiO_2'. *Phys. Rev. B* 2011, vol. 83, pp. 1–5, 235202

[125] Bekisli F., Fowler W.B., Stavola M. 'Small polaron characteristics of an OD center in TiO_2 studied by infrared spectroscopy'. *Phys. Rev. B* 2012, vol. 86, pp. 1–5, 155208.

[126] Spahr E.J., Wen L., Stavola M., *et al.* 'Giant enhancement of hydrogen transport in rutile TiO_2 at low temperatures'. *Phys. Rev. Lett.* 2010, vol. 104, pp. 1–4, 205901

[127] Stallinga P., Johannesen P., Herstrøm S., Nielsen K.B., Nielsen B.B., Byberg J.R. 'Electron paramagnetic study of hydrogen-vacancy defects in crystalline silicon'. *Phys. Rev. B* 1998, vol. 58, pp. 3842–3852

[128] Nielsen B.B., Hoffman L., Budde M. 'Si-H stretch modes of hydrogen-vacancy defects in silicon'. *Mater. Sci. Eng. B* 1996, vol. 36, pp. 259–263

[129] Ewels C.P., Öberg S., Jones R., Pajot B., Briddon P.R. 'Vacancy- and acceptor-H complexes in InP'. *Semicond. Sci. Technol.* 1996, vol. 11, pp. 502–507

[130] Zach F.X., Haller E.E., Gabbe D., Iseler G., Bryant G.G., Bliss D.F. 'Electrical properties of the hydrogen defect in InP and the microscopic

structure of the 2316 cm^{-1} hydrogen related line'. *J. Electron. Mater.* 1996, vol. 25, pp. 331–335

[131] Weinstein M.G., Song C.Y., Stavola M., *et al.* 'Hydrogen-decorated lattice defects in proton implanted GaN. *Appl. Phys. Lett.* 1998, vol. 72, pp. 1703–1705

[132] Van de Walle C.G. 'Interactions of hydrogen with native defects in GaN'. *Phys. Rev. B* 1997, vol. 56, pp. 10020–10023

[133] Bastin D., Lavrov E.V., Weber J. 'Metastable state of the $V_{Zn}H_2$ defect in ZnO'. *Phys. Rev. B* 2011, vol. 83, pp. 1–5, 195210

[134] Parmar N.S., McCluskey M.D., Lynn K.G. 'Vibrational spectroscopy of Na-H complexes in ZnO'. *J. Electron. Mater.* 2013, vol. 42, pp. 3426–3428

[135] Herklotz F., Hupfer A., Johansen K.M., Svensson B.G., Koch S.G., Lavrov E.V. 'Infrared absorption on a complex comprising three equivalent hydrogen atoms in ZnO'. *Phys. Rev. B* 2015, vol. 92, pp. 1–10, 155203

[136] Corolewski C.D., Parmar N.S., Lynn K.G., McCluskey M.D. 'Hydrogen-related complexes in Li-diffused ZnO single crystals'. *J. Appl. Phys.* 2016, vol. 120, pp. 1–5, 035702

[137] Capizzi M., Frova A. 'Optical gap of strontium titanate (deviation from Urbach tail behavior)'. *Phys. Rev. Lett.* 1970, vol. 25, pp. 1298–1302

[138] Weber G., Kapphan S., Wöhlecke M. 'Spectroscopy of the O-H and O-D stretching vibrations in $SrTiO_3$ under applied electric field and uniaxial stress'. *Phys. Rev. B* 1986, vol. 34, pp. 8406–8417

[139] T-Thienprasert J., Fongkaew I., Singh D.J., Du M.-H., Limpijumnong S. 'Identification of hydrogen defects in $SrTiO_3$ by first-principles local vibration mode calculations'. *Phys. Rev. B* 2012, vol. 85, pp. 1–5, 125205

[140] Tarun M.C., McCluskey M.D. 'Infrared absorption of hydrogen-related defects in strontium titanate'. *J. Appl. Phys.* 2011, vol. 109, pp. 1–4, 063706

[141] Varley J.B., Janotti A., Van de Walle C.G. 'Hydrogenated vacancies and hidden hydrogen in $SrTiO_3$'. *Phys. Rev. B* 2014, vol. 89, pp. 1–8, 075202

[142] Tarun M.C., Selim F.A., McCluskey M.D. 'Persistent photoconductivity in strontium titanate'. *Phys. Rev. Lett.* 2013, vol. 111, pp. 1–5, 187403

[143] Poole V.M., Jokela S.J., McCluskey M.D. 'Using persistent photoconductivity to write a low-resistance path in $SrTiO_3$'. *Sci. Rep.* 2017, vol. 7, pp. 1–6, 6659

[144] Poole V.M., Huso J., McCluskey M.D. 'The role of hydrogen and oxygen in the persistent photoconductivity of strontium titanate'. *J. Appl. Phys.* 2018, vol. 123, pp. 1–5, 161545

[145] Leitch A.W.R., Alex V., Weber J. 'Raman spectroscopy of hydrogen molecules in crystalline silicon'. *Phys. Rev. Lett.* 1998, vol. 81, pp. 421–424

[146] Hiller M., Lavrov E.V., Weber J., Hourahine B., Jones R., Briddon P.R. 'Interstitial H_2 in germanium by Raman scattering and *ab initio* calculations'. *Phys. Rev. B* 2005, vol. 72, pp. 1–4, 153201

[147] Vetterhöffer J., Wagner J., Weber J. 'Isolated hydrogen molecules in GaAs'. *Phys. Rev. Lett.* 1996, vol. 53, pp. 5409–5412

[148] Lavrov E.V., Herklotz F., Weber J. 'Identification of hydrogen molecules in ZnO'. *Phys. Rev. Lett.* 2009, vol. 102, pp. 1–4, 185502

[149] Pritchard R.E., Ashwin M.J., Tucker J.H., Newman R.C. 'Isolated interstitial hydrogen molecules in hydrogenated crystalline silicon'. *Phys. Rev. B* 1998, vol. 57, 15048–15051

[150] Shi G.A., Stavola M., Fowler W.B., Chen E. 'Rotational-vibrational transitions of interstitial HD in Si'. *Phys. Rev. B* 2005, vol. 72, pp. 1–6, 085207

[151] Huso J., Ritter J.R., Thapa D., Yu K.M., Bergman L., McCluskey M.D. 'Oxygen vibrational modes in $ZnS_{1-x}O_x$ alloys'. *J. Appl. Phys.* 2017, vol. 123, pp. 1–5, 161537

[152] Genzel L., Bauhofer W. 'Phonons in mixed crystals'. *Z. Phys. B Condens. Matter* 1976, vol. 25, pp. 13–18

[153] Chen X.K., Wiersma R., Wang C.X., *et al.* 'Local vibrational modes of carbon in GaSb and GaAsSb'. *Appl. Phys. Lett.* 2002, vol. 80, pp. 1942–1944

[154] Pritchard R.H., Davidson B.R., Newman R.C., *et al.* 'The structure and vibrational modes of H-C_{As} in passivated AlAs grown by chemical beam epitaxy'. *Semicond. Sci. Technol.* 1994, vol. 9, pp. 140–149

[155] McCluskey M.D., Haller E.E., Becla P. 'Carbon acceptors and carbon–hydrogen complexes in AlSb'. *Phys. Rev. B* 2001, vol. 65, pp. 1–4, 045201

[156] Clerjaud B., Côte D., Hahn W.-S., Ulrici W. 'Carbon–hydrogen complex in GaP'. *Appl. Phys. Lett.* 1991, vol. 58, pp. 1860–1862

[157] Davidson B.R., Newman R.C., Button C.C. 'Vibrational modes of carbon acceptors and hydrogen-carbon pairs in semi-insulating InP doped using CCl_4'. *Phys. Rev. B* 1998, vol. 58, pp. 15609–15613

[158] Najmi S., Chen X.K., Yang A., Steger M., Thewalt M.L.W., Watkins S.P. 'Local vibrational mode study of carbon-doped InAs'. *Phys. Rev. B* 2006, vol. 74, pp. 1–4, 113202

[159] Khachadorian S., Gillen R., Ton-That C., *et al.* 'Revealing the origin of high-energy Raman local mode in nitrogen doped ZnO nanowires'. *Phys. Status Solidi RRL* 2016, vol. 10, pp. 334–338

[160] Soudi A., Khan E.H., Dickinson J.T., Gu Y. 'Observation of unintentionally incorporated nitrogen-related complexes in ZnO and GaN nanowires'. *Nano Lett.* 2009, vol. 9, pp. 1844–1849

[161] Li Y.J., Zhang B., Lu W. 'Infrared absorption spectroscopy on OH–Ni complex in hydrothermally grown ZnO'. *J. Appl. Phys.* 2009, vol. 105, pp. 1–6, 093516

[162] Lavrov E.V., Weber J. 'Comment on "Infrared absorption spectroscopy on OH-Ni complex in hydrothermally grown ZnO" [J. Appl. Phys. 105, 093516 (2009)]'. *J. Appl. Phys.* 2009, vol. 106, pp. 1, 086104

[163] Gärtner F.G., Mollwo E. 'IR absorption of OH and OD centres and OH/OH, OD/OD, and OH/OD complexes in Cu-doped ZnO single crystals'. *Phys. Status Solidi B* 1978, vol. 89, pp. 381–388

[164] Parmar N.S., Corolewski C.D., McCluskey M.D., Lynn K.G. 'Potassium acceptor doping of ZnO crystals'. *AIP Adv.* 2015, vol. 5, pp. 1–7, 057107

[165] Lavrov E.V., Weber J., Börrnert F. 'Copper dihydrogen complex in ZnO'. *Phys. Rev. B* 2008, vol. 77, pp. 1–10, 155209

[166] Lavrov E.V. 'Infrared absorption spectroscopy of hydrogen-related defects in ZnO'. *Physica B* 2003, vol. 340–342, pp. 195–200

[167] Halliburton L.E., Wang L.J., Bai L.H., *et al.* 'Infrared absorption from OH^- ions adjacent to lithium acceptors in hydrothermally grown ZnO'. *J. Appl. Phys.* 2004, vol. 96, pp. 7168–7172

[168] Limpijumnong S., Zhang S.B. 'Resolving hydrogen binding sites by pressure—a first-principles prediction for ZnO'. *Appl. Phys. Lett.* 2005, vol. 86, pp. 1–3, 151910

Chapter 4
Magnetic resonance methods

Jean-Louis Cantin[1]

Since the early 1960s, electron magnetic resonance spectroscopies have been widely used to study point defects in semiconductors. Even if present in the ppm regime, point defects control the global electronic and optical properties of semiconductors. The electrical conductivity, the color, of the materials can be engineered by introducing chosen impurities or, one the contrary, can be deteriorated by detrimental adventitious defects. Electron magnetic resonance spectroscopies are quantitative techniques sensitive enough to identify these defects. The analysis of the spectra provides structural, electronic and chemical information about the defect: point symmetry, valence state, chemical nature as well as electronic and geometrical structures of the defect environment. These experimental data generally lead to the model of the defect, at the atomic scale. The first part of this chapter presents an overview of the principle and methodology of these spectroscopies in order to appreciate their contribution to modern semiconductors physics, illustrated in the second part of this chapter.

4.1 Electron spin resonance spectroscopies

The aim of this chapter is certainly not to provide a definitive theoretical treatment of electron paramagnetic resonance (EPR) but merely to give key elements to understand its application to defect studies. Further insight on the technique can be gained by examining the books and papers cited in [1–10].

Electron spin resonance (ESR) and nuclear magnetic resonance (NMR) are often presented as sister techniques as they basically share the same principle and formalism. Indeed the electron spin in ESR, or the nuclear spin in NMR, is in both cases identified via the resonant absorption of an electromagnetic wave between energy sublevels induced by an external magnetic, via the so-called Zeeman interaction. Thus, identification of the isolated electron spin carriers by ESR first relies on the detection of the resonant Zeeman interaction. ESR is generally performed using moderate magnetic field values below two Tesla, which can be

[1]Institut des Nano Sciences de Paris, Sorbonne Universite, Paris, France

accurately controlled and swept, and consequently, the transitions between the electronic Zeeman levels are induced using electromagnetic waves in the microwave domain.

Although the nuclear spin value and abundance are atom specific, rendering NMR chemically sensitive, all electrons have the same electronic spin $S = 1/2$. Furthermore, electrons are subject to the Pauli principle, which states that two electrons in the same orbital must have opposite spin projections of 1/2 and $-1/2$, respectively. Consequently, the net total spin of filled orbitals, or orbitals occupied by an even number of electrons, is generally equal to zero. Thus, although each single electron has a spin one half, most of the electrons in atoms, molecules and solids are "ESR silent," that is to say will not give rise to any ESR. ESR detects only systems containing unpaired electrons, that is to say paramagnetic species. For this reason, ESR is also referred to as EPR. This chapter focuses on the formalism required to interpretate EPR spectra and describes the experimental setup.

It is worth mentioning that, when applied to a magnetic sample, where the electron spin is magnetically coupled, the same technique is called ferromagnetic resonance (FMR). In this case, the quantum formalism presented in the following is no more necessary. Indeed, the magnetization of the sample will interact with the applied magnetic field as a giant macro spin formed by magnetic coupling of individual spin carriers, and the interaction energy is no longer quantified. FMR is thus analyzed with a phenomenological classical formalism and is used to investigate magnetic anisotropies, magnetization damping or coupling between magnetic layers.

4.1.1 What is EPR used for?

In solid state, isolated paramagnetic centers that can be studied by EPR include

- d-transition elements and f-rare earth elements which naturally have unpaired electron spins;
- free radicals, which may be radiation induced or created during chemical reactions for example;
- paramagnetic molecules, like O_2 or NO for instance;
- conduction electrons; and
- point defects like impurities, antisites and interstitials atoms, vacancies and their clusters.

Consequently, EPR is widely used to study the mechanism and kinetics of chemical or biological reactions, via the radical generated, or using spin-labeling or spin-trapping methods. This quantitative technique is also used in radiotherapy or food processing to perform radiation dosimetry. Recently, rapid and noninvasive oximetry by EPR was developed for medical purpose.

In this chapter, we will focus on the study by EPR of point defects in semiconductor materials. Indeed, impurities, vacancies, anti-sites and complexes of them, in a diamagnetic material, may exhibit a local electronic reconstruction favoring unpaired electrons, and consequently, such defects have a nonzero electron spin. Of course, point defects may exist in an $S = 0$ state and then be EPR silent. Nevertheless, in semiconductors, most of the point defects have several charge states in the gap, and generally, each of them corresponds to a different spin state. Changing the defect charge state by electrical polarization or by light irradiation is then an efficient mean to reveal and detect the defects by EPR. But of course, in these out of equilibrium conditions, the technique may not be easily quantitative.

EPR is typically three order of magnitude more sensitive than NMR. EPR probes the whole sample placed in the spectrometer. Nevertheless, the size of the sample is limited by the dimensions of the cavity, which are related to the microwave frequency, as we will see later on this chapter. The versatile spectrometers use X-Band (10 GHz) or Q-Band (40 GHz) microwave, and the maximal sample size is typically $15 \times 4 \times 4$ or $5 \times 2 \times 2$ mm^3, respectively. The sensitivity depends on the width of the resonance of the defects, but typically 10^{10} spins in the cavity would raise a detectable EPR signal. The EPR spectroscopy is thus a technique of choice to study defects with concentration as low as 10^{11}/cm^{-3}, even though in heavily doped semiconductors, the residual conductivity at low temperature may lower the sensitivity by disturbing the penetration of the microwave into the sample.

Finally, let us emphasize that EPR is a nondestructive technique, the sample being submitted to a low-power microwave (typically less than 1 W) and a moderated magnetic field (generally below 1.5 T). In the following examples, as in most cases, the semiconductor sample is crystalline, but the technique can probe polycrystalline and amorphous samples, as well as powders and nanostructures, in solution if needs be.

We will now review the formalism required to describe the interaction between the electron spin of a paramagnetic defect and an external field. This Zeeman interaction identifies the defect and gives information on the symmetry of its electronic orbitals. We will also describe the interaction between the electronic spin and the surrounding nuclear spins. These so-called hyperfine interactions inform on the atoms that make up the defect and their microscopic structure.

4.1.2 Spin Hamiltonian formalism

In EPR, the spin Hamiltonian formalism is used to describe the interactions between a system of spins and an external magnetic field. Because we are dealing with point defects into semiconductors, we will focus on the interaction of an isolated paramagnetic defect into a crystalline matrix.

The Hamiltonian corresponding to the total energy of the paramagnetic electrons in this defect is the sum of several terms.

In the following, I present the quantum formalism useful to analyze EPR spectra. I intend to give physical justification and discuss the physical meaning of the different introduced terms. I will justify specific choices in their expressions and sometimes demonstrate important results. The reader must be familiar with quantum mechanics and rudiments of linear algebra.

First, we may consider the kinetics energy of the electrons and their coulombic interactions, between them and with the nuclei of the system. The corresponding operator is

$$\sum_k \left(\frac{p_k^2}{2m} - \frac{Ze^2}{4\pi\varepsilon_0} \frac{1}{r_k} \right) + \frac{e^2}{4\pi\varepsilon_0} \sum_{jk} \frac{1}{r_{jk}}$$

The coulombic interaction with the rest of the crystal can be took into account by adding a potential V_c. This term, noted H_0, is the first term of the Hamiltonian

$$H_0 = \sum_k \left(\frac{p_k^2}{2m} - \frac{Ze^2}{4\pi\varepsilon_0} \frac{1}{r_k} \right) + \frac{e^2}{4\pi\varepsilon_0} \sum_{jk} \frac{1}{r_{jk}} + V_c$$

We now add magnetic interactions.

First, there is a magnetic coupling between the electron spin and the magnetic field produced by the nucleus around which the paramagnetic electron is orbiting. Indeed, in the electron frame, the positively charged nucleus is moving and consequently creates a magnetic field. This magnetic field depends on the angular momentum $\boldsymbol{L}$ and interacts with the magnetic dipole moment associated with the electron spin $\boldsymbol{S}$. This is the "spin–orbit coupling" H_{LS} which can be expressed

$$H_{LS} = \lambda \boldsymbol{L} \cdot \boldsymbol{S} \text{ where } = \sum_i l_i, \boldsymbol{S} = \sum_i s_i \text{ and } \lambda \text{ is spin–orbit constant.}$$

When a paramagnetic defect is placed into a uniform external magnetic field, its energy levels are shifted. This is known as the Zeeman effect. The interactions of the orbital- and spin-related magnetic momenta of the unpaired electrons with the external magnetic field are described by the operator:

$$H_B = -(\boldsymbol{\mu}_L + \boldsymbol{\mu}_S) \cdot \boldsymbol{B}$$

with

$$\boldsymbol{\mu}_L = -\beta \boldsymbol{L} \quad \text{and} \quad \boldsymbol{\mu}_S = -g_0 \beta \boldsymbol{S} \text{ where } \beta = \frac{e\hbar}{2m}$$

is the Bohr magneton and g_0 is the gyromagnetic factor (around 2.0023 for a free electron). We use here the reduced orbital and spin momenta, with dimensionless integer or half-integer eigenvalues.

In a first step, I will show how this Hamiltonian can be written using the so-called g tensor or Lande tensor. For this, I am going to use the time-independent perturbation theory and will treat $H = H_{LS} + H_B$ as a perturbation of H_0.

$$H = \lambda \boldsymbol{L} \cdot \boldsymbol{S} + \beta (\boldsymbol{L} + g_0 \boldsymbol{S}) \cdot \boldsymbol{B}$$

The first-order correction to the energy is

$$E^{(1)} = \langle \psi_n | H | \psi_n \rangle$$

where the state $|\psi_n\rangle$ corresponds to the eigenfunction of H_0 associated with the eigenvalue E_n.

If we choose (Oz) as the quantification axis of the kinetics moment then

$$\langle\psi_n|L_x|\psi_n\rangle = \langle\psi_n|L_y|\psi_n\rangle = 0.$$

The interaction with the external magnetic field cannot modify the eigenvalue of L_z which is fixed by the ligand fields, the orbital moment is said "quenched." Thus $\langle\psi_n|L_z|\psi_n\rangle = 0$, and

$$E^{(1)} = \langle\psi_n|\beta g_0 \boldsymbol{S}\cdot\boldsymbol{B}|\psi_n\rangle$$

According to the perturbation theory, the second-order correction to the energy is given by

$$E^{(2)} = -\sum_{m\neq n}\frac{\langle\psi_n|H|\psi_m\rangle\langle\psi_m|H|\psi_n\rangle}{E_m - E_n}$$

As $\langle\psi_n|\beta g_0 \boldsymbol{S}\cdot\boldsymbol{B}|\psi_m\rangle = \beta g_0 \boldsymbol{S}\cdot\boldsymbol{B}\langle\psi_n|\psi_m\rangle$, $H^{(2)}$ can be expressed

$$E^{(2)} = -\sum_{m\neq n}\frac{\langle\psi_n|\lambda\boldsymbol{L}\cdot\boldsymbol{S}+\beta\boldsymbol{L}\cdot\boldsymbol{B}|\psi_m\rangle\langle\psi_m|\lambda\boldsymbol{L}\cdot\boldsymbol{S}+\beta\boldsymbol{L}\cdot\boldsymbol{B}|\psi_n\rangle}{E_m - E_n}$$

$$E^{(2)} = -\sum_{i,j=x,y}\sum_{m\neq n}\frac{\{\langle\psi_n|L_i|\psi_m\rangle(\beta B_i+\lambda S_i)\}\{\langle\psi_m|L_j|\psi_n\rangle(\beta B_j+\lambda S_j)\}}{E_m - E_n}$$

or, more compactly

$$E^{(2)} = -\sum_{i,j=x,y}\Lambda_{ij}(\beta B_i+\lambda S_i)(\beta B_j+\lambda S_j)$$

where

$$\Lambda_{ij} = \sum_{m\neq n}\frac{\langle\psi_n|L_i|\psi_m\rangle\langle\psi_m|L_j|\psi_n\rangle}{E_m - E_n}$$

Notice that the calculations of the terms Λ_{ij} require the knowledge of the wave functions of the paramagnetic electrons and the corresponding energies E_n. In the case of defects, these calculations are quite difficult as the electronic orbitals of the defect are generally reconstructed or relaxed and thus significantly differ from the ones of the crystal.

Finally,

$$E^{(2)} = -2\beta\lambda\sum_{i,j}\Lambda_{ij}B_iS_j - \lambda^2\sum_{i,j}\Lambda_{ij}S_iS_j - \beta^2\sum_{i,j}\Lambda_{ij}B_iB_j$$

Among these three terms, the last one represents a very weak diamagnetic interaction and is negligible. The second term is independent of the external magnetic field. It is an anisotropic spin–spin interaction H_{ss}, arising from the spin–orbit coupling. This term will be further discussed later.

The first term mixes the total spin and the external magnetic field components and consequently introduces an important anisotropy in the Zeeman interaction. It can be combined with the first-order correction to the energy

$$g_0\beta \boldsymbol{B}\cdot\boldsymbol{S} - 2\beta\lambda\sum_{i,j}\Lambda_{ij}B_iS_j = \beta\sum_{i,j}g_{ij}B_iS_j,\ \text{where } g_{ij}= g_0\delta_{ij} - 2\lambda\Lambda_{ij}$$

The g_{ij} components define a matrix coupling the spin and the external magnetic field. This can be interpretated as the result of a Zeeman-type interaction with an effective magnetic moment

$$\boldsymbol{\mu} = -\beta\boldsymbol{B}\cdot\tilde{\mathbf{g}}\cdot\boldsymbol{S}$$

Note that the anisotropy of g is directly proportional to the spin–orbit coupling constant λ.

So H can be written using the so-called g tensor, or Lande tensor $\tilde{\mathbf{g}}$ whose components are g_{ij}:

$$H = \beta\boldsymbol{B}\cdot\tilde{\mathbf{g}}\cdot\boldsymbol{S} + H_{SS}$$

In the principal axis system of the g tensor, the Zeeman term of this Hamiltonian is

$$\beta\boldsymbol{B}\cdot\tilde{\mathbf{g}}\cdot\boldsymbol{S} = \beta B(\sin\theta\cos\varphi\sin\theta\sin\varphi\cos\theta)\begin{pmatrix} g_{xx} & 0 & 0 \\ 0 & g_{yy} & 0 \\ 0 & 0 & g_{zz}\end{pmatrix}\begin{pmatrix} S_x \\ S_y \\ S_z\end{pmatrix}$$

$$\beta\boldsymbol{B}\cdot\tilde{\mathbf{g}}\cdot\boldsymbol{S} = \beta B\left(g_{xx}\cos\varphi\sin\theta\, S_x + g_{yy}\sin\varphi\sin\theta\, S_y + g_{zz}\cos\theta\, S_z\right)$$

where the orientation of the magnetic field in the principal axis system is given by the polar angles θ and φ. This Hamiltonian can be expressed using an effective g value:

$$\beta\boldsymbol{B}\cdot\tilde{\mathbf{g}}\cdot\boldsymbol{S} = \beta g B S_z$$

with

$$g = \sqrt{g_{xx}^2\sin^2\theta\cos^2\varphi + g_{yy}^2\sin^2\theta\sin^2\varphi + g_{zz}^2\cos^2\theta}$$

4.1.2.1 Zero-field splitting

Now we can go further and consider the dipolar interaction between the magnetic moment of the paramagnetic electrons. Of course, this interaction only exists for systems with more than one unpaired electron, that is to say, for defects with a total spin $S \geq 1$. This interaction depends only on the electron spin, like the previous term H_{ss}, and it can thus be formally added to H_{ss} to produce the so-called zero-field splitting H_{zfs}. It is then possible to write H_{zfs} using a single tensor $\tilde{\boldsymbol{D}}$:

$$H_{zfs} = \boldsymbol{S}\cdot\tilde{\boldsymbol{D}}\cdot\boldsymbol{S}$$

It is worthwhile remembering that this term has two distinct physical origins: the spin–orbit interaction and dipolar interactions between the magnetic moment of unpaired electrons. Due to electrons indiscernibility, $\tilde{\boldsymbol{D}}$ has to be associated with a symmetrical matrix.

In the frame where $\tilde{\boldsymbol{D}}$ is diagonal, its principal values D_x, D_y and D_z are related to the spin operator by

$$H_{zfs} = D_x\boldsymbol{S}_x^2 + D_y\boldsymbol{S}_y^2 + D_z\boldsymbol{S}_z^2$$

We now fleetingly demonstrate that $D_x + D_y + D_z = 0$.

Let us consider purposely the operator:

$$H_{zfs} + \alpha\boldsymbol{S}^2$$

It is easy to see that any eigenvectors $|\varphi\rangle$ of H_{zfs} is also eigenvector of this new operator. Indeed, suppose that the eigenvalue associated to $|\varphi\rangle$ is E_f, then

$$\left(H_{zfs} + \alpha\boldsymbol{S}^2\right)|\varphi\rangle = \left\{E_\varphi + \alpha S(S+1)\right\}|\varphi\rangle$$

So the operator $H_{zfs} + \alpha\boldsymbol{S}^2$ has the same eigenvectors as H_{zfs} and has eigenvalues shifted by $\alpha S(S+1)$. As long as we deal with eigenvectors and difference of eigenvalues, the operators H_{zfs} and $H_{zfs} + \alpha\boldsymbol{S}^2$ are equivalent.

Then, it is always possible to choose

$$\alpha = -\frac{1}{3}\left(D_x + D_y + D_z\right)$$

so the trace of $H_{zfs} + \alpha\boldsymbol{S}^2$ is equal to zero. Because of the equivalence with H_{zfs}, we can always impose to the trace of $\tilde{\boldsymbol{D}}$ to be equal to zero: $D_x + D_y + D_z = 0$.

The zero-field splitting is then characterized by a symmetrical tensor $\tilde{\boldsymbol{D}}$ with $\mathrm{Tr}(\tilde{\boldsymbol{D}}) = 0$.

Note that, in the particular case of spin $S = 1/2$, the elements of $\tilde{\boldsymbol{D}}$ are equal to zero. Indeed, the diagonal elements of the matrix representing H_{zfs} in the basis $\{|m_S\rangle\}$ formed by the common eigenvectors of $\left(\boldsymbol{S}^2, \boldsymbol{S}_z\right)$ are

$$\langle m_S|D_x\boldsymbol{S}_x^2 + D_y\boldsymbol{S}_y^2 + D_z\boldsymbol{S}_z^2|m_S\rangle$$

$$= D_x\langle m_S|\boldsymbol{S}_x^2|m_S\rangle + D_y\langle m_S|\boldsymbol{S}_y^2|m_S\rangle + D_z\langle m_S|\boldsymbol{S}_z^2|m_S\rangle$$

$$= D_x\frac{1}{2}\left[S(S+1) - m_S^2\right] + D_y\frac{1}{2}\left[S(S+1) - m_S^2\right] + D_z m_S^2$$

If $S = 1/2$, $m_S^2 = \pm 1/2$ and m_S^2 has only one possible value: 1/4. Then

$$\langle m_S|D_x\boldsymbol{S}_x^2 + D_y\boldsymbol{S}_y^2 + D_z\boldsymbol{S}_z^2|m_S\rangle = \left(D_x + D_y + D_z\right)/4 = 0 \text{ because } \mathrm{Tr}\left(\tilde{\boldsymbol{D}}\right) = 0.$$

Consequently, for a defect with a single unpaired electron, and in this case only, the zero-field splitting does not affect the energy levels, and the term $\boldsymbol{S}\cdot\tilde{\boldsymbol{D}}\cdot\boldsymbol{S}$ can be omitted in the total Hamiltonian expression.

In conclusion, the Zeeman interaction, the spin–orbit interaction and the magnetic interaction between electronic spins can be described by the following Hamiltonian:

$$H = \beta \boldsymbol{B} \cdot \tilde{\mathrm{g}} \cdot \boldsymbol{S} + \boldsymbol{S} \cdot \tilde{\boldsymbol{D}} \cdot \boldsymbol{S}$$

In this expression, only the electronic spin $\boldsymbol{S}$ appears, the angular momentum is hidden in the $\tilde{\mathrm{g}}$ and $\tilde{\boldsymbol{D}}$ tensors. For this reason, this Hamiltonian is called spin Hamiltonian.

The spin value and the tensors $\tilde{\mathrm{g}}$ and $\tilde{\boldsymbol{D}}$ can be determined by analyzing the angular variation of the EPR spectrum obtained by rotating the sample into the external magnetic field. The principal values and the directions of the principal axes reflect the point symmetry of the defect. For example, a simply occupied sp^3 dangling bond has an axial (C_{3v}) symmetry and so the g tensor. An axial g tensor has only two distinct principal values, the lowest indicates the direction of the symmetry axis of the defect.

The tensors $\tilde{\mathrm{g}}$ and $\tilde{\boldsymbol{D}}$ are typical for a defect, and they identify it. Nevertheless, the knowledge of these parameters does not allow the determination of the nature of a previously unknown defect. But the detection of hyperfine interactions supplies chemical information on the defect and its environment. This usually enables the proposition of a tentative defect model which should be confirmed by theoretical calculations of the experimentally measured parameters.

4.1.3 Hyperfine interactions

The unpaired electrons of the paramagnetic defect can magnetically interact with the surrounding nuclear spins $\boldsymbol{I}$ as a nuclear magnetic dipole is associated to the nuclear spin. This magnetic coupling depends on $\boldsymbol{S}$ and $\boldsymbol{I}$ and, of course, does not exist for nuclei with a zero nuclear spin.

This interaction, called "hyperfine interaction," is generally anisotropic in crystals and is consequently written using a tensor $\tilde{\boldsymbol{A}}$. If the electronic spin interacts with different nuclear spins then the corresponding operator is a linear combination of these interactions:

$$H_{SI} = \sum_n \boldsymbol{S} \tilde{\boldsymbol{A}}_n \boldsymbol{I}_n$$

where $\tilde{\boldsymbol{A}}_n$ describes the interaction with the nuclear spin $\boldsymbol{I}_n$.

These hyperfine interactions provide information about the chemical nature of defect because both the value and the natural abundance of the nuclear spins are atom specific. When the electronic wave function of the paramagnetic spin is sufficiently delocalized, the electronic spin interacts with distant nuclei in the host crystal. These hyperfine interactions depend on the position, number and chemical nature of the atoms in the environment of the defect. They are sometimes called "super-hyperfine" interactions, the name "hyperfine interactions" being then kept for interactions with nuclear spins in the core of the defect itself.

From the multiplicity of the hyperfine lines, the nuclear spin value can be deduced: a doublet corresponds to $I = 1/2$, a triplet to $I = 1$, a sextet to $I = 5/2$ and so on. The natural abundance of interacting nuclei can be determined from the intensity ratio between the hyperfine lines and the central Zeeman resonance (arising from the electron spin interacting with nuclei with spin $I = 0$). In fact, the observational situation may be complicated by the coexistence of different isotopes with nonzero nuclear spin and also, by the presence of equivalent lattice sites in the semiconductor crystal. In this case, statistics is required to evaluate the probability of the different defect environments (isotopes on the equivalent lattices sites). Generally speaking, the larger the A values, the closer the nuclei. But spin polarization effects, for example, may strongly reduce the hyperfine interaction with the nearest neighbors.

The A values can be theoretically calculated by the linear combination of atomic orbitals or by density functional theory. The electron spin localization can be deduced from the hyperfine tensor. For instance, the s and p spin densities of a dangling bond can be deduced from the isotropic and anisotropic parts of the hyperfine tensor, giving information about the relaxation of the atom with the dangling from its lattice site.

Finally, the nuclear spins themselves interact with the external magnetic field and between them. The corresponding Hamiltonians are H_{BI} and H_{II}, respectively. These terms are essential for NMR spectroscopy and may be of importance for ENDOR (electron nuclear double resonance), but I will neglect them in the following because the nuclear magnetic moment is typically 2,000 times weaker than the electronic one.

4.1.4 Resonance

The spin Hamiltonian gives the energy splitting of the Zeeman and hyperfine sublevels. The experimental determination of this energy splitting can be performed by EPR spectroscopy via the detection of the resonant absorption of a photon. Submitting a paramagnetic defect of spin S to a controlled static external field B, the degeneracy of the ground-state level gets lifted and splits into $(2S + 1)$ Zeeman sublevels.

In the simple case of an $S = 1/2$ defect, illustrated in Figure 4.1, two Zeeman sublevels symmetrically split from the E_0 energy ground state. For a given orientation of the magnetic field with respect to the defect, they have distinct energies of $E_0 + g\beta B/2$ and $E_0 - g\beta B/2$ where g is the effective g value calculated from the g tensor. So the Zeeman splitting ΔE is proportional to the magnetic field:

$$\Delta E = g\beta B$$

By increasing the magnetic field and continuously irradiating the defect by photons of constant energy photons $h\nu$, the resonance occurs when ΔE reaches the photon energy and it follows that the resonant field B_r is

$$B_r = \frac{h\nu}{g\beta}$$

Considering typical magnetic field values from 0.1 up to 10 T, the photons are in the microwave or far infrared domains.

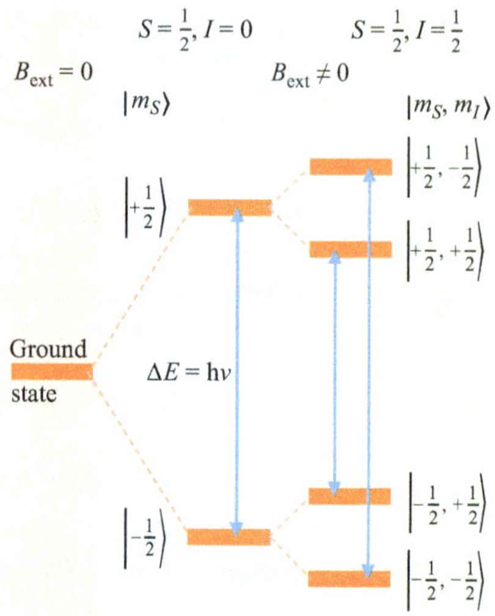

Figure 4.1 Zeeman levels scheme of an $S = 1/2$ paramagnetic defect without or with an hyperfine interaction with a nuclear spin $I = 1/2$

By measuring the angular variation of the resonant field, that is to say the values of B_r for different orientations of the sample with respect to the external magnetic field, the g tensor can be determined.

4.1.4.1 Selection rules

The selection rules are obtained by studying the conditions for which the spin Hamiltonian matrix has a nonzero element between the initial and final Zeeman level states.

The microwave has a time-dependent magnetic field B_1 which is responsible for the spin transition. The total magnetic field on the paramagnetic defect can be seen as the sum of a "static" field B which splits the energy level (Zeeman splitting) and a time-dependent field B_1 leading to inter-level transitions.

Considering that B is always much stronger than B_1, we can treat the effect of B_1 has a perturbation of the Zeeman states defined by B. Choosing the z-axis in the direction of the static field B, the Zeeman states are eigenfunctions of the Hamiltonian $g\beta\boldsymbol{S}\cdot\boldsymbol{B} = g\beta S_z \cdot B$.

The magnetic field B_1 interacts with the magnetic moment of the paramagnetic defect and the corresponding Hamiltonian is

$$H_1 = -\boldsymbol{\mu}_S \cdot \boldsymbol{B}_1 = g\beta\boldsymbol{S}\cdot\boldsymbol{B}_1$$

If B_1 is aligned with B, then $H_1 = g\beta S_z \cdot B_1$ and the matrix element coupling two Zeeman levels $|m_i\rangle$ and $|m_f\rangle$ is

$$\langle m_f | g\beta S_z \cdot B_1 | m_i\rangle = g\beta B_1 \langle m_f | S_z | m_i\rangle = g\beta B_1 m_S \langle m_f | m_i\rangle = 0$$

Thus no transition can be expected from the component of the microwave magnetic field along the direction of the external static field. To maximize transition probability, the time-dependent magnetic field B_1 has to be perpendicular to the static field B.

In this case, H_1 depends only on the operators S_x and S_y. These operators can be expressed as a function of the shift operators S_+ and S_-:

$$S_x = \frac{1}{2}(S_+ + S_-) \quad \text{and} \quad S_y = \frac{1}{2i}(S_+ - S_-)$$

Consequently, the Hamiltonian H_1 will change the value of the spin projection m_i by one unit.

For instance, let us consider B_1 along the x-axis. We have then

$$H_1 = g\beta S_x \cdot B_1 = \frac{1}{2} g\beta B_1 (S_+ + S_-)$$

So

$$\langle m_f | H_1 | m_i \rangle = \frac{1}{2} g\beta B_1 \langle m_f | S_+ + S_- | m_i \rangle$$

$$= \frac{1}{2} g\beta B_1 \{ \langle m_f | S_+ | m_i \rangle + \langle m_f | S_- | m_i \rangle \}$$

$$\langle m_f | H_1 | m_i \rangle = \frac{1}{2} g\beta B_1 \{ \langle m_f | m_i + 1 \rangle + \langle m_f | m_i - 1 \rangle \}$$

Finally, because of the orthogonality of the Zeeman states, $\langle m_f | H_1 | m_i \rangle$ is nonzero only if $|m_f\rangle = |m_i + 1\rangle$ or $|m_f\rangle = |m_i - 1\rangle$.

In conclusion, when perpendicular to the static field, the microwave magnetic field B_1 induces resonant transitions between Zeeman levels separated by one unit of spin projection quantum number. The so-called selection rule is

$$\Delta m = \pm 1$$

In the case of hyperfine interaction with a nuclear spin I, the Zeeman sublevels are splitted into $(2I + 1)$ hyperfine levels. Figure 4.1 presents the energy level configuration in the case of $I = 1/2$. The four levels do not generate four resonances in EPR because H_1 does not act upon the nuclear spin. Finally, the selection rules are

$$\Delta m_S = \pm 1 \quad \text{and} \quad \Delta m_I = 0$$

and in this case, only two resonances will be detected and a doublet builds up.

4.1.5 Transition probability: relaxation phenomena

According to time-dependent perturbation theory and the Fermi's Golden rule, the transition probability between two Zeeman levels $|m_i\rangle$ and $|m_f\rangle$ is

$$W_{m_i m_f} = \frac{2\pi}{\hbar^2} |\langle m_i H_1 m_f \rangle|^2 f(\omega - \omega_0)$$

where ω is the microwave frequency and $f(\omega - \omega_0)$ is the envelope function associated with the transition.

For a spin $S = 1/2$,

$$W_{m_i m_f} = \frac{\pi}{\hbar^2} g^2 \beta^2 B_1^2 f(\omega - \omega_0)$$

First, this expression shows that the transition probability is proportional to B_1^2 that is to say to the microwave power.

Second, we immediately find that

$$W_{m_i m_f} = W_{m_f m_i} = W$$

So the microwave-induced absorption and emission transitions have the same probability. But because, at thermal equilibrium, the numbers of spins in the upper state and lower state are not the same, the numbers of absorption and radiative emission transitions are quite different. Indeed, as the majority of the spins lie aligned with the external magnetic field in the lower energy state, most of the transitions correspond to microwave absorption transitions.

Let us mention that in the case of a paramagnetic defect submitted to a magnetic field of one Tesla or less, the Zeeman sublevels split by a few hundredth of meV only, which is much lower than the thermal energy kT, even at low temperature. Consequently, population difference in the Zeeman levels, determining the absorption intensity, is always very small.

For simplicity, let us consider once again the case of the two Zeeman states of an $S = 1/2$ defect with populations N_- and N_+, and let us introduce the total spin number N and the population difference $n = N_- - N_+$. The EPR signal is proportional to the net number of absorption transitions: Wn. The rate of microwave energy absorption is the absorbed power:

$$P_{abs} = h\nu W n$$

In fact, the populations of the Zeeman levels follow the Curie's law. At thermal equilibrium, the state populations follow the Boltzmann statistics:

$$\frac{N_+}{N_-} = e^{-\Delta E/kT}$$

where $\Delta E = g\beta B$ is the Zeeman energy splitting. When $\Delta E \ll kT$:

$$\frac{N_+}{N_-} \approx 1 - \frac{\Delta E}{kT} = 1 - \frac{g\beta B}{kT}$$

As

$$N_+ = \frac{1}{2}(N - n) \quad \text{and} \quad N_- = \frac{1}{2}(N + n)$$

this approximation leads to

$$2n = \frac{g\beta B}{kT}(N + n)$$

Because in our assumption $N \gg n$, we consider that $N + n \approx N$ and then

$$n = \frac{g\beta BN}{2kT} = n_0$$

As the electrons in the two Zeeman levels have opposite spin projection $m_S = \pm 1/2$, this population difference leads to a net magnetic moment $g\beta n_0/2$ in the direction of B. The magnetization M_0 of the sample, defined as the net magnetic moment per unit volume is

$$M_0 = \frac{g^2\beta^2 N_S}{4kT} B$$

where $N_S = N/V$ is the spin density in the volume V of the sample.

This expression directly gives the so-called static magnetic susceptibility χ_0 defined by $M_0 = \chi_0 B$:

$$\chi_0 = \frac{g^2\beta^2 N_S}{4kT}$$

This result can be generalized for a defect with spin S:

$$\chi_0 = \frac{g^2\beta^2 N_S S(S+1)}{3kT}$$

The spin resonance is detected via the microwave energy absorption by the spin system. Because the EPR signal is proportional to n, the lower the temperature, the stronger the signal. When the temperature decreases, the Curie's law shows that n increases as $1/T$, leading to a signal improvement by a factor of 75 by cooling the sample from room temperature down to 4 K.

Nevertheless, as we will see, the lowest temperature measurement is not always the more convenient. The spins in the upper state have different non-radiative ways to relax to the lower state. When the relaxation process gets slow, the lower and upper states' populations tend to equalize and, consequently, the net number of absorption falls down to zero and so the detected signal.

Like for NMR spectroscopy, the relaxation process is described by introducing two different relaxation times T_1 and T_2.

The time T_1 is called the spin–lattice relaxation time or longitudinal relaxation time. The spin–lattice relaxation describes the energy exchange process between the paramagnetic spins and the thermal energy reservoir made up of the neighboring atoms. It is the characteristic time of the exponential law followed by the magnetic moment to reach its equilibrium position, aligned with the external magnetic field. This relaxation depends on the temperature. The lower the temperature the longer T_1.

Considering the spin–lattice relaxation, the time evolution of the Zeeman level populations reads

$$\frac{dn}{dt} = \left.\frac{dn}{dt}\right|_{microwave} + \left.\frac{dn}{dt}\right|_{lattice}$$

with

$$\left.\frac{dn}{dt}\right|_{microwave} = W(N_+ - N_-) - W(N_- - N_+) = -2Wn$$

If n_0 is the thermal equilibrium value of n, then

$$\left.\frac{dn}{dt}\right|_{lattice} = (n_0 - n)\frac{1}{T_1}$$

The steady-state solution determined by $dn/dt = 0$ is

$$n = n_0\frac{1}{1 + 2WT_1}$$

When $W \gg 1/T_1$, n approaches zero and the EPR signal disappears. This is the saturation phenomenon which can be avoided by reducing the microwave power ($W \propto B_1^2$) or by increasing the measurement temperature (and so reducing T_1).

As a result, to optimize the measurements conditions of a paramagnetic defect, that is to say to maximize the n value, which increases when the temperature decreases, we often have to come to a compromise between the microwave power and the temperature because the lower the temperature the higher n_0, but the longer T_1.

The time T_2 is called spin–spin relaxation time or transverse relaxation time. The interaction between spins in the sample holds up the magnetic moment alignment with the external magnetic field. It desynchronizes the spins with each other and impacts the decay of the transverse component of the magnetic moment on its way back to equilibrium position. The characteristic time of this exponential decay is T_2 that is defined similarly to T_1 by

$$W = \frac{1}{2}\gamma^2 B_1^2 T_2$$

Consequently, the steady-state solution n can be rewritten as

$$n = n_0\frac{1}{1 + \gamma^2 B_1^2 T_1 T_2}$$

The Bloch equations, obtained in 1946 by classical physics means, describe the time evolution of the magnetization of the two-level spin system considered here. From the time evolution of the magnetization components parallel and perpendicular to the static field, Bloch has deduced the stationary solution and has introduced a complex magnetic susceptibility $\chi = \chi' + i\chi''$, where χ' and χ'' describe the dispersion and absorption phenomenon, respectively. This classical calculation leads to

$$\chi' = \frac{1}{2}\frac{\gamma M_0(\omega_0 - \omega)T_2^2}{1 + (\omega_0 - \omega)^2 T_2^2 + \gamma^2 B_1^2 T_1 T_2}$$

$$\chi'' = \frac{1}{2}\frac{\gamma M_0 T_2}{1 + (\omega_0 - \omega)^2 T_2^2 + \gamma^2 B_1^2 T_1 T_2}$$

In continuous wave (CW) EPR, the signal corresponds to the microwave power absorbed P_{abs} by the defects in sample, when the resonance condition is reached. The microwave magnetic field B_1 exerts a torque on the magnetization and the power of the corresponding moment is

$$P_{abs} = \omega B_1^2 \chi''$$

Given the ω dependence of P_{abs}, when the saturation factor $\gamma^2 B_1^2 T_1 T_2 \ll 1$, the resonance have a Lorentzian line shape.

To conclude, let us emphasize that the saturation phenomenon is sometimes useful to clear a spectrum where signals of different paramagnetic defects get superimposed. If the defects have distinct relaxation times, then by adjusting the temperature and microwave power, it is sometimes possible to saturate one signal preserving the other. This selective saturation simplifies the analysis of the spectrum.

4.1.6 Experimental setup

The detection of the resonance consists in the measurement of the absorption of the microwave energy by a given paramagnetic center in the sample when the magnetic field reaches its resonant value.

The simplest methodology would rely on a transmission measurement. This can be performed by irradiating the sample with a microwave and by measuring the transmitted intensity and then comparing it to a reference intensity coming directly from the microwave source. Unfortunately, this direct measurement, used in many optical spectroscopies, generally does not allow sufficient signal-to-noise ratio to detect the magnetic resonance of a few ppm of paramagnetic species. The first reason is that the Zeeman sublevels generated by the spectrometer magnetic field are typically split by a few hundredth of meV only, which is much lower than the thermal energy even at low temperature. Consequently, population difference in the Zeeman levels, determining the absorption intensity, is always small. The second reason is that for microwaves traveling in free space or in waveguide, both electric and magnetic components of the electromagnetic wave are present simultaneously in the same spatial region. So the absorption of the microwave energy will be dominated by electric dipole transitions which have a higher transition moment than magnetic dipole transitions, and the detection of EPR will be impossible.

In fact, the magnetic dipole transition probability depends on the square of the microwave magnetic field. To optimize the transition probability, most of the EPR spectrometers use a microwave resonator: a metallic cavity of rectangular or cylinder shape. This cavity is tailored to generate microwave standing waves (cavity modes). In this resonator, the nodes and antinodes of the electric and magnetic fields are spatially separated, and the microwave magnetic field is maximized at the cavity center, where the sample is hold. For instance, Figure 4.2 shows the spatial distribution of the magnetic and electric components of the microwave in a popular rectangular cavity using the transverse electrical mode TE_{102}. At the center of the resonant cavity, the magnetic field of the microwave is

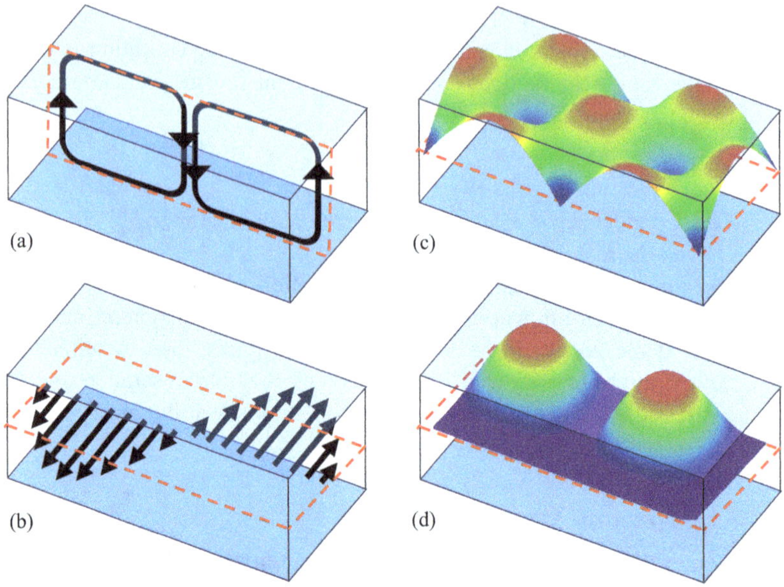

Figure 4.2 Schematic of the magnetic field lines (a) and electric field (b) in the median planes (dash lines) of the resonant TE_{102} cavity. Spatial distribution of the square of the magnetic field (c) and electric field (d) amplitudes in the cavity

maximal. To do not disturb the cavity modes, the sample holder and the sample have to be as transparent as possible to the microwave. For this reason and to do not generate adventitious paramagnetic signals, the sample holder is generally made of ultrapure quartz.

The wavelength of the microwave determines the cavity size which, in turn, imposes the maximal sample size. The microwave frequencies commonly used are 9.6 GHz (called X-Band, wavelength about 3 cm) and 35 GHz (Q-Band, wavelength about 1 cm), respectively; and the typical sample sizes are $4 \times 2 \times 12$ and $2 \times 2 \times 8$ mm^3 respectively. To maintain the standing wave pattern in the cavity, the microwave frequency has to be kept rigorously constant during the measurement.

For this reason and also because widely tunable microwave sources are not available, we use a fixed microwave energy and vary the splitting of the Zeeman levels by sweeping the external magnetic field. For each value of the external magnetic field, the resonant condition is tested by measuring the absorbed microwave power. Consequently, ESR spectra are displayed as a function of the external magnetic field values.

Because the magnetic field has to be swept during the experiment and accurately controlled to determine the g values, electromagnets are more convenient than superconducting magnets. So a large electromagnet with large magnetic coils produces homogenous magnetic field on the whole sample volume, limiting the inhomogeneous broadening of the resonance. If very high magnetic fields are required, hybrid magnets are used.

Figure 4.3 presents a scheme of a typical EPR spectrometer and a picture of the X-Band spectrometer in use at Institut des Nano Sciences de Paris is shown in Figure 4.4.

The sample is hung into the cavity. The sample and the cavity are then placed in an external magnetic field to lift the degeneracy of the Zeeman levels of the paramagnetic defects.

To perform experiments at low temperature, cryostats are used. And at Q-Band frequency, both the cavity and the sample are placed into the cryostat and cooled down, the X-Band spectrometers generally used He flow cryostats. In this case, the

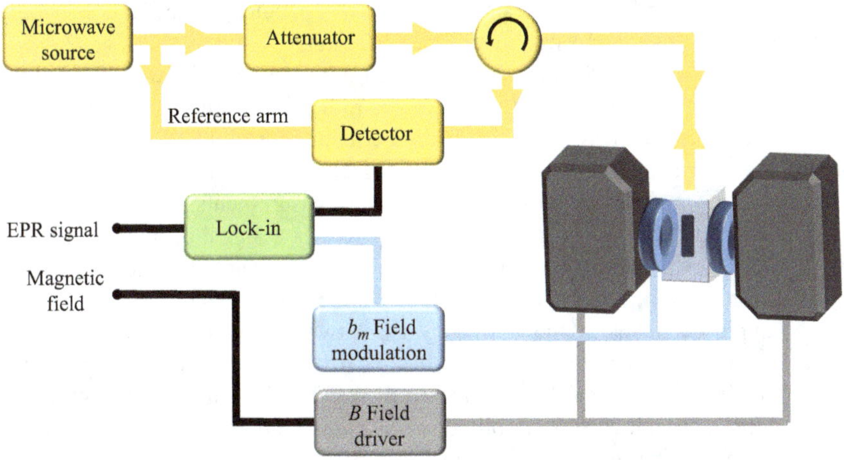

Figure 4.3 Schematic of a typical EPR spectrometer

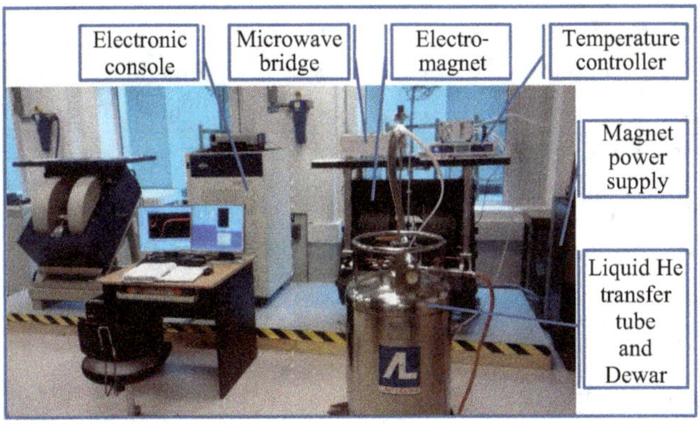

Figure 4.4 X-Band EPR spectrometer at Institut des Nano Sciences de Paris

cavity remains at room temperature and the He flows to the sample thanks to a double-walled, vacuum-insulated quartz tube inserted into the cavity.

In CW experiments described here, the microwave absorption at the resonance is detected via the variation of the Q factor of the {cavity+sample} system induced at the resonance by the variation of the amount of microwave energy stored into the cavity. To make the measurement insensitive to the power fluctuations of the microwave source, a photodiode detector compares the amount of microwave energy passing through the {cavity+sample} system to the one transmitted via a reference arm.

To increase the sensitivity of the technique and get rid of non-inter Zeeman levels related absorption, a lock-in detection is performed by modulating the external magnetic field and rejecting absorptions out of step with the magnetic field. An oscillating magnetic field b_m generated by additional Helmholtz coils is collinearly superimposed to the "static" field B. An efficient rejection is generally achieved with a modulation frequency ν_m of 100 kHz.

Nevertheless, the modulation frequency is limited by the relaxation time of the paramagnetic defect. The field modulation has to be slow compared to the relaxation time so that the spin system remains in thermal equilibrium during the passage through the resonance. This so-called slow passage condition has to be achieved to avoid saturation in order to perform quantitative measurements.

The lock-in detection of the absorption generates a signal proportional to the first derivative of the absorption line with respect to the magnetic field. Indeed, for a modulation with a peak-to-peak amplitude b_m the actual external magnetic field in the sample oscillates at the frequency ν_m between $B - b_m$ and $B + b_m$. As illustrated in Figure 4.5, for a given value B_0 of the static field, the absorption signal $D(B_0, t)$ measured by the detector oscillates. The lock-in amplifier gives the so-called EPR signal that is the amplitude of D and is proportional to the local slope of the absorption line for the static field B_0.

The line shape of the EPR resonance is the first derivative of a Lorentzian line. If inhomogeneous broadening is present, the line has a Voigt profile: a convolution of a Lorentzian line with a Gaussian one.

4.1.6.1 Choice of microwave frequency

Let us consider a sample containing two types of paramagnetic defects characterized by two effective g values g_1 and g_2 respectively. The corresponding resonant fields B_1 and B_2 are given by

$$h\nu = g_1\beta B_1 \quad \text{and} \quad h\nu = g_2\beta B_2$$

The resonances will then be separated by

$$\Delta B = B_2 - B_1 = h\nu\left(\frac{1}{g_2} - \frac{1}{g_1}\right)$$

If the values of g_1 and g_2 are close or if the resonances are broad, the two resonances may overlap and consequently may not be resolved. As ΔB linearly increases with the microwave frequency, one can think to carry the EPR

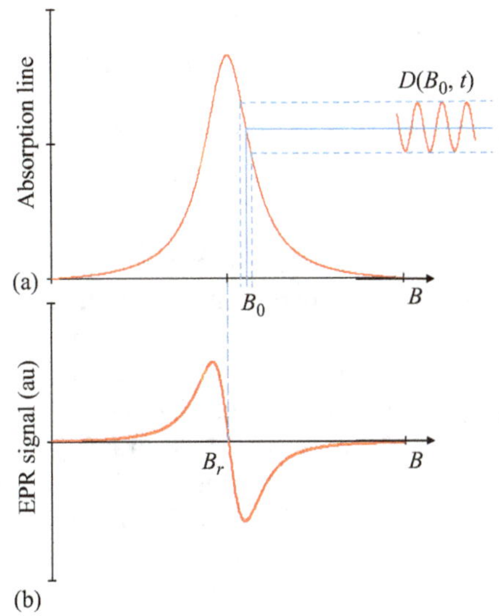

Figure 4.5 (a) Absorption line as a function of magnetic field B. Modulating the magnetic field around a given value B_0 makes the absorption to oscillate with time with an amplitude $D(B_0, t)$ proportional to the slope of the absorption line at B_0. (b) The detected signal is the amplitude of D once filtered by a lock-in detection. It reproduces the first derivative of the absorption line with respect to B. When B_0 corresponds to the resonant field B_r, $D(B_r, t) = 0$

Table 4.1 Resonant magnetic fields of a paramagnetic defect with a Lande g factor of 2 for different microwave frequencies

Band	Frequency (GHz)	Wavelength (cm)	Magnetic field ($g = 2$) (mT)
L	1	30	36
S	3	10	100
X	10	3	350
K	24	1.25	860
Q	35	0.85	1,250
W	94	0.3	3,350

experiment out with a higher microwave frequency. The commercial EPR spectrometers use microwave sources emitting in frequency bands, which are usually represented by a letter, according to radar bands standards. Table 4.1 displays for each band the mean frequency, the corresponding wavelength (in vacuum or air) and the value of the resonant field for a defect with a g value of 2.

The wavelength value gives the order of magnitude of the cavity size. Of course, the higher the frequency, the lower the wavelength. Consequently, the maximum sample size decreases when the frequency increases, from 15 × 4 × 4 mm^3 at X-Band down to 3 × 0.5 × 0.5 mm^3 at W-Band. As the spectrometer sensitivity is expected to increase with the square of the frequency, the small sample volume may not be a problem at first glance. Nevertheless, for semiconductor defect studies, the use of frequencies higher than X-Band is exceptional because of the following reasons:

- Angular variations, needed to determine the g tensor, require accurate orientation of the sample into the cavity, but the orientation of tiny crystals can be tough.
- The high-frequency microwave may not probe the whole sample. Indeed, semiconductors may have a residual conductivity even at low temperature leading to the exponential decay of the electromagnetic microwave power from the surface over a characteristic length called the skin depth. When the microwave frequency increases, the skin depth decreases as $1/\sqrt{\upsilon}$. So the higher the frequency, the smaller the skin depth. The microwave reflection on the sample surface is more important at high frequency and may render difficult or impossible the coupling of the cavity, that is to say the creation of the microwave standing waves pattern in the cavity, even for poor conductor samples.
- Strains in epitaxial films or heterostructures often induce spread in g values of defects as the local environment of the defect slightly differs from one defect to another. In this case, the width of the resonance of the defect increases with the microwave frequency, and there is no resolution enhancement.

4.1.7 Electron nuclear double resonance

In semiconductors, weak hyperfine interactions with distant nuclei are unresolved and broaden the resonance. ENDOR is the technique of choice to investigate these weak interactions. In ENDOR experiments, the EPR signal is used to detect NMR of the nuclei in interaction with the defect.

This double-resonance technique can provide more detailed information about super-hyperfine interactions over single resonance experiment. Indeed the additional restrictions imposed by quantum selection rules enable to reduce the transition redundancies observed in complex multilevel systems. Weak hyperfine couplings obscured in homogenously broadened EPR spectra can then be extracted from the spectrum albeit at the expense of a more complex experimental technique.

The detection of ENDOR is based on the partial recovery of a previously intentionally saturated EPR transition. The EPR has to be observed first, and the external magnetic field is fixed at the resonance value. The microwave power is then increased up to saturation and, as discussed earlier, the EPR signal decreases. Then, a radio frequency (RF) electromagnetic field is swept, perpendicular to both the microwave and the external magnetic fields. When resonant with nuclear sublevels, this RF field allows NMR transitions with $\Delta m_I = \pm 1$, changing the populations of the levels and thus partially desaturing the EPR resonance of the defect whose small coupling to the nucleus is within the EPR linewidth. The ENDOR spectrum is the EPR signal intensity as a function of the RF. An ENDOR line appears at the Larmor frequency of every nuclei interacting with the defect electronic spin. In addition, quadrupolar couplings can be observed by ENDOR, whereas they are forbidden, at first order, in EPR spectroscopy.

ENDOR can be carried out at X-Band frequencies using a cylindrical microwave cavity, the RF field being generated by a helix coils mounted on the quartz tube of the variable temperature Dewar inserted into the cavity. It can also be performed at Q-Band frequencies, using the standard EPR setup with an additional antenna placed near the sample to produce the RF field.

Contrary to EPR, ENDOR is not a quantitative technique because the intensity of ENDOR spectrum depends on several electrons and nuclear relaxation rates. The number of nuclei contributing to each type of hyperfine interactions cannot be directly determined. Neither the nuclear spin can be determined, but the nuclei will be unambiguously identified by their nuclear g factor measured by ENDOR.

The sensitivity of ENDOR is much reduced compared to EPR. So a high density of defects is desirable but may result in dipolar interactions or exchange mechanisms that shorten the relaxation time (T_2) and broaden the resonance, rendering the saturation impossible or the signal undetectable respectively. In the solid state, broadening due to anisotropic hyperfine interactions may also significantly complicate the ENDOR detection.

For all these reasons, ENDOR is rarely used for semiconductor defect studies. Nevertheless, an example of defect study by ENDOR is given in the next part of this chapter.

4.1.8 Pulsed spectroscopies

Pulse EPR is the most recent major development of EPR spectroscopy. By manipulating the spins with the sequences of microwave pulses, information about relaxation processes of the spin system, spin–spin distances and weakly coupled nuclei can be obtained.

In pulse EPR, the spectrum is recorded at a constant value of the external magnetic field by exciting a large frequency range simultaneously with a short high-power microwave pulse. The selectivity of the excitation is controlled by the duration of the pulse. For example, a pulse of 10 ns has a bandwidth of 100 MHz and corresponds to 3 mT.

Shorter pulse length leads to broader excitation range but, contrary to pulse NMR, it is generally not possible to excite the whole spectrum using a single microwave pulse. This strong limitation comes from the use of a resonator containing the sample. Indeed the high microwave power of the pulse remains stored in the resonator during a certain time, even if the resonators used in pulse spectrometers have a much lower Q factor than conventional CW cavities to allow rapid power dissipation. As long as high microwave power remains in the cavity, the detector of the spectrometer has to be protected, and measurements cannot be performed. This so-called dead time of the spectrometer is typically 100 ns at X-Band frequencies. It can be reduced by decreasing the Q factor of the pulse EPR resonator, but this dramatically lowers the sensitivity. Consequently, pulsed EPR is usually less sensitive than CW–EPR.

Relaxation times and spin–spin distance measurements are important in chemistry or biology to determine kinetics and complex structure of supramolecular assemblies for instance, but in contrast to g values and hyperfine coupling, they are not characteristics of isolated spins like paramagnetic defects in semiconductor. Consequently, very few defect studies benefit from pulse EPR spectroscopies. All paramagnetic defects were first evidenced by CW–EPR taking advantage of its high sensitivity and when appropriate, detailed information could be obtained by pulsed or multiple resonance techniques.

4.1.9 Optically detected magnetic resonance, electrically detected magnetic resonance

The detection of ESR, in continuous or pulsed modes, can be carried out by indirect means. This is the case of optically detected magnetic resonance (ODMR) and electrically detected magnetic resonance (EDMR). Both techniques replace the detection of the incident microwave by the detection of photoluminescence (PL) or current variation when reaching the ESR condition.

The main advantage of these detections compared to conventional one is the gain in sensitivity of several order of magnitude. This is particularly interesting for ultralow concentration of defect studies. In some case, the high sensitivity of ODMR allows single defect measurements.

Of course, ODMR is limited to photoluminescent paramagnetic defects. ODMR is worthwhile for $S = 1$ defects where microwave-induced transitions between the triplet states sublevels impact the optical properties because of spin-dependent recombination processes. Let us mention that in the case of $S > 1/2$ defects with zero-field splitting, the ODMR can be performed without applying external magnetic field. Thus, ODMR is generally performed with microwave in the 0.1–1 GHz frequency range for which tunable sources exist. In addition, thanks to its high sensitivity, ODMR can be performed without microwave resonator, and an antenna placed near to the sample may be sufficient to generate the microwave.

ODMR spectrum is the PL variation as a function of microwave frequency. Nevertheless, experimental difficulties are translated in the optical domain as the expected PL variation when passing through the resonance is typically in the 10^{-5} range.

As the ODMR signal is dependent on the external magnetic field, ODMR can be used to get information about the local magnetic field around the defect. For instance, the ODMR of single defect in nanodiamond was recently used for magnetic field mapping of magnetic domains at the nanoscale.

Because isolated defects in semiconductors are envisaged as solid-state quantum bits for quantum technologies, the ODMR technique currently knows a revival of interest as it settles the basis of individual spin manipulation.

In semiconductors, magnetic resonance can be electrically detected through the variation of their conductivity due to the opening or blocking of spin-dependent recombination or transport channels. The EDMR signal actually demonstrates the influence of the paramagnetic defects on the electrical properties of the material. As it does not depend on defect spin polarization, it has usually a higher sensitivity than conventional EPR, even though EDMR signal appears often significantly broader than the one of the same defects detected with conventional EPR. As for ODMR, EDMR experiments can be carried out without a resonator for microwave excitation. As the electrical detection requires ohmic contact and delicate electrical circuitry, this technique is not versatile and is not widely applied for defect studies.

Because of its very indirect detection of spin, EDMR is not use for unknown defect investigations. EDMR has been yet demonstrated recently with defects in SiC for instance. Nevertheless, EDMR and pulsed EDMR are currently developing to meet the technological demand of spin-based quantum information.

4.2 Illustrative examples: structural and chemical control

4.2.1 The SiC/oxide interface defects

The MOS (metal/oxide/semiconductor) structure is the key element of most of the microelectronic devices. The formation of atomically controlled semiconductor/oxide interface is of major importance for modern devices using nanometric oxide layers.

Insulating oxides may be deposited on top of the semiconductor, but when possible, the thermal oxidation of the semiconductor is undoubtedly the easiest way to grow atomically sharp interface. The semiconductor/oxide interface is complex because it operates the structural and electrical transitions between a crystalline semiconductor and an amorphous insulating material. The emblematic and historic example is SiO_2/Si interface but related interfaces: SiO_2/Si_xGe_{1-x}, HfO_2/Si, Al_2O_3/Si and Si_3N_4/Si, and more recently, SiO_2/SiC have become of high technological interest.

These interfaces share common features. The thermal oxidation of the semiconductor induces strain in the interface plane. This strain is partially released by the formation of dangling bonds, in the last unoxidized plane, at the semiconductor side.

In cubic crystals like Si, SiGe and 3C-SiC, due to sp^3 hybridization, bonds are oriented along in the four equivalent [1 1 1] directions and, as long as this hybridization is respected, dangling bonds are expected to preserve this orientation. These dangling bonds have three possible charge states, depending on the number of electrons they contain, namely: +1 (zero electron), 0 (one electron) and −1 (two electrons). They are paramagnetic with $S = 1/2$ in the neutral charge state and can be detected by EPR. They have been named “Pb defects” [11–17]. These defects pin the Fermi level at the interface between the (+/0) and (0/−) levels and consequently a large amount of the dangling bonds are in the paramagnetic state. They are the major part of the “density of interface traps” D_{it} measured by electrical measurements like capacitance–voltage measurements or charge pumping [18]. These electrically active interface defects degrade the ideal behavior of the devices.

Their identification is necessary to decide which changes in the fabrication processes will improve the device performance and reliability.

We will merely focus on the SiC/SiO_2 interface, and we are going to illustrate *g* anisotropy and its relationship to spin–orbit coupling. We will show how *g* spread can bring information on the interface stress and how hyperfine interaction can be used to study the environment of the defect. Finally, we will also illustrate the use of EPR as a quantitative technique to study the electrical passivation of these defects.

After the oxidation of 3C-SiC, a new EPR anisotropic spectrum is observed. Figure 4.6 displays the EPR spectra obtained for 3C-SiC/SiO_2 at room temperature, for three particular high-symmetry directions of the magnetic field: [0 0 1], [1 1 1]

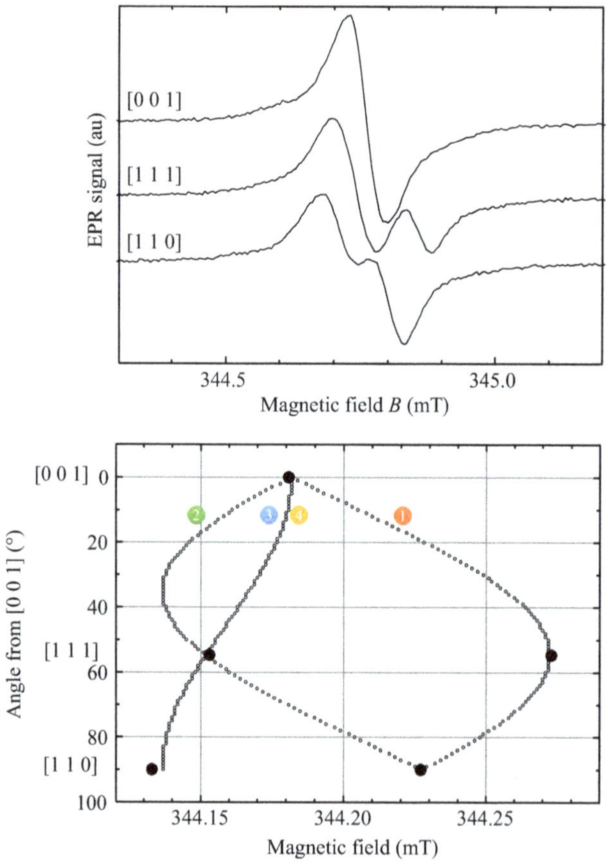

Figure 4.6 Angular variation of the X-Band EPR signal of Pb_C at 3C-SiC/SiO_2 interfaces for a rotation of the magnetic field in the $(1\bar{1}0)$ plane. Top: Room temperature EPR signals for three orientations of the magnetic field. Bottom: Angular variation of the resonance positions. The black circles correspond to the experimental positions of the resonances shown in the spectra displayed on the right

and [1 1 0]. For an arbitrary direction of the magnetic field in the $(1\,\bar{1}\,0)$ plane, the EPR spectrum is composed of three overlapping sharp resonances. This anisotropic spectrum can be fitted by only two resonances for [1 1 1] and [1 1 0] and a single line for [0 0 1] as shown in Figure 4.7.

Figure 4.6 also shows the complete angular variation of the resonant fields for a rotation of the external magnetic field in the $(1\,\bar{1}\,0)$ plane, from [0 0 1] to [1 1 0]. This angular variation is typical for a spin $S = 1/2$ defect with an axial C_{3v} symmetry. For this symmetry, the g tensor has only two distinct eigenvalues: $g_{xx} = g_{yy} = g_{\perp}$ and $g_{zz} = g_{//}$. Consequently, the g effective value expression depends only on the angle θ between the magnetic field and the trigonal axis of the defect:

$$g = \sqrt{g_{\perp}^2 \sin^2\theta + g_{//}^2 \cos^2\theta}$$

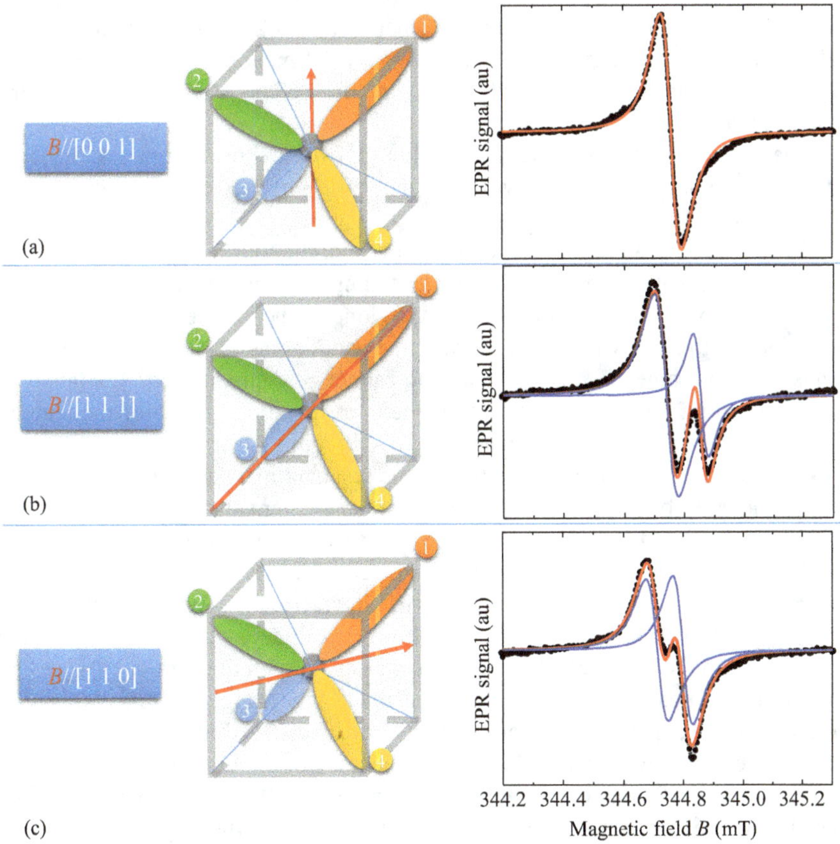

Figure 4.7 Right: Room temperature EPR spectra for three high-symmetry orientations of the magnetic field in the (1 1 0) plane: (a) B//[0 0 1], (b) B//[1 1 1] and (c) B//[1 1 0]. The simulated line is the superposition of two resonances for B//[1 1 1] and B//[1 1 0]. Left: Drawing of the four dangling bonds existing in the sample

From the highest and lowest resonant magnetic field values observed for [1 1 1] and [1 1 0], we deduce: $g_{//[1\ 1\ 1]} = 2.00238$ and $g_{\perp} = 2.00317$. As expected, the lowest g value indicates the symmetry axis of the defect.

The multiplicity of the resonances can be understood by the orientational degeneracy of equivalent defects. In the sample, dangling bonds in the four equivalent $[1\ 1\ 1]$, $[\bar{1}\ \bar{1}\ 1]$, $[\bar{1}\ 1\ \bar{1}]$ and $[1\ \bar{1}\ \bar{1}]$ directions are present. Figure 4.7 displays drawings of these four dangling bonds configurations, labeled 1–4.

When the magnetic field is along the [0 0 1] direction, Figure 4.7(a), the angle θ between the magnetic field and the axes of the four dangling bonds with the magnetic field is the same. Consequently, the four resonances of the four types of defects have the same effective g value and get superimposed, as a result, a single-line EPR spectrum is observed.

When the magnetic field rotated toward the [1 1 1] direction (Figure 4.7(b)), one type of defect has its axis aligned with the magnetic field (dangling bond labeled 1), whereas the axes of the three others (labeled 2, 3 and 4) form the same angle with the magnetic field. Consequently, two resonances are observed, with a 1:3 intensity ratio if the four orientations of the defects are equiprobable, the weaker resonance arising for $g = g_{//}$.

Finally, for the magnetic field aligned in [1 1 0] direction (Figure 4.7(c)), the dangling bonds labeled 1 and 2 will have the same g effective value, whereas the dangling bonds 3 and 4 will have $g = g_{\perp}$.

For an arbitrary direction of the magnetic field, the four defect axes make four different angles with respect to the field, leading to four effective g values and consequently four distinct resonances. But, as long as the magnetic fields lie in the $(1\ \bar{1}\ 0)$ plane (light gray plane Figure 4.7), the dangling bonds labeled 3 and 4 will always have the same effective g value and cannot be distinguished. The labels of the four dangling bonds are written near the corresponding branch of the angular variation of their respective resonance displayed Figure 4.6 (bottom).

In conclusion, the angular variation of the resonant magnetic field has allowed the determination of the g-tensor principal axis and eigenvalues.

But in the case of SiC, two types of atoms are at the interface: Si and C, and we have to attribute this defect to Si or C dangling bond. The chemical identification will be definitely provided by the hyperfine interaction, nevertheless the determined g anisotropy is in favor of a C dangling bond. Indeed, we have seen that the difference from the free electron g value is increasing with the spin–orbit coupling constant. By comparison with the ones measured for Si dangling bonds at Si/SiO_2 interface and Ge dangling bonds at $SiGe/SiO_2$ interface (see Table 4.2), the weak g anisotropy evidenced here is coherent with the smaller spin–orbit coupling of C (cf. Table 4.2) and in agreement with the assignment of this defect to a C dangling bond. Figure 4.8 compares the angular variations of these three types of dangling bond defects according to the g values given Table 4.2.

The study of the hyperfine interaction of the electron spin with the surrounding nuclear spins will give information about the chemical nature of the atom with the dangling bond. In the sample, only ^{13}C and ^{29}Si nuclei have nonzero nuclear spin. They have both a spin $I = 1/2$, and their natural abundance is 1.1% and 4.7%,

Table 4.2 Comparison of g-tensor anisotropy and spin–orbit coupling constants for three semiconductor/oxide interface defects

Atom	Spin–orbit coupling constant (cm^{-1})	Interface	Defect	g values $g_{//}...g_{\perp}$	Reference
C	29.3	SiC/SiO_2	Pb_C	2.0024…2.0032	[19]
Si	211	Si/SiO_2	Pb	2.0017…2.0089	[12]
Ge	1,450	$Si_{0.95}Ge_{0.05}/SiO_2$	Pb_{Ge}	2.0005…2.0210	[17]

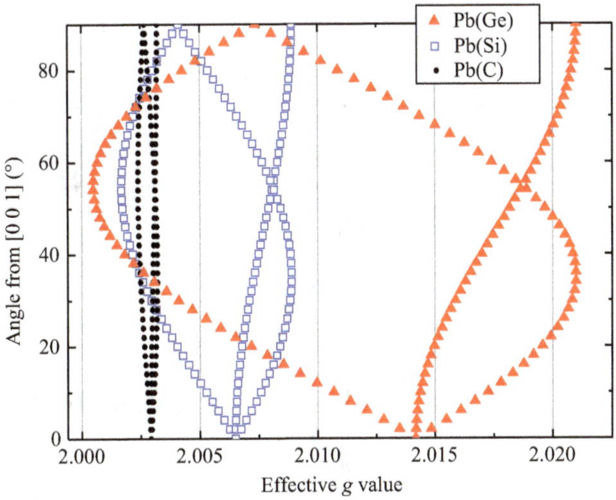

Figure 4.8 Comparison of the angular variations for the C, Si and Ge dangling bonds at SiC/SiO_2, Si/SiO_2 and $SiGe/SiO_2$ respectively

respectively (observation of hyperfine interaction with ^{17}O of 0.037% natural abundance is definitively hopeless).

Defects in hyperfine interaction with these nuclear spins 1/2 are expected to produce doublet lines instead of a single Zeeman resonance. Because of the low abundances of the nuclear spins, to detect these unlikely interactions, the most favorable magnetic field orientation is [0 0 1] for which all the resonances get added.

Figure 4.9 displays the EPR spectrum for B//[0 0 1] measured at high gain. The principal resonance is truncated. Figure 4.9 (top) shows a clearly resolved doublet, with a splitting of 1.24 mT, and a second doublet with twice this splitting. These doublets are centered on the Zeeman lines. This clearly indicates that they arise from hyperfine interactions of the defect with ligand atoms.

By numerical integration, we determine an intensity ratio between the doublet lines and the central Zeeman resonance of about 0.12 ± 0.03. The ratio would be equal to the natural abundance in the case of an interaction with a single ^{13}C or ^{29}Si nucleus. Instead, the measured ratio is about three times the natural abundance of

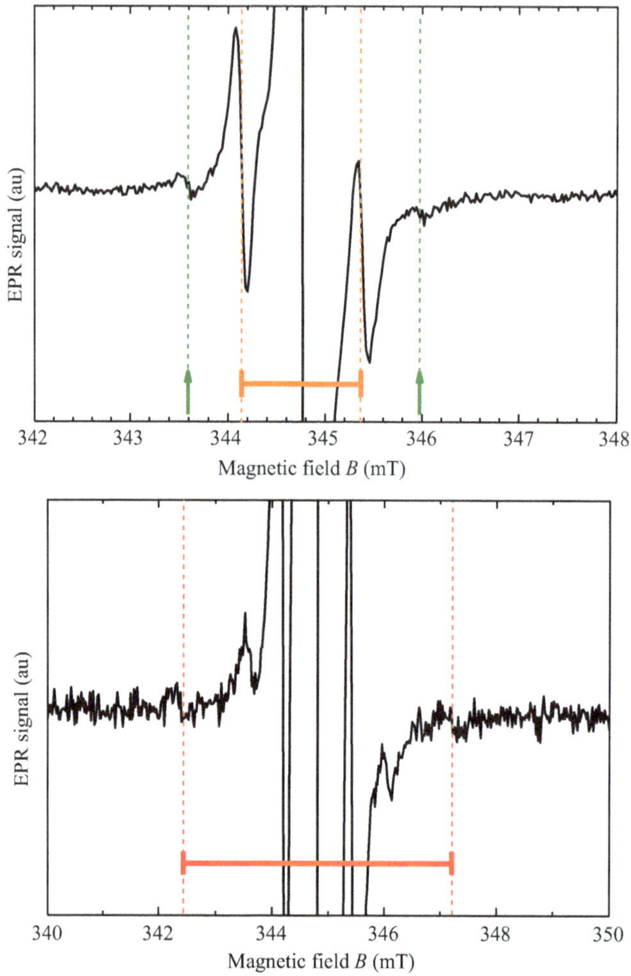

Figure 4.9 EPR spectrum of 3C-SiC/SiO$_2$ for B//[0 0 1]. Top: Zoom on the major lines attributed to the distant ^{29}Si hyperfine interactions. Bottom: The doublet lines resulting of the central hyperfine interaction with one ^{13}C nucleus

^{29}Si. We thus attribute this doublet to interaction with one ^{29}Si nucleus among three equivalent Si neighbors. This is the expected configuration for a C dangling bond defect interacting with the first nearest Si neighbors. Let us mention that the third nearest neighbors of a C atom in 3C-SiC are also three Si atoms but, as the HF interaction is decreasing with increasing ligand distance, the first assignment seems most likely.

In fact, to tidily describe the hyperfine interaction with three equivalent Si atoms, one has to consider the different statistical configurations of the three nuclei.

Given the natural abundance of ^{29}Si, the most probable configuration is three ^{28}Si, giving rise to the Zeeman central resonance, and then one ^{29}Si and two ^{28}Si, producing these doublet lines. But other much less probable configurations with two ^{29}Si and one ^{28}Si or three ^{29}Si nuclei produce low-intensity multiline spectrum characteristics for interaction with effective nuclear spin 1 or 3/2. Some of these lines spotted by arrows can be seen on the spectrum Figure 4.9 (top).

A direct identification of the chemical nature of the paramagnetic dangling bond is provided by the hyperfine interaction with the central nucleus. But for C-related defects, the observation of this doublet is difficult because its intensity is only 1% of the Zeeman resonance. But under optimized observation condition [19], we do observe the doublet shown Figure 4.9 (bottom) with a splitting of 4.8 mT. Its intensity ratio with respect to the central Zeeman line is in good agreement with the natural abundance of ^{13}C.

The hyperfine interaction is anisotropic, and the principal axes and values of the A tensor associated with this interaction are determined by analyzing the angular variation of the doublet lines. The A tensor has the same principal axes as the g tensor and the eigenvalues are $A_{//[1\ 1\ 1]} = 73 \times 10^{-4}\ \mathrm{cm}^{-1}$ and $A_{\perp} = 35 \times 10^{-4}\ \mathrm{cm}^{-1}$.

Now we will see how the hybridization and the localization of the wave function of the paramagnetic electron can be determined from these values in the frame of the Linear Combination of Molecular Orbitals (LCAO) theory. The relaxation of the defect dangling bond can then be estimated by comparing its "s" and "p" characters to the nominal ones expected from the sp^3 hybridization. The s and p spin densities can be deduced from the so-called isotropic a and anisotropic b parts of the hyperfine interaction defined by

$$A_{//} = a + 2b \quad \text{and} \quad A_{\perp} = a - b\,.$$

The isotropic part a is characteristic for the "contact" interaction arising from the electronic and nuclear wave functions overlap. As only the "s" component of the unpaired electron wave function is nonzero at the nucleus place, this isotropic part is directly proportional to the s density of the electronic wave function:

$$a = \frac{16\pi}{3} g_n\beta_n\beta|\psi_s(r=0)|^2 \approx |\psi_s(r=0)|^2 \cdot (1{,}380)\ 10^{-4}\ \mathrm{cm}^{-1}$$

The anisotropic part of the hyperfine interaction is the result of the dipolar interaction between the electronic and nuclear spins. It is proportional to the p component of the electronic wave function:

$$b = \frac{4}{5} g_n\beta_n\beta\left\langle \psi_p\left|\frac{1}{r^3}\right|\psi_p\right\rangle \approx \left\langle \psi_p\left|\frac{1}{r^3}\right|\psi_p\right\rangle \cdot (33.7)\ 10^{-4}\ \mathrm{cm}^{-1}$$

From the hyperfine tensor values measured, we deduce $a = 48 \times 10^{-4}\ \mathrm{cm}^{-1}$ and $b = 13 \times 10^{-4}\ \mathrm{cm}^{-1}$ and s and p spin densities of about 4% and 39% respectively.

The total paramagnetic electron density is only 43% at the central C atom. The electron of the dangling bond is largely delocalized and is thus able to probe its environment. The experimental s/p ratio is about 0.10, significantly smaller than the nominal value of 0.25 expected for an sp^3 hybridization. It shows that the C atom with the dangling bond does not stay at its initial tetrahedral site but relaxes down toward its three nearest neighbors. This relaxation can be further characterized by the variation of the angle θ between the dangling bond and the bond with the first Si neighbors (see Figure 4.10). It was shown that a dispersion in the θ value from its nominal of 109.47° induces a spread in the g-tensor eigenvalues [20]. The g values distribution broadens the resonances. Contrary to the case of a broadening induced by unresolved weak hyperfine interactions, this broadening is expected to linearly increase with the microwave frequency. This phenomenon is typical for interface defects and is not observed for other defects in SiC for which the resonance linewidth is frequency independent.

This spread in g values is induced by the stress in the interface plane and is a fingerprint of the interface defects [20]. It can be evidenced by comparing the EPR spectra observed at X-Band and Q-Band frequencies. Figure 4.11 displays the EPR spectra measured at 9.4 and 34 GHz for B//[1 1 1]. We observe for the high field line; a peak-to-peak linewidth of 0.07 mT, which increases to 0.21 mT, confirming the attribution of this spectrum to an interface defect.

In conclusion, we have identified by EPR the C dangling bond at the 3C-SiC/SiO_2 interface. In the last plane of unoxidized crystalline semiconductor, dangling bonds are pointing in the [1 1 1] direction, as shown in Figure 4.10. This defect exists at (1 1 1) interfaces, but it has been identified at (0 0 1) and (1 1 0) interfaces too.

The C dangling bonds have been evidenced in oxidized 4H-SiC and 6H-SiC as well [21,22]. These interface defects have recently been studied by EDMR [23]. They are clearly different from C-related defect which exist in the amorphous oxide the near-interface region and which are characterized by an isotropic spectrum [24,25].

4.2.1.1 Energy level position

In the case of SiC/SiO_2 interfaces formed with n-type SiC, we can by photo EPR directly verify the electrical activity of interface defects. Indeed, if the electrical

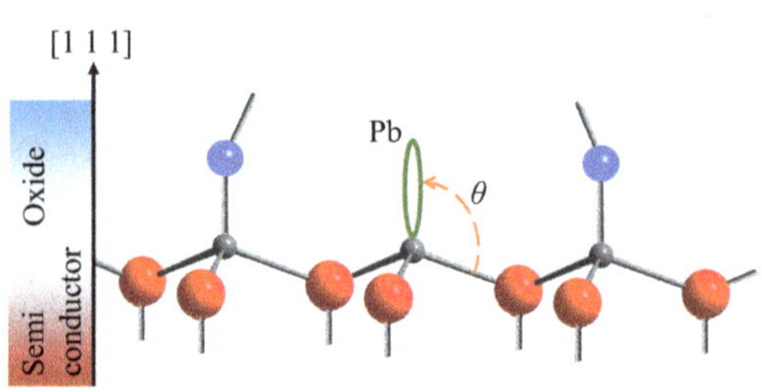

Figure 4.10 Schematic of Pb defect at oxide/(1 1 1)SiC interface. The dangling bond is represented by the green ellipse. The angle between the dangling bond axis and the others bonds is theta. The Si, O and C atoms are represented respectively by circles with increasing diameter

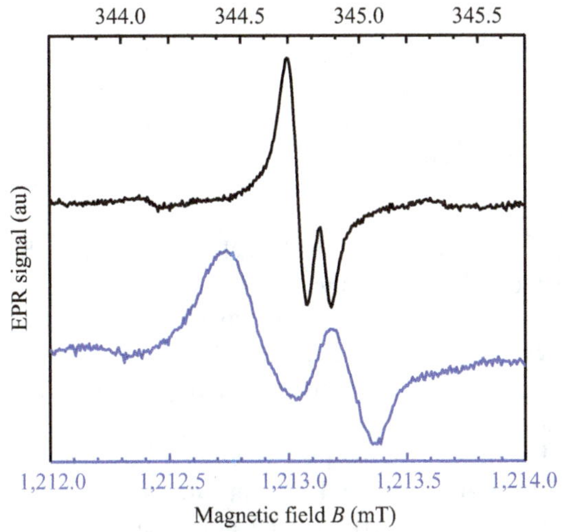

Figure 4.11 *Comparison of the EPR spectra measured at X-Band (up and top magnetic field scale) and Q-Band (down and bottom magnetic field scale) for B//[1 1 1]. The magnetic field scales are different, but the magnetic field range is 2 mT for both spectra*

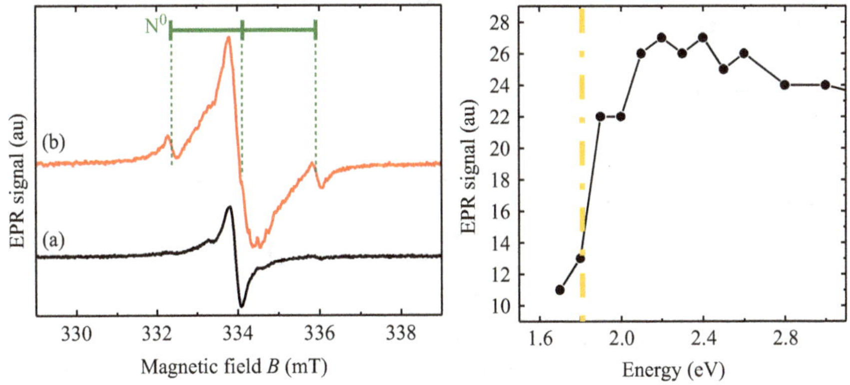

Figure 4.12 *Left: EPR spectra observed at 4 K (a) at thermal equilibrium (b) under in situ photoexcitation. The position of the EPR signal from the N neutral donor is indicated by the three dot lines. Right: Intensity of the N neutral donor as a function of the energy of excitation*

compensation of the semiconductor was not modified by the interface defects, we should observe at low temperature the neutral nitrogen N^0 donor. As shown in Figure 4.12 (left), this is not the case at thermal equilibrium. However, by in situ photoexcitation with above bandgap energy light, the typical three lines spectrum of the neutral donor N^0 is observed.

We have systematically studied as a function of light energy the intensity of the N^0 signal in order to determine the Fermi level position (Figure 4.12 (right)) [32]. A photogeneration threshold of about 1.8 eV is evidenced. Thus the Fermi level position relative to the conduction band edge is $E_c - 1.8$ eV. It corresponds to the Pb_C (−/0) level if this level pins the Fermi level.

4.2.1.2 Passivation

Taking advantage of the quantitative EPR technique, we are able to demonstrate that the Pb_C defect can be transformed into electrically inactive C–H bonds by annealing at 400°C in forming gas. This "passivation" treatment is technologically relevant as it reduces the interface trap density of SiC-based MOS devices. We have also studied the reverse mechanism occurring during vacuum annealing at temperatures ranging from 600°C up to 1,000°C.

The effect of forming annealing on the resonance line of Pb_C is illustrated in Figure 4.13 (left). After annealing at 400°C in forming gas (5% H_2, 95% N_2), the concentration of Pb_C defect, determined by double numerical integration of the resonance line, is reduced by more than one order of magnitude. The Pb_C signal can be recovered by thermal annealing in ultrahigh vacuum (UHV) at temperature higher than 800°C. The evolution of the defect concentration is displayed Figure 4.13 (right) as a function of the annealing temperature.

Considering that the variation of the defect concentration per time unit is simply proportional to the concentration of passivated centers [Pb_C-H], we get the first-order differential equation:

$$\frac{d[\mathrm{Pb_C}]}{dt} = k_d[\mathrm{Pb_C\text{-}H}]$$

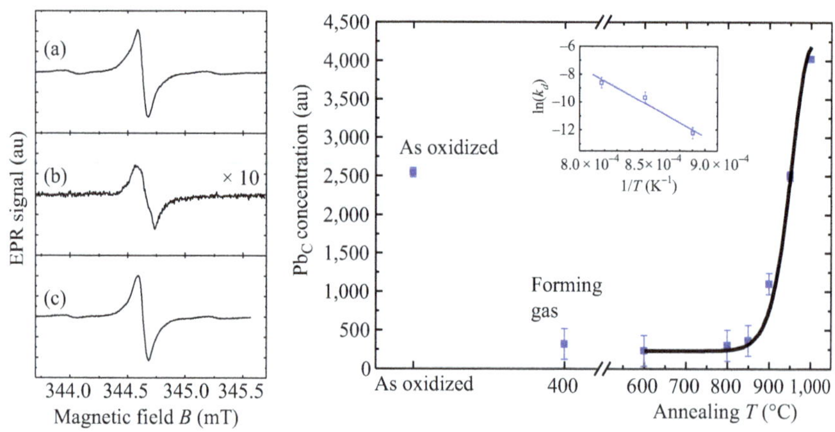

Figure 4.13 Pb_C concentration after oxidation (spectrum (a)), after forming gas annealing at 400°C (spectrum (b)) and as a function of annealing temperature in ultrahigh vacuum (spectrum (c) was observed after 1,000°C annealing). The inset shows the Arrhenius plot of the rate constants k_d of the dissociation reaction. The solid line represents the fit with a dissociation energy of 4.3 eV

The rate constant k_d is given by the Arrhenius equation: $k_d = k_0 \cdot \exp(-E_d/kT)$ where k is the Boltzmann constant, and E_d is the dissociation energy. With the Arrhenius plot of the rate constants k_d of the dissociation reaction shown Figure 4.13 (right), we determined a dissociation energy of 4.3 $\pm$ 0.2 eV [26]. The higher dissociation energy of hydrogenated C dangling bonds at SiC/SiO_2, compared to Si dangling bonds at Si/SiO_2 interfaces can be explained by the higher C–H bond energy.

4.2.2 *The N dumbbell in GaN*

The ESR techniques have shown important results in degradation studies of semiconductors exposed to high temperatures or particle irradiations for example. The primary processes of degradation always involve individual intrinsic point defects like vacancies and interstitials or clusters of them. These defects modify the electronic properties of the semiconductor and can further give rise to extended defects. An excellent recent example is provided by the identification of the nitrogen split interstitial or "dumbbell" defect in GaN. This defect is similar to carbon split interstitial observed in irradiated SiC or diamond [27]. Irradiation with high-energy particles like electrons, protons or ions, displaces lattice atoms by elastic collisions and creates primary vacancies and interstitial atoms in both Ga and N sublattices. Their diffusion and rearrangements play a crucial role in the amorphization process occurring under particle bombardments. These defects were actively studied by electrical measurements [28,29], magnetic resonance [30–33] and positron annihilation spectroscopy (PAS) [34–40].

PAS has played a key role in the identification of intrinsic defects in GaN. Contrary to EPR, PAS is not restricted to paramagnetic charge states and thus provides a complementary view of the defects. Signals attributed to gallium and nitrogen vacancies were observed by PAS [41,42], but only the interstitial Ga was clearly identified by magnetic resonance methods [43]. Curiously, experimental EPR results for the defects related to the nitrogen sublattice were missing: the N vacancy has not been identified, and the N interstitial has escaped any detection.

In GaN, the analysis of the hyperfine structure of the defects is complicated by the high concentration of nuclear spins giving rise to hyperfine interactions leading to broadened EPR lines instead of a clearly resolved structure. Indeed, there is one nitrogen isotope ^{14}N with a nuclear spin $I = 1$ with 100% of natural abundance and two Ga isotopes, ^{69}Ga and ^{71}Ga, with nuclear spin $I = 3/2$ and natural abundance of about 70% and 40%, respectively. Thus the hyperfine structure is the superposition of the two Ga-related quadruplet lines and the triplet lines of the nitrogen.

With the help of high-frequency EPR and ENDOR experiments, together with the support of theoretical calculations, a first N sublattice-related defect has been identified recently [44,45], as illustrated in the following.

Figure 4.14 shows the EPR spectra measured at 35 GHz (Q-Band) and 324 GHz of commercial freestanding 300 μm thick n-type GaN layers after irradiation with 20 meV electrons. As stated in the methodological part of this chapter, the resolution of the EPR spectroscopy is enhanced at high frequency and overlapping signals with distinct g values get resolved with the use of high frequency. But the comparison of 35 and 324 GHz spectra reveals that the overall linewidth of

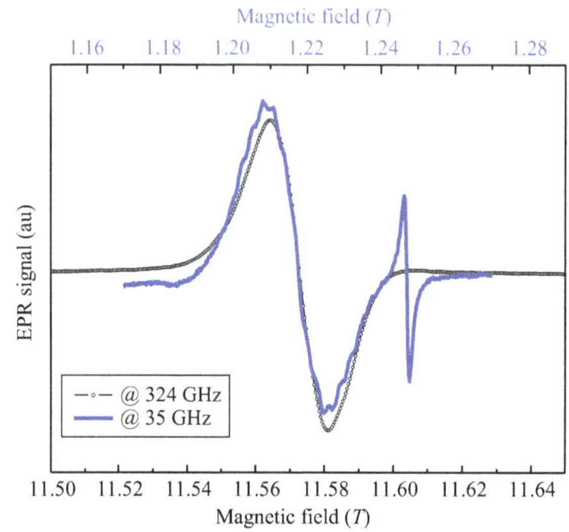

Figure 4.14 Comparison of the (N–N) spectra measured at 4 K for B//c at 35 GHz (Q-Band) and 324 GHz. The magnetic field is top (bottom) scale for the 35 GHz (324 GHz) spectrum. In the 35 GHz spectrum, the sharp resonance at about 1.248 T comes from the shallow donor

the spectrum is constant, showing that the signal arises from only one defect and that its linewidth is due to unresolved hyperfine interactions with the numerous nuclear spins in its environment.

The angular variation of the resonant field performed with high microwave frequency has allowed the determination of the spin value $S = 1/2$, the point symmetry C_1 and the g tensor of the defect: $g_{xx} = 1.9985$, $g_{yy} = 2.0016$ and $g_{zz} = 2.0036$.

ENDOR is a technique of choice to determine the weak unresolved hyperfine coupling responsible for the large EPR resonance. Because broad resonances are difficult to spot in a single ENDOR spectrum measured at a fixed magnetic field value, it is useful to perform field frequency ENDOR. This technique consists of measuring an ENDOR spectrum incrementing the magnetic field step by step through the EPR resonance. As a result, instead of a single ENDOR spectrum measured at the resonant magnetic field, the 2D ENDOR intensity plot shown Figure 4.15 is obtained, with the RF as abscissa and the applied magnetic field as ordinate.

ENDOR transitions in the 20–90-MHz range can be observed. The broadening of the ENDOR lines is a consequence of short electronic relaxation times probably due to the high defect concentration. But such a high concentration is required for ENDOR spectroscopy.

The ENDOR spectrum can be simulated by considering the hyperfine coupling with two distinct N nuclei and four distinct Ga nuclei (with two isotopes for each).

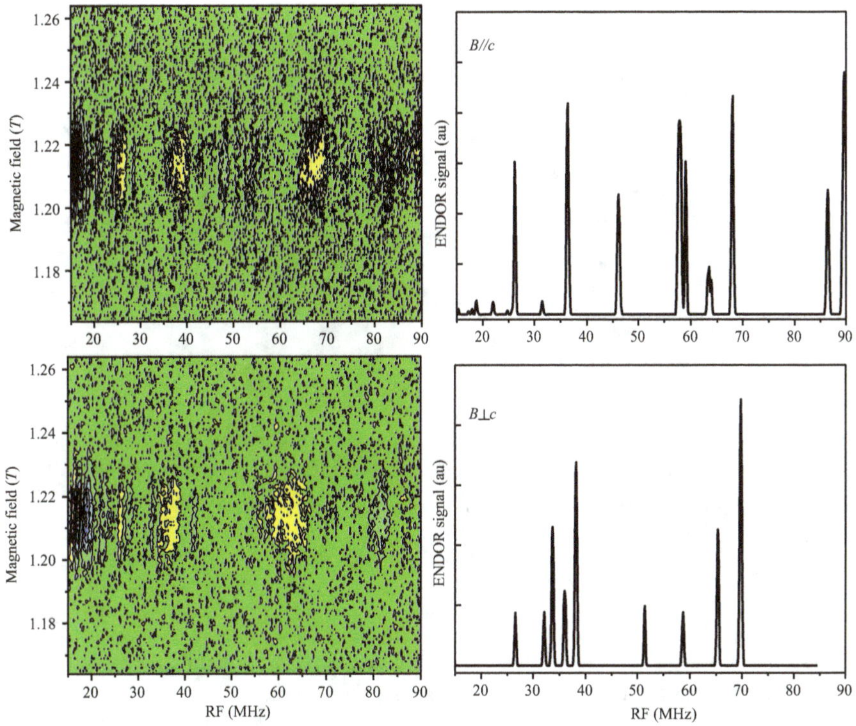

Figure 4.15 Q-Band FF-ENDOR spectrum measured for B//c and B⊥c at 6 K and simulated ENDOR spectrum at the resonant field for these two orientations

This simulation is shown in Figure 4.15 and is in good agreement with the ENDOR intensity plot. The corresponding coupling constants are N_1: 42 MHz, N_2: 31 MHz, Ga_1: 116 MHz, Ga_2: 82 MHz, Ga_3: 45 MHz and Ga_4: 50 MHz.

It is interesting to compare these hyperfine splitting to the full width of the EPR line. Let us recall that a 40 MHz hyperfine splitting corresponds to about 1.1 mT, which is an order of magnitude smaller than the 17 mT of the EPR line-width. This evidences the interest to perform ENDOR to determine hyperfine coupling unresolved by EPR.

Based on the comparison of all these experimental parameters to first principle calculations, a microscopic model of the defect can be asserted. This model is the neutral nitrogen split interstitial, that is to say a pair if N atoms located on an N site. Due to its peculiar geometry, this defect is also called N dumbbell. Figure 4.16 shows the structure of the defect.

This model describes the ground-state configuration of the defect. At temperature above 50 K, the orientation of the N dumbbell changes. Such a thermally activated reorientation is often observed for defects in semiconductors like SiC. Finally, the stable configuration at room temperature of the carbon interstitials

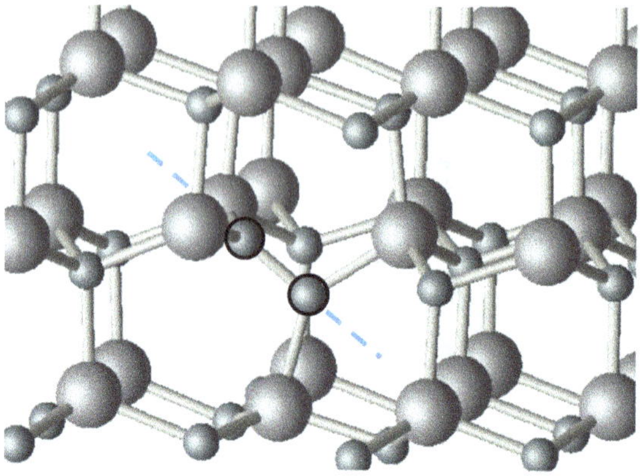

Figure 4.16 Structure of the neutral nitrogen split interstitial in GaN. Large and small spheres represent Ga an N atoms respectively. The two N atoms making up the defect are black circled, and the dotted line shows the axis of the defect

generated by the particle irradiation of GaN seems to be the nitrogen split interstitial, similarly to the case of diamond and silicon carbide.

4.3 Examples: electrical and optical activities

4.3.1 P-Type doping of GaN

EPR is well suited for the study of dopants in semiconductors. Indeed, below the Mott transition temperature, the electron or hole gets localized on the dopant atom, forming a paramagnetic center. Among the most recent doping issues for which EPR had a great insight, we can quote the p-type doping of nitrides.

In GaN for instance, p-type doping is challenging, and Mg turns out to be the most effective dopant. But the identification of substitutional Mg atoms by EPR is not straightforward. Indeed, an anisotropic EPR signal is detected in Mg-doped GaN, but the resonance attributed to Mg acceptor is broad, 10 mT large and structureless [46,47]. Its angular variation is typical for a spin $S = 1/2$ with an axial symmetry, but the g-tensor values are sample dependent and vary typically between 2.05 and 2.01 for $g_{//c}$ and 1.985 and 2.018 for $g_{\perp c}$. Figure 4.17 displays typical spectra of the Mg-related signal. Besides, this signal generally does not present any of the common features of the shallow acceptors observed in SiC and CdS for example. First, the g value anisotropy in GaN, as evaluated by the $g_{//c} - g_{\perp c}$ difference, is more than one order of magnitude smaller, even if $g_{\perp c}$ values of about 0.3 have been reported recently in thick freestanding samples [48]. Second, the usual strong angular dependence of the intensity of the EPR signal is not systematically observed.

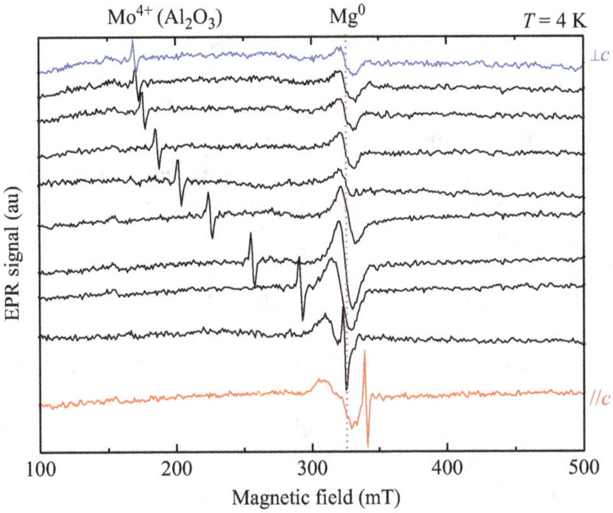

Figure 4.17 Angular variation of the Mg signal in an epitaxial 2 μm thick GaN layer grown on Al_2O_3. The Mg concentration is 1×10^{19} cm^{-3}. The spectra are measured at 4 K at X-Band, with an angular step of 10° from B⊥c (up) to B//c (down). A resonance from Mo^{4+} ions in the Al_2O_3 substrate is also observed

This puzzling situation was recently enlightened by an exhaustive study by Zvanut *et al.* [49]. The authors have shown that the spread in the *g* values is not related to a variety of defect configurations but can be attributed to local crystal fields variations. This conclusion was drawn by careful line shape analysis and comparison of Mg signal detected in samples with thicknesses from 0.4 up to 1,000 μm grew with different methods.

Two features of the Mg-related resonance were particularly monitored by rotating the sample in the magnetic field. The first one is of course the *g* anisotropy $g_{//c} - g_{\perp c}$ deduced from the angular variation of the position of the resonance. The second one is the asymmetry of the first derivative of the resonance line, as evaluated form the relative heights of the positive and negative parts of the curve. Both features are found to be sample dependent.

Interestingly, the authors show that the asymmetric line shape of the resonance can be reproduced by the superposition of a few symmetrical Lorentzian lines, each of them corresponding to an Mg acceptor with slightly different $g_{//c}$ and $g_{\perp c}$ values. The *g* values and their spread required to reproduce the EPR spectra are sample dependent. Nevertheless, the authors noticed that symmetric line shapes are observed when the *g* anisotropy is small, whereas large *g* anisotropy produces highly asymmetric EPR lines. For a given sample, the degree of asymmetry of the EPR line is not constant for all the spectra but depends on the angle between the magnetic field and the *c*-axis.

These observations turn to be in reasonable quantitative agreement with theoretical calculations of g values of defects submitted to a nonaxial crystal field [50]. In these calculations, the g values were determined by using the spin Hamiltonian formalism and the spin–orbit term, as presented in the first part of this chapter, and by adding two crystal field energies. The first one accounts for the axial distortion of the hexagonal GaN lattice. The second term is related to a crystal field perpendicular to the c-axis. The results show that while the $g_{//c}$ value remains almost unaffected, the $g_{\perp c}$ value of the center varies from 0.1 up to 2.0 when the nonaxial crystal field increases from 0.5 to 50 meV. This variation is nonlinear, and consequently, the sensitivity of the $g_{\perp c}$ value to the crystal field depends on the average crystal field value itself. For crystal field values higher than 50 meV, the $g_{\perp c}$ remains almost unchanged, equal to 2.0. Thus strong nonaxial crystal fields reduce the observed g anisotropy.

Since then, all the experimental results can be explained assuming a single Mg acceptor configuration submitted to a nonaxial crystal field and assuming that the average value and the spread of this crystal field are sample dependent. In this frame, the spread of the g values in a given sample naturally arise from the nonuniformity of this field across the sample. As the thinnest samples present the largest crystal fields, the authors suggest that the strain can be the source of this nonaxial crystal field. This gives a consistent explanation for the sample dependence of the EPR results and the spread of results in the literature.

4.3.2 *Origin of the residual conductivity of Ga_2O_3*

Ga_2O_3 has emerged as a semiconductor able to compete with SiC and GaN for power devices. In its monoclinic structure, as grown β-Ga_2O_3 has a large bandgap of 4.9 eV and a residual n-type conductivity while p-type doping is hard to achieve. The responsible for the systematic n-type behavior is a donor with the ionization energy of 16–30 meV as estimated by Hall-effect measurements [51]. As PL studies failed to detect the signal from the donor bound excitons, the electronic structure of the shallow donors in unintentionally doped β-Ga_2O_3 remains obscure.

By EPR, the first studies were performed by Gourier, Binet *et al.* [52–54], and the authors observed a narrow anisotropic resonance with linewidth of 0.03 mT and g values: $g_{xx} = 1.960$, $g_{yy} = 1.958$ and $g_{zz//b} = 1.962$, and they attributed it to the resonance of conduction electrons originating from the O vacancy (V_O), which acts as a shallow donor in many oxides. But this resonance can be observed from 4 K up to room temperature. This is quite unusual for a shallow donor since the thermal energy should, at a given temperature, allow for the ionization of the electron from the shallow defect state to the conduction band. Consequently, the paramagnetic shallow defect should transform into a nonparamagnetic state leading to the disappearance of the resonance shallow donor resonance before room temperature. This temperature behavior is, on the contrary, expected for the resonance of conduction electrons. This attribution was confirmed later by Meyer *et al.* who observed in low-doped Ga_2O_3 this signal only at high temperature [55].

Recent first-principles calculations based on density function theory of the energy levels of this defect in various configurations revealed that the oxygen vacancies are deep donors in β-Ga_2O_3 with a level at more than 1 eV below the conduction band and consequently cannot be the origin of its n-type conductivity [56].

A comprehensive EPR study was recently achieved by Son *et al.* [57] who carefully monitored the intensity of the shallow donor resonance as a function of temperature in samples with different conductivities.

The authors studied highly resistive Ga_2O_3 and confirmed that the narrow anisotropic resonance observed at room temperature disappeared at about 50 K. For lower temperature, the *Q* factor of the cavity does not significantly increase, suggesting that the conductivity of the sample is not dramatically reduced once the resonance disappeared; consequently, the resonance cannot arise from conduction electrons.

For samples with donor density below 10^{16} cm^{-3}, Son *et al.* showed that the intensity of the resonance increases from room temperature down to about 100 K. But for lower temperatures, the intensity decreases until the signal becomes undetectable at about 50 K. This unusual behavior has escaped previous studies because the narrow resonance may be easily over modulated: using a too large magnetic field modulation artificially broadens the resonance leading to a distorted estimation of its intensity.

For a classical shallow donor resonance, the lower the temperature the higher the intensity. The observation of a maximum at a relatively high temperature in the temperature dependence of the concentration of donors in the neutral paramagnetic state can only be explained if the donor has a so-called negative U. Donors of this particular type where first evidenced in III–V semiconductors and called DX centers [58]. For a donor with a negative U, the lowest energy state corresponds to a two-electron state, with a net electronic spin $S = 0$, reached after a large lattice relaxation. Thus at low temperatures, the donors lie in this diamagnetic states, and the resonance corresponding to the neutral paramagnetic state is not observed. Increasing the temperature allows the neutral state to get thermally populated and the donor resonance can be detected by EPR. When the temperature is high enough for the ionization of the neutral donor (i.e., transition from the neutral donor state to the conduction band), the intensity of the resonance decreases. The competition between these two phenomena is responsible for the existence of a maximum in the temperature dependence of the shallow donor.

From the fit of the temperature dependence, Son *et al.* determined a donor activation energy of 44–49 meV and also that the neutral state is lying at about 28 meV below the conduction band and the negative DX level at about 18 meV below the neutral state.

Because for a negative U donor, the paramagnetic state is not the lowest energy state, the paramagnetic electron may have enough energy to hop from one donor to another. This hopping mechanism renders the paramagnetic signal typical for partly delocalized electrons, but the resonance does not arise from electrons in the conduction band. At high donor concentrations, an impurity band below the conduction band is formed and reduces the activation energy to about 17 meV.

Considering on the one hand, the possible activation of the donor by high temperature annealing and on the other hand, the agreement between the concentrations of donor and Si impurities as measured by EPR and SIMS respectively, Son *et al.* propose that Si impurities act as shallow donors in Ga_2O_3, similar to Si donor in AlGaN [59]. Finally, the oxygen vacancy in Ga_2O_3 has not yet been observed by EPR. Further studies are required to identify this deep defect in Ga_2O_3, paramagnetic state of which should be revealed by EPR experiments under suited photoexcitation.

4.3.3 SiC defects as quantum bits

Quantum computing is based upon the quantum properties of matter, like discrete and entangled states. It has been shown recently that paramagnetic point defects in semiconductors may be considered as relevant elementary quantum systems in this context [60]. The control and manipulation of the spin of the defects by optical means, at room temperature, has already been achieved for the particular case of the NV defect in diamond [61]. This defect is a carbon vacancy associated with a nitrogen atom. The nitrogen is a frequent impurity in diamond which may also be introduced purposely by ion implantation for instance. The defect has strong and stable PL at room temperature and consequently is a very convenient single-photon emitter for quantum cryptography applications for example. Its spin $S = 1$ which can be manipulated by microwave pulses and optically read out because the PL intensity depends on the m_S value.

Several challenger defects have been recently identified in silicon carbide which has both fundamental and technological advantages over diamond [62,63]. On the one hand, the growth and doping of SiC are already mastered at technological and industrial levels due to its applications for high-frequency and high-power devices. On the other hand, SiC crystallizes in a variety of structures, and distinct defect configurations are expected to be relevant for quantum computing and spintronic applications. Several defects in SiC are currently the subjects of intense studies, mainly the divacancy, the silicon vacancy [64] and the NV center, which is in SiC an N dopant atom associated with a silicon vacancy [65]. The NV is not an intrinsic defect, and the position of the defect depends on the position of N atoms which can be controlled by local implantation for example. This deterministic placement is a decisive advantage for the applications of quantum technologies. In addition, the PL of NV center occurs in the near infrared region [66], a domain where the fiber optics provides the most efficient transmission, this gives this defect an edge over the others for quantum communication.

EPR has played a major role to establish the atomic structure of these defects and to study their magneto-optical properties. Like in diamond, the NV center in SiC is a spin 1 defect. Figure 4.18(a) presents the EPR spectrum of one configuration of NV in 4H-SiC. The Zeeman interaction with the electronic spin $S = 1$ removes the degeneracy of the levels according to the m_S value. If the $m_S = 1$ and the $m_S = -1$ levels split symmetrically with respect to the $m_S = 0$ level, then the selection rule $\Delta m_S = \pm 1$ allows two transitions at exactly the same resonant magnetic field, and experimentally only one resonance is observed. But when the spin–spin interaction generates a zero-field splitting, the Zeeman splitting is not symmetrical. The corresponding energy level diagram is illustrated in Figure 4.18(b).

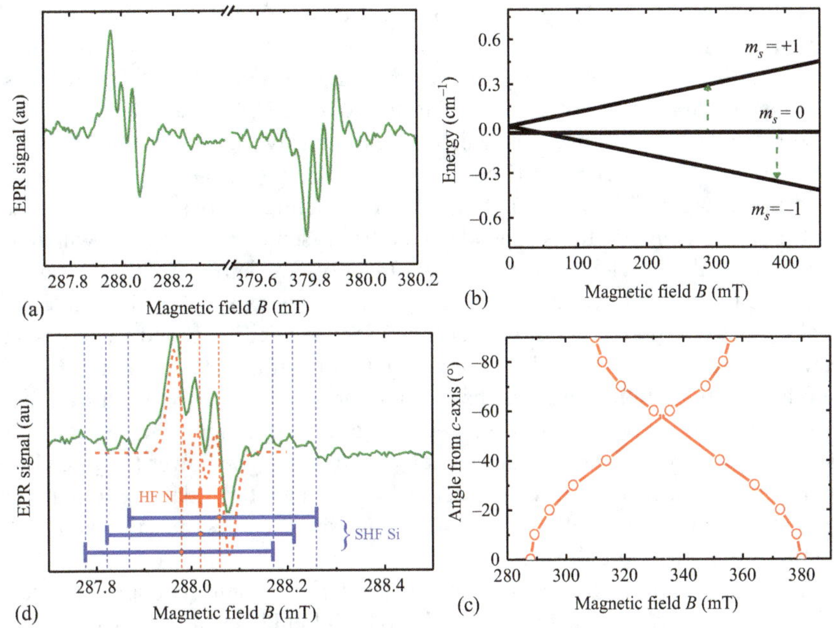

Figure 4.18 *(a) X-Band EPR spectrum of one of the axial configuration of the NV center in 4H-SiC observed for B//c at 4 K under in situ optical excitation. (b) Zeeman level energy diagram. The dash arrows show the resonant transitions at X-Band frequency. (c) Angular variation of the low field and high field lines of the NV center. (d) Low field line of the NV center. The dotted triplet line corresponds to the theoretical fit considering only the hyperfine interaction with one N nucleus (I = 1). The dash vertical lines identify the three super-hyperfine doublet lines corresponding to interaction with distant ^{29}Si*

The X-Band photons, of about 0.3 cm^{-1} of energy, produce two resonant transitions at about 300 and 400 mT which are the so-called low field and high field lines. From the angular variation of the resonant fields of these two lines displayed Figure 4.18(c), we can deduce that the defect has a C_{3v} point symmetry, an isotropic g value of 2.0030 and a spin–spin interaction D parameter, with a value of $D = 429 \times 10^{-4}$ cm^{-1}.

Figure 4.18(d) shows the low field line of the spectrum. This line is composed of three overlapping lines. This partially resolved triplet is typical for hyperfine interaction with a nuclear spin $I = 1$ of 100% abundance, which identifies the nitrogen nucleus (^{14}N, $I = 1$). The electronic spin $S = 1$ interacts with the N nucleus of the first neighbor. From the splitting of this triplet, we deduce the value of the hyperfine coupling constant $A(^{14}N) = 1.12$ MHz. By comparing the smooth theoretical triplet lines (Figure 4.18(d)) to the experimental low field line, we notice the presence in the experimental spectrum of additional doublets with low intensity, indicated by dash lines. The three doublets arise from hyperfine interaction with three equivalent Si neighbors, some of them being ^{29}Si ($I = 1/2$, 4.7% natural abundance). The corresponding hyperfine constant is $A(\mathrm{Si}_3) = 10.7$ MHz. All these experimental parameters indicate that the defect is an $\mathrm{N_CV_{Si}}$ pair. This was confirmed by theoretical calculations which unambiguously identify the NV center in 3C as well as 4H and 6H-SiC [67,68].

In the hexagonal 4H-SiC polytype, different configurations of the defect are possible. These configurations depend (1) on the crystallographic site on which the vacancy is sited (in 4H-SiC, there are two inequivalent lattice sites for Si and C atoms) and (2) on the direction of the nitrogen–vacancy axis. When the defect axis lies along the hexagonal *c*-axis, the configuration is called axial and has $\mathrm{C_{3v}}$ symmetry. So there are two possible axial configurations of the NV center in 4H-SiC. The other configurations are called basal configurations, and once again two basal configurations are expected. Because in the basal configurations, the defect axis is not a symmetry axis of the crystal, the defect symmetry gets lower ($\mathrm{D_{2d}}$), and the Zeeman level degeneracy is consequently lower than for axial configurations. Consequently, the signal intensity is spread into several lines, and the basal configurations are thus more difficult to detect.

A key feature of the defect is its optically polarizable spin state. Looking back to Figure 4.18(a), we see that the low field and high field lines have opposite phases and correspond to absorption and emission transitions respectively. This is due to the preferential polarization of the $m_S = 0$ state induced by the light excitation. The polarization rate as a function of the optical excitation can be assessed by the intensity ratio of the low-field and high-field transitions. It has been shown to be of 100% below 100 K and remains higher than 30% at ambient. This effect demonstrates the possibility to "initialize" this "quantum bit" in a defined spin state, facilitating its manipulation by microwave radiation and its optical read out for further quantum computing applications.

4.4 Summary and outlook

Over the past 50 years, the EPR spectroscopy has demonstrated its ability to elucidate the nature of point defects in semiconductors. A paramagnetic defect can be identified from the interaction of its electronic spin with a magnetic field. The atoms making up the defect and their microscopic structure can be determined from the interactions between the electronic spin and the surrounding nuclear spins. This quantitative technique has allowed the optimization of growth processes and the design of passivation treatments by monitoring the defect density in as grown or irradiated materials.

Although, in the early years, the identification of defects relied mainly on the detection of hyperfine interactions and comparison with values obtained by linear combination of atomic orbitals, sophisticated density functional theory calculations now provide not only accurate hyperfine values but also *g*-tensor values. This new and strong theoretical support is crucial to convincingly attribute a microscopic model to an EPR signal, even when well-resolved hyperfine structure is detected. The case of intrinsic defects in SiC is particularly illustrative, as numerous erroneous models have been first unwisely proposed on the basis of experimental data only.

While conventional EPR probes the whole sample volume, the high sensitivity of this technique allows for the study of monolayers and emerging 2D materials. For example, EPR was used to identify the "molecular states" in graphene [69] and to evidence electrically induced Pauli paramagnetism in this material [70]. Intrinsic defects located at grain boundaries in polycrystalline MoS_2 were also recently observed by conventional EPR. These defects, tentatively attributed to sulfur antisite defects, can be detected in one monolayer thick samples [71].

Recent experimental development, like pulse EPR, has not shown to be clearly relevant when it comes to defects in semiconductors. Nevertheless, new insights could be soon brought by coupling EPR to local microscopies, like scanning tunneling microscopy (STM). For example, a new method is currently developed which consists in measuring the spin-polarized tunnel current while irradiating the sample with a microwave and scanning the microwave frequency [72–74]. This has already enabled the detection of the Larmor resonant frequency of a single paramagnetic iron atom located at the surface of the sample, below the STM tip. This technique, called ESR–STM, combined with tip-induced atomic manipulation, might be promising for the study of fundamental interactions between paramagnetic defects in various materials, including semiconductors.

The achievement of CW–EPR spectroscopy in UHV or controlled atmosphere, carried out by the early 2000s, is certainly very interesting and insufficiently exploited to study semiconductors surface defects. The vacuum is required to stabilize surface defects like dangling bonds and to allow for the surface dangling bonds to remain unsaturated and possibly paramagnetic. To preserve the sensitivity of EPR in a UHV chamber is quite challenging due to the natural incompatibility of UHV metallic items with microwave resonant cavity systems. For the moment, UHV compatible spectrometers have been involved essentially in studies of defects and adsorbates at oxide surfaces [75–77], to study charge transfer [78] and catalytic properties [79], but they could as well be used to study semiconductors surface defects and their role in the first oxidation steps for example.

To conclude, paramagnetic defects in semiconductors have shown to be a significant interest for quantum technologies as they could form solid-state quantum bits [80]. In this frame, EPR and ODMR are techniques of choice to selectively manipulate the spins and to perform optical initialization and read out of the spin state.

References

[1] A. Abragam, B. Bleaney, "Electron Paramagnetic Resonance of Transition Ions", Clarendon Press, Oxford (1970).

[2] J.A. Weil, J.R. Bolton, "Electron Paramagnetic Resonance Spectroscopy", John Wiley Interscience, New York, NY, (2007).
[3] C.P. Poole, "Electron Spin Resonance: A Comprehensive Treatise on Experimental Techniques" (Second Edition), Interscience Publishers, New York, NY (1996).
[4] S.S. Eaton, G.R. Eaton, "Electron Paramagnetic Resonance" in Ewing's Analytical Instrumentation Handbook (Third Edition), Edited by Jack Cazes CRC Press, Boca Raton, FL, (2004).
[5] S.K. Misra, "Multifrequency Electron Paramagnetic Resonance: Theory and Applications", Wiley-VCH, Weinheim, (2011).
[6] D. Goldfarb, S. Stoll, "EPR Spectroscopy: Fundamentals and Methods", ISBN: 978-1-119-16298-8, Wiley, Chichester, (2018).
[7] A.J. Hoff, "Advanced EPR: Applications in Biology and Biochemistry", Elsevier Science Publication, Amsterdam, (1989).
[8] G. Jeschke, A. Schweiger, "Principles of Pulse Electron Paramagnetic Resonance", Oxford University Press, Oxford, (2001).
[9] R.H. Clarke, "Triplet State ODMR Spectroscopy", Wiley, New York, NY (1982).
[10] C. Boehme, H. Malissa, eMagRes **6**, 83 (2017).
[11] E.H. Poindexter, P.J. Caplan, B.E. Deal, R.R. Razouk, Journal of Applied Physics **52**, 879 (1981).
[12] J.L. Cantin, M. Schoisswohl, H.J. von Bardeleben, N. Hadj Zoubir, M. Vergnat, Physical Review B **52**, R11599 (1995).
[13] A. Stirling, A. Pasquarello, J.C. Charlier, Physical Review Letters **85**, 2773 (2000).
[14] A. Pasquarello, M.S. Hybertson, R. Car, Nature **396**, 58 (1998).
[15] J.L. Cantin, H.J. von Bardeleben, Journal of Non-Crystalline Solids **303**, 175 (2002).
[16] P. Aubert, H.J. von Bardeleben, F. Delmotte, J.L. Cantin, M.C. Hugon, Physical Review B **59**, 10667 (1999).
[17] S. Lebib, M. Schoisswohl, J.L. Cantin, H.J. von Bardeleben, Thin Solids Films **294**, 242 (1997).
[18] N.H. Thoan, K. Keunen, V.V. Afanas'ev, A. Stesmans, Journal of Applied Physics **109**, 013710 (2011).
[19] H.J. von Bardeleben, J.L. Cantin, Y. Ke, Y. Shishkin, R.P. Devaty, W.J. Choyke, Materials Science Forum **483–485**, 273 (2006).
[20] A. Stesmans, Physical Review B **48**, 2418 (1993).
[21] J.L. Cantin, H.J. von Bardeleben, Y. Shishkin, Y. Ke, R.P. Devaty, W.J. Choyke, Physical Review Letters **92**, 015502 (2004).
[22] H.J. von Bardeleben, J.L. Cantin, M. Mynbaeva, *et al.*, Electrochemical Society Proceedings **2003-02**, 39 (2003).
[23] G. Gruber, J. Cottom, R. Meszaros, *et al.*, Journal of Applied Physics **123**, 161514 (2018).

[24] P.J. Mac Farlane, M.E. Zvanut, Journal of Electronic Materials **28**, 144 (1998).

[25] P.J. Macfarlane, M.E. Zvanut, Journal of Applied Physics **88**, 4122 (2000).

[26] J.L. Cantin, H.J. von Bardeleben, Y. Ke, R.P. Devaty, W.J. Choyke, Applied Physics Letters **88**, 092108 (2006).

[27] D.C. Hunt, D.J. Twitchen, M.E. Newton, *et al.*, Physical Review B **61**, 3863 (2000).

[28] D.C. Look, D.C. Reynolds, J.W. Hemsky, J.R. Sizelove, R.L. Jones, R.J. Molnar, Physical Review Letters **79**, 2273 (1997).

[29] D.C. Look, Z.Q. Fang, B. Claflin, Journal of Crystal Growth **281**, 143 (2005).

[30] M. Linde, S. J. Uftring, G.D. Watkins, V. Härle, F. Scholz, Physical Review B **55**, R10177 (1997).

[31] C. Bozdog, H. Przybylinska, G.D. Watkins, *et al.*, Physical Review B **59**, 12479 (1999).

[32] P. Johannesen, A. Zakrzewski, L.S. Vlasenko, *et al.*, Physical Review B **69**, 045208 (2004).

[33] N.T. Son, C.G. Hemmingsson, T. Paskova, *et al.*, Physical Review B **80**, 153202 (2009).

[34] K. Saarinen, T. Suski, I. Grzegory, D.C. Look, Physica B **308–310**, 77 (2001).

[35] K. Saarinen, T. Suski, I. Grzegory, D.C. Look, Physical Review B **64**, 233201 (2001).

[36] S. Hautakangas, V. Ranki, I. Makkonen, *et al.*, Physica B **376–377**, 424 (2006).

[37] S. Hautakangas, I. Makkonen, V. Ranki, *et al.*, Physical Review B **73**, 193301 (2006).

[38] F. Tuomisto, V. Ranki, D.C. Look, G.C. Farlow, Physical Review B **76**, 165207 (2007).

[39] F. Tuomisto, Applied Surface Science **255**, 54 (2007).

[40] F. Tuomisto, K. Saarinen, T. Paskova, B. Monemar, M. Bockowski, T. Suski, Journal of Applied Physics **99**, 066105 (2006).

[41] F. Tuomisto, V. Ranki, D.C. Look, G. Farlow, Physical Review B **76**, 165207 (2007).

[42] K. Saarinen, J. Nissilä, J. Oila, *et al.*, Physica B **273–274**, 33 (1999).

[43] K.H. Chow, G.D. Watkins, A. Usui, M. Mizuta, Physical Review Letters **85**, 2761 (2000).

[44] H.J. von Bardeleben, J.L. Cantin, U. Gerstmann, *et al.*, Physical Review Letters **109**, 206402 (2012).

[45] H.J. von Bardeleben, J.L. Cantin, H. Vrielinck, *et al.*, Physical Review B **90**, 085203 (2014).

[46] E.R. Glaser, W.E. Carlos, G.C.B. Braga, *et al.*, Physical Review B **65**, 085312 (2002).

[47] B. Monemar, P.P. Paskov, G. Pozina, *et al.*, Journal of Applied Physics **115**, 053507 (2014).

[48] M.E. Zvanut, J. Dashdorj, J.A. Freitas, *et al.*, Journal of Electronic Materials **45**, 2692 (2016).

[49] M.E. Zvanut, J. Dashdorj, U.R. Sunay, J.H. Leach, K. Udwary, Journal of Applied Physics **120**, 135702 (2016).

[50] G.N. Aliev, S. Zeng, J.J. Davies, *et al.*, Physical Review B **71**, 195204 (2005).

[51] K. Irmscher, Z. Galazka, M. Pietsch, R. Uecker, R. Fornari, Journal of Applied Physics **110**, 063720 (2011).

[52] L. Binet, D. Gourier, C. Minot, Journal of Solid State Chemistry **113**, 420 (1994).

[53] E. Aubay, D. Gourier, Physical Review B **47**, 15023 (1993).

[54] D. Gourier, E. Aubay, J. Guglielmi, Physical Review B **50**, 2941 (1994).

[55] B.K. Meyer, U. Leib, A. Hofstaetter, C. Krummel, D. Kohl, Materials Science Forum **258–263**, 1473 (1997).

[56] J.B. Varley, J.R. Weber, A. Janotti, C.G. Van de Walle, Applied Physics Letters **97**, 142106 (2010).

[57] N.T. Son, K. Goto, K. Nomura, *et al.*, Journal of Applied Physics **120**, 235703 (2016).

[58] D.J. Chadi, K.J. Chang, Physical Review B **39**, 10063 (1989).

[59] X.T. Trinh, D. Nilsson, I.G. Ivanov, E. Janzén, A. Kakanakova-Georgieva, N.T. Son, Applied Physics Letters **105**, 162106 (2014).

[60] J.R. Weber, W.F. Koehl, J.B. Varley, *et al.*, Proceedings of the National Academy of Sciences **107**, 8513 (2010).

[61] M.W. Doherty, N.B. Manson, P. Delaney, F. Jelezko, J. Wrachtrup, L. Hollenberg, Physics Reports **528**, 1 (2013).

[62] W.F. Koehl, B.B. Buckley, F.J. Heremans, G. Calusine, D.D. Awschalom, Nature **479**, 84 (2011).

[63] A. Gali, Physica Status Solidi (B) **248**, 1337 (2011).

[64] G. Wolfowicz, C.P. Anderson, A.L. Yeats, *et al.*, Nature Communications **8**, 1876 (2017).

[65] H.J. von Bardeleben, J.L. Cantin, E. Rauls, U. Gerstmann, Physical Review B **92**, 064104 (2015).

[66] S.A. Zargaleh, B. Eble, S. Hameau, *et al.*, Physical Review B **94**, 060102 (2016).

[67] H.J. von Bardeleben, J.L. Cantin, A. Csóré, A. Gali, E. Rauls, U. Gerstmann, Physical Review B **94**, 121202 (2016).

[68] A. Csóré, H.J. von Bardeleben, J.L. Cantin, A. Gali, Physical Review B **96**, 085204 (2017).

[69] A. Barbon, F. Tampieri, AIMS Materials Sciences **4**, 147 (2017).

[70] N. Fujita, D. Mastumoto, Y. Sakurai, *et al.*, Nature, Scientific Reports **6**, 34966 (2016).

[71] A. Stesmans, S. Iacovo, D. Chiappe, *et al.*, Nanoscale Research Letters **12**, 283 (2017).

[72] S. Baumann, W. Paul, T. Choi, *et al.*, Science **350**, 417 (2015).

[73] F.D. Natterer, K. Yang, W. Paul, *et al.*, Nature **543**, 226 (2017).

[74] P. Willke, W. Paul, F.D. Natterer *et al.*, Science Advances **4**, eaaq **154** (2018).

[75] M. Sterrer, E. Fischbach, T. Risse, H.-J. Freund, Physical Review Letters **94**, 186101 (2005).

[76] M. Yulikov, M. Sterrer, M. Heyde, *et al.*, Physical Review Letters **96**, 146804 (2006).

[77] D. Cornu, J. Rocker, A. Gonchar, T. Risse, H.-J. Freund, Physical Review Letters **117**, 016801 (2016).

[78] G. Pacchioni, H.-J. Freund, Chemical Reviews **113** (6), 4035 (2013).

[79] R. Baran, F. Averseng, D. Wierzbicki, *et al.*, Applied Catalysis A **523**, 332 (2016).

[80] G. Wolfowicz, C.P. Anderson, A.L. Yeats, *et al.*, Nature Communications **8**, 1876 (2017).

Chapter 5

The role of muons in semiconductor research

Patrick W. Mengyan[1,2]

5.1 Introduction

Intentional or not, hydrogen (H) is a common impurity in all semiconductors that can modify the electrical, optical and magnetic properties of the host. In the early 1980s, hydrogen was experimentally determined to passivate the electrical activity of shallow donors and acceptors in silicon (Si). Further studies showed similar behaviour in other elemental and compound semiconductors (see, e.g. [1–4]). Hydrogen has since been shown to modify properties of dopants, other impurities and more extended defects over a wide range of semiconducting materials as H often forms complexes with these defects that then move the related energy levels into or out of the bandgap [5]. In addition to passivating shallow or deep donors or acceptors, H can also activate isoelectric impurities. For example, Si–H in germanium becomes a shallow acceptor [6,7].

Hydrogen has very strong reactivity and high diffusivity, so it readily interacts with other impurities (hence, it passivates defects so well). While a lot of experimental data have been collected on hydrogen-related complexes, there is very little direct information on isolated hydrogen. One example of the H data is the experimental characterisations of the *AA9* centre in silicon (Si), which have been performed by the way of electron paramagnetic resonance (EPR) and deep-level transient spectroscopy [8–11]. This remains one of the only directly studied isolated atomic H impurity centres, albeit certainly not for lack of trying.

The long-standing accepted picture of H behaviour in semiconductors is presented very thoroughly in [5].

In semiconductors, isolated atomic hydrogen can exist in a positive (H^+), neutral (H^0) or negative (H^-) charge state. Generally speaking, isolated hydrogen normally acts against the prevailing conductivity type since it is typically a negative-U centre, where U is the separation between donor and acceptor thermodynamic levels (i.e. $U = E_A(-/0) - E_D(0/+)$). This means that, in equilibrium conditions, hydrogen typically acts as an acceptor (H^-) in

[1]Department of Physics, Northern Michigan University, Marquette, MI, USA
[2]Department of Physics and Astronomy, Texas Tech University, Lubbock, TX, USA

n-type and donor (H^+) in *p*-type materials. The specific state of the impurity centre depends on how the Fermi energy of the host compares to the energy of the hydrogen defect level. The interstitial location and diffusive properties of a hydrogen impurity centre are also directly related to its charge state. Furthermore, the reactivity of the hydrogen impurity centres is also a function of both charge state and diffusivity.

While the reactivity and strong tendency of hydrogen to form complexes result in modifying the bulk electrical or optical properties of a host, making hydrogen such an interesting impurity, it is these particular interactions that can make studying the fundamental processes and mechanisms by which hydrogen interacts with the host very difficult or impossible. For instance, hydrogen is normally intentionally introduced to a material in high concentrations and often at high temperatures. This ultimately results in hydrogen associating with itself or other defects, so little to no information on isolated atomic hydrogen or systems with dilute hydrogen content can be available by these conventional means. Complications also arise relating to the H solubility properties of a host material.

In light of these complications, using positively charged muons as an indirect but experimentally accessible analogue to isolated hydrogen impurities has proven to be very productive. In fact, the vast majority of experimental data on the isolated hydrogen impurity comes from these muon-based measurements. Positively charged muons (μ^+) have the same spin and charge as a proton with a mass that is roughly nine times smaller $\left(m_\mu \approx 0.1126\,\text{amu}\right)$. When implanted into a target material, a positively charged muon can behave as a light proton or bind with an electron to form a complex known as muonium $\left(\text{Mu}^0 = \mu^+ + e^-\right)$ with properties that are very similar to that of hydrogen. Since it is the muons themselves that mimic the H impurity centres and any material can be subjected to the non-destructive muon-implantation process, these measurements neither depend (at all) on the host's solubility of hydrogen nor do they require some minimum concentration of H or the presence of specific target nuclei, unlike many other techniques, such as EPR, electron-nuclear double resonance spectroscopy (ENDOR), nuclear magnetic resonance (NMR) or infrared (IR) vibrational spectroscopy.

The aim of this chapter is to provide an introduction and overview of using muons to study defects in semiconductors for an audience with a background in material science. First is a general tutorial to relevant models and discussion of the muon-based techniques that have been important to the semiconductor field. The latter portion of this chapter highlights results from selected studies on semiconductors to demonstrate and describe some contributions that muon spin research (μSR) techniques have made to the semiconductor community in recent years. The topics I have selected are what I believe to be a good balance between general accessibility and relevance as a way to provide a fresh but rounded discussion on this topic. For detail beyond the scope of this chapter or alternative perspectives, I encourage the reader to look at some of the excellent reviews and textbooks relating to the muonium in semiconductors and μSR technique, such as [1,12–18], as well as the many provided throughout this text.

5.2 Muon spin research (μSR): the techniques

The collection of muon-based measurements relevant to molecular and material sciences is generally referred to as *muon spin research*, *MuSR* or *μSR*. Depending on the particular experimental application and configuration, more descriptive terms, such as *rotation*, *relaxation* and *resonance*, are often used. In a typical μSR experiment, 100% spin-polarised positive muons (μ^+) are implanted in a material and precess in the local environment until decaying into a positron and the associated neutrinos. The positrons are emitted preferentially aligned with the spin at the time of decay. The time evolution of the muon's spin is monitored by tracking the time and position of the emitted positrons to reveal information on the local environment for which the muon is a direct probe. There are μSR techniques that utilise negatively charged muons (μ^-) [19]; however, the vast majority of current semiconductor-based applications use the positive muon (μ^+), so any further discussion of muons and μSR will be referring to μ^+, unless otherwise explicitly noted as μ^-. It is important to note that since the μ^+ itself is the probe of the local environment, μSR does not rely on scattering or other mechanisms (e.g. application of pulsed electric or magnetic fields and optical pulses) to probe the environment of interest unlike many other available techniques (e.g. X-ray diffraction, neutron scattering, electron spin resonance, NMR and vibrational spectroscopy). Generally speaking, μ^+ is very sensitive to magnetic fields and attracted to areas of high electron density making it a viable probe for a wide range of applications such as investigating magnetism (e.g. local field distribution, ordering dynamics), superconductivity (e.g. coexistence of superconductivity and magnetism, magnetic penetration depth, coherence length), defect states in semiconductors (where muonium, $Mu^0 = \mu^+e^-$, functions as an experimentally accessible analogue to hydrogen and hydrogen complexes), H diffusion in materials ($Mu^{+,0,-}$ cf. $H^{+,0,-}$) and many more. For more detail on the technique in general and associated applications, see [16–20].

In semiconductors, muons are commonly used as an analogue for hydrogen to provide insight on the characteristics and behaviour of hydrogen impurity centres and complexes. The charged muon can also be used to probe carrier distribution and dynamics. The relevant properties of a muon and muonium as compared to hydrogen are outlined in Table 5.1. From these data, note the similarities of the electronic properties of Mu^0 and H^0; specifically, the reduced electron mass, ground state radius, binding energy and hyperfine frequency (after accounting for the muon's 3.18 times larger magnetic moment). These similarities are a result of the muon and proton having the same spin and charge as well as magnetic moment and mass that are within an order of magnitude of each other. One example of experimental data confirming similarity between the H^0 and Mu^0 centres is within 10% agreement of measured hyperfine frequencies for the bond-centred (BC) interstitial H^0 and Mu^0 centres in Si is [1,9,10,21]. A more recent direct comparison between O–H and O–Mu complexes in rutile TiO_2 has been completed where the electronic structure of each complex coupled to a nearby electron centred on a Ti^{3+}

Table 5.1 Useful properties of the muon (μ^+), proton (p^+), electron (e^-), vacuum muonium (Mu^0), vacuum hydrogen (H^0) and general hydrogenic centres [12,23]. The reduced mass for Mu^0 is calculated by $m_{Mu} = m_e/(1 + m_e/m_\mu)$. The ground state radius is $a_{Mu} = /(1 + m_e/m_\mu)a_0$ where a_0 is the Bohr radius. The binding energy is $(a_0/a_{Mu})R_y$

	Muon (μ^+)	**Proton (p^+)**	**Electron (e^-)**
Mass (m_p)	$0.1126 \approx 1/9$	1	1/1,836
Spin ($\hbar$)	1/2	1/2	1/2
Moment (μ_p)	3.18	1	657
Gyromagnetic ratio ($\gamma/(2\pi)$) (MHz/T)	135.54	42.577	28,025
Lifetime, τ (μs)	2.19709	Stable (∞)	Stable (∞)
	Muonium (Mu^0)	**Hydrogen (H^0)**	**Hydrogenic centre**
Reduced e^- mass (m_e)	0.995187	0.999456	m^*
Ground-state radius (Å)	0.531736	0.529465	$a^* = \epsilon(m_e/m^*)a_0$
Binding energy (eV)	13.5403	13.5984	$(m^*/m_e)(1/\epsilon^2)R_y$
Hyperfine frequency (MHz)	4,463	1,420	$\sim A_0/a^{*3}$

is nearly identical [22], which helps support the comparison of muonium to isolated hydrogen states in semiconductors.

Similar to hydrogen, muonium can also exist in the neutral (Mu^0), positive (Mu^+) and negative (Mu^-) charge states. Carefully note that the charged muonium states Mu^+ and Mu^- are *not* to be confused with the bare μ^+ and μ^-. The bare muon is always attracted to sites with high valence electron density (e.g. BC site in covalent semiconductors) or cation electron cloud ionic materials (e.g. oxides) as to lower its energy, just like the proton. The careful use of the notation μ^+, p^+, Mu^+ and H^+ distinguishes the first two as free particles from the last two, which are impurity (defect) centres. Additional discussion relating to the behaviour, characteristics and associated dynamics of these states continues in Section 5.3.

While the similarities of Mu and H are important, in order to properly compare the static and dynamic properties of these centres, it is critical that one also considers the differences between them. One of the biggest differences in experiments on hydrogen and muonium lies in the mass difference. Since the μ^+ is approximately nine times lighter than the proton, quantum effects such as zero-point energy and tunnelling are more important for muonium as compared to hydrogen. With careful consideration of these differences, however, the results from a study of the dynamics of the muonium centre(s) are still, at least, qualitatively comparable to that of hydrogen. In the few cases where direct comparisons can be made, it is found that the electronic properties are very similar, but barriers for motion, for instance, are higher for H than for Mu.

One key point to keep in mind is that μSR experiments almost always explore non-equilibrium states. While the properties of the Mu centres arise from the muon interactions with the host (i.e. governed by semiconductor properties, doping concentration and type, temperature and applied electric and magnetic fields), any dynamics (e.g. transitions, charge cycles) are almost certainly moving the system towards thermal equilibrium. However, at higher temperatures (e.g. above room temperature) the measured transitions, in general, represent dynamics within an equilibrium mix of states.

Other differences between muon and hydrogen experiments relate to the experimental conditions. For instance, even under the highest beam intensities, μSR experiments are well within the dilute limit; therefore, muons do not interact with other muons, and it is unlikely that enough muons interact with other point defects to be detectable. Due to the muon lifetime (2.2 μs), many muon experiments are designed to use time windows shorter than 32 μs. The short timescale and dilute nature of the muon implantation are a stark contrast to many hydrogen-based measurements where H is introduced in higher concentrations and the observations are at much longer timescales (e.g. minutes, several days or longer with respect to time after implantation), results of which may include H forming complexes with itself, the host or other defects, and hence, one is studying the longer-time history of the H centres.

Reiterating a key difference, the Mu and H experiments probe opposite ends of the reaction sequence since the Mu experiments look at the early-time history starting from the initially isolated Mu (or H) towards the impurity complexes, whereas the hydrogen experiments tend to look at the system after the H complex already formed.

5.2.1 General principles

Muons are found in nature in cosmic rays that can be detected at sea level on Earth's surface with an approximate flux of ~1 muon per square cm every minute [24]. Among many practical limitations, including the low flux, this is far too low for efficient use as a probe for condensed matter research. In practice, muons are manufactured at proton-accelerator-based facilities in several steps. The main available muon sources are *TRIUMF* (actual name) in Canada, the *Paul Scherer Institut* (PSI) in Switzerland, the *Science and Technology Facility Council ISIS Neutron and Muon Source* (STFC-ISIS) at *Rutherford Appleton Laboratory* in England and the *Muon Science Facility* (MUSE) in the *Materials and Life Science Experimental Facility* at the *Japan Proton Accelerator Research Complex* in Japan. Laboratories in South Korea and China are actively working on developing μSR facilities.

First, a proton beam (~500 to ~800 MeV, depending on the facility) from a cyclotron (TRIUMF and PSI) or synchrotron (STFC-ISIS and MUSE) interacts with a target (e.g. graphite, beryllium) that produces pions ($\pi^{\pm}$). The pion decays (τ_{π} ~26 ns) into a muon ($\mu^{\pm}$), associated neutrino (ν_{μ}) and antineutrino $(\overline{\nu}_{\mu})$.

$$\pi^{+} \rightarrow \mu^{+} + \nu_{\mu} \quad \text{and} \quad \pi^{-} \rightarrow \mu^{-} + \overline{\nu}_{\mu} \tag{5.1}$$

Since the net spin of the π^{+} is zero and within the energy range relevant to μSR, the neutrino is produced with its spin antiparallel to its momentum (i.e. negative helicity). Conservation of both linear and angular momentum requires the μ^+ to also have its spin antiparallel to its momentum. Similarly for the π^-, with the exception that the antineutrino has its spin parallel to its momentum (i.e. positive helicity), and therefore the μ^- also has positive helicity. This statement does assume that the π are at rest, which is consistent with many of the μ^+ sources in operation today. In this type of source, the μ^+ beams are sourced from the π^+ that are at rest near the surface of the production target. The other π decay mode produces much higher energy μ^+ and μ^- with negative helicity. The muon beams that use μ^+ from this near-surface π decay process are referred to as 'surface' channels. Beams for negative muon (μ^-)-based experiments require a different process because π^- and μ^- are quickly captured within the primary target; therefore, a very different μ extraction configuration is required, but it will not be discussed here since it is not as widely used for μSR work relating to defects in semiconductors. A direct result of the choice to use the pions that decay from rest near the target's surface is that the muons emerge 100% spin polarised (kinetic energy ~4.119 MeV). Once extracted from the proton target, the muons are directed down the beamline, tidied up and implanted directly into a sample. Within the first ~1 ns of entering the sample, the ~4 MeV muon beam reduces to a few eV from the initial interactions with the local environment, including ionising atoms and scattering with electrons that are along the implantation path (known as the ionisation track), followed by the capture and release of electrons (either free electrons within the system or electrons freed along the ionisation track) until the muon has a final energy within the meV range and therefore reaches the so-called stopping site at the end of the ionisation track. There is virtually no depolarisation of the muon itself during implantation as the interactions are all Coulombic and very rapid. Since almost all of the energy is transferred from the muon to the host in the early stages of the implantation process and the separation distance of the last damage and initial stopping site is separated by ~100 nm, there should be no interaction between the thermalised muon and the ionisation track (i.e. vacancies or interstitial defects produced along the ionisation track). In some systems, there may be an early-time ($\sim10^{-9}$ to $\sim10^{-7}$ s) interaction between the muon and host in the final stages of thermalisation that affect the state that is eventually formed; however, details of this process are not yet understood (e.g. see [23,25,26]). Further details on this specific aspect of implantation are beyond the scope of this chapter but ought to be considered when interpreting data.

Figure 5.1 depicts the process from muon production through thermalisation.

Once the muon is at the end of the ionisation track and sufficiently slowed, the muon interacts with the local environment and then decays after a time (scaled by the 2.2 μs lifetime) into a positron (e^+) and the associated neutrino (ν_e) and antineutrino ($\overline{\nu}_\mu$), that is

$$\mu^+ \rightarrow e^+ + \nu_e + \overline{\nu}_\mu \tag{5.2}$$

The e^+ is emitted in a direction preferential to the muon spin at the time of decay and is then tracked by detectors (e.g. scintillation) positioned around the sample.

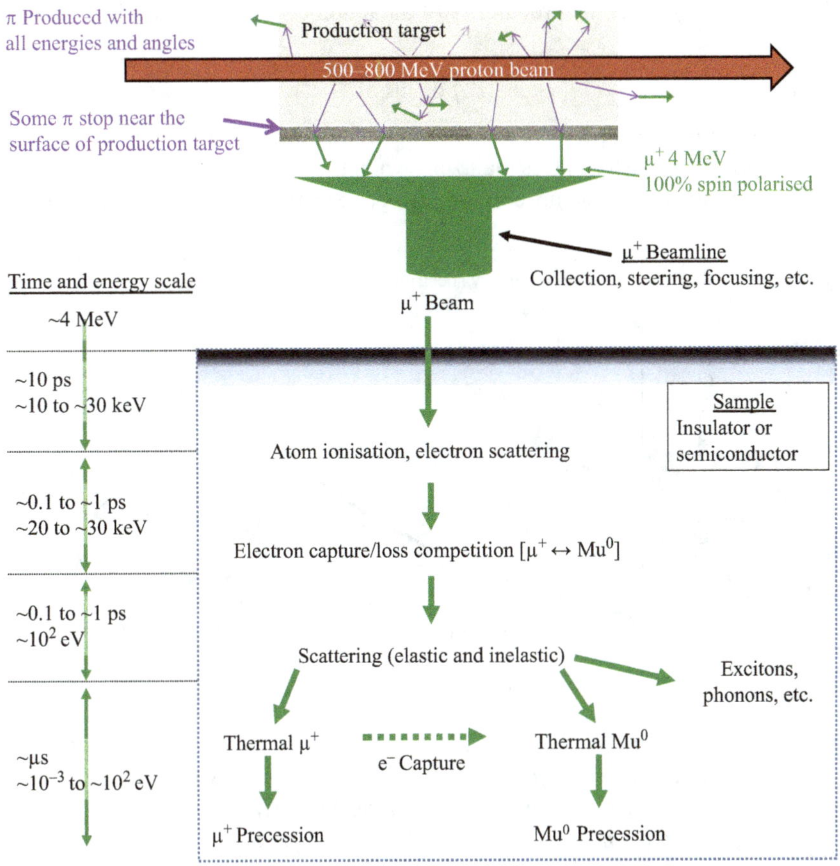

Figure 5.1 *Key steps in the production of surface muons and implantation into insulating or semiconducting samples are depicted here from the initial interaction between the proton beam and target for surface μ^+ production through the final thermalisation of the implanted muon. Typical time and energy scales are indicated for the associated processes after implantation*

An asymmetry factor (a), describing the correlation between muon spin and positron momentum, scales with the positron energy (ϵ_e) is written as

$$a(x) = \frac{2x - 1}{3 - 2x} \tag{5.3}$$

Parameterised by $x = \epsilon_e/\epsilon_{max}$ and $\epsilon_{max} = 52.83$ MeV is the maximum positron energy.

The energy spectrum ($E(x)$) is written as

$$E(x) = 2x^2(3 - 2x) \tag{5.4}$$

Both $a(x)$ and $E(x)$ relate the probability of the muon decay (P) and positron energy (ϵ_e) through

$$dP(x, \theta) = E(x)(1 + a(x)\cos(\theta))\, dx\, d(\cos(\theta)) \tag{5.5}$$

The relationships (5.3) and (5.4) are plotted in Figure 5.2 and (5.5) in Figure 5.3. Notice that the asymmetry is less than zero when the positron energy is below 50% of the maximum indicating no correlation between the positron momentum and

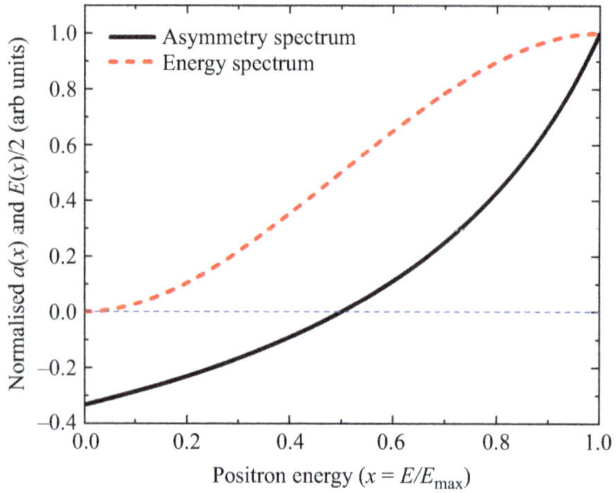

Figure 5.2 Normalised asymmetry (5.3) and energy spectrum (5.4) for positron e^+ emission from μ^+ decay. The asymmetry indicates the correlation of μ^+ spin direction and e^+ momentum. Notice that the asymmetry is less than zero when the e^+ energy is below 50% of E_{max}, indicating no correlation between μ^+ spin and e^+ momentum

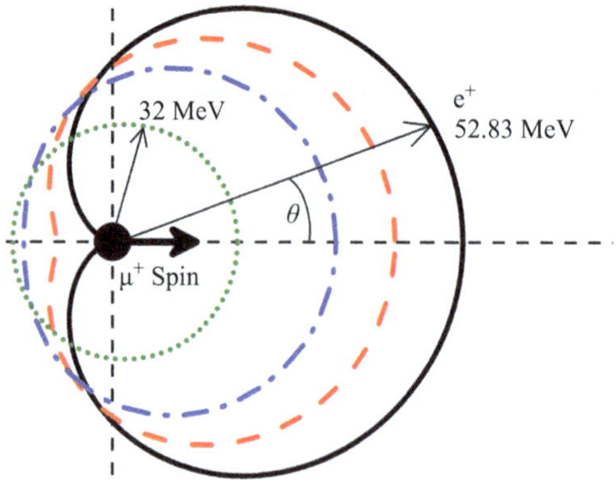

Figure 5.3 Angular distribution of e^+ emission (5.5) from μ^+ decay for 100%, 90%, 80% and 60% of the maximum e^+ energy (outer to innermost lines). The maximum e^+ energy is 52.83 MeV. Notice the near-complete isotropic distribution at 32 MeV (60%) and full anisotropy at 52.83 MeV (100%)

muon spin direction. At this limit, the angular distribution of e^+ emission is perfectly symmetric, so even if those lower energy positrons are detected (they usually are not), they do not contribute the final measurement tracking the evolution of the muon spin polarisation.

While the specific detector configurations across the variety of μSR instruments vary, they all track the time evolution of spin polarisation for an implanted muon by monitoring an ensemble of muon decay events through detection of the resulting positron. There are two modes available, time integral (TI) and time differential (TD). In the TI mode, only the total number of events (positron from muon decay) is counted in each detector. In TD mode, a clock starts upon muon implantation and is stopped when a counter detects a corresponding event (positron from muon decay). A single data point with fixed independent variables (e.g. sample temperature, applied magnetic field, illumination and applied electric field) typically include somewhere around $\sim 10^6$ to $\sim 10^8$ events (positron counts). In a particular counter, the number of recorded positrons, $N(t)$, as a function of time from implantation, t, can be expressed as

$$N(t) = N_o \exp\left(-\frac{t}{\tau_\mu}\right)(1 + Ap(t)) + B \tag{5.6}$$

where the overall normalisation constant, N_o, and time-independent background, B, cannot be experimentally determined with a single detector. The projection of the muon spin polarisation on the symmetry axis of the detector and scaling factor, $Ap(t)$, can be observed in a single detector but is not usually analysed alone. In practice, multiple counters are used so that the experimental (muon) asymmetry spectra can be calculated without needing N_o or B. The simplest case combines positron counts from a pair of counters that are 180° apart with respect to the frame of a two-dimensional rotating body, and then the experimental asymmetry, $a_o(t)$, is written as

$$a_o(t) = \frac{(N_1(t) - B_1) - (N_2(t) - B_2)}{(N_1(t) - B_1) + (N_2(t) - B_2)} \tag{5.7}$$

where the subscripts '1' and '2' index the detectors.

In order to account for differences between the physical detectors, such as counter efficiency and solid angle coverage, two adjustment parameters, α and β, can be defined

$$\alpha \equiv \frac{N_2(0)}{N_1(0)} \tag{5.8}$$

and

$$\beta \equiv \frac{A_2}{A_1} \tag{5.9}$$

In a normally functioning experimental setup, β is very close to one and independent of factors such as sample geometry and position. However, α is highly dependent on these parameters. Here, $N_{1,2}(0)$ corresponds to the number of counts in detectors '1' or '2' immediately after implantation.

Examples of typical detector arrangements are shown in Figures 5.4 and 5.5. In the two-detector configuration (Figure 5.4) with the detectors 180° apart, assuming

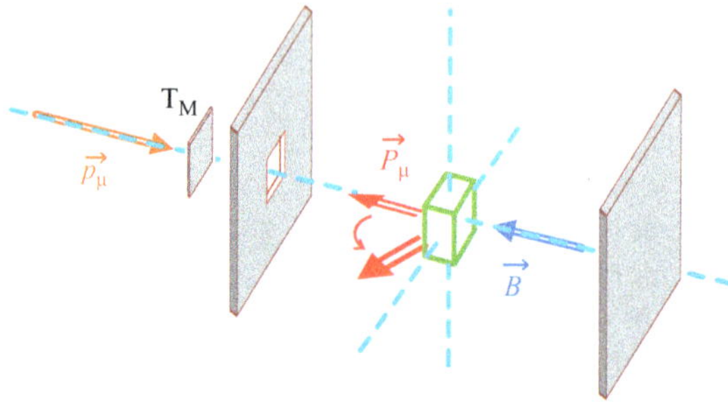

Figure 5.4 *Typical detector configuration for longitudinal (LF) and zero (ZF) applied field muon spin relaxation measurements utilising two detectors and showing the initial muon momentum $\vec{p}_\mu$ and spin $\vec{P}_\mu$. In LF, the external magnetic field $\vec{B}$ is applied parallel to the initial spin direction. In ZF, small fields are used to maintain as close to zero field as possible at the sample. A detector, T_M, is used to sense the incoming muon and start the clock. In practice, the two positron detectors may be built out of an array of several detectors whose positron counts are grouped based on placement relative to the sample*

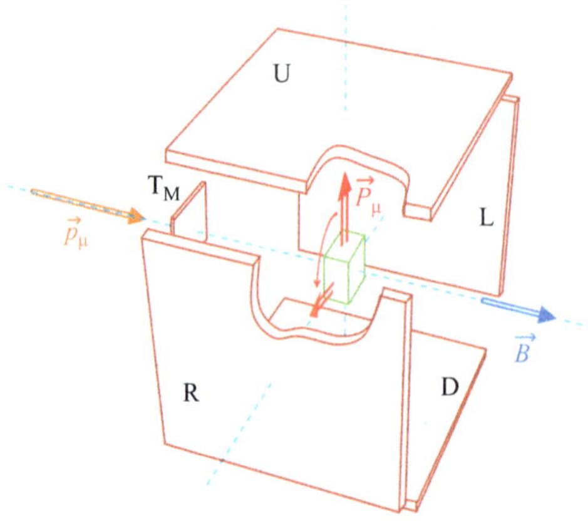

Figure 5.5 *Typical detector configuration for transverse (TF) applied field muon spin rotation measurements utilising four detectors (Up, RIght, Down and Left) and showing the initial muon momentum $\vec{p}_\mu$ and spin $\vec{P}_\mu$. In TF, the external magnetic field $\vec{B}$ is applied perpendicular to the initial spin direction. Notice the spin is rotated 90° from the momentum in this configuration. In low-TF, the LF two-detector configuration (Figure 5.4) is used with a field applied perpendicular to the initial spin direction and no spin rotation*

the typical β value of one (which is okay unless something is seriously wrong with a particular detector), and accounting for any beam-borne background, the corrected muon asymmetry, $a(t)$, can be written as

$$a(t) = \frac{N_1'(t) - \alpha N_2'(t)}{N_1'(t) + \alpha N_2'(t)} \tag{5.10}$$

A μSR time-domain spectra is prepared with this calculation for horizontally opposed detectors. A common way to calibrate for α is to apply a weak field that is perpendicular to the initial spin polarisation direction that drives the muon spin to precess, then α is adjusted so that the measured asymmetry oscillates about zero, thereby 'balancing' the detectors. An example of raw counts in two opposing counters and the corresponding corrected asymmetry is shown in the upper and lower panels of Figure 5.6, respectively.

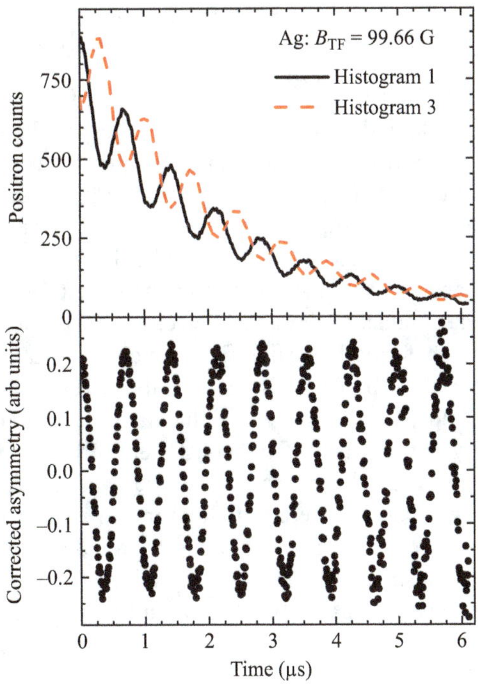

Figure 5.6 Transverse field muon spin rotation measurements on Ag (purely diamagnetic) with $B_{TF} = 9.966 \times 10^{-3}$ T (99.66 G), measured using the HiTime spectrometer at TRIUMF. Top: positron counts per individual histogram for two horizontally opposed detectors (histograms 1 and 3, solid and dashed lines) clearly showing behaviour consistent with (5.6). Bottom: data from top processed with (5.10). These data are used for calibration purposes. In this particular setup, the maximum measured asymmetry is 0.22

The four-detector configuration (Figure 5.5) has the distinct advantage of doubling the angular resolution and tracking the direction of muon spin rotation.

In practice, instruments may have a series of several physical detectors that each count incoming events but can be grouped by their relative positions about the sample so that the effective result is either the two- or four-detector configuration.

While interacting with the host, the implanted muon will precess in the effective magnetic field of the local environment ($\boldsymbol{B}_{\text{eff}}$) at the Larmor frequency $\left(\omega_\mu = \boldsymbol{B}_{\text{eff}} \times 135.54\ \text{MHz/T}\right)$ until it decays. For reference, precession frequencies for the proton, electron and neutron are 42.577, 28,025 and 29.165 MHz/T, respectively. The large magnetic moment and short lifetime enable the muon to be used as a sensitive magnetic probe and therefore detect fields down to the order of $\sim 10^{-5}$ T and sense fluctuation rates in the range of 10^4–10^{12} Hz that complement the lower ($\sim 10^{-3}$ to $\sim 10^5$ Hz) and higher ($\sim 10^8$ to $\sim 10^{13}$ Hz) ranges accessible by NMR and neutron techniques, respectively.

The effective magnetic field ($\boldsymbol{B}_{\text{eff}}$) at the muon site can be written, in general, as

$$\boldsymbol{B}_{\text{eff}} = \boldsymbol{B}_{\text{ext}} + \boldsymbol{B}_{\text{loc}} = \boldsymbol{B}_{\text{ext}} + \boldsymbol{B}_{\text{dip}} + \boldsymbol{B}_{\text{hyp}} + \boldsymbol{B}_{\text{fermi}} + \delta\boldsymbol{B}_{\text{loc}} \tag{5.11}$$

where $\boldsymbol{B}_{\text{ext}}$ is an externally applied field. $\boldsymbol{B}_{\text{dip}}$ is a sum of localised moments over the entire crystal, including site-to-site differences (i.e. the dipolar field contributions). $\boldsymbol{B}_{\text{hyp}}$ is the contribution from short-range magnetic interactions between μ^+ and local electronic moments (e.g. wave function overlap), commonly referred to as the hyperfine interaction. $\boldsymbol{B}_{\text{fermi}}$ includes contributions from the Fermi contact interaction (between μ^+ and a carrier), which can be the magnetic interaction of μ^+ and e^- spins for *s*- and *p*-materials; indirect exchange (RKKY) between μ^+ and unpaired electrons via conduction (itinerant) electrons for some *d*- and *f*-materials; or in insulators, the transferred hyperfine field which is the μ^+ and electron wave function overlap. The last term, $\delta\boldsymbol{B}_{\text{loc}}$, is included to account for fluctuations in any of the contributing field components. While fluctuating contributions to the effective field are actually part of each of the constituent terms, I explicitly include a separate fluctuation term to help emphasise that the overall effective field may be written into two parts: first, an average field from all contributions and second, contributions from any dynamics. Some systems or processes can be modelled with a core component that is static, and then any deviations from the static component can be developed as dynamics as a perturbation on top of the existing static core.

5.2.2 Polarisation functions

In any time-domain μSR experiment, the observable quantity is the time evolution of the muon spin polarisation, $P(t)$, by way of the measured asymmetry, $a(t)$ (e.g. (5.10)). In this chapter, discussion of the polarisation functions and models is limited to select examples relevant to using μSR to study defects in semiconductors, the associated results and general comments. In practice, the models created and used to interpret data have a broad range of complexity that varies based on the host system, type of experiment and, ultimately, what kind of information is of interest. Additional detail relating to other models and calculating spin polarisation functions from the spin-Hamiltonian and known field distributions is available in extensive technical detail in more specialised sources such as [12–17].

The selection of polarisation functions in this section provides a basis from which a preliminary characterisation of muon behaviour and individual states can be established. A typical μSR experiment involves collecting several time-domain (or time-integral) spectra at a series of specifically chosen parameters (e.g. temperature, magnetic field, electric field, illumination and sample or field orientations) individually characterised with an appropriate model based on polarisation functions such as those presented in this section. Then, variations in the measured spectral parameters (e.g. frequency, relaxation rate, field width, amplitude, hop rate and correlation time) as a function of controlled parameters (e.g. temperature, applied magnetic field, electric field, carrier concentration and orientation) are fit by more general models to identify and characterise states, active processes and mechanisms within the system under study. Examples of models for use in determining these general properties, such as hyperfine constants, diffusion coefficients, energy barriers, donor levels and acceptor levels, are included in Section 5.3. Section 5.4 then provides a brief discussion relating to applications of these models and some of the major contributions μSR has made to the semiconductor field over the last several years.

Perhaps the most straightforward but non-trivial case is to consider a system in which there is a single unique μ^+ stopping site within the unit cell of a particular material (meaning $\boldsymbol{B}_{\text{eff}}$ is the same for all stopping sites), that the implanted μ^+ does not move once it arrives and that $\boldsymbol{B}_{\text{eff}}$ is exactly perpendicular to the initial μ^+ spin direction. Under these conditions, the muon will precess at the Larmor frequency $\left(\omega_\mu = \boldsymbol{B}_{\text{eff}} \times 135.54\ \text{MHz/T}\right)$ until it decays. The corresponding muon spin polarisation function can be modelled as

$$P(t) = a_0 G(t) \cos(\omega t + \phi) \tag{5.12}$$

where the initial asymmetry (a_0), relaxation envelope function ($G(t)$), precession frequency (ω) and phase (ϕ) are parameters that could be fit directly to the measured $a(t)$ data. The initial asymmetry (a_0) is related to the fraction of implanted muons that are in the state described by that component of $P(t)$. The relaxation envelope function ($G(t)$) provides a measure of field inhomogeneity. The precession frequency is a direct measure of the effective field for that component, since $\omega_\mu = \boldsymbol{B}_{\text{eff}} \times 135.54\ \text{MHz/T}$. The phase ($\phi$) relates to the shift of the initial spin direction of the implanted μ^+ as compared to the initial spin direction of the μ^+ in this state. Figure 5.6 shows a real example of the corrected asymmetry from a piece of silver in a magnetic field of about 9.966×10^{-3} T (99.66 G) that is applied perpendicular to the initial spin polarisation of the implanted muons. The α parameter via (5.10) was fit so that the spectrum is balanced around zero. This non-relaxing signal corresponds to a time-independent relaxation function of $G(t) = 1$ [unit less] and nicely shows the result of a purely homogeneous field distribution. This clean μ^+ behaviour from silver demonstrates why it is often used for calibration purposes and as a way to mount a sample to provide a well-known background.

One approach to developing a spin polarisation function for a muon in a particular state is to consider how a magnetic moment interacts with a magnetic field and the distribution of effective fields (5.11) across the available muon sites throughout the material. Beginning with a general approach to a simple state, assume a promptly formed state and that the period of fluctuation ($1/\nu_B$) in the field sensed by the muon ($\boldsymbol{B}_{\text{eff}}$) is very fast compared to the lifetime of the muon (i.e. $1/\nu_B \gg \tau_\mu$ so that $\boldsymbol{B}_{\text{eff}}$ is effectively constant). With respect to the initial muon spin axis, the time evolution of the spin is

$$S_\mu(t) = S_\mu\left(\cos(\theta)^2 + \sin(\theta)^2\cos\left(\omega_\mu t\right)\right) \tag{5.13}$$

where $\omega_\mu = \gamma_\mu \boldsymbol{B}_{\text{eff}}$, the phase ($\phi$) is taken to be zero since the state is taken to be promptly formed and θ is the angle between the initial muon spin direction and $\boldsymbol{B}_{\text{eff}}$ as felt by the muon. That is, $\theta = 0°$ corresponds to the initial muon spin direction being aligned with $\boldsymbol{B}_{\text{eff}}$ (typically along $\hat{z}$) and $\theta = 90°$ corresponds to $\boldsymbol{B}_{\text{eff}}$ being exactly perpendicular to the initial muon spin direction. In this case, (5.13) reveals that components of $\boldsymbol{B}_{\text{eff}}$ that are parallel to the incoming muon spin direction do not contribute to the change in muon spin polarisation and that only components of $\boldsymbol{B}_{\text{eff}}$ that are perpendicular to the incoming muon spin direction do contribute.

By normalising (5.13) and expressing the local field as a general distribution, $\rho(\boldsymbol{B}_{\text{eff}})$, the overall spin polarisation function is written as

$$P(t) = \int \left(\cos(\theta)^2 + \sin(\theta)^2 \cos\left(\omega_\mu t\right)\right)\rho(\boldsymbol{B}_{\text{eff}})\mathrm{d}^3\boldsymbol{B}_{\text{eff}} \tag{5.14}$$

which is the statistical average over all field distributions.

A field distribution for a material containing a single unique stopping site with no μ^+ motion following thermalisation can be written as

$$\rho(\boldsymbol{B}_{\text{eff}}) = \frac{\delta(B_{\text{eff}} - B_o)}{4\pi^2 B_o^2} \tag{5.15}$$

where δ is the Dirac delta. Evaluating (5.14) with the distribution (5.15), we see

$$P(t) = \frac{1}{3} + \frac{2}{3}\cos\left(\omega_\mu t\right) \tag{5.16}$$

In a system where the dominant contribution to $\boldsymbol{B}_{\text{eff}}$ is interactions with a dense packing of nuclei surrounding a μ^+ site (concentrated dipole moments with a field that is the largest term in (5.11)), the fields are taken to have a Gaussian-shaped distribution with second moment, Δ_G, and expressed as

$$\rho(\boldsymbol{B}_{\text{eff}}) = \left(\frac{1}{\Delta_G\sqrt{(2\pi)}}\right)^3 \exp\left(-\frac{B_{\text{eff}}^2}{2\Delta_G^2}\right) \tag{5.17}$$

Averaging over the distribution yields the 'static Kubo–Toyabe' function [27]

$$P_S^{KT}(t) = \frac{1}{3} + \frac{2}{3}\left(1 - (\gamma_\mu \Delta_G t)^2\right) \exp\left(-\frac{(\gamma_\mu \Delta_G t)^2}{2}\right) \tag{5.18}$$

Figure 5.7 shows this function with a few different field distributions ($\Delta_G = \langle B_z^2 \rangle^{1/2} = ((\langle B_x^2 \rangle + \langle B_y^2 \rangle)/2)^{0.5}$ where $\hat{z}$ is taken to be along the initial muon spin polarisation) with values on the order of 10^{-4} T to low 10^{-3} T, commensurate with what is often observed in real systems. In this case, the origin of relaxation of the asymmetry is in the dephasing of the muon spin precession that is caused by the inhomogeneity in the field distribution. The minimum in (5.18) occurs at a time equal to $\sqrt{3}/(\gamma_\mu \Delta_G)$. The long-time tail of the polarisation recovers one-third of the initial polarisation (for a static state) because an average of one-third of the components of the local fields is aligned parallel to the initial muon spin, therefore not contributing to dephasing. A known limitation of the pure Gaussian-field-based approximation is in matching the data in the time range where the relaxation begins. Despite these differences, this model continually proves itself to be an excellent tool for establishing the characteristics and trends of the local-field distributions as sensed by the muons.

If an external field ($\boldsymbol{B}_{\text{ext}}$) is applied with $\theta = 0°$ (parallel to the initial spin direction, known as longitudinal field (LF)), and a magnitude that is comparable to

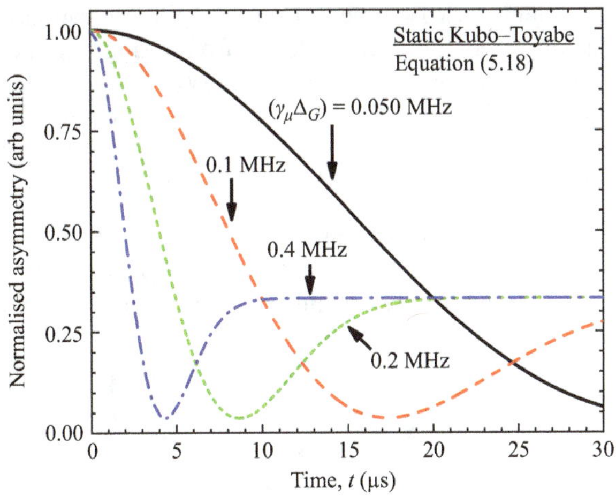

Figure 5.7 Simulated normalised asymmetry for the static Gaussian Kubo–Toyabe function (5.18) with field widths of '$\gamma_\mu \Delta G$'. Notice the minimum polarisation occurs at $t = \sqrt{3}/(\gamma_\mu \Delta G)$

the other fields that contribute to $\boldsymbol{B}_{\text{eff}}$, such as a dipolar or hyperfine fields, then the Kubo–Toyabe function for this arrangement becomes [27]

$$P^{\theta=0^\circ}(\Delta_G, B, t) = 1 - \left(\frac{2\gamma_\mu^2\Delta_G^2}{\omega_\mu^2}\right)\left(1 - \exp\left(-\frac{(\gamma_\mu\Delta_G t)^2}{2}\right)\cos(\omega_\mu t)\right) + \left(\frac{2\Delta_G^4}{\omega_\mu^3}\right)\int_0^t \exp\left(-\frac{(\Delta_G t')^2}{2}\right)\sin(\omega_\mu t')\,\mathrm{d}t' \tag{5.19}$$

where the field sensed by the muon is B and part of $\omega_\mu = \gamma_\mu B$.

If an external field ($\boldsymbol{B}_{\text{ext}}$) is applied with $\theta = 90^\circ$ (perpendicular to the initial spin direction, known as transverse field (TF)), and a magnitude that is larger than about $5\Delta_G$, the nuclear contribution ($\boldsymbol{B}_{\text{dip}}$) is dominated by the external field, resulting in the muon being sensitive to the field distribution (and variations within) along the direction of the applied field. In this case, the overall polarisation function becomes a Gaussian-damped oscillation, that is

$$P^{\theta=90^\circ}(t) = \exp\left(-\frac{(\gamma_\mu\Delta_G t)^2}{2}\right)\cos(\omega_\mu t) \tag{5.20}$$

The precession frequency is directly proportional to the perpendicular component of the effective field and scaled by the gyromagnetic ratio of the muon ($\omega_\mu = \gamma_\mu B_{\text{eff}}^{\perp}$).

If the effective field sensed by the muon has rapid fluctuations, such as from the μ^+ quickly moving among different sites or from changes in the neighbouring moments, the overall polarisation function becomes a Lorentzian-damped oscillation with relaxation rate λ, that is

$$P^{\theta=90^\circ}(t) = \exp(-\lambda t)\cos(\omega_\mu t) \tag{5.21}$$

Now considering a system with a very dilute distribution of neighbouring moments, as observed by the μ^+, with a field width parameter of σ and a local field $(\boldsymbol{B}_{i,\text{loc}} = \boldsymbol{B}_{i,\text{eff}} - \boldsymbol{B}_{\text{ext}})$ at the ith muon site is Δ_i, the distribution function can be written as

$$\rho(\Delta_i) = \frac{1}{\pi}\frac{\sigma}{\sigma^2 + \Delta_i}\bigg|\,\Delta_i = \boldsymbol{B}_{i,\text{loc}} \tag{5.22}$$

By taking a statistical average over a Gaussian distribution of possible field values, the relaxation function ($G(t)$ cf. (5.12)) is a 'static Lorentzian Kubo–Toyabe function'

$$G(t) = \frac{1}{3} + \frac{2}{3}(1 - \gamma_\mu\sigma t)\exp(-\gamma_\mu\sigma t) \tag{5.23}$$

An example of this is shown in Figure 5.8.

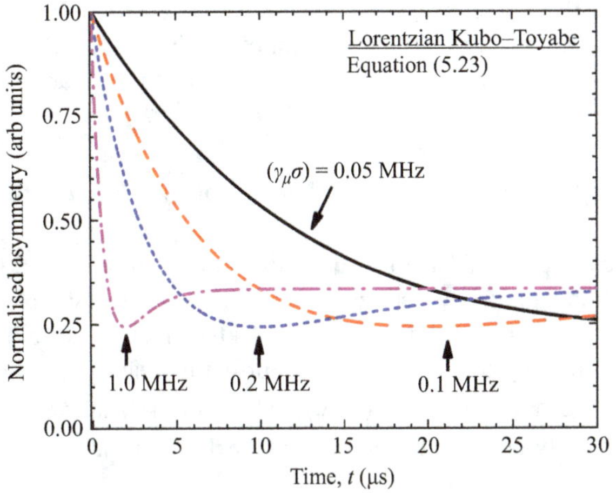

Figure 5.8 Simulated normalised asymmetry for the static Lorentzian Kubo–Toyabe function (5.23) with field widths of '$\gamma_\mu\sigma$'

If a field is applied parallel to the initial spin polarisation direction ($\theta = 0°$), the relaxation function for this dilute distribution (5.22) is [28]

$$
\begin{aligned}
G(t, \omega_\mu) = 1 &- \frac{\gamma_\mu\sigma}{\omega_\mu} j_1(\omega_\mu t)\exp(-\gamma_\mu\sigma t) \\
&- \left(\frac{\gamma_\mu\sigma}{\omega_\mu}\right)^2 (j_0(\omega_\mu t)\exp(-\gamma_\mu\sigma t) - 1) \\
&- \left(1 + \left(\frac{\gamma_\mu\sigma}{\omega_\mu}\right)^2\right)\gamma_\mu\sigma\int_0^t j_0(\omega_\mu t')\exp(-\gamma_\mu\sigma t')\,\mathrm{d}t'
\end{aligned}
\tag{5.24}
$$

where j_0 and j_1 are spherical Bessel functions.

Relaxation in the muon polarisation function is caused by inhomogeneity or changes in the field as sensed by the muon. Fluctuations in the field sensed by the muon can be caused by the ensemble of muons sampling different sites where each site has a different field (e.g. spatial inhomogeneity) or by changes in the local environment itself (e.g. changing fields from fluctuating local moments, charges attaching and releasing from the muon and the muon itself moving). In fact, detecting the difference between these two possible sources of dynamics is sometimes impossible from a single time-domain spectra. Regardless of the source, one way to work with these fluctuations is to assume that $\boldsymbol{B}_{\text{eff}}$ has a time dependence of the form

$$\boldsymbol{B}_{\text{eff}}(t) \sim \exp(-\nu t) \tag{5.25}$$

where ν is the fluctuation rate of the field as sampled by the muon, completely independent of previous or future behaviour and independent of the origin of the changes in field.

A system with concentrated moments with fields that fluctuate in time or where the muon is allowed to hop between sites can be modelled by modifying the static Kubo–Toyabe function (P_S^{KT}, (5.18)) to yield the dynamic Kubo–Toyabe function [27]

$$P_D^{KT}(t) = P_S^{KT}(t)\exp(-\nu t) + \nu \int_0^t P_S^{KT}(t')\exp(-\nu t')P_D^{KT}(t-t')\,\mathrm{d}t' \qquad (5.26)$$

Characteristic examples of this function are shown in Figure 5.9. In general, starting with a fluctuation rate near zero, (5.26) matches the static function with a full one-third recovery of the long-time tail. As the fluctuation rate increases, the one-third tail is reduced (due to dephasing). Further increase in the fluctuation rates then suppresses the dip at $t = \sqrt{3}/(\gamma_\mu \Delta_G)$ where the relaxation has an Abragam shape until eventually becoming Lorentzian (see Table 5.2).

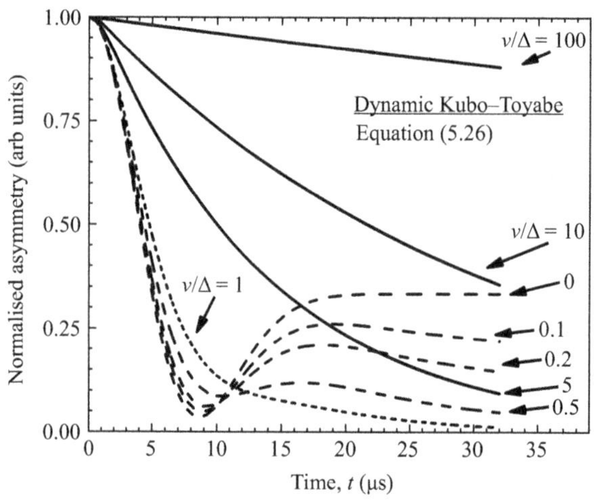

Figure 5.9 Simulated normalised asymmetry for the dynamic Gaussian Kubo–Toyabe function (5.26) with field width of $\gamma_\mu \Delta_G = 0.20$ MHz and ratio of fluctuation rate to field width (ν/Δ) indicated on the plot next to each line with the static ($\nu/\Delta = 0$) Kubo–Toyabe plotted for reference. The curves are only evaluated out to 30 μs

Table 5.2 Assumptions and resulting relaxation functions, G(t), for common analytic approximations to the dynamic Kubo–Toyabe function (5.26)

Assumption	***G(t)***	**Comment**
$\nu/\Delta < 1$	$(1/3)\exp(-(2/3)\nu t)$	Slow – KT tail affected
$1 < \nu/\Delta < 10$	$\exp(-\lambda(e^{-\nu t} - 1 + \nu t))$	Moderate – Abragam function
$\nu/\Delta > 10$	$\exp(-\lambda t)$	Fast – simple Lorentzian

Analytic approximations to (5.26) produce relaxation functions that allow for more efficient data analysis as compared to the full form that requires numerical methods to solve and use. Three common and useful approximations include assumptions that correlate the fluctuation rate (ν) and local-field distribution (Δ). Table 5.2 outlines the relevant assumption, resulting form of the relaxation function $G(t)$, and a comment on each approximation. Here, $\lambda \equiv 2\Delta^2/\nu^2$.

Note that some μSR literature report the Abragam function without the factor of two included (in λ) as written in Table 5.2 and (5.29). In the case of the Kubo–Toyabe function (5.26) and Keren function (5.29), $\boldsymbol{B}_{\text{ext}}$ is parallel to the initial spin polarisation direction (i.e. $\theta = 0° \| \hat{z}$), so fluctuations along g the $\hat{x}$ and $\hat{y}$ directions will both contribute to the muon relaxation (loss of polarisation). The form written without this factor of two corresponds to the case where $\boldsymbol{B}_{\text{ext}}$ is perpendicular to the initial spin polarisation direction (i.e. $\theta = 90°$ and hence perpendicular to $\hat{z}$) so only fluctuations along the $\hat{z}$ direction contribute to the relaxation.

Another common dynamic spin polarisation function is presented by Keren in [29] as an alternative to the Kubo–Toyabe approach (5.26). This function applies a perturbation treatment to the same dynamic input of the dynamic Kubo–Toyabe model and has an advantage of providing an analytic expression. This function incorporates static and dynamic field fluctuations and makes no assumptions regarding the instantaneous field distribution. This approximation works well if the field fluctuations are in the $\nu > \sim 10\Delta$ regime or one is only interested in the very early times of muon decay $\left(t < \sqrt{3}/\Delta\right)$. The general result is computed for $\boldsymbol{B}_{\text{ext}}$ applied along the initial spin polarisation direction of the incoming muon (i.e. $\theta = 0°$) and written as

$$P_z(t) = P_z(0)\exp(-\Gamma(t)t) \tag{5.27}$$

where $\Gamma(t)t$ has the general form

$$\Gamma(t)t = \frac{2\Delta^2}{\left(\omega_\mu^2 + \nu^2\right)^2} \cdot \left\{ \left(\omega_\mu^2 + \nu^2\right)\nu t + \left(\omega_\mu^2 - \nu^2\right)\left(1 - \exp(-\nu t)\cos\left(\omega_\mu t\right)\right) - 2\nu\omega_\mu \exp(-\nu t)\sin\left(\omega_\mu t\right) \right\} \tag{5.28}$$

and the Larmor frequency of the muon is ω_μ. In the zero-field (ZF) limit ($\boldsymbol{B}_{\text{ext}} = 0$), $\omega_\mu \to 0$ and (5.28) recovers the Abragam function since

$$\Gamma(t)t \to \frac{2\Delta^2}{\nu^2}(\exp(-\nu t) - 1 + \nu t) \tag{5.29}$$

In the faster fluctuation regime ($\nu \gg \sim 0.1\ \mu\text{s}^{-1}$), the resulting function has an exponential shape, matches the dynamic Kubo–Toyabe function and is expressed as

$$\Gamma(t)t = \frac{2\Delta^2\nu}{\omega_\mu^2 + \nu^2}t \tag{5.30}$$

In the early-time limit and when $\nu \ll 0.1\ \mathrm{ns}^{-1}$, the overall shape of the relaxation function is Gaussian and expressed as

$$\Gamma(t)t = \Delta^2 t^2 + O(t^3) \tag{5.31}$$

Since the early-time relaxation only depends on the second moment of the field distribution, it is independent of the applied field and fluctuation rate.

The Keren function, (5.27) and (5.28), does not work well when $\Delta \sim \nu$ or if Δ/ν is of a value such that the later time polarisation data begin to increase (i.e. $t > \sqrt{3}/\Delta$ is within the observable data gate). While the Keren function cannot fit data beyond $t > \sqrt{3}/\Delta$, the Kubo–Toyabe function (5.26) can recover tail extending beyond the minimum in polarisation (e.g. $t > 8.66\ \mu\mathrm{s}$ in Figure 5.9).

Thus far, only $\boldsymbol{B}_{\mathrm{dip}}$, $\boldsymbol{B}_{\mathrm{ext}}$ and fluctuations within have been considered as contributors to $\boldsymbol{B}_{\mathrm{eff}}$ and the associated spin polarisation functions. If the thermalised muon interacts with an electron, then the $\boldsymbol{B}_{\mathrm{hyp}}$ term in (5.11) may become important. In semiconductors, atomic muonium can be formed when the muon captures an electron to form the hydrogen-like (H^0), electrically paramagnetic, neutral state of muonium ($\mathrm{Mu}^0 = \mu^+e^-$), where the electron is centred on the muon. Muonium is also found to exist in the electrically diamagnetic positive (Mu^+) and negative (Mu^-) charge states, similar to H^+ and H^-.

A quantum-based density matrix approach can be used to calculate the polarisation functions for such systems. I will continue with a lighter and mostly qualitative discussion of these states and interactions while providing some of the key results to stay within the intended scope of this chapter. For more detail, I encourage the reader to spend some time with [12,15–17,29,30] that are much more specialised and provide discussion that goes well beyond the highlights.

The most straightforward muonium species occurs when there is no unpaired electron spin (e.g. Mu^+, Mu^- or any diamagnetic Mu complex) and the interactions with surrounding nuclei (e.g. dipolar, quadrupolar) are negligible. The Hamiltonian for this species in a magnetic field (B) along $\hat{z}$ is

$$\mathcal{H} = -h\tilde{\gamma}_{\mu} B I_z \tag{5.32}$$

The eigenstates of which can be labelled as $|+\rangle$ and $|-\rangle$ corresponding to the spin-up ($m_I = +1/2$) and spin-down ($m_I = -1/2$) states of the muon, where m_I is the muon magnetic quantum number. The energy eigenvalues are then $-\gamma_{\mu}B/2$ for the $|+\rangle$ and $+\gamma_{\mu}B/2$ for the $|-\rangle$ states. The resulting polarisation function for when the muon spin aligned with the applied field ($\theta = 0^\circ$) is then

$$P^{\theta=0^\circ}(t) = 1 \tag{5.33}$$

again showing that the muon polarisation is not affected by a field parallel to the muon spin direction. When the applied field is perpendicular to the initial spin direction of the muon ($\theta = 90^\circ$), the polarisation function becomes

$$P^{\theta=90^\circ}(t) = \cos(\omega_{\mu} t) \tag{5.34}$$

indicating that the precession is only in the plane that is perpendicular to the field.

These results match (5.18) and (5.20) for the bare muon if $\boldsymbol{B}_{\text{eff}}$ is perfectly homogeneous (i.e. $\Delta_G = 0 \rightarrow G(t) = 1$) or if a single muon site is considered. In the case of (5.32), only a single site is being considered. This is a good example demonstrating that the muon is only sensitive to $\boldsymbol{B}_{\text{eff}}$ and cannot necessarily delineate between different sources of a particular field by a single measurement. Often times, models that include an ensemble of time-domain spectra and changes to the shape or character of the polarisation function at various temperatures and fields are required to uncover the overall processes and mechanisms responsible for the sensed $\boldsymbol{B}_{\text{eff}}$ and hence what is happening in the material being studied.

Next, consider the case of neutral muonium Mu^0 but maintain a situation where interactions with surrounding nuclei are negligible. In semiconductors, Mu^0 is typically found to have either an isotropic or axially symmetric hyperfine interaction. For isotropic muonium, the electron Zeeman splitting is the largest term in the spin Hamiltonian. The two levels relate to the electron spin-up and spin-down states. The muon Zeeman interaction introduces an additional splitting of the electron levels that depends on whether the muon and electron spins are aligned parallel or antiparallel. The details of this treatment, including results such as the eigenvalues, eigenvectors and field dependence of these energy levels and transition frequencies, can be found in [12]. As an example, these levels for isotropic (1*s*-like) muonium in a centre with tetrahedral coordination are shown in Figure 5.10. The convention is to label states (levels) where the muon and electron

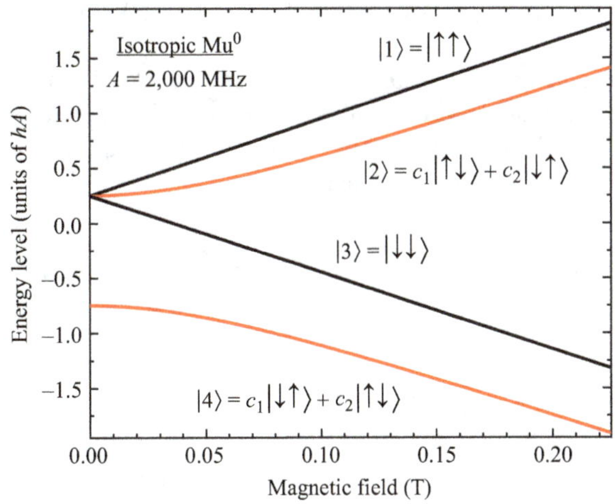

Figure 5.10 *Breit–Rabi energy levels for an isotropic muonium centre (e.g. Mu_T^0) with hyperfine constant $A = 2{,}000$ MHz. The state (level) numbers here are consistent with the discussion throughout the text. Notice that states $|1\rangle$ and $|3\rangle$ are consistent with the muon and electron magnetic quantum numbers (in the $\hat{z}$ direction) both either $+1/2$ or $-1/2$. The other two states include different mixtures of $\pm 1/2$ for both the muon and electron. See [12,32] for detail relating to the constants c_1 and c_2*

magnetic quantum numbers (in the $\hat{z}$ direction) both either $+1/2$ as $|1\rangle$ and $-1/2$ as $|3\rangle$. States $|2\rangle$ and $|4\rangle$ correspond to states with different mixtures of $\pm 1/2$ for the muon and electron.

With some work, the polarisation function for a system with field applied along the initial muon spin polarisation axis ($\theta = 0°$) is

$$P^{\theta=0°}(x,t) = \frac{1+2x^2}{2(1+x^2)} + \frac{\cos(\omega_{24}t)}{2(1+x^2)} \tag{5.35}$$

where $\omega_{24} = \omega_0\sqrt{1+x^2}$, $\omega_0 = 2\pi A_\mu$ and A_μ is the isotropic muonium hyperfine constant. For convenience, the field parameter $x \equiv B/B_0$, where the applied field is B and characteristic field $B_0 \equiv A_\mu/(\gamma_e + \gamma_\mu)$. The gyromagnetic ratio for the electron and muon are $\gamma_e = 2\pi \times 28{,}024.6$ MHz/T and $\gamma_\mu = 2\pi \times 135.54$ MHz/T, respectively. In practice, the first term of (5.35) is all that is observed since the transition frequency ω_{24} is typically too high for the experimental setup unless one is using a very high timing resolution apparatus. Qualitatively, (5.35) corresponds to the situation where the muon ensemble is initially polarised but the electrons are not; therefore, half of the ensemble is in the static $|1\rangle$ (i.e. $|++\rangle$) state, whereas the other half transitions between the mixed states ($|2\rangle$ and $|4\rangle$) at the frequency ω_{24}.

If the field is applied in a direction perpendicular to the initial muon spin polarisation ($\theta = 90°$), the polarisation function is

$$P^{\theta=90°}(x,t) = \frac{1}{2}\cdot\{\zeta\cos(\omega_{12}t) + \eta\cos(\omega_{23}t) + \zeta\cos(\omega_{34}t) + \eta\cos(\omega_{14}t)\} \tag{5.36}$$

where x is the same as in (5.35); ω_{ij} corresponds to the transition rates between $|i\rangle$ and $|j\rangle$ and whose frequency is controlled by the hyperfine interaction (Figure 5.10 shows a typical energy-level diagram for isotropic Mu^0 with these states labelled) and for convenience, ζ and η are defined as

$$\zeta = \frac{1}{2} + \frac{x}{2\sqrt{1+x^2}} \tag{5.37}$$

and

$$\eta = \frac{1}{2} - \frac{x}{2\sqrt{1+x^2}} \tag{5.38}$$

In high fields, $\zeta \to 1.0$ and $\eta \to 0$ so only the $\cos(\omega_{12}t)$ and $\cos(\omega_{34}t)$ terms in (5.36) have significant amplitude. However, at low fields, $\zeta = \eta = 1/2$, so all four terms in (5.36) are present, but only the triplet (transition rates ω_{12} and ω_{23}) is typically within an observable range. A detailed table with calculated amplitudes for each component of (5.35) and (5.36) in the two different field directions is found in Table IV of [12].

In a situation where there is a dipolar coupling between the muon and electron spin (still ignoring any coupling to nuclear moments), the hyperfine interaction of axially symmetric muonium now requires two terms. This anisotropic muonium species is often found in a BC configuration and therefore labelled as $\mathrm{Mu}^0_{\mathrm{BC}}$. Historically, this species has also been labelled as 'anomalous Mu' or Mu^*. For convenience, the hyperfine interaction is expressed in components measured parallel and perpendicular to the bond direction, $A_\parallel$ and $A_\perp$ respectively. Analytical solutions to the Hamiltonian governing this configuration are known for a few special cases of relative field directions. Thorough treatment of the general or other cases can be found in [12,16,30,31]. Continuing with the special case for which the dipolar contribution is axially symmetric, $A_\parallel$ and $A_\perp$ can be expressed in terms of an isotropic component A_{iso} and the dipolar component D

$$A_{\mathrm{iso}} = \frac{1}{3}\left(A_\parallel + 2A_\perp\right) \quad \text{and} \quad D = \frac{2}{3}\left(A_\parallel - A_\perp\right) \tag{5.39}$$

or, for convenience, rearranged and written as

$$A_\parallel = A_{\mathrm{iso}} + D \quad \text{and} \quad A_\perp = A_{\mathrm{iso}} - \frac{D}{2} \tag{5.40}$$

Considering first, a field applied along the direction of the bond $\left(\boldsymbol{B}_{\mathrm{ext}} \| A_\parallel\right)$ and hence parallel to D, the polarisation function can be written [32]

$$P^{B\|D}(x_B, t) = \frac{1 + 2x_B^2}{2(x_B^2 + 1)} - \frac{\cos(\omega_{23}t)}{2(x_B^2 + 1)} \tag{5.41}$$

With a field-scaling parameter, x_B, defined as

$$x_B = \frac{B}{B_0} = \frac{B\left(\gamma_{\mathrm{e}} + \gamma_\mu\right)}{(2A_{\mathrm{iso}} - D)/2} \tag{5.42}$$

This is similar to the result for the isotropic case, except now the decoupling field B_0 depends on both A_{iso} and D.

Now considering a field applied along a direction that is perpendicular to the bond $(\boldsymbol{B}_{\mathrm{ext}} \| A_\perp)$ and hence perpendicular to D, the polarisation function can be written

$$P^{B\perp D}(x_B, y_B, t) = \frac{x_B^2}{2(x^2 + 1)} + \frac{y_B^2}{2(y_B^2 + 1)} - \frac{\cos(\omega_{23}t)}{2(x_B^2 + 1)} - \frac{\cos(\omega_{14}t)}{2(y_B^2 + 1)} \tag{5.43}$$

which has a similar form but now has two characteristic decoupling fields, B_0 and B_1, within the field scaling parameters x_B and y_B

$$x_B = \frac{B}{B_0} = \frac{B\left(\gamma_{\mathrm{e}} + \gamma_\mu\right)}{(2A_{\mathrm{iso}} + D/2)/2} \tag{5.44}$$

and

$$y_B = \frac{B}{B_1} = \frac{B\left(\gamma_{\mathrm{e}} - \gamma_\mu\right)}{(3D/2)/2} \tag{5.45}$$

These two limiting cases demonstrate the general expectations for axially symmetric muonium. Solutions for cases where the applied field is not along one of these convenient axes are typically determined numerically. Some examples of decoupling curves for Mu^0_{BC} in Si with a few common geometries (for Zincblende systems) are plotted in Figure 5.11.

In the case of an orientationally disordered material on which μSR measurements are being conducted with the incoming muon spin polarisation direction parallel to the externally applied field ($\theta = 0°$ or LF), the polarisation of the muonium immediately after its formation (i.e. $t \rightarrow 0$ in $P(t)$) can be approximated by taking a weighted average over several of the possible angles between the applied field and principal axis of the hyperfine tensor (usually along D). The result of this approximation is expressed as [31]

$$P(x_B, y_B, t \rightarrow 0) = \frac{q + x_B^2}{1 + x_B^2} \tag{5.46}$$

with

$$q = \frac{1}{2}\frac{\langle \cos^2(\phi)\rangle + y_B^2}{1 + y_B^2} \tag{5.47}$$

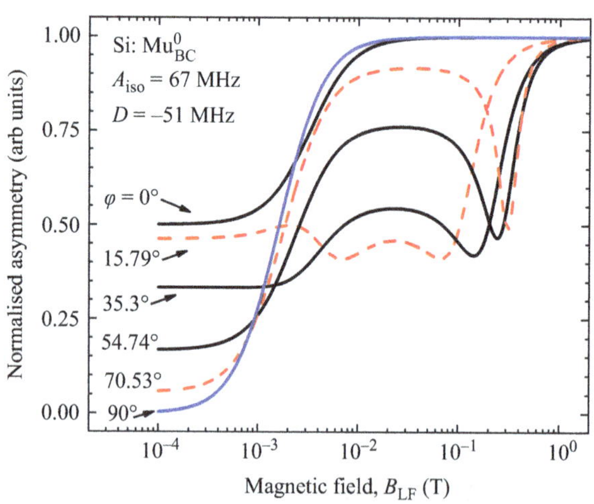

Figure 5.11 Decoupling curves for anisotropic (bond-centred) muonium in Si where $A_{iso} = 67$ MHz and $D = -51$ MHz and the BC symmetry axis is along the (1 1 1). The angle between the symmetry axis and applied magnetic field is φ. The isotropic decoupling curve is recovered when $\varphi = 0°$, where $B_{ext} \| D$ so only the isotropic component results in precession at low field. When $\varphi = 90°$, $B_{ext} \perp D$ (i.e. (5.43)), both A_{iso} and D result in precession and therefore the measured LF asymmetry goes to zero at low B_{ext}

The angle between the applied field and primary axis of D is ϕ and for a polycrystalline $\langle \cos^2(\phi) \rangle = 1/3$. The field scaling parameters x_B and y_B now include a characteristic field of the hyperfine anisotropy, B_1, and the characteristic field for the isotropic component, B_0. In this case, the field scaling parameters are

$$x_B = \frac{B}{B_0} = \frac{B(\gamma_e + \gamma_\mu)}{A_i} \tag{5.48}$$

and

$$y_B = \frac{B}{B_1} = \frac{B(\gamma_e - \gamma_\mu)}{D} \tag{5.49}$$

with B referring to the applied field, as before.

It is important to note that approximation (5.46) does not deal with the avoided level crossings (ALCs) that occur in these anisotropic cases (Figure 5.11); however, it does provide a reasonable approximation for the general shape of the decoupling curve [31].

5.2.3 *Transverse field μSR*

One of the first μSR techniques is 'transverse field muon spin rotation' (TF-μSR). In this experimental configuration, an external magnetic field ($\boldsymbol{B}_{\text{ext}}$) applied in a direction that is perpendicular to the incoming muon spin polarisation ($\theta = 90°$) drives the Larmor precession of the muon spin about the total effective field ($\boldsymbol{B}_{\text{eff}}$). The time evolution of the spin precession is followed with a set of detectors. A common detector configuration uses two orthogonal pairs (total of four) of opposing detectors in a plane perpendicular to the applied field, similar to the schematic in Figure 5.5. One way to achieve the TF orientation is to pass the muon beam through perpendicular electric and magnetic fields (Wien filter) that rotate the muon spin to the desired orientation prior to entering the sample. Another option is to design the sample environment and magnet such that $\boldsymbol{B}_{\text{ext}}$ is perpendicular to the incoming spin at the sample without rotating the muon spin. The latter option is an approach often used by instruments that are designed for ZF or LF measurements but still need some weak TF capabilities.

With the muon spin precessing about the effective field, the measured corrected asymmetry (e.g. (5.10)) can typically be modelled in the form

$$P_\mu^{\theta=90°}(t) = \Sigma_n a_n G_n(t) \cos(\omega_{n,\mu} t + \phi_n) \tag{5.50}$$

where the μ^+ in site (or state) n precesses at the Larmor frequency ($\omega_{n,\mu} = \gamma_\mu B_{n,\text{eff}}$) and thereby directly probes the net field (5.11) of that state. The asymmetry (amplitude) a_n is a measure of signal intensity that directly correlates to the probability of μ^+ persisting in state n. For example, the amplitude could provide a measure of relative site population, μ^+ vs Mu^0 fraction or a measure of the fraction of a material existing in a magnetic state. The relaxation parameter $G_n(t)$ characterises the way in which the amplitude of the measured time evolution of the precessing signal for the *n*th state is damped, which relates to the inhomogeneity of the local fields and changes therein (e.g. see Section 5.2.2). For a static centre, TF relaxation is usually the result of dephasing of the probe spins precessing at slightly different frequencies arising from slightly different local-field values at the various sites (spin–spin or T_2^{-1}, in NMR terms). In the case of rapid field fluctuations, the relaxation may be showing a true loss of polarisation due to the transitions between Zeeman energy eigenstates of the μ^+ itself (spin–lattice or T_1^{-1}, in NMR terms). The latter case is much better studied using the longitudinal applied field methods. Equations (5.20) and (5.21) show the form for a Gaussian and Lorentzian relaxing component, respectively. Phase ϕ_n refers to the apparent shift of the initial μ^+ precession within the spectra. Shifts in the phase are usually associated with decoherence of the precession from processes such as electron capture or release.

In the cases where the observed frequencies are low enough (on the order of MHz or smaller, that is $\boldsymbol{B}_{\text{eff}} < \sim 10^{-2}$ T), or spectra that contain only one or two components with sufficiently different frequencies, direct analysis of the time-domain data can be straightforward; an example is shown in Figure 5.12. Higher frequencies, or more complicated spectra, often make direct time-domain analysis

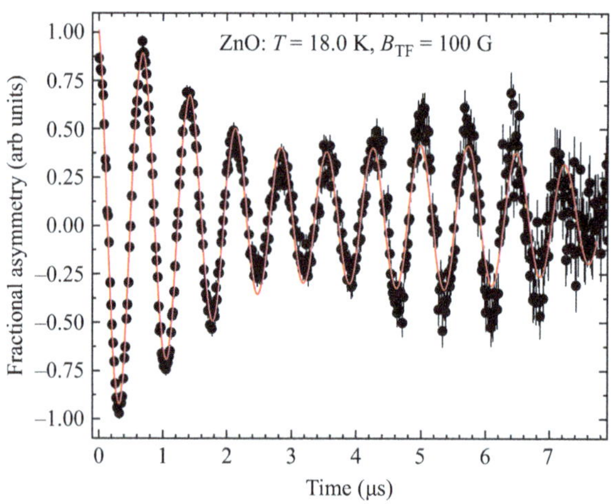

Figure 5.12 Example of low TF-μSR data on ZnO collected as part of a set of experiments on the transparent conducting oxides [110] where the corrected asymmetry is normalised by comparison to an Ag calibration. These data are fit well by a linear combination of two Lorentzian relaxing oscillations (5.50). As fit, the phase, fractional amplitude, frequency and relaxation rate for the dominant component are $24.6 \pm 0.9°$, 0.92 ± 0.02, 1.383 ± 0.001 MHz and 0.232 ± 0.007 μs^{-1}, respectively. The secondary component yields phase, fractional asymmetry, frequency and relaxation rates of $-11 \pm 5°$, 14.3 ± 0.01, 1.232 ± 0.004 MHz and 0.06 ± 0.02 μs^{-1}, respectively

impractical, if not impossible (e.g. Figure 5.13). One way to overcome some of the challenges of high-frequency oscillations (high field data) in the time-domain data is to use Fourier analysis to convert the time-domain data to a frequency domain. Tools to perform discrete Fourier transforms on data sets are readily available. Another approach that is frequently utilised involves using a discrete rotating reference frame (RRF) transformation on the time-domain data [33]. The net result of the RRF transformation is a time-domain spectrum that is repacked so that the displayed signal is the difference between the direct time-domain spectra and the rotating baseline. If, for example, there is a precession signal of $\omega_\mu = 27.1$ MHz and one sets an RRF frequency of 26.7 MHz, the transformed signal will appear to have a 0.4 MHz frequency, making it much easier to see. An example comparing the direct time-domain signal to one in an RRF is shown in Figure 5.13. Another advantage of the RRF is that if spectra contain several oscillating components with significantly different frequencies, one can isolate each component for easier fitting by choosing a frequency for the RRF that nearly matches each component.

In practice, combinations of these tools are typically used in analysing TF data. For example, one may choose to start with Fourier analysis on a particular time-domain spectra to build a preliminary model, via (5.50), by first assessing how many components with unique frequencies may exist, approximately where those frequencies are centred, the relevant line shapes (relaxation functions) and approximate fractional occupation of each state (amplitudes). Once a general idea of the general model and frequencies is established, choosing an appropriate RRF

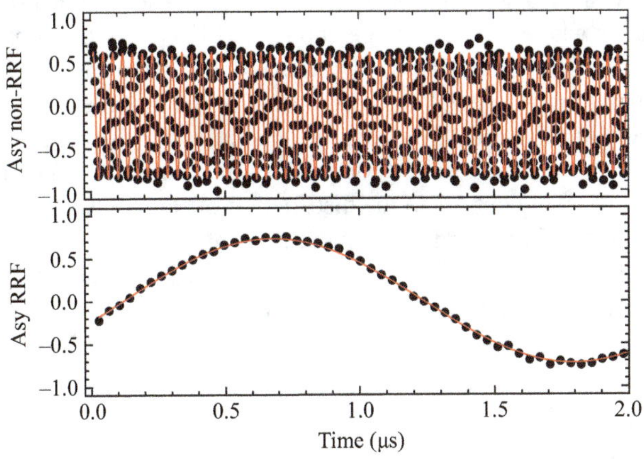

Figure 5.13 High TF μSR time-domain spectra in the (top) non-rotating lab frame and (bottom) reference frame rotating at 26.7 MHz; 0.4 MHz below the primary oscillation frequency. Here the field was set to 0.2 T, although typical high-TF measurements can go up to 7 T. Analysing data under high-field conditions is much more manageable when using the RRF

frequency and then fitting the RRF transformed time-domain data with the model can efficiently follow.

At its core, the muon's extreme sensitivity to local fields ($\sim 10^{-5}$ T) provides a variety of useful applications for the TF-μSR technique. Some examples involve probing the vortex states, magnetic penetration depth and correlation lengths within superconductors [34]; local fields, electronic structure, ordering and nuclear spin dynamics [16,17,35]. This chapter focuses on applications related to identifying and characterising $Mu^{\pm}$, Mu^0 and Mu^0-like centres and processes as an experimentally accessible analogue to hydrogen impurities in semiconductors (e.g. see Section 5.3 and [12,14–16,32,36,37]).

5.2.4 Longitudinal field μSR

Longitudinal field muon spin relaxation (LF-μSR) refers to the configuration where an external magnetic field ($\boldsymbol{B}_{\text{ext}}$) is applied in a direction parallel to the incoming muon spin polarisation ($\theta = 0°$). The time evolution of the muon spin polarisation is typically monitored with a pair of opposed counters that are along the initial spin polarisation direction, similar to what is illustrated in Figure 5.4. The set of four counters in quadrature perpendicular to the spin polarisation (as done in TF) is not normally used for LF, given that if the muon spin does not have a component that is initially perpendicular to the local field, one is unlikely to develop. Typical LF-μSR time-domain signals lack precession (especially in high fields), so the dominant feature in the measured asymmetry (e.g. (5.7) or (5.10)) is depolarisation (loss of asymmetry in the time domain and sometimes referred to as relaxation). In comparison with TF measurements, LF measurements are much more sensitive to the final state of a transition or reaction since phase coherence is not required to see a signal (unlike TF).

In conditions where the LF is large enough to effectively decouple the dipolar and hyperfine fields, any depolarisation (relaxation) in the measured asymmetry is caused by dynamic processes like muon motion or muonium state transitions that lead to exponential relaxation functions. In this case, the fluctuations in a sensed magnetic parameter are responsible for the loss of polarisation and so it is a true depolarisation (T_1^{-1}), in contrast to the distribution of static local fields resulting in dephasing $\left(T_2^{-1}\right)$ as often found in TF measurements.

As these spectra tend to lack precession $\left(\omega_\mu = 0\right)$, the measured asymmetry (e.g. (5.7) or (5.10)) for a particular temperature in the high LF regime can often be modelled in the form

$$P_\mu^{\theta=0°}(t) = \Sigma_n a_n G_n(t) \tag{5.51}$$

where the amplitude (asymmetry) for the nth state, a_n, corresponds to the signal intensity and probability of the muon being in that state. The relaxation (depolarisation) function for the nth state $G_n(t) = \exp(-\lambda_n t)$ and dependencies on experimental parameters (e.g. applied field, sample temperature, dopant type, carrier concentration and excess carrier concentration) provide significant insight into the processes responsible for the observed relaxation. Several examples of processes relevant to studying semiconductors are discussed in Section 5.3.3.

If the applied field does not dominate contributions to the effective field (5.11), then the components of $\boldsymbol{B}_{\text{eff}}$ that are perpendicular to the incoming muon spin cannot be ignored, and the depolarisation (relaxation) functions can become very complicated. For instance, in low LF for a static centre surrounded by nuclear moments, the Gaussian LF Kubo–Toyabe (5.19) is often a good approximation. Another example in the low LF regime that is appropriate for neighbouring nuclear moments with sufficiently fast dynamics or small Δ_G is (5.27) and (5.28), that is known as the LF Keren function.

LF-μSR techniques can measure behaviours such as local-field fluctuation rates (either from muon motion or from variations in the local environment itself), muonium transition rates (e.g. charge cycles, site-to-site motion and state conversion) and the strength of nuclear and electronic coupling of the μ^+ to its local environment (e.g. see [38,39]). Several of the models applicable to probing properties of semiconductors and the muonium impurity as an analogue to the hydrogen centre are discussed in Section 5.3.

5.2.5 Zero-field μSR

In zero field muon spin relaxation (ZF-μSR) measurements, the experiment is configured such that the measured magnetic field at the sample site is zero (i.e. $\boldsymbol{B}_{\text{ext}} = 0$). Instruments configured for ZF measurements typically have active field correction capabilities where the magnetic field near the sample is monitored and a dedicated set of coils are used to maintain as close to ZF as possible at the sample site. These systems are sensitive enough to offset common sources, such as the Earth's magnetic field, stray fields in nearby structures, neighbouring instruments and equipment that may be moving overhead (e.g. cranes). The detector configuration can be identical to either the two- or four-detector layout described for LF and TF measurements, respectively. The subject of the measurement and expected features drive the choice as to which configuration is best for that particular study. However, the vast majority of semiconductor-based ZF work uses the two-detector arrangement, identical to the LF setup. One may opt for an instrument where the four-detector (TF) arrangement is available if there is interest in measuring properties within an ordered magnetic phase where high timing (field) resolution is desired.

Since the μ^+ itself is the probe of the local environment and the sample under study does not need to be perturbed in any other way for the time evolution of the muon polarisation to be observed, this ZF technique is particularly well suited to look at how μ^+, Mu^+ and Mu^- behave in a material in a close to natural state, directly measure $\boldsymbol{B}_{\text{loc}}$ (e.g. directly measure dipolar fields for a static muonium centre) and characterise dynamics that result from either μ^+ ($Mu^{\pm}$) motion between sites or fluctuations in $\boldsymbol{B}_{\text{loc}}$ itself [12,27,35]. These measurements can provide insight into comparable behaviour of hydrogen centres in materials.

As done with TF and LF, models for the measured asymmetry in ZF can be built by summing polarisation (relaxation) functions, weighted by the respective amplitude for each state. The Kubo–Toyabe function (i.e. (5.18), (5.23) and (5.26)) and related approximations (e.g. Table 5.2) are frequently relevant relaxation functions when characterising the diamagnetic muonium states in semiconductors.

5.2.6 Resonance-based μSR techniques

While there are many types available, two resonance-based techniques that have been important in studying semiconductors are high-frequency AC field-driven resonance (RF-μSR) and muon level-crossing resonance (LCR). These provide additional means to study nuclear and electronic coupling of the muon to its surroundings, as well as processes in which the muon (muonium) is actively involved (e.g. transition dynamics).

5.2.6.1 AC field-driven resonance (RF-μSR)

A typical RF-μSR experiment takes the usual LF setup (muon spin, DC magnetic field and two detectors are all collinear as in Figure 5.4) and then includes a high-frequency oscillating transverse magnetic field (with frequency $\omega_{\rm rf}$) across the sample. In these experiments, the user typically maintains a fixed DC LF and sweeps $\omega_{\rm rf}$ or chooses a fixed $\omega_{\rm rf}$ and sweeps the DC LF. Either way, the muon polarisation maintains its alignment with the DC LF (maximising measured LF asymmetry) unless $\omega_{\rm rf}$ is close to the resonant condition. This would drive a precession of the muon spin, resulting in a reduction in the overall longitudinal polarisation and is therefore observed as a loss in the longitudinal asymmetry. The resonant condition occurs when $\omega_{\rm rf}$ matches one of the muonium precession frequencies (e.g. for Mu^+, $\omega_{\rm rf} = \omega_\mu = 2\pi\gamma_\mu B$).

This technique produces data that are similar to what is measured by TF techniques (e.g. state identification, electronic structure), but it carries the distinct advantage of being able to measure the final persistent state regardless of when it is formed, since RF-μSR does not require phase coherence. Recall that TF requires phase coherence and is only sensitive to states that are promptly formed.

An advantage RF-μSR has over proton resonance experiments is that because of the large magnetic moment of the muon, lower magnetic fields are required to produce significant splitting of paramagnetic Mu lines. Also, since the loss in polarisation is tracked by the positron emission and the positrons are not affected by the applied RF field, the time evolution of the muon polarisation can be tracked while the RF is being applied.

Similar to traditional resonance techniques, data collection can be done in a time resolved (TD) mode or TI mode. TD data produce the time-domain spectra for measurements with and without RF so the user has the ability to analyse individual spectra for every step. In TI mode, only the total polarisation in each detector is saved; however, the relevant information regarding the change in asymmetry with and without the RF at each field (or frequency) can be easily calculated from these totals. This change in RF asymmetry ($a_{\rm rf}$) can be calculated by looking at the difference in measured asymmetry for RF on and RF off as

$$a_{\rm rf} = \frac{N_1^{\rm ON} - N_2^{\rm ON}}{N_1^{\rm ON} + N_2^{\rm ON}} - \frac{N_1^{\rm OFF} - N_2^{\rm OFF}}{N_1^{\rm OFF} + N_2^{\rm OFF}} \tag{5.52}$$

or as

$$a_{\text{rf}} = \frac{N_1^{\text{ON}} - N_1^{\text{OFF}}}{N_1^{\text{ON}} + N_1^{\text{OFF}}} - \frac{N_2^{\text{ON}} - N_2^{\text{OFF}}}{N_2^{\text{ON}} + N_2^{\text{OFF}}} \tag{5.53}$$

where N_i^{ON} and N_i^{OFF} correspond to the total counts in detector group i with RF power on and off, respectively.

Typical on-resonance spectra as a function of swept magnetic field or ω_{rf} for spin dynamics result in Lorentzian line shapes [15,40], which can then be characterised by peak centre (resonant field or frequency), line width and amplitude at several temperatures to lead to a full characterisation of the measured states.

5.2.6.2 Avoided level-crossing resonance (ALC or LCR)

A typical avoided level-crossing resonance (ALC) measurement uses LF-μSR measurements in either TD or TI modes (depending on the facility and apparatus), where the applied LF is varied to look for quantum state mixing that occurs to avoid level crossing where two energy levels are nearly degenerate. For example, when an applied field is such that the Zeeman splitting matches a transition between energy levels of the nuclear spins, the muon and nucleus undergo simultaneous changes in levels with no change in total energy. This flip-flop of spins then produces a significant relaxation in the measured muon polarisation (in TD) and significant drop in total measured asymmetry (TI or TD).

The labels of 'level-crossing resonance' (LCR) and 'avoided level-crossing resonance' (ALC or ALCR) are both commonly used in describing this measurement and phenomena.

In the context of semiconductors and muonium studies, ALCs can be within a system including diamagnetic muonium ($\text{Mu}^{\pm}$) and neighbouring nuclei, an isolated Mu^0 ($\mu^+ + e^-$) system or Mu^0 levels affected by neighbouring nuclear spins. Here ALCs can be particularly helpful in determining locations of nearby nuclei and mapping the electronic wave function.

One of the earliest and major contributions to the semiconductor community by the way of ALCs has been in determining the physical and electronic structure of the BC muonium (axially symmetric, Mu_{BC}^0) species in silicon [21] and GaAs [41].

5.2.6.3 Quadrupolar level-crossing resonance (QLCR)

When the LCR technique is extended to systems with nuclei that have spin greater than 1/2 and large dipolar and quadrupolar moments, the Zeeman levels and nuclear quadrupolar levels of the nearby nuclei become important. Here, the condition of resonance is met when the Zeeman energy of the muon probe matches an energy difference in the nuclear quadrupolar levels of the nearby nucleus. Upon meeting the resonance condition, state mixing by the dipole–dipole interaction allows cross-relaxation (spin-flip transitions) to occur with no change in total energy. The strength of the dipolar coupling between the muon and involved nucleus controls the amplitude of the quadrupolar LCR (QLCR) spectra. The muon induces an electric field gradient responsible for the nuclear quadrupole splitting and therefore is the main contributor to the position of the resonances.

In practice, the muon Zeeman levels are tuned to match the resonance (crossing) condition by adjusting the strength of the applied LF. When the condition is met, this effect is observed as enhanced relaxation (peak in the longitudinal relaxation rate) and a dip in measured integrated LF asymmetry (polarisation). These observed features are due to the resonant transfer of polarisation (state mixing), driven by the dipole–dipole interaction between the muon and nucleus.

An example simulation of the energy levels and associated measured polarisation for QLCRs of a muon is included in Figure 5.14.

As an example, the strengths of the dipole (D_i) and quadrupole (Q_i) interactions between the muon and the ith nucleus, assuming axial symmetry about the muon–nucleus direction ($\hat{z}'$), can be expressed in the form [15,42]

$$D_i = \frac{\mu_0}{4\pi}\frac{h\tilde{\gamma}_\mu\tilde{\gamma}_{i,n}}{r_i^3} \quad \text{and} \quad Q_i = \frac{3V_{i,z'z'}eq_i}{4J_i(2J_i-1)} \tag{5.54}$$

where h is Planck's constant, $V_{i,z'z'}$ is the electric field gradient, and eq_i is the nuclear quadrupolar moment for the ith nucleus, which are well known. The dipolar term scales inversely with the cube of the distance from the ith nucleus, so typically only the nearest neighbours need to be considered.

For additional detail on this technique, see [43].

QLCR measurements can be used to identify the nearby nucleus and help determine the electric-field gradient. When used with decoupling curves or other

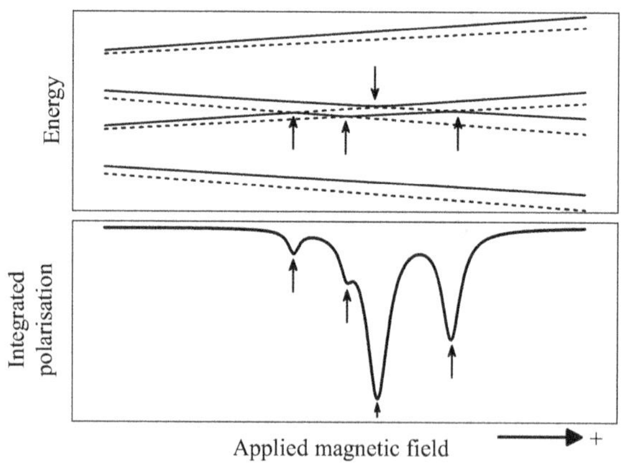

Figure 5.14 Quadrupolar level-crossing resonance Breit–Rabi diagram (top) and integrated polarisation (bottom) for a system involving a muon and spin 3/2 nucleus where the principle axis of the electric field gradient is oriented at 54.7° from the applied field and initial μ^+ spin direction. From top to bottom, the energy levels are $|+3/2,+1/2\rangle$, $|-3/2,+1/2\rangle$, $|+1/2,+1/2\rangle$, $|-1/2,+1/2\rangle$, $|+3/2,-1/2\rangle$, $|-3/2,-1/2\rangle$, $|+1/2,-1/2\rangle$ and $|-1/2,-1/2\rangle$

techniques, QLCR can also contribute to determining relative locations and establish the physical structure of the muonium states that are present.

5.2.7 Low-energy muons (LEM or LE-μSR)

The low-energy muon technique (LEM or LE-μSR) uses principles similar to that of the bulk μSR techniques already discussed, with the added ability to adjust the implantation depth (by varying incoming muon energy) between a fraction of a nanometre to several hundred nanometres instead of the tenths of millimetre implantation depths conventional μSR uses. This technique is currently unique to the Swiss Muon Source, Paul Scherrer Institut (Switzerland), see [44–46].

Some examples of work that have been utilising LE-μSR includes looking at the energy dependence of muonium states of silicon and insulators [47], thin film and near-surface states in germanium [48,49] and looking at Mu–carrier interactions as a function of depth near semiconductor interfaces to map out depletion regions [50,51].

5.2.8 Illumination and muon spectroscopy of excited states

Another tuneable control parameter for use in semiconductors is carrier concentration. Often referred to as photo-μSR, incident light on a sample undergoing otherwise routine μSR measurements can be used to probe an additional dimension of muon–carrier interactions. Historically, xenon flash lamps were the first used due to their flat power spectrum through the wavelengths relevant for the materials under study. Some of the first reported systematic measurements were on Si and Ge where TF and LF measurements were performed with and without photoexcitation to study the effect of illumination on Mu centres, including state formation and dynamics [52,53].

A technique that involves a tuned laser as the source for photoexcitation is being developed at the STFC-ISIS facility [54]. This particular system combines the high-field LF spectrometer (HIFI) with a wavelength-tuneable high-power pulsed laser system to allow for light-pump muon-probe experiments under high longitudinal magnetic fields. The pulse of light can be set to arrive at any time before, during or after the muons arrive at the sample. One example of how this has already been useful is on measuring excess carrier lifetimes in bulk Si [55]. That is, this novel technique has been able to directly access the bulk carrier lifetimes, recombination kinetics and probe processes, such as charge-state cycles, as a function of carrier injection level and temperature. The range of recombination times over which this technique can probe is from ~50 ns to over 20 ms. This particular laser system allows for much finer time resolution capabilities for the pump, as compared to previous uses with flash lamps.

An example of other ongoing efforts utilising this technique includes an attempt to directly measure Mu^0 donor and acceptor levels in 6H-SiC by matching the level with incident photon energy (effort led by J.S. Lord at STFC-ISIS). Another example involves an attempt to investigate the Mu^0 excited states in rutile TiO_2, where, based on reports of hydrogen-based IR excitation measurements, the incident light is expected to induce a transition to an excited Mu^0 state, such as a vibrational mode or other dynamics (effort led by P.W. Mengyan at NMU/TTU).

5.3 Putting muons to use in semiconductors

Muonium results provide an excellent qualitative and reasonable quantitative estimate for the behaviour and properties of isolated hydrogen impurity centres in semiconductors. Muons can also probe charge and magnetic interactions with the host on length and frequency scales that are not available by other means. It is, of course, because of the charge, gyromagnetic ratio and other similar properties to hydrogen (see Table 5.1) that the muon is so useful in this way.

General results from muon-based semiconductor experiments can be thought of as being part of three groups: (1) identification and characterisation of the muonium centre that serves as a direct analogue to hydrogen impurity centres; (2) muonium functioning as a dopant allowing for donor and acceptor levels, comparable to those of hydrogen, to be measured and (3) a general probe of charge and magnetic interactions within the host. The overall types of studies are of 'electronic structure' (e.g. surrounding electronic density, muon stopping site within the crystal and distribution of neighbouring nuclei) and 'dynamics' (e.g. muon diffusion, muon tunnelling and charge exchange).

This section combines the core information on the μSR techniques already presented with some general models to explore the implementation of these techniques in studying semiconductors.

Muon studies in semiconductors have been primarily centred on monitoring the implanted muon to provide insight to the very similar properties and behaviour of isolated hydrogen in materials. The detection of other defects or impurities in semiconductors is usually accomplished through the interactions with diamagnetic or paramagnetic muonium. A consequence of the short muon lifetime (2.2 μs) is that direct interactions between a muon and charged impurity centre require a combination of sufficiently high impurity concentration and high enough muonium mobility. That is, if the muonium (muon) does not diffuse quickly enough to encounter a defect centre before the muon decays, then the defect is undetected. If a defect centre modifies the electrical or magnetic properties of the host (e.g. charge distribution, carrier concentration and magnetic order), the effects the impurity has on the host may be detected by the muon through carrier or field interactions, even if the muon does not have a direct interaction with the defect centre itself.

Completing a full characterisation of the muonium states and behaviour provides insight into the hydrogen-like properties as well as interactions with other impurities within the host. Such a characterisation includes measuring characteristic local fields (e.g. dipolar and hyperfine fields for the diamagnetic and neutral muonium centres, respectively) and is accomplished with a combination of TF, LF, ZF and RF measurements. Refined structural details of the local environment come out of measurements such as ALCR and QLCR that probe interactions with neighbours. Dynamics in which the state participates may include motion (e.g. muonium vibrational states, tunnelling, diffusion) or fluctuations in the local field (e.g. muonium charge cycles, fluctuating neighbouring nuclei).

5.3.1 Muonium states

Similar to hydrogen, muonium can exist in a neutral paramagnetic ($Mu^0 = \mu^+e^-$) state, or as a diamagnetic state with positive (Mu^+) or negative (Mu^-) charge. In semiconductors with tetrahedral coordination, the two interstitial sites where stable or metastable states have been observed are labelled as the cage-centred tetrahedral (T) site and BC site (Figure 5.15). The T-site is in the large open cage-like region in the Zincblende lattice and typically supports the neutral (Mu_T^0) and negative muonium (Mu_T^-) states. Mu_T^0 has an atomic 1s wave function that is slightly constricted with an isotropic hyperfine parameter that is smaller than the vacuum value. For example, the hyperfine constants for Mu_T^0 in Si, Ge and diamond are 45%, 60% and 80% of the free atom value (4,463 MHz). The lowest electronic charge density is found in the T-site and therefore, it tends to be a stable location for Mu^-. The Mu^- state in the T-site is much less mobile than the Mu_T^0 or Mu^+ since the extra electron further constricts the centre. The BC site is centred in a stretched bond with axial symmetry about the bond direction. Neutral (Mu_{BC}^0) and positive (Mu_{BC}^+) charge states are typically supported in this site. Hyperfine parameters for the Mu_{BC}^0 tend to be much smaller than the T-site. For example, Si, Ge and diamond have isotropic components of $A_{iso} = 67$, 96 and 206 MHz, respectively, and dipolar components of $D = -51$, -69 and -374 MHz, respectively (i.e. (5.39) and (5.40)). All of these are on the order of a few per cent of the muonium vacuum value.

The muonium state notation used here indicates the charge state in superscript and site in subscript. For historical reasons relating to initial observations of each

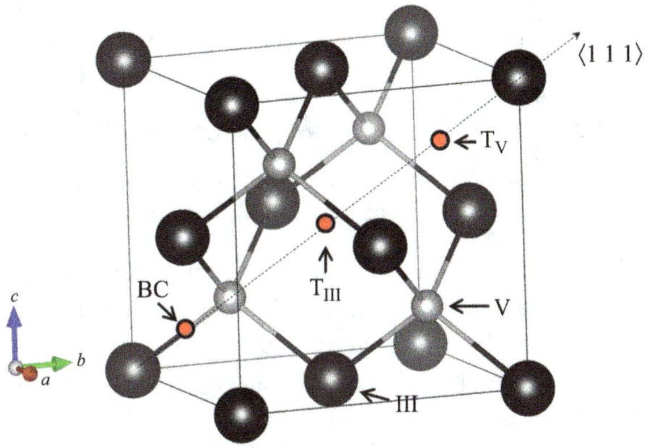

Figure 5.15 Zincblende structure with one bond-centred (BC) muonium position indicated. T_{III} and T_V are the muonium T-sites associated with group-III and group-V neighbours, respectively

neutral state, early literature refers to Mu_T^0 as 'normal muonium' denoted as Mu and Mu_{BC}^0 as 'anomalous muonium' denoted as Mu^* (first reported in 1978 [56]).

For the diamond structure, theoretical models place the lowest energy site within the cage region to be slightly offset from the geometric centre (T-site) at what is labelled as an antibonding (AB) site. The Mu_T^0 centres tend to be quite mobile and so it is possible that what has been identified as Mu_T^0 in experiments is actually an average of the symmetrically placed but offset AB sites.

There are two T-sites in the Zincblende III–V compounds. One has four group-III neighbours and the other is surrounded by group-V atoms. The one with group-III neighbours is the more stable position for Mu^- because of the partial ionic bonding characteristics within these compounds. As found in the diamond structure, Mu^0 and Mu^+ are generally most stable in the BC position, although in some cases, Mu_T^+ (near group-V neighbours) is a nearly static centre.

Sites in other common materials are discussed in Section 5.4.

5.3.2 *Identifying muonium centres*

Distinct features such as frequencies in TF measurements (Figure 5.16), a characteristic shape in LF decoupling curves (Figure 5.11), resonance peaks in RF

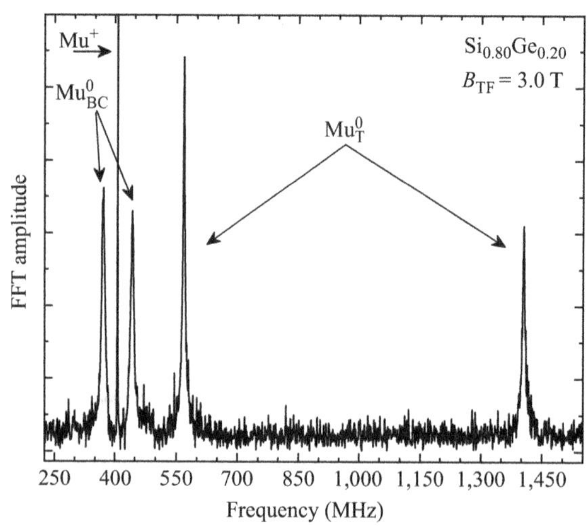

Figure 5.16 FFT of high-TF spectra showing a diamagnetic Mu (Mu^+), bond-centred (Mu_{BC}^0) and T-site (Mu_T^0) muonium in $Si_{0.80}Ge_{0.20}$ at 55 K and under a 3.0 T field applied perpendicular to the initial muon spin. So that both of the Mu_T^0 lines are visible, a combination of FFTs from the forward and backward precession directions (determined by changing the order of the 'left' and 'right' detectors in the four-detector configuration) has been used where the lower frequency Mu_T^0 line is actually −567 MHz relative to the others. The hyperfine frequency for Mu_T^0 at this temperature is therefore 1,972 ± 1 MHz

spectra and a set of frequencies (i.e. muonium triplet) in either low TF or ZF measurements all provide a means to identify the different species, basic characteristics and locations of muonium.

Although TF muon spin rotation measurements are only sensitive to states that are promptly formed, these measurements often provide the best starting point for identifying muonium centres in a material. By comparing the total measured muonium asymmetry from a sample to that of a known standard (e.g. silver), one can establish what fraction of the implanted muons are observable under those conditions. Diamagnetic muonium (Mu^+ or Mu^-) precesses at a frequency scaled by the gyromagnetic ratio of the muon. In principle, there is a slight difference in effective field between Mu^+ and Mu^- due to the two additional electrons. However, in practice, the two states are not separable based on the frequency alone. Instead, measurements relating to the relative mobility or interactions with neighbours are required to determine the charge state of the diamagnetic centre. One of the best and most cited examples such as separation is demonstrated in GaAs in [57].

In high TFs (i.e. $\omega_e \gg A_\mu$), promptly formed isotropic Mu^0 is observed as a set of two lines that are split about the diamagnetic (Larmor) frequency, where the measured frequencies are ω_{12} and ω_{34} in (5.36) and correspond to the muonium $|1\rangle \leftrightarrow |2\rangle$ and $|3\rangle \leftrightarrow |4\rangle$ transitions (see Figure 5.10), respectively. The electron Zeeman splitting is the largest term in the spin Hamiltonian. The two levels are electron spin up and spin down. The muon Zeeman interaction introduces an additional splitting of the electron levels dependent on whether or not the muon and electron spins are aligned parallel or antiparallel. The transition rate between these two states is the hyperfine constant (A_μ) and shows spin in the spin-polarisation spectra as oscillating signals as $\omega_{12} = \omega_{e\uparrow} = \omega_\mu - A_\mu/2$ and $\omega_{34} = \omega_{e\downarrow} = \omega_\mu + A_\mu/2$. In lower fields with sufficient timing resolution, the ω_{12} and ω_{23} frequencies are visible and often labelled as the muonium triplet (see Figure 5.10). Under low LF or ZF conditions and in an apparatus with sufficient timing resolution, the ω_{24} transition (between the mixed states, Figure 5.10) appears as an oscillation governed by the hyperfine frequency.

In the case of Mu^0_{BC}, the anisotropy results in the frequency of the paramagnetic muonium lines (transition frequencies between $|1\rangle \leftrightarrow |2\rangle$ and $|3\rangle \leftrightarrow |4\rangle$) having a dependence on the relative orientation of the symmetry axis and magnetic field. An example is shown for the high field limit in Figure 5.17. Full expressions and discussion of the details relating to calculations for Mu^0_{BC} amplitudes, and frequencies can be found in [12]. In high field, the precession frequencies can be approximated by

$$\omega_{12} + \omega_{34} = \left|A_\parallel \cos^2(\theta) + A_\perp \sin^2(\theta)\right| = \left|A_{\rm iso} + \frac{D}{2}\left(3\cos^2(\theta) - 1\right)\right| \tag{5.55}$$

where $A_\parallel$, $A_\perp$, $A_{\rm iso}$ and D are the parallel, perpendicular, isotropic and dipolar components of the hyperfine constant. θ is the angle between the symmetry axis and the applied field.

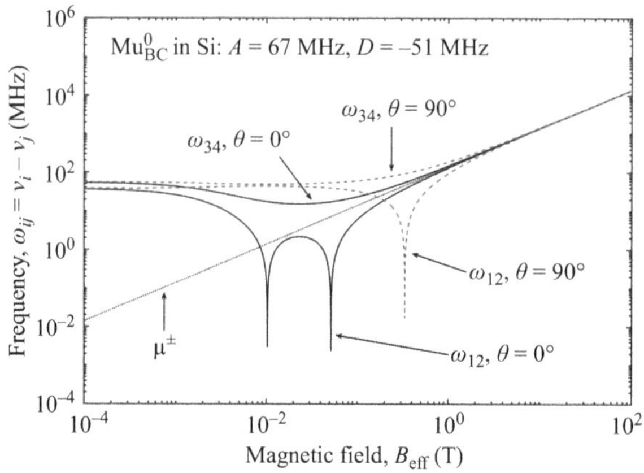

Figure 5.17 Field dependency of muonium transition frequencies for Mu^0_{BC} with the magnetic field applied parallel (0°) and perpendicular (90°) to the symmetry axis. This simulation uses parameters for Si where the isotropic and dipolar components of the hyperfine constant are 67 and −51 MHz, respectively

An example of a system in which Mu^+, Mu^0_T and Mu^0_{BC} are simultaneously observed in a high TF measurement is shown in Figure 5.16.

In materials where quadrupolar interactions of neighbouring nuclei are significant, the field dependence of the TF Gaussian rate constant (for the relevant state) can be used to characterise the interaction of the neighbour in the electric-field gradient induced by the Mu impurity. However, the QLCR approach is a much more common way to do such a study.

Characterising static Mu^0 centres that are not promptly formed, and therefore not observable in TF, requires the use of RF or LF measurements that drive a precession or otherwise do not have the phase coherence requirement that traditional TF does.

Although it can be more time consuming and not as direct as TF spectroscopic methods, using LF to measure properties of isotropic and anisotropic muonium (e.g. Mu^0_T and Mu^0_{BC}) or the $Mu^{\pm,0}$ coupling to the local environment (e.g. QLCR) can be a rich source of information on static and dynamic centres. Without the phase coherence requirement, LF techniques are more sensitive to the final state of a process or reaction sequence and can measure states that are not promptly formed (unlike TF). The time-dependent spin-polarisation function for a static and isotropic Mu^0 centre where the field is aligned with the initial spin direction of the muon is given by (5.35). It is often the case that looking at the initial relaxing asymmetry (i.e. $P(t \to 0)$) or total integrated asymmetry as a function of applied LF can be used very effectively to characterise both isotropic and anisotropic centres. An approximate general expression for an axially symmetric centre is provided in (5.46). Some examples of decoupling (repolarisation) curves for isotropic and axially symmetric centres (Mu^0_T and Mu^0_{BC}) are shown in Figure 5.11.

In practice, a complete characterisation of static centres often uses a combination of TF, LF and RF methods.

Beyond the traditionally discussed Mu^0 states where the electron is centred on the muon, some materials include a state where the muon forms an Mu^0-like complex where diamagnetic Mu is attached to an ion, such as O–Mu (cf. O–H), and then interacts with a neighbouring electron that is centred on another atom. For example, in rutile TiO_2, the muon attaches to an oxygen and is then coupled to an electron that is centred on a neighbouring Ti^{3+} ion [22]. The characterisation measurements already discussed can be used to study these states so long as proper care is taken when interpreting the results.

In metals, conduction electrons screen the Coulomb field of the muon so there is no direct binding of the electron. Without a direct binding of the muon and electron, what has historically been referred to as muonium should not be observed. However, a hyperfine interaction can still be observed, even in situations where Coulomb binding is ineffective, provided that another mechanism both localises the muon near the electron (or electron near the muon) and prevents electron spin-exchange. One example of this process occurs in the metallic $Cd_2Re_2O_7$ where a hyperfine interaction is observed, which suggests muonium. In this case, the muon is bound to a spin polaron – a nanoscale ferromagnetic droplet with giant spin, S, and electric charge e^- in which the binding electron has its spin locked to S but still couples to the muon spin through a hyperfine interaction (e.g. see [58]).

In ZF, the width of the field distribution at the muon site can be characterised by using (5.18), (5.19), (5.23) or (5.24). In ZF, the muonium hyperfine constant may show as an oscillation in the measured signal, provided the timing resolution of the apparatus is sufficiently high or the hyperfine frequency is sufficiently low.

Identifying and characterising static states is an essential step in developing an understanding of muonium behaviour in a material. In semiconductors, low-temperature states are often static and serve as an excellent starting point for studying a system.

5.3.3 Processes and dynamics involving muonium centres

In addition to characterising the local physical and electronic structure of isolated H (cf. Mu) impurity centres in a host, μSR can provide insight into the stability of these centres and their interactions with carriers, as well as reactions and processes in which isolated hydrogen may participate. In the late 1980s and early 1990s, work on elemental semiconductors (mainly Si and Ge) began to shift from identification of static centres to investigating the active processes sensed by the muon [38,40]. Since then, μSR-based semiconductor studies have continued with more complex materials and looking at all aspects of muonium in materials.

In practice, identifying and characterising the muonium (cf. isolated hydrogen) impurity centres in systems generally involves using a combination of data from the static states and dynamic states to assemble a model that describes the overall behaviour within a material. Section 5.3.2 focused on identifying and characterising static centres and states. This section highlights some of the main known dynamic processes involving muonium centres. Combining information from these sections will provide a solid basis for identifying and characterising isolated hydrogen impurities by way of μSR. While there are some good questions and important work relating to the details associated with the process by which the electron and muon interact during the muonium formation process (see, e.g. [23,25,26]), this section will focus on the overall dynamics associated with the depolarisation of the muon from site changes, ionisation and charge- or spin-exchange processes.

The muon senses dynamics as changes in the effective field that contribute to the depolarisation or dephasing of the muon. Possible sources of fluctuations in the effective field originate from processes such as the muon itself moving throughout the crystal, fluctuations in the neighbouring nuclei or interactions with charges (e.g. capture or release of or by electrons). By varying control parameters (e.g. temperature, magnetic field and carrier concentration), then analysing the spectra and the overall changes throughout each, the dominant processes responsible for the fluctuating fields can often be identified through characteristic features in the line shapes, relaxation rates and trade-offs between individual component asymmetries.

Within the scope of identifying and characterising muonium (cf. hydrogen) in semiconductors, the main processes involve one-way transitions from paramagnetic to diamagnetic muonium that have been very useful in establishing the donor and acceptor levels with respect to a host's bandgap, cyclic processes involving charge-exchange or spin-exchange, and diffusion of the Mu^0 or $Mu^{\pm}$ centres.

5.3.3.1 Transitions between neutral and charged muonium

First, consider the straightforward case of thermal ionisation of neutral muonium involving a static Mu^0 centre going to either an Mu^+ by releasing the electron or Mu^- by activated hole capture (capture electron from valence band). The easiest way to observe this particular transition is through TF-μSR or RF-μSR measurements where the Mu^0 and $Mu^{\pm}$ spectroscopic signatures are clearly visible (e.g. Figure 5.16) and a trade-off in asymmetry between the paramagnetic and diamagnetic components can be followed as a function of temperature. Recall that the amplitude corresponds to the fractional population of each state. Here, the time-domain data would be appropriately fit with an oscillating component for each visible muonium frequency in a RRF. A direct Mu_T^0 to Mu^+ one-way transition could be modelled by assuming an Arrhenius expression for carrier density and defining an ionisation rate λ_i, given by

$$\lambda_i = \sigma_i v_{e,h} \rho = \lambda_i^{(0)} \exp\left(-\frac{E_a}{k_B T}\right) \tag{5.56}$$

where σ_i is the ionisation cross section; $v_{e,h} = \sqrt{3k_B T/m_{e,h}^*}$ is the thermal velocity of the electron or hole, respectively; ρ is the carrier density at the relevant band edge; $\lambda_i^{(0)}$ is a vibrational prefactor; the temperature of the system is T; k_B is the Boltzmann constant; and E_a is the activation energy for the process. Folding this together with the appropriate weighting by muon lifetime τ_μ, expressions for the diamagnetic $a_\pm$ and paramagnetic a_0 amplitudes can be written as (see, e.g. [59])

$$a_\pm(T) = a'_\pm + a'_0 \frac{\tau_\mu^{-1}}{\lambda_i + \tau_\mu^{-1}} = a'_\pm + a'_0 \frac{f_i}{f_i + \exp(E_a/k_B T)} \tag{5.57}$$

and

$$a_0(T) = a'_0 - a'_0 \frac{\tau_\mu^{-1}}{\lambda_i + \tau_\mu^{-1}} = a'_0 - a'_0 \frac{f_i}{f_i + \exp(E_a/k_B T)} \tag{5.58}$$

where a'_0 and $a'_\pm$ are the initial ($T \rightarrow 0$ K) TF amplitudes of the paramagnetic and diamagnetic states, respectively. In this idealised example, the only two states present throughout the measured range are an Mu^0 at low temperature that converts to an Mu^+ at higher temperature. From a measurement like this, a muonium donor level could then be directly inferred from fits to the amplitude variation to extract the activation energy. Donor and acceptor levels in many materials have been inferred from such fits or models that are based on this basic process in TF measurements.

Another situation that has been the subject of studies involves active conversion between paramagnetic (Mu^0) and diamagnetic ($Mu^\pm$) muonium states. These processes are best characterised with an LF configuration. The two common possibilities for these transitions involve either a cyclic charge- or spin-exchange mechanism. The net observed effect from either mechanism amounts to a combination of Mu^0 to $Mu^\pm$ and $Mu^\pm$ to Mu^0 transitions. Charge-exchange involves an actual change in the muonium charge state by capture or release of a charge that causes variations in the effective field. In the spin-exchange process, cycles occur between a state where the muon and electron have their spins aligned parallel and another where the muon and electron spins are antiparallel. When the muon and electron spins are parallel, the system behaves like $Mu^\pm$ with a polarisation that is independent of time. When the muon and electron spins are aligned antiparallel, the resulting behaviour is similar to Mu^0 with non-oscillating and oscillating components (i.e. (5.35) for an isotropic centre).

In general, these spin-exchange and charge-exchange processes in LF result in a Lorentzian line shape and are fit with $P(t) \sim p_j(t)\exp(-t/T_{1,j})$, where p_j and $1/T_{1,j}$ are the asymmetry and measured relaxation (depolarisation) rates associated with state j, respectively. In this two-state charge-cycle model, the two states are Mu^0 and $Mu^\pm$. However, since the polarisation of the $Mu^\pm$ state is preserved along the direction of the applied LF, the polarisation is time independent, so in this case $1/T_{1,\pm} = 0$. Therefore, muon depolarisation only occurs in the Mu^0 state. In the Mu^0 state, the random electron spin direction is effectively transferred to the muon by way of the hyperfine interaction, which causes the loss of up to half of the remaining polarisation during each cycle. Therefore, the depolarisation rate is ultimately governed by how the hyperfine period compares to the lifetime of the Mu^0 state.

For simplicity, first consider an isotropic Mu^0 centre. Making no assumption about transitions out of each state, the LF dependence on the depolarisation rate ($1/T_1$) measured at a fixed temperature can be approximately modelled [38,60,61] as

$$\frac{1}{T_1} \approx \frac{1}{2}\left(\frac{\lambda_0\lambda_\pm}{\lambda_0+\lambda_\pm}\right)\left(\frac{\omega_0^2}{\lambda_0^2+\omega_{24}^2}\right) \tag{5.59}$$

where λ_0 and $\lambda_\pm$ are transition rates for $Mu^0 \rightarrow Mu^\pm$ and $Mu^\pm \rightarrow Mu^0$, respectively; $\omega_0 = 2\pi A_\mu$ with A_μ being the hyperfine constant; the muonium transition frequency between the two mixed states ($|2\rangle \leftrightarrow |4\rangle$) $\omega_{24} = \omega_0\sqrt{1+x^2}$; $x = B/B_0$ and $B_0 = A_\mu/(\gamma_e + \gamma_\mu)$. While not explicitly included, temperature dependences for this process are built into the transition rates. A brief discussion on the claimed exact expression of which (5.59) is the first term in a series expansion that is found in Section 2.8 of [32].

It is convenient to work with these cycles in either the fast or slow exchange limits (with respect to the hyperfine constant). A crossover from the fast to slow charge-cycle regime occurs when the transition frequency matches the ω_{24} rate. Recall that ω_{24} scales with externally applied field.

The so-called fast limit or rapid cycle regime describes the situation where the Mu^0 ionises before the muon polarisation can undergo one oscillation. That is to say that the Mu^0 lifetime is much smaller than the period of hyperfine oscillation ($2\pi/\omega_{24}$) or that the transition rate is much faster than the hyperfine oscillation frequency ($\lambda_0 \gg \omega_{24}$). This will be the case at sufficiently low fields. In this regime, the fractional loss of polarisation per charge exchange cycle is much smaller than the amplitude of the oscillating component in the Mu^0 polarisation function (i.e. $a_{24} = 1/(2(1+x^2))$ from (5.35)). The primary observed result is that the measured asymmetry, and therefore $1/T_1$, is independent of applied field while in this regime.

The so-called slow limit or slow cycle regime describes the situation where the mean Mu^0 lifetime is much greater than the period of hyperfine oscillation ($2\pi/\omega_{24}$) or equivalently stated as the transition rate is much smaller than the hyperfine oscillation frequency ($\lambda_0 \ll \omega_{24}$). While in the slow regime, the muon polarisation may undergo several oscillations before Mu^0 transitions to either diamagnetic state ($Mu^\pm$), and therefore, the muon may experience a fractional loss up to the amplitude of the oscillating component (i.e. $a_{24} = 1/(2(1+x^2))$ from (5.35)) each time it is in the Mu^0 state. In the very slow regime, the fractional loss with magnitude a_{24} occurs for every time the Mu^0 state is occupied. The primary observed behaviour is that since $1/T_1$ is proportional to the oscillating ω_{24} component, at large fields $1/T_1$ goes as $1/B_{ext}^2$.

Although the fundamental process responsible for change in states is different between the charge- and spin-exchange models and since the interactions are both qualitatively similar (effective Mu^0 vs $Mu^\pm$ behaviour), the resulting $1/T_1$ dependencies on the applied LF are both well approximated by the same model.

The related expression for anisotropic Mu^0 (e.g. Mu^0_{BC}) becomes much more complex; the full version is discussed in [30], but in the rapid cycle regime is approximately equivalent to the result for exchanges involving isotropic Mu^0. In the slow cycle regime, the LF depolarisation rate $(1/T_1)$ shows a sharp peak at the external field (B_{ext}) where the effective field at the muon site becomes fully transverse causing the visible LF polarisation to reduce at a maximum rate; that is,

$$B_{ext} = \frac{1}{2\gamma_\mu}\left(A_{\parallel}\cos^2(\theta) + A_{\perp}\sin^2(\theta)\right) \tag{5.60}$$

where θ is the angle between the symmetry axis and B_{ext} (as defined for Mu^0_{BC}).

In practice, the main approach members of our collaboration tend to follow when working with these data starts by fitting the individual spectra with appropriate polarisation $P(t)$ functions to determine $1/T_1$ for each temperature and field. Next, we systematically fit the LF-dependent $1/T_1$ rates at each measured temperature using (5.59) to extract a rough measure of A_μ, λ_0 and $\lambda_\pm$. An order of magnitude measure of A_μ provides a rough characterisation of the type of centre (Mu^0_T cf. Mu^0_{BC}). Once established, global fits can then be completed with the set of field-dependent curves putting an appropriate temperature-dependent constraint on A_μ (see, e.g. [12,38]).

The temperature dependencies involved with these transitions are folded into the transition rates. A very thorough discussion of some of this modelling and related processes is provided in [38–40]. As an example, ionisations can be modelled as done in (5.56); carrier capture can be modelled as

$$\lambda_c = n v \sigma_c \tag{5.61}$$

where n and v are the electron and hole densities and σ_c is the capture cross section, noting that there may be a temperature dependence to any of these parameters.

Last of this series, a combined site change and carrier capture can be written as

$$\lambda_{ac} = n v \kappa_{ac} \exp\left(\frac{E_{ac}}{k_B T}\right) \tag{5.62}$$

where κ_{ac} is an effective total process cross section for the entire transition.

5.3.3.2 Muonium in motion

One of the more straightforward dynamic processes relates to the motion of charged (diamagnetic) muonium (Mu^+ or Mu^-), where the relaxation function $G(t)$ is governed by the average field width at a site and rate of field fluctuation as sensed by the muon (e.g. see Table 5.2 and surrounding discussion). From a single time-domain spectrum, it is not possible to distinguish the difference between the two possible sources for fluctuations, either (1) a static muon sensing variations in field from fluctuating neighbouring nuclei or (2) static nuclear moments with the muon sampling several inequivalent sites. However, with some modelling, intuition and a set of complementary measurements, often times one situation can be determined to be much more likely than the other. The most common way to study diffusion of charged muonium (cf. diffusion of ionic H) has been using ZF measurements over a wide temperature range to map out the muon hop rate (ν in Table 5.2, and relevant equations). An example of the resulting hop rates from measurements on $CdGeAs_2$ is shown in Figure 5.18, along with examples of the raw data showing the characteristic dynamic Kubo–Toyabe shape.

For these types of measurements, it is common to collect both ZF and low TF data during the same experiment time period so that the total diamagnetic fraction in the material can be checked. These low TF data also provide insight into existing states and a proper α calibration (e.g. (5.10)) so that the baseline can be set. The data are then typically characterised using an appropriate set of ZF functions to construct a general model including a temperature dependence that is fit to the data and the fractional amplitude (state population), field width and fluctuation rate (hop rate) for each state. The temperature dependence of the hop rate is often fit with a set of Arrhenius functions where the prefactors and energies provide insight into the type of motion or process that is active for each state (e.g. a vibrational mode, tunnelling, local motion and global diffusion). Mu^+ and Mu^- are spectroscopically identical (in practice); however, they show very different motional properties (e.g. [57]) where Mu^+ moves much easier than Mu^-. These differences are apparent in the resulting hop rates and related prefactors and energies.

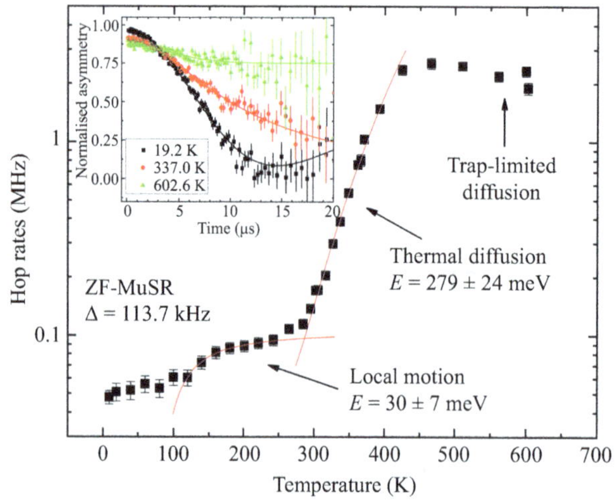

Figure 5.18 Diamagnetic Mu hop rates (via dynamic Kubo–Toyabe) in $CdGeAs_2$ showing three regimes of motion. The inset shows raw data and fits at 19 K, 337 K and 602 K where the Mu^+ is static, tunnelling between nearby T-sites and undergoing thermally activated site-to-site motion, respectively. A Δ of 113.7 kHz was used for the entire range [104]

Dynamics of diamagnetic muonium centres can also be tracked in TF by tracking the evolution of the relaxation function with temperature. The relaxation rate $\left(T_2^{-1}\right)$ is directly proportional to the fluctuation rate when the fluctuation rate is less than the muon Larmor frequency and inversely proportional when the fluctuation rate is greater than the Larmor frequency. A peak in $\left(T_2^{-1}\right)$ occurs where the fluctuation rate matches the precession frequency. The very fast fluctuation limit results in what is traditionally known as motional narrowing.

In the case of Mu_T^0 moving between sites (or effective field fluctuations at similar sites), the process of measuring the hop rates (fluctuation rates) is a little more involved. During the Mu_T^0 site-change process, the unpaired electron experiences randomly fluctuating nuclear hyperfine interactions through its successive interactions with the randomly oriented neighbouring nuclei, leading to electron spin relaxation and, through the muon–electron hyperfine interaction, muon spin relaxation. In modelling this, a fluctuating local field (δB_loc) can be accounted for by replacing the nuclear hyperfine interaction term in the Hamiltonian with the effective field nuclear hyperfine interaction field strength δ_n and a correlation time τ_c (hop rate $= 1/\tau_c$). These dynamics are best observed in the LF configuration. In the LF configuration for an isotropic Mu^0 centre (e.g. Mu_T^0) and at high fields, the spin–lattice relaxation $(1/T_1)$ is dominated by the ω_{12} transition, leading to an exponential decay of the muon polarisation function. With this time dependence added and omitting the oscillating component that is not observed on typical LF instruments, (5.35) can now be written as

$$P^{\theta=0^\circ}(x,t) = \left(\frac{1+2x^2}{2(1+x^2)}\right)\exp\left(-\frac{t}{T_1}\right) \tag{5.63}$$

With some work, the spin–lattice relaxation as a function of applied field can be characterised with (see, e.g. [62,63])

$$\frac{1}{T_1} \approx \left(1-\frac{x}{\sqrt{1+x^2}}\right)\frac{\delta_n^2\tau_c}{1+\omega_{12}^2\tau_c^2} \tag{5.64}$$

where $\omega_{12} = \pi A_\mu(1+\omega_-/\omega_+)x - \sqrt{1+x^2}$ and is the angular frequency for the transition between the muonium states $|1\rangle$ and $|2\rangle$ (see Figure 5.10). Here, $\omega_\pm \equiv (\gamma_\mathrm{e} \pm \gamma_\mu)B_\mathrm{ext}$ and $x = B/B_0$ with $B_0 = A_\mu/(\gamma_\mathrm{e}+\gamma_\mu)$, as before.

In practice, a series of TD LF measurements is made where the applied external field B_ext is varied at several different fixed temperatures. The time spectra are fit with an appropriate model based on (5.63) to extract the field and temperature dependences of the spin–lattice relaxation rates $(1/T_1)$. Next, $1/T_1$ are plotted as a function of B_ext for each fixed temperature where each LF scan is fit with (5.64) to extract the parameters of interest (e.g. the hop rate $1/\tau_c$, hyperfine constant). The temperature dependences of these parameters are useful in characterising the active states and processes. Once a full model is established, the data are often refit with the complete model to refine the parameters.

In general, a decrease in hop rate $(1/\tau_c)$ with increase in temperature (or increase in rate with decrease in temperature) is a strong indicator of quantum tunnelling that is controlled by phonon scattering. This regime is often characterised with an inverse power law temperature dependence to the hop rate $1/\tau_c = CT^{-n}$ where the power n ranges from ~3 to ~7, T is the temperature, and C is a proportionality constant (see, e.g. [64]). In some materials, a temperature-independent hop rate at low temperature suggests coherent band-like motion with a mean free path that is greater than the lattice constant and a k_BT larger than the muonium bandwidth (see, e.g. [15]). A minimum in the hop rate indicates a crossover into a thermally activated regime that is Arrhenius in character or, in some cases, trapping. Figure 5.19 is an example where Mu^0_T hop rates in $ZnGeP_2$ were measured using this method [65].

Equation (5.64) can also be used in situations where the local-field fluctuation component is due to variations in the local arrangement of atoms surrounding the Mu^0 centre (e.g. [66]). That is, instead of a time-dependent variation in the nuclear hyperfine interaction, the same effect of a fluctuating nuclear hyperfine interaction term is observed for a set of similar sites with differences in the local field for each.

5.3.3.3 Two-state bidirectional transition

To introduce this model, consider a mobile centre that alternates between a mobile and trapped state with depolarisation occurring in both states. Since this situation includes states that are mutually dependent, the model for the time evolution of the muon spin polarisation ($P(t)$) cannot be assembled with a simple weighted linear combination of states. Some specific applications of this approach have been used throughout the years (often referred to as trap and release); however, a generalised approach for this model was recently formalised by Baker [67]. This general approach allows an end user to choose separate polarisation functions that describe

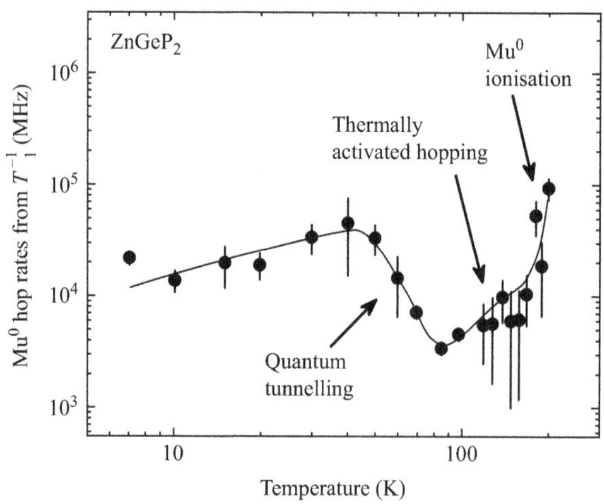

Figure 5.19 Hop rates for Mu^0 in $ZnGeP_2$ extracted from a series several LF scans at several temperatures by fitting the field dependence of the relaxation rate (T_1^{-1}) at each temperature with (5.64). [87,104]

two individual states and then make a straightforward set of substitutions to produce an appropriate model that includes rates for transitions between these states.

As an example, the first application of this generalised model was used in working on ZF data for charged muonium centres in In_2O_3, where the particular two-state model was built using a static Kubo–Toyabe function and dynamic ZF Keren function for the trapped and free states, respectively. This combination ultimately allowed for a comprehensive characterisation of the muonium centres in In_2O_3 and other related oxides (detailed discussion in [67]).

The general form of the polarisation function for this model can be written as

$$P(t) = a_1[e^{-\nu_1 t}P_1(t) + \nu_1 e^{-\nu_2 t}f(t)] + a_2[e^{-\nu_1 t}P_2(t) + \nu_2 e^{-\nu_2 t}f(t)] + \nu_1\nu_2\int_0^t \left[e^{-\nu_2(t-t')}f(t-t')(P_1(t') + P_2(t'))\right] \mathrm{d}t' \tag{5.65}$$

with

$$f(t) = \int_0^t \left[e^{-(\nu_1-\nu_2)t'}P_1(t')P_2(t-t')\right] \mathrm{d}t' \tag{5.66}$$

where P_1 and P_2 are the time-dependent polarisation functions for states 1 and 2; a_1 and a_2 are the fractional amplitudes of states 1 and 2; ν_1 and ν_2 are the inverse lifetimes of states 1 and 2 (equivalently the rate that corresponds to exiting states 1 and 2, respectively). Depending on the relevant process, temperature dependence such as Arrhenius behaviour can be included in the transition rates.

In the example application involving trap and release of centres in In_2O_3 (briefly mentioned in this section, Section 5.4.5 and thoroughly discussed in [67]), the total polarisation for the trap and release states depends on the amplitudes (fractional population) of the free and trapped states, the individual polarisation functions for the two involved states and the rates at which the centre is trapped (captured) and released (detrapping). This model does not impose any conditions or constraints on the relationship between the trapping and release rates.

In general, this two-state bidirectional transition model can be used in combination with other polarisation functions in order to account for other states that may be present but not directly involved in the transitions.

5.3.4 Measuring donor and acceptor levels

The energy associated with a paramagnetic to diamagnetic muonium conversion where there is a direct loss of an electron to the conduction band (0/+) or capture of a second electron from the valence band (0/−), as the two common examples, constitutes the muonium donor and acceptor levels. So far, the most common way to extract the energies associated with these transitions (and hence the 0/+ and 0/− levels) has been through tracking the thermal stability of the observed Mu^0 states. That is, using the relevant μSR techniques to identify and characterise the present Mu states through a broad enough temperature range such that the stable (or metastable) Mu^0 and a final stable $Mu^{\pm}$ state(s) are observed and then analysing the temperature dependences associated with the conversions of these states (see Section 5.3.3.1). This often means sweeping from cryogenic temperatures to above room temperature (in the extreme case of quartz, Mu^0 is stable above 1,200 K).

The most easily observed case occurs when a promptly formed Mu^0 goes through a direct ionisation to a promptly formed Mu^+ ($Mu^0 \rightarrow Mu^+ + e^-$). Here, the initial Mu^0 and final Mu^+ states are promptly formed and therefore observable with high TF hyperfine spectroscopy methods that show a direct trade-off between the shrinking measured asymmetry of the paramagnetic Mu^0 frequencies (satellite ω_{12} and ω_{34} lines) and growth in measured asymmetry of the corresponding diamagnetic Mu^+ frequency. Recall that the measured asymmetry for a particular state indicates the fractional population of that state. One model describing the thermal evolution of the measured asymmetry relating to diamagnetic and paramagnetic states is described by (5.57) and (5.58) and the surrounding text.

In many systems, the transition between a stable or metastable paramagnetic state to a final ionised state includes intermediate processes (e.g. repeated carrier capture and release and motion) which introduce complexity in extracting the donor or acceptor level relative to the conduction or valence band edge, respectively. With diligence in experimentally identifying the active states, processes and developing a model for the complete system, these energies (as well as many others) are still accessible.

Establishing these energy levels for muonium in a wide variety of materials has provided significant contributions to the hydrogen in semiconductor community. Also, quite notably, results of a study that used six different materials (Si, Ge, GaAs, GaP, ZnSe and 6H-SiC) where at least two distinct Mu^0 states are measured strongly support the theoretical prediction of a universal pinning level for H(+/−) [68,69]. The (+/−) level is the point at which the equilibrium charge state changes from positive to negative when crossed by the electron chemical potential (Fermi energy) and is typically about half way between the acceptor (0/−) and donor (0/+) levels. The measured Mu(+/−) level is consistently about 0.5 eV higher than the prediction for H(+/−). While the larger zero-point energy for the muon results in higher individual eigenstates for Mu, these electrically active switching (+/−) levels are actually the differences in energy between the charge states; and therefore, effects from zero-point energy in both cases should be mostly irrelevant. The source of this large difference is still an open question.

There are a few expected differences between Mu and H energies, mainly as a result of the difference in zero-point vibrational states. The Mu is nine times lighter than H, and therefore, in a harmonic model, Mu has a factor of three larger energy. This results in the lowest Mu vibrational level roughly corresponding to the first excited state for H if both Mu and H are in the same potential well. For a donor at the BC site, the additional electron is in an AB orbital, so the bond is weaker for the Mu^0_{BC} as compared to the Mu^+_{BC}. In Si, Ge and GaAs, the Mu level is observed to be ~30–35 meV deeper than the related H levels. In the case of a T-site acceptor, most hydrogen models place the lowest energy sites at four locations where each is slightly offset from the centre and shifted in a direction closer to a neighbour with very low motional barriers between these similar sites and a local maximum at the proper T-site. The heavier hydrogen is likely to localise at one of these sites, while the lighter Mu tends to localise (or at least appear to) at the central T-site.

While these thermal-dependent techniques can measure the levels with respect to a band edge, the absolute placement of the energy levels still depends primarily on the theoretical predictions of band locations.

A very recent effort using laser excitation (instead of thermal excitation) to directly measure the donor and acceptor level position within the band is underway using the laser system on the HIFI spectrometer at the STFC-ISIS facility [54,70]. In addition to positioning these levels within the band at specific temperatures, if successful, this technique may also provide another method for measuring the bandgap energy and studying interactions with carriers.

5.3.5 Effects of carrier concentrations

Interactions with charge carriers is part of what makes muonium and isolated hydrogen so interesting to study. Carrier concentration of a host is, indeed, a tuneable parameter that can provide insight into the prevailing muonium states and processes in a material. With some diligence, the muon–carrier interactions can also provide insight into processes involving the carriers themselves. Carrier concentration can be varied by traditional doping mechanisms or excess carriers can be injected by optical means such as a flash lamp or tuneable laser. Many studies have used or attempted to use means of optical excitation to delineate between positive and negative charged centres and provide additional control over charge cycles to better understand the dominant or underlying processes.

A very recent effort including the new laser system on the HIFI spectrometer at the STFC-ISIS facility [54] has been proven to be a new method to study excess carrier lifetime in bulk silicon [55]. This application demonstrates the capability of the system where depth dependence of the optical excitation and muon implantation can be controlled, and the bulk excess carrier lifetimes can be probed directly. This technique provides access to a range of recombination times (~50 ns–>20 ms), injection levels and temperatures. This technique is currently being developed, and efforts to probe other properties associated with carrier interactions in Si and several other hosts are currently underway.

5.4 Select examples of μSR contributions to the field

Over the last ~40 years, μSR has played a significant role in understanding the role and behaviour of interstitial hydrogen in semiconductors. During this time, many papers, chapters and reviews have been published on the topic. Most recently, a set of very comprehensive and easily available reports on the progress of μSR studies on semiconductors, semimetals and oxide muonics has been authored by Cox [32,36,37,71,72] along with several others in the muonium in semiconductor collaboration that have spent large portions of their careers focused on this topic.

These recent reports, along with the previous reviews, constitute a nice history of semiconductor muonics through the first generation of work that focused primarily on identification and characterisation of muonium centres. The current generation of work seems to be focused on developing a more in-depth understanding of active processes, interactions, dynamics and more complex centres in both well-known and new materials by developing new techniques and finding new ways to extract more information from the older techniques. The groundwork for many of the current studies was laid during the early phases of technique development including the extensive work on materials such as Si, Ge and GaAs. Since the recent reports are readily available, provide a comprehensive review of the first generation of work and are an accurate preview into our current work; this section will provide only a limited selection of major contributions μSR has made to the field and some highlights on selected materials. For more detail on any of these topics, I encourage the reader to pursue these reviews, the references therein and references throughout this chapter.

Some of the first major contributions μSR made to the semiconductor field were in the late 1980s, although some work dates back to 1958 [73]. What may be the first major contribution μSR made to the field appears to be the identification of Mu^0_{BC} (then labelled 'anomalous' or Mu^*) as the ground state centre in silicon. Although first reported in 1978 [74], it was not until about 10 years later that the similarities between the Mu^0_{BC} and *AA9* EPR signals in Si [8–10] identifying the H^0 ground state were realised and therefore established the electronic structure of Mu and H to be identical. Also around this time, nuclear hyperfine splittings demonstrated the BC structure in GaAs. The major efforts during this time period were spent both in developing the techniques and attempting to understand basic muonium behaviour in the elemental compounds (e.g. see [1,13–15,21,75–79]). In the early 1990s, the metastable Mu^0_T centre was linked with the highly mobile transition state for H^0 diffusion in Si [80]. In the mid-1990s, work on charge cycles and site transitions in Si, Ge and GaAs contributed a reminder to the defects in semiconductor community that equilibrium is a dynamic process [38,40,81]. In the late 1990s, clear evidence of a negatively charged ionic state was presented from work in n-type Si [82]. Next came the observation of shallow donor Mu (H) states in CdS and zinc oxide [83,84]. While similar work continued on several other semiconducting compounds, the next big contribution was an experimental confirmation of the predicted pinning of the electric switching levels H(+/−) by μSR measurements on six compounds: Si, Ge, GaAs, GaP, ZnSe and 6H-SiC [68,69]. The prediction was published in 2003 [69] and experimental confirmation published in 2008 [68]. Another contribution came with confirmation of the shallow donor state in transparent conducting oxide electrodes (In_2O_3 and SnO_2) [85]. Around this same time, the first reports for the acceptor levels of interstitial muonium in II–VI semiconductors ZnSe and ZnS appear [59]. In 2015, muon measurements on rutile TiO_2 showed identical configuration and electronic structure to that of hydrogen [22]. This direct comparison confirms the validity of muonium as an experimentally accessible analogue to the hydrogen impurity; in this case, the muonium-like centre is composed of a muon bound to an electron centred on a neighbouring Ti^{3+} (small polaron), opposed to the case of previously observed centres where the electron is centred on the muon itself. There have only been a few cases where direct comparisons between muonium and isolated hydrogen have been available. Next, a new approach to measuring the bulk carrier recombination times in Si by way of laser-based photo-μSR proves to be successful [55], introducing another class of contribution μSR may be able to make to the semiconductor community.

5.4.1 Group-IV

μSR work on group-IV tetrahedrally coordinated compounds such as Si, Ge and allotropes of carbon has been unmatched in providing original results and insight into the behaviour of isolated hydrogen impurities. These materials were among the first to establish the occupied crystallographic sites in the diamond lattice (T and BC), interaction with charge carriers, the link between site and charge states as well as the measurement of donor and acceptor levels; for example, see [38,40]. Similar work, although in much less detail, has been completed on hexagonal 4H- and 6H-SiC. In this section, only Si, Ge, SiGe alloys and SiC are highlighted.

In the diamond structured materials, muonium forms four distinct centres: Mu_T^0, Mu_T^-, Mu_{BC}^0 and Mu_{BC}^+. In Si, Mu_{BC}^0 and Mu_{BC}^+ are the ground and stable states for the neutral and positively charged centres, respectively. At the T-site, the negatively charged centre is stable, and the neutral is only metastable. The mobility for muonium formed in the T-site is quite high since it can easily diffuse through the puckered, hexagonal rings that make up the T-site caged region. Calculations for hydrogen suggest that the local energy minimum for Mu_T is actually slightly offset at the AB sites in the four directions towards each neighbour. However, measurements of the Mu_T^0 show highly isotropic hyperfine interaction, which indicates the atomic-like nature of its wave function, so either the Mu_T^0 is actually at the T-site or it is rapidly moving between the four equivalent AB sites, thereby averaging to appear as if it is at the T-site. Experimentally, the two cases are indistinguishable.

In Si, the ground state for Mu^0 is at the BC site, the stable state for Mu^+ is Mu_{BC}^+ and Mu^- is stable at the T-site. In Ge, theory predicts the two Mu^0 states (Mu_T^0 and Mu_{BC}^0) to have nearly equivalent energies with the slightly lower preference given to Mu_T^0. For both Si and Ge, most of the processes leading to transitions among the four available Mu states have been identified [38,40]. Recent work on Si using laser-based photoexcitation has made additional measurements on these processes and interactions as part of an effort to develop a new technique and to better understand these important processes in Si [55]. Similar work on Ge is underway.

An in-depth study of characteristics of muonium defect in a series of $Si_{1-x}Ge_x$ compounds further explored the defect energy levels as a function of elemental composition and, quite notably, measured the depth of an acceptor level residing in the valence band [7,86]. This ongoing effort, spanning nearly 20 years, has also provided a deeper understanding of active processes in these compounds; see, for example, [66,87].

Muonium sites and charge-state transitions have been studied in 4H- and 6H-SiC [88,89]. Donor and acceptor levels have been identified and used in part of the study that verified universal Mu(+/−) level across a number of compounds [68]. μSR measurements, combined with theoretical prediction for the three charge states of hydrogen in hexagonal SiC, place Mu sites at AB locations associated with each Si and C and possibly at the T-site associated with a silicon at the end of a short *c*-axis channel.

5.4.2 Group III–V

Studied during the same time as Si and Ge, GaAs has also been a very rich source of information for understanding interstitial hydrogen, properties of muonium and furthering the development of using μSR to study semiconductors. Since GaAs has been extensively reviewed elsewhere, these comments are kept very brief. One significant contribution made with GaAs is the complete characterisation of the Mu^- centre, including structural and motional properties (e.g. [90]). Notably, this is the first full characterisation of the Mu^- and completed with QLCR. In *n*-type GaAs, Mu^- is found in the T-site associated with Ga neighbours and a diffusion rate that is nominally ten orders of magnitude smaller than the Mu^0_T in the same system [61,62].

In the Zincblende III–V systems, the stable sites are similar to that of the diamond structure but include T-sites associated with each group-V and group-III neighbours. In GaAs, the only stable location for Mu^- is the T-site associated with Ga neighbours. While Mu^+ likely visits the T-site associated with As neighbours, evidence suggest that it is done so only as a mobile centre in GaAs. The BC site is the most stable location for both Mu^0 and Mu^+. The (0/+) donor level is placed 0.17 eV below the conduction band edge, the (0/−) acceptor level 0.60 eV above the top of the valence band and the (+/−) switching level 0.21 eV above midgap were established and reported in [91]. The behaviour in GaP is similar to that of GaAs. There are two observed sites relating to Mu^+ with three distinct regimes of motion [92]. Here, an Mu^+ charge state is supported at the BC site and the T-site that is associated with the group-V atom. At low temperatures, the Mu^+_{BC} is present, and measurements are consistent with tunnelling or local hops between the four locations about the central atom. With an increase in temperature comes the onset of dynamics that shows a detachment from the central atom and transition to global hopping among BC sites. The next step includes the group-V Mu^+_T in the motion. Interestingly, in GaAs the $Mu^+_{T,V}$ state is immediately mobile; however, in GaP, there is a temperature region where this Mu^+_T is not immediately mobile, but transitions to a mobile state at higher temperatures. The T-site that is associated with gallium neighbours hosts Mu^-, presumably with an Mu^0_T precursor [93]. GaP data show some suggestion of a mobile $Mu^0_{T,Ga}$ interacting with donors at low temperatures but stabilising to Mu^-_T by around 150 K. The activation energy for diffusive motion of the Ga T-site Mu^- is near 0.730 ± 0.040 eV and transitions out of this state via ionisation of the Mu^- above 600 K. This transition is likely associated with the transition out of the occupied state related to the acceptor level with energy of around 1.71 ± 0.32 eV; however, this could also be complicated by a charge cycle. Ultimately, the acceptor level in GaP is established to be near

0.82 ± 0.09 eV (above the valence band) and donor level 0.150 ± 0.02 eV below the conduction band. These donor and acceptor levels support the common Mu (+/−) level, although there are still several open questions relating to specific details of the higher temperature charge cycles.

Significant work on the III–V nitrides (specifically GaN, AlN and InN), including identification and characterisation of centres and dynamics, is discussed in [94–96]. These compounds are representative of the group of 2H wurtzite structured III–nitride materials with significant ionic character in these bonds. The only neutral (Mu^0) state is a shallow donor with a weakly anisotropic hyperfine constant that is close to the vacuum Mu^0 value and at unstable locations close to the group-III atom with an Mu^+ ground state. The large isotropic component (e.g. $A_{iso} = 4{,}450 \pm 436$ MHz in AlN) of the hyperfine constant indicates a lack of covalent bonding with the host and the weak dipolar contribution (e.g. $D = 167 \pm 29$ MHz in AlN) that points away from the *c*-axis suggests the centre is not in the unblocked *c*-axis channels; this site is shown in Figure 5.20. The atomic Mu^0 impurities are highly mobile and undergo at least five different dynamic processes before the ultimate ionisation. There are four distinct Mu^+ states in InN, three are static centres and the fourth shows properties characteristic of quantum-mechanical tunnelling below 300 K with field width and rate parameters consistent with motion around a single N atom. At higher temperatures, long-range thermal diffusion with a motional barrier of about 420 meV is revealed through the increasing hop rate. Above around 500 K, the dynamic behaviour transitions to diffusion-limited trapping and slow release (characterised by a trap-release model). The Mu^+ ground state is consistent with an AB site ($AB_\perp$ in Figure 5.20) where the N–Mu^+ is oriented into the channels of the 2H wurtzite structure. Site-change and motional barrier energies have been determined. In general, four sites for metastable locations of Mu^+ were observed: $AB_\perp$, $AB_\parallel$, $BC_\parallel$ and $BC_\perp$ (see Figure 5.20).

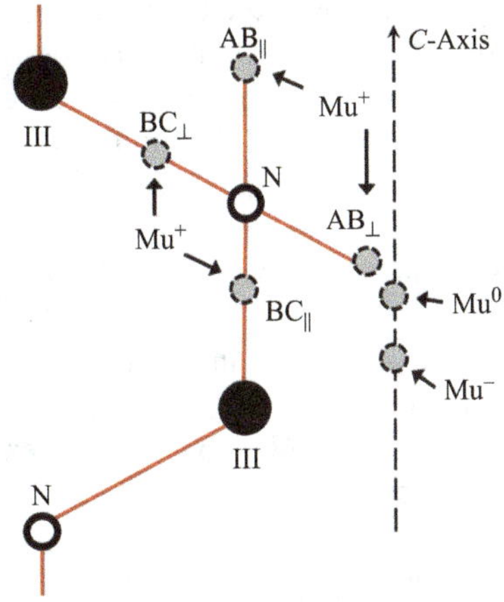

Figure 5.20 The muonium states in the III–nitride hexagonal (wurtzite) structure are shown here and discussed within Section 5.4.2 and [94–96]

5.4.3 Group II–VI

Muonium studies have been carried out in the II–VI compounds (e.g. [59,97–100]), and muonium has been observed to be stable in either a donor (0/+) configuration where it is strongly bound to an anion in the AB configuration or as an acceptor (0/−) forming in a T-site region associated with the cations. In ZnSe, two Mu^0 states with similar but distinct hyperfine interactions are observed: one, at low temperature that is very broad ($A = 3{,}283.6 \pm 0.5$ MHz) and converts to the second, which is very sharp around 30 K ($A = 3{,}454.26 \pm 0.02$ MHz). The former is assigned to muonium that is in a not yet relaxed lattice where the potential minima is slightly off the centre position and Mu tunnels among the similar off-centre sites. The dynamics presumably average out any anisotropy that ought to be present due to the off-centre positioning. The latter state is then the muonium in the centre of the T-site cage after the lattice is relaxed. Alternatively, these could correspond to a deep donor state in the T-site associated with Se (supporting Mu^0 and Mu^+) and deep acceptor in the T-site associated with Zn (supporting Mu^0 and Mu^-). In the Zincblende II–VI materials, the anion cage region is competitive with the BC site for both positively charged donor and neutral states. A shallow donor state is not observed in ZnSe. In this picture, the Zn T-site is the more stable of the two for Mu^0. With increase in temperature (between 30 K and 50 K), a transition to the Mu^- state occurs by way of delayed electron capture, but then around 60 K, the extra electron is lost. In ZnS, similar behaviour is observed with the electron capture occurring in the same temperature range, but the Mu^- state is maintained to 250 K before the single e^- ionisation. From these μSR measurements and characterisation of the acceptor state within the cation T-site on ZnSe and ZnS, electron affinities were measured to be near 3.1 eV within a few tenths of eV where the uncertainty is primarily due to uncertainty in the positioning of the band edge.

These studies find that for II–VI materials, in general, compounds with lower electron affinity (e.g. ZnSe, ZnS and ZnTe) show no shallow donor but do show an acceptor state, so in the *n*-type materials, the Fermi level is close to the conduction band, so Mu^- is formed. In materials with larger electron affinity (e.g. CdS, CdSe, CdTe and ZnO), only the donor states are within the gap, so the electron-rich acceptor (0/−) is usually unbound and therefore should not form; however, the shallow donor (0/+) states in AB sites are indeed observed. The measured isotropic and dipolar components of the hyperfine constants for Mu^0_{AB} (shallow effective-mass donor states) in these examples are $A_{iso} = 0.244 \pm 5$ and $D = 0.091 \pm 0.006$ MHz in CdS; $A_{iso} = 0.087 \pm 0.004$, $D < 0.040$ MHz in CdSe; $A_{iso} = 0.261 \pm 0.004$, $D < 0.050$ MHz in CdTe; and $A_{iso} = 0.500 \pm 0.020$, $D = 0.260 \pm 0.020$ MHz in ZnO. Interestingly, in the cubic CdTe, an isotropic Mu^0 is found with $A = 2{,}400 \pm 400$ MHz, but in the hexagonal Cd-based compounds, no isotropic Mu^0 are observed.

5.4.4 Group II–IV–V_2 chalcopyrites

Current work on the II–IV–V_2 chalcopyrite systems is not as comprehensive or extensive as their III–V and II–VI counterparts; however, electronic and magnetic features of the II–IV–V_2:Mn chalcopyrites (e.g. higher carrier mobility, band gaps from ~0.2 to ~5 eV; and above room temperature ferromagnetism) are quite appealing to the spin-based electronics community. Development of these materials for their IR active non-linear optical properties started in the early 1970s and continued through the 1990s so many properties are well known; however, they have yet to be fully optimised for their semiconducting properties. The relatively recent discovery of bulk ferromagnetism at room temperature the Mn-doped II–IV–V_2 has reignited some interest (e.g. see [101–103]).

These chalcopyrite II–IV–V_2 compounds have a structure related to their III–V Zincblende counterpart, where the group-III elements are replaced by a mix of group-II and group-IV elements that forces a doubled unit cell along the crystallographic (0 0 1) direction. These systems tend to have significant disorder in the II–IV sublattice, even in high-purity crystals, due to the minimal energy differences between the various arrangements of group-II and group-IV atoms.

A μSR study was completed on the parent and Mn-doped compounds with a goal of understanding the mechanism responsible for transferring the magnetism from the local magnetic moments to the bulk, as the existing theories are not compatible with these systems [104].

As an example, $ZnGeP_2$ is one of the chalcopyrite compounds studied. In *n*-type $ZnGeP_2$, a series of μSR measurements to characterise the muonium centres was completed to reveal a promptly formed Mu^0 centre in the T-site associated with phosphorus neighbours having an isotropic hyperfine constant $A(T \rightarrow 0) = 1{,}965.6 \pm 0.3$ MHz. This Mu^0 then proceeds to hop to the Mu^0_T associated with the II–IV sublattice, where an anisotropic hyperfine interaction is observed. From the data, it is unclear as to whether this is a two-way process; however, there is no evidence to suggest that it is not. The Mu^0 centre ionises near room temperature, leaving a surprisingly slow diffusing Mu^+. Interestingly, in the related *p*-type $CdGeAs_2$, Mu^+ is also very slow to diffuse through the bulk, even at temperatures in excess of 400 K, where traps limit the diffusion rate through much higher temperatures (Figure 5.18) [65,104].

5.4.5 Oxides

Oxide-based materials continue to show a wide range of interesting properties, such as photovoltaic activity, ferroelectric behaviour, magnetism and high-temperature superconductivity, just to name a few (see, e.g. [105,106]). Understanding the role and behaviour hydrogen plays in modifying the electrical, optical and magnetic properties of oxide-based materials has been an important topic for many years and particularly important for efficient development of materials for potential device applications. Comprehensive discussion relating to modelling the electrical activity of hydrogen with muons is found in [36,37].

One example of recent work includes a study on rutile TiO_2 that has provided additional validation of the hydrogen and muon analogy by direct comparison of hyperfine interactions in a low-temperature state where the muon and hydrogen are found to form identical configurations with nearly the same electronic structures [22]. To reiterate the importance, this is one of only a few cases where direct comparisons between the muonium and related hydrogen states exist. Measured differences are accounted for by scaling with the magnetic moments and considering differences in the zero-point motion of Mu and H. In rutile TiO_2, the low-temperature bonding configuration includes the muon attaching to one of six oxygen atoms that surround a Ti atom in a typical O–Mu (O–H) configuration and hyperfine coupled to the donor electron that is localised on the central Ti^{3+} ion. The ground state is formed by bonding to only one of the two oxygen that are in the same *a–b* plane as Ti^{3+}. There are two orientationally inequivalent sublattices, and therefore, two distinct hyperfine splittings are observed. The excited states include motion between the oxygen sites both in and out of plane with the Ti^{3+} and across the channel. Also in progress are similar studies on anatase and brookite TiO_2, the metastable phases, both of which also show significant changes to their electrical and optical properties with hydrogen incorporation.

Transparent conducting oxides (e.g. In_2O_3) have also been the focus of recent studies (work by led by Baker [67]) where a generalised model for a two-state bidirectional transition has been developed (see Section 5.3.3.3) and implemented with a trap and release picture in addition existing models to fully characterise the muonium (cf. H) behaviour in these materials. In light of the successful development and implementation of approach on In_2O_3, work on other transparent conducting oxides (e.g. Ga_2O_3, TiO_2, SnO_2 and CdO) is ongoing.

Another example of recent work that has been quite informative is on the configurations of muonium (cf. isolated hydrogen) in oxides with high-k permittivity (e.g. ZrO_2, Lu_2O_3, BeO, HfO_2 and MgO; [107,108]). The main results of this particular study include significant insight into the muonium (isolated H) configurations as well as the role of lattice dynamics in muonium formation. The two basic muonium (isolated H) configurations were determined to be an oxygen-bound donor-like configuration and an interstitial (atom-like) acceptor. The conversions between these configurations were also studied.

5.5 Final thoughts

Outlined here is a tutorial to the main principles relating to using muons to study semiconductors and a glimpse at some examples of contributions these techniques have and will continue to make to the field. μSR measurements have provided the semiconductor community with an abundance of information relating to isolated hydrogen impurities that are not easily available by other means.

The collaboration in which I am primarily involved has been working on a mix of projects that include, for example, utilising the laser system on the HIFI spectrometer at the STFC-ISIS facility to develop the MuSES technique for use in semiconductors. We are also working to understand the characteristics and behaviour of muonium (cf. H) in many of the semiconducting oxides that show a very diverse range of potential applications. Another example of recent work is that of using μSR to study a class of material (II–IV–V_2:Mn chalcopyrites) that shows promise for spin electronics due to the combination of semiconducting and bulk magnetic properties they exhibit. In this particular dilute magnetic semiconducting system, the muon serves as an experimentally accessible analogue to the isolated hydrogen impurity as well as a direct probe of the local magnetic environment.

There is a considerable amount of interesting physics in 2D materials. Even though probing near-surface interactions and properties of thin films are possible (e.g. LEM, MuSES), there is not (yet) an effective way to stop enough muons in a single monolayer to provide a feasible measurement.

Looking forward, I expect that μSR will continue to provide valuable insight on materials, especially those in which hydrogen is particularly reactive or where a probe of the local magnetic environment or local charge density could be of use.

The last few decades have been incredibly productive in this field. The ongoing efforts to utilise and further develop novel muon-based techniques to study semiconducting and many other types of systems show great promise of continuing to provide useful data to the community.

Acknowledgements

I gratefully acknowledge contributions of all members of the muonium in semiconductor collaboration, past and present. I would also like to extend a special thank you to the very hard-working individuals that keep the muon facilities operational and are continually working to improve the measurement capabilities and user accessibility. Without these dedicated individuals, none of these works would be possible. Many of the simulations in this work were completed by utilising the updated version of the software package developed by J.S. Lord [109].

References

[1] Pankove J.I., Johnson N.M. (eds.). *Hydrogen in Semiconductors*, Semiconductors and Semimetals Series, Vol. **34** (San Diego: Academic Press; 1991).

[2] Pearton S.J., Corbett J.W., Stavola M. (eds.). *Hydrogen in Crystalline Semiconductors* (Berlin: Springer-Verlag; 1992).

[3] Nickel N.H. (ed.). *Semiconductors and Semimetals*, Vol. **61** (Berlin: Academic Press; 1999).

[4] Weber J. 'Hydrogen in semiconductors: From basic physics to technology'. *Physica Status Solidi C.* 2008;**5**:535.

[5] Estreicher S.K. 'Hydrogen-related defects in crystalline semiconductors: a theorists perspective'. *Materials Science and Engineering.* 1995;**R14**:319.

[6] Carroll B.R., Lichti R.L., Mengyan P.W., *et al.* 'Spectroscopic identification of shallow muonium acceptors in $Si_{0.06}Ge_{0.94}$'. *Applied Physics Letters.* 2014;**105**:122101.

[7] Carroll B.R., Lichti R.L., King P.J.C., Celebi Y.G., Yonenaga I., Chow K.H. 'Muonium defect levels in Czochralski-grown silicon-germanium alloys'. *Physical Review B.* 2010;**82**:205205.

[8] Gorelkinskii Y.V., Nevinnyi N.N. *Soviet technical physics letters.* 1987; **13**:45.

[9] Gorelkinskii Y.V. 'Electron Paramagnetic Resonance Studies of Hydrogen and Hydrogen-Related Defects in Crystalline Silicon' in Nickel N.H. (ed.). *Semiconductors and Semimetals*, Vol. **61** (Berlin: Academic Press; 1999).

[10] Gorelkinskii Y.V., Nevinnyi N.N. 'Electron paramagnetic resonance of hydrogen in silicon'. *Physica B.* 1991;**170**:155.

[11] Bech Nielsen B., Bonde Nielsen K., Byberg J.R. 'EPR experiments on hydrogen implanted silicon crystals: Annealing properties of bond center hydrogen'. *Materials Science Forum.* 1994;**143–147**:909.

[12] Patterson B.D. 'Muonium states in semiconductors'. *Reviews of Modern Physics.* 1988;**60**:69.

[13] Cox S.F.J. 'Implanted muon studies in condensed matter science'. *Journal of Physics C: Solid State Physics.* 1987;**20**:3187.

[14] Estle T.L., Lichti R.L. 'Muonium in semiconductors and some implications for hydrogen impurities'. *Hyperfine Interactions.* 1996;**97/98**:171.

[15] Chow K.H., Hitti B., Kiefl R.F. 'μSR on Muonium in Semiconductors and its Relation to Hydrogen' in Stavola M. (ed.). *Semiconductors and Semimetals*, Vol. **51A** (San Diego: Academic Press; 1998).

[16] Schenck A. *Muon Spin Rotation Spectroscopy: Principles and Applications in Applied Physics* (Bristol: Adam Hilger; 1985).

[17] Yaouanc A., de Reotier P.D. *Muon Spin Rotation, Relaxation and Resonance: Applications to Condensed Matter* (Oxford: Oxford Press; 2011).

[18] Brewer J.H. 'Muon Spin Rotation/Relaxation/Resonance' in Trigg G.L., Vera E.S., Greulich W. (eds.). *Encyclopedia of Applied Physics*, Vol. **11** (New York: VCH; 1994).

[19] Yamazaki T., Nagamine K., Nagamiya S., *et al.* 'Negative muon spin rotation'. *Physica Scripta.* 1975;**11**:133.

[20] Brewer J.H., Fleming D.G., Crowe K.M., *et al.* 'μ+SR spectroscopy: The positive muon as a magnetic probe in solids'. *Physica Scripta.* 1974;**11**:144.

[21] Kiefl R.F., Celio M., Estle T.L., *et al.* '^{29}Si hyperfine structure of anomalous muonium in silicon: Proof of the bond-centered model'. *Physical Review Letters.* 1988;**60**:224.

[22] Vilão R.C., Vieira R.B.L., Alberto H.V., *et al.* 'Muonium donor in rutile TiO_2 and comparison with hydrogen'. *Physical Review B.* 2015;**92**:081202R.

[23] Storchak V.G., Eshchenko D.G., Brewer J.H. 'Formation dynamics of muonium centres in semiconductors – A new approach'. *Journal of Physics: Condensed Matter.* 2004;**16**:S4761.

[24] Particle Data Group. 'Review of particle physics'. *European Physical Journal C – Particles and Fields.* 1998;**3**:1.

[25] Vilão R.C., Vieira R.B.L., Alberto H.V., Gil J.M., Weidinger A. 'Role of the transition state in muon implantation'. *Physical Review B.* 2017; **96**:195205.

[26] Vilão R.C., Vieira R.B.L., Alberto H.V., *et al.* 'Barrier model in muon implantation and application to Lu_2O_3'. *Physical Review B.* 2018;**98**:115201.

[27] Hayano T.S., Uemura Y.J., Imazato J., Nishida N., Yamazaki T., Kubo R. 'Zero- and low-field spin relaxation studied by positive muons'. *Physical Review B*. 1979;**20**:850.

[28] Crook M.R., Cywinski R. 'Voigtian Kubo–Toyabe muon spin relaxation'. *Journal of Physics: Condensed Matter*. 1997;**9**:1149.

[29] Keren A. 'Generalization of the Abragam relaxation function to a longitudinal field'. *Physical Review B*. 1994;**50**:10039.

[30] Senba M. 'Anisotropic muonium atoms: Energy levels and electron spin exchange'. *Physical Review A*. 2000;**62**:042505.

[31] Pratt F.L. 'Repolarization of anisotropic muonium in orientationally disordered solids'. *Philosophical Magazine Letters*. 1997;**75**:371.

[32] Cox S.F.J. 'Muonium as a model for interstitial hydrogen in the semiconducting and semimetallic elements'. *Reports on Progress in Physics*. 2009;**72**:116501.

[33] Riseman T.M., Brewer J.H. 'The rotating reference frame transformation in MuSR'. *Hyperfine Interactions*. 1990;**65**:1107.

[34] Sonier J.E., Brewer J.H., Kiefl R.F. 'μSR studies of the vortex state in type-II superconductors'. *Reviews of Modern Physics*. 2000;**72**:076402.

[35] Keren A. 'Muons as probes of dynamical spin fluctuations: some new aspects'. *Journal of Physics: Condensed Matter*. 2004;**16**:S4603.

[36] Cox S.F.J., Lord J.S., Cottrell S.P., *et al.* 'Oxide muonics: I. Modelling the electrical activity of hydrogen in semiconducting oxides'. *Journal of Physics: Condensed Matter*. 2006;**18**:1061.

[37] Cox S.F.J., Gavartin J.L., Lord J.S., *et al.* 'Oxide muonics: II. Modelling the electrical activity of hydrogen in wide-gap and high-permittivity dielectrics'. *Journal of Physics: Condensed Matter*. 2006;**18**:1079.

[38] Lichti R.L., Cox S.F.J., Chow K.H., *et al.* 'Charge-state transitions of muonium in germanium'. *Physical Review B*. 1999;60:1734.

[39] Lichti R.L. 'Dynamics of Muonium Diffusion, Site Changes, and Charge-State Transitions' in Nickel N.H. (ed.). *Hydrogen and Semiconductors II*, Semiconductor and Semimetals Series, Vol. **61** (Berlin: Academic Press; 1999).

[40] Kreitzman S.R., Hitti B., Lichti R.L., Estle T.L., Chow K.H. 'Muon-spin-resonance study of muonium dynamics in Si and its relevance to hydrogen'. *Physical Review B*. 1995;**51**:13117.

[41] Schneider J.W., Chow K., Kiefl R.F., Kreitzman S.R., MacFarlane A. 'Electronic structure of anomalous muonium in GaP and GaAs'. *Physical Review B*. 1993;**47**:10193.

[42] Luke G.M., Brewer J.H., Kreitzman S.R., *et al.* 'Muon diffusion and spin dynamics in copper'. *Physical Review B*. 1991;**43**:3284.

[43] Cox S.F.J. 'Detection of quadrupole interactions by muon level crossing resonance'. *Zeitschrift fuer Naturforschung*. 1992;**47a**:371.

[44] Elvezio M. 'Very slow positive muons'. *Zeitschrift für Physik C Particles and Fields*. 1992;**56**(suppl 1):S243.

[45] Morenzoni E. 'Low energy muons as probes of thin films and surfaces'. *Applied Magnetic Resonance*. 1997;**13**:219.

[46] Shiroka T., Bucci C., De Renzi R., *et al.* 'Production of pulsed ultraslow muons and first μSR experiments on thin metallic and magnetic films'. *Applied Magnetic Resonance*. 2000;**19**:471.

[47] Prokscha T., Morenzoni E., Eshchenko D.G., *et al.* 'Formation of hydrogen impurity states in silicon and insulators at low implantation energies'. *Physical Review Letters*. 2007;**98**:227401.

[48] Prokscha T., Morenzoni E., Eshchenko D.G., Luetkens H., Nieuwenhuys G.J., Suter A. 'Near-surface muonium states in germanium'. *Physica B: Condensed Matter*. 2009;**404**:866.

[49] Prokscha T., Chow K.H., Stilp E., *et al.* 'Photo-induced persistent inversion of germanium in a 200-nm-deep surface region'. *Scientific Reports*. 2013;**3**:2569.

[50] Alberto H.V., Vilão R.C., Vieira R.B.L., *et al.* 'Slow-muon study of quaternary solar-cell materials: Single layers and p–n junctions'. *Physical Review Materials*. 2018;**2**:025402.

[51] Eshchenko D.G., Storchak V.G., Morenzoni E., *et al.* 'Low energy μSR studies of semiconductor interfaces'. *Physica B: Condensed Matter*. 2009;**404**:873.

[52] Kadono R., Matsushita A., Macrae R.M., Nishiyama K., Nagamine K. 'Muonium centers in crystalline Si and Ge under illumination'. *Physical Review Letters*. 1994;**73**:2724.

[53] Iwanowski M., Maier K., Major J., *et al.* 'Investigation of muon states in silicon and germanium by field-quenching and RFμSR'. *Hyperfine Interactions*. 1994;**86**:681.

[54] Yokoyama K., Lord J.S., Murahari P., *et al.* 'The new high field photo-excitation muon spectrometer at the ISIS pulsed neutron and muon source'. *Review of Scientific Instruments*. 2016;**87**:125111.

[55] Yokoyama K., Lord J.S., Miao J., Murahari P., Drew A.J. 'Photoexcited muon spin spectroscopy: A new method for measuring excess carrier lifetime in bulk silicon'. *Physical Review Letters*. 2017;**119**:226601.

[56] Fisch R., Licciardello D.C. 'Negative-U states in the gap in hydrogenated amorphous silicon'. *Physical Review Letters*. 1978;**41**:889.

[57] Chow K.H., Cox S.F.J., Davis E.A., *et al.* 'Diffusion, trapping and relaxation of Mu^+ and Mu^- in heavily-doped GaAs'. *Hyperfine Interactions*. 1997;**105**:309.

[58] Storchak V.G., Brewer J.H., Stubbs S.L., *et al.* 'Spin polarons in the correlated metallic pyrochlore $Cd_2Re_2O_7$'. *Physical Review Letters*. 2010;**105**:076402.

[59] Vilão R.C., Gil J.M., Weidinger A., *et al.* 'Acceptor level of interstitial muonium in ZnSe and ZnS'. *Physical Review B*. 2008;**77**:235212.

[60] Chow K.H., Kiefl R.F., Schneider J.W., *et al.* 'Muonium dynamics in Si at high temperatures'. *Physical Review B*. 1993;**47**:16004.

[61] Chow K.H., Hitti B., Kiefl R.F., Dunsiger S.R., Lichti R.L., Estle T.L. 'Diffusion and charge dynamics of negatively charged muonium in n-type GaAs'. *Physical Review Letters*. 1996;**76**:3790.

[62] Kadono R., Kiefl R.F., Brewer J.H., *et al.* 'Quantum diffusion of muonium in GaAs'. *Hyperfine Interactions.* 1990;**64**:635.

[63] Kadono R. 'Quantum diffusion of muonium in insulators'. *Hyperfine Interactions.* 1990;**64**:615.

[64] Storchak V.G., Prokof'ev N.V. 'Quantum diffusion of muons and muonium atoms in solids'. *Reviews of Modern Physics.* 1998;**70**:929.

[65] Mengyan P.W., Baker B.B., Lichti R.L., *et al.* 'Hyperfine spectroscopy and characterization of muonium in $ZnGeP_2$'. *Physica B: Condensed Matter.* 2009;**404**:5121.

[66] Jayarathna G.S. *Transition Dynamics for Muonium Acceptor States in Silicon Germanium Alloys* (Texas Tech University, M.Sc. Thesis: 2014).

[67] Baker B.B. *Charged Muonium Diffusion in Indium Oxide and Other Transparent Conducting Oxides* (Texas Tech University, Ph.D. Dissertation: 2015).

[68] Lichti R.L., Chow K.H., Cox S.F.J. 'Hydrogen defect-level pinning in semiconductors: The muonium equivalent'. *Physical Review Letters.* 2008;**101**:136403.

[69] Van de Walle C.G., Neugebauer J. 'Universal alignment of hydrogen levels in semiconductors, insulators and solutions'. *Nature.* 2003;**423**:626.

[70] Lord J.S., Yokoyama K., Mengyan P.W., *et al.* 'Direct spectroscopy of muon donor and acceptor levels in silicon carbide'. *STFC-ISIS Neutron and Muon source.* RB1710134, doi: 10.5286/ISIS.E.RB1710134. RB1810775, doi: 10.5286/ISIS.E.RB1810775.

[71] Cox S.F.J., Lichti R.L., Lord J.S., *et al.* 'The first 25 years of semiconductor muonics at ISIS, modelling the electrical activity of hydrogen in inorganic semiconductors and high-κ dielectrics'. *Physica Scripta.* 2013; **88**:068503.

[72] Cox S.F.J. 'The shallow-to-deep instability of hydrogen and muonium in II-VI and III-V semiconductors'. *Journal of Physics: Condensed Matter.* 2003;**15**:R1727.

[73] Swanson R.A. 'Depolarization of positive muons in condensed matter'. *Physical Review.* 1958;**112**:580.

[74] Fisch R., Licciardello D.C. 'Negative-U states in the gap in hydrogenated amorphous silicon'. *Physical Review Letters.* 1978;**41**:889.

[75] Symons M.C.R. 'Chemical aspects of muon spin rotation'. *Hyperfine Interactions.* 1984;**19**:771.

[76] Cox S.F.J., Symons M.C.R. 'Molecular radical models for the muonium centres in solids'. *Chemical Physics Letters.* 1986;**126**:516.

[77] Estle T.L., Estreicher S., Marynick D.S. 'Preliminary calculations confirming that anomalous muonium in diamond and silicon is bond-centered interstitial muonium'. *Hyperfine Interactions.* 1986;**32**:637.

[78] Estreicher S. 'Equilibrium sites and electronic structure of interstitial hydrogen in Si'. *Physical Review B.* 1987;**36**:9122.

[79] Kiefl R.F., Celio M., Estle T.L., *et al.* 'Determination of the electronic structure of anomalous muonium in GaAs from nuclear hyperfine interactions'. *Physical Review Letters.* 1987;**58**:1780.

[80] Lichti R.L., Lamp C.D., Kreitzman S.R., *et al.* 'Modern muon spectroscopic methods in semiconductor physics'. *Material Science Forum.* 1992; **83–87**:1115.
[81] Estle T.L., Chow K.H., Cox S.F.J., *et al.* 'Site stability, diffusion and charge dynamics for muonium in GaAs'. *Material Science Forum.* 1997;**258–263**:849.
[82] Hitti B., Kreitzman S.R., Estle T.L., *et al.* 'Dynamics of negative muonium in n-type silicon'. *Physical Review B.* 1999;**59**:4918.
[83] Gil J.M., Alberto H.V., Vilão R.C., *et al.* 'Novel muonium state in CdS'. *Physical Review Letters.* 1999;**83**:5294.
[84] Cox S.F.J., Davis E.A., Cottrell S.P., *et al.* 'Experimental confirmation of the predicted shallow donor hydrogen state in zinc oxide'. *Physical Review Letters.* 2001;**86**:2601.
[85] King P.D.C., Lichti R.L., Celebi Y.G., *et al.* 'Shallow donor state of hydrogen in In_2O_3 and SnO_2: Implications for conductivity in transparent conducting oxides'. *Physical Review B.* 2009;**80**:081201R.
[86] Carroll B.R., Lichti R.L., Mengyan P.W., *et al.* 'Spectroscopic identification of shallow muonium acceptors in $Si_{0.06}Ge_{0.94}$'. *Applied Physics Letters.* 2014;**105**:122101.
[87] Mengyan P.W., Lichti R.L., Celebi Y.G., *et al.* 'Muonium transitions in Ge-rich SiGe alloys'. *Physica B: Condensed Matter.* 2012;**407**:2829.
[88] Bani-Salameh H.N. *Muonium Defect Centers in Aluminium Nitride and Silicon Carbide* (Texas Tech University, Ph.D. Dissertation: 2007).
[89] Celebi Y.G., Lichti R.L., Carroll B.R., King P.J.C., Cox S.F.J. 'Muonium in 4H silicon carbide'. *Physica B: Condensed Matter.* 2009;**404**:5117.
[90] Chow K.H., Kiefl R.F., MacFarlane W.A., *et al.* 'Structure of negatively charged muonium in n-type GaAs'. *Physical Review B.* 1995;**51**:14762R; ibid 1995;**52**:16939.
[91] Lichti R.L., Bani-Salameh H.N., Carroll B.R., Chow K.H., Hitti B., Kreitzman S.R. 'Donor and acceptor energies for muonium in GaAs'. *Physical Review B.* 2007;**76**:0455221.
[92] Lichti R.L., Chow K.H., Hitti B., Davis E.A., Sjue S.K.L., Cox S.F.J. 'Motional properties of positive muonium in gallium III-V compounds'. *Physica B: Condensed Matter.* 2001;**308–310**:862.
[93] Vernon J.E., Carroll B.R., Bani-Salameh H.N., *et al.* 'Dynamics of T-site muonium states in gallium phosphide'. *Physica B: Condensed Matter.* 2009;**404**:820.
[94] Celebi Y.G., Lichti R.L., Coss B.E., Cox S.F.J. 'Dynamics of positively charged muonium centers in indium nitride'. *Physical Review B.* 2006; **74**:245219.
[95] Lichti R.L., Celebi Y.G., Cottrell S.P., Cox S.F.J., Davis E.A. 'Diffusion and trapping of Mu in the III–V nitrides'. *Journal of Physics: Condensed Matter.* 2004;**16**:S4721.

[96] Bani-Salameh H.N., Lichti R.L., Celebi Y.G., Cox S.F.J. 'Diffusion and conversion dynamics for neutral muonium in aluminium nitride'. *Physical Review B*. 2006;**74**:245203.
[97] Gil J.M., Alberto H.V., Vilão R.C., *et al.* 'Shallow donor muonium states in II–VI semiconductor compounds'. *Physical Review B*. 2001;**64**:075205.
[98] Vilão R.C., Alberto H.V., Piroto Duarte J., *et al.* 'Muonium spectroscopy in ZnSe: Metastability and conversion'. *Physical Review B*. 2005;**72**:235203.
[99] Lichti R.L., Carroll B.R., Vernon J.E., Bani-Salameh H.N., Chow K.H., Fan I. 'Possible donor and acceptor energies for Mu in ZnSe'. *Physica B: Condensed Matter*. 2009;**404**:827.
[100] Vilão R.C., Gil J.M., Weidinger A., *et al.* 'Information on hydrogen states in II–VI semiconductor compounds from a study of their muonium analogues'. *Nuclear Instruments and Methods in Physics Research A*. 2007; **580**:438.
[101] Novotortsev V.M., Kochura A.V., Marenkin S.F. 'New ferromagnetics based on manganese-alloyed chalcopyrites $A^{II}B^{IV}C^{V}_2$'. *Inorganic Materials*. 2010;**46**:1421.
[102] Cho S., Choi G., Cha G., Hong S.C., Kim Y., Zhao Y. 'Room-temperature ferromagnetism in $(Zn_{1-x}Mn_x)GeP_2$'. *Physical Review Letters*. 2002; **88**:257203.
[103] Demin R., Koroleva L., Marenkin S., *et al.* 'Room-temperature ferromagnetism in Mn-doped $CdGeAs_2$ chalcopyrite'. *Journal of Magnetism and Magnetic Materials*. 2005;**290–291**:1379.
[104] Mengyan P.W. *Magnetism in Mn-Doped Chalcopyrites* (Texas Tech University, Ph.D. Dissertation: 2014).
[105] Ramesh R., Schlom D.G. 'Whither oxide electronics?'. *Materials Research Society Bulletin*. 2008;**33**:1006.
[106] McCluskey M.D., Tarun M.C., Teklemichael S.T. 'Hydrogen in semiconductors'. *Journal of Materials Research*. 2012;**27**:2190.
[107] Vieira R.B.L. *Hydrogen in High-κ Permittivity Oxides Modelled by the Muon Analogue* (Universidade de Coimbra, Ph.D. Dissertation: 2018).
[108] Vieira R.B.L., Vilão R.C., Marinopoulos A.G., *et al.* 'Isolated hydrogen configurations in zirconia as seen by muon spin spectroscopy and ab initio calculations'. *Physical Review B*. 2016;**94**:115207.
[109] Lord J.S. 'Computer simulation of muon spin evolution'. *Physica B: Condensed Matter*. 2006;**374–375**:472 (www.isis.stfc.ac.uk/Pages/Muons-Downloads.aspx).
[110] Lichti R.L., Baker B.B., Mengyan P.W., Celebi Y.G., Catak E., Hudy L. 'Motion of positively charged muonium centres in transparent conducting oxides'. *STFC-ISIS Neutron and Muon source*. RB1120279, doi: 10.5286/ISIS.E.24088334.

Chapter 6

Positron annihilation spectroscopy, experimental and theoretical aspects

Jonatan Slotte[1], Ilja Makkonen[2] and Filip Tuomisto[2]

6.1 Introduction

Defect characterization in semiconductors is a complex process involving a plethora of properties, e.g., defect species, electrical and optical properties and thermodynamic properties, such as formation and migration energies. Hence, there is not a single experimental technique that can cover all of these properties and single-handedly contribute to the control of defects on a production line. However, when it comes to a specific type of defect, namely, vacancy defects, positron annihilation spectroscopy (PAS) is a particularly useful technique. The technique can readily be used to identify vacancy defects. Measurements can be performed both at low and elevated temperatures and consequently, the annealing, formation and migration properties of vacancy defects can be studied. In addition, also optical and electrical properties of vacancy defects can to some extent be investigated.

PAS is based mainly on two features of the positron, its ability to get trapped at open-volume defects in crystalline materials and its fate to eventually annihilate with an electron from the surrounding lattice. After a positron is implanted into a sample, it thermalizes rapidly and thereafter diffuses in the lattice as a free particle. During the diffusion process, it can get trapped at an open-volume defect, where the repulsion from the surrounding atom cores results in a potential well for the positron. Due to the reduced electron density in a vacancy, the positron lifetime will increase compared to a positron annihilating as a free particle. Furthermore, the 511 keV annihilation radiation will also be affected by the trapping of the positron, since the momentum of the annihilating electron–positron pair will be preserved in the annihilation radiation. This is observed as a Doppler shift and a change in the angular correlation of the annihilation radiation.

In addition to measuring the Doppler shift, the momentum distribution of the annihilating electron–positron pair can also be simulated within the framework of the density functional theory. The combination of experiment and theory is a very useful addition to PAS when it comes to identifying and understanding defects.

[1]Department of Applied Physics, Aalto University, Finland
[2]University of Helsinki and Helsinki Institute of Physics, Helsinki, Finland

6.2 Positrons in crystalline semiconductors

6.2.1 Positrons obtained from radioactive sources

In a traditional tabletop positron spectroscopy experiment, a radioactive β^+-source is used. The most commonly used isotope is ^{22}Na, where the emitted positron is usually accompanied by a 1.27 MeV photon from the excited daughter nucleus ^{22}Ne. This photon can be used as a timing signal in positron lifetime studies. The three-particle decay reaction (the daughter nucleus, the positron and the neutrino) results in a broad and continuous energy distribution for the positron up to certain maximum energy (E_{max} = 0.54 MeV for the ^{22}Na decay). The positron implantation profile directly from a β^+-source can be described by the empirical equation [1]

$$P(x) = e^{-\alpha x} \quad \text{where } \alpha = 16\frac{\rho\,(\mathrm{g\,cm^{-3}})}{E_{max}^{1.43}\,(\mathrm{MeV})} \tag{6.1}$$

The mean implantation depth for positrons in Si is approximately 50 μm. Hence, the so-called fast positrons obtained directly from an encapsulated source can only be used for bulk studies.

6.2.2 Slow monoenergetic positron beams for thin-layer studies

To be able to study thin layers and surfaces, a slow-positron beam is needed. Typically, the used positron energies in such a beam vary in the 0–40 keV range. To produce a low-energy positron beam, a few approaches are used. The traditional way is to use a β^+-source, and then slow down the fast positrons from the source with a moderator. This is usually a thin film of a material with a negative work function for positrons, placed in front of the source. Tungsten (W) is commonly used as moderator material. The efficiency of tungsten moderators is very low, only around 10^{-4}, which in turn limits the positron current in the beams to the fA range. Better efficiencies, up to 10^{-2}, can be achieved by using solid-state rare-gas moderators [2,3]. Various semiconductor and insulator materials have also been tested for use as the so-called field-assisted moderators [4–6]. Although some of these works show promise, field-assisted moderators have not yet found their way into routine use in slow-positron beams.

To produce beams with higher intensity, reactor-based [7–9] and accelerator-based [10,11] beams have been developed. In such positron beams, pair production is used as the source of positrons.

Irrespective of the technique used to produce the positrons for the beam, the implantation profile in a solid for monoenergetic positrons will be of the form [12,13]

$$P(x,E) = \frac{mx^{m-1}}{x_0^m} e^{-(x/x_0)^m}, \quad \text{where } x_0 = \frac{AE^n}{\rho\Gamma(1+(1/m))}. \tag{6.2}$$

m, n and A are empirical parameters and Γ is the gamma function. Commonly used empirical parameter values are $A = 4.0\ \mu\mathrm{g\,cm^{-2}}$, $m = 2$ and $n = 1.6$ [14]. As can be seen both from (6.2) and the example profiles in Figure 6.1, the width of the

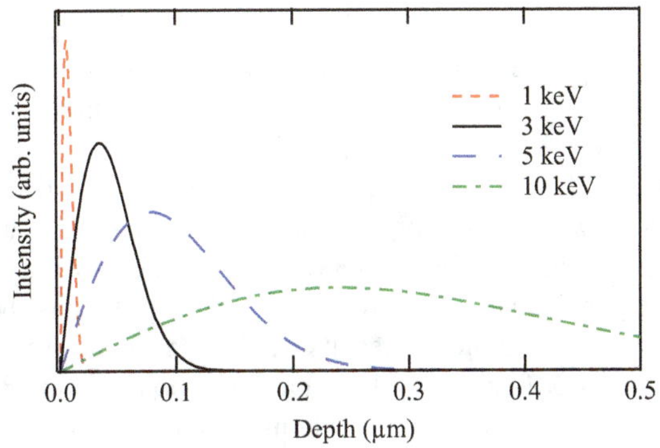

Figure 6.1 Implantation profiles for positrons with different energies in germanium

positron implantation profile is heavily dependent on the beam energy. Hence, sensitivity for the depth distribution of positron traps will quickly decline with increasing implantation energy.

6.2.3 Positron thermalization and diffusion

When a positron enters a solid, there are certain processes that will characterize the lifespan of the positron. In order of occurrence, these are thermalization, diffusion in the crystalline lattice and possible trapping into open-volume defects.

When the positron enters a crystalline solid, it will rapidly lose energy due to interactions (ionizations and core-electron excitations) with the electrons of the atoms in the host lattice. At lower kinetic energies of the positron, electron-hole excitations become important, and when the kinetic energy has decreased close to the thermal energy of the lattice, phonon scattering is the main energy loss process. Altogether the thermalization time for positrons at common measurement temperatures (30 K–700 K) is only a few picoseconds [15,16]. Hence, under normal circumstances, this is much less than the positron lifetime, and the thermalization can be omitted in the data analysis.

After thermalization, the positron will diffuse as a charged particle through the lattice. Due to its positive charge, the positron will be repelled by the positively charged nuclei, and the positron density will be concentrated to the interstitial region in the lattice. During diffusion, the most important process of interaction with the lattice is scattering from acoustic phonons, leading to a $T^{-1/2}$ dependence for the diffusion coefficient on temperature [17]. The positron diffusion length L_+ will be limited by the positron finite lifetime. In a positron-trap-free lattice, the diffusion length will be [18]

$$L_+ = \sqrt{\tau_B D_+}, \quad \text{where } D_+ = \frac{k_B T}{m^*}\tau_r. \tag{6.3}$$

In (6.3), τ_B is the positron lifetime in a positron-trap-free lattice, τ_r the relaxation time and m^* the positron effective mass. The positron diffusion length in a typical positron-trap-free semiconductor lattice is around 200 nm [19]. This value will be greatly reduced if the semiconductor contains positron-trapping defects. In layers with defect concentrations up to 10^{18} cm^{-3}, the diffusion length can be reduced by an order of magnitude.

As the positron diffusion length is two to three magnitudes lower than the width of the positron implantation profile directly from a ^{22}Na source, (6.1), positron diffusion does not have to be accounted for in an experiment with such fast positrons. However, in a beam experiment with slow positrons, the positron diffusion will more or less dominate the results of an experiment. As can be seen from Figure 6.1, the positron diffusion length will at low implantation energies be of the same order of magnitude as the width of the implantation profile.

6.2.4 Positron trapping

During diffusion, the positron will be in a delocalized Bloch state with the positron density concentrated in the interstitial region, due to repulsion from the positively charged nuclei of the host lattice atoms. The transformation from a delocalized state to a localized positron state is called trapping. In a sample containing positron traps, trapping is competing with annihilation. Hence, in order for the trapping to be observable in an experiment, it has to be fast enough, i.e., the concentration of positron traps has to be high enough.

A typical positron trap is an open-volume defect, where the missing atom lowers the repulsion for positrons. The positron senses this as a potential well, and consequently, this creates a localized state for the positron.

The parameter describing the positron trapping into a defect is the defect-trapping rate κ_D:

$$\kappa_D = \mu_D c_D \tag{6.4}$$

In (6.4), μ_D is the defect-specific trapping coefficient and c_D the concentration of the defect D. Puska *et al.* [20] studied theoretically positron trapping in semiconductors. For positively charged vacancies, μ_D is too small for trapping to occur during the short lifetime of a positron. Hence, positively charged vacancies are not detected with positrons. For neutral vacancies, positron trapping is temperature independent with a trapping coefficient of the order of $10^{14}-10^{15}$ s^{-1}. This corresponds to a value of $10^{-9}-10^{-8}$ cm^3 s^{-1}. For negatively charged vacancies, the trapping coefficient will be temperature dependent, increasing with decreasing temperature. There are two factors contributing to the temperature dependence. First, the Coulomb interaction between the negatively charged vacancy and positron leads to a $T^{-1/2}$ dependence of the trapping coefficient. Second, at low temperatures, shallow Rydberg states at the vacancy act as precursors for the final trapping into the localized state. This leads to an exponential temperature dependence for the trapping coefficient. At room temperature (RT), this transfers into roughly an order of magnitude lower detection limit for negatively charged vacancies compared to neutral vacancies.

6.2.5 The kinetic trapping model and positron lifetime spectroscopy

For lifetime studies with fast positrons, where positron diffusion does not have to be taken into account, the practical way of treating lifetime data is using the conventional trapping model introduced by West [21]. The trapping model is based on three assumptions:

1. At time $t = 0$, the positron is in a free Bloch state.
2. The positron-trapping rate κ_D is given by $\kappa_D = \mu_D c_D$, where μ_D is the defect-specific trapping rate, and c_D is the defect concentration.
3. A trapped positron may escape from a trap. The detrapping rate is denoted by δ_D.

In a conventional lifetime experiment, where the width of the positron implantation profile is several orders of magnitude wider than the positron diffusion length, the positron annihilation can be described by a set of linear rate equations:

$$\frac{dn_B}{dt} = -\left(\lambda_B + \sum_i \kappa_{D_i}\right) + \sum_i \delta_{D_i} n_{D_i} \tag{6.5}$$

$$\frac{dn_{D_i}}{dt} = -\kappa_{D_i} n_B - (\lambda_{Di} + \delta_{D_i}) n_{D_i} \tag{6.6}$$

The sum in (6.5) is over all the defects D_i. Point (1) in the trapping model gives the boundary conditions for n_B and n_{D_i},

$$n_B(0) = 1 \quad n_{D_i} = 0. \tag{6.7}$$

The probability for the positron to still exist at time t:

$$n(t) = n_B(t) + \sum_i n_{D_i}(t) = \sum_i I_i e^{-\lambda_i t} \tag{6.8}$$

The decay constants λ_i and intensities I_i ($\sum I_i = 1$) are found by solving (6.5) and (6.6). In an actual experiment, the lifetime spectroscopy measures the annihilation probability as function of time, i.e.,

$$-\frac{dn(t)}{dt} = \sum_i I_i \lambda_i e^{-\lambda_i t}. \tag{6.9}$$

By combining (6.5), (6.6) and (6.7), a simple sum rule can be found

$$-\frac{dn}{dt}\Big|_{t=0} = \lambda_B = \sum_{i=1}^{N+1} I_i \lambda_i, \tag{6.10}$$

which can be used for checking if the decomposition is compatible with the trapping model.

A useful quantity of a lifetime spectrum is the average positron lifetime τ_{av}, which is the center of mass of the measured spectrum:

$$\tau_{av} = \int_0^\infty t\left(-\frac{dn(t)}{dt}\right)dt = \sum_{i=1}^{N+1} I_i\tau_i \tag{6.11}$$

The average positron lifetime is perhaps the most important experimental quantity in a lifetime experiment, since it is independent of the trapping model. Hence, τ_{av} can be deduced from the spectrum without any knowledge on the decomposition. Furthermore, the measured positron lifetime will be a superposition of the characteristic lifetimes of the different annihilation states in a sample. Hence,

$$\tau_{av} = \eta_B\tau_B + \sum_j \eta_{D_j}\tau_{D_j}, \tag{6.12}$$

$$\tau_{av} = \frac{\lambda_B}{\lambda_B + \sum \kappa_j}\tau_B + \frac{\sum \kappa_j\tau_{D_j}}{\lambda_B + \sum \kappa_i} \tag{6.13}$$

It is worth noting that the annihilation fractions η_j are different from the lifetime intensities I_i. For details on the derivation of (6.12) and (6.13), see Appendix A2 in [22]. In a simple case where there are only two annihilation states, bulk (B) and vacancy (D), the annihilation fractions in (6.13) are

$$\eta_B = \frac{\lambda_B}{\lambda_B + \kappa_D} \quad \text{and} \quad \eta_V = \frac{\kappa_D}{\lambda_B + \kappa_D}. \tag{6.14}$$

Based on (6.13) and (6.14), κ_D needs to be at least 1% of λ_B for the detection of defects. When κ_D increases above (50–100) λ_B PAS loses its sensitivity for defects. In the case of one defect species, this phenomenon is called saturation trapping.

In a typical tabletop lifetime experiment, the event for the lifetime spectrum is the time difference between the birth photon of the positron (the 1.27 MeV γ-photon from ^{22}Ne when a ^{22}Na source is used) and the 511 keV annihilation photon. Usually, two plastic scintillators connected to photomultiplier tubes are used for the detection of the photons. The time resolution of such systems is typically 200–250 ps, which is good enough to reliably observe changes in τ_{av} of the order of 1 ps. Whether lifetime components of different defects can be decomposed from the lifetime spectrum will depend on several factors, e.g., on the lifetime difference between the components and the intensities of the different components. As a rule of thumb, separation of lifetime components is possible if $\tau_{i+1}/\tau_i > 1.3-1.5$ [18]. Typically, at most three lifetime components can be separated from a lifetime spectrum. The positron lifetime in a typical monovacancy size defect is 10%–20% above the lifetime for the positron in a trap-free sample.

6.3 Doppler broadening techniques

6.3.1 Doppler broadening spectroscopy

When a positron eventually meets its end by encountering an electron in a sample, the annihilating positron–electron pair will have momentum. This momentum will cause a Doppler shift in the annihilation radiation

$$\Delta E = \frac{1}{2} c p_L, \tag{6.15}$$

with the longitudinal momentum component p_L of the annihilating positron–electron pair in the direction of the emitted annihilation photon. This shift causes the 511 keV annihilation line to broaden accordingly.

In a Doppler broadening experiment, the energy distribution of the emitted annihilation photons is measured with an energy-sensitive high-purity Ge detector, see Figure 6.2. The energy resolution of such a detector is of the order of 1 keV at the energy of the annihilation photon. As the width of the Doppler broadened annihilation line is 2–3 keV, the resolution will heavily influence the shape of the measured spectrum. Hence, Doppler broadening results are commonly presented with the shape parameters S and W, see Figure 6.2. The low-momentum S parameter is defined as the fraction of counts ($S = N_S/N_{tot}$) in the central part of the annihilation line. The energy window is equipment dependent and typically chosen to include roughly 50% of the total counts in the annihilation line, i.e., the S-parameter value is around 0.5. In a semiconductor, the S parameter describes

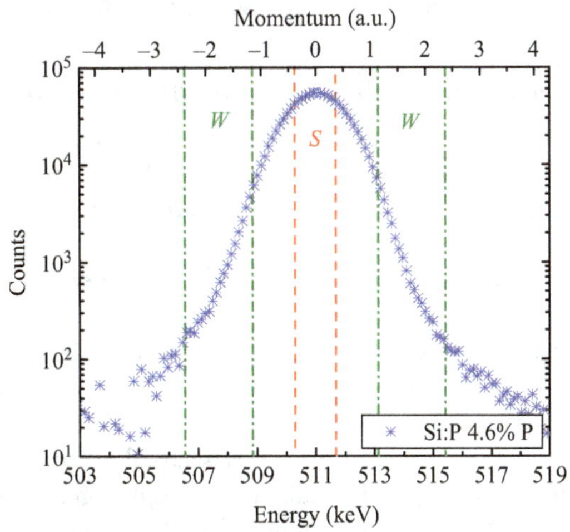

Figure 6.2 Background reduced Doppler broadening spectrum from a Si:P epilayer. Indicated in the figure are typical S and W integration windows [23]

annihilation with mainly low-momentum valence electrons. The *W* parameter is similarly defined as the fraction of counts in the high-momentum wing region of the annihilation line, i.e., it describes annihilation with mainly high-momentum core electrons. The *W*-parameter windows are chosen symmetrically on the low- and high-energy sides of the annihilation line. For a particular experimental setup, standard windows are usually used, e.g., on the beams at Aalto University, the windows S: $0 < |p| < 0.5$ a.u., W: $1.5 < |p| < 3.5$ a.u. are used as default windows. However, it is worth noting that especially when new materials are studied, it can be worthwhile to trim the *W*-parameter windows in order to obtain the best electron momentum sensitivity for a particular set of samples.

As the absolute values of the Doppler parameters are heavily dependent on the width of the energy windows and the resolution of the detector, parameters normalized to a (defect free) reference value are presented when it is feasible. This makes comparison of results measured with different equipment possible.

Similarly to the positron lifetime experiment, the measured *S* and *W* parameters are superpositions of the characteristic parameters for the different annihilation states in a sample:

$$S = \eta_B S_B + \sum_j \eta_{D_j} S_{D_j}, \quad W = \eta_B W_B + \sum_j \eta_{D_j} W_{D_j} \tag{6.16}$$

In a positron beam experiment, the annihilation fractions at different annihilation states typically change as a function of beam energy. If the superposition involves only two annihilation states, there will be a linear relation between the measured *S* and *W* parameters. Hence, e.g., the plot of the *W* parameter as a function of the *S* parameter in the *SW* plane will form a line, with a characteristic slope that depends on the involved annihilation states, see Figures 6.10 and 6.13.

6.3.2 Coincidence Doppler broadening spectroscopy

In a regular single Ge-detector measurement with a ^{22}Na source, the background in the measured spectrum is heavily influenced by the 1.27 MeV γ-photon. This limits the energy sensitivity in the experiment to approximately 511 ± 4 keV. To reduce the background and improve the sensitivity, a coincidence setup can be used. In such a setup, both annihilation photons are detected. A coincidence setup can be realized in two ways. The second detector can only be used for gating purposes; in this case, a simple and cheaper scintillator can be utilized. However, a more effective way is to use two energy-sensitive Ge detectors. In such a setup, the additional energy conservation condition ($E_{\gamma_1} + E_{\gamma_2} = 1{,}022$ keV) effectively reduces the background close to zero. As an additional bonus, the resolution of the system is improved by a factor of $\sqrt{2}$ [24]. In a so-called 2D-Doppler broadening measurement, the momentum distribution of the annihilating electron–positron pairs can be measured up to 9 keV, corresponding to a momentum range of approximately $0 < |p| < 5$ a.u. Figure 6.3 shows the complete 2D matrix, with the axes being the energy axes of the two detectors. One of the diagonals (left–right in Figure 6.3) gives the actual 2D-spectrum, the other diagonal (up–down in Figure 6.3) gives the

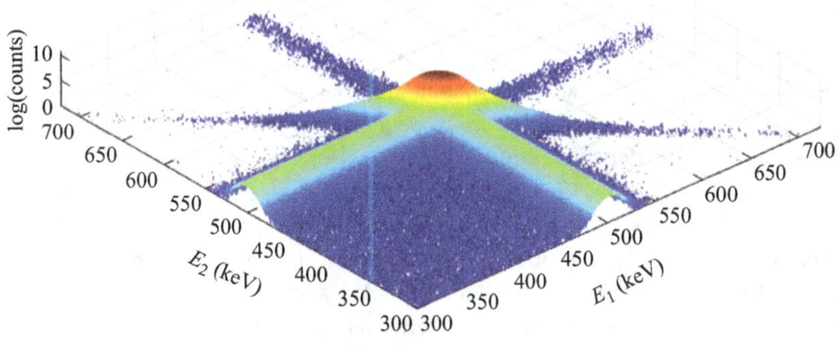

Figure 6.3 Data matrix from a 2D-Doppler experiment. The E_1 and E_2 axes denote the energy scales for the two detectors. The vertical axis is the logarithm of the detected counts in each matrix element

resolution function of the two-detector system. An added advantage of the coincidence technique is the possibility to compare experiment with theoretical simulations.

6.3.3 Supporting theory for defect identification

Combining positron annihilation experiments such as positron lifetime and especially coincidence Doppler broadening with supporting modeling improves the possibilities to really *identify* vacancy-type defects [18]. The practical modeling of positron states and annihilation in solids is always based on the so-called two-component density functional theory (DFT) for electron–positron systems [25]. Defects are usually modeled using periodic boundary conditions and supercell models.

The two-component DFT is a generalization of the very same density-functional formalism that, in addition to many other applications in materials science and chemistry, is widely used for total-energy calculations of point defects in semiconductors [26]. It is based on expressing the energy functional of the electron–positron system in the external potential due to nuclei as a functional of both the electron and positron densities. In addition to the exchange–correlation energy functional familiar from one-component DFT, one has to approximate the so-called electron–positron correlation energy functional $E_c^{e-p}[n_+, n_-]$, which represents essentially the many-body interaction energy between the positron and its screening electron cloud.

Once the atomic structure and densities of the electrons, $n_-(\mathbf{r})$, and the single positron in the system, $n_+(\mathbf{r})$, have been calculated using two-component DFT, the annihilation rate λ (the inverse of the positron lifetime τ) can be calculated as

$$\lambda = \frac{1}{\tau} = \pi r_0^2 c \int \gamma(\mathbf{r}) n_+(\mathbf{r}) n_-(\mathbf{r}) \, d\mathbf{r}, \tag{6.17}$$

where c is the speed of light and r_0 the classical radius of electron. The so-called enhancement factor $\gamma(\mathbf{r})$ (the value of the electron–positron pair correlation function at zero distance) takes into account the screening of the positron by electrons, which is a many-body effect. In general, the enhancement factor is another unknown non-local functional of the electron and positron densities, $\gamma(\mathbf{r}) = \gamma(\mathbf{r}; [n_+, n_-])$. The simplest approximation is again the local-density approximation (LDA) based on many-body calculations made for homogenous electron–positron systems, where the enhancement factor is at each point in space assumed to be the same as in a homogeneous system with densities equal to the local ones, $\gamma(\mathbf{r}) = \gamma^{hom}(n_+(\mathbf{r}), n_-(\mathbf{r}))$. For a delocalized positron, the appropriate limit is the zero-positron-density limit, and the LDA becomes a function of the electron density alone, $\gamma(\mathbf{r}) = \gamma(n_-(\mathbf{r}))$. The main LDA parametrizations used include the zero-positron density limit parametrization of Boroński and Nieminen [25] and the general two-component functional by Puska, Seitsonen and Nieminen [27]. Models beyond the LDA include mainly the semi-local generalized-gradient approximations by Barbiellini *et al.* [28–31], and the weighted-density approximation that is needed to describe the positron surface states, which always involve fully non-local electron–positron correlations [32–35].

In principle, localized positrons should be modeled using fully self-consistent two-component DFT. However, the following simpler scheme has been shown to give consistent results. (1) The electronic structure is modeled first in the absence of the positron and (2) the positron is added in the system in a postprocessing manner similarly as in the case of a delocalized positron, namely, taking the zero-positron density limits of the electron–positron correlation potential and the enhancement factor. The repulsive forces on ions due to the localized positron, which can have a considerable effect on the defect structure, can be considered within this model as well [36].

For modeling the momentum density of annihilating electron–positron pairs, we ourselves have used consistently the model by Alatalo *et al.* [37],

$$\rho(\mathbf{r}) = \sum_j \gamma_j \left| \int e^{-i\mathbf{p}\cdot\mathbf{r}} \psi_+(\mathbf{r}) \psi_j(\mathbf{r}) \, d\mathbf{r} \right|^2, \tag{6.18}$$

where the summation goes over occupied electron states, and the state-dependent enhancement factor is written as $\gamma_j = \lambda_j^{LDA} / \lambda_j^{IPM}$, where λ_j^{LDA} is the annihilation rate of state j,

$$\lambda_j = \pi r_0^2 c \int \gamma(n_-(\mathbf{r})) n_+(\mathbf{r}) |\psi_j(\mathbf{r})|^2 \, d\mathbf{r}, \tag{6.19}$$

evaluated with the LDA enhancement factor and λ_j^{IPM} the annihilation rate in the independent-particle model ($\gamma = 1$). An especially useful numerical method for supercell defect calculations is the projector-augmented wave method [38], within which it is possible to reconstruct the all-electron wave functions that are accurate up to high momenta [36,39,40].

Before comparing with experimental Doppler spectra, the three-dimensional momentum density of annihilating pairs is projected on the axis corresponding to the measurement geometry,

$$\rho(p_L) = \int\int \rho(\mathbf{p})\, dp_x\, dp_y, \tag{6.20}$$

and the resulting spectrum is convoluted with the experimental resolution function.

6.4 Temperature-dependent measurements in narrow bandgap semiconductors

As discussed earlier, negatively charged vacancies will have an increasing trapping coefficient with decreasing temperature. This will also apply to non-open-volume defects, e.g., ionized acceptors or negatively charged antisite defects in compound semiconductors. Hence, to get information on defect charge states and lifetime spectrum composition, measurements as a function of temperature can be conducted.

Furthermore, in a positron annihilation measurement, a varying measurement temperature can be used for moving the Fermi level within the bandgap of the semiconductor. In a best case scenario, this enables observation of charge-state transitions and to estimations of the charge-transition level [41,42]. PAS has also readily been utilized for point-defect annealing studies, since it is fairly easy to realize an in situ annealing setup, also in situ irradiations with subsequent in situ annealing experiments have been realized [43–45].

6.4.1 Vacancy annealing in Ge

Even though germanium has been studied for decades, defects in germanium have only been extensively investigated since the late 1990s [46]. Among the fundamental defect properties that were largely unknown or uncertain until recently were the stability of the divacancy [47–50] and the annealing/migration properties of the monovacancy [51–53]. However, the challenge with germanium has been that direct observation of point defects (with, e.g., electron paramagnetic resonance) has been difficult, and most studies have therefore relied on indirect spectroscopic tools. PAS has played a leading part in the determination of the vacancy properties of germanium, from pioneering work in the 1960s, 1970s and 1980s [54–59] to more recent work [60,61].

By performing low-temperature irradiation experiments with subsequent annealing and in situ PAS lifetime measurements, Slotte *et al.* were able to identify the monovacancy in germanium and its annealing stages [44]. Figure 6.4 shows the average positron lifetime measured at 35 K as a function of annealing temperature for a 10 MeV low-temperature proton irradiated weakly n-type Ge sample. The results showed two distinct annealing stages, at 100 K and 200 K. The first one coincided fairly well with previous reports [62–64] on the annealing of the Frenkel pair. However, no defect-specific lifetimes could be decomposed from the measured lifetime spectra since more than one positron trap was present at high enough

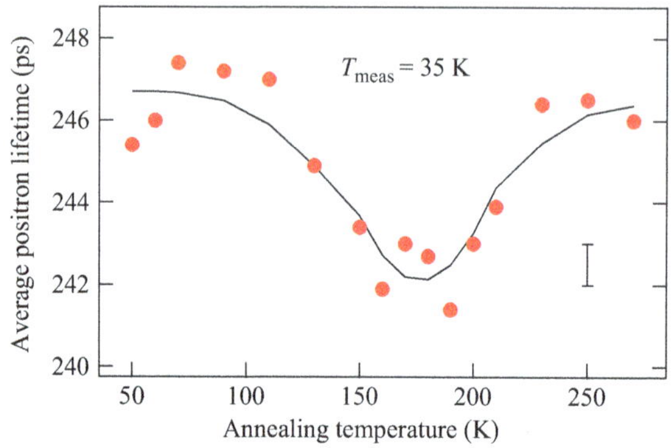

Figure 6.4 The average positron lifetime for a Ge sample irradiated at 35 K as a function of annealing temperature. The solid line is a guide to the eye. Indicated in the figure is also a typical error bar for the average lifetime. © 2011 by the American Physical Society. Reprinted, with permission, from [44]

concentration in the experiment, rendering the decomposition unsuccessful. After keeping the sample at RT for 2–4 weeks, a successful decomposition could be made, and the dominant positron trap after RT annealing was identified as the divacancy.

In an effort to identify the defect responsible for the 200 K annealing stage, a second irradiation experiment was performed at 100 K, with subsequent annealings and PAS lifetime measurements at 35 K. The result is summarized in Figure 6.5. The lifetime spectra could now be decomposed after annealings in the temperature interval 100 K–200 K, and a defect lifetime of 272 $\pm$ 4 ps could be associated with the Ge monovacancy. At 200 K, the monovacancy becomes mobile and a fraction of the monovacancies forms divacancies. This explains the increase in the defect lifetime τ_2 above 200 K and the decrease in intensity of this component.

6.4.2 Positron traps in low temperature irradiated GaSb

Undoped GaSb is p-type irrespective of the growth method [65], although the growth method seems to have an impact on the defect composition [66], furthermore n-type GaSb is heavily compensated. Both positron-lifetime and Doppler-broadening spectroscopy have been utilized for studying the cause of p-type conductivity in bulk GaSb. Ling *et al.* [67] measured Czochralski-grown undoped GaSb as function of temperature. They observed a decrease in average lifetime with decreasing temperature and interpreted this as a clear indication of positron trapping into a shallow trap. Ling *et al.* modeled the positron trapping and estimated the concentration of shallow traps to 2×10^{17} cm^{-3} and argued that the trap is the Ga_{Sb} antisite. Kujala *et al.* [68], Figure 6.6, noted that even though Ga_{Sb} plays a significant role in the p-type conductivity, the total acceptor concentration can be explained by a second Ga-related native defects, i.e., the Ga vacancy. A somewhat

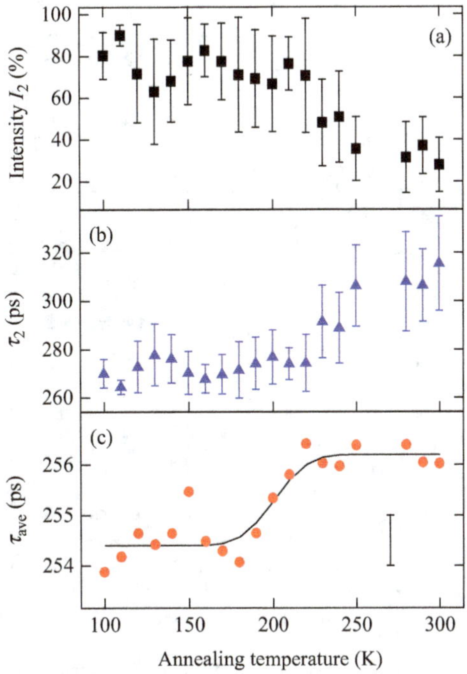

Figure 6.5 The average positron lifetime (c), defect component τ_2 (b) and τ_2 intensity (a) of a Ge sample irradiated at 100 K. The solid line is a guide to the eye. © 2011 by the American Physical Society. Reprinted, with permission, from [44]

surprising finding in [68] is that Ga_{Sb} seems to heavily influence the positron trapping up to 300 K and significantly influence positron trapping also above 300 K. Similar effects from non-open-volume defects have been observed in GaN [69].

GaSb is somewhat of a peculiarity also from a self-diffusion perspective. Bracht *et al.* [70] observed that the self-diffusion coefficients of Ga and Sb differ by three orders of magnitude. This was explained by the instability of the Sb vacancy, i.e., an Sb vacancy is transformed into a Ga vacancy and a Ga antisite defect. Hence, the concentration of vacancies to mediate self-diffusion on the Sb sublattice is several orders of magnitude lower than on the Ga sublattice, rendering a significantly lower self-diffusion coefficient for Sb compared to Ga.

Segercrantz *et al.* [45] performed both RT and low-temperature (35 K) irradiations with subsequent ex situ and in situ PAS lifetime measurements to study the balance of vacancy defects on the two sublattices in GaSb. Figure 6.7 shows the average positron lifetime in GaSb from a low-temperature irradiation experiment with subsequent in situ annealing and measurement at 35 K. The higher fluence

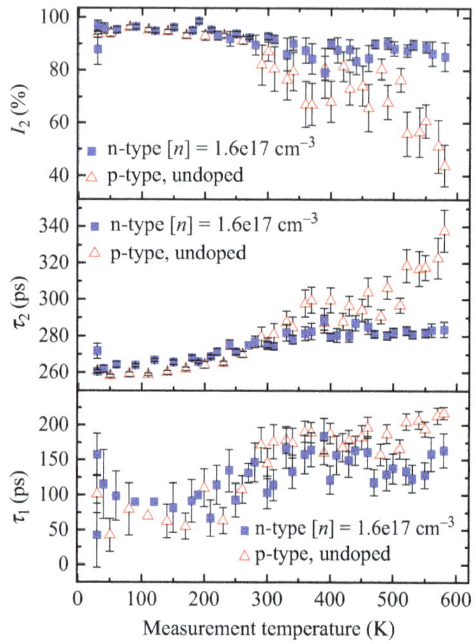

Figure 6.6 Decomposition results as a function of measurement temperature for n- and p (undoped)-type GaSb. For both samples, τ_2 decreases with decreasing measurement temperature. However, not even at 30 K does the defect component equal the expected bulk lifetime of approx. 245 ps. However, the roughly 100% intensity of τ_2 below 300 K is a clear indication that the two trap model is failing. Modified version of a figure published previously in [68]

result (7×10^{14} cm^{-2}) shows a clear annealing step at around 150 K, whereas the lower fluence results lack this step. This is explained by the differences in production rate for the two monovacancies V_{Sb} and V_{Ga} and by the significant concentration of V_{Ga} in the as-grown material. The differences in displacement energies on the two sublattices result in roughly 40% more Sb than Ga vacancies in the irradiation. A high enough fluence is needed to form enough Sb vacancies compared to Ga vacancies and make the annealing step at 200 K visible in the experiment. At lower fluences, the Ga vacancies in the as-grown material together with the Ga vacancies formed in the irradiation dominate the positron trapping. The Ga vacancy is stable well above RT, and therefore no annealing step is observed for the lower fluence sample (1×10^{14} cm^{-2}). The annealing step observed at 150 K for the higher fluence sample is associated with the annealing of the Sb vacancy. Segercrantz *et al.* estimated the activation energy for the annealing of the Sb vacancy (dashed line in Figure 6.7) and obtained a value of 0.6 ± 0.1 eV for the energy barrier.

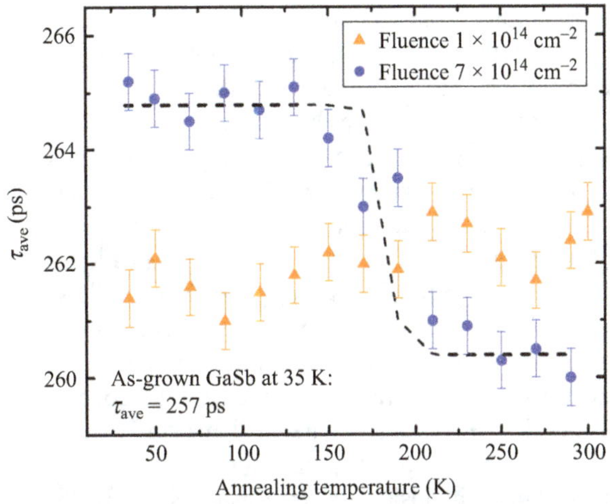

Figure 6.7 Measured τ_{av} as a function of annealing temperature. The positron measurements were performed at 35 K. The dashed curve is a fit to the data measured for the sample irradiated with a fluence of 7×10^{14} cm^{-2}. © 2017 by the American Physical Society. Reprinted, with permission, from [45]

6.5 Acceptor-like defects in wide bandgap semiconductors

Coincidence Doppler broadening measurements coupled with the theoretical modeling of positron states and annihilation at defects can at best provide very detailed information on the atomic structures of the defects trapping positrons. This section provides some examples of recent work made on GaN, its II–VI counterpart ZnO and semiconducting oxides.

6.5.1 Cation vacancies and acceptor impurities in GaN and ZnO

GaN and ZnO are among the materials whose defects have been studied the most using positron spectroscopy methods and whose results are best understood [18,71–73]. Both have a wurtzite lattice structure with a similar lattice parameter. From the positrons' point of view, these materials and their defects are extremely alike. Instead of the energy-band structure or gap, localized positrons mostly sense the atomic structure, and Ga and Zn or N and O are not that different from one another in terms of their effective sizes or atomic shell structures.

Also in these materials, the positrons are mainly sensitive to the cation vacancies that have been detected [74] and identified using coincidence Doppler broadening [75–78]. In the case of GaN (as well as AlN [79]), positrons can sense a difference between vacancies in as-grown and irradiated samples, which is attributed to the existence of Ga vacancy–oxygen complexes in the as-grown material [75].

For both GaN and ZnO, one of the main challenges has always been the effective p-type doping of the materials, a challenge that has not yet been solved for ZnO. In the case of GaN, the activation of the Mg acceptor is still very low, on the order of 1%.

Mg and Be doping of GaN as well as Na and Li doping of ZnO have been studied extensively using positrons [80–86]. The effect of these potential dopants on the positron annihilation signal is remarkably similar. Even though negative ions do trap positrons into hydrogenic states at low temperatures, in the case of Mg in GaN [82] and Na in ZnO [85], PAS does not really enable identifying the cation-site acceptors Mg_{Ga} and Na_{Zn} (although it is likely that the V_N–Mg_{Ga} complex in GaN does trap the positrons [82]). These acceptors effectively "block" positrons from trapping at a localized and strongly bound state at these cation sites, while the lowering of the Fermi lever simultaneously reduces the cation vacancy concentration. Concerning the lighter Be and Li, Be_{Ga} and Li_{Zn} in the respective materials have sufficient open volume to trap positrons [86,87]. From the positrons' point of view, they are simply large cation vacancies with a small impurity inside. Positron spectroscopy methods have been used to study the interaction of diffusing Na with traps such as Li_{Zn} and V_{Zn} [85].

The positron annihilation signal of these defects is rather close to the signal measured for defect-free bulk, however, with a slightly higher single component lifetime (<10–15 ps difference). Careful coincidence Doppler broadening experiments (Figures 6.8 and 6.9 display the so-called ratio curves, in which the spectra

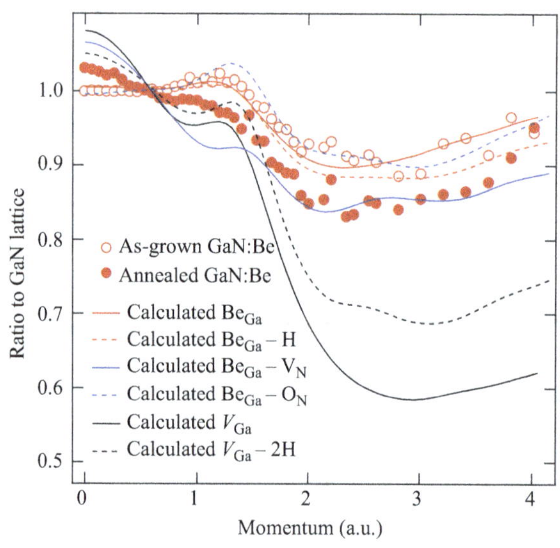

Figure 6.8 Ratio curves (the spectrum divided by a reference spectrum) obtained from coincidence Doppler broadening experiments in Be-doped Ga-polar samples (for details, see [87]) and from theoretical calculations. The comparison of experimental and theoretical data clearly shows that defects related to Be_{Ga} *are detected in Be-doped GaN. © 2017 by the American Physical Society. Reprinted, with permission, from [87]*

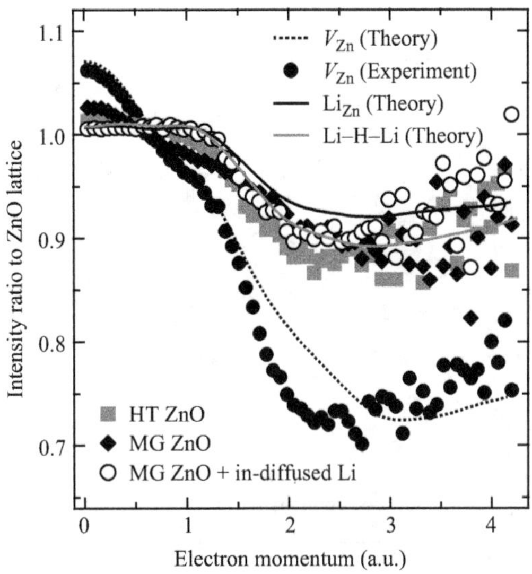

Figure 6.9 Ratio curves (the spectrum divided by a reference spectrum) from coincidence Doppler broadening measurements for as-grown hydrothermal (HT) ZnO, as-grown melt grown (MG) ZnO, and Li-doped MG-ZnO as compared to the theoretical results obtained for V_{Zn}, Li_{Zn} and the Li_{Zn}–H–Li_{Zn} complex. The experimental data obtained for a ZnO sample irradiated with oxygen ions with high concentration of V_{Zn} is shown for comparison [78]. © 2011 by the American Physical Society. Reprinted, with permission, from [86]

measured for the samples are divided by that of a defect-free bulk sample) display a signal that is distinguishable from the bulk signal but whose characteristic features are different from those of cation-vacancy-related defects [86,87]. This signal can be reproduced with modeling when Be_{Ga} in GaN and Li_{Zn} in ZnO are considered (Figures 6.8 and 6.9; and [86,87]). The modeling of a large set of candidate defects as well as past experiments have helped to rule out other candidate defects (see especially [87–89]). Figure 6.10 from [87] displays the measured GaN:Be data in the *SW* plane along with calculated defect-specific values for many candidate defects involving for instance hydrogen. The data is consistent with the Be_{Ga} defect and, furthermore, cannot be reproduced as a linear combination other annihilation states.

6.5.2 Cation and oxygen vacancies in metal oxides

There has been increased interest to study metal oxides such as transparent (semi) conducting oxides and perovskites using PAS during the past decade (see, e.g., [18,73,89] and references therein). The main complication is the often poor crystalline quality of the samples (as compared to more "traditional" elemental and compound semiconductors) and thereby the difficulty of finding suitable defect-free reference

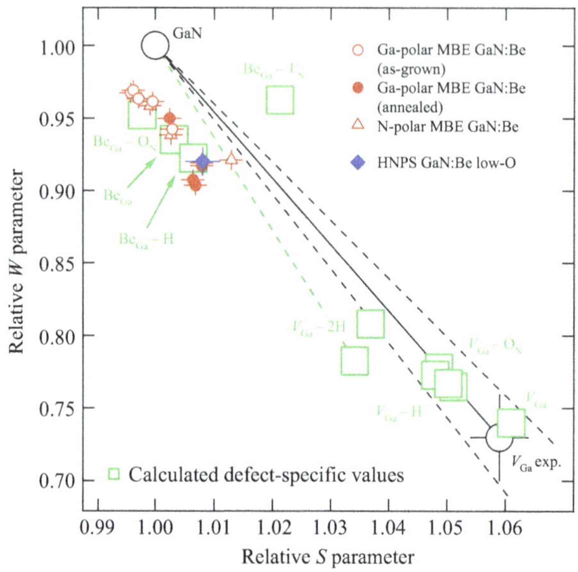

Figure 6.10 (S, W) plot with characteristic data from a variety of GaN samples. The S and W parameters are shown as normalized to those of the GaN lattice. © 2017 by the American Physical Society. Reprinted, with permission, from [87]

samples. One of the only, if not the only, well-understood system is ZnO [73]. Often also the crystal structures are complicated, especially in the case of the ternary and quaternary "complex oxides." Even binary oxides such as In_2O_3 can have a rather large supercell (40 atoms in the primitive cell, 80 atoms in the unit cell) and inequivalent cation and oxygen sites.

Doppler broadening techniques combined with accurate modeling of positron states and annihilation at defects can be a powerful tool in identifying the defects and defect complexes seen in experiments. However, the interpretation of measurement data for oxides is typically more challenging due to the lack of reference samples and the more complicated and versatile vacancy defect structures present. One typically needs to consider several batches of defected and (hopefully) defect-free samples in order to see what kinds of trends can be seen in the data for the particular material and have the ability to perform first-principles modeling in order to understand what these trends actually mean at the defect level. In the case of the "traditional" well-known semiconductors, a line in the S–W plot (a plot where W is plotted as a function of S or vice versa [90]) typically represents a mixture of two annihilation states: the delocalized bulk state at a defect-free crystal and a well-defined vacancy state. In the case of metal oxides, the origin of such a line is usually not as obvious. It can equally well represent a mixture of two different annihilation states (bulk and a vacancy or two different vacancies) or a change in vacancy complexes' size and/or structure [88]. Completely defect-free samples where positron predominantly annihilates in the delocalized bulk state are scarce and often unavailable.

In the course of our first detailed studies of metal oxides using positron annihilation supported by theory (In_2O_3 [88,91], SnO_2 [88,92] and ZnO [86,88,93]), we have learned that the sample studies often contain not only cation vacancies but also larger vacancy complexes involving both cation and oxygen vacancies. Isolated O vacancies are not expected to trap positrons due to their small open volume and positive charge. The key benefit in the defect characterization is the sensitivity of the Doppler broadening technique to the O environment around positron-trapping and annihilation sites [88].

Figure 6.11 illustrates how annihilation with O ions, more specifically their 2p shells, around cation vacancy defects dominates the high-momentum part of Doppler spectra in most systems. Chances in the local O concentration in the

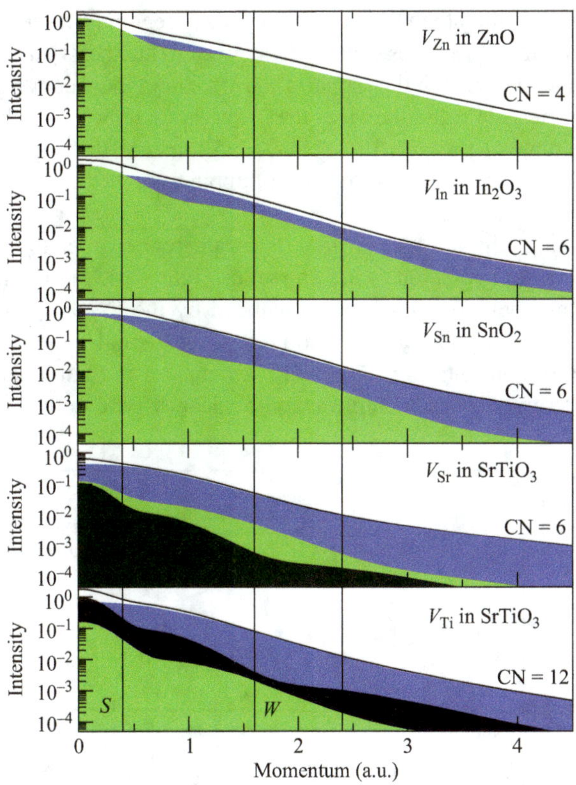

Figure 6.11 Decompositions of Doppler spectra calculated for cation monovacancies into annihilation with cations (green and black for A and B in ABO_x) and O (blue behind) displayed as superimposed segments. The calculation is based on a simple model based on superposition of free atoms in a lattice. CN denotes the coordination number of the cation in question and vertical lines mark the typical positions of the S and W windows. Simplified version of a figure published previously in [88]

vicinity of positron-trapping and annihilation sites will considerably affect the lineshape parameters S and W. Once the effect of the defect microstructure on Doppler broadening is understood through modeling, we can understand what is happening to defects in a sample series as a function of some parameter or treatment, such as annealing, surfactant concentration or irradiation dose [88].

Figure 6.12 displays an example of depth profiling of a thin film using a slow-positron beam. The samples are undoped (or unintentionally doped, uid) and Sb-doped SnO_2 thin films grown on a sapphire substrate [92,94]. The figure shows the S parameter as a function of implantation energy for a selection of SnO_2 samples. The curves exhibit a somewhat similar form within three regions: surface, oxide film and substrate. At low energies ($E < 3$ keV), the effect of surface annihilations dominates, and the S parameter gradually drops as the energy increases. For intermediate energies ($E = 5-15$ keV for the Sb-doped samples and $E = 4-7$ keV for the uid sample), positrons annihilate in the film and a plateau region is observed. This area is wider for the Sb-doped samples due to their greater thickness. At higher energies ($E > 20$ keV), positrons reach the sapphire substrate, and the S converges toward the substrate's characteristic value shown as the lower dashed line. The other dashed line shows the S parameter measured in a SnO_2 bulk crystal that has been shown to contain negligible concentrations of positron-trapping vacancy defects and is used as a reference in the following.

The SnO_2-thin-film-specific S and W parameters extracted from the depth profiles of all the studied samples are shown in Figure 6.13. For comparison with calculated values, they are shown as normalized by the (S, W) parameters of the SnO_2 bulk crystal. The measurement points fall on a single line and evolve as a function of the Sb concentration. This kind of behavior is traditionally interpreted as non-saturation trapping to a single specific vacancy-type defect, in which case,

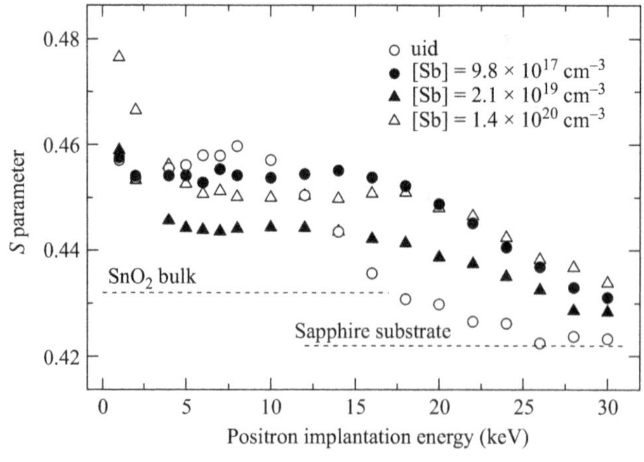

Figure 6.12 S parameter as a function of positron implantation energy for selected Sb-doped SnO_2 samples studied in [92]. The bulk SnO_2 and substrate values are marked by dash lines

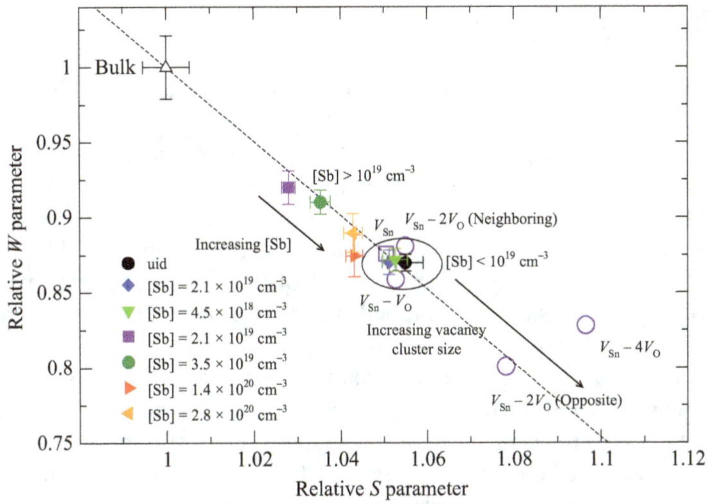

Figure 6.13 *(S, W) plot for Sb-doped SnO_2 including both experimental [92] (solid markers) and computational [88] (open markers) results. The parameters are shown as normalized to those of the SnO_2 lattice*

the location of the point on the line depends on the vacancy concentration. However, the behavior of this SnO_2 data as a function of Sb concentration is rather peculiar. The extreme S/W values (highest S and lowest W) correspond to the lowest Sb concentrations ($<10^{19}$ cm^{-3}), and there is a discontinuous behavior with increasing Sb concentration. Only at larger concentrations ($>10^{19}$ cm^{-3}), the trend can be explained by thermodynamics, i.e., the data is consistent with an increase in the cation vacancy concentration with increasing Sb concentration and increasing Fermi level. The microscopic interpretation based on computational results [88] in Figure 6.13 is such that at low-Sb concentrations, the positron trapping is in saturation and the traps are complexes formed by Sn vacancy and two or more oxygen vacancies.[1] Once the Sb concentration is increased enough, the growth quality improves (Sb acts as a surfactant), and positrons detect isolated Sn vacancies whose concentration is initially below the upper sensitivity limit and increases with increasing Sb concentration.

Even though positron annihilation is expected to be insensitive to isolated oxygen vacancies, the technique still provides an indirect way to detect them. Positrons are always very sensitive to cation vacancy defects, even if their concentration is much smaller. If there is a large enough concentration of oxygen vacancies, they will start to complex with cation vacancies. This phenomenon is possible to detect with careful Doppler broadening measurements and supporting theory.

[1]Please note that absolute S/W values are difficult to compare between experiment and theory. Instead, one should rather look at trends among a set of samples or defects considered.

6.6 Summary and outlook

In summary, PAS gives microscopic information about vacancy defects in semiconductors in the concentration range 10^{15}–10^{19} cm^{-3}. The positron lifetime is the fingerprint of the open volume associated with a defect, and it can be used to identify mono- and divacancies and larger vacancy clusters. Doppler broadening of the annihilation radiation, on the other hand, can be used to identify the nature of the atoms surrounding the vacancy. Consequently, vacancies on different sublattices of a compound semiconductor can be distinguished, and impurities associated with the vacancies can be identified. The charge state of a vacancy defect can be determined by the temperature dependence of the positron-trapping coefficient, and positron localization into Rydberg states around negative centers yields information about ionic acceptors that have no open volume. Importantly, as shown in this chapter, the methods based on positron annihilation are not restricted by the nature or physical dimensions of the semiconductor. Defects can be studied in narrow- and wide-bandgap semiconductor materials in samples of any conductivity. Bulk crystals as well as thin films can be subjected to the experiments and defects identified.

The diffusion of thermalized positrons poses challenges for performing positron annihilation studies in materials systems with reduced dimensionality. However, in certain cases, positrons can be trapped at heterointerfaces within semiconductor devices—for example, at the AlGaN/GaN interface of a high-electron mobility transistor or at the InGaN/GaN interface in a light-emitting diode. There is a considerable amount of ongoing work aiming at elucidating the positron behavior in such systems, including the trapping phenomena in other nanostructured materials. The possibility of analyzing point-like defects in detail with positrons in real device structures is around the corner.

References

[1] Brandt W, Paulin R. Phys Rev B. 1977;15(5):15.
[2] Khatri R, Charlton M, Sferlazzo P, *et al.* Appl Phys Lett. 1990;57:2374.
[3] Mills AP, Gullikson EM. Appl Phys Lett. 1986;49:11211.
[4] Beling CD, Simpson RI, Charlton M, *et al.* Appl Phys A. 1987;42(2):111.
[5] Brauer G, Anwand W, Nicht EM, *et al.* Appl Surf Sci. 1997;116:19.
[6] Al-Qaradawi IY, Sellin PA, Coleman PG. Appl Surf Sci. 2002;194(1–4):29.
[7] Hugenschmidt C, Kögel G, Repper R, *et al.* Nucl Instrum Methods Phys Res B. 2004;221:160.
[8] Schut H, van Veen A, de Roode J, *et al.* Mater Sci Forum. 2004;445–446:507.
[9] Hawari AI, Gidley DW, Xu J, *et al.* AIP Conf Proc. 2009;1099:862.
[10] Cassidy DB, Greaves RG, Deng SHM, *et al.* AIP Conf Proc. 2004;1099:866.
[11] Krause-Rehberg R, Jungmann M, Krille A, *et al.* J Phys Conf Ser. 2011;262:012003.
[12] Valkealahti S, Nieminen RM. Appl Phys A. 1984;35:51.

[13] Makhov AF. Sov Phys Solid State. 1961;2:1934.

[14] Vehanen A, Saarinen K, Hautojärvi P, *et al.* Phys Rev B. 1987;35(10):4606.

[15] Jensen KO, Walker AB. J Phys: Condens Matter. 1990;2:9757.

[16] Nissilä J, Saarinen K, Hautojärvi P. Phys Rev B. 2001;63:165202.

[17] Boev OV, Puska MJ, Nieminen RM. Phys Rev B. 1987;36(15):7786.

[18] Tuomisto F, Makkonen I. Rev Mod Phys. 2013;85:1583.

[19] Soininen E, Mäkinen J, Beyer D, *et al.* Phys Rev B. 1992;46(20):13104.

[20] Puska MJ, Corbel C, Nieminen RM. Phys Rev B. 1990;41(14):9980.

[21] West RN. In: Hautojärvi P, editor. Positrons in Solids. Heidelberg: Springer Verlag; 1979.

[22] Krause-Rehberg R, Leipner HS. Positron Annihilation in Semiconductors. Berlin: Springer; 1999.

[23] Kujala J. Aalto University Publication Series, Doctoral dissertations; 2016. p. 223.

[24] MacDonald JR, Lynn KG, Boie RA, *et al.* Nucl Instrum Methods. 1978;153:189.

[25] Boroński E, Nieminen RM. Phys Rev B. 1986;34:3820–3831.

[26] Freysoldt C, Grabowski B, Hickel T, *et al.* Rev Mod Phys. 2014;86:253–305.

[27] Puska MJ, Seitsonen AP, Nieminen RM. Phys Rev B. 1995;52:10947–10961.

[28] Barbiellini B, Puska MJ, Torsti T, *et al.* Phys Rev B. 1995;51:7341–7344.

[29] Barbiellini B, Puska MJ, Korhonen T, *et al.* Phys Rev B. 1996;53:16201–16213.

[30] Kuriplach J, Barbiellini B. Phys Rev B. 2014;89:155111.

[31] Barbiellini B, Kuriplach J. Phys Rev Lett. 2015;114:147401.

[32] Jensen KO, Walker AB. J Phys F. 1988;18(12):L277–L285.

[33] Rubaszek A. Phys Rev B. 1991;44:10857–10868.

[34] Callewaert V, Saniz R, Barbiellini B, *et al.* Phys Rev B. 2017;96:085135.

[35] Shi W, Callewaert V, Barbiellini B, *et al.* Phys Rev Lett. 2018;121:057401.

[36] Makkonen I, Hakala M, Puska MJ. Phys Rev B. 2006;73:035103.

[37] Alatalo M, Barbiellini B, Hakala M, *et al.* Phys Rev B. 1996;54:2397–2409.

[38] Blöchl PE. Phys Rev B. 1994;50:17953–17979.

[39] Ishibashi S. In: Positron Annihilation – ICPA-13. vol. 445 of Materials Science Forum. Trans Tech Publications: Zurich; 2004. pp. 401–403.

[40] Makkonen I, Hakala M, Puska MJ. J Phys Chem Solids. 2005;66(6):1128–1135.

[41] Slotte J, Saarinen K, Salmi A, *et al.* Phys Rev B. 2003;67:115209.

[42] Kuitunen K, Tuomisto F, Slotte J. Phys Rev B (BR). 2007;76:233202.

[43] Coleman PG, Burrows CP. Phys Rev Lett. 2007;98:265502.

[44] Slotte J, Kilpeläinen S, Tuomisto F, *et al.* Phys Rev B. 2011;83:235212.

[45] Segercrantz N, Slotte J, Tuomisto F, *et al.* Phys Rev B. 2017;95(18):184103.

[46] Vanhellemont J, Simoen E. J Electrochem Soc. 2007;154:H572.

[47] Peaker AR, Markevich VP, Murin LI, *et al.* Mater Sci Eng, B. 2005;124–125:166.

[48] Kolkovsky V, Petersen MC, Larsen AN. Appl Phys Lett. 2007;90:112110.

[49] Kolkovsky V, Petersen MC, Larsen AN, *et al.* Mater Sci Semicond Process. 2008;11:336.
[50] Kuitunen K, Tuomisto F, Slotte J, *et al.* Phys Rev B. 2008;78:033202.
[51] Whan RE. Phys Rev. 1965;140:690.
[52] Ershov SN, Pantaleev VA, Nagornkh SN, *et al.* Sov Phys Solid State. 1977;19:187.
[53] Shaw D. Phys Status Solidi B. 1975;72:11.
[54] Fabri G, Poletti G, Randone G. Phys Rev. 1966;151:356.
[55] Fieschi R, Gainotti A, Ghezzi C, *et al.* Phys Rev. 1968;175:383.
[56] Yeh CK, Meng HC, Lee YH, *et al.* Radiat Eff. 1974;24:95.
[57] Arifov PU, Arutyunov NY. Sov Phys Semicond. 1980;14:159.
[58] Moser P, Pautrat JL, Corbel C, *et al.* In: Jain PC, Singru M, Gopinathan KP, editors. Positron Annihilation. Singapore: World Scientific Publ. Co.; 1985.
[59] Corbel C, Moser P, Stucky M. Ann Chim (Paris). 1985;(10):733.
[60] Polity A, Rudolf F. Phys Rev B. 1999;59:10025.
[61] Slotte J, Rummukainen M, Tuomisto F, *et al.* Phys Rev B. 2008;78:085202.
[62] Mesli A, Dobaczewski L, Nielsen KB, *et al.* Phys Rev B. 2008;78:165202.
[63] Emtsev V. Mater Sci Semicond Process. 2006;9:580.
[64] Callcott TA, Mackay JW. Phys Rev. 1967;161:698.
[65] Dutta PS, Bhat HL, Kumar V. J Appl Phys. 1997;81:5821.
[66] Segercrantz N, Slotte J, Makkonen I, *et al.* Appl Phys Lett. 2014;105:082113.
[67] Ling CC, Lui MK, Ma SK, *et al.* Appl Phys Lett. 2014;85:384.
[68] Kujala J, Segercrantz N, Tuomisto F, *et al.* J Appl Phys. 2014;116:143508.
[69] Tuomisto F, Ranki V, Look DC, *et al.* Phys Rev B. 2007;76:165207.
[70] Bracht H, Nicols S, Walukiewicz W, *et al.* Nature. 2000;408:69.
[71] Saarinen K. Chapter 5 – Characterization of native point defects in GaN by positron annihilation spectroscopy. In: Manasreh O, editor. III-Nitride Semiconductors: Electrical, Structural and Defects Properties. Amsterdam: Elsevier; 2000. pp. 109–149.
[72] Tuomisto F. In: Ehrentraut D, Meissner E, Bockowski M, editors. Point Defects and Impurities in Bulk GaN Studied by Positron Annihilation Spectroscopy. Berlin, Heidelberg: Springer Berlin Heidelberg; 2010. pp. 295–316.
[73] Tuomisto F. Characterization of vacancy defects in ZnO by positron annihilation spectroscopy. In: Selim F, editor. Zinc Oxide the Future Material for Electronics: A Comprehensive Review on ZnO Physics and Defects. Kerala, India: Research Signpost; 2011.
[74] Saarinen K, Laine T, Kuisma S, *et al.* Phys Rev Lett. 1997;79:3030–3033.
[75] Hautakangas S, Makkonen I, Ranki V, *et al.* Phys Rev B. 2006;73:193301.
[76] Uedono A, Ito K, Nakamori H, *et al.* J Appl Phys 2007;102(8):084505.
[77] Uedono A, Ishibashi S, Ohdaira T, *et al.* J Cryst Growth. 2009;311(10): 3075–3079.
[78] Zubiaga A, Tuomisto F, Coleman VA, *et al.* Phys Rev B. 2008;78:035125.
[79] Mäki JM, Makkonen I, Tuomisto F, *et al.* Phys Rev B. 2011;84:081204.
[80] Saarinen K, Suski T, Grzegory I, *et al.* Phys Rev B. 2001 Nov;64:233201.
[81] Saarinen K, Ranki V, Suski T, *et al.* J Cryst Growth. 2002;246(3):281–286.

[82] Hautakangas S, Oila J, Alatalo M, *et al.* Phys Rev Lett. 2003;90:137402.
[83] Hautakangas S, Saarinen K, Liszkay L, *et al.* Phys Rev B. 2005;72:165303.
[84] Tuomisto F, Mäki JM, Zajac M. J Cryst Growth. 2010;312(18):2620–2623.
[85] Neuvonen PT, Vines L, Venkatachalapathy V, *et al.* Phys Rev B. 2011; 84:205202.
[86] Johansen KM, Zubiaga A, Makkonen I, *et al.* Phys Rev B. 2011;83:245208.
[87] Tuomisto F, Prozheeva V, Makkonen I, *et al.* Phys Rev Lett. 2017;119:196404.
[88] Makkonen I, Korhonen E, Prozheeva V, *et al.* J Phys: Condens Matter. 2016;28:224002.
[89] Tuomisto F, Rauch C, Wagner MR, *et al.* J Mater Res. 2013;28(15):1977–1983.
[90] Liszkay L, Corbel C, Baroux L, *et al.* Appl Phys Lett. 1994;64(11):1380–1382.
[91] Korhonen E, Tuomisto F, Bierwagen O, *et al.* Phys Rev B. 2014;90:245307.
[92] Korhonen E, Prozheeva V, Tuomisto F, *et al.* Semic Sci Tech. 2015;30(2): 024011.
[93] Johansen KM, Tuomisto F, Makkonen I, *et al.* Mater Sci Semicond Process. 2017;69:23–27.
[94] Korhonen E, Tuomisto F, Bierwagen O, *et al.* AIP Conf Proc. 2014;1583(1): 368–371.

Chapter 7

First principles methods for defects: state-of-the-art and emerging approaches

Elif Ertekin[1] *and Hannes Raebiger*[2]

Electronic structure theory [1,2] is a powerful tool to predict properties of defects from first principles. In principle, one needs to solve the many-particle Schrödinger equation for bulk and defect-containing materials to obtain total energies, which can then be associated with macroscopic observables via thermodynamics. In practice, the Schrödinger equation is solved at different levels of approximation, which lead to a variety of uncertainties often making it difficult to assess the degree of accuracy attainable. This chapter describes typical approaches and levels of approximation and the common sources of uncertainties encountered. On one hand, it provides a generalized framework for carrying out defect calculations. On the other hand, it is intended as a guide to experimentalists how to read theory papers, assess conclusions, and interact with theorists. Some of the common terms used throughout this chapter are briefly described in Table 7.1.

7.1 Defect levels

Before considering first-principles methods, we first give an intuitive introduction of the basic observables, *defect levels* and *charge transition levels*. A point defect D in a semiconductor host can take many forms: for example, a foreign atom, a vacancy (the absence of a host atom), or an antisite defect. When a defect D is present, it introduces a unique set of *defect levels*, one-electron states or spectral lines in the semiconductor band structure, often observed within the bandgap. Defect D may exist in various charge states $q = (\ldots, -1, 0, 1, \ldots)$, and each (D, q) has its own unique spectral lines. These spectral lines have several features analogous to the spectral lines of a free atom. For a free atom, the spectral lines describe the energy required/released when an electron is removed from an occupied orbital and added to an empty one. The same is true for a defect in a semiconductor, but in addition to adding or removing electrons to or from specific defect orbitals, the defect can now also exchange electrons with the host semiconductor, whereby carrier electrons

[1]Department of Mechanical Science and Engineering, University of Illinois at Urbana–Champaign, Champaign, IL, USA
[2]Department of Physics, Yokohama National University, Yokohama, Japan

Table 7.1 Common nomenclature and simple explanations

Term	Handwaving explanation
Defect level	Spectral levels induced by a defect, often within the band gap. In this chapter, *defect level* refers to a defect-induced one-electron *quasiparticle level* (see below, and Section 7.1). Easily confused with thermodynamic charge transition level
Quasiparticle energy	Single electron spectral energies. The energy required to add/remove an electron from a specific level (or orbital) at fixed atomic coordinates
Charge transition level	Thermodynamic activation energy for ionization of a defect. Sometimes also called a *zero-phonon* line. A difference from quasiparticle levels is that in a thermodynamic excitation, atomic coordinates in general are not fixed
Total energy	The ground state energy of an interacting many-electron system. This is the eigenvalue of many-electron Schrödinger equation (7.11), or evaluated, e.g., from (7.13) or (7.16)
Self-interaction	When the Coulomb repulsion potential energy of all electrons is calculated based on the charge density (classically), it contains a spurious term of an electron repelling itself. This is called self-interaction, which in wave-function-based methods is easily accounted for, but not so in charge-density-based methods
Finite size effect	An error introduced when describing a macroscopic system ($\sim 10^{23}$ atoms, practically infinite) within an atomic scale, finite, model (much less than 10^{23} atoms)

(or holes) may be released or captured. The activation energies of such carrier exchange are the donor or acceptor energies, or *charge transition levels*, of the given defect. It is important to distinguish which type of levels are being discussed—single particle (electron or hole) spectral energy levels, or activation energies for carrier exchange between defect and host. We begin with the spectral energy levels.

To predict or understand single electron defect levels, one can go a long way simply by considering orbital interactions, point group symmetries, and tabulated atomic orbital energies [3]. This is illustrated in Figure 7.1 for the case of cation vacancies in Cu_2O and ZnO [4]. The cation vacancy levels, i.e., the ligand orbitals of the V_{Cu}^{-} and the V_{Zn}^{2-}, can be understood as the dangling bond energies of the anion (oxygen) atoms surrounding the vacancy. The band edges of Cu_2O and ZnO can be understood to arise from the interaction of these V_{Cu}^{-} and V_{Zn}^{2-} ligand levels with the cations (Cu^{+}, Zn^{++}), as illustrated in Figure 7.1(a) and (b). The valence band maximum (VBM) can be considered as the highest occupied ligand field orbital and the conduction band minimum (CBM) as the lowest unoccupied one.

Based on such orbital-interaction diagrams, one can deduce if the vacancy dangling bond levels lie above or below the band edges using known atomic orbital energies [3]. Since the negatively charged copper vacancy V_{Cu}^{-} levels in Cu_2O (Figure 7.1(c)) are below the VBM, these readily remain filled, leaving behind a hole at the VBM edge. This suggests that the copper vacancy in Cu_2O is a shallow acceptor. By contrast, the V_{Zn}^{2-} dangling bonds lie inside the ZnO bandgap

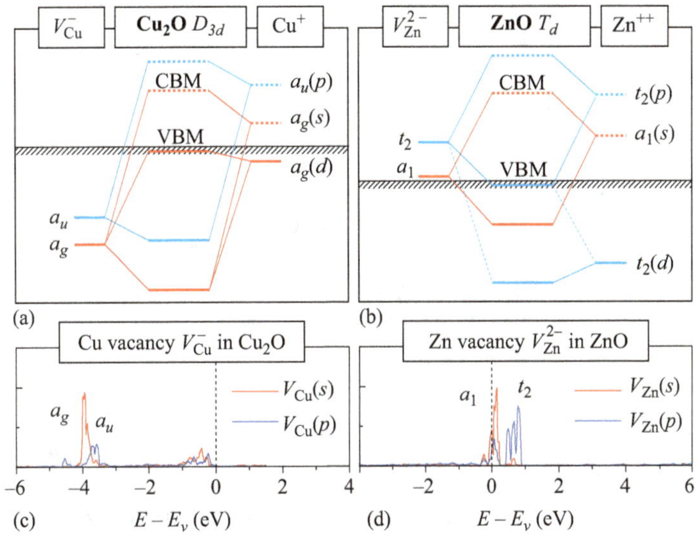

Figure 7.1 Orbital interaction diagrams for band formation of bulk (a) Cu_2O and (b) ZnO and the local density of states of the ideal cation vacancies in (c) Cu_2O and (d) ZnO. © 2007 APS. Reprinted, with permission, from [4]. Modified by permission from authors

(Figure 7.1(d)). Depopulating them leaves behind in-gap deep acceptor states. Panels (c) and (d) show the dangling bond energies of oxygen anions surrounding the vacancy site calculated from first principles, which confirm the relative energies of the ligand orbitals of the V_{Cu}^{-} and the V_{Zn}^{2-} with respect to the VBM. This simple approach can be extended to other cases of defects such as anion vacancies interacting with a cation ligand, and substitutional impurities (Figure 7.2).

When drawing such schematic diagrams, the relative energies of respective anion and cation orbitals must be known beforehand. These can be taken from atomic orbital energies [3] and the orbital interaction energies (*bonding* and *anti-bonding* states). Ultimately, the resulting defect levels (donor/acceptor levels) can be estimated by tight-binding or other model theories. However, such models require that we know a great deal about the materials, types of defects present, defect geometries, and lattice relaxations. For example, if a defect exhibits a Jahn–Teller distortion leading to a different point symmetry than that of the lattice, there will be a different set of orbitals and different resulting interactions.[1] When the necessary information is not available a priori or if more quantitative descriptions are desired, we then must tackle the many-particle Schrödinger equation and calculate the electronic structure and total energies directly from first principles.

[1]For example, both *charge neutral states* Mg and Mn impurities in GaN (cf. Figure 7.2) exhibit symmetry-lowering deformations [5,6], so for the charge neutral open shell configurations, one needs to consider different point symmetries.

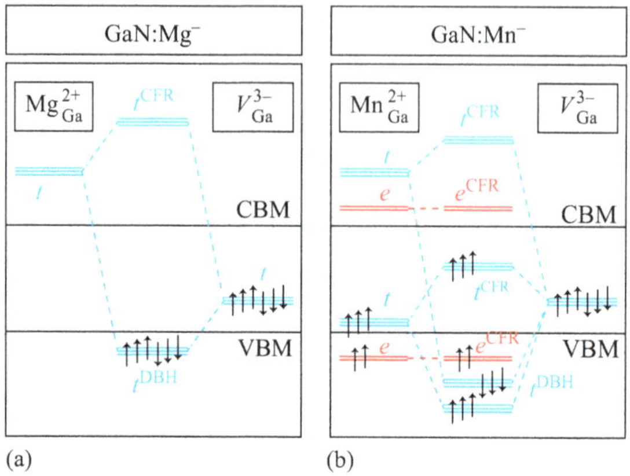

Figure 7.2 Orbital interaction diagrams for (a) Mg^- and (b) Mn^- impurities in GaN. Notice that both impurities are considered in negative charge states, *which correspond to closed shell electronic configurations.* Neutral charge states *correspond to open shell configurations, which may be prone to symmetry-breaking Jahn–Teller distortions*

This both increases the complexity of the calculation itself and necessitates linking a quantum mechanical calculation of a microscopic system with macroscopic observables.

7.2 Defect thermochemistry

7.2.1 Defect formation enthalpy

At thermodynamic equilibrium, defect formation enthalpies govern defect concentrations, defect ionization levels, and the possibility of carrier doping of a semiconductor. As formulated by Zhang and Northrup [7] in 1991, the enthalpy of formation ΔH of defect D in charge state q is

$$\Delta H_{D,q}(E_F,\mu) = (E_{D,q} - E_H) + \sum_{\alpha} n_\alpha \mu_\alpha + qE_F. \tag{7.1}$$

where $E_{D,q}$ and E_H are the total energies of the defect-containing system and the pure host, respectively. The thermodynamic environment is represented by the atomic chemical potentials μ_α for each atom of type α, by assuming that each is in equilibrium with a physical reservoir such as a gas or a bulk phase. The parameter n_α is the number of atoms of type α added to ($n_\alpha > 0$) or removed from ($n_\alpha < 0$), the reservoir to create the defect. For a defect in a charge state q, the last term describes the exchange of electrons with ($q > 0$ addition of electrons to, $q < 0$ removal of electrons from) a thermodynamic reservoir of electron chemical potential $E_F = E_v + \Delta E_F$. The Fermi energy E_F generally varies between E_v and E_c (CBM), so $\Delta E_F \in [0, E_{gap}]$. As expected, positively ionized defects ($q > 0$) have a lower formation energy when the Fermi energy is low, while negatively ionized defects ($q < 0$) have a lower formation energy when the Fermi energy is high.

Note that this formulation neglects aspects of the temperature dependence to defect formation energies (entropy terms TS) which in general include configurational, vibrational, and electronic contributions. Detailed aspects of pressure/volume dependence are also often neglected. In the most common utilization of (7.1), only the T and P dependence of the chemical potential of gas phase species such as oxygen molecules are included, as in (7.5).

7.2.2 Phase stabilities and chemical potential space

The chemical potentials of atoms μ_α are variables determined by the thermodynamic environment and depend on the growth conditions such as temperature T, vapor pressures in growth chamber, what precursor materials are used, and other environmental parameters. The chemical potential μ_α of species α can be referenced to the elemental phase $\mu_\alpha = \mu_\alpha^{\text{elem}} + \Delta\mu_\alpha$, such as the elementary solid for a metal or the molecule energy for a gas phase molecule. For all species α, the parameters $\Delta\mu_\alpha$ vary within an allowable range determined by thermodynamic constraints.

The first constraint is that the semiconductor of interest itself must be thermodynamically stable in the environment. For the binary compound A_nB_m, $\Delta\mu_A$ and $\Delta\mu_B$ must satisfy

$$n\Delta\mu_A + m\Delta\mu_B = \Delta H_f(A_nB_m), \qquad (7.2)$$

where $\Delta H_f(A_nB_m)$ is the formation enthalpy of the compound A_nB_m referenced to bulk elemental phases. This limits the allowable chemical potentials to the thick solid line indicated in Figure 7.3.

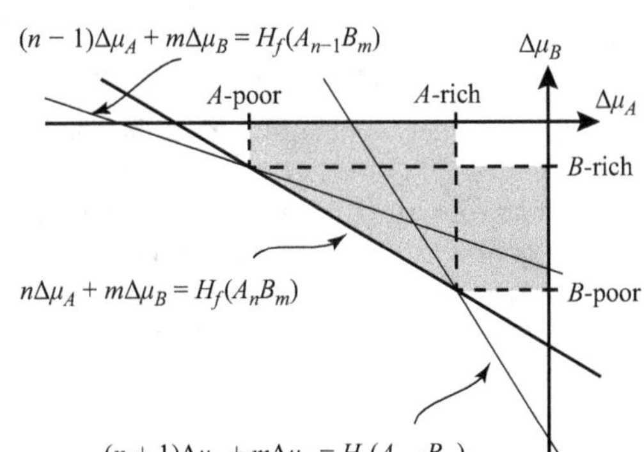

Figure 7.3 The range of $\Delta\mu_\alpha$ for A_nB_m formation. The thick solid line is the formation enthalpy for A_nB_m, and the thin solid lines for its competing phases $A_{n-1}B_m$ and $A_{n+1}B_m$. A_nB_m is stable only in the range of $\Delta\mu_\alpha$ where $H_f(A_nB_m)$ lies lower than its competing phases

The second set of constraints requires that the compound be stable with respect to the formation of possible competing phases. For instance, we require $\Delta\mu_A, \Delta\mu_B < 0$ so that the compound A_nB_m is stable with respect to the formation of the elemental A and B phases, limiting the allowable range to the quadrant shown. Finally, if other binary phases such as $A_{n-1}B_m$ and $A_{n+1}B_m$ exist, these impose additional constraints. The compound of interest should also be stable against formation of these competing binary phases. This requires that

$$(n-1)\Delta\mu_{\rm A} + m\Delta\mu_{\rm B} < \Delta H_f(A_{n-1}B_m), \tag{7.3}$$

$$(n+1)\Delta\mu_{\rm A} + m\Delta\mu_{\rm B} < \Delta H_f(A_{n+1}B_m). \tag{7.4}$$

This further constrains the allowable range for $\Delta\mu_A$, $\Delta\mu_B$ to the line segment within the gray region, as graphically shown in Figure 7.3. Thus, for binary compound A_nB_m, $\Delta\mu_\alpha$ varies between the extrema "A-rich/B-poor" and "A-poor/B-rich." The range of chemical potentials alters defect formation energies according to (7.1): vacancies of species A have lower formation energy and are more prevalent under A-poor conditions, but have a higher formation energy and are less prevalent under A-rich conditions. Thus, knowledge of chemical potential space provides insights into the native defects that are present under different thermodynamic environments.

For Cu_2O, for example, the Cu-rich/O-poor boundary is limited by Cu metal formation at $\Delta\mu_{\rm Cu} = 0$, and the Cu-poor/O-rich boundary is limited by CuO formation. The latter boundary is given $\Delta\mu_{\rm Cu}$ and $\Delta\mu_{\rm O}$ at the intersection of $2\Delta\mu_{\rm Cu} + \Delta\mu_{\rm O} = \Delta H_f({\rm Cu_2O})$ and $\Delta\mu_{\rm Cu} + \Delta\mu_{\rm O} = \Delta H_f({\rm CuO})$. The chemical potential of oxygen, $\Delta\mu_{\rm O}$, at a given temperature T and oxygen partial pressure $P_{\rm O_2}$ is given by

$$\begin{aligned}\Delta\mu_{\rm O}(T,P_{\rm O_2}) &= \frac{1}{2}\Delta\mu_{\rm O}(T,P_{\rm O_2}=P_0) + \frac{1}{2}k_BT\ln\left(\frac{P_{\rm O_2}}{P_0}\right)\\ &= \frac{1}{2}\{[H_0+\Delta H(T)] - T[S_0+\Delta S(T)]\} + \frac{1}{2}k_BT\ln\left(\frac{P_{\rm O_2}}{P_0}\right).\end{aligned} \tag{7.5}$$

The function $\Delta\mu_{\rm O}(T,P_{\rm O_2}=P_0)$ can be obtained from thermodynamic tables (see, e.g., [8]) using $\Delta H(T) = C_P(T-T_0)$ and $\Delta S(T) = C_P\ln(T/T_0)$.

For a binary compound as described here, the phase stability region is defined by a line segment (one degree of freedom). Correspondingly, for a ternary compound, the stability region is given by a planar two-dimensional area (two degrees of freedom) and for a quaternary by a volume of three-dimensional space (three degrees of freedom).

7.2.3 Defect formation enthalpy diagrams

For a fixed set of chemical potentials, the formation energies of a given defect are typically presented as a function of Fermi energy E_F, as shown in Figure 7.4. The formation energy $\Delta H_{D,q}$ for each charge state q is a line with slope q as indicated by (7.1). For each value of E_F, it is common to show only the charge state with lowest ΔH, illustrated by the broken strong line. The kinks (marked by vertical lines)

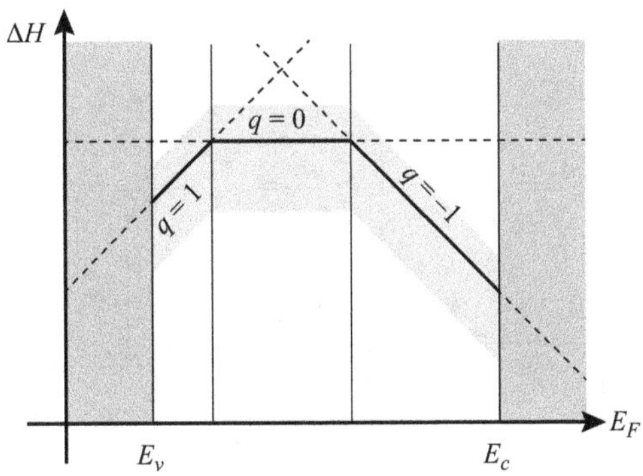

Figure 7.4 *Typical ΔH plot. Defects with charge state q have slope q when ΔH is plotted as a function of Fermi energy ΔE_F as shown by the dashed lines. The values of the Fermi energy where two lines for charge state q and q′ intersect are known as the thermodynamic charge transition levels $\varepsilon(q|q')$ (see (7.6)). The solid line indicates the lowest energy ΔH amongst all charge states q as a function of Fermi energy. When $\Delta\mu_\alpha$ is varied, ΔH for all q will rigidly shift up or down within gray area, but $\varepsilon(q|q')$ remain fixed*

where the preferred charge states change are known as *charge transition levels*, indicating values of the Fermi energy at which the most favorable charge state of the defect changes. For a change from charge state q to q', the thermodynamic charge transition level is $\varepsilon(q|q')$, given by

$$\varepsilon(q|q') = \frac{\Delta H_{D,q'} - \Delta H_{D,q}}{q - q'} - E_v \tag{7.6}$$

with respect to E_v. The absolute value of ΔH depends on growth conditions due to the chemical potentials in (7.5), which shifts the defect formation energy curves vertically (indicated by gray area above and below ΔH). However, the $\varepsilon(q|q')$ charge transition levels are not affected by changes to the $\Delta\mu_\alpha$.

7.2.4 Configuration coordinate diagrams and optical excitation

Defect charge states may change also by optical excitation or photoemission. The main difference between thermodynamic and optical processes is whether the atoms in the system have time to relax or not, as illustrated in Figure 7.5. Here, the lower curve describes the (adiabatic) potential energy of a defective system with negatively charged defect D^-, and the upper curve the same defect in charge neutral state together with an electron in the conduction band $D^0 + e$. Each of these electronic configurations has their own relaxed atomic structure, and the structural deformation from q to q' is depicted as a one-dimensional configuration coordinate (or reaction coordinate, or generalized coordinate). Vertical jumps between such

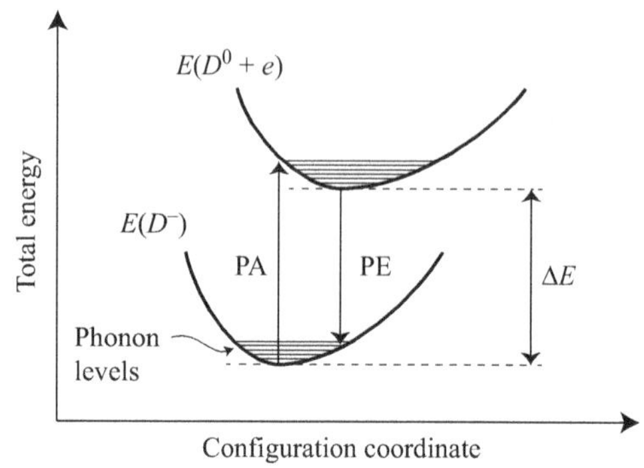

Figure 7.5 Configuration coordinate diagram for electronic excitation. The thick lines show the total energies for systems with defect D in charge state $q = -1$, and $q = 0$ with an electron donated to the CB, as a function of the configuration coordinate

potential energy curves are related to optical photoabsorption and photoemission excitations, and motion along the coordinate is quantized in phonon levels. In a thermodynamic transition between q and q', one compares total energies of *relaxed equilibrium structures*, i.e., the total energy difference ΔE of total energy minima along the configuration coordinate, and also lattice vibrations (phonons) are involved.

7.2.5 Defect equilibria and carrier concentrations

Finally, for a given thermodynamic environment $\{\mu_\alpha\}$, the equilibrium defect concentration of defect (D, q) is given by

$$c_{D,q}(E_F, \mu_i, T) = N_D \exp\left(-\frac{\Delta H_{D,q}}{k_B T}\right), \tag{7.7}$$

where N_D is the density of available crystallographic sites for the defect to form. For a given semiconductor, if the formation energy of all defects D in relevant charge states q that are expected to form have been determined (for example, all possible native defects), then we can predict the semiconductor's equilibrium Fermi energy and carrier concentration for a fixed set of chemical potentials μ_i. There is only one value of the Fermi energy E_F^{eq}, called the equilibrium Fermi energy, for which the defect concentrations and the carriers present satisfy charge neutrality for the system

$$0 = \sum_{(D,q)} q c_{D,q} + p - n \tag{7.8}$$

with carrier concentrations consistent with Fermi–Dirac statistics

$$n(T) = \int_{E_C}^{\infty} f(E, T) D_{\text{avail}}(E)\, dE, \tag{7.9}$$

$$p(T) = \int_{-\infty}^{E_V} (1 - f(E,T))D_{\text{avail}}(E)\, dE, \tag{7.10}$$

where $D_{\text{avail}}(E)$ is the available density of states. The self-consistent solution to the set of (7.7–7.10) yields E_F^{eq}, and the defect and carrier concentrations at thermodynamic equilibrium for the given environment.

7.3 Sources of uncertainty in defect formation energy calculations

Before diving deeply into the first-principles methods commonly used for the calculation of defect properties, to frame the discussion, it is insightful to provide a conceptual overview of what is needed for accurate predictions and what the typical sources of uncertainty are. Some clues to the challenges to achieving high predictive capability are revealed by considering the terms present in (7.1). Each term must be determined accurately to get a good description of the defect formation enthalpy $\Delta H_{D,q}(E_F, \mu_\alpha)$: the total energies of bulk and defect-containing supercells, the chemical potentials μ_α, and the range of allowable Fermi energies E_F.

Common sources of uncertainty that arise in the calculation of these quantities are listed in Table 7.2. These can conceptually be divided into two main categories: the accuracy of the selected computational engine for total energy calculation and finite size effects. The first refers to uncertainties arising from the selected first-principles method and its limitations; the second refers to uncertainties since defects are often simulated at concentrations much larger than those experienced in real materials. That is, even given a perfectly accurate computational engine, finite size effects would still affect the computed quantities in nonphysical ways. Whenever a set of defect

Table 7.2 Typical sources of uncertainties in first-principles defect calculations can be broadly grouped into two categories: those arising due to limitations of the selected first-principles method (selected computational engine) and those arising from finite size effects

Underlying reason	**Source of uncertainty and correction approaches**
Computational engine	*Supercell total energies*. Use higher accuracy method; apply empirical corrections
	Range of allowable chemical potentials μ_α. Use higher accuracy method; apply empirical correction
	Band gap. Use higher accuracy method, shift band edges
Finite size effects	*Elastic*. Use large supercell; estimate remaining interaction; extrapolate
	Electrostatic and potential alignment. Use large supercell; estimate remaining interaction; extrapolate
	Burstein–Moss band filling. Use large supercell; estimate remaining interaction; extrapolate

properties computed from first principles is reported, it is useful to think about how uncertainties related to the computational engine and to finite size effects have been addressed. Commonly used and emerging options for computational engines are described in Sections 7.4 and 7.5, and finite size effects are discussed in Section 7.6.

7.4 Total energies using first principles electronic structure calculations

7.4.1 Overview and many-particle Schrödinger equation

While the formulation described above is general for determining defect formation energies, charge transition levels, equilibrium Fermi energy, and carrier densities, its predictive power depends on accurately determining the quantities that appear in (7.1). In recognition of this fact, the computational first-principles modeling community has explored a variety of techniques to improve the fidelity with which these quantities can be calculated. These methods include Hartree–Fock (HF), density functional theory (DFT) in its many flavors, Green's function approaches such as the *GW* approximation (many body perturbation theory) quantum Monte Carlo (QMC), and others. By far, the most common methods lie within the framework of DFT, but nowadays one can use conventional DFT, DFT+U, hybrid DFT, and other formulations. Some approaches are listed in Table 7.3. Even experts in the field often disagree on which methods would give the most accuracy for a given problem. It is important to understand the methods available and their relative merits and shortcomings.

Table 7.3 Summary of common and emerging first-principles methodologies utilized for describing defects in semiconductors

Method	Description and merits/challenges
Hartree–Fock	Variational approach using Slater determinant ansatz for many-body wave function. No self-interaction error. Neglects electron correlation effects by definition. Biases toward overly localized description. Typical scaling $\sim CN_e^4$
DFT (LDA, GGA)	Reformulates Schrödinger equation in terms of total electron density. Approximate description of exchange and correlation (typically derived from the homogeneous electron gas). Incomplete cancelation of self-interaction biases results toward delocalized states. Typical scaling $\sim CN_e^3$
DFT (hybrid)	Incorporation of a proportion of exchange interaction in order to reduce the self-interaction error. Improves upon conventional DFT but at a sizable computational cost. Typical scaling $\sim CN_e^{3-4}$
GW	Perturbation theory to estimate electron addition and removal energies. Many different implementations (non-self-consistent vs. self-consistent, vertex corrections, etc.). Can be sensitive to initial single-particle states. Typical scaling $\sim CN_e^{4-5-6}$ depending on the approach
QMC	Direct solution approach to the many-body Schrödinger equation so few uncontrolled approximations. Large computational costs. Difficult to obtain response functions. Typical scaling $\sim(100\text{–}1{,}000)CN_e^3$

Rigorously, the total energies required for defect formation enthalpy calculations are obtained from Schrödinger's equation. In the Born–Oppenheimer approximation, the many-particle Schrödinger equation in atomic units is written as

$$\left[-\frac{1}{2}\sum_i \nabla_i^2 - \sum_\alpha \sum_i \frac{Z_\alpha}{|\boldsymbol{R}_\alpha - \boldsymbol{r}_i|} + \sum_\alpha \sum_{\beta>\alpha} \frac{Z_\alpha Z_\beta}{|\boldsymbol{R}_\alpha - \boldsymbol{R}_\beta|} + \sum_i \sum_{j>i} \frac{1}{|\boldsymbol{r}_i - \boldsymbol{r}_j|}\right] \Psi(r_1, r_2, \ldots, r_N) = E\Psi(r_1, r_2, \ldots, r_N). \tag{7.11}$$

where roman indices count electrons at $\boldsymbol{r}_i$ ($i = 1, \ldots, N$ for an N electron system) and Greek indices nuclei of atomic number Z_α at $\boldsymbol{R}_\alpha$. The first three terms operate only on a single electron, but the last term $\propto (|\boldsymbol{r}_i - \boldsymbol{r}_j|)^{-1}$ due to electron–electron Coulomb interaction renders the solution $\Psi(r_1, r_2, \ldots, r_N)$ non-separable. The problem is challenging, among other reasons, because the wave function $\Psi(r_1, r_2, \ldots, r_N)$ is highly multidimensional. There are various numerical strategies to solve (7.11), including variational methods such as HF and QMC. The task is so challenging though that most practical solution approaches, such as DFT, do not directly tackle the many-particle Schrödinger equation to solve for $\Psi(r_1, r_2, \ldots, r_N)$.

7.4.2 Hartree–Fock

HF is a technique that does tackle the many-body Schrödinger equation. It is not typically used to describe defects in semiconductors nowadays since other approaches are more accurate, but it represents a good starting point to discuss the other approaches. In HF, we approximate $\Psi(r_1, r_2, \ldots, r_N)$ as a Slater determinant composed of one-electron spin orbitals $\psi_{i\sigma}(\boldsymbol{r}_j)$

$$\Psi(r_1, r_2, \ldots, r_N) = \frac{1}{\sqrt{N!}} \begin{vmatrix} \psi_{1\sigma}(r_1) & \psi_{2\sigma}(r_1) & \cdots & \psi_{N\sigma}(r_1) \\ \psi_{1\sigma}(r_2) & \psi_{2\sigma}(r_2) & \cdots & \psi_{N\sigma}(r_2) \\ \cdots & \cdots & \cdots & \cdots \\ \psi_{1\sigma}(r_N) & \psi_{2\sigma}(r_N) & \cdots & \psi_{N\sigma}(r_N) \end{vmatrix} \tag{7.12}$$

and seek ground-state solutions by variational minimization of the total energy with respect to the parameters of Ψ, namely, $\delta E/\delta\Psi = 0$. With this *ansatz*, the total energy can be written as

$$\begin{aligned} E^{\mathrm{HF}} = {} & \sum_{i\sigma} \int \psi_{i\sigma}^*(\boldsymbol{r}) \left[-\frac{1}{2}\nabla^2 - \sum_\alpha \frac{Z_\alpha}{|\boldsymbol{R}_\alpha - \boldsymbol{r}_i|}\right] \psi_{i\sigma}(\boldsymbol{r})\, d\boldsymbol{r} \\ & + \sum_\alpha \sum_{\beta>\alpha} \frac{Z_\alpha Z_\beta}{|\boldsymbol{R}_\alpha - \boldsymbol{R}_\beta|} + \frac{1}{2} \iint \frac{\rho(\boldsymbol{r})\rho(\boldsymbol{r}')}{|\boldsymbol{r} - \boldsymbol{r}'|}\, d\boldsymbol{r}\, d\boldsymbol{r}' \\ & - \frac{1}{2} \sum_{i\sigma} \sum_{j\sigma} \iint \frac{\psi_{i\sigma}^*(\boldsymbol{r})\psi_{j\sigma}^*(\boldsymbol{r}')\psi_{i\sigma}(\boldsymbol{r}')\psi_{j\sigma}(\boldsymbol{r})}{|\boldsymbol{r} - \boldsymbol{r}'|}\, d\boldsymbol{r}\, d\boldsymbol{r}', \end{aligned} \tag{7.13}$$

where $\rho(\boldsymbol{r}) = \sum_{i\sigma}\psi^*_{i\sigma}(\boldsymbol{r})\psi_{i\sigma}(\boldsymbol{r})$ is the electron number density. Compared to (7.11), the electron kinetic energy, ion–electron Coulomb interaction, and ion–ion Coulomb interaction remain, but now the electron–electron Coulomb interaction becomes represented by two terms. The first one is called the Hartree energy and represents the classical electrostatic potential energy of a charge density interacting with itself. Note that this term represents the total charge density interacting with itself, so it also nonphysically includes the interaction of each electron with itself. This is often called *self-interaction* and will become important later.

The second term, the "exchange energy," is a quantum mechanical term that arises from the form assumed for $\Psi(r_1, r_2, \ldots, r_N)$. The use of the Slater determinant ensures that the wave function Ψ satisfies Pauli exclusion, i.e., $\Psi(\ldots, r_i, \ldots, r_j, \ldots) = -\Psi(\ldots, r_j, \ldots, r_i, \ldots)$ is antisymmetric upon swapping electron indices r_i and r_j. Conveniently, the exchange term also cancels out the nonphysical self-interaction that is contained in the classical Hartree energy.

The expression for the energy E^{HF} is minimized by solving the corresponding Euler–Lagrange equations

$$\left[-\frac{1}{2}\nabla^2 - \sum_{\alpha}\frac{Z_\alpha}{|\boldsymbol{R}_\alpha - \boldsymbol{r}_i|} + \sum_{\alpha}\sum_{\beta>\alpha}\frac{Z_\alpha Z_\beta}{|\boldsymbol{R}_\alpha - \boldsymbol{R}_\beta|} + \sum_{j\sigma}\frac{\psi^*_{i\sigma}(\boldsymbol{r}')\psi_{i\sigma}(\boldsymbol{r}')}{|\boldsymbol{r}-\boldsymbol{r}'|}\,d\boldsymbol{r}'\right.$$

$$\left. - \sum_{j\sigma}\int\frac{\psi^*_{i\sigma}(\boldsymbol{r})\psi^*_{j\sigma}(\boldsymbol{r}')\psi_{i\sigma}(\boldsymbol{r}')\psi_{j\sigma}(\boldsymbol{r})}{\psi^*_{i\sigma}(\boldsymbol{r}')\psi_{i\sigma}(\boldsymbol{r})|\boldsymbol{r}-\boldsymbol{r}'|}\,d\boldsymbol{r}'\right]\psi_{i\sigma}(\boldsymbol{r}) = \varepsilon^{\mathrm{HF}}_{i\sigma}\psi_{i\sigma}(\boldsymbol{r}) \tag{7.14}$$

called the HF equations. The eigenvalues $\varepsilon^{\mathrm{HF}}_{i\sigma}$ describe electron-removal energies and thus can be associated with spectral lines observed for optical ionization of defects in semiconductors. The last term in brackets, the exchange potential V_x when operating on $\psi_{i\sigma}(\boldsymbol{r})$ becomes

$$\begin{aligned} V_x\psi_{i\sigma}(\boldsymbol{r}) &= -\sum_{j\sigma}\int\frac{\psi^*_{i\sigma}(\boldsymbol{r})\psi^*_{j\sigma}(\boldsymbol{r}')\psi_{i\sigma}(\boldsymbol{r}')\psi_{j\sigma}(\boldsymbol{r})}{\psi^*_{i\sigma}(\boldsymbol{r})\psi_{i\sigma}(\boldsymbol{r})|\boldsymbol{r}-\boldsymbol{r}'|}\,d\boldsymbol{r}'\psi_{i\sigma}(\boldsymbol{r}) \\ &= -\sum_{j\sigma}\int\frac{\psi^*_{j\sigma}(\boldsymbol{r}')\psi_{i\sigma}(\boldsymbol{r}')}{|\boldsymbol{r}-\boldsymbol{r}'|}\,d\boldsymbol{r}'\,\psi_{j\sigma}(\boldsymbol{r}), \end{aligned}$$

which shows that (7.14) are nonlocal.

It is important to note that the selection of the form of the wave function as a Slater determinant restricts the space of functions $\Psi(r_1, r_2, \ldots, r_N)$ over which the total energy is minimized. In HF, we obtain the best possible representation of $\Psi(r_1, r_2, \ldots, r_N)$ within the search space. Consequently, the HF wave function is different from the exact one, and the HF total energy is higher than the true total energy. The degree to which $\Psi^{\mathrm{HF}}(r_1, r_2, \ldots, r_N)$ is a good description of $\Psi(r_1, r_2, \ldots, r_N)$ for a given material has been a longstanding question: for some materials, the HF description is a good approximation and it fails for many others. Materials for which the HF description is not as good are often referred to as "correlated materials," and the energy difference between the HF energy and the true energy is called the "correlation energy." Correlation effects are important: for instance, empirically we find that HF tends to recover around half of the true-binding energy for a system of atoms. The neglect of correlation results in underbinding and a typically overly localized description of electronic states (correlation is a stabilizing effect and without it, electronic orbitals tend to avoid each other). If there were no electron–electron interaction in the Hamiltonian, the HF *ansatz* would be exact: the many-body wave function would be separable. However, implementations of HF do treat the full interacting Hamiltonian. The limitation is in the *ansatz* for the wave function, which is able to include aspects of electron correlation through satisfying Pauli exclusion (exchange) but otherwise ignores correlation.

7.4.3 Density functional theory

For good reason, DFT is the workhorse method for the application of first-principles methods to defects in semiconductors. It currently offers the best available combination of predictive capability with computational tractability.

DFT is based on the works of Hohenberg and Kohn [9] who showed (i) that there is a one-to-one mapping between the ground-state many-electron wave function $\Psi(\{\boldsymbol{r}_i\})$ and the ground-state electron density $\rho(\boldsymbol{r})$, and (ii) that for a given external potential it is possible to define a universal functional for the energy $E[\rho(r)]$ where the ground-state energy is the global minimum of the functional and the density $\rho(r)$ that minimizes the functional is the ground-state density. The functional can be written as

$$E[\rho(\boldsymbol{r})] = T[\rho(\boldsymbol{r})] + E_{\text{int}}[\rho(\boldsymbol{r})] + \int V_{\text{ext}}(\boldsymbol{r})\rho(\boldsymbol{r})\, d\boldsymbol{r} + \sum_{\alpha}\sum_{\beta>\alpha} \frac{Z_\alpha Z_\beta}{|R_\alpha - R_\beta|}, \tag{7.15}$$

where $T[\rho(\boldsymbol{r})]$ and $E_{\text{int}}[\rho(\boldsymbol{r})]$ give the kinetic energy and electron–electron energy of the system of interacting electrons as a functional of the ground-state charge density. Thus, DFT offers a significant computational simplification compared with HF because it allows us to work with the total charge density $\rho(\boldsymbol{r})$ rather than the many-body wave function $\Psi(\boldsymbol{r}_1, \boldsymbol{r}_2, \ldots, \boldsymbol{r}_N)$. Numerically, we solve for a three-dimensional function rather than a $3N$ dimensional function.

The Kohn–Sham approach [10] to DFT is most widely used in practice. Here one assumes that the ground-state density of the true interacting system can be represented by that of an auxiliary (stand-in) noninteracting system for which the functional can be more easily approximated. The auxiliary system is represented as a set of fictitious, noninteracting one-electron spin orbitals $\phi_{i\sigma}(\boldsymbol{r})$ so that the electron density $\rho(\boldsymbol{r}) = \sum_{i\sigma}\phi^*_{i\sigma}(\boldsymbol{r})\phi_{i\sigma}(\boldsymbol{r})$. Then the functional can be written as

$$E^{\text{KS}} = \sum_{i\sigma}\int \phi^*_{i\sigma}(\boldsymbol{r})\left[-\frac{1}{2}\nabla^2 - \sum_{\alpha}\frac{Z_\alpha}{|\boldsymbol{R}_\alpha - \boldsymbol{r}_i|}\right]\phi_{i\sigma}(\boldsymbol{r})\, d\boldsymbol{r}$$

$$+ \sum_{\alpha}\sum_{\beta>\alpha}\frac{Z_\alpha Z_\beta}{|\boldsymbol{R}_\alpha - \boldsymbol{R}_\beta|} + \frac{1}{2}\iint \frac{\rho(\boldsymbol{r})\rho(\boldsymbol{r}')}{|\boldsymbol{r} - \boldsymbol{r}'|}\, d\boldsymbol{r}\, d\boldsymbol{r}' + E_{xc}[\rho(\boldsymbol{r})]. \tag{7.16}$$

The first term is similar to the HF kinetic energy in (7.13), but the kinetic energy of the fictitious $\phi_{i\sigma}(\boldsymbol{r})$ is different from that of the HF orbitals $\psi_{i\sigma}(\boldsymbol{r})$. The many-body electron–electron interaction is now encompassed by two terms: one that represents the electrostatic potential energy of a charge density interacting with itself similar to the Hartree energy, and the exchange–correlation energy $E_{xc}[\rho(\boldsymbol{r})]$.

The exchange–correlation energy can be thought of as correcting for "everything we get wrong" by formulating the problem in this auxiliary way. It is meant to incorporate aspects of exchange and correlation that would otherwise be missed, as well as the kinetic energy not accounted for by the first term of (Equation (7.16)). Its exact form is not known, so it must be approximated. The selection of the approximation to $E_{xc}[\rho(\boldsymbol{r})]$ gives rise to the different flavors of DFT that are commonly used today (and associated uncertainties). Common forms of the exchange–correlation energy such as the local density approximation (LDA [11,12]) and the generalized gradient approximations (GGAs [13]) are based on the exchange–correlation energy of the homogeneous electron gas, which has been numerically computed by Ceperley and Alder using QMC approaches [11]. The LDA is "local" because the exchange–correlation energy at point $\boldsymbol{r}$ with density ρ is approximated by the exchange–correlation energy of a homogeneous electron gas of density ρ. In the GGA, the exchange–correlation energy at point $\boldsymbol{r}$ depends on both ρ and the gradient $\nabla\rho$ and is called "semilocal." There are many implementations of GGAs including Perdew-Burke-Ernzerhof (PBE) [13] and Perdew-Wang (PW91) [14]. The term meta-GGA refers to exchange–correlation approximations that depend on higher order spatial derivatives of ρ, such as TPSS [15] and the recently developed "strongly constrained and appropriately normed" (SCAN [16]) meta-GGA. What is most important to note is that once a form for the exchange–correlation is selected, total energies and other computed properties depend on how well the approximation works for the material under consideration.

Applying the variational principle, one obtains again a set of Euler–Lagrange equations

$$\left[-\frac{1}{2}\nabla^2 - \sum_{\alpha}\frac{Z_\alpha}{|\boldsymbol{R}_\alpha - \boldsymbol{r}_i|} + \sum_{\alpha}\sum_{\beta>\alpha}\frac{Z_\alpha Z_\beta}{|\boldsymbol{R}_\alpha - \boldsymbol{R}_\beta|}\right.$$

$$\left. + \sum_{j\sigma}\frac{\phi^*_{i\sigma}(\boldsymbol{r}')\phi_{i\sigma}(\boldsymbol{r}')}{|\boldsymbol{r} - \boldsymbol{r}'|}\,d\boldsymbol{r}' + V_{xc}[\rho(\boldsymbol{r})]\right]\phi_{i\sigma}(\boldsymbol{r}) = \varepsilon^{\mathrm{KS}}_{i\sigma}\phi_{i\sigma}(\boldsymbol{r}) \qquad (7.17)$$

called the Kohn–Sham equations. The last term is defined as $V_{xc}[\rho(\boldsymbol{r})] = \delta E_{xc}[\rho(\boldsymbol{r})]/\delta\rho(\boldsymbol{r})$.

One important consideration from (7.15) relates to the third term, representing the electrostatic potential energy of a charge density interacting with itself. Because it is the electrostatic interaction for the total charge density, it also spuriously includes the interaction of each electron with itself. This *self-interaction* is believed by many to lie at the heart of many of the inaccuracies within LDA and PBE. Although the self-interaction is fortuitously canceled in HF, it remains in DFT. Naturally then, improvements in DFT are expected from the development of exchange–correlation approximations that reduce the self-interaction error (ideally while introducing minimal computational cost).

7.4.4 Emerging approaches—quantum Monte Carlo

DFT is formulated to obtain results not through direct solution of the many-body Schrödinger equation but through some intermediary model (the density functional in DFT). In contrast, QMC methods comprise a suite of stochastic tools that enable fundamental calculation of material properties via direct solution of the many-particle Schrödinger equation without the need for adjustable parameters. This comes at the price of now having to solve for the many-body wave function $\Psi(r_1, r_2, \ldots, r_N)$ rather than simply the charge density but offers the advantage that if we do a better job of solving the equations, we can be sure that the accuracy of the results will improve. Because of its direct treatment of electron correlation, QMC methods are among the most accurate electronic structure approaches available today and demonstrate a long and distinguished record of ground-breaking and bench-marking calculations [17–19]. However, their application to point defects in solids is relatively new. While only a few representative problems have been explored to date [20–31], they are now starting to appear more frequently in the literature.

In QMC, stochastic Monte Carlo approaches are implemented to numerically address the multiple-dimensional integrals that arise from the many-particle wave functions. A detailed description of the QMC methodology is not possible here (note some useful reviews [17,18,32,33]), however, we summarize the main methods. The *variational Monte Carlo* (VMC) framework is a straightforward implementation of the variational method, in which expectation values are evaluated using Monte Carlo integration, thus allowing the wave function to explicitly include electron correlation. Typically in VMC, wave functions can be constructed that obtain 85% of the energy due to electron correlation. In diffusion Monte Carlo (DMC), a stochastic process is used to further improve the wave function. In DMC, the wave function has no explicit form but is represented by a population of walkers. Across a range of materials, the DMC method can obtain 90%–95% of the energy due to electron correlation. Overall, the method exhibits some key advantages: direct solution approach yielding accurate results and systematically improvable treatment of correlation across a range of systems within the same theory, favorable (N^3) scaling with system size, and near-linear scalability on modern high-performance architectures. However, its computational costs remain prohibitive for wide-scale utilization, and while total energies are readily attainable, it is challenging to obtain response functions and spectral properties.

7.5 Achieving higher accuracy with density functional theory and other approaches

7.5.1 Observations

DFT is a very successful approach to predict the properties of materials from first principles. Lattice constants, binding energies, elastic constants, and other ground-state properties can be predicted with high fidelity across a wide class of material systems. DFT-based methods have greatly increased the predictive capabilities of the computational materials community for bulk material properties as well as for defects. This is most likely why nowadays it is expected that DFT predictions are good, and instances where they are not good are mostly considered anomalies. There are now, however, many known instances where DFT predicts even qualitatively incorrect behavior [34]. It is important to be aware of and understand some of the commonly observed shortcomings of DFT for materials and defect calculations, and approaches to overcome these.

The accuracy of DFT for a given material system relies on how good the description of the exchange–correlation energy is. Since this description is not perfect, DFT as implemented with local (LDA [11,12]) or semilocal (GGA [13]) functionals, does not provide quantitative predictions of defect charge transition levels or defect formation energies. These functionals are derived from the homogeneous electron gas, and there is no a priori reason to assume that the exchange–correlation energy of a homogeneous electron gas is a good description of exchange–correlation in all materials. The approximation is best for material systems with a spatially slowly varying charge density (closer to homogeneous) and may not be as good for systems with larger electron density gradients. A consequence of applying the exchange–correlation of a homogeneous electron gas to all material systems, relevant to defects, is a tendency of LDA/GGA to describe the states of the system as more delocalized (metallic) than they are in reality. The underestimated bandgap in LDA or GGA is one manifestation of the overly delocalized description: often the valence band edge is predicted to be too high and the conduction band edge too low, resulting in bandgap underestimate.

Historically, the simplest workaround has been the so-called scissors operator (Figure 7.6), whereby one cuts the band structure in the middle and shifts all occupied states down by ΔE_v, and all unoccupied states up by ΔE_c, such that

$$E_g = E_g^{\mathrm{LDA}} + \Delta E_v + \Delta E_c, \tag{7.18}$$

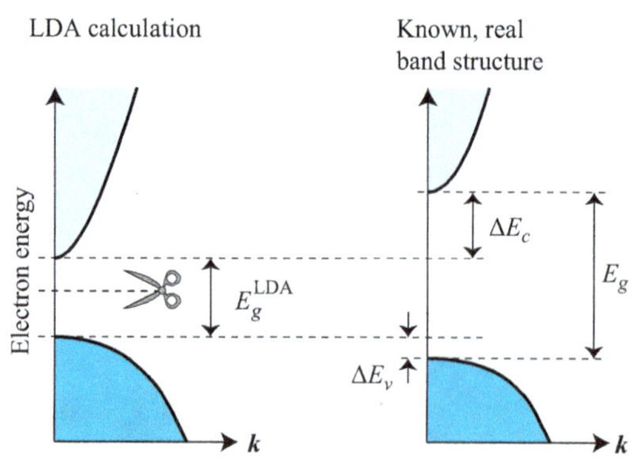

Figure 7.6 Band gap correction by scissors operator

where E_g and E_g^{LDA} are the known, correct bandgap and the LDA bandgap, respectively. These energy shifts can then be added as post-processor correction terms to total energies of defective systems as [35]

$$\Delta E_{\mathrm{scissors}} = \begin{cases} z_e \Delta E_c; & \text{for donors} \\ z_h \Delta E_v; & \text{for acceptors.} \end{cases} \tag{7.19}$$

The factor z_e (z_h) gives the number of donor electrons (acceptor holes) in the conduction band (valence band) for the system considered. It is thus assumed that *shallow* defect states shift along with the host bands, and *deep* defect states are not shifted.

Obviously, there is no unique way to determine ΔE_v and ΔE_c, and moreover, deep defect states hybridize with host band states, so shifting of the band edges should also shift deep levels somewhat. A more systematic approach is to incorporate nonlocal external potential terms $V_{\alpha,l}$ [36,37] to the Kohn–Sham Hamiltonian in (7.16). These potentials depend on the atomic species α and angular momentum l. For example, in oxides like ZnO or In_2O_3, the conduction band has mostly cation-s character, so including a large positive (repulsive) potential for cation-s states shifts the conduction band upwards, opening the gap in a similar way as the scissors operator. Once the empirical $V_{\alpha,l}$ have been determined to reproduce host band properties (bandgap, effective masses, band widths, etc.), any type of defect (deep or shallow) can be investigated, and one obtains their electronic levels self-consistently within the correct bandgap.

Another consequence of the use of the LDA or GGA description of exchange and correlation is the residual *self-interaction* error (the remaining self-interaction error once the exchange–correlation of the homogeneous electron gas is applied). This is believed to give rise to the failure to correctly predict localized states that arise in systems due to narrow bands, defects exhibiting large lattice relaxations, and deep levels, etc. The self-interaction error biases the LDA and PBE description toward delocalized states, since it artificially penalizes localization. Often, many defects are predicted to be more shallow than they actually are. Consequently, conventional DFT typically fails quantitatively, and can fail even qualitatively, to describe the localization of a defect wave function.

There are many ongoing efforts to address these shortcomings. Discussed here are (a non-exhaustive list) the generalized Koopman's approach, DFT+U, Green's function methods, hybrid functionals [38], and empirical correction schemes.

7.5.2 Generalized Koopman's theorem

Physically, the total energy of a system $E(N)$ should vary linearly with the continuous electron occupation number n_i, as shown by the green line in Figure 7.7(a). One important observation is that not only do LDA/GGA and HF not exhibit the correct linear dependence but also that they make errors of opposite sign. The variation is convex in LDA/GGA ($d^2E(n_i)/dn_i^2 > 0$, red line) and concave in HF ($d^2E(n_i)/dn_i^2 < 0$, blue line).

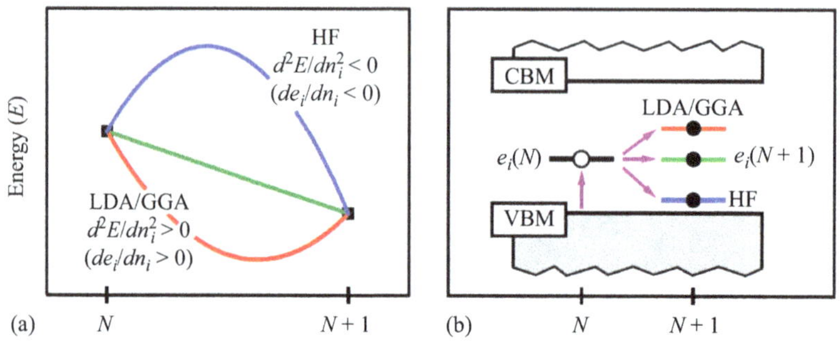

Figure 7.7 (a) In HF theory, the energy is a concave function of the occupation number n_i, whereas in DFT with the typical LDA or GGA exchange–correlation functionals, it is a convex function. The correct behavior is linear. (b) Adding an electron to the lowest unoccupied level should not shift the respective energy eigenvalue, but in HF and LDA/GGA, the eigenvalues shift as illustrated. © 2009 APS, Reprinted, with permission, from [39]

A system's ionization energy (IE), electron affinity (EA), and quasiparticle (QP) gap are given by

$$IE = E(N) - E(N-1), \tag{7.20}$$

$$EA = E(N+1) - E(N), \tag{7.21}$$

$$QP = EA - IP, \tag{7.22}$$

respectively, where $E(N)$ is the energy of the neutral system, $E(N-1)$ is the energy of the system with one electron removed, and $E(N+1)$ is the energy of the system with one electron added. In HF, Koopman's theorem [40] tells us that the eigenvalue of the highest occupied orbital $\varepsilon_i(N)$ can be associated with the IE of the system $E(N) - E(N-1) = \varepsilon_i(N)$. In DFT, Janak's theorem [41] (essentially the differential analog) tells us that $dE(n_i)/dn_i = \varepsilon_i(N)$ holds for the Kohn–Sham system, for any exchange-correlation functional.

For electron addition, if we integrate Janak's theorem from N to $N+1$, then we obtain

$$EA = E(N+1) - E(N) = \int_{x_i=N}^{x_i=N+1} \varepsilon_i(x_i)\, dx_i, \tag{7.23}$$

where $\varepsilon_i(x_i)$ indicates the how the energy eigenvalue of orbital i shifts upon continuously increasing the electron occupation. For the exact functional, since $d^2E(n_i)/dn_i^2 = 0$, then from Janak's theorem $dE(n_i)/dn_i = \varepsilon_i(N)$ is a constant and the integral evaluates to $\varepsilon_i(N)$. For real functionals that do not obey the correct linear behavior, we can conceptually write instead [42]

$$E(N+1) - E(N) = \varepsilon_i(N) + \Sigma_i + \Pi_i, \tag{7.24}$$

where $\varepsilon_i(N)$ is the energy eigenvalue before electron addition, and the terms Σ_i, Π_i denote two principal sources of deviation. The variable Π_i is the self-interaction energy after electron addition to orbital i keeping the orbitals of the neutral system fixed, and Σ_i is the energy recovery that would arise from allowing wave function relaxation.

The correct linear dependence requires that $\Sigma_i + \Pi_i = 0$, which can be considered as a condition that would be fulfilled by the exact DFT functional. In HF, $\Pi_i \equiv 0$ since there is no self-interaction, and in general $\Sigma_i < 0$, so HF eigenvalues spuriously shift downwards upon occupation (and upwards upon electron removal), Figure 7.7(b). For DFT, due to incomplete cancelation of the self-interaction energy typically $\Pi_i > 0$, and usually we observe $\Sigma_i + \Pi_i > 0$ due to residual self-interaction, causing the energy eigenvalue to increase upon electron addition. The presence of this self-interaction tends to artificially impose energetic penalties to localized orbitals, leading to a bias in DFT toward delocalized states.

Designing functionals so that they reproduce the correct linear behavior $d^2E(n_i)/dn_i^2 = 0$ can be used as a guiding principle to improving accuracy. This condition is necessary but not sufficient, as the exact functional does satisfy this condition, but there is no guarantee that any functional satisfying this condition will be good. In practice, however, it is an effective approach that tends to improve results. On the other hand, when the linearity condition is not satisfied, some inaccuracy in total energies can be expected, which in the worst case may lead to qualitatively wrong ground-state electronic structure configurations.

7.5.3 LDA+U and PBE+U

One common approach for improving the description of localized states, particularly for materials where d or f orbitals are important, is based on the LDA+U or PBE+U approach. These orbitals are relevant to many semiconductors (for instance transition metal (TM) oxides) and tend to be localized in comparison to s and p orbitals. Therefore, LDA and PBE do not properly describe them on account of the strong electron–electron interaction: they are often underbound and exhibit energies that are too high relative to orbitals derived from s and p atomic states. For the prototypical example of ZnO, the artificially high energies of the completely filled d orbitals places them too close to the VBM and induces an overly large p–d repulsion, resulting in even more severe bandgap underestimate than typically observed in LDA or PBE.

The fix in the $+U$ scheme is to apply an orbital-dependent potential that incorporates an extra Coulomb interaction U to the d orbitals [43], a penalty for specific atomic sites that depends on the on-site electronic occupation $n_{m,\sigma}$ as

$$V^U_{m,\sigma} = U\left(\frac{1}{2} - n_{m,\sigma}\right). \tag{7.25}$$

The Hubbard-U mimics a screened HF like interaction [44,45]. This potential has the effect of shifting more than half-filled orbitals to lower energies, and less than half-filled orbitals to higher energies. Thus, it drives sets of orbitals to be fully occupied or empty in order to better capture localized states. For half-filled orbitals ($n_{m,\sigma} \approx 0.5$), V^U in principle does nothing but, in practice, may lead to numerical instabilities. Hence, the approach works best for localized orbitals that are either fully occupied or empty ($n_{m,\sigma}$ close to 1 or 0). The parameter U is a variable, often treated as a fitting parameter selected to reproduce bandgaps and/or the experimental positions of the d orbitals relative to, say, the VBM. Alternatively, U can be tuned to fulfill (7.23). However, it is not reasonable to expect $+U$ to correct for all of the errors in LDA or PBE, which leaves these approaches to select U open for scrutiny.

Advantages of the approach are the relatively small computational cost beyond that of LDA or PBE itself, and the improvement of the description of such localized states. Of course, localization errors and associated errors in the bandgap are only partially corrected by $+U$, and the correction approach itself is somewhat *ad hoc*. As the Hubbard U is typically applied to localized d or f orbitals, it relies on the assumption that the other s and p orbitals are already reasonably well described. In general s, p orbitals are expected to be better described in DFT, so $+U$ may serve to reduce the relative error between different types of orbitals. It does not, however, address other sources of error inherent to LDA or PBE. For example, in ZnO where the filled 3d orbitals lie several eV below the VBM, the LDA gap of 0.7 eV opens to 1.5 eV with $U \approx 4$–5 eV, compared to the experimental gap of 3.4 eV [46].

Another disadvantage is that even p-derived or other host semiconductor orbitals tend to become localized near defects, but this effect is not well captured by LDA+U or PBE+U since $n_{(m,\sigma)} \approx 0.5$. For such systems, to correct for the self-interaction error around the defect, Lany and Zunger proposed to include a potential [39] of form

$$V^{\text{hs}}_{m,\sigma} = \lambda(1 - n_{m,\sigma}/n_{\text{host}}), \tag{7.26}$$

where n_{host} is the on-site occupancy in the unperturbed semiconductor. This has the effect of canceling the nonlinearity of the energy *vs.* occupation (Figure 7.7(a)) and correctly describes polaron formation [39].

7.5.4 *Green's function approaches: the GW method*

Basic methodology. Green's function approaches based on single-particle orbitals include the *GW* method, the random phase approximation, and the Bethe–Salpeter equation (BSE) approach. These are all methods that rely on *propagators*, which in quantum mechanics refer to functions that describe the probability amplitude for a particle to travel from one place to another in a given time (or, if Fourier transformed, for a particle to travel with a given energy and momentum). While the *GW* method is too rich in implementations to be described here in full detail, we provide a simple overview. More detailed descriptions are available in Refs [47,48].

The idea behind *GW* is to accurately determine QP energies, which are the energies associated with adding or removing electrons to a system. Unlike the other methods described in Section 7.5, the *GW* approach has a different theoretical starting point than DFT. The *GW* approximation is an approximation to calculate the *self-energy* Σ of a many-body system of electrons. The self-energy refers to the energy that a particle has due to changes that it itself causes to the environment. For instance, one contribution to Σ arises from the electrostatic potential felt by an electron due to its interactions with other electrons (screening). Since electrons have a repulsive interaction with each other, an electron will disturb other electrons around it (it will polarize them), which then changes the electrostatic potential the electron itself experiences. Practically, for the purpose of defect calculations, the *GW* approach is used to obtain (i) bandgaps and, (ii) when combined with other methods, defect charge transition levels.

In the *GW* method, the self-energy Σ is expanded in a series about a set of single-particle states in terms of two quantities: the single particle Green's function G and the screened Coulomb interaction W. The expression is usually truncated after the first term:

$$\begin{aligned}\Sigma(1,2) &= iG(1,2)W(1^+,2) - \int d3 \int d4\, G(1,3)G(3,4)G(4,2)W(1,4)W(3,2) + \cdots \\ &\approx iG(1,2)W(1^+,2)\end{aligned} \tag{7.27}$$

In this common notation, the numbers 1, 2, ... are composite indices that indicate each electron's position, spin, and the time and the superscript "+" indicates an infinitesimal forward time propagation. The two main considerations then are how to describe each component, G and W. Each is discussed briefly below.

Single particle Green's function G: The Green's function approach requires the two-point propagator $G((\boldsymbol{r},t),(\boldsymbol{r}',t'))$ for a particle from $(\boldsymbol{r},t)$ to $(\boldsymbol{r}',t')$. Peaks in the Fourier transform $\hat{G}(\omega,\boldsymbol{k})$ correspond to electron addition and removal energies, often called QP energies. The approach developed by Hedin [49] is most often used, which yields a closed system of equations that can be solved for $G(1,2)$, the nonlocal part of the self-energy $\Sigma_{xc}(1,2)$, the polarizability $P(1,2)$, screened Coulomb interaction $W(1,2)$, and parameter $\Gamma(1,2)$. Hedin's formulation also introduces a new parameter I, the irreducible scattering amplitude, whose form is unknown and must be approximated. Ideally, we would like to do everything self-consistently. Given a choice for I, we would like to start with an initial guess for G, use it to obtain other variables in Hedin's equations that depend on G, then solve Hedin's equations for the new G, and repeat the approach until G no longer changes. Since this is challenging in practice, most approaches use a single-particle approximation for G instead

$$\hat{G}(\boldsymbol{r},\boldsymbol{r}',\omega) = \sum_i \frac{\varphi_i(\boldsymbol{r})\varphi_i^*(\boldsymbol{r}')}{\omega - \epsilon_i^{QP}} \tag{7.28}$$

where ϕ_i are orbitals of some single-particle Hamiltonian. Most often, the ϕ_i are obtained from the Kohn–Sham DFT Hamiltonian, as first suggested by Hybertsen and Louie [50]. When this is done, the expression for the quasiparticle energies becomes

$$\epsilon_i^{QP} = \epsilon_i^{KS} - Re[\langle\varphi_i|V_{xc}|\varphi_i\rangle] + Re\left[\langle\varphi_i\middle|\sum\nolimits_{xc}\left(\epsilon_i^{QP}\right)\middle|\varphi_i\rangle\right] \tag{7.29}$$

where is the Kohn-Sham eigenvalue and V_{xc} is the Kohn-Sham exchange-correlation potential in (7.17), showing how the quasi-particle energies are modified from the starting Kohn-Sham eigenvalues. Since the self-energy Σ_{xc} depends on the quasiparticle energy, (7.29) should be solved iteratively.

Screened Coulomb interaction W: If we simply use for W the bare (unscreened) Coulomb interaction, we recover the standard perturbation theory expansion for the self-energy. In fact, the GW approximation with W implemented this way recovers the HF exchange energy as the self-energy. However, in real materials, the potential at some point due to the presence of an electron at another is much weaker than what is given by the bare Coulomb potential, due to screening by other electrons. Most implementations therefore go beyond the unscreened Coulomb interaction by approximating the polarizability within a noninteracting particle model

$$P(\boldsymbol{r},\boldsymbol{r}',\omega) = \sum_{i,j}\frac{\varphi_i^*(\boldsymbol{r})\varphi_j(\boldsymbol{r})\varphi_j^*(\boldsymbol{r}')\varphi_i(\boldsymbol{r}')}{\omega + \left(\epsilon_i^{QP} - \epsilon_j^{QP}\right)} \times (f_i - f_j) \tag{7.30}$$

to screen the Coulomb exchange interaction, where f_i, f_j are Fermi-Dirac occupations and the summation is taken over occupied states j and unoccupied states i. So in some sense, the GW method can be thought of as a dynamically screened HF self-energy.

Common implementations: In the most common implementation G_0W_0 (single-shot), G and the polarizability P are directly obtained from (7.28) and (7.30) using DFT QP energies. From here, W and Σ_{xc} are obtained directly from simplified forms of Hedin's Equations, and the QP energies in (7.29) are evaluated once. At the next level of sophistication GW_0, the single particle energies in (7.28) would then be updated with the new values, and the procedure repeated until self-consistency is achieved (the single-particle energies no longer change). Note that in GW_0, P is generally not updated with new single-particle energies and is only evaluated once.

Determination of bandgaps and defect charge transition levels. With this implementation, the QP gap can be obtained as the difference between the EA (the smallest electron addition energy) and the IE (the smallest electron removal energy) for the bulk material. The addition and removal energies are the poles of fourier transform $\hat{G}(\omega,\boldsymbol{k})$. There are now several systematic assessments of bandgaps determined using G_0W_0 or GW_0 starting with local or semilocal DFT orbitals which generally show promising results [51,52]. Defect charge transition levels are less straightforward to obtain since GW obtains only electron removal and addition energies (QP energies) at fixed atomic geometry. For defects, the peaks in $\hat{G}(\omega,\boldsymbol{k})$ correspond to the transition levels for defects *at fixed atomic coordinates* (see Fig. 7.5). These can be compared directly to experimental photoemission and photoabsorption spectra. To obtain the thermodynamic charge transition levels, the electron addition and removal energies need to be supplemented with atomic positions and relaxation energies obtained using some other approach. If we believe that geometries and relaxation energies are reliably obtained in DFT, we can combine these with *GW* QP energies as suggested in [53] to estimate charge transition levels. The approach improves charge transition levels over conventional DFTs, for example, giving results in good agreement with experiment for self-interstitials in silicon [53] and for *F*-center defects in magnesium oxide [27,54]. As discussed later, the results for the *F*-centers in MgO agree remarkably well with those obtained independently using QMC.

In terms of limitations, *GW* is ultimately based on perturbation theory, and like any perturbation theory, its accuracy will depend on the quality of the starting point (the description of the unperturbed system). When starting from single particle DFT Kohn–Sham orbitals, it may be less effective in cases where the starting orbitals themselves are not good. There are some cases for defects where the starting point is sufficiently poor that the method will not work: if DFT predicts a metallic state in the semiconductor to begin or if it spuriously places the defect level outside the gap. There are outstanding questions [55] to why G_0W_0 and GW_0 perform as well as they do when starting from Kohn–Sham single particle states.

7.5.5 Hybrid functionals

Hybrid functionals have become more common in the last 10 years for first-principles predictions of defect properties. Today, they represent the best compromise between accuracy and practical usability (they are available in several DFT codes), although they come at a considerable computational cost. Hybrids offer one route to reducing the self-interaction error described earlier and appear to address some of the principal shortcomings of LDA and PBE such as the tendencies toward delocalized states and the bias toward shallow defects.

The idea behind hybrid functionals is based on the consideration that DFT suffers from self-interaction errors that are fortuitously not present in HF. In HF, they are removed by the Fock exchange integral. In hybrid functionals, to reduce the self-interaction error, a proportion α of standard DFT exchange is replaced by a Fock-like exchange interaction [56] as

$$E_{xc}^{\text{hybrid}} = E_{xc}^{\text{DFT}}[\rho(\boldsymbol{r})] + \alpha(E_x^{\text{Fock}} - E_x^{\text{DFT}}[\rho(\boldsymbol{r})]) \tag{7.31}$$

where the exchange is evaluated by the Fock-like integral

$$E_x^{\text{Fock}} = -\frac{1}{2}\sum_{i\sigma}\sum_{j\sigma}\iint \frac{\phi_{i\sigma}^*(\boldsymbol{r})\phi_{j\sigma}^*(\boldsymbol{r}')\phi_{i\sigma}(\boldsymbol{r}')\phi_{j\sigma}(\boldsymbol{r})}{|\boldsymbol{r}-\boldsymbol{r}'|}\, d\boldsymbol{r}\, d\boldsymbol{r}' \tag{7.32}$$

that appears in (7.13), with the sum evaluated over occupied orbitals. It is evaluated on the DFT (not HF) orbitals.

There are many types of hybrid functionals available today. One classification is whether the hybrid incorporates a direct exchange interaction as in the integral above, or whether the interaction is screened at larger distances. The integral above is long-ranged ($\sim 1/r$), nonlocal, and numerically difficult to evaluate, introducing a sizable computational cost over LDA or PBE. An alternative approach suggested by Heyd *et al.* [57] is to truncate the long-ranged part by replacing the integral with a screened version E_{sx}^{Fock} given by

$$E_{sx}^{Fock} = -\frac{1}{2}\sum_i\sum_j f_i f_j \int\int \frac{\mathrm{erfc}(\mu|\boldsymbol{r}-\boldsymbol{r}'|)}{|\boldsymbol{r}-\boldsymbol{r}'|}\varphi_i^*(\boldsymbol{r})\varphi_j^*(\boldsymbol{r}')\varphi_i(\boldsymbol{r}')\varphi_j(\boldsymbol{r})d\boldsymbol{r}d\boldsymbol{r}' \tag{7.33}$$

where μ is called the range-separation parameter. This screening makes the interaction go to zero at large distances. For perfect metals, the exchange integral should go to zero at large distances, whereas for perfect insulators, it should remain. Screened exchange hybrid functionals such as Heyd-Scuseria-Ernzerhof (HSE06) [57], therefore, represent a middle ground between these two cases and facilitate numerical evaluation of the Fock-like integrals. Common hybrid functionals that are not screened are PBE0 [58] and B3LYP [38].

The selection of hybrid parameters such as the degree of mixing α and the range separation parameter μ then gives a degree of tunability, although these parameters should not be treated as *ad hoc* tuning parameters. The common selection of $\alpha = 0.25$ has some grounding in perturbation theory [59]. The selection of $\alpha = 0.25$ and $\mu = 0.11$ au corresponding to HSE06 empirically seems to show reasonable but not perfect performance for total energies and bandgaps for small to mid-gap semiconductors. It does not, however, work well for wide bandgap or very ionic systems (MgO, for example) for which the optimal exchange should be >0.25 and the screening is weak. Figure 7.8 shows bandgaps for several semiconductors and insulators [52] calculated using PBE0 and HSE06 hybrid functionals, the G_0W_0 approach, and experiment. Such observations have recently led to suggestions for systematic approaches to selecting hybrid functional parameters based on static dielectric constants [60] or satisfying the Koopman's theorem on specified test cases such as polarons or charge transition levels of hydrogen impurities [61]. These should be considered as guidelines, since, as always, there are no guarantees that fitting to specific test cases will result in good predictive capability for all other cases.

As the examples presented later show, the use of hybrids has improved our predictive capability for the application of first-principles methods to point defects. While there are several systematic studies and benchmarks available today, there still remain several outstanding questions to the ultimate degree of accuracy attainable using hybrid functionals. The case study of nitrogen defects in ZnO presented later in this chapter will highlight some of these questions. There are also some notable possible shortcomings. These include the slow convergence of finite size effects for shallow defects due to the long-range exchange [62]. It is also reported that typical α parameters used for the semiconductor host give an *overcorrection* for the 3d-induced in-gap states [63] for TM impurities. A systematic study of octahedral TM clusters shows that hybrid descriptions of the 3d-ligand

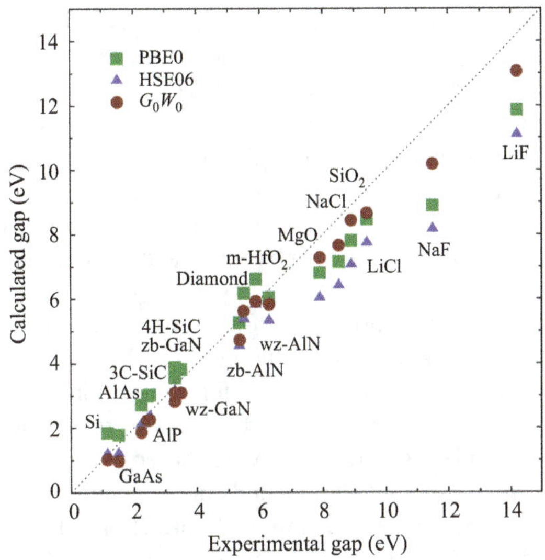

Figure 7.8 Band gaps for several semiconductors and insulators: experimental results compared to those calculated by PBE0 and HSE06 hybrid functionals and G_0W_0. © 2012 APS. Reprinted, with permission, from [52]

interactions are not in general consistent with results from variational QMC [64]. The set of available results is still growing today, and more targeted efforts are expected to reveal more insights and tendencies into the performance of hybrids as a whole.

7.5.6 Fitted chemical potentials

Finally, we address the problems in evaluating chemical potentials μ_α for accurate defect thermodynamics and describe an empirical correction scheme. The formation enthalpy of a compound, say A_nB_m, is calculated as

$$\Delta H_f(A_nB_m) = E_{\text{tot}}(A_nB_m) - n\mu_{\text{A}}^{\text{elem}} - m\mu_{\text{B}}^{\text{elem}}, \tag{7.34}$$

where E_{tot} is the total energy of the compound and μ_α^{elem} are the total energies per species α in their elemental phases. Accurate calculation of formation enthalpies ΔH_f obviously requires both accurate E_{tot} for the compound of interest and accurate μ_α^{elem} for all elements involved. However, sufficiently accurate total energies (chemical accuracy 1 kcal/mole $\approx$ 0.04 eV/atom) entail QMC or higher order quantum chemical calculations (post-HF methods), which are not practical for large/complex systems or even isolated atoms of heavy elements.

Nonetheless LDA/GGA-based calculations are observed to often give $\Delta H_{D,q}$ in reasonable agreement with experiment. The agreement is believed to be somewhat fortuitous, owing to error cancelations in the calculated total energies of the compound of interest and the elemental reference phases. That is, if one makes approximately the same error in $E_{\text{tot}}(A_nB_m)$ as in $n\mu_A^{\text{elem}} + m\mu_B^{\text{elem}}$ in (7.34), the computed formation energy will be reasonable. This is more likely to occur if all the compounds A_nB_m, A, and B themselves exhibit similar types of bonding. If they are all metals, then it is more reasonable that computed energies have a similar overall error [65]. On the other hand, this is not necessarily the case for semiconductors. For instance, obtaining the formation enthalpy for ZnO requires total energies for ZnO, metallic Zn, and gas phase O_2, and DFT does not make similar errors for all three phases.

A practical but effective approach to address the problem of incomplete cancelation of errors is to take advantage of the fact that there are systematic patterns in the incomplete error cancelation observed for conventional DFT. For example, compound semiconductors tend to be underbound with respect to elemental solid metallic phases. By tabulating and quantifying the errors in computed formation enthalpies for, say, all extant A_nB_m compounds in the chemical potential space (cf. Figure 7.3), it is possible to apply an empirical correction for the incomplete cancelation of errors. In the *fitted elemental-phase reference energies* [66,67] approach, calculated μ_α^{elem} are shifted by a constant offset $\delta\mu_\alpha^{\text{FERE}}$ as

$$\mu_\alpha^{\text{FERE}} = \mu_\alpha^{\text{LDA/GGA}} + \delta\mu_\alpha^{\text{FERE}}. \tag{7.35}$$

Here the shifts $\delta\mu_\alpha^{\text{FERE}}$ are chosen to obtain a best fit against known experimental enthalpies of formation ΔH_f^{exp} by solving the linear set of equations

$$\Delta H_f^{\text{exp}}(A_nB_m) = \Delta H_f^{\text{LDA/GGA}}(A_nB_m) - \sum_\alpha n_\alpha\mu_\alpha^{\text{FERE}}. \tag{7.36}$$

This approach, by construction, improves computed formation enthalpies. The usual questions related to the transferability of the approach should be kept in mind: the shifts are obtained by fitting to a given data set, and there are no guarantees they will apply equally well to all cases outside of the fitted data set.

7.6 Finite size effects

First principles electronic structure calculations can only deal with microscopic systems, typically of a few hundred atoms, or some thousands of atoms. Experimental samples typically are tens of orders of magnitude larger. The best way to model macroscopic defective semiconductor samples is by periodic boundary conditions, as illustrated in Figure 7.9. Here, an isolated defect, or impurity, is included in a finite supercell, typically of the order of 100 atoms. This corresponds to an impurity concentration of the order of a few percent, or $\sim 10^{21}$ cm^{-3}, which is several orders of magnitude larger than typical impurity concentrations in semiconductors. The main concern in this approach is whether or not there are spurious interactions between the defect and its periodic images, and how such interactions influence the calculated results.

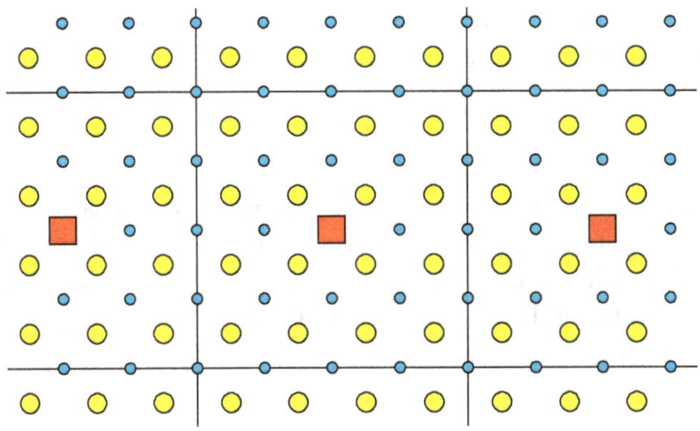

Figure 7.9 *Impurity in a periodic supercell. The blue and yellow spheres indicate host atoms, and the red square indicates an impurity. The black lines indicate supercell boundaries, beyond which the same cell is periodically repeated*

7.6.1 Charged defects

Various charge states q of a defective system are evaluated by varying the number of electrons in the system. This leads to two spurious terms in the total energy that need to be corrected for (i) error due to arbitrary shift in electrostatic potential and (ii) Coulomb repulsion of charged defect and its periodic images. These are not independent of each other and should be considered in a self-consistent manner. There is a well-established machinery [35,68–76] to evaluate the spurious errors due to finite size effects *after the fact* and amend the total energy by adding correction terms ΔE as described below.

1. *Potential alignment.* In a momentum-space formalism, to avoid divergence of the electrostatic potential V_{el} for a charged system with periodic boundary conditions, V_{el} is set as zero at reciprocal lattice vector $G = 0$ [77]. This has the effect of a constant shift ΔV_{pa} of unknown magnitude to V_{el}, which does not affect the total energies of charge neutral systems but changes the total energy by $-q\Delta V_{\text{pa}}$. Thus, for meaningful defect formation energies, one needs to add

$$\Delta E_{\text{pa}} = q\Delta V_{\text{pa}} \tag{7.37}$$

   to the total energy of the defective system. The shift ΔV_{pa} can be evaluated, e.g., by comparing the electrostatic potential at an atomic site V^{at} calculated for an atom in the pristine host $V^{\text{at}}_{\text{host}}$, and for a corresponding atom in the defective system V^{at}_{D}, which should be sufficiently far from the defect such that its potential is not perturbed by the defect. It is convenient to consider the potential of several such atoms and evaluate $\Delta V_{\text{pa}} = \overline{(V^{\text{at}}_{D} - V^{\text{at}}_{\text{host}})}$.
2. *Image charge effect.* The main source of error in charged system is due to the Coulomb interaction energy of the defect and its periodic images, which

decays very slowly ($\propto 1/r$) as supercell size increases. The image charge correction can be expressed as [69]

$$\Delta E_{\mathrm{IC}} = \frac{q^2 \alpha_{\mathrm{M}}}{2\epsilon L} + \frac{2\pi q Q}{3\epsilon L^3} + \mathcal{O}(L^{-5}). \tag{7.38}$$

The first (leading) term is the Madelung energy of point charge q screened by the dielectric constant ϵ, where α_{M} is the Madelung constant and L is the linear supercell dimension. The second term is the interaction of a point charge q with the aperiodic charge density ρ_D induced by the defect. The quantity Q is the second radial moment of this charge density

$$Q = \int_{\Omega} d\mathbf{r} \rho_D(\mathbf{r}) r^2, \tag{7.39}$$

where Ω is the volume of the supercell. Higher order terms scale as $\mathcal{O}(L^{-5})$ and are neglected. In certain limits, the monopole-quadrupole term scales as $\propto q^2/L$ instead of $\propto q/L^3$ and can be approximated as $c_{\mathrm{sh}}(1 - \epsilon^{-1})$ times the first order term, where $c_{\mathrm{sh}} \approx 1/3$ is a geometric factor that depends on the supercell shape [71]. This type of image charge correction, often also called the *Makov–Payne correction*, is very easy to evaluate, but is limited to 3D bulk systems with an isotropic dielectric response. Moreover, (7.38) implicitly assumes a negligible dipole moment for the defective system [74,78], so for polar and/or low dimensional systems, a more general scheme is desirable.

3. *Freysoldt, Neugebauer, and Van de Walle (FNV) correction.* FNV have proposed another scheme, which generalizes the above corrections (i) and (ii) [72,73,76]. Here, the correction term

$$E_{\mathrm{FNV}} = E_{\mathrm{PC}} - q\Delta V_{\mathrm{PC},q/b}\big|_{\mathrm{far}} \tag{7.40}$$

is added, where E_{PC} is the Madelung energy of a point charge (cf. first term in (7.38)) and the second term is a potential alignment like term (cf. (7.37)). Here, one evaluates the defect-induced potential

$$V_{q/b} = V_{\mathrm{D},q} - V_{\mathrm{host}}, \tag{7.41}$$

and $\Delta V_{\mathrm{PC},q/b}$ is its difference with the point charge potential $V_{\mathrm{PC},q}$,

$$\Delta V_{\mathrm{PC},q/b} = V_{q/b} - V_{\mathrm{PC},q}. \tag{7.42}$$

$\Delta V_{\mathrm{PC},q/b}\big|_{\mathrm{far}}$ is $\Delta V_{\mathrm{PC},q/b}$ evaluated far away from the defect in the supercell. This approach requires no knowledge of the microscopic screening, is applicable to any shape supercell, and also for systems of lower dimensionality [74,75].

7.6.2 Band filling for shallow defects

For shallow defects with the defect single particle level resonant in the valence or conduction bands, spurious defect–defect interactions emerge even for charge neutral defects in finite (too small) supercells. These interactions arise from artificial filling of

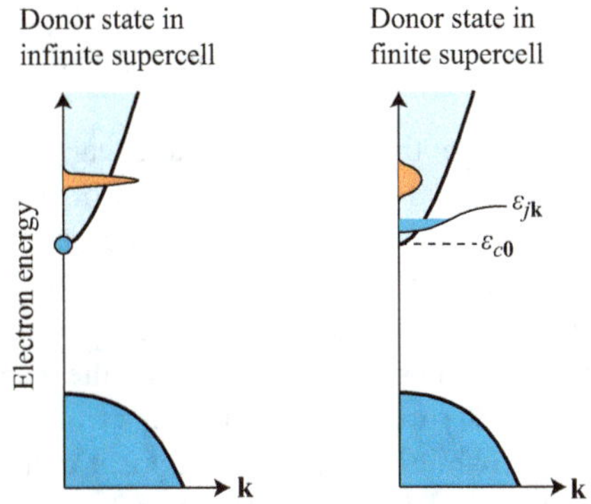

Figure 7.10 *Band filling effect. The left panel shows a donor state (orange), which is resonant inside the conduction band (CB), calculated in an infinite (sufficiently large) supercell. When the donor state has donated its electron to the CB, there is an electron precisely at the CB bottom. The right hand panel shows how (i) the donor state broadens due to defect–defect interaction and (ii) how the CB gets deformed and repelled upwards due to spurious electron–electron interactions*

the bands due to the high simulated defect density. The effect is illustrated for a shallow donor state in Figure 7.10. This finite size effect can be corrected for by adding the correction term [35]

$$\Delta E_{\mathrm{bf}} = -\sum_{j>c,\mathbf{k}} (w_{\mathbf{k}} \eta_{j\mathbf{k}} \varepsilon_{j\mathbf{k}} - \varepsilon_{c0}), \tag{7.43}$$

where $\varepsilon_{j\mathbf{k}}$ are the eigenstates for donor electrons and ε_{c0} is the CBM (assumed here to be at Γ point). The parameters $w_{\mathbf{k}}$ and $\eta_{j\mathbf{k}}$ are weights for each **k**-point for Brillouin zone integration. The same effect occurs also for acceptor type defects and is corrected for similarly.

7.7 Recent examples

7.7.1 Mg and other possible acceptors in GaN

It was experimentally first reported in 1989 that low-electron energy beam irradiation of Mg-doped GaN can give rise to modest hole concentrations $2 \times (10^{16})\ \mathrm{cm}^{-3}$ [80]. Within a few years, enhancements to growth methods resulted in practically usable p-type GaN [81]. Despite that the doping efficiency of Mg in GaN remains low, these developments ultimately led to the rise of GaN-based light-emitting diodes and other devices. It is intriguing that both to date Mg is the only known impurity that gives rise to p-type conduction in GaN, and yet it has long been a challenge for theory to understand and explain the underlying physics of how. The case study of $\mathrm{Mg_{Ga}}$ in

GaN represents an example of a system where recent developments in first-principles computational approaches have improved our understanding of the defects responsible for p-type dopability.

Following early studies in the late 1980s and early 1990s, demonstrating p-type conductivity for Mg-doped GaN, significant effort was dedicated to assess the properties of Mg_{Ga}. Theoretical [82] studies (based on conventional DFT) and experimental efforts [83] attempted to discern properties of Mg_{Ga} such as its thermal activation energy. It was common to assume that the optical emission observed at 3.3–3.5 eV (ultraviolet luminescence, UVL, near the bandgap) upon a small degree of Mg incorporation [84] arose from transitions from some shallow donor to the Mg acceptor state. Meanwhile, at higher Mg dopant concentrations, the origin of a dominant broad peak that arises in the range of 2.70–2.95 eV (blue luminescence, BL) was attributed to recombination from deep donors to the Mg acceptors. This picture of Mg_{Ga} however was not bulletproof. The identity of the shallow donor responsible for the UVL remained unclear, and strangely, annealing or electron beam irradiation intended to activate the Mg acceptors was noted to correspond to a drop in the UVL intensity [79,85,86] and the emergence of the broad BL [85–87]. These contradictory observations became the source of much discussion [88–92].

Also, given the Mg_{Ga} IE of $\sim$200 meV, large numbers of Mg dopants are needed to obtain appreciable p-type carrier concentrations. This results in the possibility of defect complexes, interaction of Mg with native defects, hydrogen, and possible self-compensation that further complicates understanding.

In 2010, Lany and Zunger used a method based on application of a localized on-site hole-state potential (7.26) to improve the description of the hole localization for acceptor defects. The magnitude of the on-site potential was selected by satisfying the generalized Koopman's condition [39] (see Sec. 7.5.2). Two possible states for Mg_{Ga} were identified (Figure 7.11): a lattice relaxed, noneffective mass state more like a deep level with the acceptor wave function largely localized to a single N atom neighboring the Mg atom, and a shallow transient state with an effective-mass-like delocalized wave function. Most critically, the use of a modified DFT designed to address the bias of conventional DFT toward delocalized states gave an interesting prediction: that cation-acceptors like Mg_{Ga} can bind holes in deep levels that are localized largely on a single neighboring N atom. In both cases, the impurity state orbitals arise from the interaction of the impurity atom Mg orbitals with the dangling bonds of the surrounding N ligands. Figure 7.11(a) shows the DOS and the charge distribution of the noneffective mass state, which is localized to one p-orbital of an N neighbor. There is a Jahn–Teller symmetry breaking, so the N atom hosting the hole has a larger bond length than the other three Ns. Figure 7.11(b) shows the DOS and the charge distribution for the shallow transient state, which exhibits a more symmetric geometry, a delocalized wave function, and was found to be only 30 meV higher in energy than the deep state. The analysis further predicted a large difference in luminescence (shallow state at 3.35 eV, deep acceptor at 2.93 eV) due to the large associated lattice relaxation for the ionized Mg_{Ga}^{-} localized state.

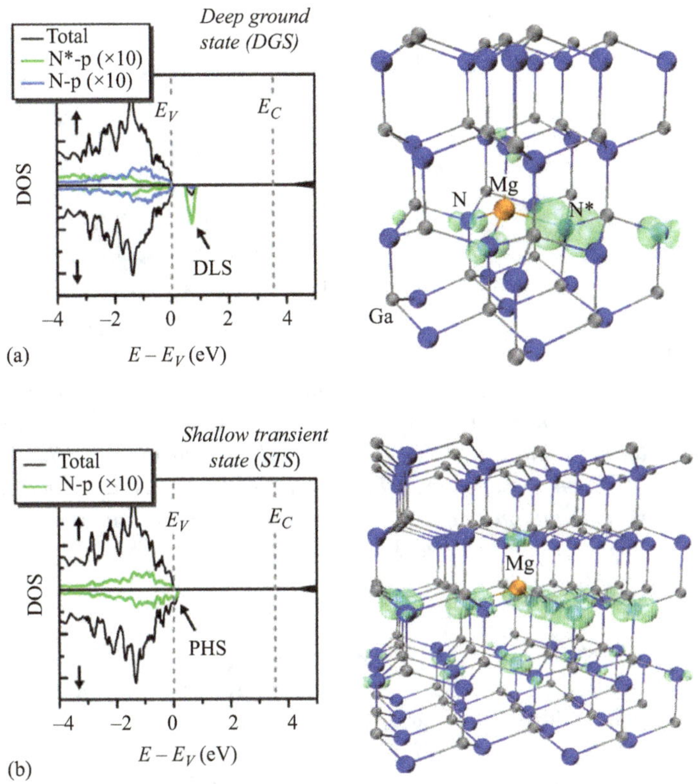

Figure 7.11 *Using an approach based on the application of an onsite potential in order to satisfy the generalized Koopman's theorem, a symmetry broken state for Mg_{Ga} can be identified. (a) The density of states and charge density of the deep level state (DLS) for the Mg_{Ga} acceptor in GaN, showing the hole charge distribution localized to a nitrogen atom N^* neighboring the Mg impurity. (b) The density of states and charge density of a shallow state, showing a delocalized hole (perturbed hole state).*

Subsequent hybrid DFT results of Lyons and Walle in 2012 [79], shown in Figure 7.12, corroborated the existence of the symmetry-broken localized state for Mg_{Ga} and predicted a $\varepsilon(-1|0)$ charge transition level at 260 meV above the VBM. They used the HSE hybrid functional with a 31% degree of mixing for the exchange, selected to give a good description of the GaN bandgap. Carrying out atomic relaxations with the hybrid functional, no metastable transient state was observed, only a highly localized hole with most of the charge on the neighboring axial nitrogen. They concluded that the existence of p-type GaN seems then to be somewhat fortuitous: despite its localized nature, Mg_{Ga}, has a relatively small IE.

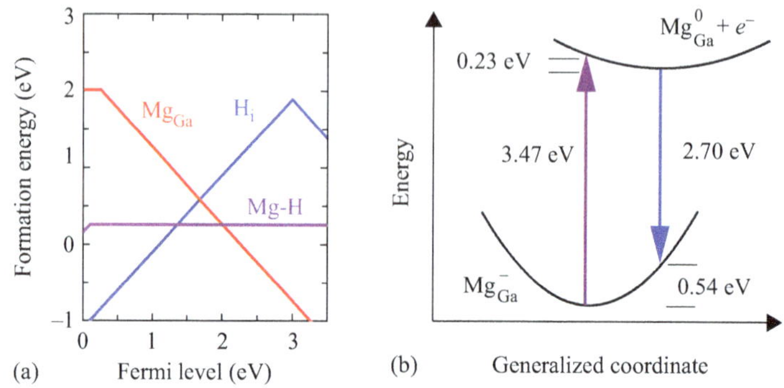

Figure 7.12 (a) Formation energy for substitutional Mg_{Ga}, hydrogen interstitial H_i, and the Mg-H complex in GaN. (b) Configuration coordinate diagram showing Frank–Condon optical transitions. In this work, the blue luminescence associated with Mg:GaN was attributed to recombination of an electron from the CBM with neutral Mg^0_{Ga}, giving an emission peak at 2.70 eV. © 2012 APS. Reprinted, with permission, from [79]

By contrast for AlN, the Al_{Ga} charge transition level occurs at 0.78 eV above the VBM, so achieving p-type conductivity is more difficult. The calculated optical emission associated with recombination of a CB electron with Mg^0_{Ga} to form Mg^-_{Ga} was calculated to be 2.70 eV; with the large deviation from the bandgap arising again from the 0.54 eV relaxation energy for the ionized Mg. Thus, the observed BL could now be attributed to transitions of CB electrons to Mg impurities, rather than from deep donors as previously surmised. The near-band-edge UVL believed to arise from the shallow donor to the supposedly shallow Mg acceptor was instead attributed to Mg-H complexes, consistent with experiments that removal of H during annealing seems to decrease the UVL intensity.

Beyond Mg_{Ga}, other possible defects that may provide p-type carriers have been analyzed extensively using first-principles. To a large extent, modern methods make it clear that most of these defects instead act as sites for carrier trapping and scattering, sources of non-radiative recombination, or simply serve as compensating acceptors. Whereas early DFT results [93] suggested that substitutional carbon on the nitrogen site C_N was a shallow, effective-mass acceptor, more recent hybrid HSE simulations instead have placed the $\varepsilon(0|-1)$ charge transition level at 0.90 eV above the VBM, indicating a deep, localized nature. These results also attribute the observed yellow emission to C_N. The deep nature can be thought of as arising from carbon p orbitals that are much higher in energy than the corresponding nitrogen p orbitals that make up the valence band. Very early on, gallium vacancies had also been considered as possible native sources of p-type conductivity in GaN. In this case, however, conventional DFT approaches already placed the $\varepsilon(-2|-3)$ transition at 1.1 eV above the VBM, which becomes further deepened to 2.09 eV using screened exchange LDA [94]. More recent results using hybrid functionals further deepen the charge transition level to 2.80 eV [85], with the difference largely arising from the identification of a different lowest energy atomic geometry using the hybrid. The case of V_{Ga} is one where the early DFT results qualitatively correctly established the deep nature of the defect, but the quantitative value of the transition level currently still continues to be refined.

7.7.2 Oxygen vacancies and other defects in metal oxides

From a first-principles perspective, defects in metal oxides are challenging due to the chemically diverse nature of metal oxides themselves, which can vary from highly ionic (MgO) to more covalent (ZnO). Since a given DFT functional cannot treat all of these systems on equal footing, the degree of accuracy exhibited by a given functional can vary from one metal oxide to another. A priori it is not always obvious which functional offers the best description of a given metal oxide. We summarize below some well-known examples highlighting the challenges associated with the application of first-principles methods to oxygen vacancies and other defects in metal oxides.

7.7.2.1 V_O in MgO

Magnesium oxide (MgO), the classic ionic binary oxide, is an insulator with a bandgap of 7.78 eV. The highly localized states and insulating nature make it challenging to obtain quantitatively correct predictions using conventional DFT-based methods. The oxygen vacancy (the F-center) is the most discussed defect in MgO. The introduction of a neutral oxygen vacancy (F^0) results in the appearance of an s-symmetry in-gap defect state with two electrons localized. The defect may become singly (doubly) ionized by release of one (two) electrons to the host, resulting in the F^+ (F^{2+}) ionized state. The paramagnetic F^+ state is active in electron paramagnetic resonance (EPR), the others are not. One puzzle to the F-center defect pertains to the very similar optical absorption energies associated with the F^0 and F^+ states (4.95 and 5.0 eV, respectively), making it difficult to distinguish one state from the other using optical techniques. First-principles modeling of optical transitions associated with F-center defects in MgO represent a success story of sorts. This example illustrates an instance where two independent many-body techniques (G_0W_0+BSE [54] and DMC [27]) have achieved quantitative estimates of absorption and emission in remarkable agreement with each other and, to a large extent, also with experiment.

Figure 7.13 summarizes the results for the optical absorption and emission peaks obtained for the case of exchange of electrons with the conduction band using both approaches. These are obtained as vertical (Franck–Condon) transitions on configuration coordinate diagrams. In both cases, 64 atom supercells were used and geometries were obtained using conventional DFT (LDA or PBE), while the higher level theory was used to obtain the excitation energies. The correspondence in supercell size and (presumably) defect geometries helps to make the comparison of results more direct. The G_0W_0 results are obtained using single-particle orbitals obtained from a hybrid scheme that incorporates 25% Fock exchange with 75% LDA exchange and correlation. The raw electron addition and removal energies obtained with G_0W_0 overestimate optical absorption energies for F^0, F^+ in

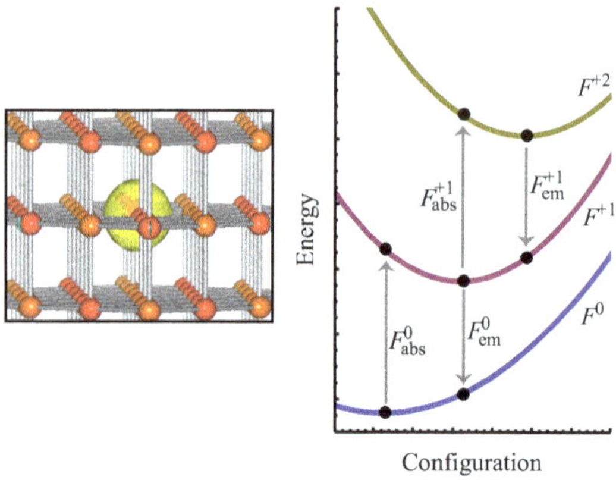

	G_0W_0+BSE	DMC	Experiment
F^{0}_{abs}	4.95	5.0(1)	5.00
F^{+1}_{abs}	4.92	5.1(1)	4.95
F^{0}_{em}	3.6	3.8(1)	3.1–3.2
F^{+1}_{em}	3.4	3.5(1)	2.3–2.4

(Energies in eV)

Figure 7.13 The oxygen vacancy (F-center) in magnesium oxide, the classic ionic binary oxide. The optical absorption and emission peaks as obtained from G_0W_0 + BSE in [54] and diffusion Monte Carlo in [27] are in very good agreement with each other. © 2013 APS. Adapted, with permission, from [27]

comparison to experiment; better agreement is achieved if the electron-hole interaction (exciton) is taken into account via the BSE [95,96]. Meanwhile, the absorption and emission energies obtained by DMC match the G_0W_0+BSE results remarkably well. The DMC-computed energies directly represent excited state many body wave functions that inherently include the exciton and thus the electron-hole interaction. It is also notable that both techniques obtain an optical emission peak at around 3.4–3.5 eV for the F^{+1} optical emission, larger than the traditional experimental assignment of 2.3–2.4 eV. The authors of [54] attribute the experimentally observed lower emission energy to transitions involving exchange with holes in the valence band; these transitions, however, were not computed in DMC.

7.7.2.2 V_O in ZnO

Interest in native defects in ZnO largely centers on understanding the unintentional n-type conductivity, which historically was attributed to the presence of oxygen vacancies [99]. Early studies (2000–10) from several groups [46,100–105] reporting first-principles calculations of native defects in ZnO exhibit a large numerical spread and substantial quantitative discrepancies. For example, for the oxygen vacancy, the reported $\varepsilon(+2|0)$ charge transition level varied from as low as 0.2 eV [106] to as high as 2.5 eV [107] above the VBM. Unlike MgO which is a classic binary ionic oxide, zinc oxide exhibits a more covalent nature. It also suffers from a severe underestimation of the bandgap energy when conventional functionals such as LDA and GGA are used, which yield gaps of ~0.75 eV in comparison to the experimental value of 3.44 eV. This has dramatic effects on calculated defect levels and formation energies, and consequently, the large spread in reported values results from the different approaches to address problems associated with the bandgap error.

More recently, approaches that yield more accurate band edge positions have consolidated results in a more narrow range. The oxygen vacancy in ZnO is one case where there remains some uncertainty in the quantitative position of the defect charge transition levels and defect formation energies, although agreements have been achieved. Figure 7.14 summarizes results for V_O in ZnO using screened exchange hybrid functionals [97,98] carried out in 2008 and 2010, respectively. These works were followed up in 2015 by QMC calculations [29] which, remarkably enough, placed the charge transition levels at a very similar position, giving confidence in set of results. The overall correspondence between the charge transition levels obtained using different methods is quite good. On the other hand, the defect formation energies themselves are about 0.4–0.5 eV higher in magnitude in QMC compared to hybrid functionals for reasons that remain unclear.

Based on these more recent studies, we can now conclude that the oxygen vacancy V_O is a donor defect that can have a reasonably low formation energy in an oxygen poor environment. However, it is a deep donor with the $\varepsilon(+2|0)$ charge transition level appearing $\sim$1 eV below the conduction band edge and hence cannot be the cause of the unintentional n-type conductivity of ZnO. Most analyses further agree that V_O is a negative-U center, with the $+1$ charge state always metastable; this is consistent with the difficultly in observing the paramagnetic $+1$ state in EPR. The negative-U behavior arises from large lattice relaxations of the

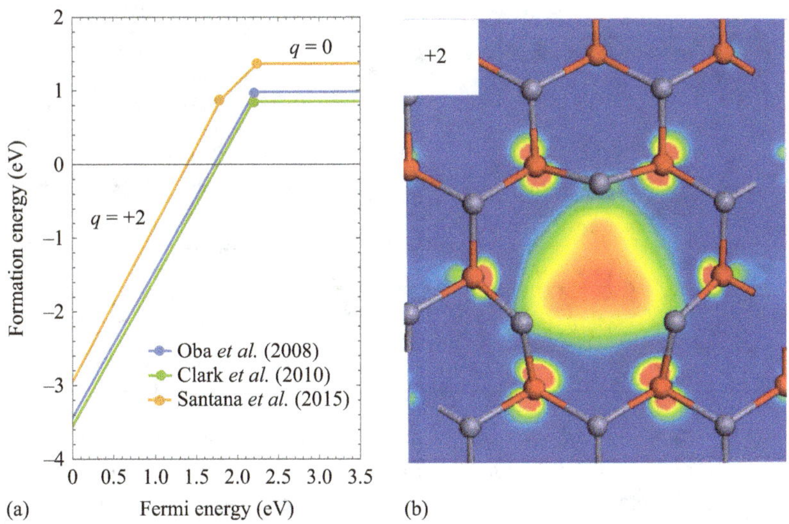

Figure 7.14 (a) Comparison of results for V_O in zinc oxide according to hybrid functional DFT calculations [97,98] and diffusion Monte Carlo calculations [29] are in quite good agreement. (b) The charge density of the $q = +2$ charge state of the oxygen vacancy. © 2010 APS. Reprinted, with permission, from [98]

+2 charge state that reduces its formation energy. (The QMC results in Figure 7.14 predict a narrow range of Fermi energies for which the $q = +1$ state is stable, but this probably lies within the stochastic error of the results and is not conclusive.) Questions remain about the ultimate source of n-type conductivity in ZnO then, which have been attributed to several possible factors including interstitial and/or substitutional hydrogen and others [108].

7.7.2.3 N_O in ZnO

The analysis of nitrogen defects N_O in ZnO has been a longstanding challenge for first-principles modeling. The interest has largely been driven by the technological interest in achieving p-type ZnO, which remains elusive, but would open the door to ZnO-based optoelectronic devices. Substitutional nitrogen at the oxygen site had long been considered one of the best candidates for this purpose [120]. Nitrogen, in the same row of the periodic table as oxygen, contains one fewer valence electron; it was believed that N_O therefore may act as an effective p-type dopant.

Figure 7.15 summarizes recent first-principles results for the $\varepsilon(0|-1)$ charge transition level for N_O, in comparison to recent experimentally reported values. Even early computational results from Park *et al.* [115] in 2002 obtained using the LDA predict a charge transition level $\varepsilon(0|-1)$ around 0.4 eV above the VBM,

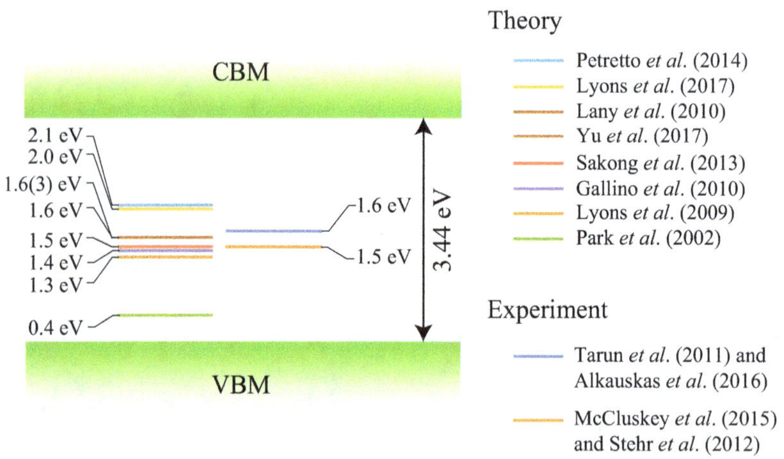

Figure 7.15 Summary of computational and experimental results for the acceptor transition level $\varepsilon(0|-1)$ for N_O in ZnO. Computational results from Petretto et al. (2014) [109], Lyons et al. (2017) [110], Lany et al. (2010) [111], Yu et al. (2017) [28], Sakong et al. (2013) [112], Gallino et al. (2010) [113], Lyons et al. (2009) [114], and Park et al. (2002) [115]. Experimental results from Tarun et al. (2011) [116], Alkauskas et al. (2016) [117], McCluskey et al. (2015) [118], and Stehr et al. (2012) [119]

suggesting that N_O is not shallow. In 2009, Lyons *et al.* [114] carried out Heyd, Scuseria, and Ernzerhof hybrid functional calculations using an exchange mixing parameter of $\alpha = 36\%$, selected in order to reproduce the experimental bandgap of ZnO. They obtained a much deeper value of $\varepsilon(-1|0) = 1.3$ eV above the VBM. Lany *et al.* [111] in 2010 used a generalized Koopman's approach and reported $\varepsilon(-1|0) = 1.6$ eV above the VBM. Interestingly, they predicted a symmetry broken defect geometry around the neutral N_O, with an associated relaxation energy of 0.8 eV playing a large role in the deepening of the charge transition level. Also in 2010, Gallino *et al.* [113] reported $\varepsilon(0|-1) = 1.4$ eV, with little geometry relaxation, obtained using the hybrid B3LYP functional. These values appearing in a consolidated range around 1.3–1.6 eV began to give confidence in the deeper nature of N_O. It is, however, interesting to note that the reported values were obtained in slightly different manners and with questions remaining about the role of symmetry breaking lattice distortions on the deepening. In any case, the results created an emerging consensus of the deep nature of the defect. This motivated new photoluminescence emission experiments [116,117] in 2011 that measured the level at 1.46 eV. Photo EPR measurements around a similar time frame [118,119] report a transition level at $\varepsilon(0|-1) = 1.6$ eV.

While these values appeared to narrow the range of the charge transition level to around mid-gap, the work of Sakong *et al.* [112] in 2013 predicted charge transition levels of $\varepsilon(-1|0) = 0.97$ and 1.46 eV, showing a discrepancy of 0.5 eV when mixing parameters of $\alpha = 25\%$ (standard HSE06) and $\alpha = 36\%$ (to reproduce the bandgap) are used. This highlights some of the challenges associated with the use of hybrids. Interestingly, more recent first-principles results appear to further deepen the defect level, which reopens some questions. The work of Petretto *et al.* [109] reports a much deeper level at 2.1 eV using HSE06; in this case, the deepening can be attributed to the selection of mixing parameter $\alpha = 45\%$. Their value also includes an estimate of finite size corrections. In 2017, Lyons *et al.* [110] reported a deepened value of $\varepsilon(-1|0) \approx 2.0$ eV again using HSE with $\alpha = 36\%$. The reasons for the difference from their previous value [114] of 1.3 eV are the symmetry breaking lattice distortion (0.45 eV) and the incorporation of a FNV finite size effect correction (0.25 eV). Thus, there are now some indications that hybrid functional descriptions accounting for finite size effects may overly deepen the transition level in comparison to the experimental results. Also, different methods predict different relaxation energies associated with the distorted neutral N_O charge state. A recent DMC analysis places the charge transition level at $\varepsilon(0|-1) \approx 1.0 \pm 0.3$ eV when 72-atom supercells are used (without the electrostatic finite size effect). This result falls on the lower end of typically reported hybrid density functional results for the same size supercells. On the other hand, the relatively large error bar on the DMC result makes it difficult to make definitive conclusions.

In summary, while there is fair agreement on the deep nature of N_O, efforts to obtain the precise position of the charge transition level from first-principles are not fully consistent with each other. After over a decade of scrutiny, in many ways N_O in ZnO still highlights some of the existing challenges for first-principles methods.

7.7.3 Transition metal impurities

TM impurities typically yield multiple deep levels that trap mobile carriers and thus often are undesired contaminants in semiconductor electronics. However, they exhibit a variety of interesting optical and magnetic properties. Moreover, TM impurities exhibit remarkably flexible hybridization in various bonding environments, and TMs can replace a host cation in nearly any semiconductor [121,122]. Over the past few decades, the prospect of making conventional semiconductors ferromagnetic [123] has sparked considerable interest [124,125]. Moreover, 3d TM impurities often persist as undesired contaminants, so it is important to know their basic characteristics. They also offer a good testing ground for the theoretical methods described earlier, as electronic structure theory often struggles for 3d-ligand systems in many ways [1,34].

Defect level formation for a 3d TM impurity (here, Mn in GaN) is illustrated by the orbital interaction diagram shown in Figure 7.2(b). The 3d orbitals in a ligand field split into several symmetry-broken levels and always induce multiple orbitals in semiconductor bandgap. These 3d-derived in-gap orbitals usually lead to deep impurity levels that are strongly localized on the 3d impurity atom. Remarkably, these 3d-induced deep gap states typically exhibit multiple stable charge states [126], and thus can be multiply ionized, which implies that a single 3d impurity can trap multiple charge carriers. These multiple charge states are facilitated by a self-regulating response (SRR), whereby the 3d impurity can partially undo (or strengthen) its hybridization with the host material in order to accommodate extra carriers localized at the defect site [126]. This SRR mechanism is illustrated by an orbital interaction diagram in Figure 7.16. For charge state q (Figure 7.16(a)), the bonding orbital formed upon TM-host interaction is mostly 3d like, i.e., a crystal field resonance (CFR), and the anti-bonding one is mostly host-like, i.e., a dangling-bond hybrid (DBH). When adding an electron to the system, it localizes on the in-gap impurity-derived orbital, which repels all other orbitals on the TM site, inducing an upward shift of all TM-derived orbitals (Figure 7.16(b)). Such orbital shift completely changes the hybridization, and in the extreme case, a previously CFR orbital is converted into a DBH orbital. Thus electrons previously localized on the TM CFR leak into the host, maintaining a local charge neutrality around the TM site [126]. The ensuing multiple TM charge states have distinctly different magnetic and optical properties, different bond lengths to the ligands, and even different symmetries due to Jahn–Teller distortions.

Figure 7.17 illustrates the CFR orbitals for a variety of 3d TM impurities in In_2O_3 and ZnO, calculated from first principles [127]. The figure also illustrates how GGA's dramatic underestimation of the oxide bandgaps would lead to most TM levels to incorrectly lie inside the conduction band.[2] The chemical trend in orbital energies follows that of the isolated atoms [3], but the 3d orbital is split into e and t orbitals. Notice that the ordering of these e and t orbitals is opposite in the

[2]This calculation was carried out applying the nonlocal external potential correction together with PBE+U [128].

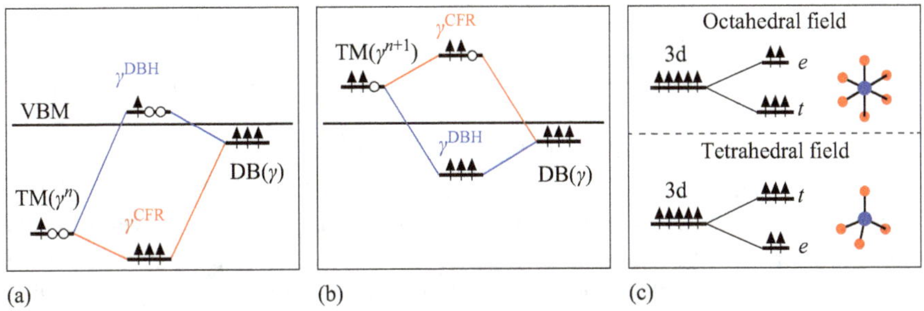

Figure 7.16 TM–host interaction orbital interaction diagram. (a) Charge state $q = 0$. (b) Charge state $q = -1$. (c) Crystal field splitting in octahedral and tetrahedral fields. In each panel, the TM (3d derived) orbital with irreducible representation γ (here of 3-fold symmetry) interacts with host dangling bond DB of same representation, forming bonding and anti-bonding states in the center of the panels, further identified as the crystal field resonance (CFR) and dangling-bond hybrid (DBH) based on their dominant character. (a)–(b) © 2008 Nature. Reprinted, with permission, from [126]

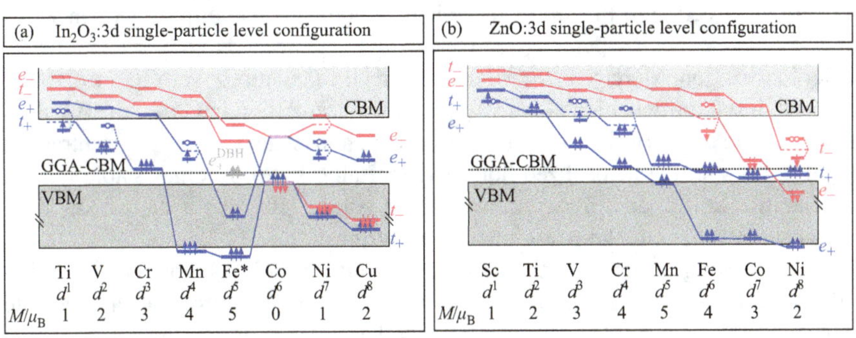

Figure 7.17 One-electron crystal-field resonance orbitals for charge neutral TM impurities in (a) In_2O_3 and (b) ZnO calculated from first principles. © 2009 APS. Reprinted, with permission, from [127]

octahedral and tetrahedral sites for In_2O_3 and ZnO. The heavier 3d elements are stable in a low-spin configuration and lighter elements in a high-spin configuration, but most TM impurities have a nonzero magnetization for their charge neutral state. The electronic configuration is typically described as d^n, where n is the number of TM valence electrons minus cation valence. While the d shell is partially occupied for all TM impurities shown in Figure 7.17, many TM impurities have a closed e or t subshell. For example, Cr in In_2O_3 has a full t^3 shell as its highest occupied orbital, and Fe in In_2O_3 has full t^3e^2 shells. Such closed shell configurations imply that the particular TM element is an isovalent impurity, and such isovalent TMs typically have very high solubilities. On one hand, high solubility implies that these TM elements being ubiquitous in nature, also end up as contaminants in semiconductors, and on the other hand, it offers the prospect of very condensed impurity doping.

Assuming a similar trend for TM impurities in GaAs as shown above for In_2O_3 and ZnO, one would assume Mn assumes a d^4 configuration.[3] Instead, Mn prefers a d^5 configuration and introduces a carrier hole in the host, which, however, remains bound to Mn in the d^5 + hole configuration [128]. This weak binding of the hole eluded first principles calculations for decades, which led to a *Battle of the Bands* between proponents of the d^5 + hole and d^4 interpretations [129]. Early first-principles calculation offered an impurity band involving Mn in the d^4 configuration [130], and DFT combined with tight-binding or dynamical mean field theories suggested a d^5 + hole scenario with a delocalized hole state [131,132]. Moreover, none of these early works could reproduce experimentally observed features [133,134]. Self-interaction corrected DFT finally was able to describe the *weakly bound* hole, as well as the delocalized hole state in the d^5 + hole configuration [135], which also are well in agreement with experiment [133,134]. The scenario is similar to the hole state bistability around the Mg acceptor in GaN [5], albeit the hole is not strongly localized on a single anion site but weakly localized over some tens of host atoms. But, in order to observe such weak localization, not only a self-interaction correction to fulfill the generalized Koopmans' condition (Section 7.5.2) but also a large enough supercell (up to 500 atoms) was needed to observe the correct carrier localization. Moreover, these large supercell calculations were able to capture the metal–insulator transition along varying Mn concentration in GaAs: the (weakly) localized hole state (insulator state) only is ground state for low concentrations ($\lesssim$0.5% Mn), and the delocalized hole state (metallic state) becomes stable for Mn concentration $\gtrsim$1%, and for concentrations around $\sim$1%, the system exhibits a superposition state, i.e., the phase transition is critical.

It is tempting to compare the above bistability of the d^5 + hole configuration with the acceptors introduced by Mg into GaN, which also introduces a hole state strongly bound on a single anion orbital. Another hole-binding impurity in GaN is the Zn acceptor [136]. What is common for Mg and Zn is that in the +II oxidation state, they have a closed-shell configuration ($3s^0$ for Mg, and $3d^{10}$ for Zn), and further oxidizing the impurity to +III breaks the closed shell; in such cases, it is often favored to introduce a hole in the host semiconductor rather than to break a closed shell. Mn in GaN exhibits similar behavior [6], where the d^5 seems to behave like a closed shell configuration. The same calculation further shows that, indeed, also Fe in GaN prefers to keep its d^5 shell intact in GaN when holes are doped from an external source. When probed for bistabilities, these deep d^5 centers exhibit multifarious hole configurations, where the hole can be trapped on crystallographically inequivalent anion sites, or exhibit various degrees of delocalization (holes delocalized along 1D chain, or within 2D plane), and the type of

[3] Ga in GaAs is trivalent, and Mn has 7 valence electrons, so one would expect Mn replacing Ga to assume the d^n configuration with $n = 7 - 3 = 4$.

localization can be controlled by axial strain [6]. Thus, in epitaxial samples, substrate strain is expected to play a key role in carrier localization.

The computational difficulty of modeling TM impurities in semiconductors faces the same problems as discussed above for defects in wide gap materials (GaN, ZnO, etc.), namely, the problem in predicting bandgaps and carrier localization around defects. In other words, the 3d levels can be misplaced with reference to the host bands (Figure 7.17), and/or polaron formation is not correctly described. However, for TM impurities, neither the hybrid functionals [63] nor the *GW* approximation [137] offer a consistent systematic improvement, so one should take special care to assure that known experimental benchmarks are met. In addition to the problems pertaining to computational engine, because of the high solubility of TM impurities, one also needs to take into consideration defect–defect interactions, as well as cluster and complex formation.

7.7.4 Defect clustering and aggregation

When the defect concentration is high, or defects are very mobile, defect clustering and/or complex formation becomes important. Assuming the defects are immobile and introduced randomly (e.g., by ion implantation), two or more impurities may by chance occupy neighboring lattice sites. The probability for any given defect to be part of a cluster is $P = 1 - (1 - c)^{\mathrm{CN}}$, where c is the defect concentration and CN the coordination number. For Si, c is very small and CN = 4, so P will be small. For compound semiconductors, however, CN is the coordination of the respective sublattice, which, e.g., in the zinc-blende or wurtzite structures is 12. For TM impurities, c may be of several atomic percent, so, e.g., for $c = 5\%$, $P = 1 - 0.95^{12} = 0.46$, which is too large to be neglected. Similar probability expressions for specific clusters (dimer, trimer, etc.) in different lattice types were given by Behringer [138], showing that the brunt of such random clusters can be expected to be dimers, but also a significant fraction of trimers can be expected. Of course also the assumption of defects being immobile is not valid. For example, copper is notorious for its high diffusivity and aggregation in silicon or germanium [139–141], and interstitial hydrogen tends to migrate around and passivate existing defects [79,142,143].

When clusters and complexes are present, the defect levels cannot be assumed as simple superpositions of their constituent individual defects, as evidenced by Figures 7.12 and 7.18. This is because (i) point symmetry of each defect and (ii) the microscopic screening of the defect electronic orbitals change upon cluster or complex formation. Instead, one has to calculate the formation enthalpies, defect levels, etc. (see Section 7.2) for each cluster and complex individually. Nonetheless, as shown for isolated defects in Section 7.1 even without full first principles calculation, one can obtain a basic understanding of defect complex spectra by simple phenomenological models [144]. When two identical defects occupy neighboring sites, for such dimer one should expect twice the amount of gap levels compared with the respective isolated defect. This can be reasoned as a Heitler–London type interaction, where each gap level splits into bonding and antibonding levels [145–148]. Moreover, the point symmetry of each individual defect in the dimer is in general

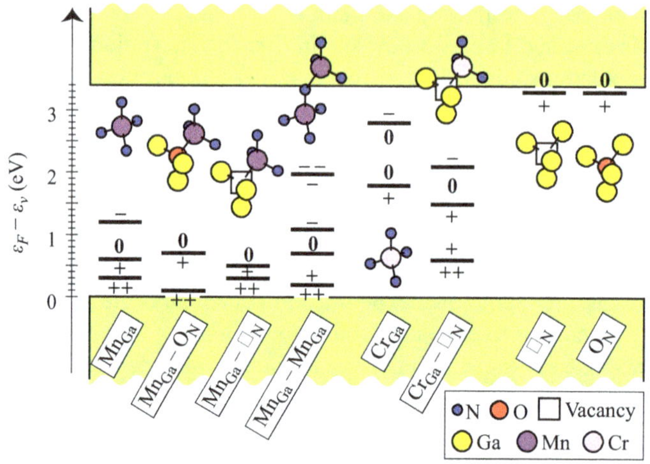

Figure 7.18 Charge transition levels of several defect complexes and their constituent defects in GaN. (Notice that here □ denotes a vacancy.) © 2010 APS. Reprinted, with permission, from [144]

lower than for the isolated defect, which may lead to even further splitting. Similar orbital-interaction models can also be used for different defect species on the same sublattice. For defects on different sublattices, there is necessarily lowering of symmetry, which lifts possible degeneracies of defect levels. Moreover, the two defects form a chemical bond, which obviously is different to the typical impurity–ligand interaction, which causes significant shifting of the defect orbitals, described by the SRR theory described earlier (see Section 7.3) [144].

To assess stability of a cluster or complex $D_n I_m$ in structural configuration ξ and charge state q, we define the binding energy

$$\begin{aligned} E^b_{D_nI_m,\xi,q} &= \Delta H_{D_nI_m,\xi,q} - n\Delta H_{D,q_D} - m\Delta H_{I,q_I} \\ &= E_{D_nI_m,\xi,q} - nE_{D,q_D} - mE_{I,q_I} + (n+m-1)E_H + (q - nq_D - mq_I)E_F. \end{aligned} \tag{7.44}$$

Because there is no exchange of atoms with external reservoir, the element chemical potentials cancel out, and E^b only depends on E_F. Notice, however that, at any given E_F, the complex and its constituent defects all have different charge states, and the ensuing E^b is piecewise linear with kinks corresponding to the charge transition energies of the complex and its constituent defects, as illustrated in Figure 7.19. For the special case that $q - nq_D - mq_I = 0$, (7.44) reduces to its widely used form, $E^b_{D_nI_m,\xi} = nE_D - mE_I + (n+m-1)E_H$, but as seen, e.g., in Figure 7.12, $q - nq_D - mq_I = 0$ is true for a specific range of E_F, but not for the entire gap. Figure 7.19(b) shows binding energies calculated for some defect complexes in GaN, which gives us further insight to E^b. The constant regions of E^b correspond to the regions where $q - nq_D - mq_I = 0$; the Mn$_{Ga}$–Mn$_{Ga}$ repulsion corresponds to two negatively charged Mn$_{Ga}$ impurities forming a doubly negative dimer, and the attractive interaction between Mn$_{Ga}$ and V_N or O$_N$

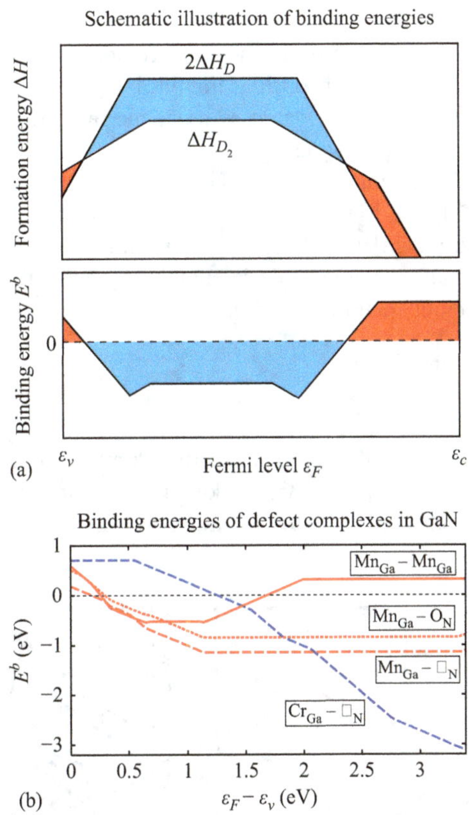

Figure 7.19 *(a) Illustration of the evaluation of binding energy for dimer D_2 composed of two defects D. In the region where ΔH for D_2 is lower than for D, th0065 binding energy is negative, illustrated the filled blue region. Positive E^b (red filled region) means that the individual defects are lower in energy. (b) Binding energies of some defect complexes in GaN. (Notice that here □ denotes a vacancy.) © 2014 AIP. Reprinted, with permission, from [149]*

corresponds a negatively charged Mn_{Ga} and positively charged V_N or O_N forming a neutral complex, and the respective E^b are fairly close to the Coulomb potential screened by the GaN dielectric constant. Thus, one may be tempted to approximate E^b for two charged impurities forming a complex simply by $q_1q_2/\epsilon R$, but this would completely neglect any contribution to E^b due to changes in hybridization and/or lattice relaxations, which can be even larger than the point ion Coulomb interactions, manifested by the strong binding of two charge neutral Mn_{Ga} forming a charge neutral dimer.

For very condensed doping, defects may aggregate into clusters [148,149], form extended defects [150], or lead to spinodal decomposition [125]. Such condensed doping can also occur due to intrinsic defects. TiO_2 is an extreme example of a non-stoichiometric oxide, which can be so deficient of oxygen that it is better described by the formula Ti_nO_{2n-1}. For large $n > 37$, these compounds are stable in the rutile phase with the oxygen-deficiency accommodated by point defects, typically oxygen vacancies V_O and interstitial titanium Ti_i [151,152]. For $4 < n \leq 37$, the oxygen deficiency is accommodated by a change in structure, namely, a phase transition into the so-called Magnéli phases [153], which can be described as a stack of rutile TiO_2 layers and corundum Ti_2O_3 layers. (Notice that Ti_2O_3 is not a stable phase alone.) These Magnéli phases can also be described as rutile TiO_2, where point defect forms ordered phases, but it is not immediately obvious whether one should accommodate the oxygen deficiency by V_O or Ti_i formation, that is, one has to make a choice between the reactions

$$nTiO_2 + V_O \rightarrow Ti_nO_{2n-1} \tag{7.45}$$

$$(2n-1)TiO_2 + Ti_i \rightarrow 2Ti_nO_{2n-1}. \tag{7.46}$$

Choosing the latter leads to the more interesting result. Corundum Ti_2O_3 and the Magnéli phases exhibit the same electronic doping behavior as rutile TiO_2, i.e., there is a quadruple donor transition close to the rutile conduction band edge (Figure 7.20).

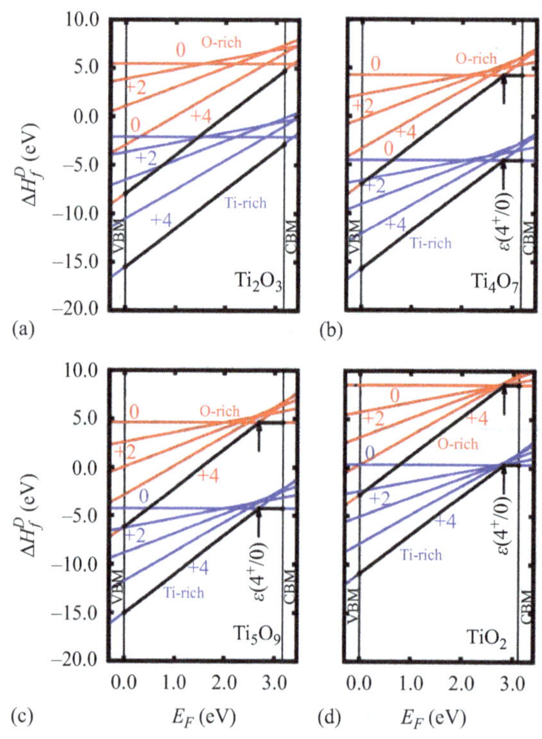

Figure 7.20 Formation energy diagrams for Ti_i formation in (a) corundum Ti_2O_3, (b) Magnéli phase Ti_4O_7, (c) Magnéli phase Ti_5O_9, and (d) rutile TiO_2. (a)–(c) are calculated using the HSE hybrid [154], (d) is calculated using screened exchange [152]. © 2016 NPG. Reprinted, with permission, from [154]

Thus, despite being pure crystallographic phases of their own with nominally zero defects, the Ti_nO_{2n-1} Magnéli phases behave as if they contained the very large concentration of $n/2$ *pseudodefects* with a quadruple donor transition [154].

7.8 Summary and outlook

In this chapter, we have outlined the thermodynamic concepts that connect microscopic observables with macroscopic systems for defects in semiconductors and summarized the basic formulations used for describing defects from first principles. First-principles approaches for defects have evolved substantially in the last 10 years, and our predictive capability is at its highest level yet. We have seen enormous successes, and first principles now regularly provides insights and explanations for observations made in experiment, or even makes predictions a priori. On the other hand, we are not yet at the level of demonstrating true quantitative predictive capability in all cases every time. In spite of being called *first principles*, ab initio, or *parameter free*, various of approximations and simplifications are carried out along the way, and often results are still justified a posteriori by comparison to experiment. Refinement of approaches and techniques to address uncertainties in calculated total energies, bandgaps, and finite size effects is an ongoing and active area of development. This is an evolving field where new approaches emerge and old ones are improved. Continued efforts at systematic analysis across methods will play a critical role in unveiling trends and identifying the next steps to move the community forwards toward more quantitative predictions.

References

[1] Slater JC. The Self-Consistent Field for Molecules and Solids: Quantum Theory of Molecules and Solids, vol. 4. New York: McGraw-Hill; 1974.

[2] Szabo A, Ostlund NS. Modern Quantum Chemistry: Introduction to Advanced Electronic Structure Theory. 1st ed. Mineola: Dover Publications, Inc.; 1996.

[3] Desclaux JP. Relativistic Dirack–Fock expectation values for atoms with $Z = 1$ to $Z = 120$. At Data Nucl Data Tables. 1973;12(4):311–406.

[4] Raebiger H, Lany S, Zunger A. Origins of the p-type nature and cation deficiency in Cu_2O and related materials. Phys Rev B. 2007;76(4):045209.

[5] Lany S, Zunger A. Dual nature of acceptors in GaN and ZnO: The curious case of the shallow MgGa deep state. Appl Phys Lett. 2010;96(14):142114. Available from: https://doi.org/10.1063/1.3383236.

[6] Raebiger H, Bae S, Echeverra-Arrondo C, *et al.* Control of hole localization in magnetic semiconductors by axial strain. Phys Rev Mater. 2018;2(2): 024402.

[7] Zhang SB, Northrup JE. Chemical potential dependence of defect formation energies in GaAs: Application to Ga self-diffusion. Phys Rev Lett. 1991;67:2339–2342. Available from: https://link.aps.org/doi/10.1103/PhysRevLett.67.2339.

[8] Weast R, Astle M. CRC Handbook of Chemistry and Physics. 60th ed. Boca Raton, FL: CRC Press; 1979.

[9] Hohenberg P, Kohn W. Inhomogeneous electron gas. Phys Rev B. 1964; 136(3B):B864–B871.

[10] Kohn W, Sham LJ. Self-consistent equations including exchange and correlation effects. Phys Rev. 1965;140(4A):A1133–A1138.

[11] Ceperley DM, Alder BJ. Ground state of the electron gas by a stochastic method. Phys Rev Lett. 1980;45:566–569. Available from: https://link.aps.org/doi/10.1103/PhysRevLett.45.566.

[12] Perdew JP, Zunger A. Self-interaction correction to density-functional approximations for many-electron systems. Phys Rev B. 1981;23:5048–5079. Available from: https://link.aps.org/doi/10.1103/PhysRevB.23.5048.

[13] Perdew JP, Burke K, Ernzerhof M. Generalized gradient approximation made simple. Phys Rev Lett. 1996;77:3865–3868. Available from: https://link.aps.org/doi/10.1103/PhysRevLett.77.3865.

[14] Wang Y, Perdew JP. Spin scaling of the electron-gas correlation energy in the high-density limit. Phys Rev B. 1991;43:8911–8916. Available from: https://link.aps.org/doi/10.1103/PhysRevB.43.8911.

[15] Tao J, Perdew JP, Staroverov VN, *et al.* Climbing the density functional ladder: nonempirical meta–generalized gradient approximation designed for molecules and solids. Phys Rev Lett. 2003;91:146401. Available from: https://link.aps.org/doi/10.1103/PhysRevLett.91.146401.

[16] Sun J, Remsing RC, Zhang Y, *et al.* Accurate first-principles structures and energies of diversely bonded systems from an efficient density functional. Nat Chem. 2016;8:831–836.

[17] Grossman JC. Benchmark quantum Monte Carlo calculations. J Chem Phys. 2002;117(4):1434–1440. Available from: https://doi.org/10.1063/1.1487829.

[18] Petruzielo FR, Toulouse J, Umrigar CJ. Approaching chemical accuracy with quantum Monte Carlo. J Chem Phys. 2012;136(12):124116. Available from: https://doi.org/10.1063/1.3697846.

[19] Foulkes WMC, Mitas L, Needs RJ, *et al.* Quantum Monte Carlo simulations of solids. Rev Mod Phys. 2001;73:33–83. Available from: https://link.aps.org/doi/10.1103/RevModPhys.73.33.

[20] Alfè D, Gillan MJ. Schottky defect formation energy in MgO calculated by diffusion Monte Carlo. Phys Rev B. 2005;71:220101. Available from: https://link.aps.org/doi/10.1103/PhysRevB.71.220101.

[21] Hood RQ, Kent PRC, Reboredo FA. Diffusion quantum Monte Carlo study of the equation of state and point defects in aluminum. Phys Rev B. 2012;85:134109. Available from: https://link.aps.org/doi/10.1103/PhysRevB.85.134109.

[22] Hood RQ, Kent PRC, Needs RJ, *et al.* Quantum Monte Carlo study of the optical and diffusive properties of the vacancy defect in diamond. Phys Rev Lett. 2003;91:076403. Available from: https://link.aps.org/doi/10.1103/PhysRevLett.91.076403.

[23] Parker WD, Wilkins JW, Hennig RG. Accuracy of quantum Monte Carlo methods for point defects in solids. Phys Status Solidi (B). 2011;248(2):267–274. Available from: https://onlinelibrary.wiley.com/doi/abs/10.1002/pssb.201046149.

[24] Leung WK, Needs RJ, Rajagopal G, *et al.* Calculations of silicon self-interstitial defects. Phys Rev Lett. 1999;83:2351–2354. Available from: https://link.aps.org/doi/10.1103/PhysRevLett.83.2351.

[25] Batista ER, Heyd J, Hennig RG, *et al.* Comparison of screened hybrid density functional theory to diffusion Monte Carlo in calculations of total energies of silicon phases and defects. Phys Rev B. 2006;74:121102. Available from: https://link.aps.org/doi/10.1103/PhysRevB.74.121102.

[26] Ertekin E, Winkler MT, Recht D, *et al.* Insulator-to-metal transition in selenium-hyperdoped silicon: Observation and origin. Phys Rev Lett. 2012;108:026401. Available from: https://link.aps.org/doi/10.1103/PhysRevLett.108.026401.

[27] Ertekin E, Wagner LK, Grossman JC. Point-defect optical transitions and thermal ionization energies from quantum Monte Carlo methods: Application to the *F*-center defect in MgO. Phys Rev B. 2013;87:155210. Available from: https://link.aps.org/doi/10.1103/PhysRevB.87.155210.

[28] Yu J, Wagner LK, Ertekin E. Fixed-node diffusion Monte Carlo description of nitrogen defects in zinc oxide. Phys Rev B. 2017;95:075209. Available from: https://link.aps.org/doi/10.1103/PhysRevB.95.075209.

[29] Santana JA, Krogel JT, Kim J, *et al.* Structural stability and defect energetics of ZnO from diffusion quantum Monte Carlo. J Chem Phys. 2015;142(16):164705. Available from: https://doi.org/10.1063/1.4919242.

[30] Shin H, Luo Y, Ganesh P, *et al.* Electronic properties of doped and defective NiO: A quantum Monte Carlo study. Phys Rev Materials. 2017;1:073603. Available from: https://link.aps.org/doi/10.1103/PhysRevMaterials.1.073603.

[31] Saritas K, Ming W, Du MH, *et al.* Excitation energies of localized correlated defects via quantum Monte Carlo: A case study of Mn^{4+}-doped phosphors. J Phys Chem Lett. 2019;10(1):67–74. Available from: https://doi.org/10.1021/acs.jpclett.8b03015.

[32] Wagner LK, Ceperley DM. Discovering correlated fermions using quantum Monte Carlo. Rep Prog Phys. 2016;79(9):094501. Available from: http://stacks.iop.org/0034-4885/79/i=9/a=094501.

[33] Kolorenč J, Mitas L. Applications of quantum Monte Carlo methods in condensed systems. Rep Prog Phys. 2011;74(2):026502. Available from: http://stacks.iop.org/0034-4885/74/i=2/a=026502.

[34] Zunger A, Lany S, Raebiger H. The quest for dilute ferromagnetism in semiconductors: Guides and misguides by theory. Physics. 2010;3:53.

[35] Persson C, Zhao YJ, Lany S, *et al.* n-Type doping of CuInSe2 and CuGaSe2. Phys Rev B. 2005;72(3):035211.

[36] Lany S, Raebiger H, Zunger A. Magnetic interactions of Cr-Cr and Co-Co impurity pairs in ZnO within a band-gap corrected density functional approach. Phys Rev B. 2008;77(24):241201.

[37] Raebiger H, Lany S, Zunger A. Control of ferromagnetism via electron doping in In_2O_3: Cr. Phys Rev Lett. 2008;101(2):027203.
[38] Becke AD. A new mixing of Hartree–Fock and local density-functional theories. J Chem Phys. 1993;98(2):1372–1377. Available from: https://doi.org/10.1063/1.464304.
[39] Lany S, Zunger A. Polaronic hole localization and multiple hole binding of acceptors in oxide wide-gap semiconductors. Phys Rev B. 2009;80(8): 085202.
[40] Koopmans T. Über die zuordnung von wellenfunktionen und eigenwerten zu den einzelnen elektronen eines atoms. Physica. 1934;1:104–113.
[41] Janak J. Proof that $\partial E/\partial n_i = \varepsilon$ in density-functional theory. Phys Rev B. 1978;18(12):7165–7168.
[42] Zunger A, Freeman AJ. Ground- and excited-state properties of LiF in the local-density formalism. Phys Rev B. 1977;16(6):2901–2926.
[43] Dudarev SL, Botton GA, Savrasov SY, *et al.* Electron-energy-loss spectra and the structural stability of nickel oxide: An LSDA+U study. Phys Rev B. 1998;57(3):1505–1509.
[44] Hubbard J. Electron correlations in narrow energy bands. Proc R Soc A. 1963;276(1365):238–257.
[45] Anisimov VI, Zaanen J, Andersen OK. Band theory and Mott insulators: Hubbard UU instead of Stoner II. Phys Rev B. 1991;44(3):943–954.
[46] Janotti A, Van de Walle CG. Native point defects in ZnO. Phys Rev B. 2007;76:165202. Available from: https://link.aps.org/doi/10.1103/PhysRevB.76.165202.
[47] Bechstedt F, Fuchs F, Kresse G. Ab-initio theory of semiconductor band structures: New developments and progress. Phys Status Solidi (B). 2009;246(8):1877–1892. Available from: https://onlinelibrary.wiley.com/doi/abs/10.1002/pssb.200945074.
[48] Aryasetiawan F, Gunnarsson O. The GW method. Rep Prog Phys. 1998;61(3): 237. Available from: http://stacks.iop.org/0034-4885/61/i=3/a=002.
[49] Hedin L. New method for calculating the one-particle Green's Function with application to the electron-gas problem. Phys Rev. 1965;139:A796–A823. Available from: https://link.aps.org/doi/10.1103/PhysRev.139.A796.
[50] Hybertsen MS, Louie SG. Electron correlation in semiconductors and insulators: Band gaps and quasiparticle energies. Phys Rev B. 1986;34:5390–5413. Available from: https://link.aps.org/doi/10.1103/PhysRevB.34.5390.
[51] Shishkin M, Kresse G. Self-consistent *GW* calculations for semiconductors and insulators. Phys Rev B. 2007;75:235102. Available from: https://link.aps.org/doi/10.1103/PhysRevB.75.235102.
[52] Chen W, Pasquarello A. Band-edge levels in semiconductors and insulators: Hybrid density functional theory versus many-body perturbation theory. Phys Rev B. 2012;86:035134. Available from: https://link.aps.org/doi/10.1103/PhysRevB.86.035134.
[53] Rinke P, Janotti A, Scheffler M, *et al.* Defect formation energies without the band-gap problem: Combining density-functional theory and the *GW*

approach for the silicon self-interstitial. Phys Rev Lett. 2009;102:026402. Available from: https://link.aps.org/doi/10.1103/PhysRevLett.102.026402.

[54] Rinke P, Schleife A, Kioupakis E, *et al.* First-principles optical spectra for *F* centers in MgO. Phys Rev Lett. 2012;108:126404. Available from: https://link.aps.org/doi/10.1103/PhysRevLett.108.126404.

[55] Shaltaf R, Rignanese GM, Gonze X, *et al.* Band offsets at the Si/SiO_2 interface from many-body perturbation theory. Phys Rev Lett. 2008; 100:186401. Available from: https://link.aps.org/doi/10.1103/PhysRevLett.100.186401.

[56] Becke AD. Density-functional thermochemistry. III. The role of exact exchange. J Chem Phys. 1993;98(7):5648–5652. Available from: https://doi.org/10.1063/1.464913.

[57] Heyd J, Scuseria GE, Ernzerhof M. Hybrid functionals based on a screened Coulomb potential. J Chem Phys. 2003;118(18):8207–8215. Available from: https://doi.org/10.1063/1.1564060.

[58] Perdew JP, Ernzerhof M, Burke K. Rationale for mixing exact exchange with density functional approximations. J Chem Phys. 1996;105(22):9982–9985. Available from: https://doi.org/10.1063/1.472933.

[59] Adamo C, Barone V. Exchange functionals with improved long-range behavior and adiabatic connection methods without adjustable parameters: The mPW and mPW1PW models. J Chem Phys. 1998;108(2):664–675. Available from: https://doi.org/10.1063/1.475428.

[60] Chen W, Miceli G, Rignanese GM, *et al.* Nonempirical dielectric-dependent hybrid functional with range separation for semiconductors and insulators. Phys Rev Mater. 2018;2:073803. Available from: https://link.aps.org/doi/10.1103/PhysRevMaterials.2.073803.

[61] Miceli G, Chen W, Reshetnyak I, *et al.* Nonempirical hybrid functionals for band gaps and polaronic distortions in solids. Phys Rev B. 2018;97:121112. Available from: https://link.aps.org/doi/10.1103/PhysRevB.97.121112.

[62] Bang J, Sun YY, Abtew TA, *et al.* Difficulty in predicting shallow defects with hybrid functionals: Implication of the long-range exchange interaction. Phys Rev B. 2013;88:035134. Available from: https://link.aps.org/doi/10.1103/PhysRevB.88.035134.

[63] Ivády V, Abrikosov IA, Janzen E, *et al.* Role of screening in the density functional applied to transition-metal defects in semiconductors. Phys Rev B. 2013;87(20):205201.

[64] Droghetti A, Alfe D, Sanvito S. Ground state of a spin-crossover molecule calculated by diffusion Monte Carlo. Phys Rev B. 2013;87(20):205114.

[65] Wolverton C, Ozolins V. First-principles aluminum database: Energetics of binary Al alloys and compounds. Phys Rev B. 2006;73(14):144104.

[66] Lany S. Semiconductor thermochemistry in density functional calculations. Phys Rev B. 2008;78(24):245207.

[67] Stevanović V, Lany S, Zhang X, *et al.* Correcting density functional theory for accurate predictions of compound enthalpies of formation: Fitted elemental-phase reference energies. Phys Rev B. 2012;85(11):115104.

[68] Leslie M, Gillan M. The energy and elastic dipole tensor of defects in ionic-crystals calculated by the supercell method. J Phys C: Solid State. 1985;18(5): 973–982.

[69] Makov G, Payne MC. Periodic boundary conditions in ab initio calculations. Phys Rev B. 1995;51(7):4014–4022.

[70] Lany S, Zunger A. Assessment of correction methods for the band-gap problem and for finite-size effects in supercell defect calculations: Case studies for ZnO and GaAs. Phys Rev B. 2008;78(23):235104.

[71] Lany S, Zunger A. Accurate prediction of defect properties in density functional supercell calculations. Modell Simul Mater Sci Eng. 2009; 17:084002.

[72] Freysoldt C, Neugebauer J, Van de Walle CG. Fully ab initio finite-size corrections for charged-defect supercell calculations. Phys Rev Lett. 2009; 102:016402. Available from: https://link.aps.org/doi/10.1103/PhysRevLett.102.016402.

[73] Freysoldt C, Neugebauer J, Van de Walle CG. Electrostatic interactions between charged defects in supercells. Phys Status Solidi (B). 2010;248(5): 1067–1076.

[74] Kumagai Y, Oba F. Electrostatics-based finite-size corrections for first-principles point defect calculations. Phys Rev B. 2014;89(19):195205.

[75] Komsa HP, Pasquarello A. Finite-size supercell correction for charged defects at surfaces and interfaces. Phys Rev Lett. 2013;110(9):095505.

[76] Chen W, Pasquarello A. Correspondence of defect energy levels in hybrid density functional theory and many-body perturbation theory. Phys Rev B. 2013;88(11):115104.

[77] Ihm J, Zunger A, Cohen M. Momentum-space formalism for the total energy of solids. J Phys C: Solid State. 1979;12:4409.

[78] Dabo I, Kozinsky B, Singh-Miller NE, *et al.* Electrostatics in periodic boundary conditions and real-space corrections. Phys Rev B. 2008;77(11): 115139.

[79] Lyons JL, Janotti A, Van de Walle CG. Shallow versus deep nature of Mg acceptors in nitride semiconductors. Phys Rev Lett. 2012;108: 156403. Available from: https://link.aps.org/doi/10.1103/PhysRevLett.108.156403.

[80] Amano H, Kito M, Hiramatsu K, *et al.* P-type conduction in Mg-doped GaN treated with low-energy electron beam irradiation (LEEBI). Jpn J Appl Phys. 1989;28(12A):L2112. Available from: http://stacks.iop.org/1347-4065/28/i=12A/a=L2112.

[81] Nakamura S, Mukai T, Senoh M, *et al.* Thermal annealing effects on P-type Mg-doped GaN films. Jpn J Appl Phys. 1992;31(2B):L139. Available from: http://stacks.iop.org/1347-4065/31/i=2B/a=L139.

[82] Van de Walle CG, Neugebauer J. First-principles calculations for defects and impurities: Applications to III-nitrides. J Appl Phys. 2004;95(8):3851–3879. Available from: https://doi.org/10.1063/1.1682673.

[83] Gotz W, Johnson NM, Walker J, *et al.* Activation of acceptors in Mg-doped GaN grown by metalorganic chemical vapor deposition. Appl Phys Lett. 1996;68(5):667–669. Available from: https://doi.org/10.1063/1.116503.

[84] Monemar B, Paskov PP, Pozina G, *et al.* Properties of the main Mg-related acceptors in GaN from optical and structural studies. J Appl Phys. 2014;115(5): 053507. Available from: https://doi.org/10.1063/1.4862928.

[85] Lyons JL, Alkauskas A, Janotti A, *et al.* First-principles theory of acceptors in nitride semiconductors. Phys Status Solidi (B). 2015;252(5):900–908. Available from: https://onlinelibrary.wiley.com/doi/abs/10.1002/pssb.201552062.

[86] Miceli G, Pasquarello A. Self-compensation due to point defects in Mg-doped GaN. Phys Rev B. 2016;93:165207. Available from: https://link.aps.org/doi/10.1103/PhysRevB.93.165207.

[87] Lyons JL, Van de Walle CG. Computationally predicted energies and properties of defects in GaN. NPJ Comput Mater. 2017;3(1):12.

[88] Eckey L, von Gfug U, Holst J, *et al.* Photoluminescence and Raman study of compensation effects in Mg-doped GaN epilayers. J Appl Phys. 1998;84(10): 5828–5830. Available from: https://doi.org/10.1063/1.368853.

[89] Reshchikov MA, Yi GC, Wessels BW. Behavior of 2.8- and 3.2-eV photoluminescence bands in Mg-doped GaN at different temperatures and excitation densities. Phys Rev B. 1999;59:13176–13183. Available from: https://link.aps.org/doi/10.1103/PhysRevB.59.13176.

[90] Oh E, Park H, Park Y. Excitation density dependence of photoluminescence in GaN:Mg. Appl Phys Lett. 1998;72(1):70–72. Available from: https://doi.org/10.1063/1.120647.

[91] Kaufmann U, Schlotter P, Obloh H, *et al.* Hole conductivity and compensation in epitaxial GaN:Mg layers. Phys Rev B. 2000;62:10867–10872. Available from: https://link.aps.org/doi/10.1103/PhysRevB.62.10867.

[92] Hautakangas S, Saarinen K, Liszkay L, *et al.* Role of open volume defects in Mg-doped GaN films studied by positron annihilation spectroscopy. Phys Rev B. 2005;72:165303. Available from: https://link.aps.org/doi/10.1103/PhysRevB.72.165303.

[93] Neugebauer J, Van de Walle CG. Gallium vacancies and the yellow luminescence in GaN. Appl Phys Lett. 1996;69(4):503–505. Available from: https://doi.org/10.1063/1.117767.

[94] Gillen R, Robertson J. A hybrid density functional view of native vacancies in gallium nitride. J Phys: Condens Matter. 2013;25(40):405501. Available from: http://stacks.iop.org/0953-8984/25/i=40/a=405501.

[95] Onida G, Reining L, Rubio A. Electronic excitations: density-functional versus many-body Green's-function approaches. Rev Mod Phys. 2002;74:601–659. Available from: https://link.aps.org/doi/10.1103/RevModPhys.74.601.

[96] Fuchs F, Rödl C, Schleife A, *et al.* Efficient $\mathcal{O}(N^2)$ approach to solve the Bethe–Salpeter equation for excitonic bound states. Phys Rev B. 2008;78:085103. Available from: https://link.aps.org/doi/10.1103/PhysRevB.78.085103.

[97] Oba F, Togo A, Tanaka I, *et al.* Defect energetics in ZnO: A hybrid Hartree–Fock density functional study. Phys Rev B. 2008;77:245202. Available from: https://link.aps.org/doi/10.1103/PhysRevB.77.245202.

[98] Clark SJ, Robertson J, Lany S, *et al.* Intrinsic defects in ZnO calculated by screened exchange and hybrid density functionals. Phys Rev B. 2010; 81:115311. Available from: https://link.aps.org/doi/10.1103/PhysRevB.81.115311.

[99] Neumann G. In: Kaldis E, editor. Current Topics in Materials Science, vol. 7. Amsterdam: North-Holland Publishing Company; 1981.

[100] Lany S, Zunger A. Anion vacancies as a source of persistent photoconductivity in II–VI and chalcopyrite semiconductors. Phys Rev B. 2005; 72:035215. Available from: https://link.aps.org/doi/10.1103/PhysRevB.72.035215.

[101] Lany S, Zunger A. Dopability, Intrinsic conductivity, and nonstoichiometry of transparent conducting oxides. Phys Rev Lett. 2007;98:045501. Available from: https://link.aps.org/doi/10.1103/PhysRevLett.98.045501.

[102] Erhart P, Klein A, Albe K. First-principles study of the structure and stability of oxygen defects in zinc oxide. Phys Rev B. 2005;72:085213. Available from: https://link.aps.org/doi/10.1103/PhysRevB.72.085213.

[103] Zhang SB, Wei SH, Zunger A. Intrinsic n-type versus p-type doping asymmetry and the defect physics of ZnO. Phys Rev B. 2001;63:075205. Available from: https://link.aps.org/doi/10.1103/PhysRevB.63.075205.

[104] Oba F, Nishitani SR, Adachi H, *et al.* Ab initio study of symmetric tilt boundaries in ZnO. Phys Rev B. 2001;63:045410. Available from: https://link.aps.org/doi/10.1103/PhysRevB.63.045410.

[105] Lany S, Zunger A. Assessment of correction methods for the band-gap problem and for finite-size effects in supercell defect calculations: Case studies for ZnO and GaAs. Phys Rev B. 2008;78:235104. Available from: https://link.aps.org/doi/10.1103/PhysRevB.78.235104.

[106] Lee EC, Kim YS, Jin YG, *et al.* Compensation mechanism for N acceptors in ZnO. Phys Rev B. 2001;64:085120. Available from: https://link.aps.org/doi/10.1103/PhysRevB.64.085120.

[107] Janotti A, Van de Walle CG. Oxygen vacancies in ZnO. Appl Phys Lett. 2005;87(12):122102. Available from: https://doi.org/10.1063/1.2053360.

[108] Janotti A, Van de Walle CG. Hydrogen multicentre bonds. Nat Mater. 2006;6(1):44–47.

[109] Petretto G, Bruneval F. Comprehensive ab initio study of doping in bulk ZnO with group-V elements. Phys Rev Appl. 2014;1:024005. Available from: https://link.aps.org/doi/10.1103/PhysRevApplied.1.024005.

[110] Lyons JL, Varley JB, Steiauf D, *et al.* First-principles characterization of native-defect-related optical transitions in ZnO. J Appl Phys. 2017;122(3): 035704. Available from: https://doi.org/10.1063/1.4992128.

[111] Lany S, Zunger A. Generalized Koopmans density functional calculations reveal the deep acceptor state of No in ZnO. Phys Rev B. 2010;81:205209. Available from: https://link.aps.org/doi/10.1103/PhysRevB.81.205209.

[112] Sakong S, Gutjahr J, Kratzer P. Comparison of density functionals for nitrogen impurities in ZnO. J Chem Phys. 2013;138(23):234702. Available from: https://doi.org/10.1063/1.4810862.

[113] Gallino F, Di Valentin C, Pacchioni G, *et al.* Nitrogen impurity states in polycrystalline ZnO. A combined EPR and theoretical study. J Mater Chem. 2010;20:689–697. Available from: http://dx.doi.org/10.1039/B915578C.

[114] Lyons JL, Janotti A, Van de Walle CG. Why nitrogen cannot lead to p-type conductivity in ZnO. Appl Phys Lett. 2009;95(25):252105. Available from: https://doi.org/10.1063/1.3274043.

[115] Park CH, Zhang SB, Wei SH. Origin of p-type doping difficulty in ZnO: The impurity perspective. Phys Rev B. 2002;66:073202. Available from: https://link.aps.org/doi/10.1103/PhysRevB.66.073202.

[116] Tarun MC, Iqbal MZ, McCluskey MD. Nitrogen is a deep acceptor in ZnO. AIP Adv. 2011;1(2):022105. Available from: https://aip.scitation.org/doi/abs/10.1063/1.3582819.

[117] Alkauskas A, McCluskey MD, Van de Walle CG. Tutorial: Defects in semiconductors—Combining experiment and theory. J Appl Phys. 2016;119(18):181101. Available from: https://doi.org/10.1063/1.4948245.

[118] Stehr JE, Hofmann DM, Meyer BK. Electron paramagnetic resonance and photo-electron paramagnetic resonance investigation on the recharging of the substitutional nitrogen acceptor in ZnO. J Appl Phy. 2012;112(10): 103511. Available from: https://doi.org/10.1063/1.4765729.

[119] McCluskey MD, Corolewski CD, Lv J, *et al.* Acceptors in ZnO. J Appl Phys. 2015;117(11):112802. Available from: https://doi.org/10.1063/1.4913827.

[120] McCluskey MD, Jokela SJ. Defects in ZnO. J Appl Phys. 2009;106(7): 071101. Available from: https://doi.org/10.1063/1.3216464.

[121] Clerjaud B. Transition-metal impurities in III–V compounds. J Phys C: Solid State. 1985;18(19):3615–3661.

[122] Zunger A. Electronic structure of 3d transition-atom impurities in semiconductors. In: Seitz F, Turnbull D, editors. Solid State Phys, vol. 39. Orlando, FL: Academic; 1986. pp. 275–464.

[123] Ohno H. Making nonmagnetic semiconductors ferromagnetic. Science. 1998;281(5379):951–956.

[124] Dietl T, Ohno H. Dilute ferromagnetic semiconductors: Physics and spintronic structures. Rev Mod Phys. 2014;86(1):187–251.

[125] Dietl T, Sato K, Fukushima T, *et al.* Spinodal nanodecomposition in semiconductors doped with transition metals. Rev Mod Phys. 2015;87(4): 1311–1377.

[126] Raebiger H, Lany S, Zunger A. Charge self-regulation upon changing the oxidation state of transition metals in insulators. Nature. 2008;453(7196): 763–766.

[127] Raebiger H, Lany S, Zunger A. Electronic structure, donor and acceptor transitions, and magnetism of 3d impurities in In_2O_3 and ZnO. Phys Rev B. 2009;79(16):165202.

[128] Kaufmann U, Schneider J. Optical and ESR spectroscopy of deep defects in III-V semiconductors. Adv Solid State Phys. 1980;20:87–116.
[129] Samarth N. Ferromagnetic semiconductors: Battle of the bands. Nat Mater. 2012;11(5):360–361.
[130] Mahadevan P, Zunger A, Sarma D. Unusual directional dependence of exchange energies in GaAs diluted with Mn: Is the RKKY description relevant? Phys Rev Lett. 2004;93(17):177201.
[131] Mašek J, Máca F, Kudrnovský J, *et al.* Microscopic analysis of the valence band and impurity band theories of (Ga,Mn)As. Phys Rev Lett. 2010;105(22): 227202.
[132] Di Marco I, Thunström P, Katsnelson MI, *et al.* Electron correlations in $Mn_xGa_{1x}As$ as seen by resonant electron spectroscopy and dynamical mean field theory. Nat Commun. 2013;4:2645.
[133] Yakunin A, Silov A, Koenraad PM, *et al.* Spatial structure of an individual Mn acceptor in GaAs. Phys Rev Lett. 2004;92(21):216806.
[134] Richardella A, Roushan P, Mack S, *et al.* Visualizing critical correlations near the metal-insulator transition in $Ga_{1-x}Mn_xAs$. Science. 2010;327(5966): 665–669.
[135] Bae S, Raebiger H. Critical metal-insulator transition due to nuclear quantum effects in Mn-doped GaAs. Phys Rev B. 2016;94(24):241115.
[136] Droghetti A, Pemmaraju CD, Sanvito S. Predicting d^0 magnetism: Self-interaction correction scheme. Phys Rev B. 2008;78(14):140404.
[137] Lany S. Band-structure calculations for the 3d transition metal oxides in G W. Phys Rev B. 2013;87(8):085112.
[138] Behringer RE. Number of single, double, and triple clusters in a system containing two types of atoms. J Chem Phys. 1958;29:537.
[139] Fuller CS, Struthers JD, Ditzenberger JA, *et al.* Diffusivity and solubility of copper in germanium. Phys Rev. 1954;93(6):1182–1189.
[140] Dash WC. Copper precipitation on dislocations in silicon. J Appl Phys. 1956;27(10):1193–1195.
[141] Struthers JD. Solubility and diffusivity of gold, iron, and copper in silicon. J Appl Phys. 1956;27(12):1560–1560.
[142] Neugebauer J, Van de Walle CG. Hydrogen in GaN: Novel aspects of a common impurity. Phys Rev Lett. 1995;75(24):4452–4455.
[143] Varley JB, Janotti A, Singh AK, *et al.* Hydrogen interactions with acceptor impurities in SnO_2: First-principles calculations. Phys Rev B. 2009;79(24): 245206.
[144] Raebiger H. Theory of defect complexes in insulators. Phys Rev B. 2010;82(7): 073104.
[145] Yamamoto T, Katayama-Yoshida H. Materials design for the fabrication of low-resistivity p-type GaN using a codoping method. Jpn J Appl Phys. 1997;36:180–183.
[146] Yamamoto T, Katayama-Yoshida H. Effects of oxygen incorporation in p-type AlN crystals doped with carbon species. Physica B: Condens Matt. 1999;273–274:113–115.

[147] Raebiger H, Ayuela A, Nieminen R. Intrinsic hole localization mechanism in magnetic semiconductors. J Phys: Condens Matter. 2004;16(41):L457–L462.

[148] Raebiger H, Ayuela A, von Boehm J, *et al.* Clustering of Mn in (Ga,Mn)As. J Magn Magn Mater. 2005;290:1398–1401.

[149] Raebiger H, Nakayama H, Fujita T. Control of defect binding and magnetic interaction energies in dilute magnetic semiconductors by charge state manipulation. J Appl Phys. 2014;115(1):012008.

[150] Nakayama H, Fujita T, Raebiger H. Magnetic properties and stability of quasi-one-dimensional Cr chains embedded in (Zn,Cr)Te. Appl Phys Express. 2013;6(7):073006.

[151] Janotti A, Varley JB, Rinke P, *et al.* Hybrid functional studies of the oxygen vacancy in TiO_2. Phys Rev B. 2010;81(8):085212.

[152] Lee HY, Clark SJ, Robertson J. Calculation of point defects in rutile TiO_2 by the screened-exchange hybrid functional. Phys Rev B. 2012;86(7):075209.

[153] Andersson S, Collén B, Kuylenstierna U, *et al.* Phase Analysis studies on the titanium-oxygen system. Acta Chem Scand. 1957;11:1641–1652.

[154] Padilha ACM, Raebiger H, Rocha AR, *et al.* Charge storage in oxygen deficient phases of TiO_2: Defect Physics without defects. Sci Rep. 2016;6:28871.

Chapter 8

Microscopy of defects in semiconductors

Fabien C.-P. Massabuau[1], *Jochen Bruckbauer*[2], *Carol Trager-Cowan*[2] *and Rachel A. Oliver*[1]

8.1 Introduction

In the study of defects in semiconductors, many techniques provide access to the structure or properties of ensembles of defects. Microscopy techniques, however, allow us to access the structure and properties of individual defects. Whilst in one sense, this can be a weakness, leading to results which lack statistical significance, the ability to image defects at high resolution can provide clarity of understanding and a wealth of information, which is extraordinarily valuable. It is said that 'seeing is believing', and microscopy represents the single most direct method to 'see' defects, or rather to image them, since the resolution of modern microscopes extends right down to the atomic scale, far beyond the power of human sight. Increasingly, microscopy techniques also provide access to the nanoscale resolution of the properties of defects and material in their close vicinity, allowing direct correlation of structure and properties.

There are any number of microscopy techniques that can be useful in addressing defects in semiconductors. Here, we choose to focus on three main families: scanning probe microscopy, scanning electron microscopy and transmission electron microscopy. We note that other powerful microscopy techniques, particularly atom probe tomography, are addressed elsewhere in this book. In this chapter, we will first address the basic principles of our selected microscopy techniques (Section 8.2): it is vital that the reader who either chooses to use microscopy in their own research on defects or relies on microscopy data from the literature should understand the physical principles underlying image formation and hence the strengths and (perhaps more importantly) limitations of the techniques. In our discussions of image formation, we will elucidate the mechanisms by which defects are typically imaged in each technique. Then, in the latter part of the chapter (Section 8.3), we describe some key examples of the application of microscopy to semiconductor materials, addressing both point and extended defects and both two-dimensional (2D) and three-dimensional (3D) materials.

[1]Department of Materials Science and Metallurgy, University of Cambridge, Cambridge, UK
[2]Department of Physics, SUPA, University of Strathclyde, Glasgow, UK

8.2 Basic principles of the microscopy techniques

8.2.1 Scanning probe microscopy

8.2.1.1 Atomic force microscopy

In its most basic form, atomic force microscopy (AFM) is a tool for measuring surface morphology at the nanoscale. Hence, where defects intersect a surface and affect the surface topography, AFM can be used to identify them. Whilst it is possible to achieve atomic resolution imaging in AFM in ultra-high vacuum, for semiconductors, this is more usually achieved using scanning tunnelling microscopy (STM) (described in Section 8.2.1.3). Here, we will address the application of AFM in air to image nanoscale (but not atomic level) surface features, related to defects. Examples of defects which influence the surface topography include dislocations, which necessarily create a pit where they intersect the surface, in order that the surface tension may balance the dislocation line tension. Dislocation imaging in AFM is one of the examples we address later in this chapter, in Section 8.3.1.2. Here, we address the principles of AFM and the basic image formation mechanism [1].

AFM employs a sharp tip with end radius typically less than 10 nm, suspended from the end of a microcantilever. In the simplest realisation of the technique, a piezoelectric scanner is used to scan the tip over the surface in contact with the sample, and where it encounters a change in the surface topography, the cantilever deflects. This deflection is monitored by reflecting a laser beam from the backside of the cantilever onto a photodetector. If the path length followed by the laser beam is long, nanoscale motion of the cantilever is amplified into larger changes in the position of the laser beam at the photodetector, and hence the surface morphology can be measured. This very simple implementation of AFM, similar to the operation of a record player, is rarely used, and indeed even when the technique was first invented [2], a feedback circuit was employed to control the deflection of the cantilever and hence the force on the surface. The feedback circuit acts to continuously adjust the height of the cantilever, aiming to achieve a fixed cantilever (or equivalently laser spot) deflection. This fixed deflection value is the 'set point' for the feedback circuit. The cantilever height is adjusted by changing the voltage applied to the z-piezo – a piezoactuator oriented to provide motion in the z-direction (i.e. normal to the surface). These voltages are then converted to heights using calibration data. The accuracy of AFM is hence only as good as the calibration data it relies on, and the calibration should usually be checked approximately every 6 months. The instrumentation used in AFM is illustrated schematically in Figure 8.1. Note that the figure shows a system in which the tip is scanned and the sample held stationary. This configuration allows large samples (such as full semiconductor wafers) to be scanned. However, many microscopes, particularly those used for smaller samples, keep the tip stationary and achieve relative motion between the tip and sample by moving the sample using a piezoactuator.

The vertical resolution of a well-calibrated atomic force microscope is limited in theory by the ability to control the z-motion via the application of voltage to the z-piezo.

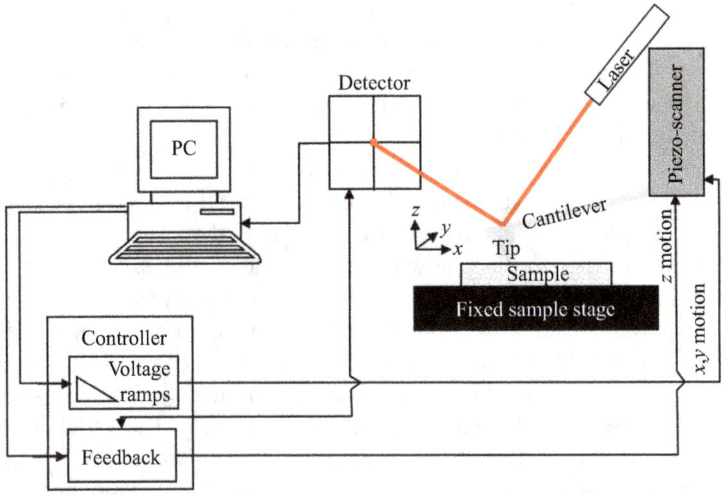

Figure 8.1 Schematic illustration of an atomic force microscope for a tip-scanning system

In practice, the vertical resolution is actually determined by noise, and the noise floor of typical commercial microscopes is often less than 0.5 Å, allowing the imaging of atomic-scale features such as monolayer steps. The horizontal resolution is in general determined by the radius of the apex of the tip used for imaging. Since the forces acting on the tip include fairly long-range van der Waal's interactions, even where portions of the tip away from the very apex do not directly contact the surface, they still influence the tip deflection, and hence the resolution. Hence, the creation and preservation of a sharp tip is paramount if high-resolution imaging is to be achieved.

When an atomic force microscope is operated, as described earlier, with the tip in constant contact with the surface, this is known as 'contact mode'. The constant rubbing of the tip over the surface tends to wear the tip, blunting its sharp apex (or sometimes damaging the sample), so many users prefer to employ an 'intermittent contact' mode instead to reduce tip wear and hence improve resolution. The most common intermittent contact mode is the so-called tapping mode. In tapping mode, the tip is oscillated at a frequency close to its resonant frequency, usually by a small piezo element on the tip holder. When the tip is close enough to the surface to make contact at the bottom of each oscillation, the presence of the surface limits or damps the amplitude of oscillation, which then varies linearly with tip-sample separation. Hence, the amplitude of oscillation may be used as the input parameter to a feedback circuit: a constant amplitude setpoint is maintained by varying the voltage applied to the *z*-piezo so that the tip traces out a contour of constant surface height. As in contact mode, calibration data is used to convert the voltages applied to the *z*-piezo to heights, forming the topography imaging. The amplitude of vibration is also affected by a second mechanism even when the tip does not contact the surface: the natural frequency of cantilever oscillation is a function of the force gradient experienced by the cantilever. Since the frequency of the driving oscillation is fixed, changes in the natural frequency can take the vibration on or off resonance, changing the amplitude of vibration. This effect has less influence on tapping mode imaging than the damping of the amplitude but is used in 'non-contact' mode to image the surface whilst applying minimal forces to the tip and sample.

Non-contact imaging is not usually an effective way to achieve high resolution in air, since most samples imaged in the air are covered by a layer of fluid from atmospheric humidity, and non-contact imaging requires the tip to remain beyond this fluid layer, which tends to blur the outlines of surface features. To achieve higher lateral resolutions in air, alternative approaches have recently been developed with the aim of controlling and minimising the force on the tip to maintain its sharpness. An example of this is 'PeakForce' tapping mode [3], in which rather than oscillating the tip close to resonance, the tip is oscillated off resonance by the main z-piezo and its deflection is measured throughout each oscillation cycle. The variation in force during the oscillation cycle is then estimated, and the highest force experienced during the period in which the tip is in contact with the surface is monitored. This is the 'peak force' and is used as the feedback parameter so that the cantilever traces a contour over the surface at which this peak force is constant. There is increasing evidence that by maintaining low peak forces throughout imaging, blunting of the tip is minimised, facilitating high lateral resolutions in imaging, but the influence of materials properties, rather than topography, on this mode is not yet fully understood.

8.2.1.2 Electrical characterisation techniques in AFM

The potential uses of the AFM tip extend far beyond topographic imaging, with many opportunities available to measure the properties of materials. Characterisation of electrical properties generally requires a conductive tip, and this can be achieved by coating a standard silicon tip in a metallic layer. It is important that the metallic layer is robust enough to avoid it wearing away during imaging, and PtIr layers are often used for this purpose. An alternative is to use all-metallic pyramid structures on the end of metal or silicon cantilevers, but whilst such structures overcome problems with coating wear, these tips are more likely to deform plastically, becoming blunt. For some applications, particularly scanning spreading resistance microscopy (SSRM) of Si, tips coated with or fully formed from highly doped diamond are used, providing a hard, mechanically stable imaging probe.

SSRM is part of a family of techniques, which also includes conductive AFM (C-AFM) and tunnelling AFM (TUNA). In these techniques, the tip is usually kept in direct contact with the sample surface, a bias is applied between the tip and the sample and the resulting current flow is monitored. The techniques differ mainly in the nature and sensitivity of the amplifiers used to measure the current, with SSRM employing a logarithmic amplifier to access the resistivity of a broad range of samples, whilst C-AFM and TUNA employ linear amplifiers, with TUNA addressing smaller currents than C-AFM. Very recently, PeakForce TUNA [4] and other similar techniques have allowed currents to be measured in intermittent contact modes during the contact phase of the oscillation cycle, allowing current maps of very fragile materials to be built up.

These current flow techniques are applicable to any material with sufficient conductivity to carry a measurable current. More specific to semiconductors is scanning capacitance microscopy (SCM). SCM is again a contact mode technique. Assuming that in air, a semiconductor surface has a native oxide present, the metal coating on the surface, the native oxide and the semiconductor itself form a metal-oxide-semiconductor capacitor (see Figure 8.2(a)). The capacitance of such a device varies with applied voltage in a way that is dependent on the local carrier concentration. Consider the application of a negative bias to a tip in contact with an n-type semiconductor. The electrons are repelled from the tip, effectively increasing the thickness of insulating material between the metallic tip and the underlying semiconductor, and the capacitance of the device is reduced. If the device was left to equilibrate, minority carriers would accumulate at the semiconductor oxide interface, and the capacitance would then be controlled just by the oxide thickness. However, if the bias applied to the tip varies at high frequency, there is insufficient time for this accumulation to occur. (The relevant *C–V* curves are illustrated schematically in Figure 8.2(b).) Under these circumstances, the amplitude of the

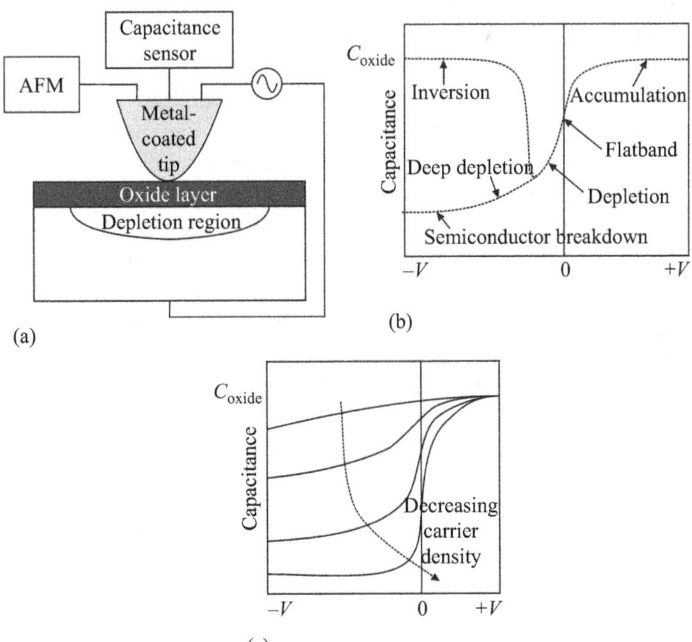

Figure 8.2 Schematic illustrations of (a) the instrumentation for SCM, (b) C–V curves for an n-type semiconductor for low-frequency bias changes (leading to inversion) and high frequency bias changes (leading to deep depletion) and (c) C–V curves for n-type semiconductors with different carrier densities for high frequency bias changes. C_{oxide} is the capacitance of the oxide layer at the semiconductor surface

change in capacitance with applied voltage is a function of the carrier concentration in the semiconductor, and hence depends on the dopant concentration. For the same applied change in voltage, a larger depletion region forms in a lightly doped semiconductor than a highly doped semiconductor, causing a larger change in capacitance. (The variation in the *C–V* curves with dopant concentration is illustrated schematically in Figure 8.2(c).) The rate of change of capacitance with voltage is monitored by a gigahertz frequency capacitance sensor and is used as a measure of the carrier density in the material. The phase of the capacitance response relative to the drive voltage provides information as to whether depletion occurs during the positive or negative half of the drive cycle, hence determining whether the material is doped n- or p-type.

A typical SCM data set thus consists of two images: (1) a phase image in which n-type material is seen as black and p-type as white. Non-semiconducting material (insulating or metallic) does not deplete and no fixed phase can be measured, so such regions appear grey and noisy. (2) An amplitude image in which highly doped material is seen as having lower amplitude than lightly doped material, typically represented by a lower brightness in a greyscale image. It should, however, be noted that undoped (or non-semiconducting) material shows no change in capacitance with voltage and thus yields no signal (seen as an even lower brightness in a greyscale image).

SCM and SSRM are very widely used in the study of semiconductors. Perhaps, slightly less common is electrical force microscopy and its more quantitative variant Kelvin probe force microscopy (KPFM). These techniques usually employ a lift mode – in which the topography of the sample is recorded during an initial scan line, and then the tip is raised through a fixed height and rescans the same trajectory during a second scan line during which electrical property data is collected. The advantage of this approach is that during the topography scan, the tip's oscillation is strongly affected by forces arising from contact between tip and the sample. In the lift scan, however, forces arising from the potential difference between a biased tip and the sample may be more easily detected, allowing measurement of the variation in potential across the sample surface.

As with topography, all electrical AFM techniques have resolution limited by the finite size of the tip. Furthermore, the sample topography can profoundly influence the electrical data, since changes in the tip-sample interaction area as the tip passes over varying topographic features alter the electrical interaction. C-AFM provides a simple example: if the tip-sidewall runs up against a 3D island forming a large contact area, an increased current can flow, even though the properties of the material may be unchanged relative to an adjoining area. Similar issues apply to the other techniques that are mentioned here and to myriad other methods which have been developed to probe electrical materials in AFM. This short section does not provide an exhaustive account, but rather a flavour of the available options. For further details, the reader should consult [5]. One interesting recent development worthy of mention is the integration of light sources with AFM set-ups to allow measurement either in darkness or under illumination. This approach is most commonly used to assess the photoconductivity of solar cell materials in C-AFM or

TUNA [6] but is equally applicable to other electrical AFM techniques and provides a wealth of new opportunities for the characterisation of optoelectronic materials.

8.2.1.3 Scanning tunnelling microscopy

The most common application of STM in the study of semiconductors is in the study of the atomic structure of clean (i.e. free from native oxide) semiconductor surfaces in ultra-high vacuum. In this context, the termination of extended defects at the surface may be observed in exquisite detail [7]. Point defects in the surface structure are frequently observed and, whilst defects seen following surface cleaning or material deposition may not be representative of bulk defects, insight into defects in the bulk may be achieved by careful interpretation of images of cleaved surfaces [8]. Specially prepared samples are sometimes required, for example material grown at lower temperatures, to ensure a sufficient density of defects to allow the relevant features to be found and identified. Where shallow subsurface point defects influence the electronic structure of the overlying surface layers, they may be detected, in addition to defects immediately at the surface. In 2D materials, the distinction between surface and bulk defects is essentially meaningless, since there is no 3D bulk material. Hence, STM may find increasing application in assessing point defects in such novel structures.

Unlike AFM, STM does not utilise a deflecting cantilever. Instead, a very sharp metallic needle forms the tip and is scanned over the sample surface, in very close proximity but with a few Ångstrom vacuum gap between the two. Under the influence of an applied bias, electrons may tunnel across the gap, but the magnitude of the tunnelling current is exponentially dependent on the distance between the tip and the sample. Hence, the contribution of atoms away from the very apex of the tip to imaging is much smaller than in AFM, and it is this exponential dependence that allows STM to be used routinely for atomic resolution imaging, with the tunnelling current flowing predominately between a single atom at the sample surface and a single atom at the tip apex. Similar to AFM, a feedback circuit may be employed, with a z-piezo being used to maintain a fixed tunnelling current. The voltages applied to the z-piezo form a map of the topography of the sample, with the tip tracing out a contour of constant tunnelling current, equivalent to a constant tip-sample distance. This simple description immediately begs the question: what do we mean by a constant tip-sample distance in this situation? Simple models of the tip-sample interaction indicate that a better description of an atomic resolution STM image at constant tunnelling current is a contour of constant local density of states (LDOS) at the Fermi level of the sample surface [9]. Whilst the original models from which this understanding is derived are known to be incomplete, they provide a useful starting point for the discussion of STM data and illustrate why – for example – a subsurface impurity atom which may have little or no impact on the surface *height* can nonetheless impact the STM data via its impact on the local electronic structure.

Within our understanding of STM images as mapping the LDOS, we can also understand the opportunities offered by imaging at different tip biases. For a tip positively biased relative to the sample, electrons flow from the occupied states of the sample into the unoccupied states of the tip. Reversing the polarity of the bias instead accesses the unoccupied states of the sample. When imaging compound semiconductors, the electronegativity difference between the metallic and non-metallic species can allow images of the occupied states of the sample to be dominated by the electron-rich non-metallic sites, whereas images of the unoccupied states primarily address the metallic sites. This ability to distinguish between the metallic and non-metallic sublattices can allow the identification of, for instance, which sublattice accommodates a particular amphoteric dopant atom; for example in highly Si-doped GaAs, Si_{Ga} donors can be observed but so can Si_{As} acceptors, Si donor–Ga vacancy complexes (Si_{Ga}–V_{Ga}) and planar Si clusters on (1 1 1) planes [10], all of which act as compensating defects reducing the overall electron concentration which can be achieved (Figure 8.3).

Additional insights into the impact of defects on the electronic structure can be gained using scanning tunnelling spectroscopy (STS): the recording of site-specific current–voltage (I–V) or dI/dV–V curves. Under conditions where electrons are injected from the occupied states of the tip into the unoccupied states of the sample, peaks in the dI/dV–V curve can (at a basic level) be interpreted as corresponding to energies with a high LDOS. More detailed interpretation of such curves can be challenging, and this approach is less widely applied than voltage-dependent imaging, but as early as 1993, this technique was successfully used to reveal donor levels associated with bulk antisite defects in GaAs. Later studies address a wide range of samples [8], including 2D materials such as graphene [11].

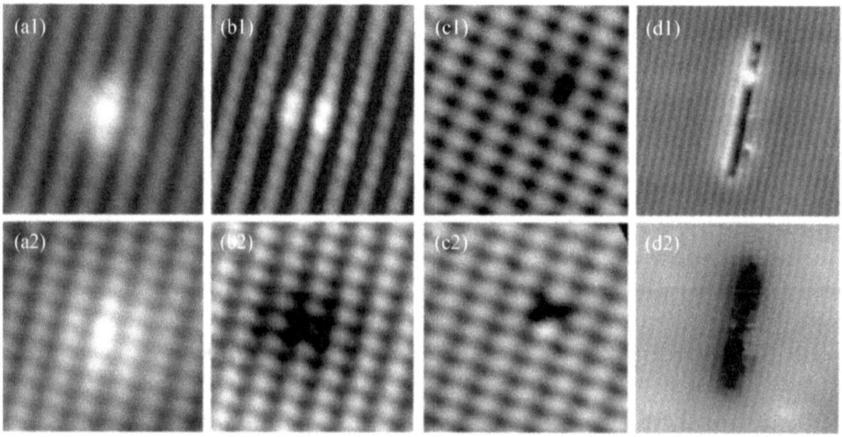

Figure 8.3 STM images of defects at cleavage surfaces of Si-doped as-grown GaAs bulk crystals. Top row: occupied states images, recorded at biases from −2.0 to −2.2 V. Bottom row: unoccupied states images, recorded at biases from +1.4 to +1.5 V. (a1) and (a2) show a SiGa donor, (b1) and (b2) a SiAs acceptor, (c1) and (c2) a SiGa–Ga-vacancy complex and (d1) and (d2) the intersection line of a planar Si cluster. © 1996 American Physical Society. Reprinted, with permission, from [10]

8.2.2 Scanning electron microscopy

8.2.2.1 Brief introduction to SEM

In a scanning electron microscope (SEM), an image is formed when the intensity of the resultant electrons, X-rays, light or current is monitored as a focussed electron beam is rastered over the surface of the sample. In the SEM, the energy of the electron beam is usually selected from a range of energies spanning from 100 eV to 30 keV [12–15]. Figure 8.4 shows a schematic of an SEM and also illustrates the

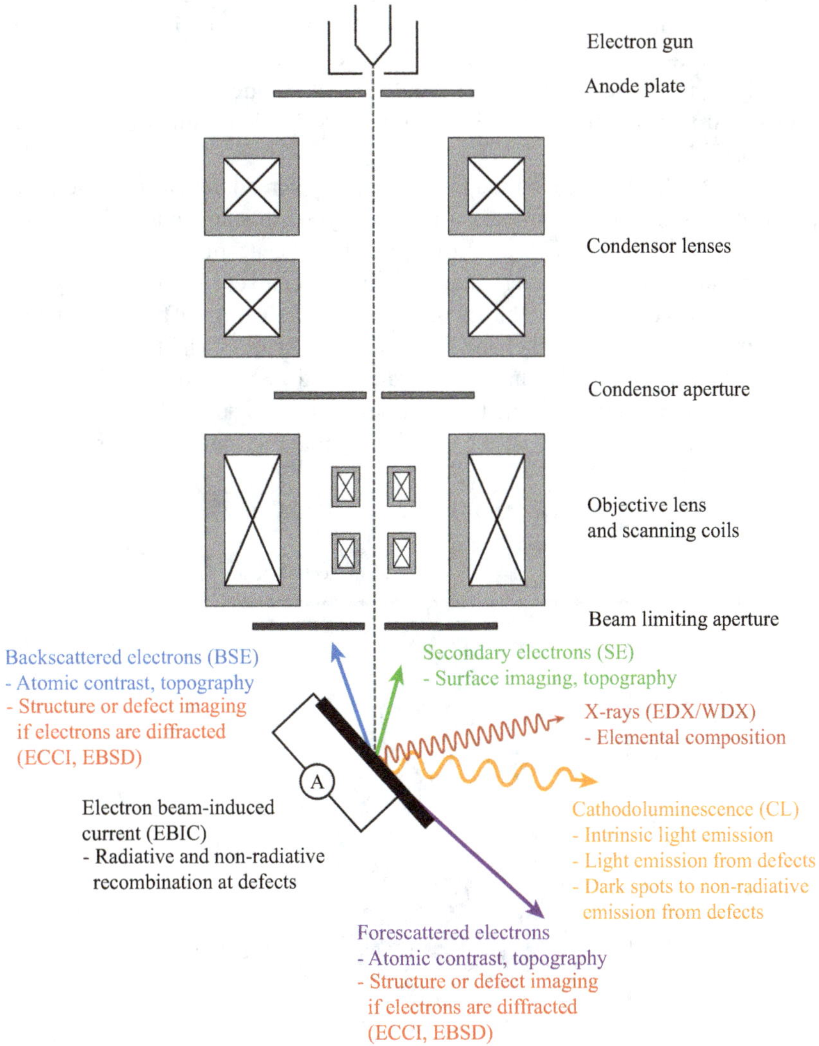

Figure 8.4 Schematic of an SEM and an illustration of the range of signals which can be obtained

range of signals which can be used to generate images and highlights those signals which can reveal defects present in the sample.

Resolution

The spatial resolution of the images generated in the SEM depends on the electron beam spot size that is dependent on the size of the electron beam source, the electron energy, the beam current and electron optical aberrations, in particular chromatic (due to the range of energies emitted by the electron source) and spherical aberrations and astigmatism. It also depends on the interaction volume; that is the volume within the sample from which the signal is obtained. The interaction volume depends on the signal being measured as displayed in Figure 8.5, the atomic number and density of the material under investigation, the energy of the incident beam and the angle of the incident beam. For light and current-based measurements, resolution may also be determined by carrier diffusion [16,17].

The size of the electron source and the energy spread depends on the type of electron gun used. Table 8.1 summarises the types of guns available and their typical properties [13–15,18]. For imaging defects, a small source size, low energy spread, low beam divergence and high beam current/beam brightness are required. In addition, for the imaging of defects using diffraction imaging, usually referred to as electron channelling contrast imaging (ECCI), a beam with low divergence is required. This will be discussed in Section 8.2.2.2. An SEM with a Schottky emitter is the instrument best suited for imaging and analysing defects. The resolution achievable for different imaging modes will be discussed later in this section.

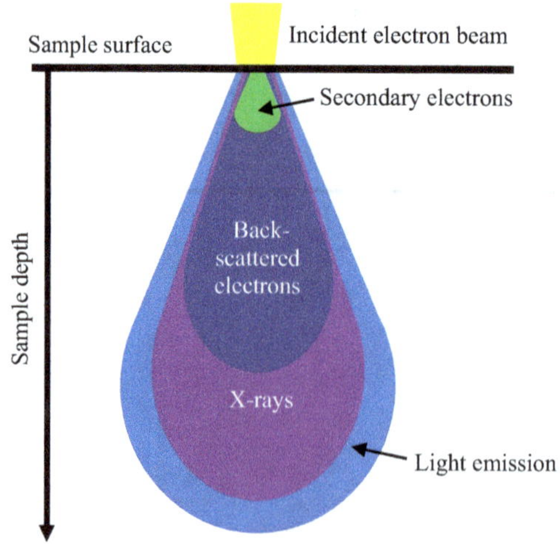

Figure 8.5 Illustrating the signals detected in the SEM and their interaction volume within the sample

Table 8.1 Summary of electron guns used in scanning electron microscopes [13–15,18]

Electron gun type	Source size (μm)	Energy spread (eV)	Beam current (nA) (at 20 keV)	Beam brightness ($A/cm^2/str$) (at 20 keV)
Tungsten hairpin thermionic emitter	≈20	≈3	≈1,000	10^5
LaB_6 thermionic emitter	≈10	≈2	≈1,000	10^6
Cold cathode field emitter	$\approx 5 \times 10^{-3}$	≈0.3	≈20	10^8
Schottky emitter	$\approx 20 \times 10^{-3}$	≈0.8	100–500	10^8

Signals used to generate images in the SEM

The signals that are used to generate images in the SEM are illustrated in Figure 8.4 and are described next. Figure 8.5 schematically illustrates the interaction volume for each of these signals.

Secondary electrons

The interaction of the electron beam with the sample gives rise to the emission of secondary electrons (SEs) through ionisation of the sample under study. SEs are low-energy electrons and are defined as those electrons with energy less than 50 eV. When the intensity of SEs is monitored as the beam is scanned over the sample, a topographic image of the surface of the sample is formed. A topographic image is produced because the intensity of the SE signal is dependent on the orientation of the sample surface with respect to the incident electron beam and to the position of the SE detector. There is a higher yield of SEs from a surface inclined with respect to the electron beam when compared to that of a flat surface, and regions of the sample facing the detector will appear brighter than those facing away. Because of their low energy, SEs have a high probability of reabsorption in the sample. Therefore, only those generated close to the surface are emitted. SE imaging is, therefore, an extremely surface-sensitive imaging mode (i.e. the interaction volume is small) and images with a resolution of order 1 nm can be obtained in an SEM with a cold cathode field emitter or Schottky emitter. Figure 8.6 shows an SE image revealing the topography of a nanoscale GaN pyramid structure. Note that the edges of the pyramids are bright which is due to more SE emission from an edge when compared to a flat surface.

SEs are typically detected by an Everhart–Thornley detector. This detector comprises an electron collector, which is positively biased at a voltage of around 200 V to attract the SEs, and a scintillator biased at around 10 keV to convert the electrons to light which is coupled to a photomultiplier tube. SEs may also be detected by an in-lens detector where the SEs are collected through the objective lens of the SEM.

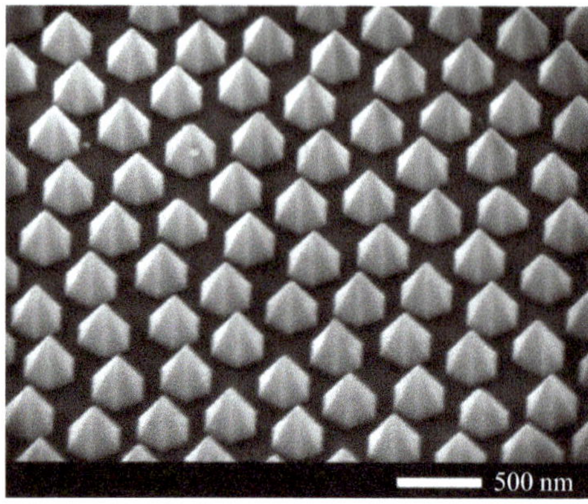

Figure 8.6 Secondary electron image of an array of GaN-based nanoscale pyramids. © 2012. Reprinted, with permission, from [19]

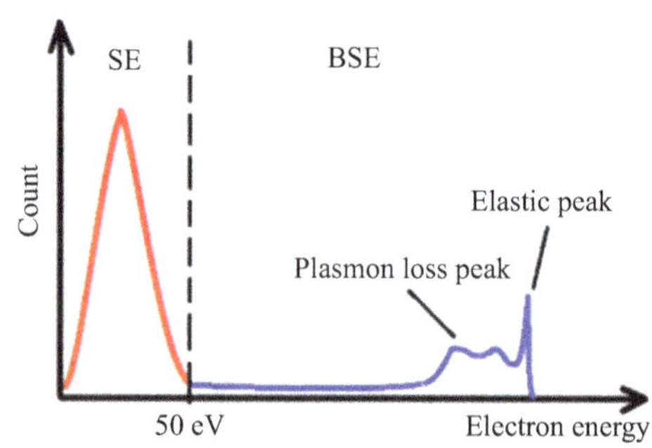

Figure 8.7 Schematic of the electron energy spectra. © 2016. Reprinted, with permission, from [20] under Creative Commons Attribution (CC BY) license: http://creativecommons.org/licenses/by/4.0/

Backscattered electrons

Backscattered electrons (BSEs) are those which have been elastically or inelastically scattered from the sample. Their energy can range from that of the incident beam down to that of the SEs. Figure 8.7 illustrates the energy distribution of SEs and BSEs in the SEM. The backscattered yield is dependent on the atomic number and density of the material under study, so atomic contrast can be provided by this imaging mode. The backscattered yield is also dependent on the inclination of the sample surface so BSEs can also be used to provide an image of surface topography. As BSEs generally come from deeper within the sample than SEs, BSE images generally exhibit a lower resolution than SE images.

BSEs may also be detected by an Everhart–Thornley detector. In this case, the electron collector is negatively biased (at around −50 V) to repel the lower energy SEs. BSEs may also be detected by an electron sensitive diode(s) placed underneath the final SEM lens (pole piece). This is usually a four-quadrant diode: if the signals from all the quadrants are added together, compositional contrast dominates; if the quadrants are operated in adjacent pairs and the signal from one pair subtracted from the other, topographic contrast dominates.

Where the electrons incident on the sample undergo diffraction and an image is formed by monitoring the intensity of the BSEs as the beam is rastered over the sample, the contrast exhibited in this image is related to the sample's structural properties. This imaging modality is commonly referred to as ECCI. Where the electrons exiting the sample undergo diffraction, and the resultant electron backscatter diffraction (EBSD) patterns (EBSPs) are recorded at each point in the raster scan; information derived from each EBSP can be used to produce a map revealing the sample's structural properties. This mapping technique commonly referred to as EBSD mapping. These techniques are discussed in the next sections.

8.2.2.2 Electron channelling contrast imaging

Electron channelling contrast images are produced when a sample is positioned so that a crystal plane or planes in the sample are at, or are close to, the Bragg angle. That is, the electrons entering the sample are diffracted and it is this that provides the dominant structural contrast in the resulting image. Electrons diffracted on exiting the sample can contribute to the structural contrast, but this component is highly dependent on the position of the detector. Any deviation in crystallographic orientation or in lattice constant due to local strain may then be revealed by variation in the contrast in the electron channelling image constructed by monitoring the intensity of backscattered or forescattered electrons as the electron beam is scanned over the sample. Extremely small changes in orientation and strain are detectable, revealing, for example, low-angle tilt and rotation boundaries and atomic steps and enabling extended defects such as dislocations and stacking faults to be imaged [21–30]. ECCI can provide similar information on defects as the transmission electron microscope (TEM) (see Sections 8.2.3 and 8.3.1.4) where the defects either thread to the surface or lie within around 50 nm of the surface.

For successful imaging of the structural properties of materials using ECCI, all that is required is that the sample has a reasonably smooth and clean surface. For the imaging of metal surfaces, this usually requires careful sample polishing to produce a high-quality surface [31], this is also the case for semiconductor wafers cut from bulk crystals. However, for most epitaxially grown semiconductor thin films, no surface preparation is required. The spatial and depth resolution of ECCI is of the order of tens of nanometres. Figure 8.8 illustrates the two geometries, namely the backscatter and forescatter geometries, which are used to acquire ECCI micrographs. The backscatter geometry (Figure 8.8(a)) has the advantage that this geometry does not require a high tilt of the sample and, therefore, a significant correction of the image to account for tilt is not required. This geometry also allows the easiest imaging of large samples, for example full semiconductor wafers. The forescatter geometry (Figure 8.8(b)) has the advantage that images exhibit better signal-to-noise with respect to the backscatter geometry due to the increase in the intensity of BSEs. The detector used to detect the BSEs is generally an electron-sensitive diode. The key to

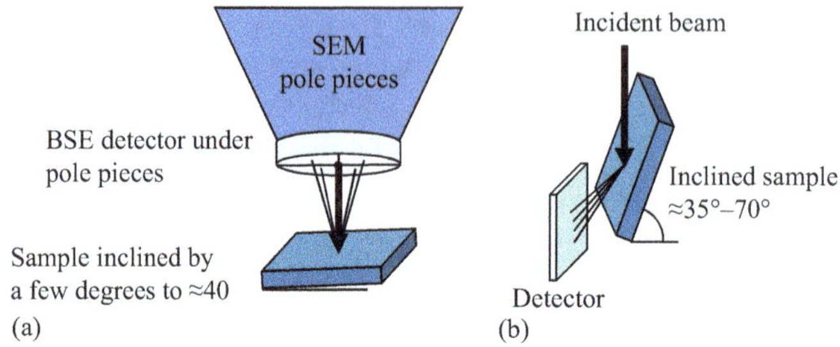

Figure 8.8 Illustrating the (a) backscatter and (b) forescatter geometries for the acquisition of ECCI micrographs

Figure 8.9 (a) ECCI micrograph of a GaN thin film showing threading dislocations. Inset: an enlarged image of three dislocations. The different grey levels in the image indicate areas of different orientation. © 2014. Reprinted, with permission, from [32]. (b) ECCI micrograph of a c-plane GaN thin film showing threading dislocations and atomic steps. © 2007. Reprinted, with permission, from [22]. (c) ECCI micrograph of an m-plane GaN thin film showing stacking faults (one highlighted by the ellipse). Partial dislocations can also be seen at the ends of the stacking faults (image courtesy of G. Naresh-Kumar)

the acquisition of good quality ECCI micrographs is the use of a good amplification system.

Figure 8.9 shows electron channelling contrast images from nitride semiconductor thin films. All the images (and the inset of Figure 8.9(a)) show threading dislocations (TDs) as spots with the black–white (B–W) contrast related to the strain field associated with the dislocations. Figure 8.9(b) shows an ECCI micrograph revealing atomic steps. Figure 8.9(c) shows stacking faults and associated partial dislocations on either end of the stacking fault. A simple geometric procedure has been developed to identify a given TD in c-plane nitride thin films as either edge-, screw- or mixed-type by exploiting differences in the direction of the B–W contrast between two ECCI micrographs acquired under quasi-two-beam conditions from two symmetrically equivalent crystal planes. Crystal planes were chosen for which the diffraction vectors ($\mathbf{g}_1$ and $\mathbf{g}_2$) were at 120° to each other, where the **g**-vector was determined through the acquisition of electron channelling patterns (ECPs) [25].

An ECP is obtained when changes in the BSE intensity are recorded as the angle of the incident electron beam is changed relative to the surface of a single crystal area of the sample. When a backscattered image is acquired at low magnification, as the beam is scanned over the sample, it changes its angle with respect to the surface of the sample (in our case by around $\pm 2.5°$), allowing an ECP to be obtained [25]. Note that, this method of acquiring an ECP is only possible if the scanned area of the sample (of order 5 mm × 5 mm in size) is smooth and of the same crystallographic orientation. Alternatively, if 'beam rocking' electron optics are available in the SEM, an ECP can be acquired from an area ranging from of order 10 μm × 10 μm to 500 nm × 500 nm in size [33], such ECPs are often referred to as selected area ECPs (SAECPs or SACPs). When the beam changes its angle with respect to the sample, different planes of the crystal satisfy the Bragg condition, giving rise to the appearance of overlapping bands of bright and dark lines (Kikuchi lines) superimposed on the image of the sample; an ECP from a GaN thin film is shown in Figure 8.10. The ECP is related to a 2D projection of the crystal structure, with the Kikuchi lines corresponding to different planes in the crystal. Comparing the ECP with kinematical and/or dynamical electron diffraction simulations allows the pattern to be indexed, i.e. the planes in the ECP can be identified. The plane or planes which intersect the centre of the ECP, usually referred to as the pattern centre, are those from which the incident electron beam is diffracted. In the example shown in Figure 8.10, the incident electron beam was diffracted from one of the $\{1\ 1\ \bar{2}\ 0\}$ planes, that is a **g**-vector of $\langle 1\ 1\ \bar{2}\ 0\rangle$ was selected.

In principle, if the diffraction condition can be selected, defects can be identified by investigating their visibility as a function of the diffraction condition, similar to the technique used to identify defects from diffraction images in the TEM (see Section 8.3.1.4). However, as ECCI micrographs from semiconductors are usually acquired in a plan-view geometry, i.e. from the top of the as-grown surface of the sample, and the collected signal is backscattered as opposed to transmitted electrons, the relaxation of the surface strain associated with a defect plays a significant role with regard to its visibility and this must be taken into account when attempting to identify defects with ECCI.

As mentioned earlier, TDs in c-plane GaN can be identified by exploiting differences in the direction of the B–W contrast between two ECCI micrographs

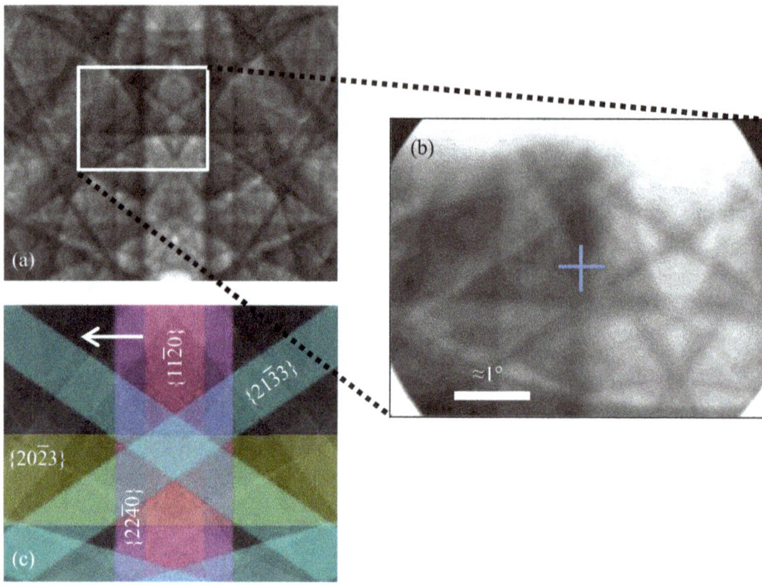

Figure 8.10 *(a) Dynamical simulation of an electron channelling pattern (ECP) from a GaN thin film. Electron beam energy is 30 keV, sample tilt ≈40° (courtesy of Aimo Winkelmann). (b) Experimental ECP. The blue cross marks the pattern centre (PC). (c) Kinematical simulation of the ECP with some indexed planes highlighted (produced with ESPRIT DynamicS (Bruker Nano) software). In this case, the PC intersects one of the {1 1 $\bar{2}$ 0} planes so a* **g**-*vector of* ⟨*1 1 $\bar{2}$ 0*⟩ *was selected*

acquired under quasi-two-beam conditions from two symmetrically equivalent crystal planes. As a result of surface relaxation, the B–W contrast direction is perpendicular to the **g**-vector for a pure screw dislocation [25]. For an edge dislocation, the direction of the B–W contrast generally depends on its Burgers vector and not on the **g**-vector. Hence, for edge dislocations, we expect the direction of the B–W contrast to either reverse or remain the same when the diffraction conditions are changed. The degree of contrast observed for an edge dislocation, that is its visibility, is dependent on the selected diffraction condition. If the **g**-vector and the Burgers vector of the edge dislocation are orthogonal, the visibility of the edge dislocation will be strongly reduced and this can help with the identification of the dislocation (see Section 8.3.1.4). If the diffraction conditions are selected so that the contrasts observed for both edge and screw dislocations are similar, mixed dislocations may be identified as their B–W contrast direction changes in a manner intermediate to that of the edge and screw dislocations. In practice, it is difficult to achieve exact two-beam conditions; however, it is possible to get sufficiently close to allow quantitative analysis of the resultant ECCI micrographs. Where it is not possible to acquire an ECP, but it is possible to identify the TDs as either an edge dislocation or one with a screw component (i.e. a screw or a mixed dislocation) by comparing two ECCI micrographs acquired for the sample placed in at least two orientations. If, for a given dislocation, the B–W contrast direction 'flips' (rotates by 180°), or stays the same, on comparison of the two images, then it is likely to be an edge dislocation. The B–W contrast direction for a screw or mixed dislocation will be observed to change but by an angle other than 180°. In practice, the direction of the B–W for each dislocation should be compared for at least three images to unambiguously identify a dislocation as either an edge dislocation or one containing a screw component.

Simulations of dislocation contrast may also help the understanding of the contrast exhibited by dislocations observed in ECCI micrographs. For example, Carnevale *et al.* [28] used simulations to support their analysis of misfit dislocations in thin films of GaP grown on Si, while Picard *et al.* have simulated the contrast for inclined threading edge and screw dislocations in $SrTiO_3$ [34] and Pascal *et al.* have simulated contrast for threading edge dislocations in GaN [35].

The conditions required to image defects using ECCI are quite stringent: a high brightness, small beam diameter (nanometres), high current (nanoamps or higher) and low divergence (a few mrad) electron beam are required. As discussed earlier, such conditions are met in an SEM equipped with a Schottky gun. ECCI micrographs are generally acquired at electron beam energies of 20–30 keV to obtain the best signal-to-noise ratio. The performance of the diodes used to detect BSEs reduces for lower electron beam energies [21]. An amplifier system specifically designed to optimise the contrast and brightness of the ECCI micrographs is also invaluable for obtaining good images. Lastly, ideally, the SEM should be equipped with beam rocking to enable the acquisition of SAECPs. The acquisition of EBSPs (discussed in the next section) can also help with the determination of diffraction conditions [31].

Until recently, ECCI was mostly used to investigate the structural properties of metals and geological materials, but its use for the characterisation of defects in semiconductors is steadily expanding. To date, ECCI has been used to image defects in nitride semiconductors [22–25], $Si_{1-x}Ge_x$ [36], SiC [37], GaAs [27], GaP [28,29], $GaAs_yP_{1-y}$ [29] and GaSb [30]. For more information on ECCI, the following papers provide informative reviews of the technique [21,30,36,37].

8.2.2.3 Electron backscatter diffraction

In EBSD, the electron beam is incident on a sample which typically is tilted at an angle near to 70°. The primary electrons (typically with energies between 10 and 30 keV) are inelastically scattered through high angles forming a diverging source of electrons which can be diffracted as they leave the sample. A simple description for the formation of an EBSP presumes that those electrons that satisfy the Bragg condition for a given plane need to have directions on two so-called Kossel cones relative to both the upper and lower surfaces of that plane, see Figure 8.11(a). The intersections of these Kossel cones with the detector screen approximately mark the edges of the Kikuchi bands, which thus have a width of about twice the Bragg angle, with the projection of the relevant lattice plane in the centre of the Kikuchi band.

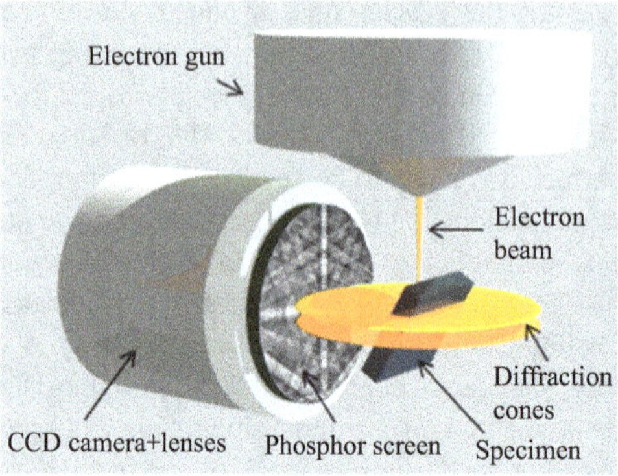

(a)

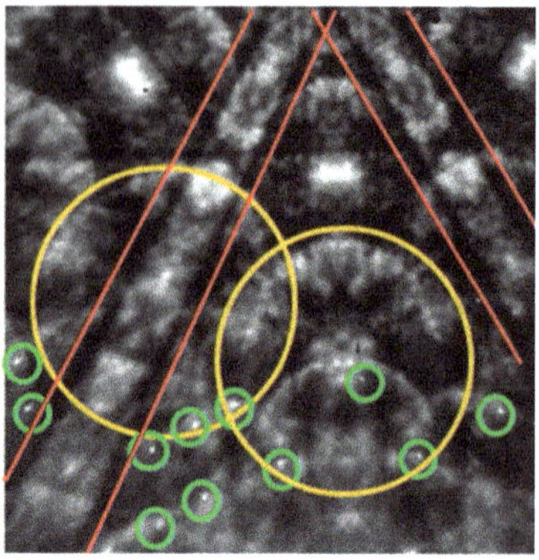

(b)

Figure 8.11 *(a) Illustration of the EBSD detection geometry and a conventional EBSD detector. (b) An EBSP from a GaN thin film acquired at an energy of 5 keV and a probe current of ≈1.5 nA. The red lines outline a pair of Kikuchi bands, the large yellow circles indicate high-order Laue zone (HOLZ) rings and the small green circles highlight reflection high-energy electron (RHEED) spots. © 2015. Reprinted, with permission, from [38] under Creative Commons Attribution (CC BY) license: http://creativecommons.org/licenses/by/4.0/*

An EBSP consists of a large number of overlapping Kikuchi bands and is related to a 2D, gnomonic projection of the crystal structure, see Figure 8.11(b). The Kikuchi lines appear as almost straight lines, because the Kossel cones are very shallow, as the Bragg angle is of order 1°. Each Kikuchi band is effectively a trace of the lattice plane from which it is formed; an EBSP can thus provide a direct measurement of the angles between lattice planes and directions in a sample's crystal structure.

The EBSP is generally detected by an electron sensitive phosphor or scintillator screen and a charge-coupled device (CCD) or complementary metal-oxide-semiconductor (CMOS) camera [39] (see Figure 8.11(a)), although there have been recent developments of direct electron cameras [38,40]. The EBSP shown in Figure 8.11(b) was acquired at 5 keV from a GaN thin film using energy-filtered direct electron detection. Direct electron detection allows high-quality EBSPs to be acquired at low electron beam energies and pattern acquisition down to 3 keV is achievable [38].

Rotations of a crystal (e.g. tilts or twists in a thin film) will produce corresponding rotations of the EBSP, while strain will produce a distortion of the EBSP. Therefore, if EBSPs are acquired from a grid of points on a sample, it is possible to analyse these EBSPs to produce maps of tilts, twists and strains. For a more detailed discussion of the EBSD technique, see [21,39,41].

Recent developments of EBSD analysis techniques, where experimental EBSPs are compared with either a reference experimental pattern or simulated pattern using cross-correlation, have greatly improved the sensitivity of EBSD and its applicability to the study of semiconductors. Sensitivities of less than 1 part in 10,000 to changes in tilt, in-plane rotation and strain are now possible [42]. Once such sensitivity is achievable, it is then possible to calculate the density of dislocations necessary to accommodate the observed misorientations and strain, that is the geometrically necessary dislocations (GNDs) [43]. For an example of what can be achieved with such cross-correlation high-resolution EBSD (cross-correlation HR-EBSD), Figure 8.12(a) shows an EBSD-derived twist map for a patterned epitaxially overgrown GaN thin film [44]. Figure 8.12(b) shows the distribution of GNDs derived from the measurement of tilt, twist and strain.

As for ECCI, EBSD has mostly been used to investigate the structural properties of metals and geological materials; however, with the development of cross-correlation HR-EBSD, its use for the characterisation of semiconductors is steadily expanding. For example, it has been used to study crystal orientation and/or strain in Si [45], Ge [46], SiGe [47] and nitride semiconductors [22,44,48,49]. A detailed description of this technique for the study of InAlN thin films and its correlation with ECCI can be found in [49].

The cross-correlation of experimental EBSPs with simulated patterns has also enabled EBSD to reveal the presence of polytypes, e.g. wurtzite and zincblende GaN in the same sample [50], antiphase domains [51] or inversion twinning [52]. For example, Figure 8.13(a) shows an EBSD-derived map of a GaP thin film containing antiphase domains [51]. Antiphase domains arise in zincblende-type crystals due to their non-centrosymmetric nature. The zincblende crystal structure is not invariant to a rotation of 90° around the [0 0 1] direction, because the arrangement of the anions and cations in the (1 1 1) plane is inverted relative to that of the (1 $\bar{1}$ 1) plane. This is illustrated in Figure 8.13(b). Figure 8.13(a) shows an EBSD-derived map from a GaP thin film revealing domains (green and blue) that are rotated by 90° relative to each other around the [0 0 1] surface normal, corresponding to an interchange between the [1 1 1] and [1 $\bar{1}$ 1] directions. The orientation of each point of the map was obtained

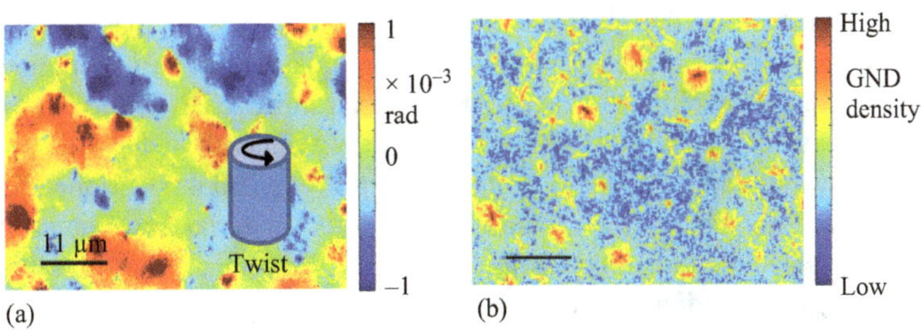

Figure 8.12 (a) A cross-correlation HR-EBSD map of twist from a patterned epitaxially overgrown GaN thin film. (b) GND distributions derived from the measurement of tilt, twist and strain using cross-correlation HR-EBSD. Maps courtesy of A. Vilalta-Clemente, A. J. Wilkinson and G. Naresh-Kumar

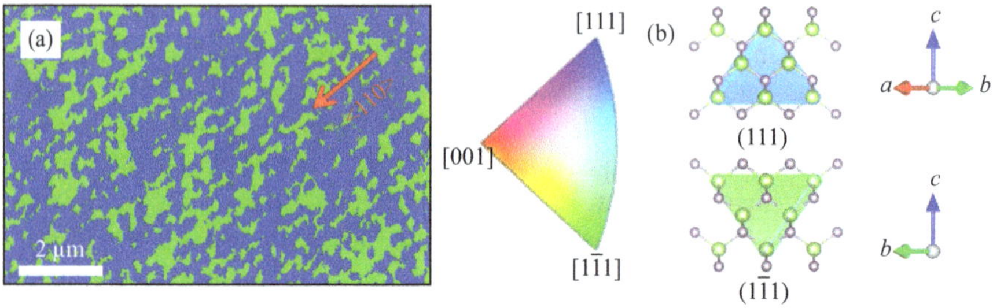

Figure 8.13 (a) EBSD-derived map revealing regions exhibiting different orientations of the zincblende lattice with respect to the [0 0 1] direction in a GaP thin film (antiphase domains). The antiphase domains are identified from cross-correlation of each experimental pattern in the map with dynamical simulations. © 2017. Reprinted, with permission, from [51] under Creative Commons Attribution (CC BY) license: http://creativecommons.org/licenses/by/4.0/. (b) Illustrating the different stacking of the Ga and phosphorous atoms in the GaP lattice for the (1 1 1) and (1 $\bar{1}$ 1). The images were generated using VESTA 3 [53]

from cross-correlation of the acquired EBSP with simulated patterns. The alignment of the antiphase domains along the ⟨1 1 0⟩ direction is due to a small miscut of the Si substrate [51].

8.2.2.4 Cathodoluminescence

Cathodoluminescence (CL) is the process of light generation by an impinging beam of electrons on a material [54]. CL reveals the influence of crystal structure, composition and strain on intrinsic luminescence and/or reveals defect or dopant-related luminescence. The primary electrons lose energy through impact ionisation, which is an inelastic process, generating electron–hole pairs. These pairs can then recombine radiatively generating light emission. It is important to note that in the case of semiconductors, the electron energy is a few orders of magnitude larger than the band gap, which makes it possible to even look at wide band gap materials, such as diamond, aluminium nitride (AlN) or boron nitride (BN). There are, however, short wavelength limits imposed on the wavelength detection, such as absorption by water vapour at a wavelengths shorter than 200 nm (the spectrometer and detector are not inside the vacuum system), the material and quantum efficiency of the detector and other optical elements (e.g. absorption of ultraviolet (UV) light by glass windows or lenses).

The previously mentioned interaction volume (Section 8.2.2.1) defines the lateral dimension and the depth at which the electron beam generates electron–hole pairs. After generation, these electron–hole pairs can diffuse before recombination through non-radiative or radiative processes. Their diffusion length increases the volume where the CL signal is generated compared with the volume within which the electron–hole pairs are formed. In GaN [55,56], the diffusion length of the charge carriers is about 50–60 nm, whereas in GaAs, it can be several microns dependent on the doping concentration [57,58]. The knowledge of the size of the interaction volume and which electron beam parameters influence it are crucial for CL imaging when it comes to depth information and spatial resolution. The interaction volume in a solid can be simulated using Monte Carlo methods, for example, CASINO [59] or NIST DTSA-II (https://www.cstl.nist.gov/div837/837.02/epq/dtsa2/index.html).

The spatial resolution is mostly influenced by the size of the electron beam spot, the excitation volume and the carrier diffusion length in the studied material [16]. The highest spatial resolution for bulk material will, therefore, be ultimately limited by the diffusion length. The diffusion length, though, is a bulk property. Confinement within nanostructures, such as quantum wells (QWs), small surface features or defects, further limits the diffusion of the generated carriers by either increasing or decreasing the probability of radiative or non-radiative recombination. This allows features several orders of magnitude smaller than the diffusion length to be observed; for example, 20 nm spaced QWs in GaAs have been resolved [60] or peak shifts in the emission from InGaN/GaN QWs over an $\approx$10 nm distance have been measured [61]. Other resolution-limiting effects include surface contamination by carbon, charging of the sample, mechanical vibrations and electrical noise [60].

Conventional CL imaging either records a panchromatic or monochromatic CL image. For panchromatic CL, the total luminescence intensity is measured, whereas for monochromatic CL, the intensity within a specific wavelength interval is chosen. By contrast, CL hyperspectral imaging is a more powerful technique in which an entire luminescence spectrum is recorded at each pixel in an image created by a scanning electron beam [62,63]. The resultant multidimensional data set can be treated mathematically to extract images describing subsets of the recorded data. For example, numerical peak fitting can produce 2D CL images of the energy peak position, peak area and peak half-width illustrating the spatial luminescence behaviour of the sample under investigation. Another mathematical tool is principal component analysis, a multivariate statistical analysis technique, which is useful for obtaining sample information when dealing with low signal-to-noise spectra [64]. An alternative way to build up a hyperspectral CL data set is by recording a large set of CL intensity images at different wavelengths, where the wavelength is selected by using a monochromator and the light is detected by a photomultiplier tube [65].

TDs that give rise to non-radiative recombination are observed as dark spots in a CL image. If the excitation volume is sufficiently small, the size of the dark spots is determined by the minority carrier diffusion length [32]. Figure 8.14 shows a CL intensity image from GaN where dislocations are revealed as dark spots.

Figure 8.15 shows the CL data obtained through peak fitting from an InGaN/GaN multiple QW (MQW) structure. The surface exhibits a high density of V-shaped pits (V-defects), common in this type of structure, caused by TDs propagating to the surface and opening up into a pit at one of the QW interfaces. Apart from the previously mentioned non-radiative recombination at the TDs reducing the CL intensity, the peak energy map exhibits a blueshift, where a pit is located

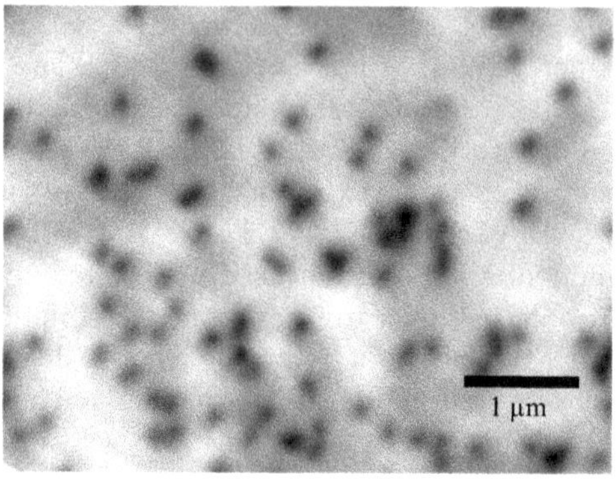

Figure 8.14 *CL intensity image of the near band-edge emission in a GaN layer. The dark spots are related to non-radiative recombination at threading dislocations*

compared with the surrounding planar areas. This blueshift may be caused by a change in InN content or thinner QWs on the facets of the V-pits [61]. The energy shift around the pits is most likely related to strain variations caused by the presence of the dislocations.

Other examples for CL imaging of defects include stacking faults [66,67] and misfit dislocations in GaN [68], dislocations in diamond [69] and GaInP [70] and grain boundaries in silicon [71].

8.2.2.5 Electron beam-induced current

Electron beam-induced current (EBIC) describes a technique where a measureable current is generated when a high-energy electron beam is incident on a sample and electron–hole pairs are created. These electron–hole pairs can undergo three mutually exclusive processes. They can recombine either radiatively (producing CL) or non-radiatively, or they can be swept apart towards electrical contacts and produce a current, which is EBIC [12]. The last process can only occur if there is an internal electric field present in the sample, such as in p–n or Schottky diodes or light-emitting diodes (LEDs), leading to carrier drift towards the contacts. This current can be measured using a current amplifier. Although EBIC is used as a standalone technique for the investigation of recombination properties of defects, it can be combined with CL to provide complementary information [71,72]. The generation of electron–hole pairs can be described by the generation rate G that is a combination of the non-radiative and radiative recombination rates R_{NR} and R_R, respectively, and the measured current I_{EBIC} as follows:

$$G = R_{NR} + R_R + I_{EBIC}/q.$$

This indirectly provides information on non-radiative recombination processes since the CL provides a measure of the radiative recombination and the EBIC

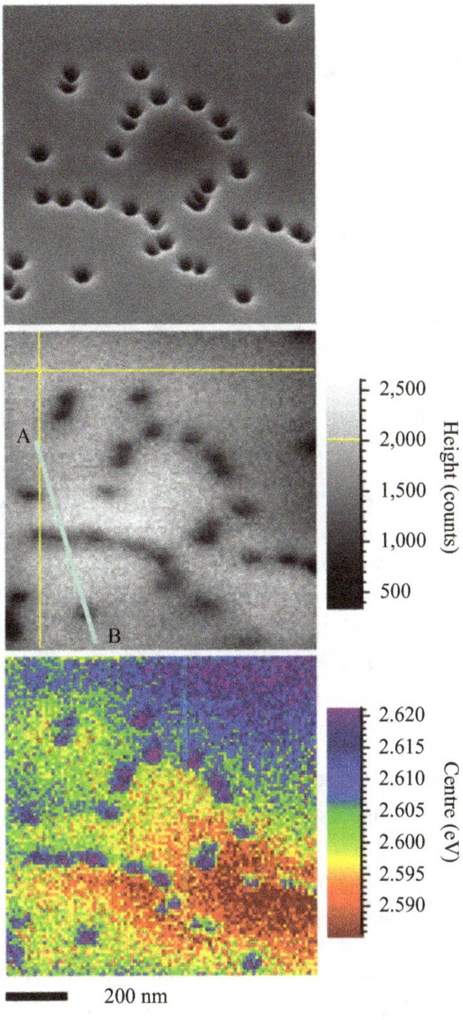

Figure 8.15 SE micrograph (top), CL intensity (middle) and peak energy image (bottom) of the MQW emission of an InGaN/GaN MQW structure. © 2012. Adapted and reprinted, with permission, from [19]

directly measures the current [73]. The resolution of EBIC is similar to CL as discussed in Section 8.2.2.4 since the same electron–hole pairs are involved in this process. However, knowledge of the structure and in particular the depth of the space charge region (caused by the internal electric field) are important for the choice of beam conditions which in turn has an influence on the resolution [17].

An example of EBIC recorded together with the CL of a blue-emitting LED consisting of InGaN/GaN MQWs is shown in Figure 8.16 [74]. The CL intensity image

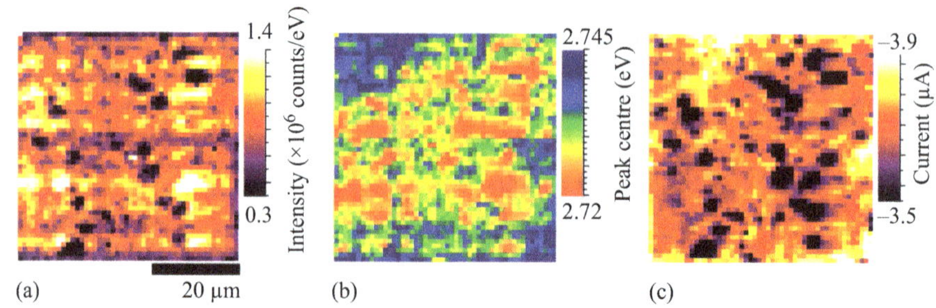

Figure 8.16 (a) CL intensity and (b) MQW peak energy images and (c) EBIC image of an InGaN/GaN LED. © 2015. Adapted and reprinted, with permission, from under Creative Commons Attribution (CC BY) license: http://creativecommons.org/licenses/by/4.0/

and EBIC image both exhibit dark spots which coincide at the same position. Since the dark spots in the CL image appear in the same location of the corresponding EBIC image, they are not caused by carrier escape from the junction but rather non-radiative recombination at defects, possibly clusters of point defects in the MQW region.

Other examples for EBIC include the investigation of grain boundaries in silicon [72], stacking faults in 4*H*-SiC [75], measurement of the diffusion length in GaN [76,77] and β-Ga_2O_3 [78], recombination behaviour of dislocations in bulk GaN [79] and solar cells [80].

8.2.2.6 Energy- and wavelength-dispersive X-ray spectroscopy

Energy- and wavelength-dispersive X-ray spectroscopy (EDX/WDX) are two similar techniques for measuring X-rays, which only differ in how the same emitted X-rays are detected. When a high-energy electron beam strikes a material, it can cause the ejection of an inner shell electron. This hole can then be filled through relaxation of an electron from a higher energetic state. The energy the electron has to lose can result in the emission of a characteristic X-ray. The X-ray energies are specific to the atomic structure of an element and its energy levels [81]. The detection of these X-rays, therefore, allows the elemental identification that is used for compositional analysis of materials [13].

In WDX, the X-rays emitted from the material are dispersed to different angles depending on their energy using a diffracting crystal according to Bragg's law. The detector, generally a gas proportional counter or scintillation counter, only detects X-rays of one energy at a time and the angle of diffraction has to be changed in order to record a whole X-ray spectrum, similar to a monochromator for light detection. In EDX, the X-rays are directly measured using a solid-state silicon detector. The X-rays impinging on the silicon generate electron–holes pairs, the number of which is proportional to the X-ray energy. WDX has a much higher energy resolution compared with EDX, but it is more time consuming to generate an entire X-ray spectrum. EDX on the other hand provides an almost instantaneous X-ray spectrum, but with lower resolution and more noise due to the electronics.

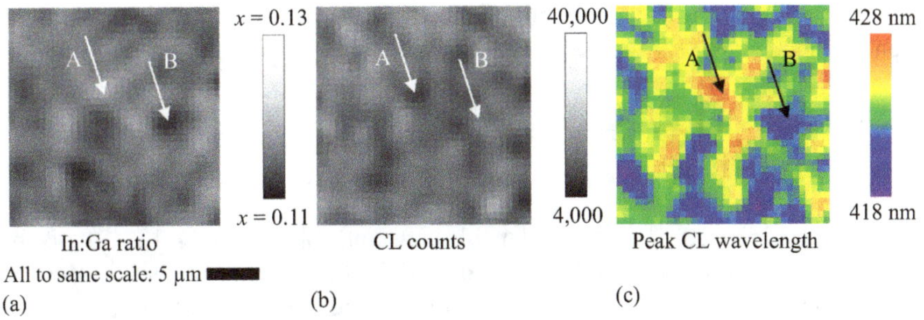

Figure 8.17 *(a) In:Ga ratio calculated from In and Ga WDX maps, (b) CL intensity and (c) peak wavelength maps acquired simultaneously from an InGaN epilayers. © 2003. Modified and reprinted, with permission, from [83]*

A comprehensive review of X-ray spectroscopy and its principles can be found in Goldstein *et al.* [13].

Quantitative measurements of the chemical composition of identified elements are generally carried out by WDX in an electron probe microanalyser (EPMA). That is an instrument similar to an SEM, but a dedicated machine with numerous WDX spectrometers containing different diffracting crystals in order to cover a range of X-ray energies. For quantitative analysis, the results of the sample under investigation have to be compared against standards of known composition and matrix corrections need to be applied taking additional effects (e.g. material density, absorption, energy loss) into account that could lead to errors in the results [12]. The spatial resolution mainly depends on the size of the excitation volume, which is defined by the electron beam energy. This means that there is a trade-off between the highest spatial resolution and the necessary, minimum beam energy for excitation of X-rays [82].

Figure 8.17(a) shows an example of a composition image of the In:Ga ratio, which was calculated from WDX elemental images of indium and gallium for an InGaN epilayer [83]. The CL was simultaneously recorded and CL images of the intensity and peak wavelength of the InGaN near band-edge emission are displayed in Figure 8.17(b) and (c), respectively. Local changes in composition are directly evidenced by an energy shift and intensity change since the band gap is a function of alloy composition as observed in points A and B corresponding to indium-rich and poor regions, respectively.

8.2.3 Transmission electron microscopy

8.2.3.1 Brief introduction to transmission electron microscopy

Electron-matter interaction in a TEM

A TEM, akin to an SEM, is based on the interaction of a beam of electrons accelerated to a high voltage (typically 80–400 kV for a TEM) within a sample. Using a relativistic approach, the wavelength of electrons accelerated to a high voltage is given by the formula.

$$\lambda = \frac{h}{\sqrt{2m_0eV\left(1+\frac{eV}{2m_0c^2}\right)}}$$

where $h = 6.626 \times 10^{-34}$ Js is the Planck's constant, $m_0 = 9.109 \times 10^{-31}$ kg is the rest mass of the electron, $e = 1.602 \times 10^{-19}$ C is the elementary charge, V is the accelerating voltage and $c = 2.998 \times 10^8$ m/s is the speed of light in a vacuum. This expression shows that the wavelength, and consequently the theoretical resolution of the microscope, decreases with increasing voltage. For a conventional TEM, with an accelerating voltage of 80–400 kV, the wavelength of the electrons ranges between 1.6 and 4.2 pm which seems well-suited for the observation of structures at the atomic scale. Nevertheless, it should be noted that in practice, it is not the electron wavelength that limits the resolution of the TEM, but instead aberrations in the electromagnetic lenses. In fact, the best microscopes currently used for atomic resolution images employ a rather low voltage (i.e. 60–200 kV), but some (or several) lens aberrations are corrected.

An electron, accelerated to a high voltage, will interact with the atoms constituting the sample in several ways, either by interaction with the atomic nucleus or with the electron cloud. The interaction of the incident electron with the nucleus is usually considered to be elastic (i.e. resulting in no loss of energy for the incident electron), whilst interaction with the electron cloud is usually inelastic (i.e. the incident electron transmits part of its energy to the material). Each of these interactions gives rise to different signals, which contain information about the material, and which are outlined in Figures 8.18 and 8.19. One needs to bear in mind that in a TEM, the material exposed to the incident beam consists of a thin foil, as opposed to the bulk material typically imaged in an SEM, and the electrons used in imaging

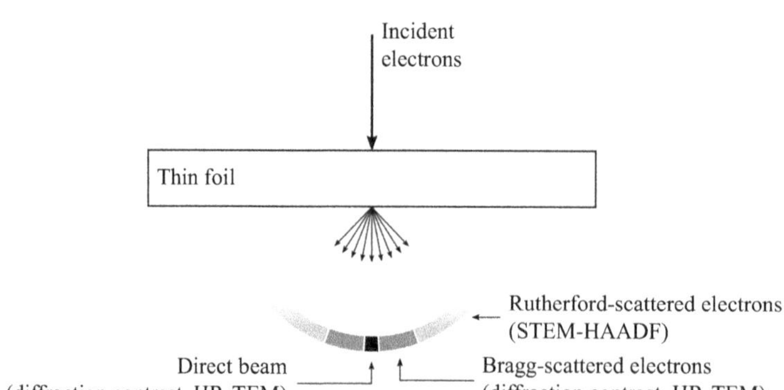

Figure 8.18 Diagram of the signals generated (and imaging techniques using these signals) by the elastic interaction between a high-energy electron beam and a thin sample

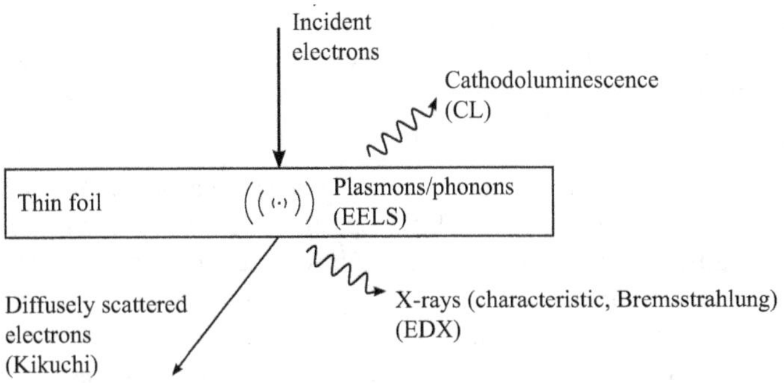

Figure 8.19 Schematic of the signals generated by the inelastic interaction between a high-energy electron beam and a thin sample

are transmitted through that thin foil. Hence, the signals relevant to a TEM differ from those outlined for SEM as described in Figure 8.4 in Section 8.2.2.1.

Elastic interaction of an incident electron with the material arises from the interaction of the electrons with the nucleus of the atoms. We will see later that this interaction is not *strictly* elastic. The incident electron trajectory is then deflected when approaching the nucleus. However, different scattering mechanisms result in an incident electron being deflected to different scattering angles. Hence, the scattering cross section, σ, which is the probability of a scattering event occurring, is related to the scattering angle. Because the TEM foil is thin, some of the incident electrons are not scattered and a direct, unscattered beam is transmitted through the sample. Scattering at low angles (15–50 mrad) is dominated by Bragg scattering – which is the same phenomenon used in X-ray diffraction and ECCI as described in Section 8.2.2.2. Elastic scattering at such low angles (15–50 mrad) is usually coherent (i.e. the electrons are in phase). Bragg scattering is controlled by the structure of the sample and results in a diffraction pattern. These electrons will be used for diffraction contrast and high-resolution transmission electron microscopy (HR-TEM) imaging, as we will describe later. Scattering at higher angles (>50 mrad) is usually incoherent and follows the Rutherford scattering rule $\sigma_{Ruth} = 1.62 \times 10^{-24}(Z^2/E_0^2)\,\mathrm{cotan}^2(\theta/2)$, where Z is the atomic number of the scattering atoms, E_0 is the incident electron energy and θ is the deviation angle. Since σ_{Ruth} is proportional to Z^2, Rutherford scattering is exploited in scanning transmission electron microscopy with a high-angle annular dark-field detector (STEM-HAADF), which provides Z-contrast images as described in Section 8.2.3.3.

Inelastic scattering corresponds to the interaction of the incident electron with the electron clouds of the sample atoms. The incident electron trajectory is also deviated, but at much lower angles (<15 mrad) than the deviations caused by the nucleus. The energy transferred to the material can result in various effects, which are summarised in Figure 8.19. Of relevance in this chapter are the diffusely scattered electrons and X-rays. Diffuse scattering results in the formation of Kikuchi bands, which are very useful when orienting crystalline samples in the TEM. X-rays are generated by electronic transitions in the atoms and are used for EDX spectroscopy. Other signals that could be of interest, but not discussed here, are CL and plasmons and phonons which result in energy losses used in electron energy loss spectroscopy (EELS).

Beam damage

The high-energy electrons used in a TEM can damage the sample. Depending on the type of interaction of the electron with the specimen (elastic or inelastic), different damage mechanisms can occur. Atomic displacement, beam sputtering, specimen heating, specimen charging, structural damage, mass loss and hydrocarbon contamination have been reported in different types of materials [84]. In this section, we will pay special attention to the mechanisms of atomic displacement and beam sputtering as they are less straightforward, and therefore more difficult to identify.

Atomic displacement originates from the elastic interaction between an incident electron and the nucleus of an atom in the sample. This interaction, in spite of being considered elastic (as we mentioned earlier), results in a transfer of energy from the electron to the nucleus. This transferred energy, small compared to the electron energy, is given by the formula.

$$E = E_{\text{max}} \sin^2\left(\frac{\theta}{2}\right)$$

where E is the energy transferred to the nucleus (in eV), E_{max} is the maximum energy that can be transferred corresponding to a 180° electron deviation (in eV) and θ is the incident electron trajectory deviation. The maximum energy E_{max} can be written as a function of the incident electron energy E_0 as follows [84,85]:

$$\begin{aligned} E_{\text{max}} &= E_0 \frac{E_0 + 2m_0c^2}{E_0 + (1 + (m_0/M))^2(Mc^2/2)} \approx 2E_0 \frac{E_0 + 2m_0c^2}{Mc^2} \\ &\approx E_0 \frac{1.02 + E_0 10^{-6}}{465.7A} \end{aligned}$$

where m_0, M and A are the rest mass of the electron, the nucleus mass and the atomic mass number of the nucleus, respectively. E_{max} is generally rather low – of the order of a few eV – but can reach several tens of eV for light nuclei (e.g. B, C, N, O) and/or high beam energy. Figure 8.20 illustrates the variation of E_{max} with E_0 for atoms that compose common semiconductor materials. If the transferred energy exceeds the displacement energy of the nucleus, E_d, the atom can be moved from its initial location in the crystal, thus creating a Frenkel pair (i.e. a vacancy and an interstitial) in the crystal. The displacement energy E_d depends strongly on the nucleus itself, on the nature and strength of the bond of this nucleus in the crystal and on the crystallographic direction in which the nucleus is to be extracted. The values of displacement E_d can often be obtained using molecular dynamics simulations – a rough order of magnitude would be in the range 10–100 eV.

Atomic displacement can also happen through sputtering. Beam sputtering involves atoms close to the sample surface so that a displaced nucleus can be ejected from the sample. Given that the nucleus does not have to be relocated to an

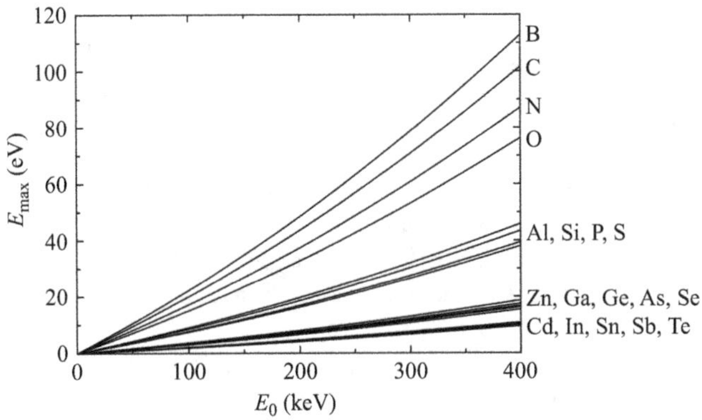

Figure 8.20 Maximum energy transferred to the sample E_{max} as a function of the incident beam energy E_0 for a range of common atoms in semiconductors

interstitial site in the crystal, this mechanism requires much less energy than atomic displacement into the 'bulk' crystal. In such a case, the displacement energy E_d can be expressed as a function of the sublimation (i.e. transition from the solid phase to the gas phase) energy E_{sub} of the material. Egerton *et al.* [86] showed that a reasonable approximation of E_d can be given by $E_d = (5/3)E_{sub}$. Given that the sublimation energy of most materials is of the order of a few eV, it appears quite likely that most samples are susceptible to beam sputtering in the TEM. Sputtering will need to be particularly carefully considered in TEM studies of the emerging 2D materials.

In both cases – atomic displacement and beam sputtering – one can see that it is possible to avoid electron beam damage by reducing the acceleration voltage of the microscope below the knock-on voltage. If this is not possible – if the gun does not allow such a low voltage, or if the lower voltage results in a poor image contrast – one can try to reduce the radiation dose by limiting the specimen exposure time. Another potential solution, assuming that beam sputtering is the dominant beam damage mechanism, would be to coat the exit surface of the TEM foil with a heavy element [84].

Sample preparation

In order to conduct a TEM experiment, the sample has to meet two key requirements: the sample must be electron transparent and the sample must fit in the TEM sample holder. These conditions imply that the sample must be thinner than 200 nm approximately and that the sample must be mounted on top of a 3.05-mm diameter ring. Therefore, one of the important parts of a TEM experiment is sample preparation. Different methods exist for preparing samples (mechanical polishing, focussed ion beam (FIB), electropolishing, chemical etching, ultramicrotome, etc.) depending on the material properties. The most commonly applied in studies of semiconductor defects are mechanical polishing and FIB. Box 8.1 outlines a typical procedure for sample preparation by mechanical polishing, with some tips to help ensure successful results.

Box 8.1: Typical procedure for preparation of TEM samples by mechanical polishing, with some tips and tricks

- Using a diamond scribe and/or a diamond saw, cut two rectangles of typical size 3 mm × 1.5 mm from the original sample.
  Tip: cut along specific crystallographic directions – this will facilitate finding the zone-axis during the TEM experiment.
- Stick the top surfaces of the two rectangles to each other using an epoxy resin.
  Tips: (1) Make the glue line as thin as possible to avoid damaging the front surface during the ion milling step. (2) Make sure that the sample can sit on a support stub on its own with the glue line perpendicular to the support. This ensures that the electron beam will be almost parallel to a zone-axis once in the TEM (some microscopes do not allow for large sample tilts).
- Mount the sample on a stub using thermoplastic wax.
- Grind the first side of the sample using SiC or diamond abrasive discs until a flat surface is obtained (take special care to avoid any chipping near the glue line).
  Tips: (1) Gradually decrease the grain size of the disc. (2) Make sure that for a given disk the sample thickness does not fall below ~2–3 times the grain size of the disk – this will ensure that there will be no damage from grinding in the final foil.
- Polish the first side of the sample using a cotton wheel in a dimpler until no scratches are visible in an optical microscope.
- Flip the sample and repeat the grinding/polishing procedure until the sample thickness reaches 20–40 μm approximately.
  Tips: Use a P400 disc (average grain 35 μm) down to a sample thickness of 80–100 μm, then P600 (average grain 25.8 μm) down to 50–70 μm and finally P800 (average grain 21.8 μm) down to 40–60 μm. Final polishing down to 20–40 μm is made with a cotton wheel in the dimpler.
- Stick a 3.05-mm TEM ring onto the sample.
- Mill the sample using an ion milling system until a hole appears in the sample. The ion milling system uses two Ar^+ ion guns operating between 0.1 and 6 kV accelerating voltage with an incident angle up to $\pm 10°$.
  Tips: (1) Position the sample with the glue line on the rotation axis of the ion mill stage. (2) Operate the system at 5–6 kV until colourful fringes start appearing around the glue line, then decrease the voltage down to 3 kV until a hole forms near the glue line, finally, decrease the voltage down to 0.1–2 kV for 15 min to clean the surface from high-energy ion mill damage.

Preparation by FIB involves putting the sample in an SEM equipped with an electron gun and a Ga^{3+} ion gun, as well as a source for Pt deposition – other elements (often C, W) can also be used. The electron gun is usually operated at fairly low voltage (5 kV) and is used for imaging, while the ion gun operates at much higher voltage (often 30 kV) and is used for milling. Imaging with the ion gun is also possible but one has to minimise the exposure to the beam as it damages the sample (as could be expected from a beam whose primary function is to mill the sample). In essence, FIB aims to excavate a thin slice out of the bulk sample, and a typical procedure, with some helpful tips, is given in Box 8.2.

Box 8.2: Typical procedure for preparation of TEM samples using FIB, with some tips and tricks

- Deposit a protective Pt layer on the surface of the sample in order to protect it from the high-energy ion beam.
  Tip: To minimise damage, first deposit Pt using the electron gun, then with the ion gun.
- Using the ion gun for removing material, and the electron gun for monitoring, mill two ditches, one on each side of the Pt layer.
- Now that the sides of the sample slice are exposed, make a cut underneath – this will separate the slice from the bulk sample.
- Extract the slice from the bulk sample using a sharp needle-shape probe, and transfer the sample to a TEM grid.
- Gradually thin down the sample slice from both sides gradually reducing the ion beam current (in order to remove previous beam damage) until electron transparency thickness (~200 nm).
  Tip: For best results (to make a sample suitable for atomic-scale imaging), perform a final cleaning of the sample using the ion gun at a lower voltage, or using a separate ion milling system (such as the system mentioned for mechanical preparation in Box 8.1) at 0.1–1 kV.

Transmission electron microscopy equipment

Although the exact architecture of a TEM will depend on its mode of operation (e.g. parallel beam, convergent beam), most TEMs are built with a common arrangement. Figure 8.21 presents the basic architecture for a TEM operating in parallel beam illumination, and each component is described next.

Electron gun: the most common types of electron guns in recent TEMs are the LaB_6 filament and the field emission gun (FEG). Although the emission mechanisms are different for these two types of guns – thermionic (LaB_6) or field emission (FEG) – the main idea is that an electron source is placed at the cathode with respect to two anodes (or Wehnelt (i.e. a 'grid' which acts as an electrostatic lens)

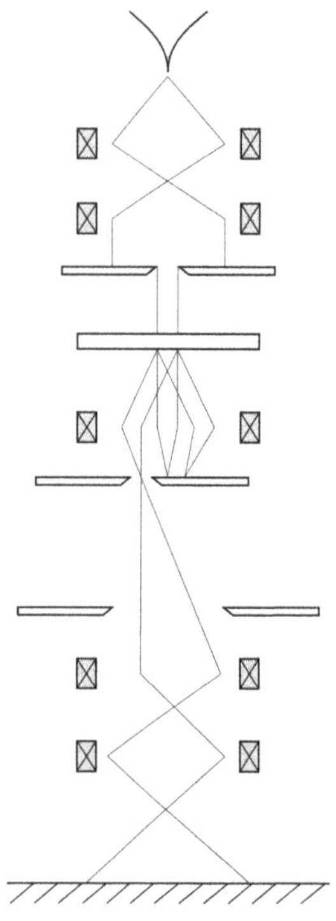

Figure 8.21 Schematic of a parallel beam TEM. To illustrate the principle behind diffraction contrast imaging, here an objective aperture is positioned around one of the diffracted beams, thus forming a dark-field image (in practice, the objective aperture is centred on the optic axis and the incident beam is tilted)

plus anode in the case of a thermionic gun). The electrons are then extracted from the source (cathode) and accelerated by the anode to the operating voltage of the microscope. An undeniable advantage of FEGs over LaB_6 filaments is that electrons emanate from a much smaller region of the cathode, leading to a smaller, brighter and more coherent source.

Condenser lens 1: the first condenser lens is used to demagnify (or magnify in the case of a FEG) the electron source.

Condenser lens 2 (C2): the second condenser lens is important in that it determines the area of exposure of the specimen. By underfocussing C2, the sample is illuminated by a quasi-parallel beam. By focussing C2, the sample is illuminated by a convergent beam (in some microscopes, a lens called the 'upper objective lens' is inserted between C2 and the sample to further decrease the convergence angle). A parallel incident beam is used to achieve a coherent electron beam – which gives rise to sharp diffraction patterns. When the electron beam is made convergent, the electron probe can be rastered across the surface (hence the appellation scanning transmission electron microscopy or STEM) with a resolution close to the probe size (related to the convergence angle). In aberration-corrected microscopes, the C2 lens is followed by aberration correctors that are essentially a set of additional multipole lenses whose functionality is to correct a specific type of aberration – e.g. astigmatism, spherical aberration, chromatic aberrations.

Condenser aperture: along with the second condenser lens, an aperture can be inserted in order to control the illumination of the specimen as well as the convergence angle of the beam.

Sample: the sample is mounted in the sample holder which can be translated in the three spatial directions and tilted around the two axes perpendicular to the optic axis.

Objective lens: an objective lens is placed after the sample in such a way that the sample is in the focal plane of the lens. The distance between the sample and the objective lens is called 'eucentric height'.

Objective aperture: the objective aperture is located in the focal plane of the objective lens. Consequently, the objective aperture is placed where the diffraction pattern of the sample forms. This aperture is used to select a particular set of diffraction conditions (and therefore a particular set of information about the sample), thus enabling particular imaging conditions which will be described later.

Selected area aperture: the selective area aperture is placed in the image plane of the objective lens. It allows one to virtually select which part of the sample is illuminated.

Intermediate and projector lenses: the intermediate and projector lenses are used to select which mode, between imaging and diffraction, will be observed on the viewing screen. When imaging mode is selected, the lenses are adjusted so that the viewing screen is in the image plane of the combination of the two lenses. Similarly, when diffraction mode is selected, the lenses are adjusted so that the viewing screen is in the focal plane of the combination of the two lenses. Using two lenses enables the magnification of the image or diffraction pattern.

Viewing screen: the electrons thus selected can finally hit the phosphor screen leading to observable fluorescence. To record an image, the viewing screen can be lifted to allow the beam to hit a photographic film or a camera.

8.2.3.2 Techniques for diffraction contrast imaging of defects – bright field, dark field, weak beam dark field

Diffraction contrast imaging is achieved by viewing the diffraction pattern of the specimen and using the objective aperture to select only one particular diffraction spot. Hence, only the planes diffracting the beam in this particular way will appear on the viewing screen. Depending on the beam going through the objective aperture (either the direct beam or a scattered beam) and on the sample orientation allowing the excitation of different diffraction conditions, this type of imaging is qualified as bright-field imaging, dark-field imaging or weak beam dark-field (WBDF) imaging. These imaging modes are mainly used for the characterisation of defects as illustrated in Figure 8.22. The application of this technique to the characterisation of dislocations will be further discussed in Section 8.3.1.4. In theory, diffraction contrast

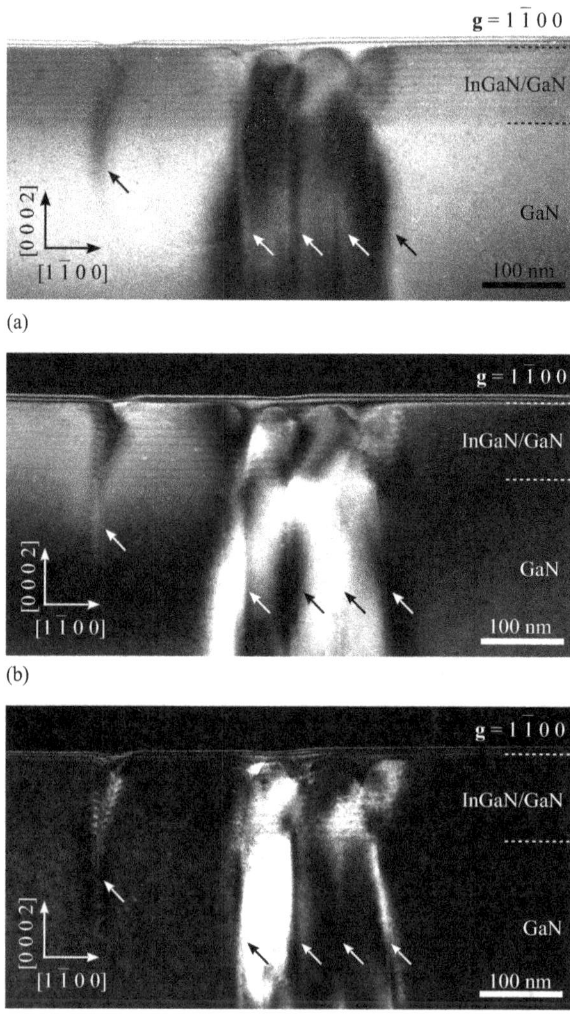

Figure 8.22 Bright field (a), dark field (b) and WBDF (c) TEM images of an InGaN/GaN QW structure observed along the $\langle 1\ 1\ \bar{2}\ 0\rangle$ *zone-axis using* $\mathbf{g} = 1\ \bar{1}\ 0\ 0$*. Dislocations in each image are indicated by an arrow*

imaging is possible regardless of the sample orientation, as long as a diffraction pattern can be obtained. In practice, the contrast of the image can be strongly increased if the sample is tilted away from the zone-axis, along the Kikuchi band associated with the selected reflection. This results in the sample being in 'two-beam conditions', where only the direct beam and one diffracted beam are strongly excited. It is also recommended to select a deviation parameter $\boldsymbol{s} > 0$ in order to further improve the contrast. (*s* is also called the excitation error and describes, in the reciprocal space, how far away the system is from the Bragg conditions.)

Bright-field imaging is obtained when an objective aperture is used to select only the direct beam. Given that the direct beam contains the information about the planes that do not diffract the incident beam, all the material which does not diffract will appear bright, whilst regions of material where the beam is scattered (e.g. defects) will appear dark (Figure 8.22(a)).

Dark-field imaging is obtained when an objective aperture is used to select a diffracted beam. One issue with dark-field imaging as can be observed in Figure 8.21 is that the selected beam is off-axis and therefore subject to aberrations by the intermediate and projector lenses. In most TEMs, this is solved by tilting the incident beam adequately so that the diffracted beam is kept on-axis. When the sample is in two-beam conditions, the majority of electrons not contributing to the direct beam are contributing to this particular diffracted beam. Therefore, a dark-field picture results in the negative of a bright-field picture, with all the diffracting planes appearing brighter (Figure 8.22(b)).

WBDF imaging occurs when the same conditions as for dark field are fulfilled, but the sample is tilted slightly further (perpendicularly to the Kikuchi band) to increase the deviation parameter **s**. In that case, the diffracted beam **g** becomes weak, and the higher order diffraction spot 3**g** (or 5**g**) becomes stronger. Like in dark field, an objective aperture is put around **g**. The resulting image is similar to a dark-field picture except that only the planes that are strongly diffracting appear (Figure 8.22(c)). This technique allows us to have a better resolution when observing dislocations, for example, only the planes closest to the core will appear. Therefore, dislocations appear much thinner, i.e. dislocation densities in high dislocation density films can be counted more easily and their Burgers vectors found.

8.2.3.3 Techniques for compositional analysis near defects – high-angle annular dark-field STEM, EDX

High-angle annular dark-field STEM

We mentioned earlier that high-angle scattering (>50 mrad) of the electron beam by the nuclei of atoms is dictated by Rutherford scattering, with the cross section proportional to Z^2. STEM-HAADF is built around this effect. By using an eponymous annular detector that collects electrons scattered at a high angle, a *Z*-contrast image is generated as the convergent probe is scanned across the sample. The term dark field originates from the fact that only diffracted electrons contribute to the image. STEM-HAADF is therefore suited for the study of compositional variations across the sample, where regions of higher average atomic number appear brighter than regions of lower average atomic number (as illustrated by the bright contrast from InGaN QWs in a GaN matrix in Figure 8.23). However, STEM-HAADF cannot be straightforwardly used to quantify the composition. Other techniques (e.g. EDX) often embedded in the same microscope are usually used for this purpose. One also has to bear in mind that the thickness of the TEM foil will affect the image contrast – thinner regions appear darker than thicker regions – and that strain

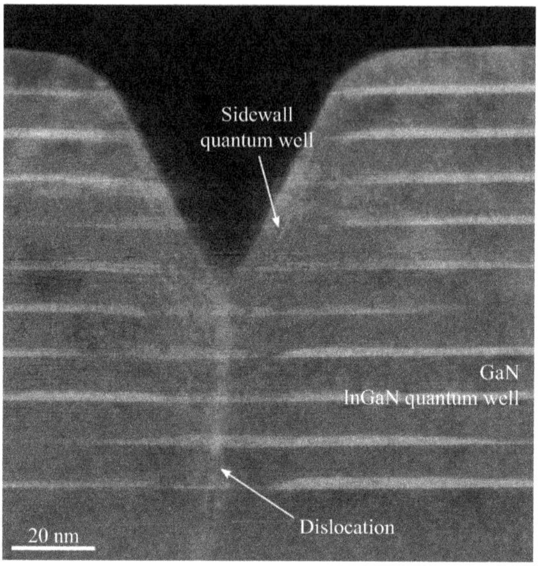

Figure 8.23 STEM-HAADF image of a dislocation in an InGaN/GaN MQW sample. The QWs appear brighter due to the greater atomic number of InGaN over GaN. QWs can also be seen forming on the sidewalls of the pit formed at the termination of the dislocation. The dislocation is also visible indicating strain-related contrast

can also influence the contrast if the detector does not collect solely electrons diffracted at a sufficiently large angle (illustrated by contrast arising at a dislocation in Figure 8.23).

Energy-dispersive X-ray spectroscopy

Unlike STEM-HAADF that provides essentially qualitative compositional information, EDX is in essence quantitative. Provided that sufficiently high signal can be detected, the atomic species, present in the sample and their fraction, can be determined. The atomistic mechanism for the generation of the EDX signal is explained in Section 8.2.2.6. Here, we add some comments about typical quantification routes used in transmission electron microscopy. Quantification of the proportion of each species can be obtained by analysis of the intensity of the peaks relating to different elements in the X-ray spectrum. Most transmission electron microscopy analysis software employs the Cliff–Lorimer method [87]. EDX can, therefore, be a useful technique to investigate compositional fluctuations at defects, for example, in order to investigate segregation phenomena. However, a common mistake is to overestimate the capabilities of the technique. For example, as we explicitly mentioned, the signal needs to be sufficiently high to be able to extract information from a spectrum, which can be more difficult to achieve when the signal arises from a small volume in a thin TEM foil. Although the systems are continuously being improved, minute amounts of impurities in the material, as is the case for semiconductor dopants, will not be detected. Unfortunately, trying to increase the EDX signal often results in an increase of the electron dose in the sample, which can damage the sample. Furthermore, the Cliff–Lorimer analysis does rely on a calibration factor, termed the Cliff–Lorimer factor, which can be derived from standard samples, but which can vary with the foil thickness and electron beam energy among other factors. This can limit the accuracy of the quantification unless treated with great care.

8.2.3.4 Techniques for high-resolution imaging of defects – high-resolution transmission electron microscopy and scanning transmission electron microscopy

HR-TEM

In the previous section, we have seen that diffraction contrast arises from selecting a single diffraction spot using the objective aperture. Thus, only the information conveyed by the selected diffraction condition – a specific set of planes in Bragg's condition – will contribute to the image. However, if one uses a larger objective aperture, or no aperture at all, several diffraction spots can contribute to the image. For example, in a configuration where two (**0g** and **g**) or three beams (−**g**, **0g** and **g**) are allowed through the aperture, one can visualise on the viewing screen a set of B–W fringes resulting from the interference of these beams. Although these fringes cannot be directly interpreted as atomic planes (those that relate to the **g** vector), the fringe spacing relates to the plane spacing. If more diffraction spots are selected, then the superposition of interference fringes from several **g** vectors results in interference 'spots', that one can relate to the crystal structure of the sample. HR-TEM is obtained when the microscope is set up in order to make these interferences visible on the screen. Let us emphasise this again: although it can be very tempting to associate atoms with the interference spots on the screen, one must always be aware that this is technically *incorrect*. Moreover, the image displayed on the viewing screen is a convolution of the interference pattern and of the contrast transfer function of the objective lens (i.e. the ability of the objective lens to transfer certain spatial frequencies). This contrast transfer function is far from being constant for all frequencies and can dramatically affect the image contrast just by changing the focus of the microscope. As a general rule of thumb, HR-TEM should mainly be used qualitatively, typically to observe the undefected crystal or identify the presence of a defect. For more quantitative interpretations, such as identifying the atomic arrangement at the core of a dislocation, one must compare the experimental image with simulations. An additional interesting analysis technique in HR-TEM is geometric phase analysis [88], which allows one to map the distribution of strain across the image – for example to identify the tensile and compressive regions of a dislocation.

High-resolution STEM

With the advent of aberration correctors, high-resolution STEM (HR-STEM) is perhaps currently the best-suited technique for the study of defects at the atomic

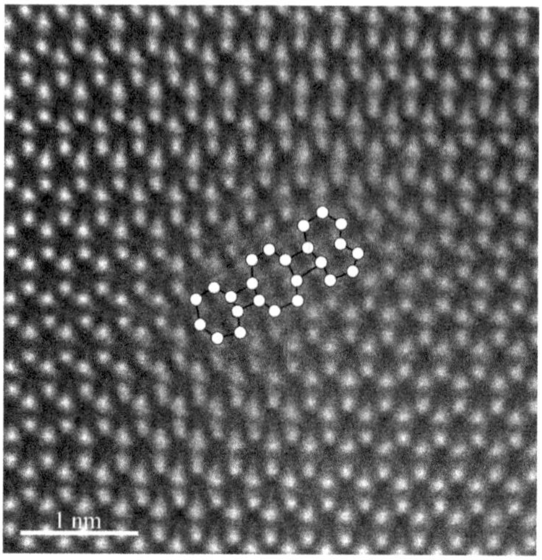

Figure 8.24 HR-STEM image of the core structure (here double 7/4/8/4/9 atom rings) of a mixed dislocation in AlGaN observed along [0 0 0 1]

scale. HR-STEM is based on the same idea as STEM-HAADF where the contrast is due to Rutherford scattering – i.e. proportional to Z^2. The main difference is that aberration correctors allow for smaller probe size and therefore improved resolution, down to the atomic scale. A crucial difference with the previous technique (HR-TEM) is that in HR-STEM, the images can be directly interpreted in terms of atoms. You still have to bear in mind that the typical TEM sample is not atomically thin, hence an 'atom' in the image correspond in fact to an atomic column. This is, therefore, the technique of choice to study atomic arrangement around defects (see Figure 8.24).

8.3 Examples of the application of microscopy to semiconductor materials

8.3.1 Dislocation densities and Burgers vectors in GaN

8.3.1.1 Motivation for the application of microscopy to study dislocation types and densities in GaN

The nitride family of materials is perhaps unique amongst the conventional compound semiconductors in the extent to which nitride devices are robust to the presence of high densities of extended defects, particularly dislocations. This allows the use of heteroepitaxial layers grown on highly mismatched substrates in the fabrication of both electronic and optoelectronic devices in GaN, with dislocation densities in working LEDs and high electron mobility transistors (HEMTs) frequently in excess of 10^9 cm^{-2}. However, dislocations do impact the performance of nitride devices, particularly laser diodes [89], and this has provided an impetus for the development of techniques both for the reduction of dislocation densities in heteroepitaxial materials and for the growth of bulk GaN with dislocation densities down to ~10^4 cm^{-2} [90]. Provision of feedback for the development of the materials growth has thus required the availability of techniques which allow quantification of dislocation densities ranging over five orders of magnitude. In typical (0 0 0 1) GaN heteroepitaxial layers, the basis of most devices, three types of TDs are observed: pure edge, or 'a-type' (Burgers vector, $\mathbf{b} = 1/3\langle 1\ 1\ \bar{2}\ 0\rangle$), pure screw, or 'c-type' ($\mathbf{b} = \langle 0\ 0\ 0\ 1\rangle$), and mixed, or 'c+a type' ($\mathbf{b} = 1/3\langle 1\ 1\ \bar{2}\ 3\rangle$). Quantification of the proportion of dislocations of each type can aid understanding of the mechanisms of both dislocation generation and reduction. Information on these mechanisms can also be deduced from the spatial distribution of dislocations across the surface of a sample. Microscopy provides the means to access all of these parameters of the dislocation distribution in the nitrides, but different methods are more appropriate to different dislocation density regimes.

8.3.1.2 AFM methods for dislocation density quantification

As was mentioned in Section 8.2.1.1, the termination of a dislocation at a sample surface necessarily results in the formation of a small surface pit, to achieve an equilibrium of forces between the dislocation line tension and the surface tension. The formation of such pits or 'dimples' was predicted by Frank as early as 1951 [91], who also noted that dislocations with a screw component intersecting a close-packed surface must be associated with a step edge emanating from the dislocation termination. As grown GaN surfaces exhibit dislocation pits associated with all types of TDs. Step edge terminations are indeed seen at screw and mixed dislocations. This allows AFM imaging of the dislocations and discrimination between dislocations with and without a screw component. However, the relevant pits are small in lateral extent (typically 20–30 nm for mixed dislocations and less than 10 nm for edge dislocations) and also shallow (less than a nanometre). Pits associated with edge dislocations in particular are often only visible when tip sharpness is carefully preserved. (Whilst in general blunt tips limit the lateral resolution rather than the vertical resolution, a blunt tip cannot penetrate into a narrow pit, or may not penetrate its full depth. In the latter case, the ability to detect the dislocation depends on the relative magnitudes of the depth to which the tip penetrates and the background noise.) A variety of wet etching methods have been applied to increase the size of dislocation-related pits at GaN surfaces (and are reviewed in [92]), but in many cases, the resulting pits are sufficiently large that for samples with dislocations densities in excess of 10^8 cm^{-2} (which includes most common materials for LEDs and HEMTs), pits tend to merge making detailed quantification difficult.

A more successful approach for (0 0 0 1)-oriented GaN grown by the most common technique, metal-organic vapour-phase epitaxy (MOVPE), is the application of a surface treatment using SiH_4 and NH_3 in the deposition system immediately post-growth. This can be straightforwardly used to increase the diameters of pits associated with edge and mixed dislocations to approximately 45 and 60 nm, respectively, allowing them to be easily imaged in AFM [92]. Tapping mode images of a sample before and after surface treatment are shown in Figure 8.25. Edge dislocations may be distinguished from mixed dislocations not only by the

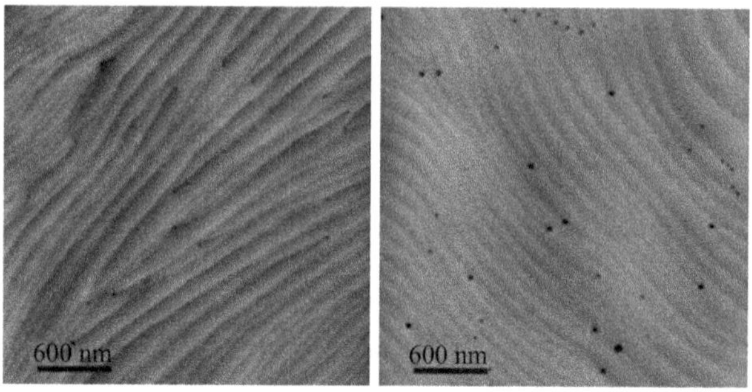

(a) (b)

Figure 8.25 Images of a (0 0 0 1) GaN surface before (a) and after (b) treatment with SiH_4 and NH_3 to increase the size of the dislocation pits.

smaller pit size but also because the treatment leaves terraces distinguishable at the surface so that step terminations at mixed dislocations can be identified. The identification of pure screw dislocations by this method has not been thoroughly explored since most MOVPE-grown materials contain very few pure screw dislocations. This method has been applied to samples with dislocation densities ranging from ~10^5 cm^{-2} [93] to 5×10^9 cm^{-2} [94], although for the lowest dislocation densities, it is very slow, since imaging must be carried out at a sufficient resolution to locate the small pits over very large areas. It has been extensively used to explore spatial distributions of dislocations in GaN [95] and to relate them to defect formation mechanisms. It has also been used to locate dislocations terminating at the surface of large (few micron height) 3D islands, including on semi-polar facets (facets inclined to (0 0 0 1)) [92]. It has been used on films grown in non-polar orientations [96] and has in this context been suggested to have some efficacy for the identification of stacking faults as well as dislocations. However, it should be noted that the specific surface treatment outlined here is not generalisable to other semiconductors and indeed fails on strained AlGaN surfaces, although it is effective on relaxed AlGaN [97]. Nonetheless, it provides a useful example of the level of detail which basic topographic AFM can provide in materials where sufficient surface topography exists or can be generated at the surface terminations of extended defects.

8.3.1.3 SEM methods for dislocation density quantification

One of the advantages of ECCI in the SEM is the ability to acquire images rapidly (in the order of minutes) from large areas of the surface of a sample (of order 10 μm × 10 μm). For dislocations which thread through the sample and reach the surface, dislocation densities can be determined by simply counting the number of dislocations in the image. There are a number of different techniques for obtaining a count of the dislocations ranging from a simple manual count, to the use of appropriate plugins in image processing packages such as ImageJ [98], to the development of bespoke image processing software exploiting machine-learning techniques. If the diffraction condition(s) can be determined, it is in principle also possible to identify the dislocation type as described in Section 8.2.2.2. Using the procedure described in Section 8.2.2.2, Naresh-Kumar *et al.* [25] identified 300 TDs in a GaN thin film. Approximately 51% were identified as pure edge dislocations (Burgers vectors $1/3\langle 1\ 1\ \bar{2}\ 0\rangle$), ≈42% were mixed dislocations (Burgers vectors $1/3\langle 1\ 1\ \bar{2}\ 3\rangle$) and pure screw dislocations accounted for ≈7% of the total dislocation density (Burgers vectors $\langle 0\ 0\ 0\ 1\rangle$).

While ECCI is an excellent technique for quantification of defects which reach the surface, it is harder to quantify dislocations below the surface due to the surface sensitivity of the ECCI technique and the difficulty in determining the precise sampling depth of the technique. However, it is possible to obtain quantitative data on dislocations which lie below the surface, as illustrated for example by the paper by Carnevale *et al.* [99] on misfit dislocations in GaP lying at up to around 100 nm from the surface.

The ability to image large areas of a sample makes it possible to probe the statistical distribution of the dislocations threading to the surface of a sample. Ripley's K-function [100,101] is one of the most widely used methods to analyse point pattern data providing information on the variation of the density of points under investigation, and it determines whether they are clustered (under-dispersed), random (dispersed) or regular (over-dispersed).

Ripley's function produces concentric circles with increasing radii around each identified point in an image and counts the number of neighbouring points that are contained within each circle. This process is repeated for each point within the image with the cumulative total for a given radius r computed over a range of values of r to produce the distribution function $K(r)$. $L(r)$ is a function derived from $K(r)$ which is easier to interpret and is simply the deviation between the measured $K(r)$ and a point distribution of the same density but exhibiting complete spatial randomness (CSR) [102].

To determine the statistical significance of any deviation from CSR, any deviation can be evaluated using simulated significance intervals by computing the lowest and highest value of $L(r)$ for each radius r to define the lower and upper bounds of a significance envelope. A Monte Carlo method is generally employed for these simulations. If the empirical L-function lies above the upper bound of the CSR envelope, this implies that there is a higher density of points than would be expected under CSR and there is clustering. Similarly, if the empirical L-function lies below the lower bound, this indicates that there is a lower density of points than would be expected under CSR, and thus the points are over-dispersed.

Figure 8.26(a) shows an ECCI micrograph from a GaN thin film where the dislocations are highlighted by red circles. This point pattern has then been analysed using $L(r)$, and this is shown in Figure 8.26(b). The deviation of experimental $L(r)$ outside the 90% significance envelope indicates a statistically significant over dispersion of dislocation for a radius of ≲150 nm; this is likely due to the mutual repulsion of neighbouring dislocations due to their associated strain fields [104,105] and to limits imposed by spatial resolution which is ≈70 μm for this

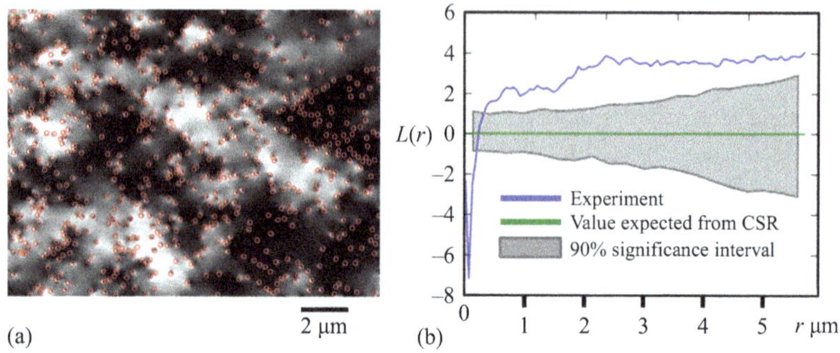

Figure 8.26 *(a) ECCI micrograph from a GaN thin film with dislocations highlighted with red circles (b) L(r) for the dislocation distribution shown in (a), showing the L(r) determined from experiment, L(r) expected from complete spatial randomness (CSR = 1) and a significance interval of 90% (dislocations identified and statistical analysis undertaken using in-house software developed by Simon Kraeusel). © 2014. Adapted and reprinted, with permission, from [103]*

image. There is also statistically significant clustering of dislocations for a radius of $\gtrsim$500 nm. Examination of Figure 8.26 shows that the dislocations are arranged in long lines, in particular, along boundaries between regions exhibiting different grey scales in the image. The changes in grey scale indicate changes in orientation between subgrains in the film. The clustering indicated by the $L(r)$ is probably due to these lines of dislocations.

For more information on a statistical analysis of the distribution of dislocations, the reader is referred to the work of Moram *et al.* who have published detailed studies on the use of spatial analysis techniques for the analysis of defects observed in AFM and CL images, see for example [105,106]. Note that Moram *et al.* employed the AFM techniques for dislocation identification described in Section 8.3.1.2.

8.3.1.4 TEM methods for dislocation density quantitation

In Section 8.2.3.2, we saw that dislocations can be observed using a TEM, and we will discuss this in more detail shortly. Probably the most 'accurate' way to obtain the density of dislocations is by using plan-view imaging. In GaN devices, the dislocations that matter the most are often those which thread towards the surface. These will be seen end-on in a plan-view geometry. The density of dislocations will, therefore, be obtained by directly counting how many dislocations intersect the foil over a given area. Here, the word 'accurate' above was purposely written with quotation marks because the area of material is relatively small (typically less than 1 μm^2), making the density thus obtained statistically less reliable than if it was obtained with other techniques (e.g. AFM or SEM). Nevertheless, its reliability is still typically better than density values obtained from cross-sectional TEM observation.

However, cross-sectional imaging *can* provide a density of dislocation, but over a much smaller area of materials hence reducing the statistical strength of the measurement. Moreover, in this geometry, most dislocations will run parallel to the TEM foil. Therefore, in order to obtain a density of dislocations in cross section, one has to count the number of dislocations at a certain position in the foil, but the thickness of the TEM foil is also required. This last value can be obtained through additional TEM analysis, typically by comparing an experimental convergent beam electron diffraction pattern with a simulated one. EELS can also be used for this purpose, by comparing the zero-loss peak with the total intensity of the low-loss spectrum. As such, cross-sectional TEM imaging is not the ideal technique to obtain a reliable density of dislocations. The real strength of the approach is that one can visualise the behaviour of dislocations throughout the thickness of the film – which is information that cannot be obtained by plan-view imaging or any surface techniques. For example, in Figure 8.27, not only can we see that a high density of dislocations is generated near the substrate/film interface (bottom of the picture) but also we see that the dislocations interact with each other within *approximately* 2 μm from the interface, hence reducing greatly the number of dislocations that thread to the surface (top of the image).

$\boldsymbol{g} \cdot \boldsymbol{b}$ *analysis*

Let us consider further why dislocations generate contrast in the TEM: we will see that not only can dislocations be imaged in the TEM but that the technique also allows one to determine the Burgers vector of the dislocations. A coherent electron beam will be diffracted by the crystal lattice. However, a dislocation generates a strain field that distorts the crystal locally. The overall idea is that if the set of planes that diffracts the beam also happens to be locally distorted due to the

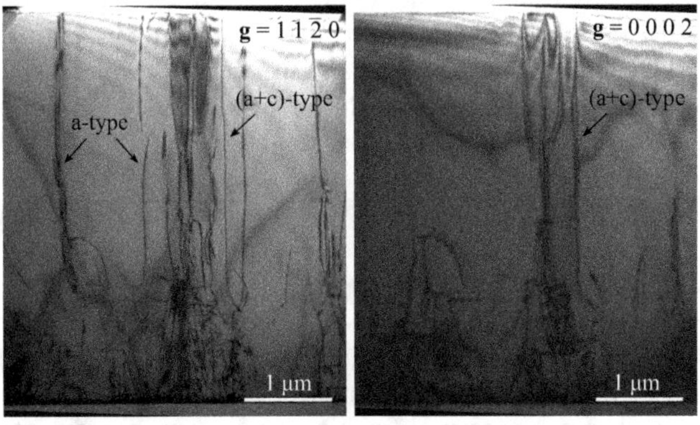

Figure 8.27 Bright-field images of dislocations in a GaN film grown on sapphire, observed using the diffraction conditions $\mathbf{g} = 1\ 1\ \bar{2}\ 0$ *(left) and* $\mathbf{g} = 0\ 0\ 0\ 2$ *(right)*

presence of the dislocations, then the contrast will arise. Given that the lattice distortion is the most pronounced near the core of the dislocation, this means that one can observe using a TEM the dislocation line inside the sample (see Figure 8.27). Additionally, the lattice distortion will be affected by the dislocation's Burgers vector. Hence, by investigating which diffraction conditions give rise to contrast, one can identify the Burgers vector of individual dislocations. Experimentally, dislocation observation is achieved by performing one of the diffraction contrast techniques described in Section 8.2.3.2 (bright field, dark field or WBDF) and selecting a **g** vector that satisfies $\boldsymbol{g} \cdot \boldsymbol{R} \notin \mathbb{Z}$. When $\boldsymbol{g} \cdot \boldsymbol{R} \in \mathbb{Z}$, the dislocation will be out-of-contrast, this is why this is often referred to as the 'invisibility criterion'. $\boldsymbol{R}$ denotes the displacement field of the dislocation, which represents the distortion of the lattice, and takes the form (for a dislocation in an isotropic medium):

$$\boldsymbol{R}(r,\phi) = \frac{1}{2\pi}\left(\boldsymbol{b}\phi + \frac{1}{4(1-\nu)}(\boldsymbol{b}_e + \boldsymbol{b} \times \boldsymbol{u}(2(1-2\nu)\ln r + \cos 2\phi))\right)$$

where $\boldsymbol{b}$ is the Burgers vector of the dislocation, $\boldsymbol{b}_e$ is the edge component of the Burgers vector, $\boldsymbol{u}$ is the line direction of the dislocation, ν is the Poisson's ratio and r and ϕ are the polar coordinates. We can see that for dislocations, the expression is rather complicated as $\boldsymbol{g} \cdot \boldsymbol{R}$ will depend on $\boldsymbol{g} \cdot \boldsymbol{b}$ and $\boldsymbol{g} \cdot \boldsymbol{b} \times \boldsymbol{u}$. A simplification of the invisibility criterion is that the defect will be out-of-contrast when $\boldsymbol{g} \cdot \boldsymbol{b} = 0$, bearing in mind that some residual contrast may arise from the term $\boldsymbol{g} \cdot \boldsymbol{b} \times \boldsymbol{u}$. In Figure 8.27, this criterion was used to differentiate edge ($\boldsymbol{b} = \boldsymbol{a} = 1/3[1\ 1\ \bar{2}\ 0]$) to mixed ($\boldsymbol{b} = \boldsymbol{a} + \boldsymbol{c} = 1/3[1\ 1\ \bar{2}\ 3]$) dislocations: edge dislocations will be visible when $\boldsymbol{g} = 1\ 1\ \bar{2}\ 0$ (left image) but not when $\boldsymbol{g} = 0\ 0\ 0\ 2$ (right image), while mixed dislocations will be visible under both conditions.

Core identification by HR-STEM

Alternatively, HR-STEM in plan-view geometry can be employed to identify the Burgers vector of dislocations. Dislocations are, therefore, seen end-on and the resolution of the technique allows one to observe the atomic arrangement near the dislocation. The Burgers vector of the dislocation can, therefore, be obtained by drawing a Burgers circuit around the dislocation (see Figure 8.28). Bear in mind that you will obtain the projection of the Burgers vector in the viewing plane so that any component of the Burgers vector parallel to the zone axis will not contribute to the Burgers circuit. Hence, unfortunately, Burgers circuit drawn around edge and mixed dislocations will appear similar, and no component of the Burgers vector of a screw dislocation will be visible.

However, the real asset of HR-STEM is that one can identify the core configuration of the dislocation, as depicted in Figure 8.28. The core configuration of the dislocation not only depends in part on the dislocation type but also can reveal other information about the material, such as the amount of strain, atomic segregation, etc. However, to be able to draw meaningful conclusions, this approach requires the identification of a statistically significant number of defects per sample. Even so, the problem of projection persists and the distinction between edge

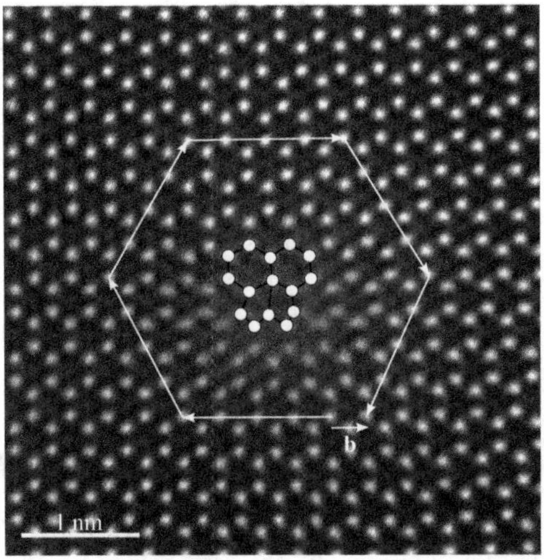

Figure 8.28 HR-STEM image of the core structure (here double 5/6 atom rings) of a mixed dislocation in AlGaN observed along [0 0 0 1]. The white arrows indicate a Burgers circuit around the core

and mixed dislocations can be achieved by observing the Eshelby twist [107]. The Eshelby twist is an apparent rotation of the lattice that is caused by the relaxation of stresses at the free surface. This is due to the screw component of a dislocation. Hence, mixed and screw dislocations will exhibit an Eshelby twist, while edge dislocations will not, which can be used to discriminate them in HR-STEM images.

8.3.1.5 Intercomparison of the various methods

It is possible to achieve accurate quantification of dislocation densities and Burgers vectors using a scanning probe microscope, an SEM or a TEM. Indeed, careful comparisons have been made of the dislocation densities recorded using AFM on SiH_4-treated surfaces and plan-view TEM imaging on the same sample, and the results are consistent between the different techniques [92]. Similarly, comparisons have been made between the results of ECCI and AFM on SiH_4-treated surfaces, not just on the same sample but on the same defects in the same sample, and again good agreement has been achieved [32]. However, the different techniques have different strengths and weaknesses, and not all samples will yield accurate results using all techniques. One factor that limits each technique is the fact that separate dislocation-related features tend to overlap as the dislocation separation is reduced. In ECCI, it is possible to discriminate between TDs if they are more than about 80 nm apart. For AFM, using the SiH_4 treatment, this spacing is somewhat lower, with mixed dislocations separated by about 50 nm and edge dislocations separated by about 40 nm being distinguishable [92]. For AFM on untreated surfaces, using a very sharp tip, it is possible to reduce these values further, but at some cost in terms of the reliability of the measurement, since tip blunting can quickly make it impossible to resolve the tiny edge dislocation pits.

Pragmatically, AFM on treated surfaces and ECCI remain effective as a mean of both measuring dislocation density and assessing Burgers vectors up to a dislocation density of mid 10^9 cm^{-2}. In high-resolution plan-view STEM, where the individual dislocation cores can be imaged, the minimum spacing at which separate dislocations can be resolved is clearly far smaller. It should be noted, however, that in cross-sectional TEM imaging, even when the WBDF technique is used to minimise the width of the dislocation-related features, projection effects may cause contrast from dislocations separated from one another in the through-sample-thickness direction to overlap, limiting the effectiveness of this approach in quantification at very high dislocation densities.

Although HR-STEM provides the ultimate resolution for samples with very high dislocation densities, increasingly the GaN used in technological applications has a much lower concentration of defects. When assessing materials grown by hydride vapour-phase epitaxy [108] (with dislocation densities less than 10^8 cm^{-2}) or ammonothermal growth [93] (with dislocation less than 10^6 cm^{-2}), the small area typically sampled by plan-view TEM imaging can mean that not one single dislocation is found in the field of view. In these low dislocation density samples, SEM-based methods such as ECCI or CL present a huge advantage in being able to assess a large area quickly and easily. AFM on treated samples can address large areas [109] (e.g. 30 μm × 30 μm), although this approach is very slow. It is also very reliant on the SiH_4 treatment. This can be applied as a post-growth treatment to materials grown by other techniques but that reduces the convenience of this approach. Overall, the availability of a range of techniques to suit the requirements of different experiments is rather useful, with either ECCI or AFM being particularly useful for fast-turnaround estimation of dislocation density, with minimal sample preparation, and STEM being well suited to more in-depth studies of the mechanisms by which dislocations interact.

8.3.2 Imaging of the impact of dislocations on materials properties

8.3.2.1 The interaction of dislocations with doping

In considering the interaction of dislocations with doping in gallium nitride, two essentially separate issues arise: (1) methods for the reduction of the dislocation density lead to local changes in intentional or unintentional dopant incorporation so that variations in dislocation density are often also related to variations in doping density. (2) Dopants, whether intentional or unintentional, may segregate to dislocation cores or to strained regions adjacent to dislocation cores. The two issues overlap to some extent where the dopant segregation affects the movement or bending of dislocations, hence influencing dislocation density-reduction mechanisms. However, we shall address separately how microscopy can contribute to understanding of each of these issues, and in so doing provide short cases studies on (for the former) the application of SCM in understanding impurity incorporation and (for the latter) the application of TEM techniques to understand the influence of impurity species on dislocation cores.

Many methods for the reduction of dislocation densities in the heteroepitaxial growth of GaN involve the formation of 3D islands which are then coalesced to produce a 2D film. Dislocations tend to bend towards the inclined surfaces of the islands in order to shorten their line-length in the crystal and hence reduce the total energy. When the islands coalesce, bent dislocations may then meet and either annihilate or react producing a new TD. Either process may reduce the dislocation density. The formation of 3D islands may be induced by altering the growth conditions during the initial stages of growth on the foreign substrate [94]. Alternatively, a GaN film with a high dislocation density may first be grown and then a mask of some sort used to prevent further 2D growth and induce 3D growth on the unmasked areas. Such a mask may be self-assembled – for example incomplete layers of SiN_x deposited *in situ* during MOVPE growth are commonly used – or may be deposited and deliberately patterned *ex situ* to provide a regular array of masked and unmasked regions [110]. The most common variant on the latter process is the 'epitaxial lateral overgrowth' or ELOG method in which a GaN layer is patterned with stripes of an amorphous material, usually SiN_x or SiO_2, which acts as a mask [111]. In one popular variant, facetted ELOG (FACELOG), the V:III ratio and temperature are reduced to encourage the growth of facetted stripes in the unmasked regions, forming triangular prisms of GaN usually aligned along $[1\ \bar{1}\ 0\ 0]$. Dislocations tend to bend towards the inclined side facets to shorten their line length in the crystal. Once the growth of the triangular prism structures is complete, the V:III ratio and temperature are increased, and a flux of a magnesium precursor may be added to the growth environment. All these changes tend to encourage lateral growth, overgrowing the mask stripes, so that the prisms of GaN coalesce to form a flat film. At the coalescence boundaries, dislocations meet and may annihilate.

Figure 8.29(a) shows a schematic of this growth process and may be compared to the TEM image in Figure 8.29(b) which reveals the impact of the facetted growth and lateral overgrowth on the dislocation structure [109]. Figure 8.29(c) and (d) shows SCM images of the same sample, providing the phase and amplitude of the SCM signal, respectively [112]. The SCM phase data shows that the GaN triangular prisms formed in the initial growth stage are n-type (appearing black), whilst the regions where lateral coalescence has occurred are p-type (appearing white). The latter point might seem unsurprising since an Mg precursor was employed to enhance the lateral growth rate. However, the p-type material has not been deliberately annealed, and so the acceptors are expected to be compensated by hydrogen. However, a clear signature of p-type conductivity is seen. The SCM amplitude data reveals that there are variations in the concentration of the electrons in the n-type region, with detailed analysis [112] suggesting that during growth, there was a higher concentration of n-type dopant incorporation on semi-polar inclined $\{1\ 1\ \bar{2}\ 2\}$ facets than on the $(0\ 0\ 0\ 1)$ facet. The donor in this case is unknown: SCM does not give information about the chemical nature of the dopant. It may be Si (from the mask) or oxygen (from the growth environment).

The observation that n-dopants are incorporated unintentionally on semi-polar inclined facets is ubiquitous in SCM studies of GaN samples in which dislocation

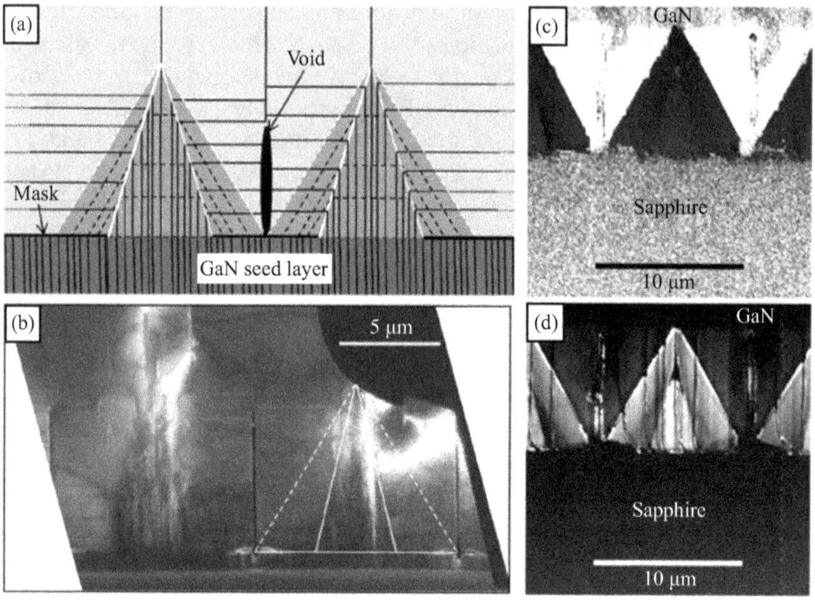

Figure 8.29 Dislocations and unintentional doping in GaN grown by a FACELOG technique: (a) schematic, in which dislocations are shown as thin dark lines. (b) (0 0 0 2) WBDF TEM image in which the white lines on the right-hand half of the image correspond to the white lines in (a). (c) SCM phase image in which black regions indicate n-type doping and white regions indicate p-type doping. (d) SCM amplitude image which provides information on the carrier concentration variation across the structure. (All images are cross sections.) (a and b) © 2010. Reprinted, with permission, from [109]. (c and d) © 2009 Reprinted, with permission, from [112]

density reduction is achieved using a transition from 3D to 2D growth [110]. Whilst unintentional doping in GaN was, in the early days of heteroepitaxial growth studies, often attributed to point defects such as nitrogen vacancies [113], later work showed that it is largely attributable to oxygen contamination, and SCM studies have shown that the majority of this contamination is confined to specific spatial regions in which semi-polar facets are present during growth, with the uptake of oxygen being far lower on both polar and non-polar facets. Overall, these studies illustrate that SCM is useful not only in identifying and quantifying unintentional (and indeed intentional) doping but also in deriving its origins and its relationship to both point and extended defects.

AFM-based techniques including SCM [114], C-AFM [115] and surface photovoltage microscopy [115] have also been used to address the electrical properties of dislocation cores in materials doped with different species. However, to understand the impact of doping on dislocation core structure, TEM techniques are far more appropriate. As shown in Section 8.2.3.3, HR-STEM-HAADF imaging may be used to resolve the core structure of dislocations. The structure of dislocation cores in undoped GaN has been identified by this approach: a-type dislocations typically form a 5/7-atom ring structure as illustrated in Figure 8.30(a)–(c) [116]. This experimental finding is consistent with theoretical predictions of the stable dislocation structure made using both density functional theory [122] and multiscale modelling [123] approaches. Neither n- nor p-doping leads to changes in the observed structure of a-type dislocations [116,124]. For (a+c)-type dislocations, on the other hand, both dissociated and undissociated dislocation cores are seen, with approximately equal frequency in undoped material. The undissociated cores exhibit a double 5/6-atom ring core structures, illustrated in Figure 8.30(d)–(f). A variety of dissociated core structures have been observed experimentally, one example of which (a 7/4/8/4/9 atom ring) is shown in Figure 8.30(g)–(i). Cores of different dissociated length may be observed, with the number of 4/8-atom rings increasing as the core length increases. Whilst Si-doping has little effect on these core structures [124], the addition of Mg-doping led to a much reduced occurrence of dissociated cores which has been

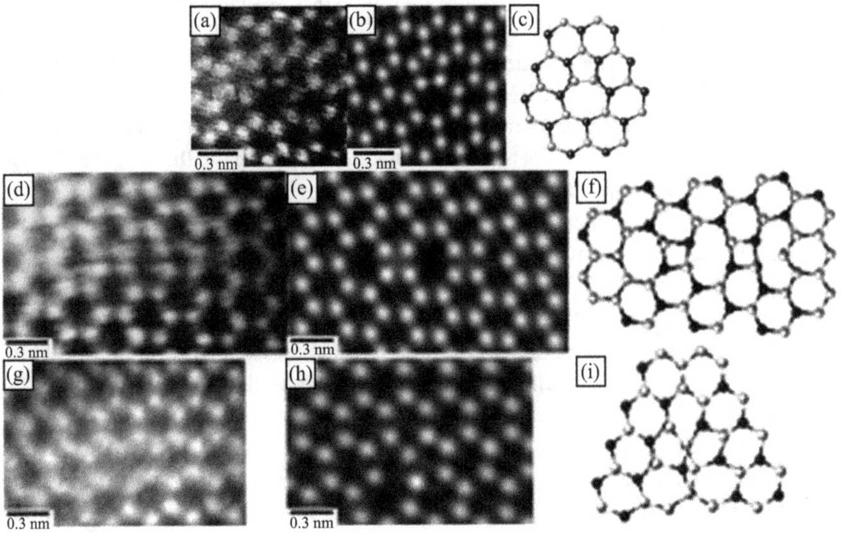

Figure 8.30 Typical STEM-HAADF images of (a) an a-type dislocation core, (d) an (a+c)-type dissociated core with a 7/4/8/4/9 atom ring structure and (g) a double 5/6-atom (a+c)-type undissociated core. Parts (b), (e) and (h) are corresponding simulated STEM-HAADF images for (a), (d) and (g), respectively. Similarly (c), (f) and (i) show the relevant core structures for each row obtained from molecular dynamics simulations, in which the white spheres represent Ga atoms and the dark spheres represent N atoms. © 2013 American Physical Society. Reprinted, with permission, from [116]

attributed to the segregation of Mg to the tensile region adjacent to the core, which reduces the dislocation line energy and hence the driving force for dissociation [116]. These studies, the conclusions of which are supported by consistent findings from molecular dynamics simulations, illustrate the power of STEM-HAADF in identifying not only the core structure but also variations in the local chemical composition (in this case by inference rather than direct observation) which can profoundly affect the impact of the dislocations on the local electrical properties.

8.3.2.2 Combining ECCI and CL to investigate the influence of defects on light emission

Although, as was explained in Section 8.3.1.1, GaN-based optoelectronic devices are remarkably robust to the presence of TDs, it was shown, as early as 1997, by a comparison of AFM and CL images that dislocations in GaN can indeed act as non-radiative recombination centres [125]. TDs can have various effects on a material's optical and electrical properties leading to other recombination channels besides near band-edge recombination or act as scattering centres affecting the carrier transport. The influence of the dislocation type on the optical properties, including whether they are radiative or non-radiative centres for recombination, shows strong ambiguity in the literature. Sugahara *et al.* have found that all types of dislocations in homo- and heteroepitaxial GaN act as non-radiative recombination centres for minority carriers as seen in transmission electron and CL images [126]. Hino *et al.* on the other hand discovered evidence in AFM, SEM and photoluminescence (PL) measurements that only dislocations with a screw component (screw- and mixed-type) act as centres for non-radiative recombination, whereas edge-type does *not* behave that way [127]. The opposite was reported by Albrecht *et al.*, who found evidence in thick homoepitaxial GaN layers that only dislocations with an edge-component (edge- and mixed-type) are recombination active, meaning that only edge- and mixed-type dislocations are centres for non-radiative recombination and screw-type dislocations have *no* impact on the luminescence, i.e. produce no dark spots in CL image [128]. More recently, Medvedev *et al.* showed that screw-type dislocations introduced through scratching or indentation exhibit *bright* luminescence, which is shifted by 0.3 eV from the regular GaN band-edge emission peak even at room temperature [129].

Besides influencing the optical properties of GaN-based devices, dislocations can also impact on their electrical performance. High current densities and low effective Schottky barrier heights were observed in areas close to dislocations having a screw component in an n-doped GaN film [130]. Dangling bonds near edge-type dislocations in AlGaN form acceptor-like states, which reduce the free carrier concentration and mobility with increasing edge dislocation density [131]. The same authors also correlate a decrease of PL intensity with increasing edge-type dislocation density to these dislocations being centres for non-radiative recombination. Furthermore, a reduction of TD density showed a drastic decrease of reverse-biased leakage current in GaN p–n junctions [132].

To further investigate, the optical behaviour of dislocations CL, ECCI and AFM imaging was performed on the same micron scale area of a Si-doped GaN epilayers as displayed in Figure 8.31(a)–(c), respectively [32].

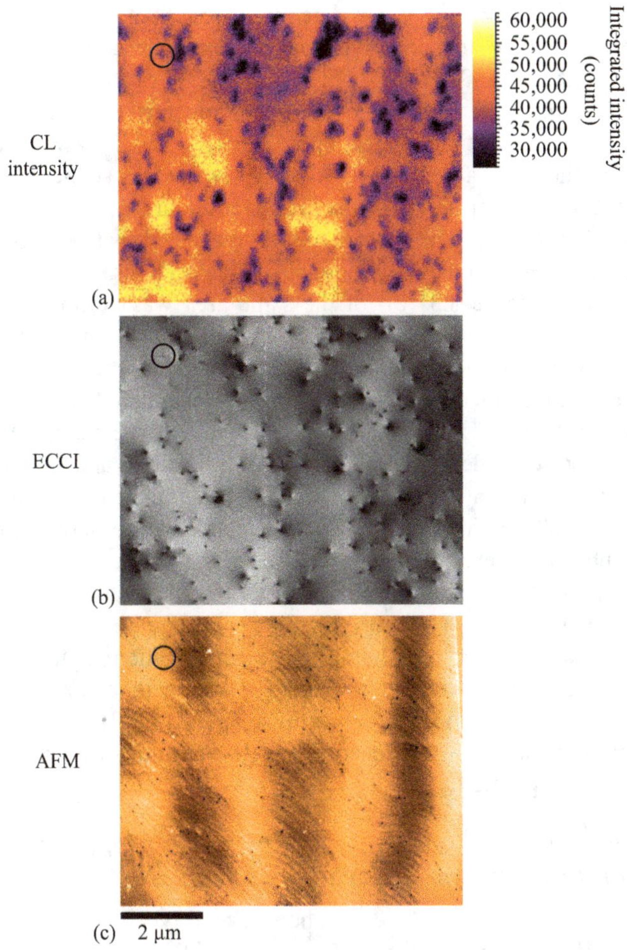

Figure 8.31 (a) CL intensity of the GaN band-edge emission, (b) ECCI and (c) AFM images from the same micron scale area of a Si-doped GaN epilayers. © 2013. Reprinted, with permission, from [32]

The CL intensity image of the GaN near band-edge emission shows a large number of dark spots, associated with non-radiative recombination at TDs [125,126]. They have varying diameter and appear in clusters of two or more dislocations in close proximity, which can lead to an underestimation of the TD density. The CL measurement was performed at a relatively high current (5 kV, ≈6 nA) in order to excite an intense enough GaN band-edge emission due to carbon contamination from the previously performed ECCI measurement on the same area leading to larger observed black spots.

From several ECCI images, it was determined that 60% of the dislocations are edge-type, fewer than 2% are screw-type and the remainder are mixed-type with a total TD density of $(5.1 \pm 0.4) \times 10^8$ cm^{-2}. It is possible to discriminate between TDs in the ECC image in Figure 8.31(b) if they are further than approximately 80 nm apart. For the investigated area, approximately 750 TDs were found using the ECCI technique, with about 400 TDs being resolved as single dislocations in the CL image. Careful comparison of the images in Figure 8.31(a) and (b) shows a *one-to-one* correlation of the dark spots in the CL image and spots with B–W contrast in the electron channelling image for those single dislocations.

AFM imaging (Figure 8.31(c)) of the same area was performed to confirm that all dislocations were visible in the ECCI micrograph for the selected channelling conditions. The AFM studies utilised the SiH_4 treatment described in Section 8.3.1.2. Careful comparison identified isolated dark spots corresponding to single dislocations in the CL image to be TDs of both edge- and mixed-type in the AFM image. Therefore, it can be concluded that both pure edge dislocations and TDs with a screw component act as centres for non-radiative recombination in the investigated Si-doped GaN layer. The results are inconclusive about pure screw dislocations acting as non-radiative recombination centres, because their density (<2% of the total dislocation population) is not high enough to identify single and isolated screw TDs for a comparative measurement.

Other examples of combining luminescence information from CL with structural information from ECCI are the investigation of growth temperate of GaInP on TDs [70], grain boundaries in CdTe thin-film solar cells [133] or the investigation of recombination behaviour of dislocations in III–V photovoltaic materials using EBIC instead of CL [134].

8.3.3 A multi-microscopy example: structure, properties and interactions of dislocations in InGaN

In the previous subsections, we have seen how several types of information can be acquired *within the same* microscope (topography and SCM in AFM, ECCI and CL in the SEM). In this section, we expand the idea of correlative microscopy by describing how various properties of the materials can be correlated using multiple *separate* microscopes. We refer to this approach as 'multi-microscopy', and the use of AFM alongside ECCI and CL in Section 8.3.2.2 provides the first illustration of its usefulness. We will further illustrate this approach by linking the properties of dislocations in InGaN using AFM, CL in the SEM and TEM imaging, with a focus here on the practical aspects of how the correlative experiment can be achieved.

8.3.3.1 Practical aspects

Probably, the most important aspect of the multi-microscopy approach is that the identification of the same region is done with the help of one or several 'markers'. It is particularly helpful to have markers at different length scales which can significantly reduce the time required to find the same feature (here, a specific dislocation) in all the different microscopes. Here, we employed, as a micro-scale marker, a TEM grid but one could also use an intentional scratch or even unintentional debris on the surface. Our nanoscale markers here are a natural feature of our sample: trench defects. The sample preparation can obviously be adjusted depending on the requirements of the experiment (especially based on which microscopes are to be used). Figure 8.32 shows an example for the study of TDs in InGaN using AFM, CL in the SEM and TEM imaging (in that specific order).

The sample is first cut to a 3-mm square – in anticipation of a future TEM sample preparation. Cleaning of the sample surface is highly recommended beforehand to prevent having particles between the sample surface and the grid which will be stuck to it. We used a 300 square mesh copper grid (mesh size *approximately* 50 μm) as is usually employed for TEM sample preparation. In order to avoid damaging and bending the grid during manipulation with tweezers, a 2-mm hole copper ring is stuck onto the grid prior to any manipulation of the grid. The set-up is then stuck to the surface of the sample by putting four small drops of glue at the edges of the set-up and positioned so that the grid (not the copper ring) is in contact with the sample surface. It is important to put only small drops of glue outside the grid, and not below it, in order to avoid having the glue spreading under the grid, which would increase the grid-sample distance and eventually also cover the regions of interest. It is crucial that the grid-sample distance remains as low as possible, otherwise the AFM tip may not be able to reach the sample through the mesh. Once the grid is securely in place, AFM measurements may be performed.

It is recommended to perform AFM observation of the sample before CL in the SEM to avoid sample contamination due to electron beam exposure. Given that the measurements performed in AFM are often tip-dependent, you should use the same fresh tip to record all the images during a single AFM session. In the context of the study of dislocations in InGaN, AFM was used to record morphological properties, which are, for each dislocation, the size of the pit terminating the dislocation – which is indicative of the dislocation type [92] – and the distance to the nearest neighbouring dislocation. This is obviously non-exhaustive and any information that can be recorded by AFM (e.g. capacitance, conductivity) can also be acquired. For accurate measurements on small defect-related features, typical image sizes are small (1–3 μm). Once the requisite small-scale images have been recorded within a particular grid square, a larger scan (10–20 μm) should be taken that will facilitate the identification of the same region in the next microscope(s), in this case, the CL in the SEM.

In the SEM, the first requirement is to find the same region which was analysed by AFM, using the micron-scale markers and overview scan. In this example, CL in the SEM was used to record the optical properties – intensity and emission wavelength – of the dislocations previously analysed by AFM and compared to that of the surrounding defect-free material. Again, an overview SEM image of the whole 50 μm square should be recorded at the end of the experiment to help with the identification of the approximate position of the region of interest within the square during TEM sample preparation.

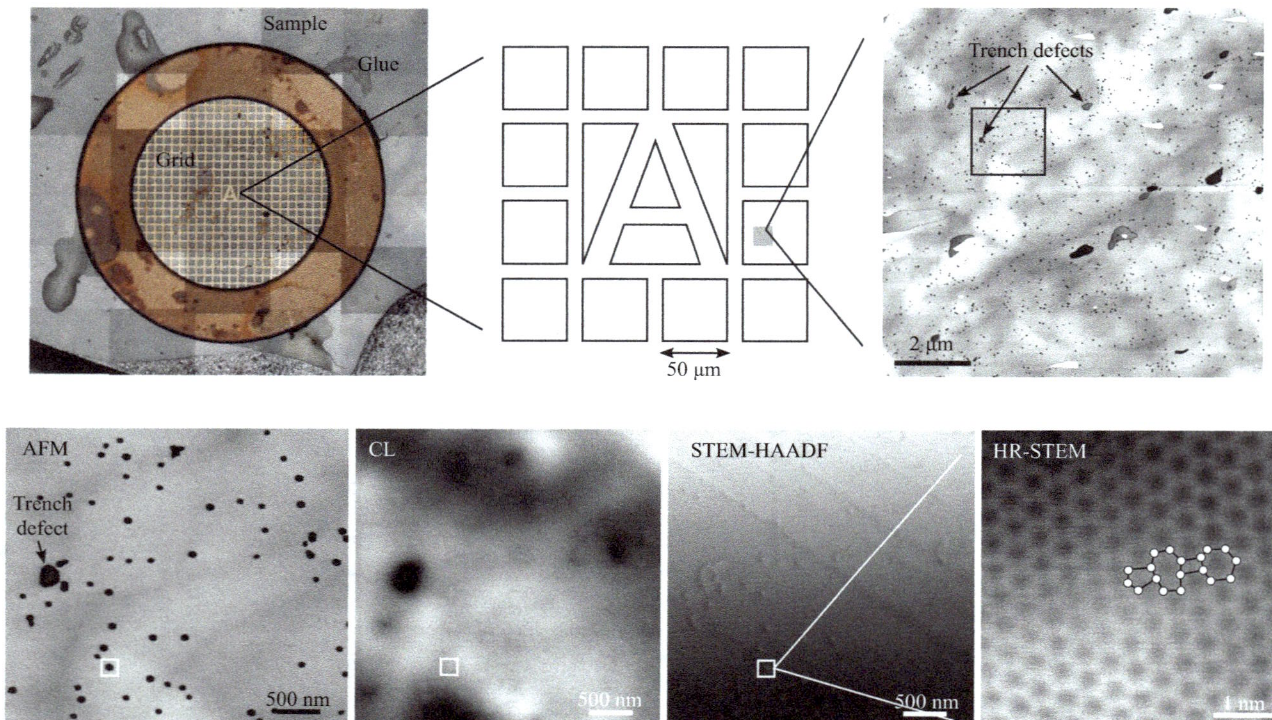

Figure 8.32 *(Top) Illustration of the multi-microscopy set-up. (Bottom from left to right) AFM, CL in the SEM, STEM-HAADF of the same micrometre-scale region of material, and HR-STEM image the dislocation indicated by a white square*

Our approach to the preparation of the sample for plan-view TEM imaging in this multi-microscopy example follows the standard mechanical polishing preparation procedure (as outlined in Box 8.1 in Section 8.2.3.1), although there are a couple of points to note. First, the grid-sample holder has to be reinforced with more glue, otherwise it won't sustain the mechanical stress applied during mechanical grinding. Glue can be placed anywhere on the grid, as long as the regions of interest are avoided. The second point, perhaps the most critical, has to do with the ion milling. Here, it is crucial to position the centre of rotation of the ion mill as close as possible to the region of interest. The idea is to minimise the time spent milling once the hole has formed – otherwise it will blunt the edges of the hole, and sputtering of the grid to the sample may occur. One should be extremely careful and stop the ion milling when the hole forms. After the formation of the hole, it is recommended to use a low-voltage (e.g. 1–3 kV) ion beam in single modulation mode in order to enlarge the hole towards the region of interest. Initial TEM imaging is needed to check whether the region of interest is electron-transparent. If not, the gentle ion milling step must be repeated. It is highly unlikely that the whole region of interest will be suitable for TEM analysis at the same time (the foil thickness will vary across the region of interest). Hence, several iterations of gentle ion milling and TEM observation will be required in order to be able to analyse as many defects as possible. In this example, the atomic configuration of the dislocation core was obtained by TEM imaging – as illustrated in Figure 8.32.

8.3.3.2 Results

The extensive data set thus generated can now be analysed to extract potential correlations between the morphological, optical and structural properties of each dislocation. Figure 8.33 illustrates the multi-microscopy approach, whereby we can obtain general trends over the entire data set involving data from several microscopes. Here, we correlate the intensity ratio (which is the CL emission intensity at the dislocation compared to that of the surrounding defect-free material) which we obtained by CL, to the distance to the nearest neighbour which we measured by AFM. The data set is also separated into edge and mixed dislocations that were identified by analysis of the AFM data. Interpretation of such trends (along with other trends not shown here) enabled us to highlight the segregation of In atoms near the core of dislocations, which leads to the bright spot seen in the CL image in Figure 8.32, in contrast to the dark spots seen where dislocations intersect MQWs and form V-pits in Section 8.2.2.4. More details about the analysis can be found in [135].

The strength of the multi-microscopy approach is also the ability to extract individual data points from the trend and analyse them further. For example, in Figure 8.33, two mixed dislocations with similar properties (low-intensity ratio and small distance to the nearest neighbour) have been selected and analysed in the TEM. Despite their similarities in AFM and CL, their core configurations are intrinsically different – one is dissociated (left image) and the other one is undissociated (right image). This highlights that the core structure of a dislocation (in InGaN) does not dominate its optical properties.

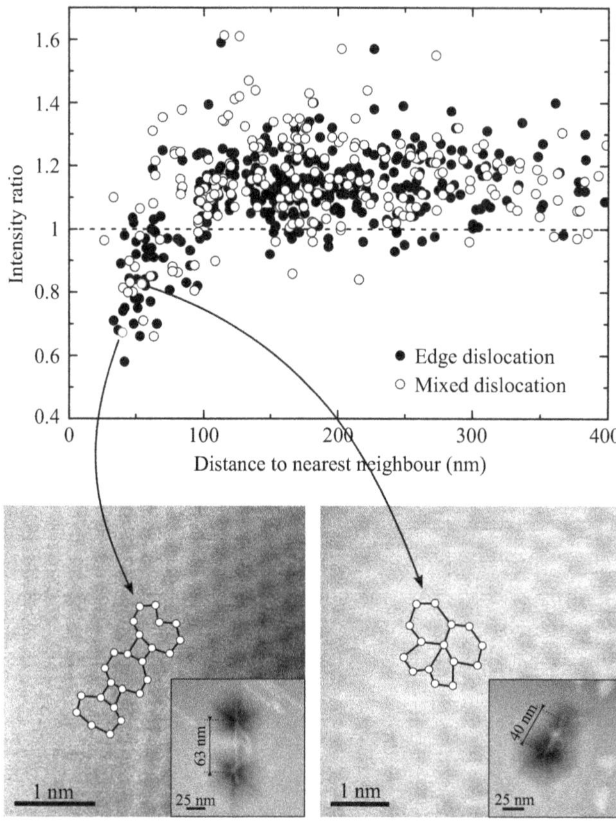

Figure 8.33 (Top) Plot of the intensity ratio (measured by CL) versus distance to the nearest neighbour (measured by AFM) as a function of the dislocation type (obtained by AFM). (Bottom) HR-STEM images of the core of the dislocations corresponding to the data points highlighted in the top graph. In inset, zoom-out STEM-HAADF image of the dislocation

8.3.4 Imaging defects in 2D materials

8.3.4.1 Addressing the electrical properties of defects in 2D transition metal dichalcogenides by scanning probe microscopy

Layered materials such as $NbSe_2$ and WSe_2 which can be mechanically exfoliated to produce 2D semiconductor samples have long been attractive to researchers using STM to study defects since cleavage provides a straightforward means of producing clean, atomically flat surfaces [136]. As the popularity of 2D materials has increased, so has the breadth of scanning probe microscopy studies, expanding to encompass not only mechanically exfoliated samples but also materials grown by chemical vapour deposition [137]. Furthermore, a wide range of scanning probe microscopy techniques has been applied, including C-AFM (which can resolve atomic scale periodicity in samples imaged under a dry nitrogen atmosphere [138]), KPFM and differential surface photovoltage microscopy [137]. STM remains, however, an extremely powerful approach to the identification of defects in 2D materials and the assessment of their impact on electronic structure. A wide variety of atomic-scale features is observed in STM images, believed to be linked to point defects and/or impurities, which are influenced (as is usual) by both topographic and electronic contributions, which complicate image interpretation [136,139]. The identification of the defect is somewhat more straightforward for dislocations, since at low-angle grain boundaries (for example), the misorientation of the atomic planes in adjacent regions provides a clear signature of the nature of the defects at the boundary, independent of the atomic resolution images of the defects themselves.

Huang *et al.* [117] imaged dislocations at low-angle grain boundaries in WSe_2 and observed an array of dislocations characterised by bright 'butterfly features' – i.e. pairs of bright lobes located symmetrically on either side of the boundary (see Figure 8.34). These butterfly features had significant apparent height when the boundary was imaged at a negative tip bias (Figure 8.34(a)–(c)), but with a positive tip bias, no such protrusion was observed (Figure 8.34(d)). This suggested that the apparent protrusion was actually an electronic feature, linked to redistributions of the LDOS at the defects, rather than out-of-plane lattice deformations. STS revealed the presence of states in the band gap between the Fermi level and the conduction band minimum. These unoccupied states will be imaged at negative tip

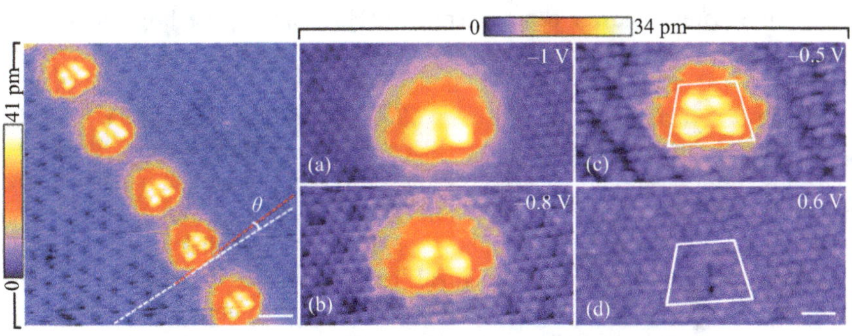

Figure 8.34 STM images of a low-angle grain boundary in WSe_2. Leftmost image: Overview of a boundary with a misorientation angle, θ, of 4.5°, displaying an array of butterfly features corresponding to dislocations. Scale bar: 2 nm. (a)–(d) Images of a particularly butterfly feature recorded under different tip biases. Note that, the apparent protrusion disappears when the tip is positively biased relative to the sample, accessing the occupied states of the sample. Scale bar: 1 nm. © 2016 American Chemical Society. Reprinted (adapted), with permission, from [117]

bias, when the electrons tunnel into them and explain the bright features observed in imaging under these conditions.

A number of possible dislocation core structures were proposed and the projected density of states was simulated with density functional theory. A 4/6 atom ring core structure was identified which allowed good agreement between simulated and experimental STM images and also consistency with the STS data in terms of the gap states. Low-angle grain boundaries subjected to external strain were then modelled based on these structures, and the models suggested that strain can be used to modify the gap states and hence the electrical impact of the grain boundaries on devices.

It is important to note here that the comparison of images predicted by atomistic modelling with real experimental data was key to identification of the defect structure, and there are clear parallels to the identification of dislocation core structures in GaN using STEM-HAADF data (Section 8.3.2.1), despite the very different natures of both the imaging technique and the material under study. Equally, the use of STS and voltage-dependent imaging to identify features linked to unoccupied states is a similar approach to that which was applied to the understanding of point defects in GaAs in Section 8.2.1.3.

8.3.4.2 ECCI examination of graphene nucleation on C-face silicon carbide

Investigating defects in 2D films using ECCI is an enticing prospect. To obtain images in such thin films requires that the penetration depth of the electron beam, and therefore the energy of the incident electron beam, be set to as low as is possible when acquiring an ECCI micrograph. It has been shown that the channelling contrast is greater at lower beam energies [21]; however, the brightness of the electron source decreases with beam energy and this, coupled with the decreasing performance of the electron-sensitive diodes with energy [21], means that in practice, the lowest energies used to acquire ECCI micrographs are of order 10 keV.

In spite of this, attempts have been made to analyse 2D films with ECCI. One interesting result has been published by Hite *et al.* [118] where the initial stages of epitaxial graphene growth were studied by characterisation of graphene formed in localised areas on C-face 6*H*-SiC substrates. Their SEM studies revealed that the initial growth begins with small, hexagonal ‘basins’ recessed into the SiC substrate. These graphene-covered basins are produced through the sublimation of Si on the formation of the graphene. ECCI revealed that the graphene growth nucleated around threading screw dislocations in the SiC substrate, see Figure 8.35.

Future developments in SEM and electron detector performance will lead to many more SEM studies of 2D materials.

8.3.4.3 Dislocation and point defect identification in h-BN by transmission electron microscopy and future outlook using machine learning

Hexagonal boron nitride (h-BN) is a wide band gap semiconductor (*approximately* 6 eV) that exhibits single photon emission in the visible and near UV part of the

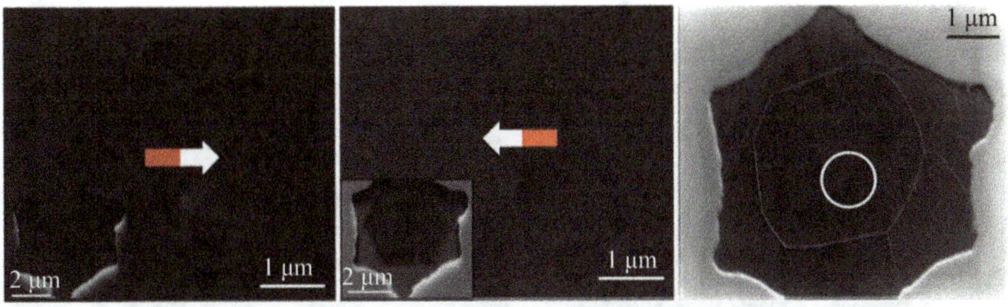

(a) (b) (c)

Figure 8.35 *(a) Magnified ECCI image of a screw dislocation in the centre of a graphene-covered basin (GCB), with an arrow indicating the direction of the dark-to-light contrast. The inset shows an image of the entire GCB. (b) Same GCB, only tilted to the opposite side of the Bragg angle (0.4° difference) – behaviour that confirms the nature of the threading screw dislocation. The arrow indicates that the dark-to-light contrast has changed direction. (c) Secondary electron image of the same GCB shows giraffe stripes and no features where the threading screw dislocation (TSD) is located (circled), proving that the signal at the TSD location is not a morphological feature. © 2011 American Chemical Society. Reprinted, with permission, from [118]*

spectrum [140,141]. This emission is currently thought to occur at defect sites but no direct correlation between a specific defect type and emission property has so far been demonstrated. This is indeed a challenging task. The first step is to demonstrate an ability to identify the structure and chemistry of such defects – using the TEM.

2D materials are of course atomically thin. This indicates that all kinds of atomic-scale defects stretch through the entire (single atom) thickness of the material. This property is particularly interesting for the study of point defects, as in 3D materials, these are extraordinarily difficult to identify by TEM imaging due to their small size comparative to the thickness of the TEM foil. The main difficulty for imaging atomic-scale defects in 2D materials is beam damage – especially for those made of light element, i.e. graphene and h-BN. As we have seen in Section 8.2.3.1, the maximum energy transferred can be significant in such materials. Moreover, owing to the atomically thin nature of the materials, the displacement energy is much lower than for bulk materials. For example, they would be more prone to beam sputtering than to atomic displacement. Hence, when one wants to observe 2D materials in the TEM, one must employ low voltages (80 kV or even 60 kV) and extremely low electron dose (implying using a TEM instead of STEM).

Just to name a few examples, aberration-corrected HR-TEM has been used to observe the atomic structure of grain boundaries in chemical vapour-deposited h-BN, which consist of a succession of 5 and 7 atom rings (Figure 8.36) [119].

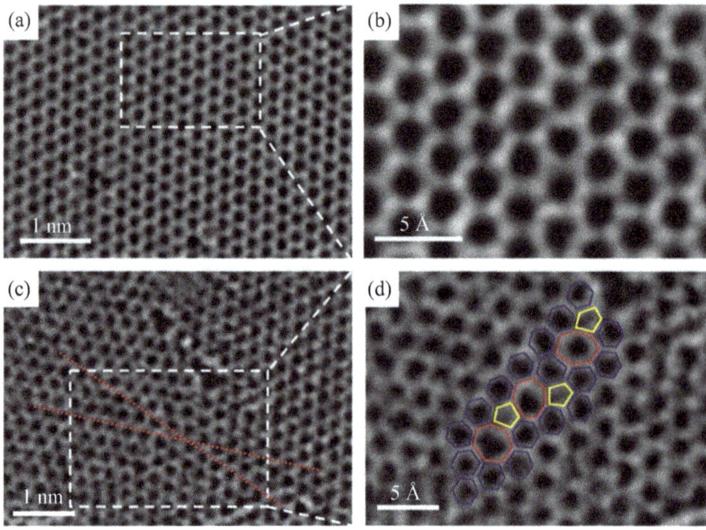

Figure 8.36 Aberration-corrected HR-TEM images of h-BN monolayer containing (a and b) no defect, (c and d) a grain boundary. © 2013 American Chemical Society. Adapted, with permission, from [119]

In another study, a single B vacancy in bilayer h-BN has been observed and shown to result in interlayer bonding (Figure 8.37) [120].

The increasing access to aberration-corrected microscopes will undoubtedly result in a larger volume of data of materials at the atomic scale (not necessarily limited to 2D materials). However, manual identification of the atomic structure of defects is time-consuming and requires an experienced user. In 2018, we are now at the advent of artificial intelligence, machine learning and neural networks, which could make the defect identification task automated, faster and (with proper training of the algorithm) unbiased and more reliable than if it were done by a human. Machine learning for image recognition is embedded everywhere in our daily life and has now been applied to identify the structure and composition of defects in the TEM [142] and STEM [121]. These algorithms can be trained from simulated images of simple defects. Even though it was trained on basic defects in graphene, a deep learning algorithm was able to identify more complex defects such as 3-fold and 4-fold coordinated Si impurities in experimental images of graphene (Figure 8.38). Without further training, the algorithm was also able to identify defects in other materials. These studies allow us to envision a future where microscopy (not only TEM imaging) of defects in semiconductors would be nearly fully automated.

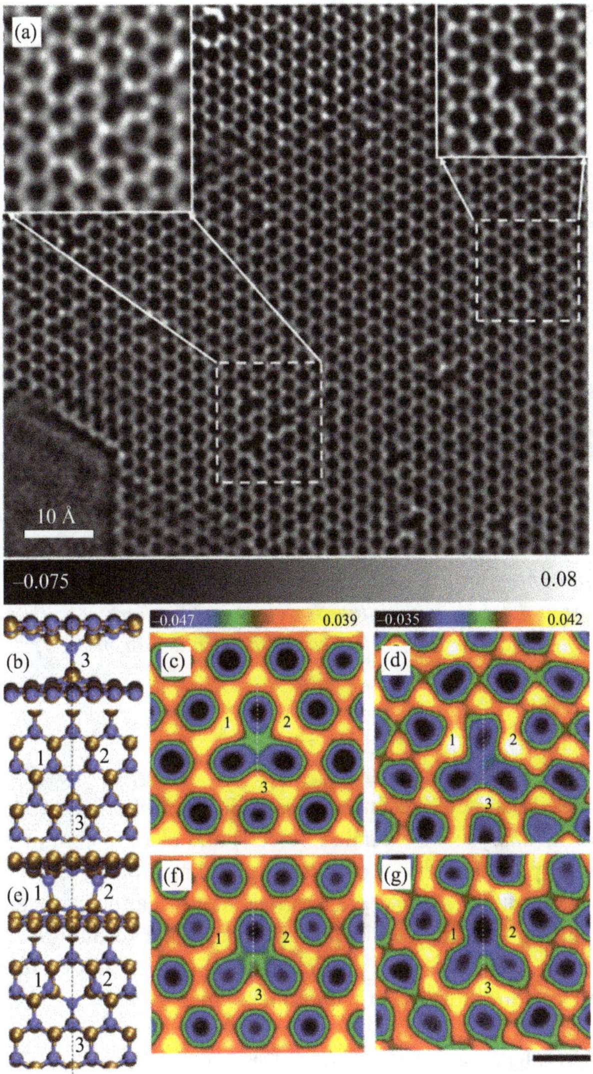

Figure 8.37 *(a) Aberration-corrected HR-TEM images of h-BN bilayer containing B vacancies, (b) atomic structure model, (c) image simulation and (d) experimental image of vacancy with one interlayer bond. (e–g) Same for vacancy with two interlayer bonds. © 2011 American Physical Society. Reprinted, with permission, from [120]*

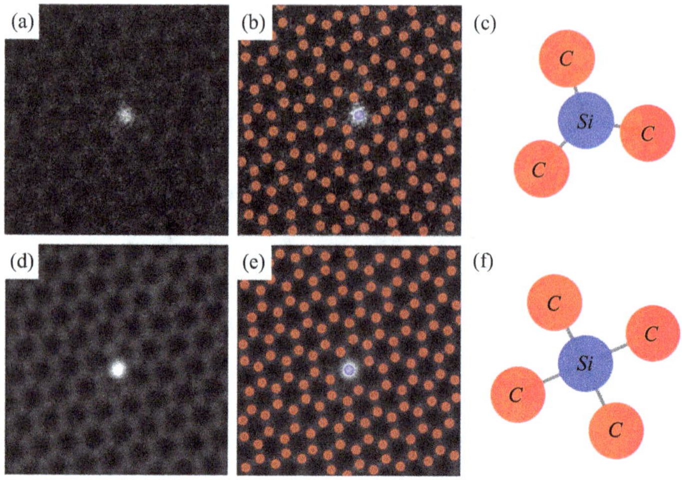

Figure 8.38 *(a) Aberration-corrected HR-STEM image of 3-fold coordinated Si defect in graphene. (b) Output from machine-learning approach for the image in (a), and (c) representation of the defect. (d and e) Same for a 4-fold coordinated Si defect in graphene. © 2017 American Chemical Society. Adapted, with permission, from [121]*

8.4 Conclusions and outlook

This chapter has illustrated the enormous usefulness of microscopy techniques, particularly for the analysis of extended defects, and also, in certain circumstances for the analysis of point defects. It has also highlighted some recurring challenges, which include appropriate sample preparation, identification of artefacts, damage to the sample during imaging, achieving an appropriate resolution to address any particular problem and achieving sufficient sampling to allow statistically significant conclusions to be made in techniques which sometimes image only very limited areas.

Despite these challenges, microscopy techniques play a vital role in the characterisation of defects in semiconductors and as they continue to develop, new opportunities will arise. Although this chapter has been necessarily selective, focussing on some of the current key techniques, new methodologies are continuously being developed, and we note a number of emerging trends. In the TEM sphere, we are increasingly seeing the integration of techniques for the characterisation of (opto)electronic properties of materials directly into the electron microscope. One exciting example of this is the application of CL in the TEM, where the limited interaction volumes attainable due to the use of thin film samples aid the achievement of excellent resolution [143]. Similarly, SEM-based techniques are becoming ever more sophisticated, with CL experiments in the SEM being extended to allow time-resolved as well as time-integrated experiments [144]. In scanning probe microscopy, the range of techniques to allow the electrical characterisation of semiconductors at the nanoscale is ever-expanding and may increasingly be integrated with mechanical property measurements, even for very stiff materials (such as most semiconductors). Hence, nanoscale studies of semiconductors are extending to embrace piezotronic and optopiezotronic devices, where mechanical stimuli are used to manipulate the electronic and optoelectronic properties of nanostructures [145].

All of these endeavours that link materials properties to materials structure have the potential to generate vast data sets. Hence, parallel developments in machine learning and big data techniques will be vital in the future to maximise the information and understanding which can be extracted from microscopy data.

Acknowledgements

The authors would like to thank G. Naresh-Kumar, Ben Hourahine and Aimo Winkelmann for their support during the writing of this chapter.

References

[1] Eaton P., West P. 'Atomic Force Microscopy'. Oxford: *Oxford University Press*. 2010.
[2] Binnig G., Quate C.F. 'Atomic force microscope'. *Physical Review Letters*. 1986;**56**(9):930.
[3] Timoney P., Zhang X.X., Vaid A., *et al.* 'Device level 3D characterization using PeakForce AFM'. Article Number: 97781A in Sanchez M.I., Ukraintsev V.A. (eds.) Metrology, Inspection, and Process Control for Microlithography XXX. Proceedings of SPIE Vol. 9778. Bellingham: SPIE; 2016.
[4] Desbief S., Hergué N., Douhéret O., *et al.* 'Nanoscale investigation of the electrical properties in semiconductor polymer–carbon nanotube hybrid materials'. *Nanoscale.* 2012;**4**(8):2705.
[5] Oliver R.A. 'Advances in AFM for the electrical characterization of semiconductors'. *Reports on Progress in Physics*. 2008;**71**:076501.
[6] Coffey D.C., Reid O.G., Rodovsky D.B., Bartholomew G.P., Ginger D.S. 'Mapping local photocurrents in polymer/fullerene solar cells with photoconductive atomic force microscopy'. *Nano Letters.* 2007;**7**(3):738.
[7] Cox G., Szynka D., Poppe U., *et al.* 'Scanning tunneling microscopy of crystal dislocations in gallium arsenide'. *Physical Review Letters*. 1990; **64**(20):2402.
[8] Feenstra R.M., Woodall J.M., Pettit G.D. 'Observation of bulk defects by scanning tunneling microscopy and spectroscopy: Arsenic antisite defects in GaAs'. *Physical Review Letters*. 1993;**71**(8):1176.
[9] Tersoff J., Hamann D.R. 'Theory and application for the scanning tunneling microscope'. *Physical Review Letters*. 1983;**50**(25):1998.
[10] Domke C., Ebert P., Heinrich M., Urban K. 'Microscopic identification of the compensation mechanisms in Si-doped GaAs'. *Physical Review B*. 1996;**54**(15):10288.

[11] Just S., Zimmermann S., Kataev V., Büchner B., Pratzer M., Morgenstern M. 'Preferential antiferromagnetic coupling of vacancies in graphene on SiO_2: Electron spin resonance and scanning tunneling spectroscopy'. *Physical Review B*. 2014;**90**(12):125449.

[12] Holt D.B., Joy D.C. (eds.). 'SEM Microcharacterization of Semiconductors'. Cambridge, MA: *Academic Press*. 1989.

[13] Goldstein J., Newbury D., Joy D., *et al.* 'Scanning Electron Microscopy and X-ray Microanalysis'. 3rd edn. New York: *Springer*. 2007.

[14] Reimer L. 'Scanning Electron Microscopy: Physics of Image Formation and Microanalysis'. 2nd edn. New York: *Springer*. 1998.

[15] Zhou W., Wang Z.L. (eds.). 'Scanning Microscopy for Nanotechnology: Techniques and Applications'. Berlin: *Springer Science & Business Media*. 2007.

[16] Edwards P.R., Martin R.W. 'Cathodoluminescence nano-characterization of semiconductors'. *Semiconductor Science and Technology*. 2011;**26**(6): 064005.

[17] Yakimov E.B. 'Investigation of electrical and optical properties in semiconductor structures via SEM techniques with high spatial resolution'. *Journal of Surface Investigation. X-ray, Synchrotron and Neutron Techniques*. 2012; **6**(6):887–9.

[18] Jeol. 'Scanning Electron Microscope A to Z: Basic Knowledge for Using the SEM'. https://www.jeol.co.jp/en/applications/pdf/sm/sem_atoz_all.pdf.

[19] Edwards P.R., Jagadamma L.K., Bruckbauer J., *et al.* 'High-Resolution Cathodoluminescence Hyperspectral Imaging of Nitride Nanostructures'. *Microscopy and Microanalysis*. 2012;**18**:1212–19.

[20] Wan Q., Masters R.C., Lidzey D., *et al.* 'Angle selective backscattered electron contrast in the low-voltage scanning electron microscope: Simulation and experiment for polymers'. *Ultramicroscopy*. 2016;**171**:126–38.

[21] Wilkinson A.J., Hirsch P.B. 'Electron diffraction based techniques in scanning electron microscopy of bulk materials'. *Micron*. 1997;**28**(4):279–308.

[22] Trager-Cowan C., Sweeney F., Trimby P.W., *et al.* 'Electron backscatter diffraction and electron channeling contrast imaging of tilt and dislocations in nitride thin films'. *Physical Review B*. 2007;**75**(8):085301.

[23] Picard Y.N., Twigg M.E., Caldwell J.D., Eddy Jr C.R., Mastro M.A., Holm R.T. 'Resolving the Burgers vector for individual GaN dislocations by electron channeling contrast imaging'. *Scripta Materialia*. 2009;**61**(8):773–6.

[24] Hite J.K., Mastro M.A., Eddy Jr C.R. 'Approach for dislocation free GaN epitaxy'. *Journal of Crystal Growth*. 2010;**312**(21):3143–6.

[25] Naresh-Kumar G., Hourahine B., Edwards P.R., *et al.* 'Rapid nondestructive analysis of threading dislocations in wurtzite materials using the scanning electron microscope'. *Physical Review Letters*. 2012;**108**(13):135503.

[26] Schulze A., Strakos L., Vystavel T., *et al.* 'Non-destructive characterization of extended crystalline defects in confined semiconductor device structures'. *Nanoscale*. 2018;**10**(15):7058–66.

[27] Callahan P.G., Haidet B.B., Jung D., Seward G.G., Mukherjee K. 'Direct observation of recombination-enhanced dislocation glide in heteroepitaxial GaAs on silicon'. *Physical Review Materials*. 2018;**2**(8):081601.

[28] Carnevale S.D., Deitz J.I., Carlin J.A., *et al*. 'Rapid misfit dislocation characterization in heteroepitaxial III–V/Si thin films by electron channeling contrast imaging'. *Applied Physics Letters*. 2014;**104**(23):232111.

[29] Yaung K.N., Kirnstoetter S., Faucher J., *et al*. 'Threading dislocation density characterization in III–V photovoltaic materials by electron channeling contrast imaging'. *Journal of Crystal Growth*. 2016;**453**:65–70.

[30] Picard Y.N., Kamaladasa R., De Graef M., *et al*. 'Future prospects for defect and strain analysis in the SEM via electron channeling'. *Microscopy Today*. 2012;**20**(2):12–16.

[31] Zaefferer S., Elhami N.N. 'Theory and application of electron channelling contrast imaging under controlled diffraction conditions'. *Acta Materialia*. 2014;**75**:20–50.

[32] Naresh-Kumar G., Bruckbauer J., Edwards P.R., *et al*. 'Coincident electron channelling and cathodoluminescence studies of threading dislocations in GaN'. *Microscopy and Microanalysis*. 2014;**20**:55.

[33] Mansour H., Guyon J., Crimp M.A., Gey N., Beausir B., Maloufi A.N. 'Accurate electron channeling contrast analysis of dislocations in fine grained bulk materials'. *Scripta Materialia*. 2014;**84**:11–14.

[34] Picard Y.N., Liu M., Lammatao J., Kamaladasa R., De Graef M. 'Theory of dynamical electron channeling contrast images of near-surface crystal defects'. *Ultramicroscopy*. 2014;**146**:71–8.

[35] Pascal E., Hourahine B., Naresh-Kumar G., Mingard K., Trager-Cowan C. 'Dislocation contrast in electron channelling contrast images as projections of strain-like components'. *Materials Today: Proceedings*. 2018;**5**(Issue): 14652–61.

[36] Crimp M.A., Simkin B.A., Ng B.C. 'Demonstration of the $g \cdot b \times u = 0$ edge dislocation invisibility criterion for electron channelling contrast imaging'. *Philosophical Magazine Letters*. 2001;**81**(12):833–7.

[37] Naresh-Kumar G., Thomson D., Nouf-Allehiani M., *et al*. 'Electron channelling contrast imaging for III-nitride thin film structures'. *Materials Science in Semiconductor Processing*. 2016;**47**:44–50.

[38] Vespucci S., Winkelmann A., Naresh-Kumar G., *et al*. 'Digital direct electron imaging of energy-filtered electron backscatter diffraction patterns'. *Physical Review B*. 2015;**92**(20):205301.

[39] Schwartz A.J., Kumar M., Adams B.L., Field D.P. (eds.). 'Electron Backscatter Diffraction in Materials Science'. New York: Springer. 2009.

[40] Wilkinson A.J., Moldovan G., Britton T.B., Bewick A., Clough R., Kirkland A.I. 'Direct detection of electron backscatter diffraction patterns'. *Physical Review Letters*. 2013;**111**(6):065506.

[41] Wilkinson A.J., Britton T.B. 'Strains, planes, and EBSD in materials science'. *Materials Today*. 2012;**15**(9):366–76.

[42] Wilkinson A.J., Meaden G., Dingley D.J. 'High-resolution elastic strain measurement from electron backscatter diffraction patterns: New levels of sensitivity'. *Ultramicroscopy*. 2006;**106**(4–5):307–13.

[43] Littlewood P.D., Britton T.B., Wilkinson A.J. 'Geometrically necessary dislocation density distributions in Ti–6Al–4V deformed in tension'. *Acta Materialia*. 2011;**59**(16):6489–500.

[44] Vilalta-Clemente A., Naresh-Kumar G., Nouf-Allehiani M., *et al.* 'High-resolution electron backscatter diffraction in III-nitride semiconductors'. *Microscopy and Microanalysis*. 2015;**21**(S3):2217–18.

[45] Vaudin M.D., Gerbig Y.B., Stranick S.J., Cook R.F. 'Comparison of nanoscale measurements of strain and stress using electron back scattered diffraction and confocal Raman microscopy'. *Applied Physics Letters*. 2008;**93**(19):193116.

[46] Dingley D.J., Wilkinson A.J., Meaden G., Karamched P.S. 'Elastic strain tensor measurement using electron backscatter diffraction in the SEM'. *Journal of Electron Microscopy*. 2010;**59**(S1):S155–63.

[47] Villert S., Maurice C., Wyon C., Fortunier R. 'Accuracy assessment of elastic strain measurement by EBSD'. *Journal of Microscopy*. 2009; **233**(2):290–301.

[48] Ishido T., Matsuo H., Katayama T., Ueda T., Inoue K., Ueda D. 'Depth profiles of strain in AlGaN/GaN heterostructures grown on Si characterized by electron backscatter diffraction technique'. *IEICE Electronics Express*. 2007;**4**(24):775–81.

[49] Vilalta-Clemente A., Naresh-Kumar G., Nouf-Allehiani M., *et al.* 'Cross-correlation based high resolution electron backscatter diffraction and electron channelling contrast imaging for strain mapping and dislocation distributions in InAlN thin films'. *Acta Materialia*. 2017;**125**:125–35.

[50] Liu R., Schaller R., Chen C.Q., Bayram C. 'High internal quantum efficiency ultraviolet emission from phase-transition cubic GaN integrated on nanopatterned Si (1 0 0)'. *ACS Photonics*. 2018;**5**(3):955–63.

[51] Naresh-Kumar G., Vilalta-Clemente A., Jussila H., *et al.* 'Quantitative imaging of anti-phase domains by polarity sensitive orientation mapping using electron backscatter diffraction'. *Scientific Reports*. 2017;**7**(1):10916.

[52] Winkelmann A., Nolze G., Himmerlich M., Lebedev V., Reichmann A. 'Point-group sensitive orientation mapping using EBSD'. Proceedings of the 6th International Conference on Recrystallization and Grain Growth; Pittsburgh, USA, Jul 2016. Cham: Springer. pp. 281–6.

[53] Momma K., Izumi F. 'VESTA 3 for three-dimensional visualization of crystal, volumetric and morphology data'. *Journal of Applied Crystallography*. 2011;**44**(6):1272–6.

[54] Yacobi B.G., Holt D.B. 'Cathodoluminescence microscopy of inorganic solids'. New York: *Plenum Press*. 1990.

[55] Chichibu S., Wada K., Nakamura S. 'Spatially resolved cathodoluminescence spectra of InGaN quantum wells'. *Applied Physics Letters*. 1997; **71**:2346.

[56] Sugahara T., Sato H., Hao M., *et al.* 'Direct evidence that dislocations are non-radiative recombination centers in GaN'. *Japanese Journal of Applied Physics*. 1998;**37**:L398.

[57] Ettenberg M., Kressel H., Gilbert S.L. 'Minority carrier diffusion length and recombination lifetime in GaAs:Ge prepared by liquid-phase epitaxy'. *Journal of Applied Physics*. 1973;**44**:827.

[58] Casey H.C., Miller B.I., Pinkas E. 'Variation of minority-carrier diffusion length with carrier concentration in GaAs liquidphase epitaxial layers'. *Journal of Applied Physics*. 1973;**44**:1281.

[59] Drouin D., Couture A.R., Joly D., Tastet X., Aimez V., Gauvin R. 'CASINO v2.42—A fast and easy-to-use modelling tool for scanning electron microscopy and microanalysis users'. *Scanning*. 2007;**29**:92.

[60] Norman C.E. 'Challenging the spatial resolution limits of CL and EBIC'. *Solid State Phenomena*. 2001;**19**:78–9.

[61] Bruckbauer J., Edwards P.R., Wang T., Martin R.W. 'High resolution cathodoluminescence hyperspectral imaging of surface features in InGaN/GaN multiple quantum well structures'. *Applied Physics Letters*. 2011;**98**:141908.

[62] Christen J., Grundmann M., Bimberg D. 'Scanning cathodoluminescence microscopy: A unique approach to atomic-scale characterization of heterointerfaces and imaging of semiconductor inhomogeneities'. *Journal of Vacuum Science and Technology B*. 1991;**9**:2358.

[63] Martin R.W., Edwards P.R., O'Donnell K.P., *et al.* 'Cathodoluminescence spectral mapping of III-nitride structures'. *Physica Status Solidi (A)*. 2004;**201**:665.

[64] Edwards P.R., Sleith D., Wark A.W., Martin R.W. 'Mapping localized surface plasmons within silver nanocubes using cathodoluminescence hyperspectral imaging'. *Journal of Physical Chemistry C*. 2011;**115**:14031.

[65] Thonke K., Tischer I., Hocker M., *et al.* 'Nanoscale characterisation of semiconductors by cathodoluminescence'. *IOP Conference Series Materials Science and Engineering*. 2014;**55**:012018.

[66] Tischer I., Feneberg M., Schirra M., *et al.* 'I_2 basal plane stacking fault in GaN: Origin of the 3.32 eV luminescence band'. *Physical Review B*. 2011;**83**:035314.

[67] Lähnemann J., Jahn U., Brandt O., Flissikowski T., Dogan P., Grahn H.T. 'Luminescence associated with stacking faults in GaN'. *Journal of Physics D*. 2014;**47**:423001.

[68] Hsu P.S., Young E.C., Romanov A.E., *et al.* 'Misfit dislocation formation via pre-existing threading dislocation glide in semipolar heteroepitaxy'. *Applied Physics Letters*. 2011;**99**:081912.

[69] Martineau P., Gaukroger M., Khan R., Evans D. 'Effect of steps on dislocations in CVD diamond grown on {0 0 1} substrates'. *Physica Status Solidi (C)*. 2009;**6**:1953.

[70] Mukherjee K., Reilly C.H., Callahan P.G., Seward G.G.E. 'Recombination activity of threading dislocations in GaInP influenced by growth temperature'. *Journal of Applied Physics*. 2018;**123**:165701.

[71] Lee W., Chen J., Chen B., Chang J., Sekiguchi T. 'Cathodoluminescence study of dislocation-related luminescence from small-angle grain boundaries in multicrystalline silicon'. *Applied Physics Letters*. 2009;**94**:112103.

[72] Knab M., Hocker M., Felser T., *et al.* 'EBIC investigations on polar and semipolar InGaN LED structures'. *Physica Status Solidi (B)*. 2016; **253**:126.

[73] Wallace M.J., Edwards P.R., Kappers M.J., *et al.* 'Bias dependence and correlation of the cathodoluminescence and electron beam induced current from an InGaN/GaN light emitting diode'. *Journal of Applied Physics*. 2014;**116**:033105.

[74] Wallace M.J., Edwards P.R., Kappers M.J., *et al.* 'Effect of the barrier growth mode on the luminescence and conductivity micron scale uniformity of InGaN light emitting diodes'. *Journal of Applied Physics*. 2015;**117**:115705.

[75] Chen B., Chen J., Sekiguchi T., *et al.* 'Electron-beam-induced current study of stacking faults and partial dislocations in Schottky diode'. *Applied Physics Letters*. 2008;**93**:033514.

[76] Bandić Z.Z., Bridger P.M., Piquette E.C., McGill T.C. 'Electron diffusion length and lifetime in p-type GaN'. *Applied Physics Letters*. 1998;**73**:3276.

[77] Yakimov E.B. 'What is the real value of diffusion length in GaN?'. *Journal of Alloys and Compounds*. 2015;**627**:344.

[78] Yakimov E.B., Polyakov A.Y., Smirnov N.B., *et al.* 'Diffusion length of non-equilibrium minority charge carriers in β-Ga_2O_3 measured by electron beam induced current'. *Journal of Applied Physics*. 2018;**123**:185704.

[79] Yakimov E.B., Polyakov A.Y., Lee I.-H., Pearton S.J. 'Recombination properties of dislocations in GaN'. *Journal of Applied Physics*. 2018;**123**: 161543.

[80] Hamer P., Tweddle D., Martin T., Wilshaw P., Hiscock M., Lindsay J. 'EBIC-Enabled NanoManipulators – investigating dislocations in mc-solar cells'. *Microscopy and Microanalysis*. 2017;**23**:1426.

[81] Moseley H.G.J. 'The high-frequency spectra of the elements'. *Philosophical Magazine*. 1914;**27**:703.

[82] Friel J.J., Lyman C.E. 'X-ray mapping in electron-beam instruments'. *Microscopy and Microanalysis*. 2006;**12**:2.

[83] Edwards P.R., Martin R.W., O'Donnell K.P., Watson I.M. 'Simultaneous composition mapping and hyperspectral cathodoluminescence imaging of InGaN epilayers'. *Physica Status Solidi (C)*. 2003;**0**:247.

[84] Egerton R.F., Li P., Malac M. 'Radiation damage in the TEM and SEM'. *Micron*. 2004;**35**:399–409.

[85] Banhart F. 'Irradiation effects in carbon nanostructures'. *Reports on Progress in Physics*. 1999;**62**:1181–221.

[86] Egerton R.F., McLeod R., Wang F., Malac M. 'Basic questions related to electron-induced sputtering in the TEM'. *Ultramicroscopy*. 2010;**110**:991–7.

[87] Cliff G., Lorimer G.W. 'The quantitative analysis of thin specimens'. *Journal of Microscopy*. 1975;**103**:203–7.

[88] Hytch M.J., Snoeck E., Kilaaas R. 'Quantitative nanoscale characterisation by electron microscopy'. *Ultramicroscopy.* 1998;**74**:131.

[89] Perlin P., Suski T., Leszczyński M., *et al.* 'Properties of InGaN blue laser diodes grown on bulk GaN substrates'. *Journal of Crystal Growth.* 2005; **281**(1):107.

[90] Dwilinski R., Doradzinski R., Garczynski J., *et al.* 'Excellent crystallinity of truly bulk ammonothermal GaN'. *Journal of Crystal Growth.* 2008; **310**(17):3911.

[91] Frank F.C. 'Capillary equilibria of dislocated crystals'. *Acta Crystallographica.* 1951;**4**:497.

[92] Oliver R.A., Kappers M.J., Sumner J., Datta R., Humphreys C.J. 'Highlighting threading dislocations in MOVPE-grown GaN using an in situ treatment with SiH4 and NH3'. *Journal of Crystal Growth.* 2006; **289**(2):506.

[93] Oehler F., Zhu T., Rhode S., Kappers M.J., Humphreys C.J., Oliver R.A. 'Surface morphology of homoepitaxial c-plane GaN: Hillocks and ridges'. *Journal of Crystal Growth.* 2013;**38**:12.

[94] Das Bakshi S., Sumner J., Kappers M.J., Oliver R.A. 'The influence of coalescence time on unintentional doping in GaN/sapphire'. *Journal of Crystal Growth.* 2009;**311**(2):232–7.

[95] Moram M.A., Oliver R.A., Kappers M.J., Humphreys C.J. 'The spatial distribution of threading dislocations in gallium nitride films'. *Advanced Materials.* 2009;**21**(38–39):3941.

[96] Moram M.A., Johnston C.F., Kappers M.J., Humphreys C.J. 'Measuring dislocation densities in nonpolar a-plane GaN films using atomic force microscopy'. *J. Physics D—Applied Physics.* 2010;**43**(5):055303.

[97] Ketteniss N., Oliver R.A., McAleese C., Kappers M.J., Zhang Y., Humphreys C. J. 'The role of strain in controlling the surface morphology of $Al_xGa_{1-x}N$ following in situ treatment with SiH_4 and NH_3'. *Applied Surface Science.* 2008;**254**(7):2124.

[98] Rueden C.T., Schindelin J., Hiner M.C., *et al.* 'ImageJ2: ImageJ for the next generation of scientific image data'. *BMC Bioinformatics.* 2017;**18**(1):529.

[99] Carnevale S.D., Deitz J.I., Carlin J.A., *et al.* 'Rapid misfit dislocation characterization in heteroepitaxial III-V/Si thin films by electron channeling contrast imaging'. *Applied Physics Letters.* 2014;**104**(23):232111.

[100] Ripley B.D. 'Modelling spatial patterns'. *Journal of the Royal Statistical Society, Series B (Methodological).* 1977;**1**:172–212.

[101] Diggle P. 'A kernel method for smoothing point process data'. *Applied Statistics.* 1985;**1**:138–47.

[102] Besag J. 'Contribution to the discussion on Dr Ripley's paper'. *Journal of the Royal Statistical Society.* 1977;**39**:193–5.

[103] Trager-Cowan C., Naresh-Kumar G., Allehiani N., *et al.* 'Electron channeling contrast imaging of defects in III-nitride semiconductors'. *Microscopy and Microanalysis.* 2014;**20**:1024.

[104] Gmeinwieser N., Schwarz U.T. 'Pattern formation and directional and spatial ordering of edge dislocations in bulk GaN: Microphotoluminescence spectra and continuum elastic calculations'. *Physical Review B*. 2007;**75**(24):245213.

[105] Moram M.A., Ghedia C.S., Rao D.V., *et al.* 'On the origin of threading dislocations in GaN films'. *Journal of Applied Physics*. 2009;**106**(7):073513.

[106] Moram M.A., Gabbai U.E., Sadler T.C., Kappers M.J., Oliver R.A. 'The use of spatial analysis techniques in defect and nanostructure studies'. *Journal of Electronic Materials*. 2010;**39**(6):656–62.

[107] Lozano J.G., Yang H., Guerrero-Lebrero M.P., *et al.* 'Direct observation of depth-dependent atomic displacements associated with dislocations in gallium nitride'. *Physical Review Letters*. 2014;**113**(13):135503.

[108] Richter E., Stoic T., Zeimer U., Netzel C., Weyers M., Tränkle G. 'Si doping of GaN in hydride vapor-phase epitaxy'. *Journal of Electronic Materials*. 2013;**42**(5):820.

[109] Bennett S.E., Holec D., Kappers M.J., Humphreys C.J., Oliver R.A. 'Imaging dislocations in gallium nitride across broad areas using atomic force microscopy'. *Review of Scientific Instruments*. 2010;**81**(6):063701.

[110] Zhu T., Oliver R.A. 'Unintentional doping in GaN'. *Physical Chemistry Chemical Physics*. 2012;**14**(27):9558–73.

[111] Gibart P. 'Metal organic vapour phase epitaxy of GaN and lateral overgrowth'. *Reports on Progress in Physics*. 2004;**67**(5):667–715.

[112] Sumner J., Oliver R.A., Kappers M.J., Humphreys C.J. 'Scanning capacitance microscopy studies of unintentional doping in epitaxial lateral overgrowth GaN'. *Journal of Applied Physics*. 2009;**106**(10):104503.

[113] Ilegems M., Montgomery H.C. 'Electrical properties of n-type vapor-grown gallium nitride'. *Journal of Physical Chemistry of Solids*. 1973;**34**(5):885–95.

[114] Hansen P.J., Strausser Y.E., Erickson A.N., *et al.* 'Scanning capacitance microscopy imaging of threading dislocations in GaN films grown on (0 0 0 1) sapphire by metalorganic chemical vapor deposition'. *Applied Physics Letters*. 1998;**72**(18):2247–9.

[115] Simpkins B.S., Yu E.T., Chowdhury U., *et al.* 'Local conductivity and surface photovoltage variations due to magnesium segregation in p-type GaN'. *Journal of Applied Physics*. 2004;**95**(11):6225–31.

[116] Rhode S., Horton M., Kappers M.J., *et al.* 'Mg doping affects dislocation core structures in GaN'. *Physical Review Letters*. 2013;**111**(2):025502.

[117] Huang Y.L., Ding Z., Zhang W., *et al.* 'Gap states at low-angle grain boundaries in monolayer tungsten diselenide'. *Nano Letters*. 2016;**16**:3682–8.

[118] Hite J.K., Twigg M.E., Tedesco J.L., *et al.* 'Epitaxial graphene nucleation on C-face silicon carbide'. *Nano Letters*. 2011;**11**(3):1190–4.

[119] Gibb A.L., Alem N., Chen J.-H., *et al.* 'Atomic resolution imaging of grain boundary defects in monolayer chemical vapor deposition-grown hexagonal boron nitride'. *Journal of the American Chemical Society*. 2013;**135**:6758–61.

[120] Alem N., Yazyev O.V., Kisielowski C., *et al.* 'Probing the out-of-plane distortion of single point defects in atomically thin hexagonal boron nitride at the picometer scale'. *Physical Review Letters*. 2011;**106**:126102.

[121] Ziatdinov M., Dyck O., Maksov A., *et al.* 'Deep learning of atomically resolved scanning transmission electron microscopy images: Chemical identification and tracking local transformations'. *ACS Nano.* 2017;**11**:12742–52.

[122] Lee S.M., Belkhir M.A., Zhu X.Y., Lee Y.H., Hwang Y.G., Frauenheim T. 'Electronic structures of GaN edge dislocations'. *Physical Review B.* 2000;**61**(23):16033.

[123] Lymperakis L., Neugebauer J., Albrecht M., Remmele T., Strunk H.P. 'Strain induced deep electronic states around threading dislocations in GaN'. *Physical Review Letters.* 2004;**93**(19):196401.

[124] Rhode S.L., Horton M.K., Fu W.Y., *et al.* 'Dislocation core structures in Si-doped GaN'. *Applied Physics Letters.* 2015:**107**(24):243104.

[125] Rosner S.J., Carr E.C., Ludowise M.J., Girolami G., Erikson H.I. 'Correlation of cathodoluminescence inhomogeneity with microstructural defects in epitaxial GaN grown by metalorganic chemical-vapor deposition'. *Applied Physics Letters.* 1997;**70**:420.

[126] Sugahara T., Sato H., Hao M., *et al.* 'Direct evidence that dislocations are non-radiative recombination centers in GaN'. *Japanese Journal of Applied Physics.* 1998;**37**:L398.

[127] Hino T., Tomiya S., Miyajima T., Yanashima K., Hashimoto S., Ikeda M. 'Characterization of threading dislocations in GaN epitaxial layers'. *Applied Physics Letters.* 2000;**76**:3421.

[128] Albrecht M., Weyher J.L., Lucznik B., Grzegory I., Porowski S. 'Nonradiative recombination at threading dislocations in n-type GaN: Studied by cathodoluminescence and defect selective etching'. *Applied Physics Letters.* 2008;**92**:231909.

[129] Medvedev O., Vyvenko O., Ubyivovk E., *et al.* 'Intrinsic luminescence and core structure of freshly introduced a-screw dislocations in n-GaN'. *Journal of Applied Physics.* 2018;**123**:161427.

[130] Brazel E.G., Chin M.A., Narayanamurti V. 'Direct observation of localized high current densities in GaN films'. *Applied Physics Letters.* 1999;**74**:2367.

[131] Chen K.X., Dai Q., Lee W., *et al.* 'Effect of dislocations on electrical and optical properties of n-type $Al_{0.34}Ga_{0.66}N$'. *Applied Physics Letters.* 2008;**93**:192108.

[132] Kozodoy P., Ibbetson J.P., Marchand H., *et al.* 'Electrical characterization of GaN p–n junctions with and without threading dislocations'. *Applied Physics Letters.* 1998;**73**:975.

[133] Stechmann G., Zaefferer S., Schwarz T., *et al.* 'A correlative investigation of grain boundary crystallography and electronic properties in CdTe thin film solar cells'. *Solar Energy Materials and Solar Cells.* 2017;**166**:108.

[134] Nay Yaung K., Kirnstoetter S., Faucher J., *et al.* 'Threading dislocation density characterization in III–V photovoltaic materials by electron channelling contrast imaging'. *Journal of Crystal Growth.* 2016;**453**:65.

[135] Massabuau F.C.-P., Chen P., Horton M.K., *et al.* 'Carrier localization in the vicinity of dislocations in InGaN'. *Journal of Applied Physics.* 2017;**121**:013104.

[136] Prodan A., Marinković V., Gril R., *et al.* 'Scanning tunneling microscopy of defects in $NbSe_2$'. *Journal of Vacuum Science & Technology B.* 2000; **18**(1):60–3.

[137] Almadori Y., Bendiab N., Grévin B. 'Multimodal Kelvin probe force microscopy investigations of a photovoltaic WSe_2/MoS_2 type-II interface'. *ACS Applied Materials & Interfaces.* 2018;**10**:1363–73.

[138] Bampoulis P., Sotthewes K., Siekman M.H., Zandvliet H.J.W. 'Local conduction in $Mo_xW_{1-x}Se_2$: The role of stacking faults, defects, and alloying'. *ACS Applied Materials & Interfaces.* 2018;**10**:13218–25.

[139] Addou R., Wallace R.M. 'Surface analysis of WSe2 crystals: Spatial and electronic variability'. *ACS Applied Materials & Interfaces.* 2016;**8**:26400–6.

[140] Bourrellier R., Meuret S., Tararan A., *et al.* 'Bright UV single photon emissionat point deffectsin *h*-BN'. *Nano Letters.* 2016;**16**:4317–21.

[141] Tran T.T., Bray K., Ford M.J., Toth M., Aharonovich I. 'Quantum emission from hexagonal boron nitride monolayers'. *Nature Nanotechnology.* 2016;**11**:37–41.

[142] Madsen J., Liu P., Kling J., *et al.* 'A deep learning approach to identify local structures in atomic-resolution transmission electron microscopy images'. *Advanced Theory and Simulations.* 2018;**1**:1800037.

[143] Griffiths J.T., Zhang S., Rouet-Leduc B., *et al.* 'Nanocathodoluminescence reveals mitigation of the stark shift in InGaN quantum wells by Si doping'. *Nano Letters.* 2015;**15**(11):7639.

[144] Liu W., Carlin J.-F., Grandjean N., Deveaud B., Jacopin G. 'Exciton dynamics at a single dislocation in GaN probed by picosecond time-resolved cathodoluminescence'. *Applied Physics Letters.* 2016;**109**(04):042101.

[145] Whiter R.A., Boughey C., Smith M., Kar-Narayan S. 'Mechanical energy harvesting performance of ferroelectric polymer nanowires grown via template-wetting'. *Energy Technology.* 2018;**6**(5):928.

Chapter 9

Three-dimensional atomic-scale investigation of defects in semiconductors by atom probe tomography

Lorenzo Rigutti[1] *and Didier Blavette*[1]

Laser-assisted atom probe tomography (APT) is the only analytical microscope able to map out the spatial distribution of chemical species in the three dimensions of space at the atomic scale. We show here its capability in the investigation of clustering and segregation of solute elements to defects in semiconductors, including dislocations, stacking-faults (SFs), interfaces and grain boundaries. The principles of APT techniques are introduced with a particular emphasis on the importance of quantitativity of composition measurements, spatial resolution and three-dimensional (3D) character of APT analyses. Limitations in terms of sensitivity and precision of composition measurements are discussed and confronted to performances of secondary ion mass spectrometry (SIMS). Salient information brought up by APT is illustrated through various examples (quantum dots (QDs), oxides, microelectronic devices, etc.). Whenever possible, APT composition profiles are compared to SIMS results. The use of other complementary techniques (transmission electron microscopy (TEM), optical spectroscopy) for a correct interpretation of APT information on defects is also discussed.

9.1 Introduction

Few analytical microscopes are able to image and characterize materials quantitatively both at the atomic scale and in the three dimensions of space. Electron microscopy tomography is an impressive instrument. However, it is not yet able to reach the ultimate scale. In the particular field of microelectronics, SIMS has been the reference tool for dopant profiling in semiconductors. However, SIMS faces its ultimate limits in terms of spatial resolution for latest generation transistors, and this technique is unable to provide 3D images. In this landscape, APT has gained a unique place.

[1]UNIROUEN, INSA Rouen, CNRS, Groupe de Physique des Matériaux, Normandie Université, 76000 Rouen, France

Because of both its high spatial resolution (0.1 nm in depth and a few tenths of a nm at the sample surface) and 3D imaging capability, APT plays an increasing role in particular in nanoscience including nanoelectronics. APT is the only technique able to map out the distribution of chemical elements in the three dimensions of space at the atomic scale. The analysed volume is typically $50 \times 50 \times 200$ nm^3.

The early developments of APT in the 1990s have been an outstanding event. The first prototypes were designed at Oxford and at the University of Rouen in France (the tomographic atom probe) [1,2]. The latter instrument has been marketed by CAMECA. Later, a new breakthrough was carried out with the design of the local electrode atom probe by Kelly *et al.* [3]. This innovation made it possible to drastically increase analysis rates and mass resolution. This instrument was marketed by IMAGO, a company founded in 1998. IMAGO has now joined CAMECA-AMETEK that is now the only company to release APT. Because of its ultimate spatial resolution (atomic planes can be imaged) and its quantitativity in composition measurements (down to light elements including hydrogen), APT has played a major role in materials science, in particular in the investigation of clustering and the segregation of solute atoms to crystal defects. The local composition in a small selected region of the analysed volume (a few nm^3) is simply derived from the number of atoms of each species. A major advantage of APT is its quantitativity combined with its high spatial resolution that makes it possible to quantitatively analyse both very small precipitates (a few nm in diameter) solute-enriched crystal defects (interfaces, grain boundaries (GBs), SFs, dislocations, etc.). APT was the first instrument able to image in 3D at the atomic-scale Cottrell atmospheres that are the segregation of 'impurities' to dislocation lines [4,5].

With the implementation of ultra-fast pulsed laser to the instrument, a new breakthrough has been achieved [6]. Previously limited to metals, semiconductors as well as oxides and other dielectrics can now be analysed routinely. This challenging innovation has made APT a unique approach for both the academic world and industry [7–11]. A recent monography written as a practical guide for APT users has been recently published [12].

The latter decade has witnessed numerous salient results using APT [9,13]. Let us mention the snow-plow effect exhibited for Pt in NiSi silicides used in contacts of Metal-oxide-semiconductor field effect transistors (MOSFET) or the clustering of boron at the nanoscale in heavily implanted silicon [8,14,15]. Besides, the characteristic size of latest generation nanotransistors (30 nm) is now below the APT field of view (50 nm) so that the 3D distribution of dopants in the device is now achievable [16].

In this chapter, the role of APT in the investigation of extended defects and solute segregation in semiconductors is discussed on the basis of several salient studies mainly carried out in our laboratory (Groupe de Physique des Matériaux) and dealing with one-dimensional (1D) (dislocations), two-dimensional (2D) (interfaces, SFs, GBs) and 3D defects (clusters, QDs). The principles of APT are first presented including a discussion of limitations in terms of spatial resolution and quantitativity. Results and performances are also compared to those of SIMS.

9.2 Atom probe tomography

9.2.1 Generalities and history

APT is an extension in three dimensions of the '1D' atom probe invented in the late 1960s by Müller *et al.* [17]. This first generation of atom probe did not provide any 3D images and it was only possible to get concentration profiles in a small selected region (1 nm) of the sample. Its principle is based on the field evaporation of surface atoms of the specimen (a sharply pointed needle, $R \sim 50$ nm) and the elemental identification of field evaporated ions by time-of-flight (t-o-f) mass spectrometry. With the development of 3D atom probe (APT) in the 1990s, 1D atom probes have no longer been used.

Atom probe has been always combined with field ion microscopy (FIM), which has been the first atomic resolution microscopy technique. FIM is based on the field ionization of a rare gas near the tip surface [18]. Ion beams originating from protruding surface atoms (high field) received on a screen make it possible to get a magnified image (magnification $G = 10^6$) of the specimen surface resolved at the atomic scale.

Formerly limited to metals or good conductors, the implementation of an ultra-fast pulsed laser (300 fs up to a few tens of ps) to the instrument has opened up APT to semiconductors and oxides [6,19]. Kellogg and Tsong were the first to implement a pulsed laser beam on a 1D atom probe [20]. Laser pulses give rise to very rapid thermal pulses that promote the field evaporation of surface atoms. This innovation has made APT a very powerful approach in nanosciences, in particular for the investigation of microelectronics materials, nanowires (silicon based or metallic) and magnetic multilayers for spintronic, including tunnel junctions containing highly resistive oxide barriers [21].

9.2.2 Principles and performances

The principle of the tomographic atom probe is based on the pulsed field evaporation of surface atoms as ions and their chemical identification by t-o-f mass spectrometry. The position of atoms at the sample surface is derived from the impact coordinates onto a time-resolved position-sensitive detector (PSD). A simple point projection is involved (Figure 9.1). The in-depth investigation of the sample is provided by the layer-by-layer evaporation of the specimen. A typical volume of $50 \times 50 \times 200$ nm^3 (10^7–10^8 atom) can be analysed in a few hours. The related analysis rate is close to 1 layer/s.

The high electric field (F) required to field evaporate surface atoms (a few tens of V/nm) is obtained by applying a high electric voltage (V_0) to the specimen prepared in the form of a sharply pointed needle. The tip radius (R) is close to 30 nm. Whereas electrochemical techniques were formerly used to prepare tips from bulk materials, ion milling (focused ion beam, FIB) is now a routine procedure to prepare sharply pointed needles from a large variety of samples including multilayers, nanotransistors, nanopowders and buried interfaces [22–24].

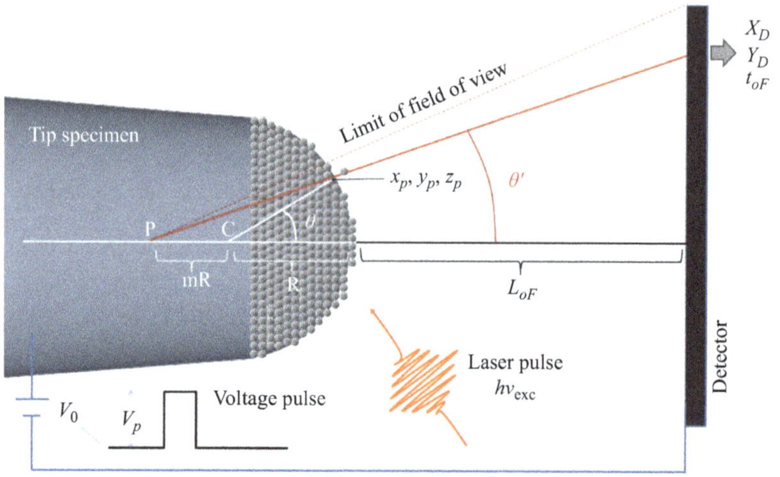

Figure 9.1 The principle of atom probe tomography. In laser-assisted instruments, electric pulses are replaced by laser pulses

The electric field F created at the tip surface is simply given by $F = V_0/\beta R$ where β is called the *field factor* ($\beta = 1$ for an ideal sphere). The field factor β takes on values between 2 and 8, depending on the shank angle and on the electrostatic environment of the specimen. As shown in Figure 9.1, the specimen has the shape of a truncated cone terminated by a spherical cap. Consequently, the tip radius slightly increases in proportion as the material is field evaporated. Because of the shank angle, a slow increase of the applied voltage is necessary in order to keep constant the applied field and to compensate for the increase of the tip radius. Consequently, V is automatically increased during the analysis in such a way to maintain the detection rate constant (typically 0.01 ion/pulse).

The evaporation rate (ion/s) is thermally activated and is given by

$$\Phi = \Phi_0 \exp(-Q_0(1 - F/F_e)/kT) \tag{9.1}$$

Q_0 is the zero-field activation energy for the surface atoms to field evaporate. In a first approximation, the overall activation energy decreases linearly with the applied field (F, (9.1)). F_e is the theoretical evaporation field for which the activation energy is null and $\Phi = \Phi_0$. Increasing the tip temperature (T) or the applied field (F) promotes field evaporation.

The chemical identity of atoms is derived from the t-o-f of field-evaporated ions. High voltage pulses V_p, superimposed to the DC voltage V_0, lead to the correlated field evaporation of surface atoms as ions. Pulses provide the start time for the timer. Emitted ions are repelled by the positively polarized surface and accelerated up to a detector that provides the stop time from which is derived the t-o-f of ions. The electric mode is generally used for metals. For bad conductors (semiconductors, oxides), electric pulses are replaced by sharp laser pulses (0.3–30 ps).

Laser pulses lead to rapid heat pulses that promote field evaporation of surface atoms. The repetition rate (pulses/s) is of the order of 100 kHz in latter generations of APT.

Field evaporated ions are detected on a single particle PSD (channel plates). For typical flight paths (L ~ 0.1–0.5 m), t is of the order of a few hundreds of nanoseconds. The mass to charge ratio of ions (M/n) is derived from the conservation of ion energy:

$$ne(V_0 + V_p) = 1/2Mv^2 \tag{9.2}$$

where $v = L/t$ is the asymptotic ion velocity far from the tip surface and n is the charge state. Ions are generally evaporated as singly charged ions but they are subsequently post-ionized in the close vicinity (nm) of the tip surface. Ionization states are between $n = 1$ and $n = 4$ and increase with the applied field. The mass spectrum related to the analysis of AlGaN/GaN quantum wells (QWs) is provided in Figure 9.2(a). Here, an electric field close to 25 V/nm was derived from the charge state ratio of Ga (2% of Ga^{2+} in Figure 9.2(a)) using the post-ionization theory [25]. Note that the vertical scale is given as the logarithm of the number of detected atoms so that to enhance the dynamics. Al (27 amu) is detected as both doubly and triply charged ions as well as AlN^{2+} molecular ions. The two isotopes of Ga (69 and 71 amu) are clearly displayed in the one plus charge state. GaN molecular ions are also detected. N field evaporates as doubly charged ions

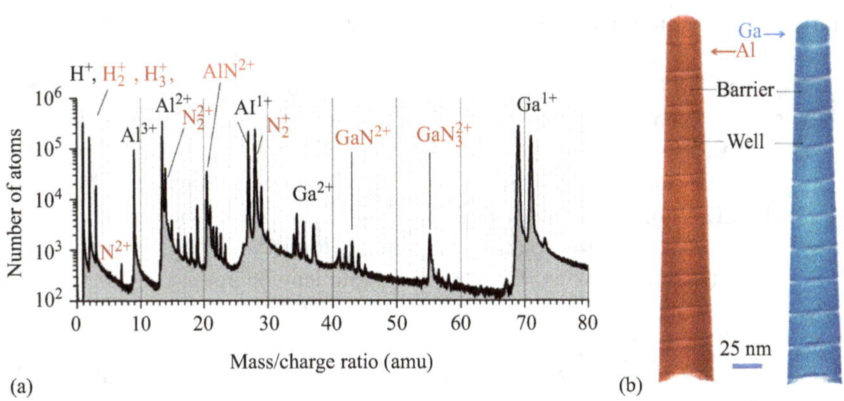

Figure 9.2 (a) Mass spectrum related to the analysis of GaN/AlGaN quantum wells. The tip temperature was 40 K. Laser pulses of 400 fs with a wavelength of 243 nm were used (pulse energy/pulse = 1 nJ, focalized in a light spot 60 µm in diameter leading to an intensity of 0.2 W/µm²). (b) 3D reconstruction of GaN/AlGaN quantum wells by atom probe tomography (initial tip radius R ~ 30 nm). Al-enriched barriers (stack of 11 bi-layers) alternate with Ga-enriched wells. The cone shape of the reconstructed volume is due to the shank angle α of the tip analysed (here α ~ 3.5°). the initial radius of the tip before depth analysis was 300 nm. Analysis conditions are given in Figure 9.3

but also as N_2 molecules detected in both the 1+ and 2+ charge states. Note the presence of hydrogen ions that mainly come from the analysis chamber. Hydrogen is present even under ultra-vacuum pressure conditions (a few 10^{-11} Torr).

The mass resolution (full width half maximum, $M/\Delta M > 1{,}000$) is generally high enough to separate most of elements in spectra. However, peak overlaps may occur whenever mass-to-charge (m/n) ratios of different ions are nearly identical (e.g. $^{14}N_2{}^+$ and $^{28}Si^+$ or $^{16}O_2{}^+$ and $^{64}Zn^{2+}$), causing potential ambiguities in the attribution of a detected event to a given species. For bad heat conductors, the thermal pulse can last over a long time after the laser pulse. This leads to a tailing edge in mass peaks (Figure 9.2(a)). This may be quite problematic for chemical species with close m/n ratios. The presence of this correlated background in mass spectra also limits the detection sensitivity of trace elements according to the location of the related peak. The sensitivity is obviously better for m/n values far from major peaks with long tails. In these regions, a non-correlated background remains. Increasing the mass resolution leads to a much better sensitivity because the width of the mass window ($\delta m/n$) to select elements can thus be reduced. The background noise is then reduced in proportion. The ultimate detection limit is a few tens of ppm (i.e. 0.001 at.%) [26] but may be for certain elements and/or materials not better than 100 ppm.

Under proper conditions (tip temperature below 80 K, low DC field), the ionization rate is the same for every chemical species. This ensures the quantitativity of concentration measurements. In a binary AB alloy, a too high base temperature or a too high DC field (too low laser energy for a given detection rate) leads to the preferential evaporation of less cohesive atoms (e.g. B atoms) between pulses. A part of B atoms is thus not detected in coincidence with evaporation pulses leading to an underestimation of B concentration [27].

The use of a time-resolved PSD makes it possible to position ion impacts and to calculate the position from which atoms originate at the tip surface (Figure 9.1). Eventually, 3D reconstructions, such as the one of the GaN/AlGaN multi-QW set shown in Figure 9.2(b), are obtained. The detection efficiency of channel plates used in the detector is close to 60%. Last improvements increase the efficiency up to 90%. Delay lines are now most often used in detectors to locate ion impacts. PSDs generally fail to assign correct impact positions for multiple events in particular for ions which both have the same m/n and are close to each other on the detector. Small detection rates are therefore generally required (0.01 ion/pulse) in order to reduce multiple events that may bias both positions and composition data. Recent improvements were brought about to detectors in order to increase their performance [28,29].

A simple point projection, close to a stereographic projection, is involved in the reconstruction of the analysed volume (Figure 9.1). The magnification (G) is given by

$$G = L/(m+1)R \tag{9.3}$$

where m defines the position of the centre of projection (CP $= mR$ with $m \sim 0.6$). For $L = 10$ cm and $R = 30$ nm, the magnification reaches $G = 2 \times 10^6$.

The position of atoms at the tip surface (x_p) is roughly derived from that of impacts (X_D) using the equation $x_p = X_D/G$. A resolution of 200 μm on the detector leads to 0.1 nm on the tip surface. More details on the reconstruction procedure may be found elsewhere [30].

Because of the shank angle, R increases with depth (z). This results in a decrease of the magnification (G, (9.3)) in proportion as the tip is field evaporated. Thus the analysed area ($A = D^2/G^2$ with D the detector diameter) increases with z and the reconstructed volume has the shape of a cone as shown in Figure 9.2. The analysed area in Figure 9.2(b) increases from 25 up to 45 nm. For $L = 10$ cm and $D = 8$ cm, the field of view is close to 45°.

The third coordinate, namely the depth scale (z), is derived from the expression of the elemental volume. For δn ions detected, this volume simply writes

$$\delta n \Omega = QA\delta z \tag{9.4}$$

where Ω is the atomic volume and δz is the elemental depth analysed related to the detection of δn ions. This latter expression is only valid for a small value of δz (i.e. small δn) for which $A(R)$ can be considered as constant. By summing elemental depth δz, z is finally calculated.

The maximum depth that can be explored with 3DAP is a few hundreds of nm. It is in theory limited by the maximum voltage (V) that can be applied to the tip. Let us remind that V is automatically raised with z in proportion to the increase of R so as to maintain the detection rate constant. A small shank angle would make it possible in principle to analyse a larger depth. However, the main limiting factor is due to the tip fracture caused by the high electrostatic pressure that is exerted on samples. Tip fracture generally occurs before reaching the maximum voltage (~20 kV). One indirect advantage of laser pulses is that sample fractures are much less frequent because of both the lower field applied (no electric pulses superimposed), and the absence of cyclic stress due to electric pulses, increasing therefore the life time of tips.

Whereas the depth resolution of APT reaches the ultimate limit of a single atomic plane, the lateral resolution (parallel to the sample surface) is far from being as good. In single-phase materials, the resolution reaches 0.3 nm. The atomic resolution is, therefore, not achieved in the three dimensions of space so that imaging the crystal lattice is generally impossible in 3D [31,32]. Unfortunately, it is not technology (detector) but physics of field evaporation that limits the lateral resolution. Aberrations in the ion trajectories very close to the tip surface are the main source of deterioration of the spatial resolution. Atomic simulations of ion trajectories have clearly indicated that it is the subtle movements of ions leaving the surface (tip) that are responsible for the dispersion of impact positions on the detector [33,34]. These aberrations depend on the short-range neighbourhood of the atom leaving the surface and on the local atomic structure and arrangement [35,36]. In two-phase materials, ion trajectory aberrations are more pronounced close to interfaces between phases that have different evaporation fields due to their different cohesion energies. This leads to a degraded spatial resolution (>1 nm) that makes more tricky the measurement of the composition of nano-clusters. Composition bias may interfere for sizes below 2–3 nm when phases have evaporation fields that differ significantly. Several approaches have been developed to deal with this issue [37–39]. In addition to this, field-induced surface migration of atoms may also deteriorate performances [40].

9.2.3 APT versus SIMS

In order to assess the performance of APT, test samples were investigated and results were compared to depth profiles provided by SIMS analysis. Test samples consisted in a stack of four thin SiGe layers (thickness close to 10 nm) with increasing Ge concentration (5, 10, 15, 18 at.% of Ge, Figure 9.4). The APT profile was built in the computer by moving a thin slice (thickness ~1 nm) perpendicularly to layers. The direct comparison of depth profiles provided by APT (laser-assisted wide angle TAP, LaWaTAP and CAMECA) and SIMS shows a very good agreement (Figure 9.4). Compositions expressed as atomic fractions ($X = n/N$ with n, the number of Ge ions detected and N, the number on detected ions per sampling slice) are observed to be quantitative and the depth scale is shown to be well calibrated. Note that concentrations per unit of volume are simply written as $C = X/\Omega$ with Ω, the mean atomic volume.

APT measurements are however subjected to much larger fluctuations than SIMS (Figure 9.3). Local composition fluctuations δX are mainly due to the limited number of ions (N) contained in each sampling slice:

$$\delta X = 2\sigma \tag{9.5}$$

with σ, the standard deviation as given by

$$\sigma = \sqrt{(X(1 - X)(1 - Q)/N)} \tag{9.6}$$

with Q, the detector efficiency ($Q \sim 0.6$).

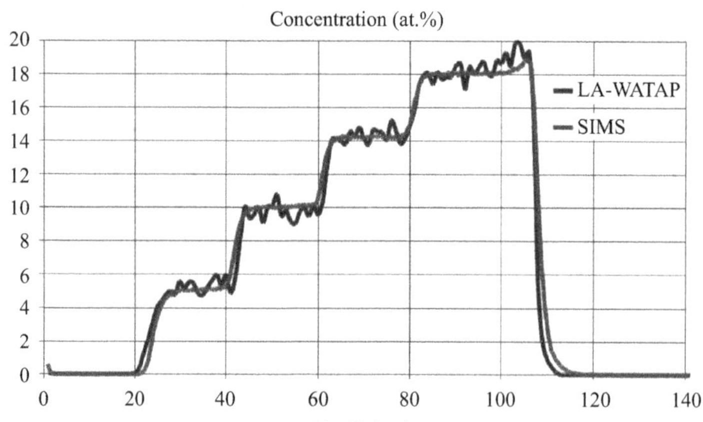

Figure 9.3 Comparison of concentration profiles related to the analysis of a step-like SiGe multilayer (Maya) using a laser-assisted wide angle tomographic atom probe (LA-WATAP, dark fluctuated profile) and SIMS (grey smooth line). Courtesy CAMECA

APT has the drawback of its advantage: due to the small analysed area (typically 30 nm), profiles are subjected to larger statistical fluctuations due to sampling errors (σ). However, APT has a higher spatial resolution that makes it possible to get concentrations on much smaller volumes. This makes APT able to reveal the existence of composition modulations at the nm-scale related for instance to solute-enriched clusters, the size of which may be a few nm. The 3D atomic map of boron atoms in silicon samples containing boron deltas (ultra-thin boron-rich layers) is provided in Figure 9.4 [19]. Four deltas appear in this reconstruction. The related depth profile taken perpendicular to boron layers is compared to SIMS profile in Figure 9.5. Again a good agreement is observed. As expected, the boron layers are separated by 18 nm and a peak boron concentration around 10^{21} boron atoms/cm^3 (~2 at.%) is measured for each delta. Let us also mention the larger steepness of APT profiles compared to SIMS. Interfaces appear more abrupt in APT profiles, illustrating therefore the higher depth resolution of APT. However, the base level of boron between deltas is lower in SIMS profiles. APT has a lower sensitivity compared to SIMS. As shown in Figure 9.5, the background in this particular experiment was rather large (~10^{19} at./cm^3). Depending on the material and the chemical element, sensitivity better than 10^{18} at./cm^3 (<20 ppm) can today be reached in last generations of APTs.

Laser-assisted APT has shown to be a powerful approach for the investigation of dopant distribution in implanted silicon. The implantation profiles as provided by APT and SIMS related to boron in implanted silicon aged 1 h at 600 °C are provided in Figure 9.6. The high implantation dose (5×10^{15} at./cm^2, boron ions of 10 keV), close to that of ultra-shallow junctions in last-generation nano-transistors, exceeds the solubility limit of boron in Si leading to the nucleation of

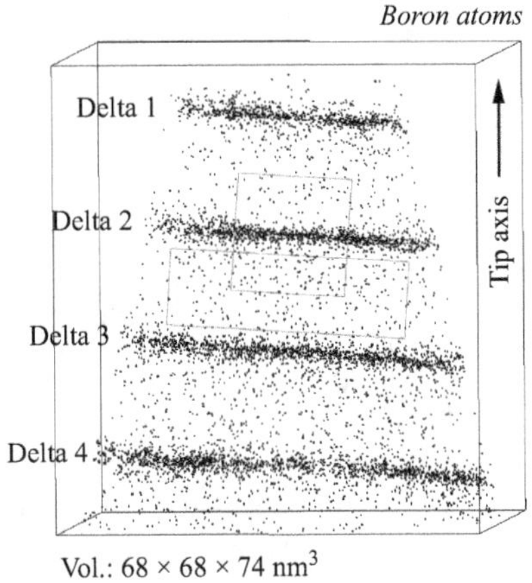

Figure 9.4 3D reconstruction of boron distribution in silicon samples. For clarity, only boron atoms are showed. Four boron deltas are exhibited. Layers are separated by 19.3 nm. Courtesy E. Cadel. © 2009 American Institute of Physics. Adapted, with permission, from [19]

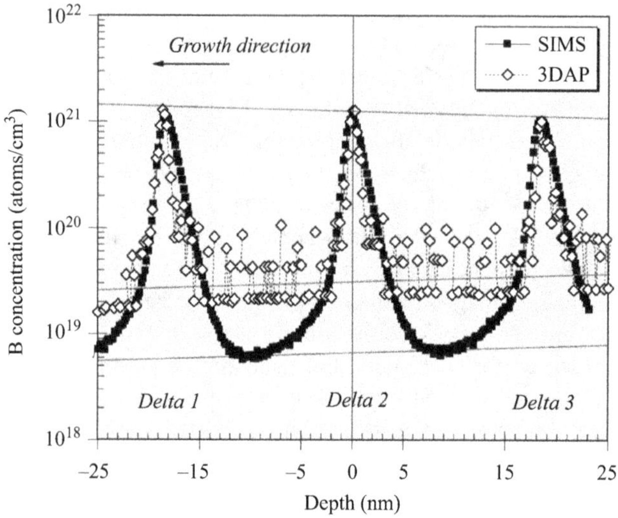

Figure 9.5 *Concentration profile of boron derived from 3D map (Figure 9.7). A thin slice 0.2 nm thick was moved in a direction perpendicular to the three first boron deltas. The surface area of the sampling box is 25 nm × 25 nm. SIMS profile is superimposed to that of APT. Courtesy E. Cadel. © 2009 American Institute of Physics. Adapted, with permission, from [19]*

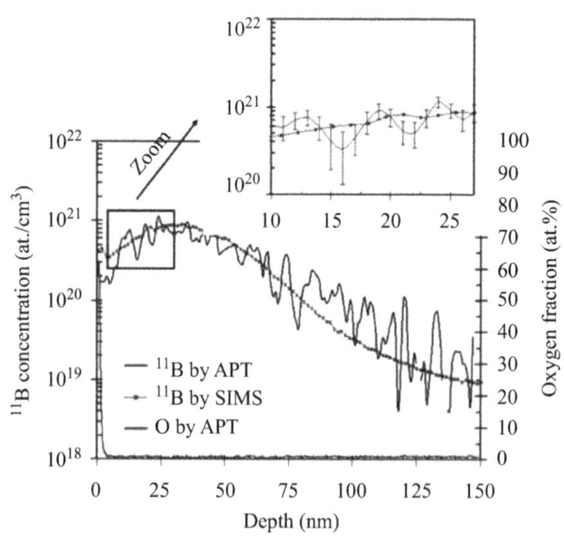

Figure 9.6 *Boron (^{11}B) concentration profile as given by SIMS and APT. APT profile reveals the presence of a native oxide at the sample surface. APT depth profiles were obtained by moving a small box (2 nm thick and 12 nm wide) through the analysed volume in a direction perpendicular to the sample surface that was exposed to boron implantation. Courtesy O. Cojocaru*

boron-enriched clusters. The zoom of Figure 9.6 reveals concentration modulations due to the presence of very small platelet-shaped boron clusters (not shown here) containing about 50 Si+B atoms with a boron content close to 7 at.% [41]. Similar BICs (boron interstitial clusters) were observed by TEM by Cristiano *et al.* [42]. A fairly good agreement between SIMS and APT depth profiles is again observed. The maximum concentration (9×10^{20} at./cm^3), close for both techniques, is detected at a depth of 35 nm in both APT and SIMS profiles (implantation peak). Significant differences are however apparent. Statistical fluctuations appear in the APT profile. Besides, the APT profile gives a boron concentration of 2×10^{20} at./cm^3 at the sample surface, whereas the SIMS profile shows a concentration twice larger (4.2×10^{20} at./cm^3). It is thought that SIMS overestimates boron concentration close to the surface due to matrix effects and the presence of a silicon oxide at the surface (Figure 9.6). Moreover, the presence of native oxide might also have an influence.

9.3 Three-dimensional defects

In this section, we present the application of APT to the study of different categories of 3D extended defects, such as inclusions forming a well-separated phase with respect to the host matrix (nanocrystals (NCs) and QDs) and clustering impurities.

9.3.1 Nanocrystals, quantum dots

NCs and QDs can be considered as 3D functional defects. The latter can be introduced by different means (epitaxial growth or annealing of metastable phases) in order to obtain specific optical properties, such as the emission of single photons for quantum cryptography [43]. Typically, electronic states can be confined within such structures and the energy of radiative transitions can be tailored by varying their morphological and compositional parameters. The case study we report in this section is a correlative analysis of the optical properties of a set of GaN QDs grown within an AlN matrix – their micro-photoluminescence (μPL) spectra were drawn from atom probe specimens (tips) – and their structural properties, assessed by a combination of electron tomography (ET) and APT. The application of electron microscopy is important for different reasons: it gives access to the crystal symmetry of the systems and provides a framework for the optimization of APT reconstructions. It also provides further complementary structural information, as it will be detailed in the following. These systems have been synthesized by molecular beam epitaxy in a regime in which Ga and Al interdiffusion is thermodynamically suppressed.

The reconstructed positions of the Ga atoms extracted from an APT analysis of a GaN/AlN multilayer containing such QDs are shown in Figure 9.7. Part (a) reports the whole analysed volume: 27 GaN-rich layers, also called 'wetting layers' in the Stranski–Krastanov growth mode, are visible. QDs lie on these wetting layers and can be better recognized as 3D islands in the zoom provided in Figure 9.7(b).

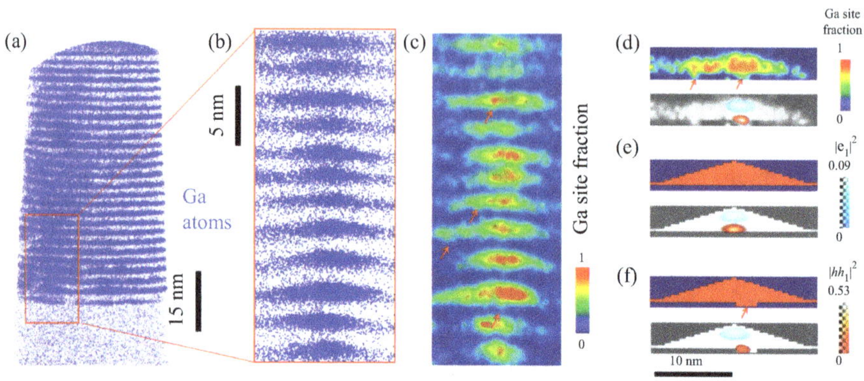

Figure 9.7 Atom probe tomographic study of a GaN/AlN multilayer containing QDs. (a) Reconstructed positions of the Ga atoms within the whole probed volume. (b) Reconstructed positions of the Ga atoms within a 0.5 nm thick slice corresponding to a cross section of a selected stack of QDs. (c) Ga site fraction map within the volume displayed in (b). The red arrows point to interface monolayer fluctuations at the bottom of several QDs. (d) Cross-section of an APT reconstructed compositional map related to a single QD (top), the corresponding electron and heavy hole ground state probability densities (bottom) are shown in the following. Interface fluctuations at the base of the QD in the site fraction maps are highlighted by red arrows. Parts (e) and (f) display the same distributions as in (d), but in (e) the sample is simulated as a hexagonal pyramid according to the morphological parameters derived by STEM tomography. In (f), a cylindrical monolayer fluctuation in the growth direction with 2 nm diameter has been added. Note that in this last case the hole state localizes at the fluctuation. © 2017 American Chemical Society. Adapted, with permission, from [45]

This zoom shows a 0.5 nm thick slice approximately cut through the centre of a QD stack, highlighting the vertical correlation of QDs. Figure 9.7(c) displays the Ga site fraction, defined as the local fraction of Ga atoms over the total number of Ga and Al atoms $N_{Ga}/(N_{Al}+N_{Ga})$, within the same volume defined in (b). The map indicates a compositional interface gradient. However, this apparent compositional gradient is a consequence of the limited lateral resolution of the APT technique, estimated here at ~2 nm. It was previously discussed that in APT, the lateral resolution is significantly worse than the in-depth resolution, as the latter is generally not influenced by surface phenomena such as short-range diffusion and local aberration effects, particularly important at interphase boundaries. Furthermore, the lateral resolution is particularly affected because of the strong mismatch in the binding energies of GaN and AlN, which translates into a significant difference of the evaporation fields: GaN is significantly easier to evaporate than AlN. The in-depth resolution is of the order of 0.25 nm, corresponding to an atomic monolayer in the polar direction of the crystal [19]. Interestingly, Figure 9.7(c) and the close-up reported in Figure 9.8(a) show interface fluctuations at the bottom GaN/AlN interface of several QDs, as pointed out by the red arrows. These features are probably related to the strain conditions leading to the QD nucleation during the growth and can possibly originate from steps on the AlN barrier surface, or from extended defects (dislocations) propagating through the structure [44]. Their amplitude is approximately that of one or two atomic monolayers (1 ML = 0.25 nm).

The correlation between optical and structural properties of these QDs can be assessed by different approaches. The basic idea is using the structural information in order to calculate the electron and hole states within a suitable model. In the present case and in the case illustrated in Section 9.4, the scanning transmission electron microscopy (STEM) and APT information was used to build the band landscapes of the system under study. Subsequently, the Schrödinger equation was solved for electron and holes states within a continuum model (the effective mass approximation). If the electron–hole (e–h) states are directly calculated over all compositional maps obtained by APT, as shown in the bottom panel of Figure 9.7(d), the agreement between the statistical distributions of the μPL energies (Figure 9.8(a)) and of the energies obtained by this calculation (Figure 9.8(b)) is poor. This is the result of the fairly large compositional gradient observed due to APT artefacts. The latter leads to an overestimation of the bandgap energy within the QDs. A much better agreement (Figure 9.8(c) and (d)) is obtained by calculating the e–h states within simpler, truncated hexagonal structures as shown in Figure 9.7(e), in which the geometrical parameters are obtained by ET analysis. However, APT still yields information which is not accessible by ET, namely the presence of interface fluctuations at the lower interface of the QD. From the optoelectronic point of view, these features may induce carrier localization, i.e. a higher degree of confinement and a diminution of the photoluminescence (PL) energy – an important effect for the optical properties. In fact, if one takes into account the presence of such fluctuations and calculates the e–h states within volumes such as the one depicted in Figure 9.7(f), the agreement between the expected distribution of transition energies (Figure 9.8(c) and (d)) and the distribution of the PL energies further improves [45]. This case study points out several advantages and drawbacks of APT. On the one hand, it yields the possibility of imaging 3D NCs [46,47] or QDs, with a spatial resolution close to the lattice constant in the direction of the analysis. However, it must be kept in mind that the lateral resolution may be insufficient for the application of models aiming to correlate structural and functional properties. In general, the application of complementary techniques (in this case STEM in ET mode) can provide the missing structural information.

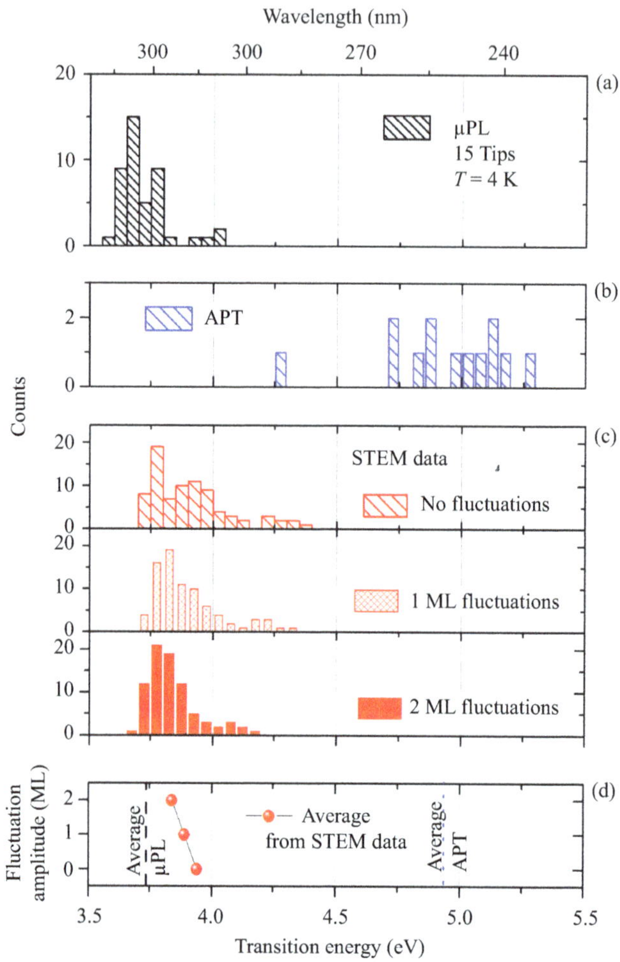

Figure 9.8 *Correlative APT-ET-μPL study of the optical and structural properties of GaN/AlN quantum dots. Histograms of (a) μPL peak energies, (b) transition energies calculated over the QD compositional distribution measured by APT, (c) transition energies calculated within pyramidal QDs as derived from ET, with the addition of a bottom interface fluctuation of 0 ML (top), 1 ML (middle) and 2 ML (bottom). The average transition energies calculated from STEM data are plotted in (d) as a function of the interface fluctuation amplitude, where they are compared with the average μPL peak energy and with the average energy calculated based on the APT composition maps. © 2017 American Chemical Society. Adapted, with permission, from [45]*

9.3.2 Impurity clusters

Atom probe allows studying the behaviour of doping impurities, which are a key element of defect engineering in electronics. As the sensitivity of APT is in the range of 10 ppm, it is possible to assess impurities in concentrations roughly above the 10^{17} cm^{-3} threshold. Gallium nitride represents a particularly interesting case study. In the mainstream technology, doping is obtained in GaN with Mg and Si impurities for the p-type and the n-type, respectively. The detection of Si represents a major problem, because the mass peak of $^{28}Si^{+}$ and $^{28}Si^{2+}$ overlaps with $^{14}N_2^{+}$ and $^{14}N^{+}$ peaks at 28 and 14 amu, respectively. The other Si isotopes, present in much lower quantity (8% of the total amount of Si), may also be hidden by nitrogen hydride molecular ions related to residual H in the analysis chamber. However, n-type doping of GaN(Si) does not present particular technical challenges. On the other hand, p-type doping was up to the early 1990s, the main challenge in nitride technology. Only through the discovery of the procedures for Mg activation, it became possible to develop the nowadays well-affirmed branch of nitride optoelectronics. Mg impurities act as acceptors when they substitute Ga atoms, but they are characterized by relatively high activation energy (0.2 eV). Furthermore, only a part of Mg is activated, possibly according to its configuration within the lattice matrix: at high concentrations, Mg may occupy interstitial sites, where it would not act as an acceptor. Atom probe was applied to this problem: although the spatial precision is not sufficient to discriminate between interstitial and substitutional configurations. In fact, APT can assess whether Mg tends to form clusters. The study performed by Bennett *et al.* [48] first reported on the Mg clustering behaviour in GaN/AlGaN superlattice. Other studies followed [49,50], but no systematic study of the dependence of Mg clustering on concentration and annealing parameters has been carried out yet. A precise quantification of the actual amount of activated Mg has not been addressed either. We report in Figure 9.9 the analysis of the p-type region of GaN/InGaN-based light-emitting diode. Figure 9.9(a) shows the distribution of Mg impurities within a reconstructed subvolume. In part (b), the isosurfaces (i.e. the surfaces surrounding regions in which the atomic fraction of Mg is higher than a certain threshold – in this specific case set at 3 at.%) highlight the regions containing higher Mg concentrations. The visual inspection of these images is not sufficient to conclude about clustering, as fortuitous clusters can occur even for random impurity distributions.[1] In the following, we illustrate two methods for quantitative assessment of clustering.

The first method is based on volume sampling and is known as frequency distribution analysis [51,52]. The application of this method is reported in Figure 9.9(c). The distribution of an atomic species i can be sampled in volume elements containing a given number of atoms N_{bin}. For each of these elements, the

[1]There is a common abuse of language which tends to confuse a 'uniform', or 'homogeneous' distribution with a 'random' distribution. The distribution of an atomic species i within a lattice cannot be uniform or homogeneous – with the exception of the scarcely interesting cases where the species i constitutes either 0% or 100% of the atoms – as there will always be statistical fluctuations, and the fraction of solute atoms depends on the position and size of the sampling volume.

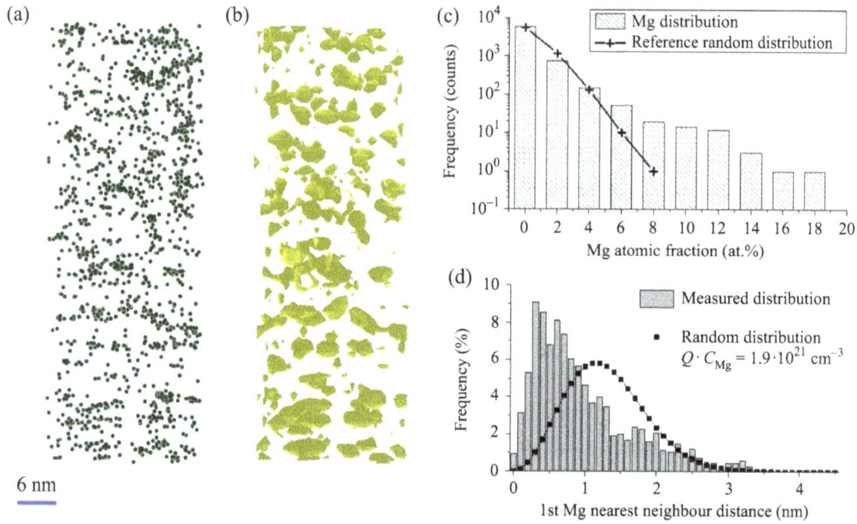

Figure 9.9 Assessment of Mg clustering in p-type GaN. (a) Spatial distribution of Mg atoms in the reconstructed subvolume of the p-type region of a GaN/InGaN light-emitting diode. (b) Isosurfaces surrounding the regions where the Mg fraction is higher than 3 at.%. (c) Frequency distribution analysis of the Mg fraction (bars) compared with the reference binomial (crosses) corresponding to a random distribution of the impurities. (d) First nearest neighbour (1NN) distance distribution extracted from the experimental data (bars) compared with the 1NN distance distribution expected for a random solution with the same concentration C_{Mg}

atomic fraction of the species i is determined as $x_i = n_i/N_{bin}$. This allows for testing the frequencies of site fraction values versus the binomial law describing the probability of finding the atomic fraction x_i within a bin:

$$P_0\left(x_i = \frac{n_i}{N_{bin}}\right) = P(n_i; \langle x_i \rangle, N_{bin}) = \binom{N_{bin}}{n_i} \langle x_i \rangle^{n_i} (1 - \langle x_i \rangle)^{N_{bin}-n_i} \tag{9.7}$$

where $\langle x_i \rangle = N_i/N_{atoms}$ is the average atomic fraction, N_i and N_{atoms} being the total number of atoms of the species i and the total number of atoms, respectively, detected within the whole sampling volume V. The average concentration of the species i is given by $C_i = N_i/(QV)$. In the data shown in Figure 9.9, the average fraction of Mg is $\langle x_i \rangle = 0.9\%$ and its concentration is $C_{Mg} = 3 \times 10^{20}$ cm^{-3}. Figure 9.9(c) reports the binomial distribution (crosses) and the measured distribution of Mg (bars) from the volume shown in Figure 9.9(a) and (b) with sampling bins containing $N_{bin} = 50$ atom. The measured distribution contains more contributions with high Mg content than the binomial distribution, indicating that

Mg is not randomly distributed but forms clusters. The conformity of a measured distribution to a binomial distribution was assessed by a chi^2 test.

The second method [53,54] consists in computing the first nearest neighbour (1NN) distances between solute atoms (Mg here) and in comparing the experimental 1NN distance distribution with the theoretical distribution that is expected for a random distribution of solute atoms:

$$P_0^{1NN}(r) = 4\pi Q C_i r^2 \exp\left[-\frac{4}{3}\pi Q C_i r^3\right] \quad (9.8)$$

where Q is the detection efficiency of the Mg impurities and C_i is the average concentration of solute atoms per unit of volume. The distance distribution of Mg 1NN measured for the dataset shown in Figure 9.9(a) and (b) is reported in part (d) of the same figure (bars). The measured distribution is significantly different from the random distribution (squares) and shows that Mg 1NNs tend to be closer together than predicted for a random solid solution. A behaviour similar to that of Mg in GaN has been assessed for other highly concentrated impurities, such as Boron in Si [41]. A number of other methods are available for the assessment and for advanced analysis of clustering [55].

9.4 Two-dimensional defects

9.4.1 Segregation to stacking faults

APT provides unique information on segregation effects to 2D defects. However, the investigation of planar defects poses specific metrology issues. We shall illustrate this aspect through a first example dealing with CdSe:ZnTe multilayers that could be promising candidates for photovoltaic devices. TEM investigations that were carried out on the same material revealed the presence of numerous {1 1 1} SFs [56]. APT images shown in Figure 9.10 reveal that these SFs are enriched in Zn. Such a phenomenon is called Suzuki effect [57]. A zoom of a segregated SF ($11 \times 11 \times 17$ nm^3) is shown in Figure 9.10(c). The Zn atoms observed in the CdSe layer away from the fault are attributed to noise counts. The related concentration profiles of Zn, Cd and Se shown in Figure 9.10(d) were built perpendicular to the SF. The increase of Zn signal is correlated with a decrease in the Cd signal that is of the same amplitude (~8%). Zn atoms are, therefore, thought to be in substitution to Cd atoms in the ordered CdSe structure. The Zn enrichment spreads over a width (w) close to 6 nm that is much larger than the actual width of SFs (~$d_{1\,1\,1}$). This is due to the spatial resolution of APT that is limited by local trajectory aberrations that affect the ion trajectory close to SFs. Such aberrations are caused by local magnification effects due to differences in tip curvatures related to the presence of regions of different evaporation fields (see Section 9.2). Zn-enriched SFs in CdSe are thought to have a higher evaporation field (F) than CdSe. This high-field (F) region leads to the formation of a protuberance at the tip surface of smaller radius $R = V/\beta F$ that accommodates the higher evaporation field.

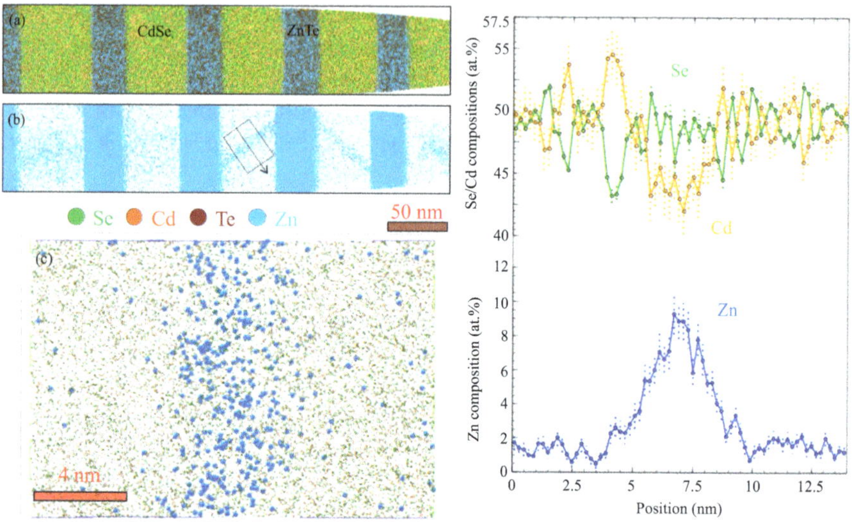

Figure 9.10 (a) and (b) 3D APT reconstruction of the CdSe–ZnTe superlattice showing the distribution of Zn, Te, Cd and Se atoms. (c) Zoom of the Zn distribution close to the enriched fault displayed in (b). Concentration profiles of Zn, Cd and Se as derived from the sampling volume in (c) and taken perpendicular to the stacking fault. Courtesy B. Bonef

This leads to the so-called defocusing effects in the ion trajectories that enlarge the apparent thickness of the enriched zone in the reconstructed image (Figure 9.10(c)). As the amount of atoms is preserved (here Zn), the observed amplitude of the concentration peak is smaller (9 at.% of Zn, Figure 9.10(d)). The peak concentration of Zn in the SF is therefore underestimated. However, a correction based on the conservation of the amount of solute elements in the segregated region (Zn) may be brought about. The integral of the concentration profile, which is proportional to the total number of solute atoms segregated to the defect, is indeed not affected by trajectory overlaps. Assuming that the concentration peak (Figure 9.10(d)) has a Gaussian shape (width $w \sim 4\sigma$ with σ the standard deviation of the Gaussian distribution), elemental considerations show that the real concentration C^* can be derived from the measured concentration C as following:

$$C^* = \frac{\sigma\sqrt{2\pi}}{\delta} \tag{9.9}$$

where δ is the actual width of the [1 1 1] SF ($d_{1\,1\,1} = 0.5$ nm) and $\sigma = w/4$ is equal to 1.5 nm according to Figure 9.10(d). This correction is based on the transformation of the experimental Gaussian profile into a square wave of height C^* and width δ. It leads to C^* that equals 66 at.%. The defect is then almost saturated in Zn atoms (ideally 100% of Zn).

Such biases would be minimized if the SF was parallel to the tip surface. In such a situation, the spatial resolution is excellent as only the depth resolution contributes and the latter is much better than the lateral resolution (cf. Section 9.2). Unfortunately this experiment would require to prepare an APT sample oriented 60° relative to the [0 0 1] direction, which is rather challenging.

9.4.2 Stacking faults in heterostructures

SFs can significantly influence the optical properties of heterostructures [58]. Their effect can be assessed by correlative microscopy methods in which APT plays an important role [59]. Figure 9.11(a) shows a SEM image and a scheme of a GaN microwire containing a system of 20 InGaN QWs. The stem of the wire is oriented along the crystal [0 0 0 1] direction, while the QWs lie on the $\{1\,\overline{1}\,0\,0\}$ planes. In these systems, it is possible to fabricate APT specimens conserving part of their optical emission properties. Figure 9.11(b) is a microscopic image displaying the PL emitted by a tip specimen under excitation with a 244 nm laser at cryogenic temperature ($T = 4$ K). The relative spectrum, shown in Figure 9.11(c), consists of a series of narrow lines. This suggests that carriers localize and recombine radiatively within a strongly inhomogeneous system. Some of the structural properties of the system could be verified by STEM, as reported in Figure 9.11(d). Here, two QWs are visible as the bright-contrasted bands lying horizontally in the figure. The interfaces of the QWs appear as quite sharp on the core side and somehow diffuse on the surface side (on the top and bottom part of the image, respectively). High-resolution imaging, shown in Figure 9.11(e), also reveals the presence of SFs lying on the {0 0 0 1} planes (basal SFs of type I_1, consisting in one single violation of the hexagonal stacking sequence).

The STEM-high-angle annular dark field (HAADF) image of a specimen tip subsequently analysed by APT is shown in Figure 9.12(a), while the APT 3D reconstruction of the positions of the In atoms is shown in Figure 9.12(b). Note that the comparison between STEM and APT makes it possible to perform as an accurate 3D reconstruction. The APT analysis completes the STEM observations. APT can show, for instance, that the diffuse contrast on the surface side of the QW interface in the STEM image of Figure 9.11(d) is related to a compositional gradient rather than to a rough QW interface, as displayed in Figure 9.12(c). Another important observation is that the In in-plane distribution within QWs is not a random distribution [51]. This can be qualitatively visualized either from the 3D reconstruction, as shown in Figure 9.12(b) or by calculating the 2D In/(In+Ga) concentration averaged over the thickness of a rectangular box containing the whole QW, as shown in Figure 9.12(d). The actual In III-site fraction can be then correctly measured by sampling nm-sized volume elements in 3D. The measurements indicate that the peak In III-site fraction is around 20%, which is significantly larger than what could be obtained by simple 1D profiling. In a 1D profile, indeed, the composition of In-poor and In-rich regions averages out. Furthermore, it can be proved that the distribution of In in the QWs not illuminated by the electron beam in the STEM but imaged by APT is similar to the QWs

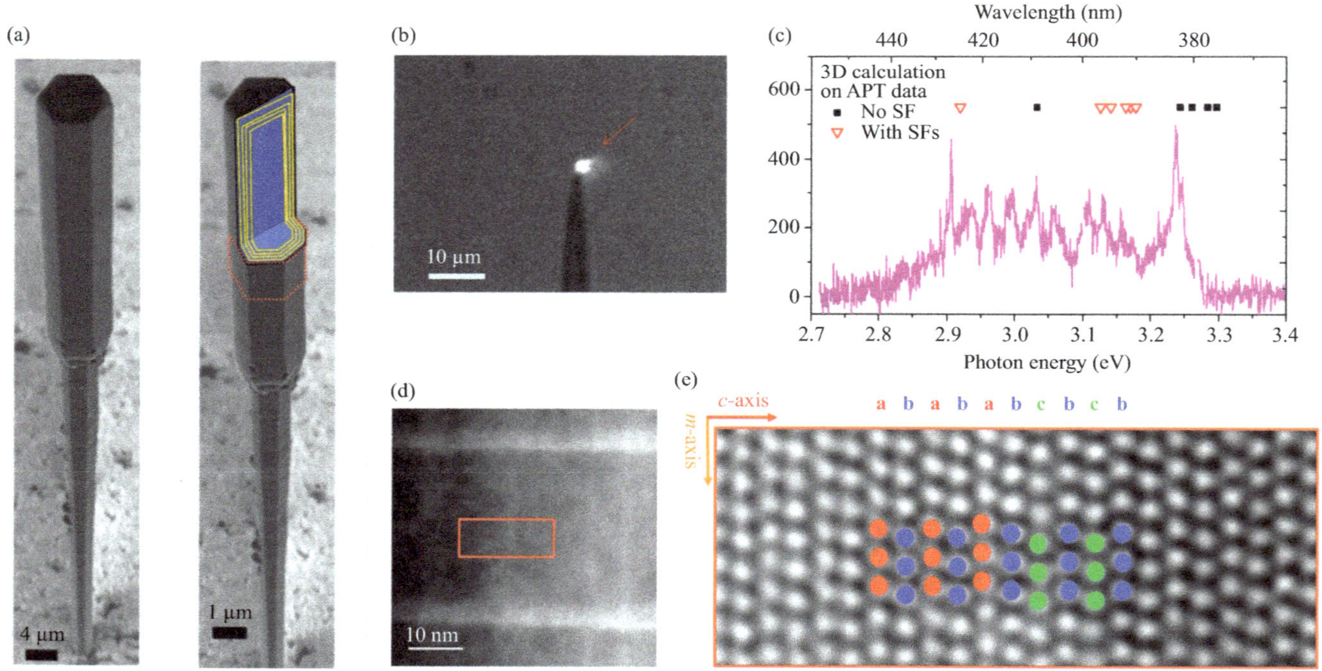

Figure 9.11 *(a) SEM image a GaN microwire containing radial InGaN QWs on its sidewalls; the right-hand side image shows a scheme of the multishell quantum well system, with the red dashed line approximately indicating the region from which APT tips were extracted. (b) Optical microscopic image of the PL (highlighted by the red arrow) emitted by a specimen mounted on a tungsten tip under excitation with a 244 nm cw laser at 4 K. (c) µPL spectrum of an APT tip specimen. (d) STEM-HAADF image recorded on a specimen showing two QWs (horizontal bright-contrasted bands) and several vertical bright-contrasted bands, which are shown in the zoom in (e) to be basal stacking faults. © 2014 American Chemical Society. Adapted, with permission, from [59]. © 2016 American Institute of Physics. Adapted, with permission, from [60]*

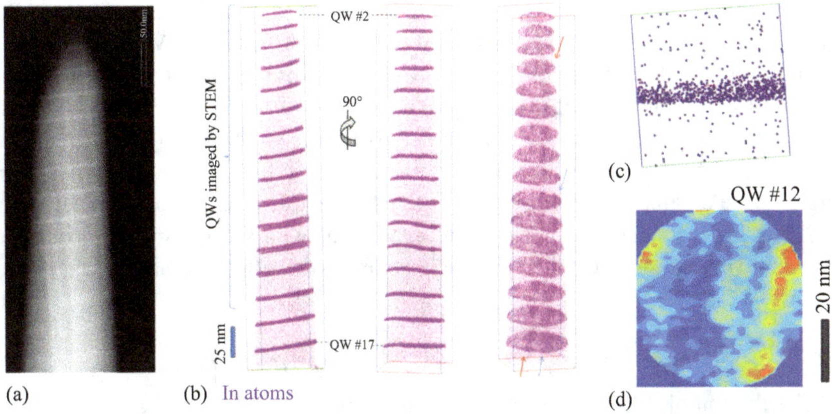

Figure 9.12 (a) HAADF-STEM image of the tip milled out of nano-chunk C: the bright contrast nearly horizontal lines correspond to InGaN quantum wells inside a GaN matrix. (b) APT 3D reconstruction of 16 out of 20 QWs contained in the tip. Only 10% In atoms are shown for clarity; the arrows point to the position of two series of In-rich regions, most likely related to basal SFs propagating through the multi-QW structure. (c) Close-up of a region of QW #4. (d) 2D density plot of the InN fraction calculated in a rectangular box containing QW #12. © 2014 American Chemical Society. Adapted, with permission, from [59]

imaged by STEM, indicating that the exposure to the electron beam was sufficiently short not to induce an artificial In rearrangement within the system. In-rich bands are defined along the *a*-direction [11–20] of the crystal. Finally, the In-rich bands propagate through the whole QW systems, as highlighted by the arrows in the rightmost image of Figure 9.12(b) [59]. The direction of the In-rich regions and their propagation along the {0 0 0 1} planes correlates well with the presence and with the orientation of the SFs [60].

The effect of the non-uniform In distribution and that of the SFs assessed by the correlative STEM and APT analysis can be taken into account for the calculation of the expected optical transition energies of the QW system. In In-rich regions, the local bandgap is lowered. For this reason, carriers tend to localize in these potential energy minima. A similar effect results from the presence of SFs. A type I_1 basal SF (Figure 9.11(e)) can be considered as an inclusion with cubic symmetry in the GaN (or InGaN) hexagonal lattice. The bandgap of hexagonal GaN is $E_g = 3.45$ eV, while SFs in GaN have characteristic emission lines at 3.41 eV. The calculations of the PL transition energies have been performed within a continuum model (the effective mass approximation), solving the Schrödinger equation for electrons and holes within APT-reconstructed volumes. As such volumes only contain information about the 3D chemical distribution of the system, a further set of calculations was performed by adding the effect of the SF on the potential energy landscape: an SF crossing a QW in correspondence of an In-rich region is assumed to lower the bandgap by additional $\Delta E = 50$ MeV. There is some uncertainty in the literature concerning the actual alignment between the conduction band Γ-point energies of the hexagonal matrix and of the SF, but the variation of this parameter can be shown to yield only small variations in the calculated transition energy values. In this way, two simulated sets of expected transition energies were obtained, as displayed in Figure 9.11(c). The set calculated neglecting the SFs is in reasonable but not optimal agreement with the PL spectrum of a tip specimen. The agreement becomes much better if the effect of SFs is taken into account [60].

9.4.3 Intergranular segregation of dopants in silicon

The distribution of dopants within latter generations of MOSFET transistors (metal oxide semiconductor–field effect transistor) implemented in ultra-large scale integrated devices is of utmost importance. Even if the dopant level does not exceed the solubility limit (no supersaturation), dopants such as B, P and As may segregate to 2D-defects also leading to partial electrical deactivation. In latest generations MOS transistors, the gate is made of a heavily doped poly-crystalline silicon layer to ensure good electric conductivity. APT investigations revealed that both As and P segregate to GBs in the gate (Figure 9.13) [61]. APT and SIMS depth-profiles (Figure 9.13(b)) again show good agreement. For small grain sizes, such a pronounced intergranular segregation reduces the dopant concentration in grains (i.e. the concentration of electrically active carriers) and in consequence the

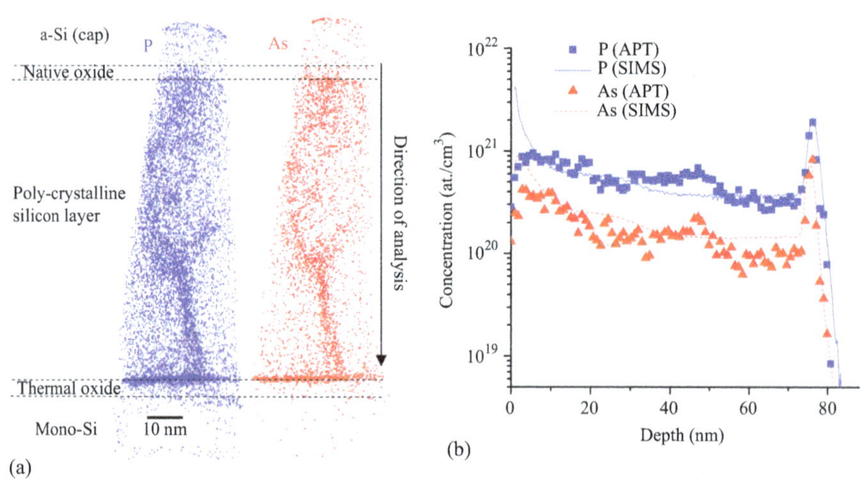

Figure 9.13 (a) 3D map of As and P atoms in an implanted and annealed (1,000 °C, 1.5 s) poly-Si sample (gate of MOSFET transistor). The segregation of As and P at grain boundaries is clearly evidenced. Silicon atoms are not represented for the sake of clarity. (b) SIMS and APT concentration profiles of As and P in the poly-Si layer for the same sample. Courtesy S. Duguay. © 2010 American Institute of Physics. Adapted, with permission, from [61]

overall conductivity of the poly-Si layer. This alters the ability of the gate to properly control the electron transport in the channel of MOSFETs. The segregation of As and P to both GBs and SiO_2/Si interfaces is also clearly evidenced in Figure 9.13(a). From such images, the dopant concentration within grains and at GBs can be measured and corrections may be brought in the same way as for CdSe–ZnTe superlattices previously discussed.

9.4.4 Segregation of dopants to gate interfaces in MOSFET transistors

Figure 9.14 illustrates the capability of APT to analyse and reconstruct part of a p-MOS nano-device in order to reveal the distribution of boron atoms after ion

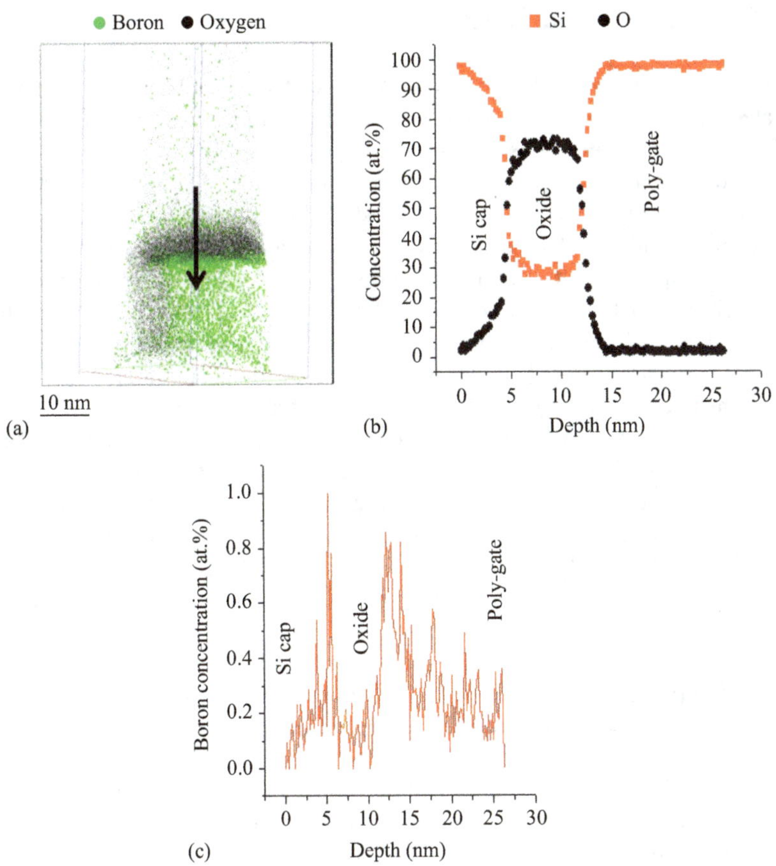

Figure 9.14 (a) APT image of the gate of a PMOS transistor. Only oxygen atoms (black) and boron atoms (green) are represented for clarity. (b) Si and O concentration profiles through the oxide surrounding the gate. (c) Related boron concentration profile showing segregation to interfaces on both sides of the oxide. Courtesy S. Duguay. © Cambridge University Press. Reproduced, with permission, from [62]

implantation [62]. The sample was protected prior FIB milling by an amorphous Si (a-Si) cap. As shown in 3D reconstruction (Figure 9.14(a)), an oxide was present between the a-Si cap and the boron-enriched gate (heavily implanted silicon) of the transistor. The silicon oxide region (gate oxide) is clearly exhibited in the depth profile depicted in Figure 9.14(b) and identified as SiO_2 (66 at.% of O). The depth profile in Figure 9.14(c) shows that boron is subjected to large concentration fluctuations. In particular, boron is observed to segregate to interfaces on both sides of the gate oxide (Si-cap/oxide and oxide/poly-gate). The atomic fraction of B in the poly-Si gate is close to 0.2 at.% compared to 0.8 at.% at interfaces. Similar interfacial segregations of dopants (As) were observed in high-k metal stacks of MOSFETs [63].

9.5 One-dimensional defects

Because of its high spatial resolution, APT is able to exhibit the presence of decorated dislocations. Such solute-enriched clouds that form around dislocation lines are called Cottrell atmospheres [64]. Such atmospheres were evidenced and composition characterized in various metallic materials. Atomic planes are very often hard to image in the vicinity of dislocations so that no crystallographic information is available. The dislocation can be detected only because of its decoration by impurities. Revealing the crystallography of the dislocation was achieved for the first time in boron-doped FeAl ordered alloys. The (0 0 1) atomic planes were imaged, and the Burgers circuit was drawn revealing the presence of an edge (0 0 1) dislocation [4]. Boron was shown to segregate along (0 0 1) edge dislocations at the expense of Al and local concentrations were measured.

In semiconductors, it is much more challenging to visualize lattice planes. However, decorated dislocations can be revealed. APT could provide evidence for As, B and Ni impurity segregation in Si [5,65,66]. In one of these experiments, illustrated in Figure 9.15, silicon was implanted with a high boron dose (5×10^{15} at./cm^2) at 30 keV and further annealed at 950 °C for 30 s. TEM images (Figure 9.15(a)) revealed the presence of a high density of dislocation loops (~10^{11}/cm^2) distributed around the projected range of implanted boron atoms. As shown in Figure 9.15(b), boron enrichment is observed along a dislocation loop having up to 50 nm in diameter. The segregation energies of boron atoms to these defects were derived from the theory of Cottrell and were estimated to be ~0.35 eV.

Finally, it is worth mentioning that the application of an electric field to an APT specimen may strongly perturb dislocations. As detailed further in Section 9.6, a mechanical stress develops as a consequence of the surface electric field. The resolved stress acting on a dislocation line may exceed the Peierls–Nabarro limit, and the dislocation can move within the specimen. This phenomenon is illustrated in Figure 9.16, which is a series of FIM micrographs recorded in sequence during the evaporation of a GaN tip specimen. In the first image (top), the crystal does not exhibit any evidence of the presence of dislocations within the limits of the spatial resolution and of the field of view. In the image in the middle, the edges of the

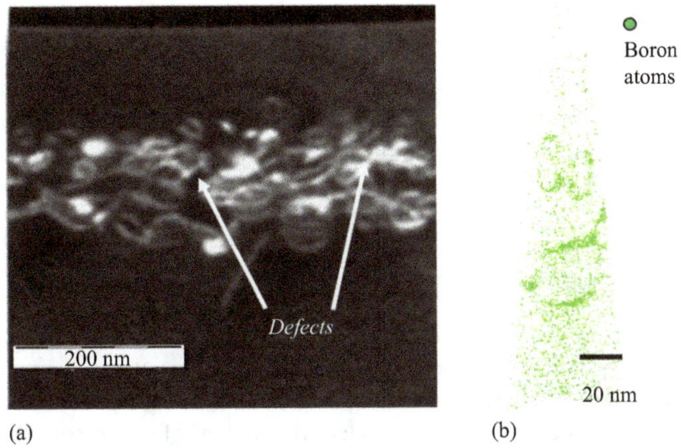

Figure 9.15 APT reconstructions of an Si sample implanted with ^{11}B at 5×10^{15} at./cm^2 with an energy of 30 keV and annealed at 950 °C for 30 s under N_2 revealing the formation of a boron enriched Cottrell atmosphere and other extended enriched defects: (a) TEM image evidencing dislocation loops, and (b) reconstruction of the APT analysis showing the decoration of extended defects (dislocation loops) by boron atoms. Courtesy S. Duguay. © 2014 American Institute of Physics. Adapted, with permission, from [65]

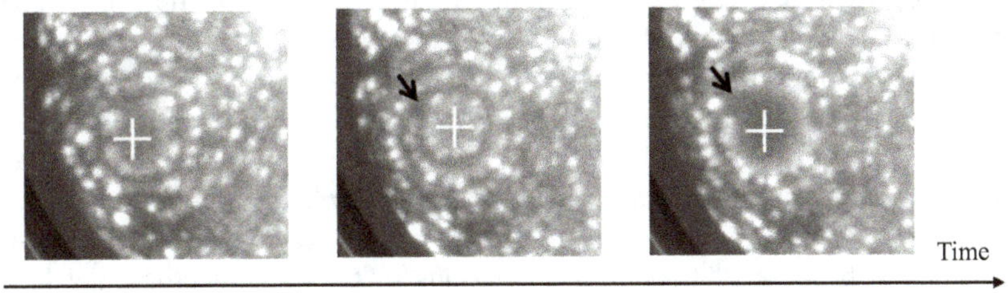

Figure 9.16 Sequence of FIM images recorded during the evaporation of a GaN tip, showing that a dislocation (indicated by a black arrow) moves towards the [0 0 0 1] pole of the crystal (white cross)

upmost terrace on the [0 0 0 1] pole of the crystal close on themselves, while those of the terrace below it do not close and form a spiral. Finally, in the last image at the bottom, the edges of the upmost terrace do not close but form a spiral. This situation corresponds to the movement of a dislocation with a component of the Burgers vector parallel to the [0 0 0 1] direction towards the [0 0 0 1] pole of the crystal, where it is presumably at equilibrium [67]. The study of field-induced plastic deformation in field-emission semiconductor tip specimens is an interesting but still relatively unexplored subject, due to the difficulty of imaging and analysing the movement of individual dislocations.

9.6 Point defects

The study of individual point defects (intrinsic defects or impurity-related) by APT is an extremely challenging task, mostly because of insufficient spatial resolution and detection efficiency of the technique. However, several approaches may allow overcoming these limitations in a certain number of relevant cases, as it will be detailed in the following.

9.6.1 Three-dimensional field ion microscopy

Because of its atomic resolution, FIM has been an attractive instrument in the study of point defects in crystals. By field evaporation of surface atoms, the material can be in-depth explored layer-by-layer. Computer-aided reconstruction of the explored volume may then be generated. This has led to a revival of the instrument, referred to as 3D FIM [68]. Point defects can be investigated in the three dimensions of space. However, the field-induced displacement of surface atoms during field evaporation limits the spatial resolution. The full 3D reconstruction of the crystal remains highly challenging [69].

Vacancies appear in dark contrast in FIM micrographs and can be imaged more easily on low-index terraces where atoms are individually imaged (e.g. (0 1 2) in cubic systems such as gold). The spatial distribution of vacancies was for instance investigated in FeAl metallic alloys [70]. However, this is thought to be much more challenging in semiconductors because of the lower resolution of FIM images in bad conductors, as shown in Figure 9.16. As regards solute atoms (impurities, dopants), FIM images may exhibit contrast if their evaporation field is different from that of solvent atoms. Bright (resp. dark) contrast is observed for high (resp. low) evaporation field chemical species. It is worth mentioning that there is no way to make the difference between darkly imaged impurities and vacancies. In addition, because of the limited resolution of FIM, it remains almost impossible to identify the nature, substitutional or interstitial, of the observed impurity [71].

In contrast to FIM, APT cannot image vacancies. There are at least two reasons for this. First, because of trajectory aberrations in the ion trajectory, the lateral resolution is larger than atomic spacing and the crystal cannot be reconstructed in 3D (see Section 9.1). The second reason comes from the detection efficiency that is not equal to unity (typically 60%). Even if we were able to get an atomic-resolution image of the crystal, the large fraction of atoms that cannot be imaged (40%) would appear as vacancies. APT remains an attractive instrument for the investigation of the spatial distribution of impurities, as reported in Section 9.3. However, again it is impossible to identify the type of site, substitutional or interstitial.

9.6.2 In situ photoluminescence

Despite the strong limitations of APT for the analysis of single-point defects, μPL applied to APT specimens represents a promising approach to overcome them, at least in the case of optically active defects. This has been recently demonstrated in a study of colour centres in diamond as illustrated in Figure 9.17 [72]. The basic

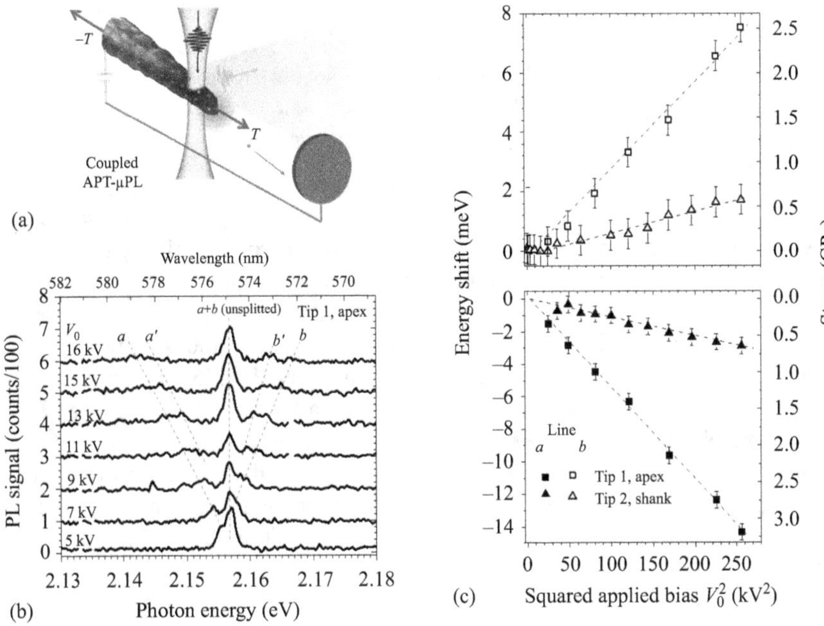

Figure 9.17 *(a) Scheme of μPL spectroscopy carried out in situ within an atom probe: the laser beam excites colour centres within a nanoscale needle and triggers field ion evaporation when placed on the apex of the tip. (b) μPL spectra of an NV^0 complex placed close to the apex of a diamond nano-needle, recorded at different applied biases V_0. (c) The energy splitting of the a, b components of the emission line scales linearly with the square of the applied voltage and allows measuring the stress active on the defect. © 2017 American Chemical Society. Adapted, with permission, from [72]*

idea, depicted in Figure 9.17(a), is combining μPL and APT within the same instrument [73]. The same pulsed laser used for triggering the ion evaporation can thus be simultaneously used for exciting optical transitions. The PL spectra can then be recorded. Several interesting results could already be achieved even without the application of the APT itself. As shown in Figure 9.17(b), the μPL spectrum of the neutral nitrogen-vacancy (NV^0) complex exhibits a dependence on the voltage V_0 applied to the tip. The appearance of discrete spectral lines (marked as a, a', b, b' in the plot) indicates that single defects are detected by the laser beam that was placed next to the tip apex (the position of the laser was controlled through the ion detection rate). Furthermore, the energy splitting of the spectral components of the PL spectrum of NV^0 is sensitive to the applied voltage. In particular, they exhibit a linear dependence on the square of the applied voltage V_0, as reported in Figure 9.17(c) for the a, b components visible in the spectra of part (b) and for those measured from another specimen. This dependence can be related to the stress state

σ at the defect, related to Maxwell stress σ_{apex} induced by the field at the apex of the tip ($\sigma = A\sigma_{apex}$ with the constant $A < 1$, because the stress decreases at a certain distance from the apex due to the increase of the specimen cross section):

$$\Delta E_{a,b} = k_{a,b}\sigma = k_{a,b}A\sigma_{apex} = k_{a,b}A\frac{\varepsilon_0 F^2}{2} = k_{a,b}A\frac{\varepsilon_0 V_0^2}{2(\beta R)^2} \quad (9.10)$$

where k_a, k_b are the proportionality constants, ε_0 is the vacuum permittivity, β and R have been defined in Section 9.1. Comparing the energy splitting with the data available from the literature on the piezospectroscopic behaviour of NV^0 [74], it becomes possible to measure the stress active on the defect [72]. Even more, by polarization-resolved spectroscopy it is possible to access the symmetry of the defect (in the case of NV^0, it is a trigonal C_{3v} symmetry) [74,75]. These results are important in perspective, because with a fully operational APT combining μPL it will be possible to correlate the PL spectra of single-colour centres with their chemical signature as obtained by APT, and to monitor the evolution of the spectrum as the defect approaches the surface of the analysed specimen. The same interesting perspectives may concern point defects within larger structures, for instance inside QWs, QDs or NCs.

9.7 Conclusion

Laser-enhanced APT has been shown to be a unique approach to investigate segregation to crystal defects and clustering in semiconductors [76–82]. The technique can be applied to virtually all 3D semiconductor systems, whatever the structure, crystalline or a-Si. The investigation of 2D objects (e.g. graphene) remains a very challenging topic today, as this would require to place a 2D system at the apex of a nanometric tip. Compared to SIMS, which is a key tool in microelectronics, APT has many advantages (3D imaging, spatial resolution) but it has also several shortcomings or drawbacks (tedious preparation of specimen, fracture of tips under the high electrostatic stress applied, lower mass resolution, statistics and sensitivity compared to SIMS). Instruments are in fact more complementary than concurrent. Statistical fluctuations in APT are larger because of the smaller volume analysed compared to SIMS. The analysed area of SIMS (100 μm wide except in nano-SIMS), 10^6 times larger than that of APT (100 nm wide) combined with its higher mass resolution (less noise in the narrower selection mass window), leads to much better sensitivity (1 appm versus 10–100 ppm for APT). Even if the low ionization efficiency of SIMS (between 0.1% and 1%) compared to APT (ionization rate = 1, detection efficiency $Q = 0.5$) reduces the difference of collected ions (N), statistics remain much better in SIMS analysis (a factor close to 30). However, this advantage is counterbalanced by the higher lateral resolution of APT and its unique 3D imaging capability.

Recent developments in TEM during the last decade have been very impressive and demonstrate that ET at a comparable scale is now available. Material containing finely dispersed precipitates a few nm in size can, for instance, be imaged in 3D using both techniques applied on the same specimen. Such a correlative approach can be of great help to asses APT image distortions, correct images and eventually obtain the most possibly accurate 3D reconstructions [45,83]. Some projects have been launched in order to combine APT and TEM instruments. The ultimate challenge would be to perform TEM observations on specimens during their analysis by APT, obtaining real-time images of them while being evaporated. The high field applied to the tip generates a high stress within the specimen that could also be used to study the dynamics of movement of defects (vacancies, interstitials, dislocations, etc.). Finally, further insight can be provided and new perspective can be disclosed by correlative and in situ optical spectroscopy directly performed on APT tip specimens.

Acknowledgements

The authors greatly acknowledge CEA-INAC, CEA-LETI, STMicroelectronics and D. Mangelinck (IM2NP) for very fruitful scientific collaborations, F. Vurpillot for his help in APT reconstructions, L. Mancini, E. Di Russo, L. Venturi, A. Vella, O. Cojocaru, R. Estivill, S. Duguay and E. Cadel for APT experiments and interpretations.

References

[1] Cerezo A, Godfrey TJ, Smith GDW. Application of a position-sensitive detector to atom probe microanalysis. Rev Sci Instrum. 1988;59(6):862–6.

[2] Blavette D, Bostel A, Sarrau JM, Deconihout B, Menand A. An atom probe for three-dimensional tomography. Nature. 1993;363(6428):432–5.

[3] Kelly TF, Larson DJ. Local electrode atom probes. Mater Charact. 2000; 44(1):59–85.

[4] Blavette D, Cadel E, Fraczkiewicz A, Menand A. Three-dimensional atomic-scale imaging of impurity segregation to line defects. Science. 1999;286(5448):2317–19.

[5] Thompson K, Flaitz PL, Ronsheim P, Larson DJ, Kelly TF. Imaging of arsenic Cottrell atmospheres around silicon defects by three-dimensional atom probe tomography. Science. 2007;317(5843):1370–4.

[6] Gault B, Vurpillot F, Vella A, *et al.* Design of a femtosecond laser assisted tomographic atom probe. Rev Sci Instrum. 2006;77(4):043705.

[7] Hoummada K, Tellouche G, Blum ID, Portavoce A, Mangelinck D. Arsenic clustering during formation of the transient Ni silicide. Scr Mater. 2012; 67(2):169–72.

[8] Cojocaru-Mirédin O, Mangelinck D, Hoummada K, *et al.* Snowplow effect and reactive diffusion in the Pt doped Ni–Si system. Scr Mater. 2007; 57(5):373–6.

[9] Inoue K, Yano F, Nishida A, *et al.* Dopant distribution in gate electrode of n- and p-type metal-oxide-semiconductor field effect transistor by laser-assisted atom probe. Appl Phys Lett. 2009;95(4):043502.

[10] Koelling S, Richard O, Bender H, *et al*. Direct Imaging of 3D Atomic-Scale Dopant-Defect Clustering Processes in Ion-Implanted Silicon [Internet]. 2013 [cited 2018 Apr 11]. Available from: https://pubs.acs.org/doi/abs/10.1021/nl400447d.

[11] Han B, Takamizawa H, Shimizu Y, *et al*. Phosphorus and boron diffusion paths in polycrystalline silicon gate of a trench-type three-dimensional metal-oxide-semiconductor field effect transistor investigated by atom probe tomography. Appl Phys Lett. 2015;107(2):023506.

[12] Lefebvre W, Vurpillot F, Sauvage X. Atom Probe Tomography: Put Theory Into Practice. London: Academic Press; 2016. 418 pp.

[13] Inoue K, Yano F, Nishida A, *et al*. Three dimensional characterization of dopant distribution in polycrystalline silicon by laser-assisted atom probe. Appl Phys Lett. 2008;93(13):133507.

[14] Cojocaru-Mirédin O, Mangelinck D, Blavette D. Nucleation of boron clusters in implanted silicon. J Appl Phys. 2009;106(11):113525.

[15] Blavette D, Wang H, Bonvalet M, Hüe F, Duguay S. Atom-probe tomography study of boron precipitation in highly implanted silicon. Phys Status Solidi A. 2013;211(1):126–30.

[16] Grenier A, Duguay S, Barnes JP, *et al*. Three dimensional imaging and analysis of a single nano-device at the ultimate scale using correlative microscopy techniques. Appl Phys Lett. 2015;106(21):213102.

[17] Müller EW, Panitz JA, McLane SB. The atom-probe field ion microscope. Rev Sci Instrum. 1968;39(1):83–6.

[18] Müller EW. Das Feldionenmikroskop. Z Für Phys. 1951;131(1):136–42.

[19] Cadel E, Vurpillot F, Lardé R, Duguay S, Deconihout B. Depth resolution function of the laser assisted tomographic atom probe in the investigation of semiconductors. J Appl Phys. 2009;106(4):044908.

[20] Kellogg GL, Tsong TT. Pulsed-laser atom-probe field-ion microscopy. J Appl Phys. 1980;51(2):1184–93.

[21] Blavette D, Al Kassab T, Cadel E, *et al*. Laser-assisted atom probe tomography and nanosciences. Int J Mater Res. 2008;99(5):454–60.

[22] Larson DJ, Cerezo A, Juraszek J, Hono K, Schmitz G. Atom-probe tomographic studies of thin films and multilayers. MRS Bull. 2009;34(10):732–7.

[23] Miller MK, Russell KF, Thompson K, Alvis R, Larson DJ. Review of atom probe FIB-based specimen preparation methods. Microsc Microanal. 2007;13(6):428–36.

[24] Estivill R, Audoit G, Barnes J-P, Grenier A, Blavette D. Preparation and analysis of atom probe tips by xenon focused ion beam milling. Microsc Microanal. 2016;22(3):576–82.

[25] Kingham DR. The post-ionization of field evaporated ions: A theoretical explanation of multiple charge states. Surf Sci. 1982;116(2):273–301.

[26] Koelling S, Li A, Cavalli A, *et al*. Atom-by-atom analysis of semiconductor nanowires with parts per million sensitivity. Nano Lett [Internet]. 2016 Dec 21 [cited 2017 Jan 26]. Available from: http://dx.doi.org/10.1021/acs.nanolett.6b03109.

[27] Rigutti L, Mancini L, Hernández-Maldonado D, *et al.* Statistical correction of atom probe tomography data of semiconductor alloys combined with optical spectroscopy: The case of $Al_{0.25}Ga_{0.75}N$. J Appl Phys. 2016;119(10): 105704.
[28] Da Costa G, Vurpillot F, Bostel A, Bouet M, Deconihout B. Design of a delay-line position-sensitive detector with improved performance. Rev Sci Instrum. 2005;76:013304-1-013304-8.
[29] Da Costa G, Wang H, Duguay S, Bostel A, Blavette D, Deconihout B. Advance in multi-hit detection and quantization in atom probe tomography. Rev Sci Instrum. 2012;83:123709-1-123709-6.
[30] Bas P, Bostel A, Deconihout B, Blavette D. A general protocol for the reconstruction of 3D atom probe data. Appl Surf Sci. 1995;87–88:298–304.
[31] Vurpillot F, Renaud L, Blavette D. A new step towards the lattice reconstruction in 3DAP. Ultramicroscopy. 2003;95:223–9.
[32] Moody MP, Gault B, Stephenson LT, Haley D, Ringer SP. Qualification of the tomographic reconstruction in atom probe by advanced spatial distribution map techniques. Ultramicroscopy. 2009;109(7):815–24.
[33] Vurpillot F, Bostel A, Menand A, Blavette D. Trajectories of field emitted ions in 3D atom-probe. Eur Phys J Appl Phys. 1999;6(2):217–21.
[34] Vurpillot F, Bostel A, Blavette D. The shape of field emitters and the ion trajectories in three-dimensional atom probes. J Microsc. 2001 Dec 24; 196(3):332–6.
[35] Waugh AR, Boyes ED, Southon MJ. Investigations of field evaporation with a field-desorption microscope. Surf Sci. 1976;61(1):109–42.
[36] Vurpillot F, Bostel A, Blavette D. Trajectory overlaps and local magnification in three-dimensional atom probe. Appl Phys Lett. 2000;76(21): 3127–9.
[37] Oberdorfer C, Schmitz G. On the field evaporation behavior of dielectric materials in three-dimensional atom probe: A numeric simulation. Microsc Microanal. 2011;17(1):15–25.
[38] Vurpillot F, Cerezo A, Blavette D, Larson DJ. Modeling image distortions in 3DAP. Microsc Microanal. 2004;10(3):384–90.
[39] De Geuser F, Lefebvre W, Danoix F, Vurpillot F, Forbord B, Blavette D. An improved reconstruction procedure for the correction of local magnification effects in three-dimensional atom-probe. Surf Interface Anal. 2007; 39(2–3):268–72.
[40] Gault B, Müller M, La Fontaine A, *et al.* Influence of surface migration on the spatial resolution of pulsed laser atom probe tomography. J Appl Phys. 2010;108(4):044904.
[41] Cojocaru-Mirédin O, Cadel E, Vurpillot F, Mangelinck D, Blavette D. Three-dimensional atomic-scale imaging of boron clusters in implanted silicon. Scr Mater. 2009;60(5):285–8.
[42] Cristiano F, Hebras X, Cherkashin N, Claverie A, Lerch W, Paul S. Clusters formation in ultralow-energy high-dose boron-implanted silicon. Appl Phys Lett. 2003;83(26):5407–9.

[43] Holmes MJ, Choi K, Kako S, Arita M, Arakawa Y. Room-temperature triggered single photon emission from a III-nitride site-controlled nanowire quantum dot. Nano Lett. 2014;14(2):982–6.

[44] Rouviere JL, Simon J, Pelekanos N, Daudin B, Feuillet G. Preferential nucleation of GaN quantum dots at the edge of AlN threading dislocations. Appl Phys Lett. 1999;75(17):2632–4.

[45] Mancini L, Moyon F, Hernàndez-Maldonado D, *et al.* Carrier localization in GaN/AlN quantum dots as revealed by three-dimensional multimicroscopy. Nano Lett [Internet]. 2017 Jun 27 [cited 2017 Jul 10]. Available from: http://dx.doi.org/10.1021/acs.nanolett.7b01189.

[46] Roussel M, Talbot E, Gourbilleau F, Pareige P. Atomic characterization of Si nanoclusters embedded in SiO_2 by atom probe tomography. Nanoscale Res Lett. 2011;6(1):164.

[47] Shinde D, Arnoldi L, Devaraj A, Vella A. Laser-material interaction during atom probe tomography of oxides with embedded metal nanoparticles. J Appl Phys. 2016;120(16):164308.

[48] Bennett SE, Ulfig RM, Clifton PH, *et al.* Atom probe tomography and transmission electron microscopy of a Mg-doped AlGaN/GaN superlattice. Ultramicroscopy. 2011;111(3):207–11.

[49] Bennett SE, Saxey DW, Kappers MJ, *et al.* Atom probe tomography assessment of the impact of electron beam exposure on $In_xGa_{1-x}N$/GaN quantum wells. Appl Phys Lett. 2011;99(2):021906.

[50] Larson DJ, Prosa TJ, Olson D, *et al.* Atom probe tomography of a commercial light emitting diode. J Phys Conf Ser. 2013;471(1):012030.

[51] Galtrey MJ, Oliver RA, Kappers MJ, *et al.* Three-dimensional atom probe studies of an $In_xGa_{1-x}N$/GaN multiple quantum well structure: Assessment of possible indium clustering. Appl Phys Lett. 2007;90(6):061903.

[52] Stephenson LT, Moody MP, Liddicoat PV, Ringer SP. New techniques for the analysis of fine-scaled clustering phenomena within atom probe tomography (APT) data. Microsc Microanal. 2007;13(6):448–63.

[53] Philippe T, Gruber M, Vurpillot F, Blavette D. Clustering and local magnification effects in atom probe tomography: A statistical approach. Microsc Microanal. 2010;16(5):643–8.

[54] Philippe T, De Geuser F, Duguay S, *et al.* Clustering and nearest neighbour distances in atom-probe tomography. Ultramicroscopy. 2009;109(10):1304–9.

[55] Lefebvre-Ulrikson W. Chapter Ten – Correlative Microscopy by (Scanning) Transmission Electron Microscopy and Atom Probe Tomography. In: Atom Probe Tomography [Internet]. Academic Press; 2016 [cited 2017 Jul 10]. pp. 319–51. Available from: http://www.sciencedirect.com/science/article/pii/B9780128046470000103.

[56] Bonef B, Grenier A, Gerard L, *et al.* High spatial resolution correlated investigation of Zn segregation to stacking faults in ZnTe/CdSe nanostructures. Appl Phys Lett. 2018;112(9):093102.

[57] Cadel E, Fraczkiewicz A, Blavette D. Suzuki effect on {0 0 1} stacking faults in boron-doped FeAl intermetallics. Scr Mater. 2004;51(5):437–41.

[58] Jönen H. Highly efficient light emission from stacking faults intersecting nonpolar GaInN quantum wells. Appl Phys Lett. 2011;99(1):011901.

[59] Rigutti L, Blum I, Shinde D, *et al.* Correlation of microphotoluminescence spectroscopy, scanning transmission electron microscopy, and atom probe tomography on a single nano-object containing an InGaN/GaN multiquantum well system. Nano Lett. 2014;14(1):107–14.

[60] Mancini L, Hernández-Maldonado D, Lefebvre W, *et al.* Multi-microscopy study of the influence of stacking faults and three-dimensional In distribution on the optical properties of m-plane InGaN quantum wells grown on microwire sidewalls. Appl Phys Lett. 2016;108(4):042102.

[61] Duguay S, Colin A, Mathiot D, Morin P, Blavette D. Atomic-scale redistribution of dopants in polycrystalline silicon layers. J Appl Phys. 2010; 108(3):034911.

[62] Blavette D, Duguay S. Atom probe tomography in nanoelectronics. Eur Phys J – Appl Phys [Internet]. 2014 Oct [cited 2018 Apr 11];68(1). Available from: https://www.cambridge.org/core/journals/the-european-physical-journal-applied-physics/article/atom-probe-tomography-in-nanoelectronics/1076ED0C57BC9765515705D90578F281.

[63] Estivill R, Juhel M, Gregoire M, Grenier A, Delaye V, Blavette D. Atomic scale investigation of arsenic segregation in high-k metal gate stacks. Scr Mater. 2016;113:231–5.

[64] Cottrell AH, Bilby BA. Dislocation theory of yielding and strain ageing of iron. Proc Phys Soc Sect A. 1949;62(1):49.

[65] Duguay S, Philippe T, Cristiano F, Blavette D. Direct imaging of boron segregation to extended defects in silicon. Appl Phys Lett. 2010;97(24):242104.

[66] Hoummada K, Mangelinck D, Gault B, Cabié M. Nickel segregation on dislocation loops in implanted silicon. Scr Mater. 2011;64(5):378–81.

[67] Pashley DW. The direct observation of imperfections in crystals. Rep Prog Phys. 1965;28(1):291.

[68] Vurpillot F, Danoix F, Gilbert M, Koelling S, Dagan M, Seidman DN. True atomic-scale imaging in three dimensions: A review of the rebirth of field-ion microscopy. Microsc Microanal. 2017;23(2):210–20.

[69] Katnagallu S, Gault B, Grabowski B, Neugebauer J, Raabe D, Nematollahi A. Advanced data mining in field ion microscopy. ArXiv171210245 Phys [Internet]. 2017 Dec 29 [cited 2018 Apr 11]. Available from: http://arxiv.org/abs/1712.10245.

[70] Paris D, Lesbats P, Levy J. An investigation of the distribution of vacancies in an ordered Fe·Al alloy by field ion microscopy. Scr Metall. 1975;9(12):1373–8.

[71] Danoix F, Vurpillot F. Chapter Three – Basics of Field Ion Microscopy. In: Atom Probe Tomography [Internet]. Academic Press; 2016 [cited 2017 Jul 10]. pp. 73–95. Available from: http://www.sciencedirect.com/science/article/pii/B9780128046470000036.

[72] Rigutti L, Venturi L, Houard J, *et al.* Optical contactless measurement of electric field-induced tensile stress in diamond nanoscale needles. Nano Lett. 2017;17(12):7401–9.

[73] Rigutti L, Vella A, Vurpillot F, *et al.* Coupling atom probe tomography and photoluminescence spectroscopy: Exploratory results and perspectives. Ultramicroscopy. 2013;132:75–80.

[74] Davies G. Dynamic Jahn-Teller distortions at trigonal optical centres in diamond. J Phys C Solid State Phys. 1979;12(13):2551.

[75] Grazioso F, Patton BR, Delaney P, Markham ML, Twitchen DJ, Smith JM. Measurement of the full stress tensor in a crystal using photoluminescence from point defects: The example of nitrogen vacancy centers in diamond. Appl Phys Lett. 2013;103(10):101905.

[76] Sha G, Ringer SP. Effect of laser pulsing on the composition measurement of an Al–Mg–Si–Cu alloy using three-dimensional atom probe. Ultramicroscopy. 2009;109(5):580–4.

[77] Torres KL, Thompson GB. Grain boundary enrichment in the FePt polymorphic A1 to L10 phase transformation. Ultramicroscopy. 2009;109(5): 606–11.

[78] Gruber M, Oberdorfer C, Stender P, Schmitz G. Laser-assisted atom probe analysis of sol–gel silica layers. Ultramicroscopy. 2009;109(5):654–9.

[79] Nishimura A, Nogiwa K, Otobe T, *et al.* Influence of laser irradiation condition on a femtosecond laser-assisted tomographic atom probe. Ultramicroscopy. 2009;109(5):467–71.

[80] Qiu Y, Cristiano F, Huet K, *et al.* Extended defects formation in nanosecond laser-annealed ion implanted silicon. Nano Lett. 2014;14(4):1769–75.

[81] Barnes JP, Grenier A, Mouton I, *et al.* Atom probe tomography for advanced nanoelectronic devices: Current status and perspectives. Scr Mater. 2018;148:91–7.

[82] Estivill R, Juhel M, Servanton G, *et al.* Boron atomic-scale mapping in advanced microelectronics by atom probe tomography. Appl Phys Lett. 2017;110(25):252105.

[83] Arslan I, Marquis EA, Homer M, Hekmaty MA, Bartelt NC. Towards better 3-D reconstructions by combining electron tomography and atom-probe tomography. Ultramicroscopy. 2008;108(12):1579–85.

Chapter 10
Ion-beam modification of semiconductors

Kai Nordlund[1] *and Sergei. O. Kucheyev*[2]

In this chapter, we give an overview of the basic physics and experimental method behind the conventional ion-beam modification of semiconductors. In particular, we describe the basic understanding of the keV ion implantation process used for doping Si and the fascinating amount of materials physics involved in the path from the initial implantation to the dopant activation. We also describe two recent developments related to the physics of implantation: the mechanisms of surface ripple formation and time-resolved experiments used to shed new light on defect migration and interaction processes.

10.1 Introduction

Since the 1930s, several technical approaches have been developed that can accelerate ionized atoms to kinetic energies much higher than the milli-eV ones they have at thermodynamic equilibrium around room temperature [1]. Initially, they were largely developed for basic research in nuclear and particle physics but have since then found also a wide range of other research and industrial applications. Currently, ion accelerators are in routine use for material characterization [2,3] and increasingly also for cancer treatment [4]. The accelerators come in many sizes and shapes, ranging from the 27-km circumference of the large hadron collider at CERN [5,6] to ion guns [7] sometimes the size of just around 10 cm [8].

From a monetary and society impact point of view, the by far most significant practical application of ion beams is in the industrial manufacturing of Si chips [9,10]. An estimate from 2012 was that the value of the annual sales of any kind of industrial accelerators was 2.2 billion USD, out of which roughly two-thirds were for ion implanters [10]. However, the physics of radiation damage processes, occurring during ion-beam processing of semiconductor chips, is the same as in other related fields of radiation effects in nuclear ceramics (which are, in fact, wide-bandgap semiconductors) and electronics used in radiation environments of space, accelerators, and nuclear reactors.

[1]Department of Physics, University of Helsinki, Helsinki, Finland
[2]Lawrence Livermore National Laboratory, Livermore, CA, USA

The aim of this chapter is to present such basic physics underlying ion-beam modification of semiconductors, as well as some recent developments in this field. We do this in the following order. In Section 10.2, we present the theoretical and experimental background of ion irradiation of materials, which applies for any solid material. In Section 10.3, we then focus on the ion-induced damage production in semiconductors under typical keV irradiation conditions and also present the physical effects occurring during the postimplantation annealing. Then, in the last sections, we present recent progress in two aspects of ion implantation: surface ripple formation (Section 10.4) and a new approach to measure the defect migration kinetics (Section 10.5).

10.2 Theoretical and experimental background

10.2.1 Theory of ion stopping and energy deposition

When an energetic ion travels in anything else except perfect vacuum, its motion will be slowed down by collisions with the nuclei and electrons in the system. The first step toward understanding effects of ions on materials is understanding the nature of this slowing down, since this also determines the range of materials modification and the implantation depth.

The theory of ion slowing down in materials has been studied systematically since the 1910s, when Niels Bohr wrote the first paper on the topic [11]. The basic theory was well established by the 1970s, when Lindhard *et al.* had compiled a comprehensive description of the stopping power of energetic ions in materials [12]. This theory, naturally developed in conjunction with experimental developments and verification, describes the ion slowing down with the stopping power S, which is divided into two independent parts, the nuclear and electronic stopping power:

$$S(E) = \frac{dE}{dr} = S_n(E) + S_e(E), \tag{10.1}$$

where dE denotes energy loss over a short path length dr which the ion traverses in the material. S has thus units of force and can indeed be considered an average frictional force that slows down ion movement.

Both theory and experiments indicate that the two quantities S_n and S_e are indeed well separable. The former depends on interatomic collisions described by screened Coulomb potentials and the latter on several electronic excitation mechanisms. Although damage formation mechanisms due to S_n and S_e are fundamentally different, recent studies show that there is a narrow energy interval in which both S_e and S_n can act synergistically to enhance damage production [13–15]. At very high energies, also nuclear reactions and other nuclear effects such as *bremsstrahlung* can contribute to the stopping with a separate term S_{nr}; however, for typical ion implantation keV and MeV energies, these effects are negligible.

The relation between nuclear and electronic stopping power is illustrated in Figure 10.1. At high energies, the energy transfer nuclear collision cross section

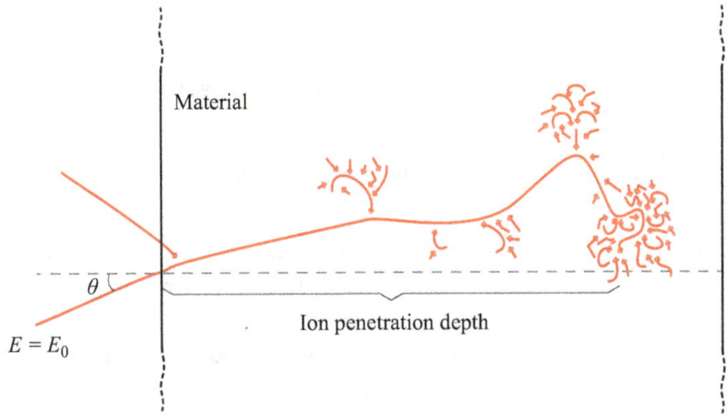

(a)

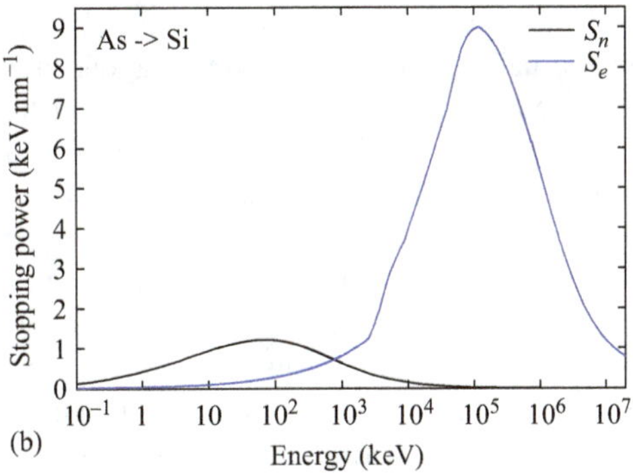

(b)

Figure 10.1 *(a) Schematic figure of a typical ion implantation process. Initially the ion moves ballistically, i.e., changes direction only in occasional strong collisions with nuclei. However, during this stage, it does at the same time lose continuously energy by "collisions" with electrons. Since the electron mass is many orders of magnitude lower than ion masses, this does not cause a change in the movement direction of the ion, however. After it has lost enough energy, the ion starts colliding more frequently with atoms, leading to a dense set of collisions, a "collision cascade." (b) Stopping power: the reason to the behavior depicted in part (a) derives from the slowing down force, usually called the stopping power S. At high energies, the electronic stopping power dominates in all materials. When the ion has slowed down enough, the nuclear stopping power starts to dominate, resulting in the set of dense collisions*

decreases strongly, hence the electronic stopping power dominates at the highest energies. For ions, heavier than the proton, the nuclear stopping power dominates at the lowest energies (as a curiosity, we note that contrary to protons, antiprotons have a stronger nuclear stopping also at the lowest energies [16]).

If the stopping power would not be determinate (i.e., no stochastic effects were at play) and ions would travel in a straight path, one could obtain the penetration range R of ions in a material in a straightforward manner by integrating the stopping power as

$$R = \int_0^{E_0} \frac{dE}{S(E)} \quad \text{if and only if ion travels in straight path} \tag{10.2}$$

However, since nuclei and electrons are discrete objects, it is clear that the stopping is in fact a stochastic quantity. Hence, the range is not a singular value but a distribution of ranges. This broadening is known as the straggling of the stopping and ion range distributions. For the range distribution, the straggling is simply defined as the standard deviation of the range distribution of the individual ranges around the mean range $\bar{R}$.

The straggling of the electronic stopping power has a fairly accurate theory [17–19]. It is, however, usually significant only for light ions like protons or very high energies [17,20]. On the other hand, the straggling of the nuclear stopping power is strong at keV energies due to the finite number of interatomic collisions and the possibility of very large deflections of the ion in a few low impact parameter collisions. This leads to a strong broadening of the range distribution; in many cases, the straggling of the range distribution is of comparable magnitude to the mean range. A few typical range distributions are illustrated in Figure 10.2. The results show that due to stochastic nature of the collisions, the distribution is in all cases wide, with the width of the distribution being of comparable magnitude to the mean range.

The basic picture of ion implantation given earlier is significantly complicated by the possibility that in crystalline materials ions may be focused along low-index crystal directions. If this happens, they may penetrate deeper than expected from considerations of collision with atoms in random directions. This effect, known as ion channeling, has been known since the 1960s and can lead to major changes in the penetration depths [23–25]. It is illustrated for the case of 10 keV Si ions implanted at Si (0 0 1) surfaces in Figure 10.2(b).

In semiconductor processing, it is common to attempt to avoid the channeling by tilting the sample in a "random" direction with respect to the ion beam. For Si, e.g., a tilt of $\theta = 7^\circ$ (polar rotation angle off (0 0 1) surface normal), $\phi = 27^\circ$ (azimuthal rotation angle off (1 0 0) direction) is often considered a "good non-channeling direction." However, as illustrated in Figure 1.2(b), even in this case, some ions can be scattered into a channeling direction, leading to a "secondary channeling tail." The choice $\theta = \phi = 20^\circ$ recommended in [22] shows clearly less—but still some—channeling tail. Since the electrical properties of semiconductors are highly sensitive to even small dopant concentrations, the possibility of having such secondary channeling tails needs to be considered in planning ion beam doping of semiconductors.

The possibility to end up in a channeling direction can be systematically analyzed with so-called channeling maps, which show the mean range or nuclear energy deposition as a function of crystal or implantation direction [22]. An example of such maps is shown in Figure 10.2 for the case of 10 keV Si ions in Si. The maps show that in the Si crystal, the $\langle 1\ 1\ 0 \rangle$, $\langle 2\ 1\ 1 \rangle$, $\langle 1\ 1\ 1 \rangle$, and $\langle 1\ 0\ 0 \rangle$ are

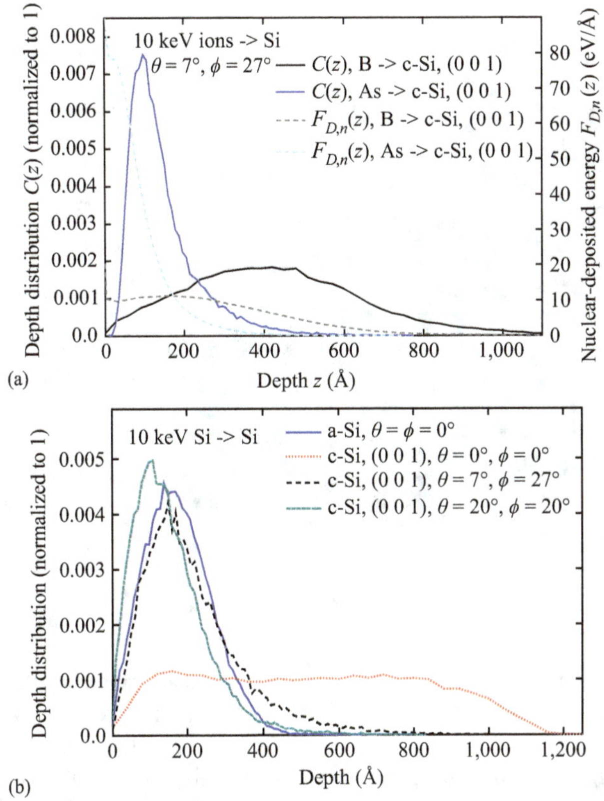

Figure 10.2 *(a) Ion range (depth) distributions C(z) for 10 keV B and As ions in Si. Also shown with dashed lines are the nuclear energy deposition $F_{D,n}$ distributions. The calculations were done for the good non-channeling direction $\theta = 7°, \phi = 27°$. Data and figure are original for this chapter. (b) Ion range distributions C(z) for 10 keV Si ions in Si. Distributions are shown for amorphous Si calculated both with the molecular dynamics range calculation method MDRANGE [21]. Also shown are distributions for crystalline Si at 300 K for perpendicular impact in the (0 0 1) crystal direction, and the non-channeling directions $\theta = 7°$, $\phi = 27°$, and $\theta = \phi = 20°$ [22]. In all cases in (a) and (b), the range profile was obtained from a simulation of 30,000 implanted ions at room temperature. The a-Si was modeled based on a 3-nm cell with random atom coordinates as described in [22]. Data and figure are original for this chapter. (c) Map of mean ranges as a function of implantation angle for 10 keV Si ions in Si. The red colors show strong channeling directions and blue weak (no) channeling. The strong variations show the importance of ion channeling effects. © 2016. Reprinted, with permission, from [22]*

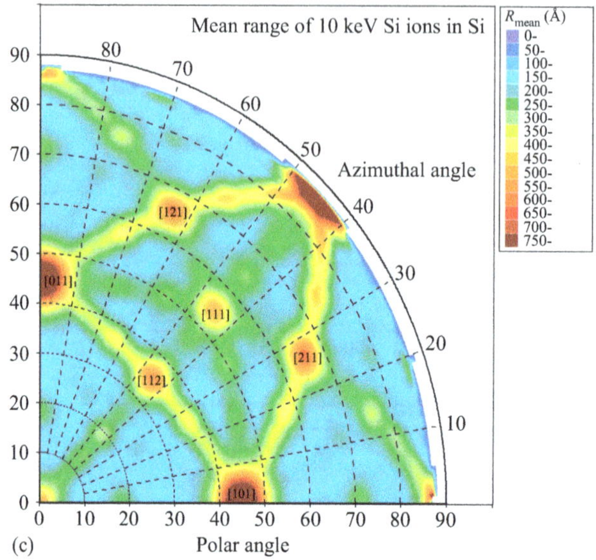

Figure 10.2 (Continued)

strong channeling directions, in this order. The mean range in the channeling direction can be five times larger than that in a "random" direction. Moreover, the figure makes it evident that there are also numerous planar channeling planes between the strong axial channeling ones. Due to this, the probability of a random crystal orientation to be a channeling direction is actually quite high (estimated as 20%–60% depending on case in [22]), and care has to be taken if complete avoidance of channeling is required. This can be accomplished by adding a Si ion implantation step that amorphizes the near-surface layer before subsequent implants with dopants.

In the absence of diffusion effects, the ion range distribution determines directly the final concentration depth dependence $C(z)$ of dopants. However, as evident from Figure 1.1(a), the ions collide several times before they stop. Hence, the energy deposition to the material does not follow the same distribution as the penetration depth. The damage production can be quantified by the nuclear and electronic energy depositions: $F_{D_n}(z)$ and $F_{D_e}(z)$, which naturally are distributed closer to the surface than the penetration depths.

For keV energy ions and recoils in semiconductors (and metals), practically all damage is produced by the nuclear collisions. At higher energies, culminating in the so-called swift heavy ion regime, also electronic excitations can produce damage in several semiconductors [14,26–29], but the discussion of these effects is beyond the scope of the current chapter. The nuclear energy deposition profile $F_{D_n}(z)$ (see Figure 10.2(a)) can be used as a good measure of the depth distribution of ballistic displacements causing primary lattice damage. We proceed with discussions on damage production in Section 10.3.

10.2.2 Basics of ion implantation experiments

In this section, we briefly discuss the equipment typically used in ion implantation experiments. For more information on ion implantation hardware, the reader is referred to one of the numerous books and reviews on this topic [1,2,10].

Different types of experiments on ion-beam modification of semiconductors require different accelerators. For example, sputtering and ripple formation experiments are typically done with high-current sputter guns (with energies typically limited to ~50 keV) rather than conventional implanters. In their design, these ion guns are more like small-area dry etching machines rather than implanters. Conventional low- and medium-energy ion implantation involves ion energies of 1–300 keV, and air insulated implanters are typically used in this case. High-current implanters for ion energies below ~10 keV typically feature a deceleration lens. High-energy machines capable of accelerating voltages $\gtrsim$300 kV require an insulating gas inclosure, which has historically been done with SF_6. However, since SF_6 is a very potent greenhouse gas (~24,000 times worse than CO_2), there has been a move to replace SF_6 with more environmentally friendly candidates. A recent review on this topic can be found in [30].

Implanters used in chip manufacturing are typically designed to perform a limited number of implant types but do it numerous times with minimum maintenance, in a high-yield process. These machines should also be compatible with the rest of the relevant device processing equipment in the fab. As a result, accelerators for semiconductor chip manufacturing are rarely used for research in academia and national laboratories, where preference is given to accelerators with a different architecture. Research-grade implantation machines are capable of performing different types of implants, with maximum flexibility of the ion type, energy, dose, dose rate, and target temperature and positioning. As a result, different manufacturers typically supply implanters for the industry and research institutions.

All accelerators have the same basic elements. These include the ion source, the acceleration stage, tools for mass separation and beam manipulation by focusing, rastering, or pulsing, tools for beam diagnostics, and the target chamber (also often referred to as the "endstation"). In all implantation experiments, there are also several common key considerations. These include accurate dosimetry, beam characteristics, and control of sample temperature during irradiation. Good vacuum conditions are paramount for ion-beam processing. Residual gases cause beam neutralization, contributing to dosimetry errors. Moreover, ion irradiation results in the breakdown and polymerization of residual hydrocarbons adsorbed on the sample surface.

Another important consideration in ion accelerators is the choice of the ion source. Both negative and positive ion sources are used. Negative ion sources are required for tandem accelerators used for high-energy implants. Sputtering-based negative ion sources could also be used for low-energy (air insulated) machines as such sources are well studied and can produce ions from almost the entire periodic table (except for inert gas ions) but with limited ion currents. Hence, all relatively high-current instruments ($\gtrsim$100 μA) employ positive ion sources. For implantation, the ion species available, source intensity, and current stability are of primary importance. Another critical parameter is the cost of source operation, which is typically related to the source maintenance schedule. Similar to the other aspects of the implanter architecture, machines for chip manufacturing are equipped with sources capable of reliable, high-performance operation with one or few species, while accelerators for research tend to use sources that are more flexible and suitable for either generating a wide range of species or beams with extremely stable currents required, for example, for studies of defect-interaction dynamics.

In almost all practical cases, ion beams are rastered to achieve the uniformity of ion fluence (which, in the semiconductor community, is exclusively referred to as ion dose). In this case, conditions of beam focusing and rastering influence the instantaneous dose rate and, hence, damage formation efficiency in regimes with pronounced dynamic annealing (DA). We discuss these dynamic defect-interaction effects in Section 10.5.

Accurate dosimetry (which could probably be called "fluencimetry" outside the semiconductor field) is not trivial, particularly if ions with different masses, energies, dose rates, sample areas, impact angles, and sample temperatures are used. It is generally a good practice to regularly verify the dosimetry calibration, particularly after any modifications to the sample holder and/or target chamber are made. Most implanters use either external or internal Faraday cup systems. Both have their pros and cons. The internal Faraday cup refers to the case when the sample holder is a part of the Faraday cup assembly. In contrast, in an external Faraday cup arrangement common for systems with rastered beams, several (typically four) Faraday cups are placed in front of the sample holder. Excellent reviews of various approaches to achieve accurate dosimetry can be found in [31–33].

Also important for implantation is the control of sample temperature during implantation. Since implantation is done in vacuum, for sample temperatures, when radiative heat transfer is ineffective (i.e., at temperatures $\lesssim 450^\circ$C), samples need to be well thermally anchored to a heat sink (i.e., the sample holder) in experiments with large dose rates and/or energies when thermal load is significant. In most practical cases, this is done by attaching the samples to a temperature-controlled sample holder with some thermally conductive glue. This is important since mechanically clamped samples have an effectively three-point thermal contact to the holder and, hence, can experience substantial heating in experiments with high power beams. For experiments involving liquid helium temperatures, different but fairly standard practices for cryostat design and sample mounting are used. These have recently been summarized in an excellent book by Ekin [34].

10.3 Damage production and annealing

In this section, we overview the most important stages of the scientific understanding of all the stages that occur during a prototypical sequence of ion implantation and annealing under semiconductor chip manufacturing. The stages discussed are schematically illustrated in Figure 10.3.

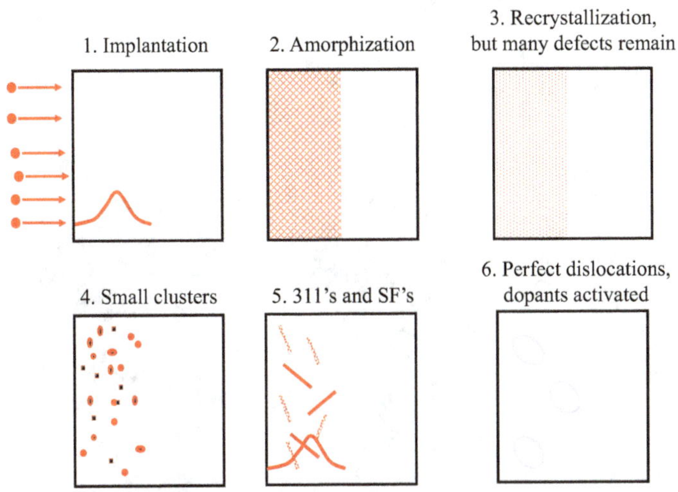

Figure 10.3 Schematic figure of the main stages of Si processing during Si chip manufacturing. The stages in the schematic are discussed in subsections in the text as follows: 1: Sections 10.3.1 and 10.3.2, 2: Section 10.3.3, 3: Section 10.3.4, and 4–6: Section 10.3.5

10.3.1 Low-fluence primary damage

As discussed in Section 10.2.1, the incoming ion produces damage via nuclear scattering events with sample atoms. These recoils in turn collide with other atoms, leading to a sequence of scattering events known as a collision cascade.

The nature of the damage produced in such cascades in Si is illustrated in Figure 10.4. The illustration is based on results from molecular dynamics (MD) computer simulations of atom motion, a widely employed method to theoretical study the primary damage in all classes of materials [35,37–42]. The data shows how a low-energy 100 eV recoil produces a few point defects or nearby Frenkel pairs. This kind of damage corresponds somewhat to the one that the Kinchin–Pease equation or a binary collision approximation (BCA) would predict to be formed. However, already at the energy of 1,000 eV, one sees that the damage takes a rather complicated form, which cannot be assigned to point defect damage in a meaningful way. A heavier atom recoil such as Xe can already at the rather moderate energy of 3,000 eV produces a continuous disordered region, corresponding to those observed experimentally in transmission electron microscopy (TEM), e.g., due to heavy Bi ion irradiation [43]. Also diffuse X-ray scattering (DXS) studies give direct experimental evidence that medium-mass ion irradiation produces clustered defects [44]. These disordered regions are often called amorphous zones. In Si, they are well visible only for heavy ions, while in other semiconductors, such as Ge and GaAs, heavy ion irradiation conditions lead to the formation of large amorphous pockets easily visible by TEM [45–47].

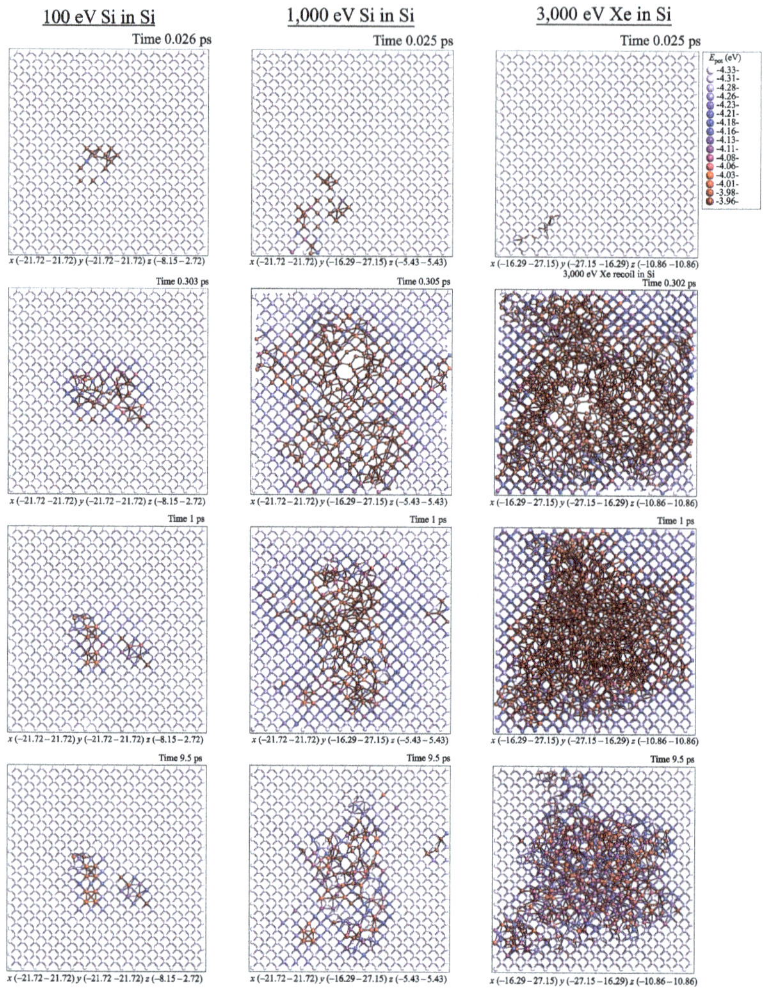

Figure 10.4 Illustration from molecular dynamics (MD) simulations of how different recoil energies and recoil atom types produce different types of damage in Si. In each case, a recoil energy was given to a lattice atom (in the case of 3,000 eV Xe, one Si atom was replaced by Xe) in a non-channeling direction, and the ensuing cascade was simulated following the principles described in [35]. The figures show a 4.4 × 4.4 × 1.1 nm (integer multiples of the Si unit cell size of 0.543 nm) region with the most damage. The actual simulation cell was much larger to fully contain the collision cascade. Chemical bonds are drawn for atoms within a distance of 0.3 nm from each other. The colors/gray scale show the potential energy of the atoms as given by the Stillinger–Weber interatomic potential [36], with light atoms having a potential energy near equilibrium and darker ones those in an excited collisional state or a final defective state (bottom row). After the 9.5 ps shown in the bottom frames, the system state did not evolve anymore on MD timescales. Original data and figure made for this chapter

10.3.2 Defect migration and interaction

MD simulations suggest that the primary damage production at around room temperature and below is essentially temperature independent. After this athermal stage, which lasts of the order of 10 ps [48], the produced damage typically evolves further due to thermally activated processes. This longer term migration includes point defect migration and interaction that could lead to either the recombination of vacancies and interstitials or point defect clustering. Thermally activated defect migration during irradiation is in principle the same as the migration of thermally generated point defects extensively investigated in the previous studies of dopant diffusion.

The defect migration in common semiconductors, especially Si, has been studied extensively for a long time [49], and we do not attempt a complete review of this topic here. We just mention a couple of key observations. The most important one is that almost all studies of defect diffusion are indirect and require often speculative models to interpret experimental results. For example, it is not obvious which point defects are responsible for the appearance of specific peaks in deep-level transient spectroscopy (DLTS), electron paramagnetic resonance (EPR), or optical absorption or luminescence spectra. A notable exception is the isolated vacancy in Si which can be detected by EPR studies. Such EPR studies have been carried over many decades and have shown that isolated vacancies in Si produced by electron irradiation are mobile well below room temperature. As expected, the vacancy mobility is affected by the vacancy charge state, which is determined by the position of the Fermi level, which in turn depends on the doping of the material [50]. DLTS and DXS studies have given similar conclusions [51,52].

The activation energy (E_a) for the migration of neutral vacancies in Si is 0.4 eV, while an E_a of 0.1 eV is typically assumed for the migration of interstitials in Si. The situation with interstitials in Si is much less clear since they cannot be directly observed by EPR. The description of more complex point defect complexes falls entirely into the realm of theoretical studies. For example, density-functional theory simulations suggest that also the di-interstitial and tri-interstitial in Si are highly mobile [53].

Moreover, comparisons of electron and ion-irradiated Si indicate that the onset of defect mobility is different for these two defect generation methods [44,54]. This could be attributed to the electron excitation during the former kind of irradiation inducing the defect mobility [44]. This is, however, consistent with the previous EPR studies revealing that the defect formation and clustering behavior in Si bombarded with ions that produce collision cascades are much more complex than in Si irradiated with high-energy electrons that produce isolated Frenkel pairs. More recent studies of damage in detectors also hint that it is possible that in fact most defects in Si produced by any ion irradiation are not simple point defects but small defect complexes involving 10–20 atoms [55]. A comparison of electron and ion irradiation results is not straightforward simply because the stable defects formed in these two cases are not the same.

Damage formed in cascades can also undergo annealing by thermal activation or electronic excitation. In GaAs, amorphous pockets recrystallize spontaneously even at room temperature under electron irradiation in the transmission electron microscope [47]. In Si and Ge, it appears that at least the larger damage amorphous pockets are stable at room temperature. They, of course, will experience recrystallization at elevated temperatures [38]. Recrystallization could also be induced by low-energy recoils [56] or electron irradiation [57]. Due to such processes, the material will not amorphize above a certain amorphization temperature for any irradiation dose [38]. The material reaches a state when processes of defect production and annihilation are balanced. The exact maximum amorphization temperature, however, is a function of the ion mass, energy, and the dose rate since these parameters determine the balance between defect production and annihilation. In recent years, kinetic Monte Carlo modeling has proven to be able to reproduce and explain many of these defect migration phenomena [58,59].

A new approach to study the postproduction defect mobility in semiconductors is reviewed in Section 10.5.

10.3.3 High-fluence damage overlap and amorphization

To achieve the kinds of dopant concentrations typically desired for Si chip manufacturing, ion fluences are needed where collision cascades spatially overlap numerous times before the total dose is reached. Since even single ions can produce significant amorphous-like damage, it appears obvious that during prolonged irradiation, the damage can buildup until the material amorphizes. This buildup of lattice defects can be measured by different techniques such as Rutherford backscattering/channeling (RBS/C), x-ray diffraction (XRD), electron microscopy, Raman spectroscopy, or electrical and other optical measurements, with RBS/C being the most common since it allows to rapidly evaluate depth profiles of displaced atoms [60–63].

A typical case of damage buildup in Si as seen by RBS/channeling is shown in Figure 10.5. This case is for self-ions, so impurity buildup is not an issue in interpreting the results. At the lowest fluence of 5×10^{13} Si ions cm^{-2}, a low defect concentration of a few percent is visible in the signal. This corresponds to a situation with only little damage overlap. As the fluence increases, the damage level raises rapidly and reaches the almost fully “amorphous” level at the fluence of 5×10^{14} Si ions cm^{-2}. Note, however, that ion channeling cannot unambiguously determine the amorphous state. Channeling spectra from polycrystalline and amorphous samples are the same [65]. Above this fluence, the amorphous layer keeps growing deeper into the sample.

The evaluation of the “damage overlap” fraction is not trivial as it requires the knowledge the defect configuration after cascade thermalization. A recent work to describe the cascade overlap within a model that takes into account the fractal nature of collision cascades has been reported by Wallace *et al.* [66].

The amorphization mechanism of Si and the structure of amorphous Si have been studied very extensively. For a long time, there were attempts to describe the amorphization with relatively simple models such as homogeneous or heterogeneous amorphization [67–69]. The former means a process where point defects overlap until they reach a critical dose, at which the material collapses into an amorphous state. The

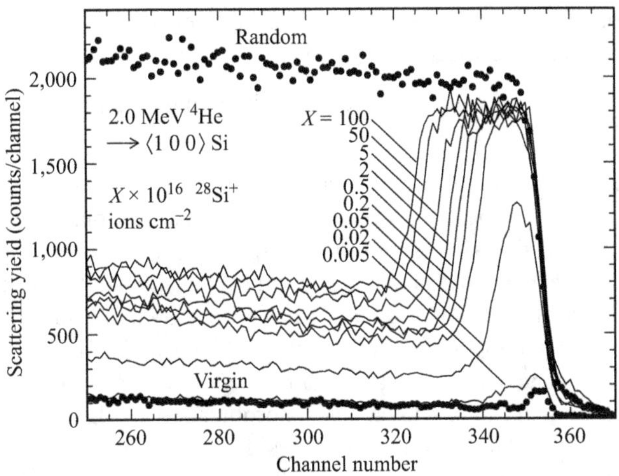

Figure 10.5 Damage buildup in Si: RBS/channeling spectra for 2 MeV He ions incident on Si(1 0 0) samples implanted with ions in the fluence range 5×10^{13}–1×10^{18} 50 keV ^{28}Si ions cm^{-2}. Also shown in the figure is the random-oriented spectrum and the spectrum in the ⟨1 0 0⟩ channeling direction for a virgin (unimplanted) sample. © 1995 American Physical Society. Reprinted, with permission from the authors, from [64]

latter means one where amorphous pockets are created independently of each other, and amorphization occurs just as buildup of these. Homogeneous amorphization is also sometimes called "defect-stimulated" and heterogeneous "direct impact." However, taking together a wide range of experimental and computer simulation observations, one can realize that the buildup of damage and amorphization of Si, at least at room temperature and above, is a much more complex process that proceeds via defect migration and interaction. We can list the following eight different mechanisms (including selected references that provide evidence for each):

- 1(A). Direct impact amorphization to produce amorphous pockets [43,45]
- 2(A). Defect-stimulated buildup [44,70]
- 3(A). Cascade-induced amorphous zone growth at amorphous–crystalline interfaces [71]
- 4*(A). Cascade-induced partial recrystallization at amorphous–crystalline interfaces [72,73]
- 5*(A). Subthreshold recoil-induced recrystallization [71]
- 6*(T). Interstitial-vacancy recombination or point defect clustering via thermal defect migration [59]
- 7*(T). Thermal annealing of damage [74]
- 8*(T). Recrystallization of amorphous–crystalline interfaces (including those surrounding amorphous pockets) by mobile defects [75,76].

Moreover, if the irradiation condition involves high levels of electronic excitations, there is at least one more mechanism active:

- 9*(A) Electronic excitation-induced bond breaking leading to recrystallization [46,57,77].

In the list, the specification "(A)" means the process is athermal, i.e., would occur even at 0 K, while (T) means it occurs by thermal activation. The asterisks "*" indicate mechanisms that slow down the amorphization.

All of these mechanisms can act at the same time and can be quite significant: e.g., the subthreshold recoil mechanism 5* has been shown to be able to slow down amorphization at low temperatures by roughly a factor of 2 [71].

The one firm conclusion that can be drawn from these observations is that it is a wild goose chase to try to pin down "*the* amorphization mechanism of silicon" for a broad range of irradiation conditions. It is a complex process that involves a balance of several mechanisms acting at the same time, both for and against amorphization.

The structure of amorphous Si (a-Si), whether produced by irradiation or other means, has also been studied very extensively [78–82]. The studies agree on that well-relaxed a-Si has an average coordination number around or slightly below 4.0 [79,83], i.e., that the material does have some dangling bonds [81]. However, it has long been clear that the results on the structure of a-Si depend on the material preparation conditions. A recent study where the same characterization techniques were used on several kinds of differently synthesized, a-Si firmly showed that there are statistically significant differences between samples made in different ways [84].

The amorphization behavior of other semiconductors can be even more complex than that for Si. In particular, radiation damages in GaN and its alloys with InN and AlN have been studied extensively. For example, the work by Wendler *et al.* shows a very complicated three-stage amorphization behavior even at 15 K [85]. MD computer simulations can describe the initial damage buildup stage but fail to reproduce the latter two stages [86]. Experiments show that the latter two stages involve some sort of complex dislocation structures; however, there is no good understanding of how these form and evolve. At higher temperatures, damage buildup in GaN is even more complicated, exhibiting very efficient DA, strong cascade density effects, ion-beam-induced material decomposition with the formation of N_2 bubbles and the development of gas-bubble-nucleation-driven ion-beam-induced porosity, and the strong influence of the sample surface that acts as the preferential nucleation site for amorphization and defect recombination [87,88].

Peculiar damage accumulation behavior has also been observed in $Al_xGa_{1-x}As$, $Al_xGa_{1-x}N$, and $In_xGa_{1-x}N$ alloys [89–101]. Experiments show that pure AlAs is quite resistant to amorphization [89], while GaAs amorphizes readily. This is somewhat surprising since both materials, and their alloys, have exactly the same zincblende crystal structure. The amorphization behavior changes drastically by a relatively small change in the Al content x from 0.5 to 0.75 [90,91]. Moreover, studies of irradiation of AlAs/GaAs interfaces show surprising damage buildup asymmetries [92–94].

10.3.4 Recrystallization of amorphous state

In Si processing, the ion implantation of course introduces dopants at the desired depths. However, the desired electronic functionality requires that Si be in a high-quality crystalline state, with dopants electrically activated. Hence, the amorphous zones, amorphous layers, and point defect clusters produced by the implantation process need to be recrystallized or turned electrically inactive. Typically, this is achieved by one of two high-temperature annealing stages: either one can use a furnace anneal with temperatures around 700°C for a long time or a high-power flash lamp anneal, typically referred to as rapid thermal annealing (RTA). The latter heats up the surface layer to 1,000°C for a short time up to several seconds [80,102]. The latter approach dominates in industry because of its speed and better control of dopant out-diffusion, which obviously is an advantage in production lines.

Especially during RTA anneals, the recrystallization of Si is quite rapid. It can be readily measured with RBS/channeling experiments, similar to the ones used to measure the defect buildup [103]. This recrystallization is well understood from a range of MD computer simulations [59,72,73,104,105], which show how thermal activation allows reforming bonds at the interface back into the crystalline state. However, the crystal left after this kind of recrystallization often leaves after it a high concentration of defects. We discuss next the evolution of these defects after the base lattice is in principle fully recrystallized, but the annealing still keeps the Si hot, allowing for thermally activated defect migration that can reduce the defect concentration.

Another approach to annealing implant damage in semiconductors is laser-assisted annealing, which was studied extensively in the 1970s and 1980s before the RTA technique became established. There has been renewed interest in laser annealing in recent years, with demonstrations of various nonequilibrium states by an educated choice of annealing parameters. For more details on recent progress with laser annealing, the reader is referred to recent articles [106–108].

10.3.5 Defect evolution during recrystallization and dopant activation

Immediately after the recrystallization, the Si lattice has a high concentration of remaining small interstitial clusters [80,109,110]. The small di-interstitial (i2) and tri-interstitial (i3) are apparently quite mobile [53]. However, the tetra-interstitial (i4) is believed to be quite stable, as well as the larger i8 one [110]. They are hence believed to be an important intermediate state during the annealing. On further annealing, however, these clusters are expected to destabilize and emit free interstitial into the lattice.

Isolated interstitials are highly mobile in Si and tend to coalesce to form so-called 311 defects that have been extensively studied by TEM [110–113]. These are rod-like defects that lie on 311 crystal planes and are specific to tetrahedral semiconductors. They are quite stable because all Si atoms involved in them are 4-fold-coordinated, i.e., have the same sp^3 coordination as Si atoms in a perfect crystal [114–117], see Figure 10.6. In spite of their perfect coordination state, they

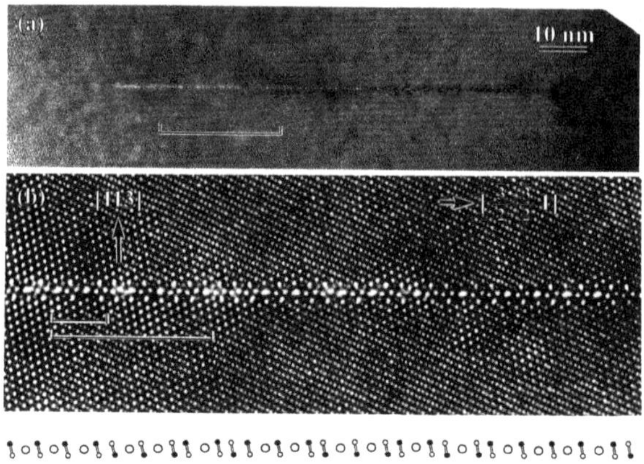

Figure 10.6 Structure of the rod-like 311 defects as seen in TEM images. © 1991. The Japan Society of Applied Physics. Reprinted with permission from [115]

are not desirable for the electronic functionality. Hence, a lot of research effort has focused on ways to remove these defects by subsequent annealing steps or by controlling the defect state of as-implanted samples by either performing low-temperature amorphizing (pre)implants or doing implantation at elevated temperatures (so-called hot implants) [118,119].

These stages of annealing can also affect the positions of the dopant atoms. This is an especially large effect for boron dopants, which have been found to diffuse much more during the annealing than they should as calculated from the equilibrium diffusion coefficient of B in Si. This effect, known as transient-enhanced diffusion, was a major problem for further Si chip miniaturization in the 1990s and was hence studied intensively. It became rather soon clear that the extra B diffusion was driven by Si interstitials, i.e., the B atom was bound to fast-migrating interstitials [120].

The source of these interstitials remained unclear for some time. Eaglesham *et al.* stated that the 311 defects were the source of the interstitials [111]. Somewhat later, however, Claverie *et al.* showed that the process is even more complex and is in reality an Ostwald ripening process, where also intermediate boron clusters play a role [121–125].

After the 311 defects emit interstitials, these in turn start coalescing into stacking faults. On even further annealing, these unfault into perfect dislocation loops [126,127]. These perfect loops are, as the only remaining defect in the system, finally acceptable for the electrical operation. The dopant atoms should have after this annealing ended up in substitutional sites, leading to the desired electrical functionality of the processed component.

10.4 Ripple formation on semiconductors

In this section, we consider the topic of spontaneous nanoscale ripple formation on semiconductor surfaces by ion irradiation. We first summarize the experimental background on the topic, and then proceed to discuss the recent paradigm change in the theoretical explanation of why ripples form. We emphasize that we do not attempt here to make a comprehensive review of the topic (Web of Science search gives more than 1,000 references with the keywords "ion beam" and "ripple"). Other recent references provide a more complete review of this topic [128,129].

10.4.1 Experimental background

When initially smooth surfaces are subjected to prolonged ion irradiation, they sometimes change from a flat to a rough morphology. In some cases, the change is into a well-ordered wave-like morphology, often called surface ripples. This was first reported by Cunningham *et al.* in 1960 in metals [130], see Figure 10.7. In 1962, it was reported that for the same irradiation condition, well-ordered ripples on glass can form either perpendicular or parallel to the angle of the incoming ion beam with respect to the surface normal [131], see Figure 10.8. Note that, the terminology in the field can be somewhat confusing. While visually it seems that the ripples on the left panel of Figure 10.8 are perpendicular to the incoming beam, the ripple formation theory deals with the wave vector of the ripples. Hence, the ripples of the kind shown in the left panel of Figure 10.8 are called parallel mode ripples.

During conventional high-dose semiconductor processing as in dry etching, the formation of such ripples can be a nuisance, and for a long time, the effect was generally rather avoided than desired. In common semiconductors such as Si, it does not usually occur at or near perpendicular impact, and hence this was not a large problem. However, as nanoscience became a rage in the 1990s, interest into the ripple-like features heightened again, as they are typically on the nanoscale, and are an example of self-organized surface feature formation.

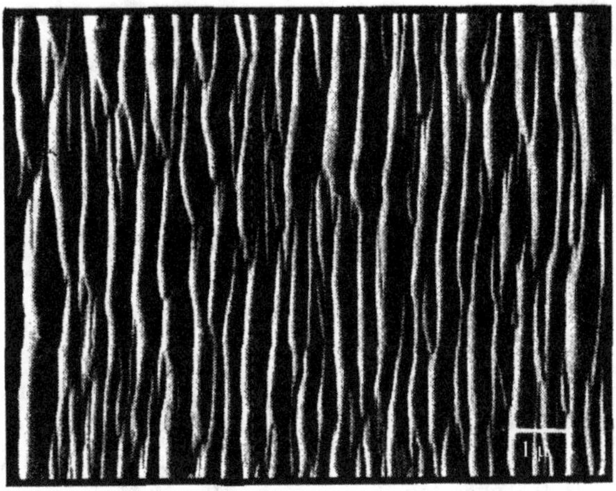

Figure 10.7 Ripples on Au formed by 8 keV Ar bombardment. © 1960 American Institute of Physics. Reprinted, with permission, from [130]

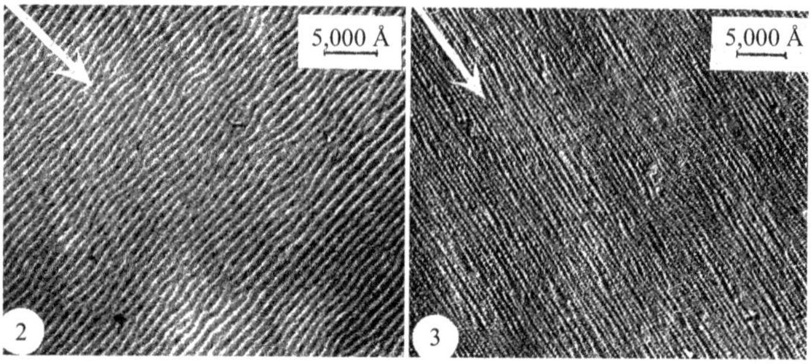

Figure 10.8 *Ripples on glass formed by ion bombardment from an ion source producing mixed ions accelerated with a 4 kV voltage. The arrow shows the incoming beam direction. Left (part 2): ion incidence angle 60°, right (part 3): ion incidence angle 10°. © 1962 French Academy of Sciences. Reprinted, with permission, from [131]*

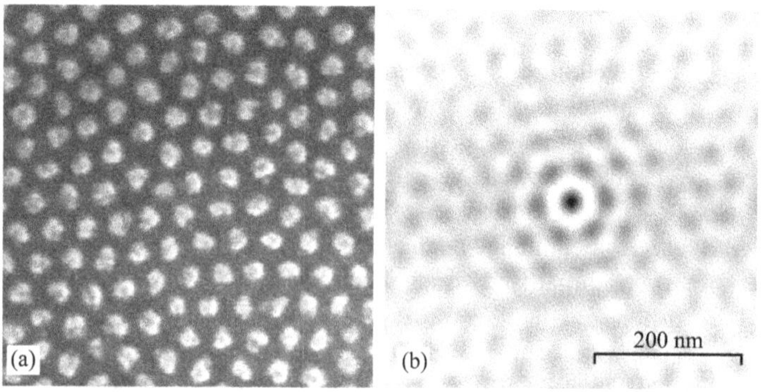

Figure 10.9 *(a) Scanning electron microscopy image of quantum dots formed on GaSb by 420 eV Ar irradiation. (b) Autocorrelation of the same image. The clear hexagonal symmetry in the autocorrelation pattern shows that the dots have a strong short-range order with hexagonal symmetry of neighboring dots around each site. (a) © 1999. American Association for the Advancement of Science. Reprinted, with permission, from [132]*

From a potential nanotechnology application point of view, particularly interesting was the observation that compound semiconductors may under some irradiation conditions reorganize into quantum dots with a high degree of short-range order, see Figure 10.9. Taken together, this and other studies have shown that ion irradiation of surfaces can be used to achieve a very rich gamut of different kinds of surface morphologies [133].

10.4.2 Theoretical models

The ion-beam processing leading to ripple formation is almost invariably done in the ion energy regime where the irradiation also leads to sputtering, i.e., the removal of atoms from the surface [134–136]. Hence, it was initially natural to search for the explanation from sputtering.

The original theoretical explanation for ripple formation was provided by Sigmund, who in 1973 considered the possible difference between an ion impact point and the point of maximum atom emission [137,138], see Figure 10.10(a). These calculations show that that the sputtering yield may actually be lower from the top of a ridge and higher at the bottom of a valley on a surface. In case an irradiation condition has this feature, it naturally leads to any high feature becoming higher and a low feature becoming deeper during prolonged irradiation. In other words, there will be a self-organized roughening of the surface. Moreover, since the difference between ion impact point and maximum emission has for any given irradiation energy and ion a well-defined length scale, one can from this simple argument also deduce that there is a preferential length scale for the features.

The work of Sigmund was extended by Bradley and Harper, who in 1982 presented a comprehensive differential equation theory that allows predicting the ripple wavelength for a given irradiation condition [139]. Importantly, it gives an

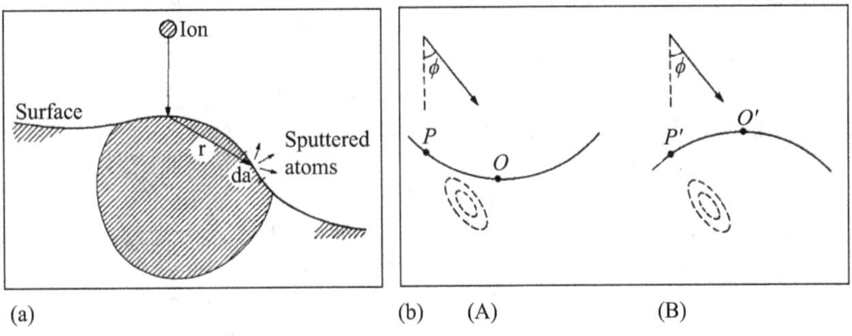

Figure 10.10 Original illustrations of the sputter-dominated ripple formation mechanism. (a) Sigmund showed in 1973 that the maximum sputtering yield does not necessarily occur at the point of impact, but away from it. This may have the consequence that the sputtering yield may be lower from the top of a ridge and higher at the bottom, leading to a surface instability that can lead to ripple formation. (b) Illustration of the dependence of energy deposition on incidence angle by Bradley and Harper [139]. Based on the Sigmund model, Bradley and Harper showed that at inclined incidence, the nuclear deposited energy is higher at the troughs (A) than the crests (B), leading to the ripple formation at inclined angles of incidence.

explicit description of the dependence of ripple wavelength and orientation (perpendicular or parallel) on the incoming ion angle.

By the arguments presented so far, it would seem that the ripples would grow in height indefinitely. However, it is rather clear that surface diffusion can play an opposing, smoothening effect. This was included already in the Bradley–Harper theory as a fourth-order derivative of the surface height [139]. We note that even in the absence of thermally activated surface diffusion, very sharp surface features are inherently unstable [140] and hence would smoothen out spontaneously.

There have been numerous extensions and further developments of the Bradley–Harper theory, to include higher order (nonlinear) displacement terms, white noise, random surface shapes, viscous relaxation, and other effects [129,141–143].

Parallel to this line of work, however, other studies made it clear that ion irradiation also leads to major other athermal atom redistribution effects, ion beam mixing [48,144–146]. Moreover, studies of single ion impacts on surfaces showed that even a single ion can produce massive surface modification, sometimes displacing thousands of atoms at the surface to form crater-like morphologies [147–152], see Figure 10.11(a). This is in contrast to the sputtering yields, which for keV ions even at most are a few tens of atoms [134,136]. Hence, it is natural to ask how much the atom displacements contribute to the sputtering.

The first model to systematically consider the role of the atom displacements in ripple formation was developed by Carter and Vishnyakov [154], who extended the Bradley–Harper differential equation model with an "atomic drift" term motivated by the momentum transfer from the ions and used this as a smoothing term to motivate why perpendicular irradiation does not lead to ripple formation in Si. This was further supported by Moseler *et al.*, who showed by MD simulation that in amorphous materials, the beam can be considered to produce a "downwards drift" of material [155]. However, the idea of momentum transfer-induced drift does not account for the fact that experimentally observed surface craters have rims also behind and on the sides of the ion impact point [147,151]. In experiments and MD simulations, the craters have been observed in the wide energy range of 1–400 keV [152], i.e., the same one where most ripples are formed.

The work toward including cratering effects to understand ripple formation started in 2008, when Kalyanasundaram *et al.* showed that even for very low energies, one can describe the surface height change as an "average crater" and introduced the concept of a crater function, that is, a mathematical description of the height change $h(x, y)$ around the ion impact point, see Figure 10.11(b).

In 2011, Norris *et al.* then developed a mathematical formalism to incorporate the crater functions into a differential equation form similar to the Bradley–Harper one [156]. This allowed to analyze both sputtering and crater-function-related atom redistribution contributions to ripple formation in a unified manner. By parameterizing both sputtering and crater function contributions from MD simulations, the Norris formalism made it possible to analyze the relative contributions of each to ripple formation. The results indicated that under a typical ripple formation condition of 250 eV Ar bombardment of Si, the atom redistribution contribution dominates over the sputtering one, see Figure 10.12.

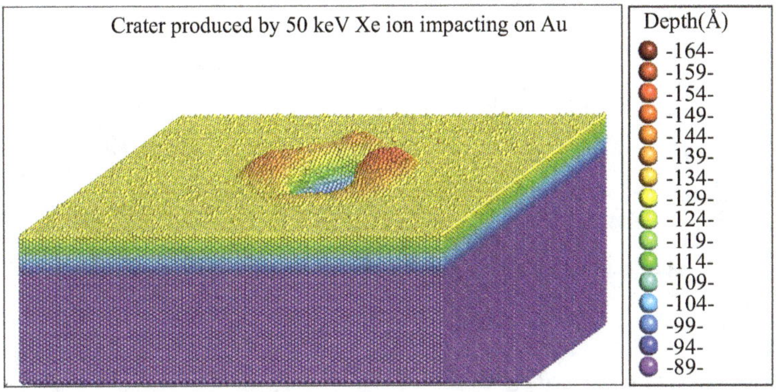

(a)

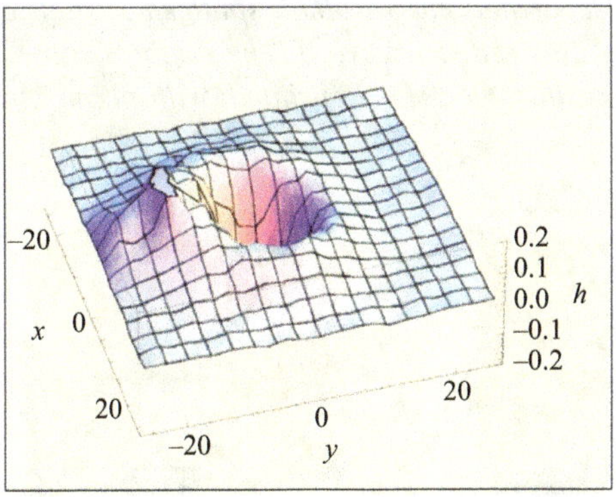

(b)

Figure 10.11 (a) A typical crater shape produced by 50 keV Xe irradiation of Au. Original work for this chapter following principles used in [152]. The colors show the height of the atoms above the surface. (b) Average crater function for 500 eV Ar irradiation of Si. © 2008. Reprinted, with permission, from [153]

One crucial reason for why the redistribution contribution is so significant is that in an amorphous material, there is a very large number of atoms with small (less than interatomic separation) displacements. Their contribution is comparable or larger than that from far-traveling ballistically displaced atoms. This is also one reason why the redistribution was not considered earlier: the BCA approach cannot account for such small displacements [157].

Since this work, the inclusion of crater functions has become common in the analysis of ripple formation. Harrison and Bradley have pointed out that the crater function approach should be extended to include crater shapes at curved surfaces [158], and Nietiadi *et al.* did indeed calculate such functions [159]. Hofsäss has expanded the formalism with additional higher order terms [129], and Hossain *et al.* used the crater function asymmetry to explain ripple orientation [160].

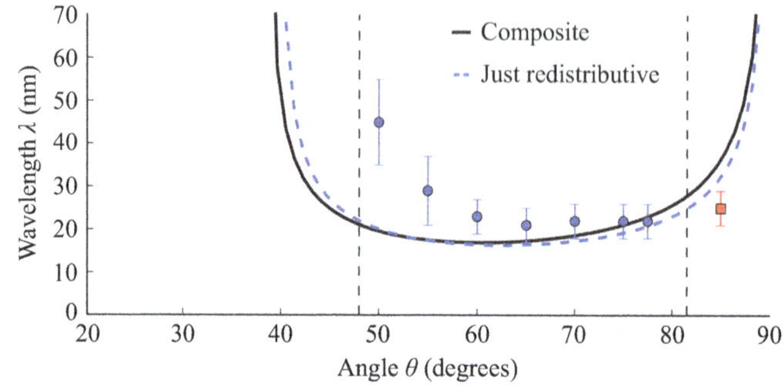

Figure 10.12 Ripple wavelengths predicted by the formalism of Norris et al., showing also separately the results if sputtering is left out of the model (curve "Just redistributive"). The symbols show experimental results. © 2011. Reprinted, with permission from [156]

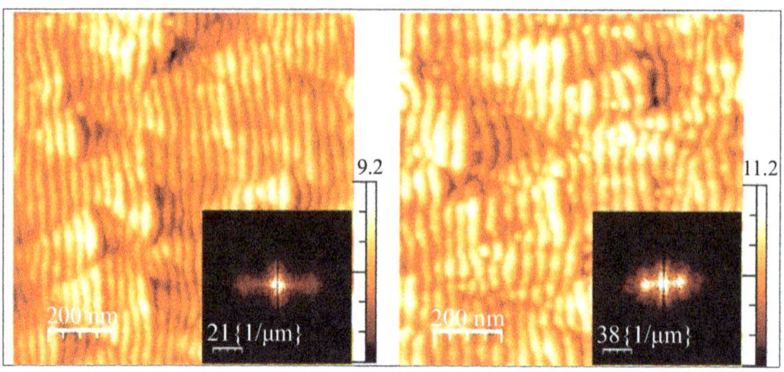

Figure 10.13 Ripples formed by 30 eV Ar irradiation of Si, at incidence angles of 65° and 70° off-normal. © 2018. Reprinted, with permission from [161]

Most recently, a combined experimental and simulation study showed that one can observe ripple formation on Si even at the very low energy of 30 eV [161], see Figure 10.13. At this energy, the sputtering yield is very small, which provides strong evidence that at least for low energies and amorphous materials, sputtering is indeed dominated by atom redistribution.

The situation for crystalline materials is, however, less clear. Silicon in these studies was considered an amorphous material since, even if it was initially crystalline, it has been certainly fully amorphized by the time the ion fluences are so high that ripples are observed [162]. However, in metals that do not amorphize, there cannot be the large number of very small displacements observed in amorphous materials [157]. Hence, the redistributive contribution must be smaller, and ripple formation may indeed be dominated by sputtering. Resolving this balance needs further work comparing metals and semiconductors under similar irradiation conditions.

10.5 Time-resolved experiments to probe defect evolution

10.5.1 Dynamic annealing

As explained in Section 10.3, the formation of stable radiation damage in materials is a dynamic phenomenon. For most irradiation environments relevant to ion-beam processing of semiconductors, the damage buildup proceeds via defect migration and interaction processes that occur over time scales of $\gtrsim$1 ps, after the rapid stages of formation and thermalization of collision cascades. These relatively long-term defect migration and interaction processes are typically referred to as DA.

Defect migration and interaction could lead to point defect annihilation or, in turn, to clustering and the growth of various extended lattice defects, such as dislocations, stacking faults, planar defects, nonstoichiometric inclusions, amorphous zones, bubbles, and voids during irradiation. Such defect interaction is often the reason why lattice disorder experimentally observed after irradiation significantly departs from predictions based only on collisional processes. The DA typically manifests experimentally as a dependence of radiation damage on irradiation conditions.

Both theoretical and experimental studies of DA are challenging. As discussed next, the mobile defects dominating DA often have lifetimes of $\gtrsim 10^{-6}$ s. Hence, the MD method cannot be readily applied, and kinetic Monte Carlo and rate-theory approaches are often used instead [163,164]. These latter techniques require explicit assumptions about the defect interaction processes involved and can be challenging due to our limited understanding of the actual (both stable and metastable) defect configurations and reaction pathways in most materials. This applies to even crystalline Si, which is perhaps the most extensively studied material [165]. There is also limited experimental data available that can be used to critically assess and benchmark DA models. For example, damage buildup and dose-rate data often cannot differentiate between physically different models [118]. Due to such inherent complexity, DA has remained a subject of on-going research for many decades [66,87,119,165–220]. In this section, we discuss both traditional dose-rate effect (DRE) and more recent pulsed beam experiments aimed at better understanding and predicting DA in semiconductors.

10.5.2 Traditional dose-rate effect experiments

In regimes with strong DA, damage formation intricately depends on irradiation conditions. These include ion dose (or fluence), target temperature, ion energy, mass, and the dose rate (which is also often referred to as beam flux). Most radiation damage studies in semiconductors have focused on understanding dose and temperature dependencies [201], while experiments specifically probing the DA regime have traditionally focused on measurements of dependencies of radiation damage on sample temperature, the dose rate, and the density of collision cascades. Most DA studies in the past have, however, involved measurements of the DRE. Such a DRE has been known for a very long time [202], and several irradiation regimes characterized by a spectacularly strong DRE have been documented [171,203,204]. Despite that, systematic studies of the DRE are relatively scarce since they are more technically challenging than damage buildup studies. Indeed, DRE experiments require an accelerator facility with stable beam currents, accurate dosimetry, and capability of broad beam (rather than rastered) irradiation [165].

The DRE reflects *inter*-cascade (rather than *intra*-cascade) DA processes, caused by the interaction of mobile defects originating in different collision cascades created in close proximity of each other. The DRE is complicated by the fact that the efficiency of DA depends on both the density of mobile defects (determined not only by the dose rate but also by ion mass and energy) and their rates of diffusion and interaction, which are typically temperature dependent. This defect interaction involves a convolution of both spatial (such as the effective defect diffusion length, L_d) and temporal (i.e., the lifetime of the mobile point defects, τ) parameters of DA. Attempts to understand DA from flux effect data have been somewhat cumbersome [165,168,194,205] since explicit assumptions are required to extract τ and L_d from experimental data by modeling defect-interaction processes.

The magnitude of the DRE also depends on the dose regime since mobile defects tend to interact with preexisting defect structures in the lattice. An irradiation regime particularly relevant to ion implantation is that when relatively high doses are delivered, resulting in the amount of stable lattice disorder of $\gtrsim$1 at.%, as measured by a channeling-based method. Numerous previous studies of the DRE in this important dose regime have been reported for various semiconductors, including Si [202,205,206], Ge [169,204], GaAs [169,203], InP [207], GaN [87], 4H-SiC [208–210], and 6H-SiC [211]. These reports [87,169,202–211] have demonstrated that the damage formation efficiency typically increases with increasing dose rate.

Here, we describe the methodology of DRE experiments with an example of a recent systematic study of 3C-SiC (relevant to both electronics [221] and nuclear [222] technologies) by a combination of ion channeling and TEM [194]. It illustrates the intriguing complexity of the DRE in SiC: it is pronounced and highly nonlinear (with saturation at high dose rates), and its magnitude strongly depends on both ion dose and temperature.

Figure 10.14 shows damage buildup curves for 3C-SiC bombarded at 100°C, 150°C, and 200°C with 500 keV Ar ions with two different dose rates of 1.1×10^{13} and 4.2×10^{13} cm^{-2} s^{-1}. Figure 10.14 shows that all six damage buildup curves are sigmoidal, which is in agreement with the other damage buildup studies of 3C-SiC in this temperature range [118,193,212,213]. The sigmoidal shape suggests nucleation-limited (i.e., stimulated) defect accumulation. The solid and dashed lines in Figure 10.14 show results of fitting the buildup with a phenomenological stimulated amorphization model from [118], providing an excellent fit for all the cases.

It is further seen from Figure 10.14 that, for low doses, below some critical values denoted by arrows, the damage level is essentially independent of the dose rate, while for larger doses, damage buildup becomes dose-rate-dependent.

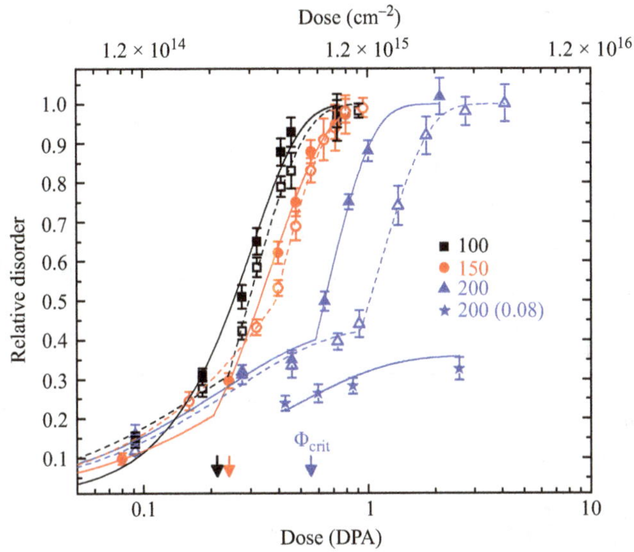

Figure 10.14 *Dose dependencies of relative bulk disorder in 3C-SiC bombarded with 500 keV Ar ions at three different temperatures (given in the legend in units of °C) with two different dose rates of (closed symbols) 4.2×10^{13} cm^{-2} s^{-1} and (open symbols) 1.1×10^{13} cm^{-2} s^{-1}. Star symbols show data points for bombardment at 200 °C with a dose rate of 8×10^{11} cm^{-2} s^{-1}. Solid and dashed lines are results of fitting with a stimulated amorphization model from [118]. Arrows mark critical doses (Φ_{crit}) at different temperatures for a dose rate of 4.2×10^{13} cm^{-2} s^{-1}. Reprinted, with permission, from [194].*

A similar effect of a reduced DRE at low doses has also been reported for room-temperature bombardment of Si with 40 keV ^{14}N ions [223] and of GaAs with 100 keV ^{28}Si ions [171]. Such a dose dependence of the DRE magnitude indicates that *inter*-cascade DA, giving rise to the DRE, is more efficient for conditions with preexisting lattice defects. The dose rate strongly influences the regime of lattice amorphization at larger doses, rather than the buildup of point defect clusters in the low-dose regime.

Figure 1.14 (star symbols) also shows a dependence for bombardment at 200°C with a much lower dose rate of 8×10^{11} cm^{-2} s^{-1}. It reveals that, with decreasing dose rate, a pronounced DRE is also present even for low doses. Hence, the weak DRE at low doses occurs only in a certain dose rate range. This effect is better illustrated in Figure 10.15(a), which summarizes dose-rate dependencies for different sample temperatures in the range of 25°C–200°C. Each of these dose-rate dependencies was measured with the total dose kept constant, and doses at different temperatures were different and chosen so that the relative damage level for the

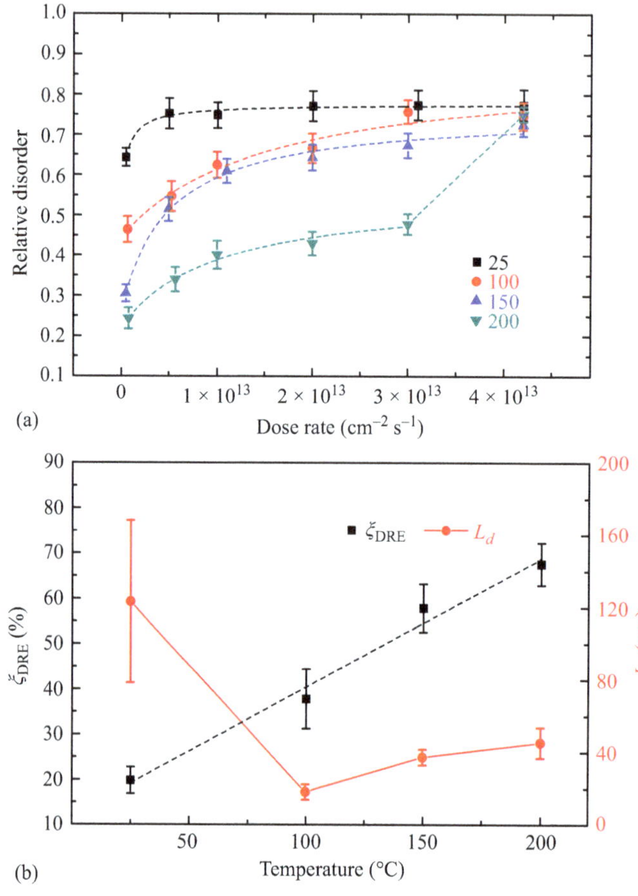

Figure 10.15 *(a) Dose-rate dependence of relative disorder in 3C-SiC bombarded at different temperatures (given in the legend in units of °C) with 500 keV Ar ions to total doses of 0.29, 0.36, 0.75, and 0.82 DPA for 25°C, 100°C, 150°C, and 200°C, respectively. Dashed lines are results of fitting with a second-order decay model described in the text. (b) Temperature dependencies of (left axis) the magnitude of the dose-rate effect (ξ_{DRE}) and (right axis) the effective defect diffusion length. The dashed line in (b) is a linear fit to the data. Reprinted, with permission, from [194]. © 2017 American Institute of Physics*

highest dose rate used was ~0.75. Figure 10.14(a) reveals a monotonic increase of damage with increasing dose rate over the entire temperature range investigated. For temperatures ≤150°C, these dependencies are strongly sublinear and saturate for large dose rates. Such a saturation effect is more pronounced at lower temperatures. Figure 10.15(a) also shows that, for the 200°C case, the saturation is observed only in the dose rate range of $\sim(1-3) \times 10^{13}$ cm^{-2} s^{-1}, and damage exhibits a strong increase to ~0.75 on further increasing dose rate from 3×10^{13} to 4.2×10^{13} cm^{-2} s^{-1}.

The DRE reflects *inter*-cascade DA. As discussed earlier, the dominant DA process could often be described by the following two defect-interaction parameters [165]: a characteristic lifetime of mobile defects (τ) and the effective defect diffusion length (L_d). In this (admittedly simplified) scenario, after ballistic generation and thermalization of a collision cascade, mobile defects experience DA over a time scale of $\sim\tau$ and diffuse over distances of $\sim L_d$. The rate of ion impacts onto L_d-defined areas scales with the dose rate (F) as $\left(1/t_{\text{impact}}\right) \simeq 4L_d^2F$. The interaction of mobile defects leading to the DRE occurs when two or more ions impact onto L_d-defined areas over the lifetime of mobile defects (τ). At very low-dose rates, $t_{\text{impact}} \gg \tau$, and the interaction of mobile defects generated in different collision cascades is suppressed. In this case, damage buildup proceeds via ballistic and intra-cascade DA processes. With increasing dose rate, t_{impact} decreases, the probability of ion impacts into L_d-defined areas over the time scale of $\sim\tau$ increases, and the DRE is observed.

The sublinear shape of DRE dependencies from Figure 10.15(a) is consistent with the second-order decay, which is also in agreement with pulsed beam results described as follows [187]:

$$n\left(t_{\text{impact}}\right) = n_\infty + \frac{n(0) - n_\infty}{1 + \left(t_{\text{impact}}/\tau\right)}, \tag{10.3}$$

where n is the level or relative disorder, and n_∞ is relative disorder for $t_{\text{impact}} \gg \tau$. Dashed lines in Figure 10.15(a) show results of fitting $n(F)$ dependencies with τ values taken from the pulsed beam measurements of [189] and L_d as a fitting parameter. The resultant temperature dependence of L_d is plotted in Figure 10.15(b) (right axis).

Figure 1.15(b) (left axis) shows the temperature dependence of the DRE magnitude, defined as

$$\xi_{\text{DRE}} = \frac{n(F_{\max}) - n(F_{\min})}{n(F_{\max})}, \tag{10.4}$$

where $F_{\max} = 4.2 \times 10^{13}$ and $F_{\min} \sim 5 \times 10^{11}$ cm^{-2} s^{-1} are the maximum and minimum F values used. This reveals a close-to-linear dependence, quantifying the above statement that the DRE becomes stronger with increasing sample temperature. Extrapolation of this linear dependence suggests that the DRE will vanish at temperatures lower than ~ -45°C.

The above results (Figures 1.14 and 1.15) clearly indicate that DA processes in 3C-SiC are thermally activated. At higher temperatures, both intra- and inter-cascade defect recombination processes are expected to become more efficient due to increased defect mobility. Thus, at higher temperatures, a smaller fraction of mobile point defects is available to form the stable lattice defects that are measured after irradiation. Data from Figure 10.15 shows that an increase in temperature results not only in increased defect recombination (leading to lower damage) but also in a more efficient defect clustering for large dose rates (resulting in larger damage). In addition, TEM analysis reported in [194] has revealed larger defect clusters for irradiation with larger dose rates, indicating that the dose rate influences both the concentration and type of stable lattice defects. Such knowledge of dose-rate dependencies is important for understanding DA. Additional insights can be obtained from experiments with pulsed ion beams that we discuss in the next section.

10.5.3 Pulsed-ion-beam method

In this section, we summarize DA studies with the pulsed-ion-beam method recently developed for accessing the dynamic regime of defect accumulation [66,165,185, 186,190]. This pulsed-beam method has by now been applied to measure τ and L_d for Si, Ge, GaAs, 3C-SiC, and 4H-SiC at different temperatures and defect generation rates [66,165,185–193,195,196]. These results have tremendously clarified the previous estimates of τ, ranging over many orders of magnitude even for Si.

The inset in Figure 10.16 shows a schematic of the time dependence of the dose rate during pulsed-beam irradiation, defining the additional pulsing parameters

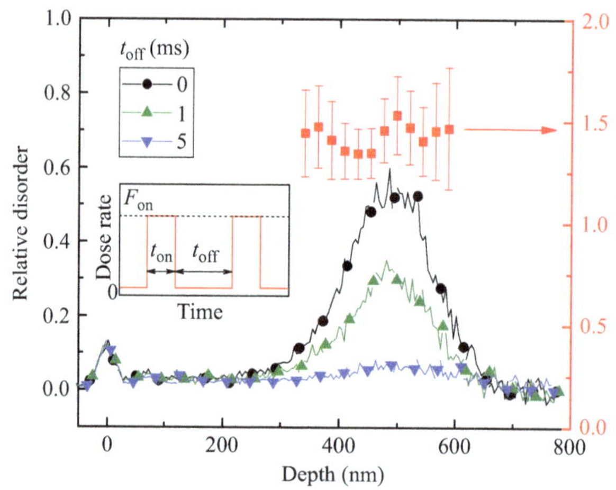

Figure 10.16 Left axis: selected depth profiles of relative disorder in Si bombarded at 100°C with a pulsed beam of 500 keV Ar ions with $F'_{on} = 1.8 \times 10^{13}$ cm^{-2} s^{-1}, $t'_{on} = 1$ ms, a total ion dose of 8.75×10^{14} cm^{-2}, and different t_{off} values given in the legend. For clarity, only every tenth experimental point is depicted. Right axis: the depth dependence of the dynamic annealing time constant (τ) measured as described in the text. The inset is a schematic of the time dependence of the instantaneous beam flux for pulsed beam irradiation, defining t'_{on}, t_{off}, and F'_{on}.

(t'_{on}, t_{off}, and F'_{on}). In such experiments, the total ion dose is split into a train of equal pulses, and the measurement involves ion irradiation of a series of specimens with all except one of the irradiation parameters fixed (either t_{off} or t'_{on}). Irradiation is followed by ex situ measurements of the level of stable disorder by, for example, ion channeling or other characterization techniques. Parameters τ and L_d are evaluated based on the analysis of experimental dependencies of the level of stable disorder on t_{off} and t'_{on}, respectively [165,185–188,191]. Moreover, a depth dependence of these parameters can be obtained in cases of depth-resolved damage measurements.

Each pulsed-beam measurement for any given combination of the target material, temperature, ion mass, energy, and F'_{on} starts with the selection of the total ion dose, so that DA in the damage level of interest can be studied. Such a selection of the total ion dose can sometimes be done based on the literature on damage buildup. However, for regimes with pronounced DA, damage typically strongly and nonlinearly depends on the irradiation parameters. Hence, the first set of experiments in any pulsed-beam series is typically the measurement of the traditional damage buildup curve, which is the dependence of damage on ion dose for a continuous (rather than pulsed) beam with $F = F'_{on}$, as, for example, shown in Figure 10.14.

The robustness of the pulsed-beam method has further been evaluated in a systematic study [165] of the dependence of τ and L_d in Si at room temperature on the doping level, ion dose, and the defect generation rate defined by F'_{on}, t'_{on}, and the density of collision cascades. Results have shown that the average density of collision cascades influences not only the efficiency of damage accumulation and annealing but also the defect-interaction dynamics. Moreover, a nontrivial dependence of τ on the choice of parameters of the active part of the beam duty cycle (i.e., F'_{on} and t'_{on}) [165] suggests an important role of the instantaneous concentrations of mobile defects in defect dynamics. Next, we give examples of pulsed-beam measurements in different semiconductors.

10.5.3.1 Measurement of defect relaxation time constants

Figure 10.16 (left axis) shows representative depth profiles of relative disorder (measured by ion channeling) in Si bombarded with 500 keV Ar ions with three different t_{off} values (given in the legend), and all the other parameters kept constant. It is seen that, with increasing t_{off}, while the damage level in the first ~250 nm from the sample surface remains unchanged, the intensity of the bulk disorder peak (which we will refer to as n) dramatically decreases. This observation suggests different dynamic mechanisms of bulk and surface disordering. It is consistent with several other depth-resolved studies of defect-interaction dynamics in Si [165,185,186,191].

Figure 10.17 (bottom axis) shows the $n(t_{off})$ dependence for Si at 100°C under 500 keV Ar ion bombardment. The monotonic $n(t_{off})$ decay revealed is related to the interaction of defects generated in different pulses (and, hence, in different cascades). As the beam is pulsed off the target, the concentration of mobile defects decreases via DA with a characteristic time constant τ. For irradiation with

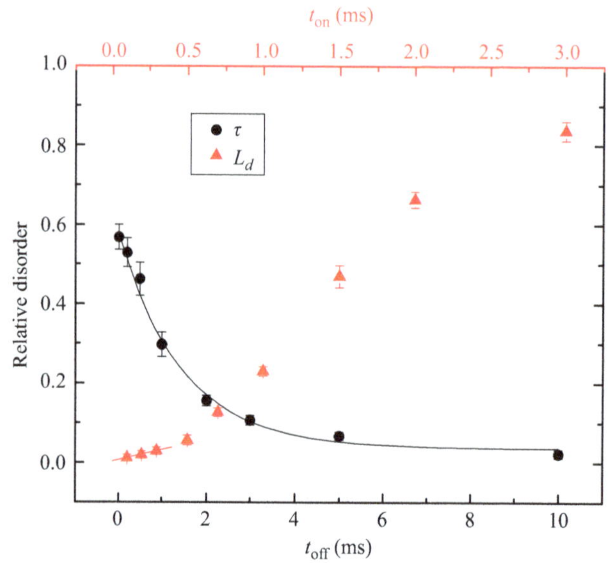

Figure 10.17 *Relative bulk disorder in Si bombarded with a pulsed beam of 500 keV Ar ions (bottom axis) as a function of t_{off} with a fixed $t'_{on} = 1$ ms and $F'_{on} = 1.8 \times 10^{13}$ cm^{-2} s^{-1} and (top axis) as a function of t'_{on} with a fixed $t_{off} = 10$ ms and $F'_{on} = 10^{14}$ cm^{-2} s^{-1}. Fitting curves (bottom axis) with the first-order (i.e., exponential) decay and (top axis) with linear equations are shown by solid lines. © 2017. Elsevier. Reprinted, with permission, from [192]*

$t_{off} \gg \tau$, DA processes have essentially decayed in time intervals between individual ion pulses.

The time constant of the dominant DA process (τ) can be quantitatively evaluated by analyzing such experimental $n(t_{off})$ dependencies. For example, τ can be obtained by fitting $n(t_{off})$ dependencies with either the first order,

$$n(t_{off}) = n_\infty + (n(0) - n_\infty)e^{-t_{off}/\tau_1}, \tag{10.5}$$

or the second order,

$$n(t_{off}) = n_\infty + \frac{n(0) - n_\infty}{1 + (t_{off}/\tau_2)}, \tag{10.6}$$

decay equations, where n_∞ is relative disorder for $t_{off} \gg \tau_{1,2}$. The "1" and "2" subscripts refer to whether the fitting was done with the first or second-order decay equation, respectively. Data from Figure 10.17 is best fitted with the first-order decay equation. Figure 10.16 (right axis) gives an example of depth-resolved τ_1 measurements. It shows that the τ_1 is essentially independent of depth at the ion end of range region for these irradiation conditions.

10.5.3.2 Measurement of defect diffusion lengths

The L_d is evaluated based on the experimental $n(t'_{on})$ dependence, with all the other irradiation parameters kept constant. In these measurements, it is important to maintain $t_{off} \gg \tau$ in order to suppress the interaction of mobile defects generated in different pulses [165,186]. The $n(t'_{on})$ dependence reflects the interaction of mobile defects created in different cascades within the same pulse. An example of the measurement of the $n(t'_{on})$ dependence is illustrated in Figure 10.17 (top axis) for the case of Si bombarded at 100°C with 500 keV Ar ions. It is seen that damage increases monotonically with increasing t'_{on}.

Two different approaches to extract the L_d from $n(t'_{on})$ data have been reported [66,165,188,190]. Both are based on the analysis of $n(t'_{on})$ dependencies for low t'_{on} values (i.e., for low ion dose increments per pulse), when the average lateral distance between neighboring ion impacts in each pulse is larger than L_d. This corresponds to the onset of the interaction of mobile defects generated in adjacent (sub)cascades within each pulse. With increasing dose per pulse $(F'_{on}t'_{on})$, the average distance between individual collision cascades for each pulse decreases, resulting in a more efficient interaction of mobile defects generated in different cascades.

In the first approach to evaluate L_d, the complexity of cascade formation and subsequent DA is approximated by the following scenario [165,186,190]. At any given depth in the plane perpendicular to the sample surface, each ion creates a damage zone with a circular area (A) with a radius $L_d + R_{ballistic}$, where $R_{ballistic}$ is the radius of the ballistic cascade, which can be estimated with a binary collision code. Measurements reported in [165] have shown that $L_d \gg R_{ballistic}$. Ion impacts obey Poisson statistics as they are random and uncorrelated. Hence, there is a finite probability (with a Poisson parameter of $AF'_{on}t'_{on}$) of multiple ion impacts into L_d-defined areas. For $t'_{on} < \tau$, the majority of defect relaxation occurs during the passive (t_{off}) rather than active (t'_{on}) portion of the beam duty cycle. Hence, the average density of ion-beam-generated mobile point defects after each pulse can be expressed as $\rho_{displacement} \propto 1 + 4L_d^2 F'_{on}t'_{on}$. If we assume that, for such low t'_{on} values, the efficiency of the stable damage formation scales with the displacement density, then $n \propto 1 + 4L_d^2 F'_{on}t'_{on}$. This linear equation is used to fit the initial (linear) portion of the $n(t'_{on})$ dependence, with the L_d as a fitting parameter. For example, fitting the data from Figure 10.17 (top axis) gives $L_d = 78 \pm 14$ nm.

10.5.3.3 Measurements of activation energies and cascade density effects

In the second approach to L_d evaluation from $n(t_{on})$ data proposed in [66], cascades are not approximated as circular for any given depth slice but are analyzed with a model that accounts for their fractal nature.

Numerous previous experiments have shown that, even for cases when defect production by electronic excitation is negligible, the damage production efficiency and DA can depend strongly on the mass and energy of bombarding particles and, hence, on the geometry of collision cascades [224,225]. Average cascade densities are determined by the target material, ion mass, energy, and the distance from the sample surface. With increasing ion or target mass, the cross section of elastic scattering increases, leading to larger volumetric densities of atomic displacements within cascades [183,224,225]. As a result, for a given target material and ion energy, light ions tend to create dilute collision cascades, while heavier ions create denser cascades [183,224,225].

The influence of the three-dimensional shape of collision cascades on radiation damage processes is one of the most complex aspects of radiation defect physics. In fact, even the definition of the cascade density is not straightforward and has been debated over the past several decades [66,88,214–216,220,224,226–228]. In the 1980s, significant progress toward understanding geometry of collision cascades was made by Winterbon *et al.* [229,230], Cheng *et al.* [231], and Rossi *et al.* [232]. They [229–232] showed that the distribution of displacements within a collision cascade could be described within the framework of fractal geometry [233]. A number of subsequent theoretical studies were focused on computations of fractal dimensions of (sub)cascades at different stages of their evolution and on understanding limitations of the fractal approach [234–241].

Despite the accumulated evidence [88,229–232,234–241] that, in many practical cases, collision cascades are, in fact, fractals, the correlation of the fractal parameters of cascades with experimental observations of a pronounced effect of ion mass and energy on the DA efficiency has remained elusive. A recent work reported in [66] links the fractal parameters of collision cascades to experimental data on radiation damage buildup. Their results have shown that, at least for Si and SiC bombarded with 500 keV ions with masses ranging from ^{4}He to ^{238}U, collision cascades are indeed mass fractals over a relatively large length scale range of $\sim 0.5-30$ nm, with fractal dimensions (D) of $\sim$1–2. Furthermore, the results [66] did not reveal any evidence of multifractality (for given ion mass and depth from the surface), significantly simplifying the mathematical treatment of cascade geometry [230–232,235–240]. Fractal parameters depend on ion mass, energy, and the depth from the surface. The maximum cascade density often occurs at depths larger than those of the maxima of the profiles of nuclear energy loss and implanted ions. This could explain several previous observations of a shift of the bulk defect peak to larger depths in irradiation regimes with strong DA [118,119,217–219,242,243].

The ion mass dependence of D is shown in Figure 10.18 (left axis) for ion masses from ^{4}He to ^{238}U. It reveals that, for all the ion masses considered, cascades are indeed mass fractals (i.e., $D < 3$), with D in the range of $\sim 0.8-1.7$ at depths corresponding to maxima of the vacancy generation profile. It is also seen that D monotonically increases with increasing ion mass and exhibits two well-defined close-to-linear regimes for masses above and below that of Ne.

In [66], the collision cascade density ρ was defined as the average local density of lattice vacancies within the sphere with a radius of R_c at a given depth. Figure 10.18 (right axis) shows the mass dependence of ρ for $R_c = 10$ nm. It is seen from Figure 10.18 (right axis) that ρ monotonically increases with ion mass and follows a similar dependence to that of D with two close-to-linear regimes with very different slopes, intersecting at a mass close to that of Ne.

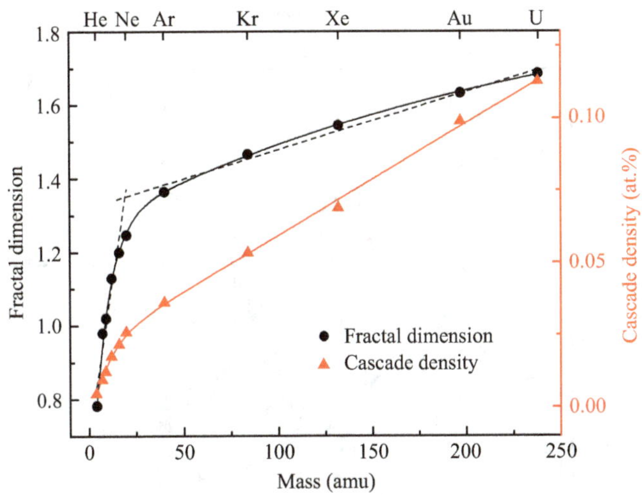

Figure 10.18 *Ion mass dependence of (left axis) the fractal dimension and (right axis) the cascade density (for $R_c = 10$ nm) for cascades created by 500 keV ions in SiC at their respective maxima of the vacancy generation profile. Dashed lines are to guide the eye. From [66]*

This fractal model has also been used to analyze results of pulsed-ion-beam measurements for determining L_d as the relevant length scale over which fractal collision cascade density is averaged in the analysis of the dependence of the DA efficiency on ion mass [66]. This analysis has yielded an L_d of ~10 nm for 3C-SiC at 100°C and ~20 nm for Si at 25°C. For both materials, L_d decreases slightly with increasing ion mass (i.e., the cascade density), which can be attributed to an increased density of defect traps in denser cascades.

In [196], an analysis of cascade densities was further extended to Si bombarded in the temperature range from ~−30°C to 210°C with ions in a wide range of masses, from ^{20}Ne to ^{129}Xe, creating collision cascades with different densities. They have demonstrated that the complexity of the influence of irradiation conditions on several parameters of defect-interaction dynamics can be reduced to a deterministic effect of a single parameter, the average cascade density, calculated by taking into account the fractal nature of collision cascades.

Figure 10.19 shows results from [196] in an Arrhenius plot of the DA rate (defined as τ^{-1}), and with kT having the usual meaning. It provides evidence that DA processes are thermally activated in the entire range of ion masses studied. This is in qualitative agreement with a number of other studies of DA in Si with continuous ion beams [197,198,200,244]. It is also seen from Figure 10.19 that, for each ion species, the DA rate exhibits two well-defined Arrhenius regions with different activation energies (E_as) where different DA mechanisms dominate. The low-T E_a is ~0.1 eV for all the ion species studied, while the high-T E_a is ~0.4 eV

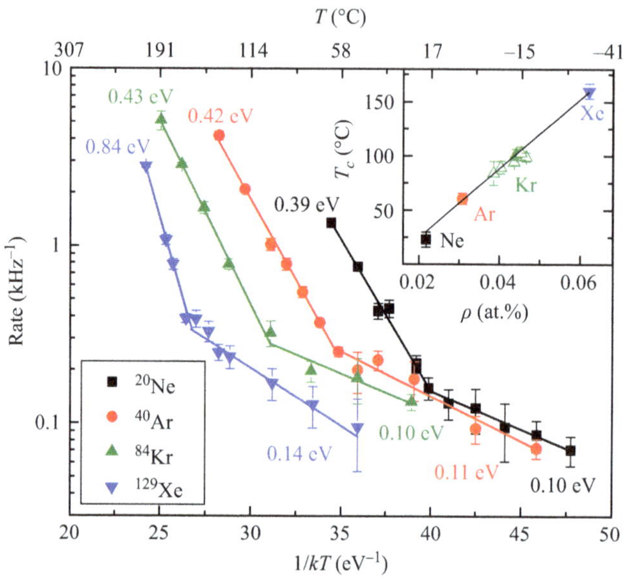

Figure 10.19 Arrhenius plot of the DA rate for Si bombarded with pulsed beams of Ne, Ar, Kr, and Xe ions. Straight lines show results of linear fitting, revealing the low- and high-T activation energies shown. The inset shows the critical temperature (T_c), separating the two DA regimes, as a function of the average cascade density (ρ) at depths corresponding to maxima of vacancy generation profiles. Open symbols for Kr ions are T_c values at different depths between ~150 and 400 nm. © 2018. American Physical Society. Reprinted, with permission, from [196]

for Ne, Ar, and Kr, and it doubles for the case of Xe. Figure 10.19 also shows that, for each ion mass, the two Arrhenius regimes are separated by a critical temperature (T_c), which increases with increasing ion mass and, hence, the cascade density. This effect is better illustrated in the inset in Figure 10.19, revealing a linear dependence of T_c on the cascade density. These results have clearly demonstrated a crucial role of the collision cascade density and can be used to predict radiation defect dynamics in Si.

The physical meaning of these E_a values has been clarified by rate-theory modeling in [191]. It has attributed the high-T E_a of 0.4 eV to the rate-limiting migration of vacancies, which is consistent with a number of previous studies of vacancy diffusion in Si [245–247]. The low-T E_a of 0.1 eV has been associated with the rate-limiting formation of divacancies as the main channel of vacancy annihilation [191]. The fact that DA for bombardment with different ion species, generating cascades in a wide range of densities, is characterized by the same low-T E_a of 0.1 eV (Figure 10.19) strongly suggests that the rate-limiting DA processes are cascade-density-independent in this temperature regime.

In contrast, Figure 10.19 shows that, for the high-T regime, E_a doubles on increasing ion mass from Kr to Xe. This effect has been attributed to the fact that ρ is depth dependent. Closer to the sample surface, away from the maximum of ballistic displacements, the ion energy is higher, and the cross section for elastic collisions is lower, leading to lower cascade densities. This is illustrated by Figure 10.20 (left axis, closed symbols), which compares depth profiles of the cascade density for 500 keV Kr and Xe ions. It shows that the shape of such profiles is similar to that of nuclear energy loss distributions. The maximum cascade density occurs at ~300 and 180 nm for Kr and Xe ions, respectively, and it rapidly decreases toward the surface. The vertical dashed line at ~45 nm in Figure 10.20 shows the depth at which the cascade density for Xe is equal to the maximum of that for Kr. If the change in the high-T E_a observed for Xe ions is a cascade-density-related effect, E_a is expected to decrease to ~0.4 eV at depths $\lesssim$45 nm.

Figure 10.20 (right axis, open symbols) tests this hypothesis and shows depth dependencies of the high-T E_a for Kr and Xe ions. The E_a for Kr is approximately constant in the entire depth range studied (~100–400 nm), while, in the same depth range, T_c for the same Kr ions decreases with decreasing ρ toward the surface, as shown by open symbols in the inset of Figure 10.19. In contrast to the case of Kr, Figure 10.20 (right axis) shows that the high-T E_a for Xe ions decreases from ~0.95 eV at ~300 nm (which is well beyond the maximum of the bulk relative disorder at the maximum of the vacancy generation profile, at 180 nm) to

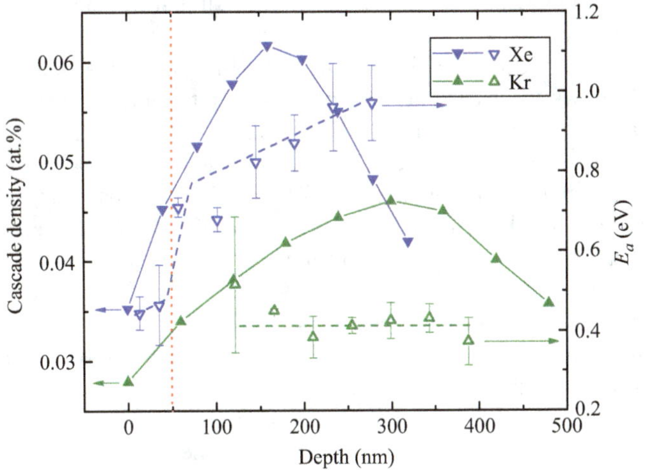

Figure 10.20 Depth dependencies of (left axis, closed symbols) the average cascade density and (right axis, open symbols) the high-T E_a in Si bombarded with 500 keV Kr or Xe ions, as indicated in the legend. The vertical dashed line denotes the depth at which the cascade density for Xe ions equals the peak cascade density for Kr ions. Dashed lines through E_a dependencies are to guide the eye. © 2018. American Physical Society. Reprinted, with permission, from [196]

~0.7 eV at ~100 nm. At shallower depths, where the cascade density $\lesssim$0.045 at.%, the high-T E_a for Xe reduces more rapidly to ~0.45 eV.

Hence, for cascades with densities $\lesssim$0.045 at.%, DA appears to be rate-limited by the migration of Frenkel pair defects as discussed earlier [191]. For denser cascades, a different high-T DA mechanism characterized by a larger and cascade-density-dependent E_a becomes dominant. This high-cascade-density regime has been associated with the onset of collective phenomena of the formation of thermal spikes [224]. Such spikes lead to the formation of zones of heavily disordered or amorphous material. The new thermally activated DA mechanism with a high E_a could involve processes of defect interaction with the defective zones formed in spikes. Specific DA mechanisms in the regime of dense collision cascades are not clear at the moment, and these cascade-density-driven defect-interaction phenomena call for future systematic studies, both experimental and theoretical.

10.5.4 Comparison of radiation defect dynamics in different semiconductors

A comparison of radiation defect dynamics in different semiconductors studied by the pulsed-beam method has recently been given in [195]. This study has included Si, Ge, GaAs, and 4H-SiC and revealed that both the high-temperature activation energy and the temperature below which DA becomes negligible scale with the melting point. Interestingly, the DA behavior of GaAs, which is a prototypical material with strong DA [172], is qualitatively similar to that for elementary semiconductors such as Si and Ge and different from that for another compound semiconductor, 4H-SiC, all studied by the pulsed-beam method. As an example, Figure 10.21 shows the Arrhenius plot of the DA rate, defined as τ^{-1}, for GaAs bombarded in the temperature range of 20°C–100°C with pulsed beams of 500 keV Xe ions. Similar to the case of Si and Ge, two well-defined Arrhenius regimes are clearly seen, one below and one above a $T_c \sim 60$°C. Note that T_c coincides with the temperature when the temperature dependence of the DA efficiency saturates for GaAs, Si, and Ge [190,191]. Linear fitting of the data in Figure 10.21 gives two activation energies of $E_a^{HT} = 0.62 \pm 0.09$ eV and $E_a^{LT} = 0.11 \pm 0.01$ eV, above and below a T_c of 60°C, respectively. Superscripts "LT" and "HT" refer to low and high temperature, respectively. Figure 10.21 also shows the three data points measured for GaAs bombarded at 20°C, 40°C, and 60°C with 500 keV Ar ions. It reveals that both E_a^{HT} and T_c depend on ion mass (and, hence, the cascade density) as the Arrhenius plots for the case of Ar ion bombardment is offset to lower Ts with E_a^{HT} of 0.45 ± 0.07 eV. This is consistent with results for Si (Figure 10.19).

Also plotted in Figure 10.21 are DA rates for Si, Ge, and 4H-SiC bombarded with 500 keV Ar ions. It is seen from Figure 10.21 that E_a^{HT} and the cutoff temperature of DA (T_0, at which DA efficiency becomes negligibly small) scale with the melting point. This is better illustrated by Figure 10.22. The scaling of T_0 with the melting point is not unexpected. Indeed, DA is related to defect mobility, which is expected to increase when temperature approaches the material's melting point. The physics behind a decrease in E_a^{HT} with increasing melting point is less

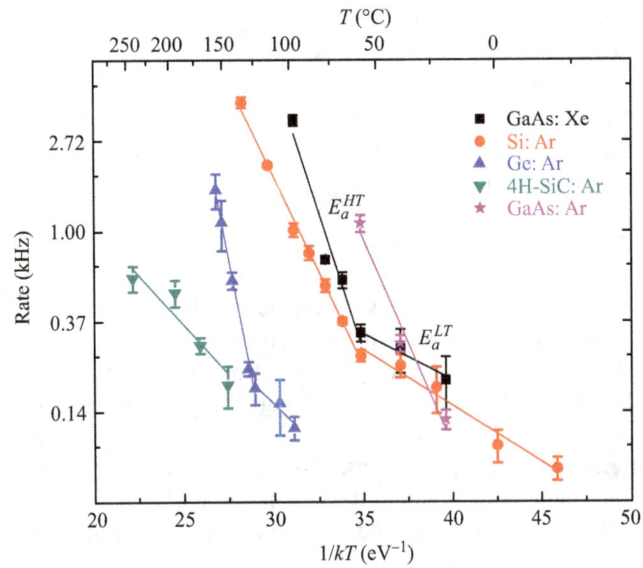

Figure 10.21 Arrhenius plot of the DA rate for four different materials bombarded with either 500 keV Ar or 500 keV Xe ions, as indicated in the legend. AIP Publishing. Reprinted, with permission, from [195]

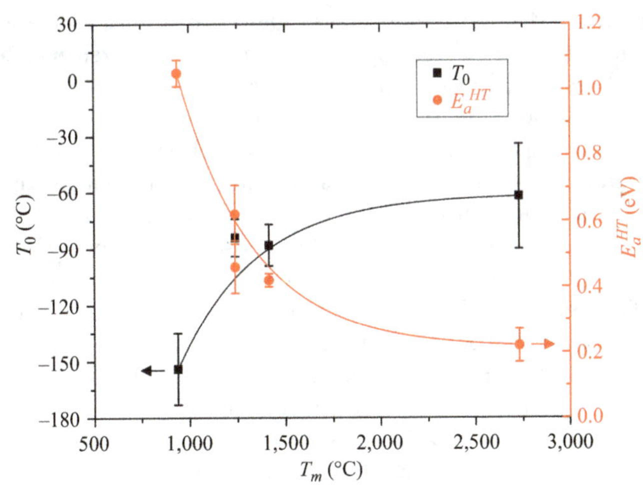

Figure 10.22 Dependence of (left axis) the cutoff temperature of DA (T_0) and (right axis) the high-temperature activation energy of DA (E_a^{HT}) on the melting temperature for the four materials studied by the pulsed beam method. Solid lines are to guide the eye. AIP Publishing. Reprinted, with permission, from [195]

clear. It could be related to the fact that the melting point reflects the bond energy in the material, given that DA processes are expected to be more efficient in materials with larger energy bonds resulting in a larger energy gain upon defect annihilation and the recovery of broken or distorted bonds. These results deserve future modeling effort to correlate the E_a values measured with specific defect migration and interaction processes.

10.6 Conclusions

In conclusion, in this chapter, we have reviewed the basic physics of ion implantation of materials, with emphasis on semiconductors. We have discussed the concepts of ion stopping and energy deposition into electronic and nuclear subsystems, radiation defect processes resulting from such electronic and nuclear energy loss events, and the influence of ion channeling on energy deposition. We have also briefly reviewed general aspects of ion irradiation experimentation, emphasizing the importance of controlling radiation parameters such as ion mass, energy, the angle of ion-beam incidence, the time dependence of the dose rate, the total dose, and sample temperature.

Throughout this chapter, examples for Si have been chosen to illustrate the wide range and complexity of physical phenomena of defect formation and evolution under ion bombardment and subsequent processing. We have also provided examples of two aspects of ion implantation where major progress has been made recently. In the first example on ripple formation in semiconductors, we have described how, in the last approximately 10 years, the dominant explanation to ripple formation has changed from one of sputtering to a redistribution-dominated one. The review of the second example, time-resolved measurements, has shown how pulsed-ion-beam experiments combined with a suitable analysis could be used to shed new light on defect migration and interaction processes.

Despite many years of extensive research, it is clear that there is still ample room for research needed to fill in current gaps in our understanding of radiation damage processes. Radiation damage physics is remarkably rich and complex. Radiation damage phenomena are inherently mesoscopic and dynamic rather than static in nature. They span length scales from atomic to macroscopic and time scales from femtoseconds to years. Defect-interaction dynamics still remains one of the least understood aspects of radiation damage. More work is needed to understand how defect-interaction dynamics influences damage accumulation, its thermal stability, and dopant activation in semiconductor devices, including low-dimensional systems where the presence of surfaces, interfaces, and a large surface-to-volume ratio could further strongly influence defect formation. We hope to see future experimental work in this area with pulsed ion beams, direct in situ time-resolved diagnostics, and pump-and-probe techniques. This also brings up the need for better diagnostics to provide spatially- and time-resolved data on defect evolution in order to benchmark various current and future theories and models.

Finally, it should be mentioned that the field of ion-beam modification of semiconductors is not isolated from the other research communities working on radiation effects in materials for advanced nuclear energy systems, superconductors, materials for nuclear stockpile stewardship, a large family of ion assisted thin-film deposition methods, and even organic semiconductors that are (intentionally or not) exposed to ion bombardment during various fabrication steps or their subsequent use. A closer interaction between these research fields by sharing experimental, theoretical, and modeling approaches and techniques would be mutually beneficial.

Acknowledgments

We acknowledge the numerous discussions and collaboration over the years, which have given us the knowledge instrumental to write this chapter. The work was partly funded by the Academy of Finland in the MESIOS project. We also acknowledge computational resources from the Finnish Grid and Cloud Infrastructure (persistent identifier urn:nbn:fi:research-infras-2016072533). The work at LLNL was performed under the auspices of the US DOE by LLNL under Contract DE-AC52-07NA27344.

References

[1] E. Wilson and E. J. N. Wilson, *An Introduction to Particle Accelerators* (Oxford University Press, Oxford, UK, 2001).
[2] J. A. Leavitt, L. C. McIntyre, Jr., and M. R. Weller, in *Handbook of Modern Ion Beam Analysis*, edited by J. R. Tesmer and M. Nastasi (Materials Research Society, Pittsburgh, Pennsylvania, 1995).
[3] L. C. Feldman and J. W. Mayer, *Fundamentals of Surface and Thin Film Analysis* (North-Holland, New York, NY, 1986).
[4] M. d'Avila Nunes, *Hadron Therapy Physics and Simulations* (Springer, New York, NY, 2014).
[5] L. Evans and P. Bryant, J. Instrum. 3, S08001 (2008).
[6] The Large Hadron Collider, https://home.cern/topics/large-hadron-collider.
[7] R. Hill and P. W. M. Blenkinsopp, Appl. Surf. Sci. 231, 936 (2004).
[8] Example Images of Small Ion Sources, http://www.casetechnology.com/source.html.
[9] J. W. Mayer and S. S. Lau, *Electronic Materials Science For Integrated Circuits in Si and GaAs* (MacMillan, New York, NY, 1990).
[10] R. E. Hamm and M. E. Hamm, *Industrial Accelerators and Their Applications* (World Scientific, New Jersey, 2012).
[11] N. Bohr, Philos. Mag. 25, 10 (1913).
[12] J. Lindhard, M. Scharff, and H. E. Schiøtt, Kgl. Danske Vid. Selskab, Mat. Fys. Medd. Kong. Dan. Vid. Selsk. 33 (14), 1 (1963).
[13] M. Backman, F. Djurabekova, O. H. Pakarinen, *et al.*, J. Phys. D: Appl. Phys. 45, 505305 (2012).
[14] L. Thomé, A. Debelle, F. Garrido, *et al.*, Nucl. Instrum. Methods Phys. Res., Sect. B307, 43 (2014).

[15] W. J. Weber, E. Zarkadoula, O. H. Pakarinen, *et al.*, Sci. Rep. 5, 7726 (2015).
[16] K. Nordlund, D. Sundholm, P. Pyykkö, D. Martinez Zambrano, and F. Djurabekova, Phys. Rev. A 96, 042717 (2017).
[17] E. Bonderup and P. Hvelplund, Phys. Rev. A 4, 562 (1971).
[18] A. Luukkainen and M. Hautala, Radiat. Eff. 59, 113 (1982).
[19] Q. Yang and R. J. MacDonald, Nucl. Instrum. Methods Phys. Res., Sect. B 83, 303 (1993).
[20] Y. Martynenko and M. Y. Nagel, Plasma Phys. Rep. 38, 996 (2012).
[21] K. Nordlund, Comput. Mater. Sci. 3, 448 (1995).
[22] K. Nordlund, F. Djurabekova, and G. Hobler, Phys. Rev. B 94, 214109 (2016).
[23] M. T. Robinson and O. S. Oen, Appl. Phys. Lett. 2, 30 (1963).
[24] E. V. Kornelsen, F. Brown, J. A. Davies, B. Domeij, and G. R. Piercy, Phys. Rev. 136, 849 (1964).
[25] D. S. Gemmell, Rev. Mod. Phys. 46, 129 (1974).
[26] C. Trautmann, S. Klaumünzer, and H. Trinkaus, Phys. Rev. Lett. 85, 3648 (2000).
[27] M. Toulemonde, W. Assmann, C. Dufour, A. Meftah, F. Studer, and C. Trautmann, Mat. Fys. Medd. Kong. Dan. Vid. Selsk. 52, 263 (2006).
[28] P. Kluth, C. S. Schnohr, O. H. Pakarinen, *et al.*, Phys. Rev. Lett. 101, 175503 (2008).
[29] M. E. Toimil-Molares, Beilstein J. Nanotechnol. 3, 860 (2012).
[30] H. Katagiri, H. Kasuya, H. Mizoguchi, and S. Yanabu, IEEE Trans. Dielectr. Electr. Insul. 15, 1424–1429 (2008).
[31] C. M. McKenna, Radiat. Eff. 44, 93 (1979).
[32] P. L. F. Hemment, Dosimetry techniques, Radiat. Eff. 44, 31 (1979).
[33] D. M. Jamba, Nucl. Instrum. Methods 189, 253 (1981).
[34] J. W. Ekin, *Experimental Techniques for Low-Temperature Measurements* (Oxford University Press, Oxford, 2006).
[35] K. Nordlund, M. Ghaly, R. S. Averback, M. Caturla, T. Díaz de la Rubia, and J. Tarus, Phys. Rev. B 57, 7556 (1998).
[36] F. H. Stillinger and T. A. Weber, Phys. Rev. B 31, 5262 (1985).
[37] H. M. Urbassek and K. T. Waldeer, Phys. Rev. Lett. 67, 105 (1991).
[38] M.-J. Caturla, L. A. M. T. Díaz de la Rubia, and G. H. Gilmer, Phys. Rev. B 54, 16683 (1996).
[39] R. E. Stoller, Comprehensive Nuclear Materials, Vol. 1, p. 293 (2012).
[40] K. Nordlund and F. Djurabekova, J. Comput. Electron. 13, 122 (2014), invited review paper for special issue on device modelling.
[41] A. A. Leino, F. Djurabekova, and K. Nordlund, Eur. Phys. J. B 87, 242 (2014).
[42] J. Polvi, P. Luukkonen, K. Nordlund, T. T. Järvi, T. W. Kemper, and S. B. Sinnott, J. Phys. Chem. B 116, 13932 (2012).
[43] M. O. Ruault, J. Chaumont, J. M. Penisson, and A. Bourret, Philos. Mag. A 50, 667 (1984).

[44] P. Partyka, Y. Zhong, K. Nordlund, R. S. Averback, I. K. Robinson, and P. Ehrhart, Phys. Rev. B 64, 235207 (2002).

[45] M. W. Bench, I. M. Robertson, and M. A. Kirk, Nucl. Instrum. Methods Phys. Res., Sect. B 59/60, 372 (1991).

[46] I. Jencic, M. W. Bench, I. M. Robertson, and M. A. Kirk, J. Appl. Phys. 78, 974 (1995).

[47] M. W. Bench, I. M. Robertson, M. A. Kirk, and I. Jenčič, J. Appl. Phys. 87, 49 (2000).

[48] K. Nordlund, S. J. Zinkle, A. E. Sand, *et al.*, J. Nucl. Mater. 512, 450 (2018).

[49] U. M. Gösele and T. Y. Tan, MRS Bull. XVI, 42 (1991).

[50] G. D. Watkins, in *Defects and Diffusion in Silicon Processing*, Vol. 469 of MRS Symposium Proceedings, edited by T. Díaz de la Rubia, S. Coffa, P. A. Stolk, and C. S. Rafferty (Materials Research Society, Pittsburgh, 1997), p. 139.

[51] J. L. Benton, S. Libertino, P. Kringhøj, D. J. Eaglesham, and J. M. Poate, J. Appl. Phys. 82, 120 (1997).

[52] P. Ehrhart and H. Zillgen, in *Defects and Diffusion in Silicon Processing*, Vol. 469 of MRS Symposium Proceedings, edited by T. Díaz de la Rubia, S. Coffa, P. A. Stolk, and C. S. Rafferty (Materials Research Society, Pittsburgh, 1997), p. 175.

[53] S. K. Estreicher, M. Gharaibeh, P. A. Fedders, and P. Ordejón, Phys. Rev. Lett. 86, 1247 (2001).

[54] K. Kyuno, D. C. Cahill, R. S. Averback, J. Tarus, and K. Nordlund, Phys. Rev. Lett. 83, 4788 (1999).

[55] E. Holmström, M. Hakala, and K. Nordlund, Phys. Rev. B 82, 094102 (2010).

[56] M. J. Caturla, T. Díaz de la Rubia, and G. H. Gilmer, J. Appl. Phys. 77, 3121 (1995).

[57] J. Frantz, J. Tarus, K. Nordlund, and J. Keinonen, Phys. Rev. B 64, 125313 (2001).

[58] L. Pelaz, L. A. Marqués, M. Aboy, J. Barbolla, and G. H. Gilmer, Appl. Phys. Lett. 82, 2038 (2003).

[59] L. Pelaz, L. A. Marqués, and J. Barbolla, J. Appl. Phys. 96, 5947 (2004).

[60] L. C. Feldman, J. W. Mayer, and S. T. Picraux, *Materials Analysis by Ion Channeling* (Academic, New York, NY, 1982).

[61] O. W. Holland, J. Narayan, and D. Fathy, Nucl. Instrum. Methods Phys. Res., Sect. B 7/8, 243 (1985).

[62] P. J. M. Smulders and D. O. Boerma, Nucl. Instrum. Methods Phys. Res. 29, 471 (1987).

[63] S. Zhang, K. Nordlund, F. Djurabekova, Y. Zhang, G. Velisa, and T. S. Wang, Phys. Rev. E 94, 043319 (2016).

[64] K. Nordlund, J. Keinonen, E. Rauhala, and T. Ahlgren, Phys. Rev. B 52, 15170 (1995).

[65] S. Zhang, O. H. Pakarinen, M. Backholm, F. Djurabekova, K. Nordlund, J. Keinonen, and T. S. Wang, J. Phys.: Condens. Matter 30, 015403 (2018).

[66] J. B. Wallace, L. B. Bayu Aji, L. Shao, and S. O. Kucheyev, Sci. Rep. 7, 17574 (2017).
[67] G. F. Cerofilini, L. Meda, and C. Volpones, J. Appl. Phys. 63, 4911 (1988).
[68] S. U. Campisano, S. Coffa, V. Rainieri, F. Priolo, and E. Rimini, Nucl. Instrum. Methods Phys. Res., Sect. B 80/81, 514 (1993).
[69] D. N. Seidman, R. S. Averback, P. R. Okamoto, and A. C. Baily, Phys. Rev. Lett. 58, 900 (1987).
[70] L. J. Lewis and R. M. Nieminen, Phys. Rev. B 54, 1459 (1996).
[71] J. Nord, K. Nordlund, and J. Keinonen, Phys. Rev. B 65, 165329 (2002).
[72] B. Weber, D. M. Stock, and K. Gärtner, Nucl. Instrum. Methods Phys. Res., Sect. B 148, 375 (1999).
[73] K. Gärtner and B. Weber, Nucl. Instrum. Methods Phys. Res., Sect. B 180, 274 (2001).
[74] M. W. Bench, I. M. Robertson, and M. A. Kirk, in *Proceedings of the 46th Annual Meeting of the Electron Microscopy Society of America*, edited by G. W. Bailey (San Francisco Press, San Francisco, CA, 1988), p. check.
[75] S. Cannavò, M. G. Grimaldi, E. Rimini, G. Ferla, and L. Gandolfi, Appl. Phys. Lett. 47, 138 (1985).
[76] K. Saarinen, P. Hautojarvi, J. Keinonen, E. Rauhala, and J. Raisanen, Phys. Rev. B 43, 4249 (1991).
[77] F. Spaepen and D. Turnbull, in *Laser Annealing of Semiconductors*, edited by J. M. Poate and J. W. Mayer (Academic Press, New York, NY, 1982), Chap. 2, pp. 15–42.
[78] K. Laaziri, S. Kycia, S. Roorda, *et al.*, Phys. Rev. Lett. 82, 3460 (1999).
[79] K. Laaziri, S. Kycia, S. Roorda, *et al.*, Phys. Rev. B 60, 13520 (1999).
[80] E. Chason, S. T. Picraux, M. Poate, *et al.*, J. Appl. Phys. 81, 6513 (1997).
[81] S. Roorda, R. A. Hakvoort, A. van Veen, P. A. Stolk, and F. W. Saris, J. Appl. Phys. 72, 5145 (1992).
[82] S. Roorda, Nucl. Instrum. Methods Phys. Res., Sect. B (1999), proceedings of the IBMM98 conference, in press.
[83] J. S. Custer, M. O. Thompson, D. C. Jacobson, *et al.*, Appl. Phys. Lett. 64, 436 (1994).
[84] E. Holmstrm, B. Haberl, O. Pakarinen, *et al.*, J. Non-Cryst. Solids 438, 26 (2016).
[85] E. Wendler, A. Kamarou, E. Alves, K. Gärtner, and W. Wesch, Nucl. Instrum. Methods Phys. Res., Sect. B 206, 1028 (2003).
[86] J. Nord, K. Nordlund, and J. Keinonen, Phys. Rev. B 68, 184104 (2003).
[87] S. O. Kucheyev, J. S. Williams, C. Jagadish, J. Zou, and G. Li, Phys. Rev. B 62, 7510–7522 (2000).
[88] S. O. Kucheyev, A. Y. Azarov, A. I. Titov, P. A. Karaseov, and T. M. Kuchumova, J. Phys. D: Appl. Phys. 42, 085309 (2009).
[89] A. Gaber, H. Zillgen, P. Ehrhart, P. Partyka, and R. S. Averback, J. Appl. Phys. 82, 5348 (1997).
[90] H. H. Tan, C. Jagadish, J. S. Williams, J. Zou, D. J. H. Cockayne, and A. Sikorski, J. Appl. Phys. 77, 87 (1995).

[91] P. Partyka, R. S. Averback, D. V. Forbes, J. J. Coleman, and P. Ehrhart, J. Appl. Phys. 83, 1265 (1998).

[92] J. L. Klatt, R. S. Averbach, D. V. Forbes, and J. J. Coleman, Phys. Rev. B 48, 17629 (1993).

[93] B. A. Turkot, B. W. Lagow, I. M. Robertson, *et al.*, J. Appl. Phys. 80, 4366 (1996).

[94] E. Wendler, B. Breeger, and W. Wesch, Nucl. Instrum. Methods Phys. Res., Sect. B 175, 83 (2001).

[95] S. O. Kucheyev, J. S. Williams, J. Zou, S. J. Pearton, and Y. Nakagawa, Appl. Phys. Lett. 79, 602–604 (2001).

[96] S. O. Kucheyev, J. S. Williams, J. Zou, *et al.*, J. Appl. Phys. 92, 3554–3558 (2002).

[97] S. O. Kucheyev, J. S. Williams, J. Zou, *et al.*, Appl. Phys. Lett. 80, 787–789 (2002).

[98] S. O. Kucheyev, J. S. Williams, J. Zou, and C. Jagadish, J. Appl. Phys. 95, 3048–3054 (2004).

[99] M. Fialho, S. Magalhaes, M. P. Chauvat, P. Ruterana, K. Lorenz, and E. Alves, J. Appl. Phys. 120, 165703 (2016).

[100] L. M. Zhang, W. L. Jiang, A. Dissanayake, *et al.*, J. Appl. Phys. 119, 245704 (2016).

[101] D. N. Faye, E. Wendler, M. Felizardo, *et al.*, J. Phys. Chem. C 120, 7277 (2016).

[102] K. Ohdaira, N. Tomura, S. Ishii, and H. Matsumura, Electrochem. Solid-State Lett. 14, H372 (2011).

[103] J. Linnros and G. Holmén, J. Appl. Phys. 62, 4737 (1987).

[104] L. A. Marqués, M.-J. Caturla, H. Huang, and T. Díaz de la Rubia, Mater. Res. Soc. Symp. Proc. 396, 201 (1994).

[105] C. Krzeminski, Q. Brulin, V. Cuny, E. Lecat, E. Lampin, and F. Cleri, J. Appl. Phys. 101, 123506 (2007).

[106] B. P. Bob, A. Kohno, S. Charnvanichborikarn, *et al.*, J. Appl. Phys. 107, 123506 (2010).

[107] S. F. Lombardo, S. Boninelli, F. Cristiano, *et al.*, Mater. Sci. Semicond. Process. 62, 80 (2017).

[108] W. J. Yang, J. Mathews, and J. S. Williams, Mater. Sci. Semicond. Process. 62, 103 (2017).

[109] J. Kim, F. Kirchhoff, W. G. Aulbur, J. W. Wilkins, F. S. Khan, and G. Kresse, Phys. Rev. Lett. 83, 1990 (1999).

[110] N. E. B. Cowern, G. Mannino, P. A. S. F. Roozeboom, *et al.*, Phys. Rev. Lett. 82, 4460 (1999).

[111] D. J. Eaglesham, P. A. Stolk, H.-J. Gossmann, and J. M. Poate, Appl. Phys. Lett. 65, 2305 (1995).

[112] N. Arai, S. Takeda, and M. Kohyama, Phys. Rev. Lett. 78, 4265 (1997).

[113] K. Arakawa, K. Saitoh, H. Mori, and K. Ono, Nucl. Instrum. Methods Phys. Res., Sect. B 206, 76–80 (2003), iBMM 2002 conference proceeding, submitted for publication.

[114] A. Bourret, in *Microsc. Semicond. Mater. Conf.*, No. 87 in Inst. Phys. Conf. Ser., edited by A. G. Cullis and P. D. Augustus (IOP Publishing, Bristol, 1987).
[115] S. Takeda, Jpn. J. Appl. Phys. 30, L639 (1991).
[116] A. Parisini and A. Bourret, Philos. Mag. A 67, 605 (1992).
[117] K. Nordlund, T. H. Metzger, A. Malachias, *et al.*, J. Appl. Phys. 98, 073529 (2005).
[118] J. B. Wallace, L. B. Bayu Aji, T. T. Li, L. Shao, and S. O. Kucheyev, J. Appl. Phys. 118, 105705 (2015).
[119] D. Bachiller-Perea, A. Debelle, L. Thome, and M. Behar, J. Nucl. Mater. 478, 268–274 (2016).
[120] B. Colombeau, N. E. B. Cowern, F. Cristiano, *et al.*, Appl. Phys. Lett. 83, 1953 (2003).
[121] A. Claverie, B. Colombeau, B. de Mauduit, *et al.*, Appl. Phys. A 76, 1025 (2003).
[122] M. Aboy, L. Pelaz, L. A. Marqués, L. Enriquez, and J. Barbolla, J. Appl. Phys. 94, 1013 (2003).
[123] F. Cristiano, N. Cherkasin, X. Hebras, *et al.*, Nucl. Instrum. Methods Phys. Res., Sect. B 216, 46 (2004).
[124] M. Aboy, L. Pelaz, L. A. Marqués, *et al.*, Appl. Phys. Lett. 86, 031908 (2005).
[125] M. Aboy, L. Pelaz, E. Bruno, S. Mirabella, and S. Boninelli, J. Appl. Phys. 110, 073524 (2011).
[126] D. J. Eaglesham, P. A. Stolk, H.-J. Gossmann, T. E. Haynes, and J. M. Poate, Nucl. Instrum. Methods Phys. Res., Sect. B 106, 191 (1995).
[127] D. Luebbert, J. Arthur, M. Sztucki, T. H. Metzger, P. B. Griffin, and J. R. Patel, Appl. Phys. Lett. 81, 3167 (2002).
[128] R. Cuerno, L. Vazquez, R. Gago, and M. Castro, J. Phys.: Condens. Matter 21, (2009), special issue on ripple formation.
[129] H. Hofsäss, Appl. Phys. A 114, 401–422 (2013).
[130] R. L. Cunningham, P. Haymann, C. Lecomte, W. J. Moore, and J. J. Trillat, J. Appl. Phys. 31, 839 (1960).
[131] M. Navez, C. Sella, and D. Chaperot, C. R. Acad. Sci. Paris 254, 240 (1962).
[132] S. Facsko, T. Dekorsy, C. Koerdt, *et al.*, Science 285, 1551 (1999).
[133] F. Frost, B. Ziberi, A. Schindler, and B. Rauschenbach, Appl. Phys. A 91, 551 (2008).
[134] R. Behrisch (ed.), *Sputtering by Particle Bombardment I* (Springer, Berlin, 1981).
[135] P. Sigmund, Nucl. Instrum. Methods Phys. Res., Sect. B 27, 1 (1987).
[136] Y. Yamamura and H. Tawara, At. Data Nucl. Data Tables 62, 149 (1996).
[137] P. Sigmund, Phys. Rev. 184, 383 (1969).
[138] P. Sigmund, J. Mater. Sci. 8, 1545 (1973).

[139] R. M. Bradley and J. M. E. Harper, J. Vac. Sci. Technol., A 6, 2390 (1988).
[140] J. Frantz, M. Rusanen, K. Nordlund, and I. T. Koponen, J. Phys.: Condens. Matter 16, 2995 (2004).
[141] R. Cuerno and A.-L. Barabási, Phys. Rev. Lett. 74, 4746 (1995).
[142] M. A. Makeev and A.-L. Barabasi, Nucl. Instrum. Methods Phys. Res., Sect. B 222, 316 (2004).
[143] E. Chason, T. M. Mayer, B. K. Kellerman, D. T. McIlroy, and A. J. Howard, Phys. Rev. Lett. 72, 3040 (1994).
[144] B. M. Paine and R. S. Averback, Nucl. Instrum. Methods Phys. Res., Sect. B 7/8, 666 (1985).
[145] C. P. Flynn and R. S. Averback, Phys. Rev. B 38, 7118 (1988).
[146] S.-J. Kim, M.-A. Nicolet, R. S. Averback, and D. Peak, Phys. Rev. B 37, 38 (1988).
[147] K. L. Merkle and W. Jäger, Philos. Mag. A 44, 741 (1981).
[148] W. Jäger and K. L. Merkle, Philos. Mag. A 57, 479 (1988).
[149] M. Ghaly and R. S. Averback, Phys. Rev. Lett. 72, 364 (1994).
[150] M. Ghaly, K. Nordlund, and R. S. Averback, Philos. Mag. A 79, 795 (1999).
[151] R. C. Birtcher and S. E. Donnelly, Nucl. Instrum. Methods Phys. Res., Sect. B 148, 194 (1999).
[152] E. Bringa, K. Nordlund, and J. Keinonen, Phys. Rev. B 64, 235426 (2001).
[153] N. Kalyanasundaram, M. Ghasisaeidi, J. B. Freund, and H. T. Johnson, Appl. Phys. Lett. 92, 131909 (2008).
[154] G. Carter and G. Vishnyakov, Phys. Rev. B: Condens. Matter 54, 17647 (1996).
[155] M. Moseler, P. Gumbsch, C. Casiraghi, A. C. Ferrari, and J. Robertson, 309, 1545 (2005).
[156] S. A. Norris, J. Samela, C. S. Madi, *et al.*, Nat. Commun. 2, 276 (2011).
[157] L. Bukonte, F. Djurabekova, J. Samela, K. Nordlund, S. A. Norris, and M. J. Aziz, Nucl. Instrum. Methods Phys. Res., Sect. B 297, 23 (2013).
[158] M. P. Harrison and R. M. Bradley, Phys. Rev. B 89, 245401 (2014).
[159] M. L. Nietiadi and H. M. Urbassek, Appl. Phys. Lett. 103, 113108 (2013).
[160] M. Z. Hossain, K. Das, J. B. Freund, and H. T. Johnson, Appl. Phys. Lett. 99, 151913 (2011).
[161] A. Lopez-Cazalilla, A. Ilinov, K. Nordlund, *et al.*, J. Appl. Phys. 123, 235108 (2018).
[162] T. K. Chini, F. Okuyama, M. Tanemura, and K. Nordlund, Phys. Rev. B 67, 205403 (2003).
[163] R. S. Averback and T. Díaz de la Rubia, Solid State Phys. 51, 281 (1998).
[164] A. A. Kohnert, B. D. Wirth, and L. Capolungo, Comput. Mater. Sci. 149, 442 (2018).
[165] J. B. Wallace, S. Charnvanichborikarn, L. B. Bayu Aji, M. T. Myers, L. Shao, and S. O. Kucheyev, J. Appl. Phys. 118, 135709 (2015).

[166] C. R. Musil, J. Melngailis, S. Etchin, and T. E. Haynes, *J. Appl. Phys.* 80, 3727 (1996).
[167] E. Wendler, W. Wesch, and G. Götz, Nucl. Instrum. Methods Phys. Res., Sect. B 52, 57 (1990).
[168] G. Carter, J. Appl. Phys. 79, 8285 (1996).
[169] T. E. Haynes and O. W. Holland, Appl. Phys. Lett. 59, 452 (1991).
[170] T. E. Haynes and O. W. Holland, Nucl. Instrum. Methods Phys. Res. Sect. B 59/60, 1028 (1991).
[171] T. E. Haynes and O. W. Holland, Appl. Phys. Lett. 58, 62–64 (1991).
[172] R. A. Brown and J. S. Williams, J. Appl. Phys. 81, 7681 (1997).
[173] N. A. G. Ahmed, C. E. Christodoulides, and G. Carter, Radiat. Eff. 52, 211 (1980).
[174] M.-O. Ruault, J. Chaumont, and H. Bernas, Nucl. Instrum. Methods 209/210, 351 (1983).
[175] M.-J. Caturla, T. Díaz de la Rubia, L. A. Marqués, and G. H. Gilmer, Phys. Rev. B 54, 16683 (1996).
[176] H. Hensel and H. M. Urbassek, Phys. Rev. B 57, 4756 (1998).
[177] K. Nordlund, M. Ghaly, R. S. Averback, M. Caturla, T. Díaz de la Rubia, and J. Tarus, Phys. Rev. B 57, 7556 (1998).
[178] M. Posselt, L. Bischoff, and J. Teichert, Appl. Phys. Lett. 79, 1444 (2001).
[179] Y. Guerin, G. S. Was, and S. J. Zinkle, MRS Bull. 34, 10 (2009).
[180] H. M. Naguib and R. Kelly, Radiat. Eff. 25, 1 (1975).
[181] M. T. Robinson, J. Nucl. Mater. 216, 1 (1994).
[182] J. S. Williams, Rep. Prog. Phys. 49, 491 (1986).
[183] R. S. Nelson, in *Radiation Damage and Defects in Semiconductors*, Conf. Series Number 16, edited by J. E. Whitehouse (Institute of Physics Publishing, 1973), p. 140.
[184] B. G. Svensson, C. Jagadish and J. S. Williams, Phys. Rev. Lett. 71, 1860 (1993).
[185] M. T. Myers, S. Charnvanichborikarn, L. Shao, and S. O. Kucheyev, Phys. Rev. Lett. 109, 095502 (2012).
[186] S. Charnvanichborikarn, M. T. Myers, L. Shao, and S. O. Kucheyev, J. Phys.: Condens. Matter 25, 162203 (2013).
[187] J. B. Wallace, L. B. Bayu Aji, L. Shao, and S. O. Kucheyev, Appl. Phys. Lett. 106, 202102 (2015).
[188] L. B. Bayu Aji, J. B. Wallace, L. Shao, and S. O. Kucheyev, J. Phys. D: Appl. Phys. 49, 19LT01 (2016).
[189] L. B. Bayu Aji, J. B. Wallace, L. Shao, and S. O. Kucheyev, Sci. Rep. 6, 30931 (2016).
[190] J. B. Wallace, L. B. Bayu Aji, L. Shao, and S. O. Kucheyev, Sci. Rep. 7, 13182 (2017).
[191] J. B. Wallace, L. B. Bayu Aji, A. A. Martin, S. J. Shin, L. Shao, and S. O. Kucheyev, Sci. Rep. 7, 39754 (2017).

[192] J. B. Wallace, L. B. Bayu Aji, L. Shao, and S. O. Kucheyev, Nucl. Instrum. Methods Phys. Res., Sect. B 409, 347–350 (2017).
[193] L. B. Bayu Aji, J. B. Wallace, and S. O. Kucheyev, Sci. Rep. 7, 44703 (2017).
[194] L. B. Bayu Aji, J. B. Wallace, and S. O. Kucheyev, J. Appl. Phys. 121, 235106 (2017).
[195] L. B. Bayu Aji, J. B. Wallace, and S. O. Kucheyev, J. Appl. Phys. 124, 025701 (2018).
[196] J. B. Wallace, L. B. Bayu Aji, L. Shao, and S. O. Kucheyev, Phys. Rev. Lett. 120, 216101 (2018).
[197] P. J. Schultz, C. Jagadish, M. C. Ridgway, R. G. Elliman, and J. S. Williams, Phys. Rev. B 44, 9118 (1991).
[198] R. D. Goldberg, J. S. Williams, and R. G. Elliman, Nucl. Instrum. Methods Phys. Res., Sect. B 106, 242 (1995).
[199] R. D. Goldberg, J. S. Williams, and R. G. Elliman, Phys. Rev. Lett. 82, 771 (1999).
[200] J. Linnros, B. Svensson, and G. Holmen, Phys. Rev. B 30, 3629 (1984).
[201] W. Wesch and E. Wendler, *Ion Beam Modification of Solids: Ion–Solid Interaction and Radiation Damage* (Springer, Berlin, 2016).
[202] F. L. Vook and H. J. Stein, Radiat. Eff. 2, 23–30 (1969).
[203] R. A. Brown and J. S. Williams, Phys. Rev. B 64, 155202 (2001).
[204] G. Holmén, P. Högberg, and A. Burén, Radiat. Eff. 24, 39–44 (1975).
[205] A. I. Titov, A. Y. Azarov, L. M. Nikulina, and S. O. Kucheyev, Phys. Rev. B 73, 064111 (2006).
[206] O. W. Holland, S. J. Pennycook, and G. L. Albert, Appl. Phys. Lett. 55, 2503–2505 (1989).
[207] U. G. Akano, I. V. Mitchel, and F. R. Shepherd, Appl. Phys. Lett. 62, 1670–1672 (1993).
[208] Y. Zhang, W. J. Weber, W. Jiang, C. M. Wang, V. Shutthanandan, and A. Hallén, J. Appl. Phys. 95, 4012–4018 (2004).
[209] J. Slotte, K. Saarinen, M. S. Janson, *et al.*, J. Appl. Phys. 97, 033513 (2005).
[210] A. Y. Kuznetsov, J. Wong-Leung, A. Hallén, C. Jagadish, and B. G. Svensson, J. Appl. Phys. 94, 7112–7115 (2003).
[211] M. Posselt, L. Bischoff, J. Teichert, and A. Ster, Appl. Phys. Lett. 93, 1004–1013 (2003).
[212] W. J. Weber, N. Yu, and L. M. Wang, J. Nucl. Mater. 253, 53–59 (1998).
[213] A. Debelle, L. Thome, D. Dompoint, *et al.*, J. Phys. D: Appl. Phys. 43, 455408 (2010).
[214] A. I. Titov, S. O. Kucheyev, V. S. Belyakov, and A. Y. Azarov, J. Appl. Phys. 90, 3867 (2001).
[215] A. I. Titov, V. S. Belyakov, and S. O. Kucheyev, Nucl. Instrum. Methods Phys. Res., Sect. B 194, 323 (2002).

[216] A. I. Titov, P. A. Karaseov, A. Y. Azarov, and S. O. Kucheyev, Nucl. Instrum. Methods Phys. Res., Sect. B 267, 2701–2704 (2009).
[217] E. Wendler, T. Bierschenk, W. Wesch, E. Friedland, and J. B. Malherbe, Nucl. Instrum. Methods Phys. Res., Sect. B 268, 2996–3000 (2010).
[218] S. Charnvanichborikarn, M. T. Myers, L. Shao, and S. O. Kucheyev, Scr. Mater. 67, 65–68 (2012).
[219] S. Charnvanichborikarn, M. T. Myers, L. Shao, and S. O. Kucheyev, J. Appl. Phys. 114, 213512 (2013).
[220] A. I. Titov, A. Y. Azarov, L. M. Nikulina, and S. O. Kucheyev, Nucl. Instrum. Methods Phys. Res., Sect. B 256, 207–210 (2007).
[221] A. Fissel, Phys. Rep. 379, 149–255 (2003).
[222] L. L. Snead, T. Nozawa, Y. Katoh, *et al.*, J. Nucl. Mater. 371, 329–377 (2007).
[223] A. I. Titov and G. Carter, Nucl. Instrum. Meth. Phys. Res. Sect. B 119, 491–500 (1996).
[224] D. A. Thompson, Radiat. Eff. 56, 105 (1981).
[225] G. W. Kinchin and R. S. Pease, Rep. Prog. Phys. 18, 1 (1955).
[226] H. L. Heinisch and B. N. Singh, Philos. Mag. A 67, 407–424 (1993).
[227] S. O. Kucheyev, J. S. Williams, A. I. Titov, G. Li, and C. Jagadish, Appl. Phys. Lett. 78, 2694–2696 (2001).
[228] A. Y. Azarov, A. I. Titov, P. A. Karaseov, and A. Hallen, Nucl. Instrum. Methods Phys. Res., Sect. B 267, 1247–1250 (2009).
[229] K. B. Winterbon, Radiat. Eff. 60, 199–207 (1982).
[230] K. B. Winterbon, H. M. Urbassek, P. Sigmund, and A. Gras-Marti, Phys. Scr. 36, 689–692 (1987).
[231] Y. T. Cheng, M. A. Nicolet, and W. L. Johnson, Phys. Rev. Lett. 58, 2083–2086 (1987).
[232] F. Rossi, D. M. Parkin, and M. Nastasi, J. Mater. Res. 4, 137–143 (1989).
[233] B. B. Mandelbrot and R. Pignoni, *The Fractal Geometry of Nature* (W.H. Freeman, New York, NY, 1983).
[234] J. C. Moreno-Marin, U. Conrad, H. M. Urbassek, and L. Gras-Marti, Nucl. Instrum. Methods Phys. Res., Sect. B 48, 404–407 (1990).
[235] F. Kun and G. Bardos, Phys. Rev. E 50, 2639–2645 (1994).
[236] F. Kun and G. Bardos, Phys. Rev. E 55, 1508–1513 (1997).
[237] J. Martan, The influence of the interaction potential on the fractal dimension of the collision cascade, Vacuum 63, 529–533 (2001).
[238] L. Luneville, D. Simeone, and G. Baldinozzi, Nucl. Instrum. Methods Phys. Res., Sect. B 266, 2702–2706 (2008).
[239] D. Simeone, L. Luneville, and Y. Serruys, Phys. Rev. E 82, 011122 (2010).
[240] E. Antoshchenkova, L. Luneville, D. Simeone, R. E. Stoller, and M. Hayoun, J. Nucl. Mater. 458, 168–175 (2015).
[241] A. De Backer, A. E. Sand, K. Nordlund, *et al.*, Subcascade formation and defect cluster size scaling in high-energy collision events in metals, EPL 115, 26001 (2016).

[242] S. Pellegrino, L. Thome, A. Debelle, S. Miro, and P. Trocellier, Nucl. Instrum. Methods Phys. Res., Sect. B 327, 103–107 (2014).
[243] F. Garrido, L. Vincent, L. Nowicki, G. Sattonnay, and L. Thome, Nucl. Instrum. Methods Phys. Res., Sect. B 266, 2842–2847 (2008).
[244] A. Kinomura, J. S. Williams, and K. Fujii, Phys. Rev. B 59, 15214 (1999).
[245] G. D. Watkins, Mater. Sci. Semicond. Process. 3, 227 (2000).
[246] P. G. Coleman and C. P. Burrows, Phys. Rev. Lett. 98, 265502 (2007).
[247] Y. Shimizu, M. Uematsu, and K. M. Itoh, Phys. Rev. Lett. 98, 7 (2007).

Chapter 11

Characterizing defects with ion beam analysis and channeling techniques

Lino Pereira[1], Andre Vantomme[1] and Ulrich Wahl[1,2]

This chapter deals with the use of ion beam analysis (IBA) techniques, in particular in channeling geometry, to study defects in semiconductors. After a tutorial (Section 11.1) introducing the basic principles of IBA and channeling techniques, selected examples of their use to characterize defects (e.g., lattice location of dopants and implantation damage) are described in Section 11.2. Finally, Section 11.3 consists of a brief outlook into future developments and applications.

11.1 Tutorial

In Section 11.1.1, we introduce the basic principles of IBA techniques that are commonly used to study defects in semiconductors: Rutherford backscattering spectrometry (RBS), elastic recoil-detection analysis (ERDA), proton (or particle)-induced X-ray emission (PIXE) and nuclear reaction analysis (NRA). While the principles described in Section 11.1.1 form the basis of IBA to determine the composition of materials, in Section 11.1.2, we introduce the implementation of these techniques in channeling geometry, which allows characterizing the structure of defects at the atomic scale. We also introduce electron emission channeling (EC), a related technique that is specifically developed to determine the lattice location of implanted ions in their host lattice.

11.1.1 Ion beam analysis

The most commonly used IBA techniques, and specifically those introduced here (RBS, ERDA, PIXE and NRA), are typically based on the use of monoenergetic ions, with kinetic energies of the order of MeV. These ions, in the form of collimated beams, impinge onto the sample being analyzed, where they are scattered and excite a variety of other particles and radiation. The IBA technique is defined by which particle or radiation is detected. RBS is based on detecting the same

[1]Instituut voor Kern- en Stralingsfysica, KU Leuven, Leuven, Belgium
[2]Centro de Ciências e Tecnologias Nucleares, Instituto Superior Técnico, Universidade de Lisboa, Lisbon, Portugal

particles (ions) that make up the initial beam, after being elastically backscattered by the nuclei in the sample under scrutiny. If the energy that is transferred from the incident ion to the target nucleus is sufficiently large, the recoiling target ions can also be detected: this constitutes ERDA. The incident ions also interact with the electrons of the atoms composing the target and in particular result in ionization. PIXE is based on the detection of X-rays emitted upon decay of a higher energy electron into the vacancy. For sufficiently high energies, the distance of closest approach between the incident ion and the target nucleus is small enough to allow nuclear reactions to occur. Depending on the incident ion and target species, and the energies involved, a wide range of particles and radiation (γ or α particles, protons, neutrons, etc.) can result. NRA encompasses the various techniques based on their detection. In this section, we introduce key physical parameters associated with atomic collisions, i.e., kinematic factor, cross section, stopping power and related observables, i.e., type, amount and energy of the detected particles. Although we will focus somewhat on RBS (of particular relevance for the remainder of the chapter), the physical mechanisms and quantities are also relevant for ERDA, PIXE and NRA. The understanding of these parameters and observables allows using IBA techniques to determine the composition (depth-resolved, in some cases) of the target materials. In Section 11.1.2, we go beyond these more standard applications of IBA techniques, introducing their implementation in channeling geometries. These *channeling versions* of IBA techniques provide information on the crystallinity of the host and on the position of (impurity) atoms in a crystalline lattice, allowing us, in particular, to study the structure of defects with sub-Å precision.

11.1.1.1 Atomic collisions

As an ion from the incident beam collides with an atom in the target, an amount of kinetic energy is transferred, which depends on the masses of the colliding particles. Assuming conservation of energy and momentum, we can derive the ratio of the energy of the incident particle before (E_0) and after (E_1) collision [1]:

$$\frac{E_1}{E_0} = \left[\frac{\left(M_2^2 - M_1^2 \sin^2\theta\right)^{1/2} + M_1\cos\theta}{M_1 + M_2}\right]^2 \tag{11.1}$$

where M_1 and M_2 are the masses of the incident and target particles, respectively, and θ is the scattering angle (Figure 11.1). This ratio is known as the kinematic factor. Similarly, the recoil energy E_2 of the initially stationary target atom is given by [1]:

$$\frac{E_2}{E_0} = \frac{4M_1M_2}{\left(M_1 + M_2\right)^2}\cos^2\phi \tag{11.2}$$

where ϕ is the recoil angle (Figure 11.1). For a known beam energy E_0, measuring the energy E_1 of the backscattered ion (RBS) or E_2 of the recoiled particle (ERDA) detected at a well-defined angle with respect to the initial beam direction allows to

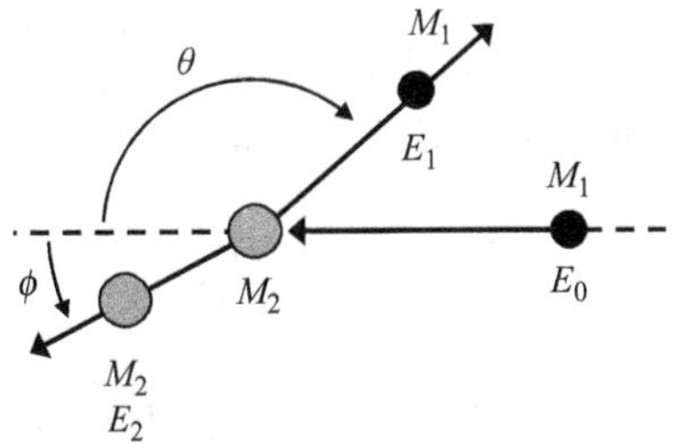

Figure 11.1 Schematic representation of a collision between an incident ion (with mass M_1 and initial energy E_0) and a target atom (with mass M_2 and initially at rest). After the collision, the incident ion is scattered at an angle θ with energy E_1, whereas the target atom recoils at an angle ϕ with energy E_2

identify the target composition (via M_2). With PIXE and NRA, the elements composing the target are identified based on the energy of the characteristic X-rays (PIXE) and of the products of the nuclear reaction (NRA).

While the energy is a signature of which elements compose the target, information on their concentration is provided by the number of detected particles. Converting the number of detected particles to target concentration requires knowledge of the cross section for that particular interaction, in addition to parameters defined by the experimental apparatus (beam, detector, etc.). Taking RBS as a representative example, we can derive the expression for the cross section considering the force between the colliding particles. Since the distance of closest approach between colliding particles is typically smaller than the radius of the electronic orbitals, and larger than the nuclear radius (not the case of NRA), the collision is governed by the Coulomb interaction between two bare nuclei with charges Z_1 (incident ion) and Z_2 (target atom). Taking the target atom as stationary during the collision, the problem reduces to calculating the scattering cross section $\sigma(\theta)$ for central force scattering, which gives the Rutherford cross section [1]:

$$\sigma_{\mathrm{RBS}}(\theta) = \left(\frac{Z_1 Z_2 e^2}{4E}\right)^2 \frac{1}{\sin^4(\theta/2)} \tag{11.3}$$

i.e., a measure of the probability for an incident ion with energy E to be scattered to an angle θ with respect to its initial direction. This is the expression originally derived by Rutherford, which together with the experiments by Geiger and Marsden, established at the beginning of the twentieth century our current picture of the atom. A more accurate expression is derived considering the two-body scattering (i.e., without assuming that the target atom remains at rest), although the correction is small for $M_1/M_2 \ll 1$ ratio, as is typically the case for RBS (with a light ion beam, such as He). At lower energies (larger distance of closest approach), a correction is also required to account for electron screening. With increasing energy (smaller distance of closest approach), deviations from the Rutherford cross section also emerge, as the effects of the nuclear force (i.e., not just Coulomb) become important. A similar derivation as done for (11.3), though keeping track of the recoil angle ϕ of the target particle, gives the ERDA cross section [2]:

$$\sigma_{\mathrm{ERDA}}(\phi) = \left(\frac{Z_1 Z_2 e^2}{2E}\right)^2 \left(\frac{M_1 + M_2}{M_2}\right)^2 \frac{1}{\cos^3 \phi} \tag{11.4}$$

Calculating the cross section for PIXE is somewhat more complex, as it involves multiple steps (ionization, radiative emission, self-absorption in the target, etc.). With NRA, the cross section is strongly dependent on the type of reaction (inelastic scattering, formation of a compound nucleus, etc.) and on the nuclei involved (incident and target) [3]. As a result, it exhibits a non-monotonous behavior characterized by ample resonances. A notation of ${}^{A}_{Z}\mathrm{X}(a,b){}^{A'}_{Z'}\mathrm{Y}$ is conventionally used, where ${}^{A}_{Z}\mathrm{X}$ and ${}^{A'}_{Z'}\mathrm{Y}$ are the target and newly formed nuclei, respectively (elements X and Y, with atomic numbers Z and Z', mass numbers A and A'); a and b are the bombarding and emitted particles, respectively. For example, ${}^{11}_{5}\mathrm{B}(\mathrm{p},\alpha){}^{8}_{4}\mathrm{Be}$ corresponds to a reaction in which a beam of protons (p) is used to irradiate a sample containing ${}^{11}_{5}\mathrm{B}$ atoms that are transmuted into ${}^{8}_{4}\mathrm{Be}$, with the emission of an α particle.

11.1.1.2 Energy loss

Earlier, we introduced how IBA techniques provide quantitative elemental analysis (which elements are present and in what amount). In this section, we describe how that information is in fact depth resolved for techniques such as RBS, ERDA and some implementations of NRA (not PIXE). The energy of the detected charged particle (i.e., excluding photons) carries a signature of the depth where it was scattered (RBS), recoiled (ERDA) or emitted (NRA), because charged particles lose kinetic energy as they traverse a solid. This energy loss occurs via two processes: electronic energy loss (energy transfer to the electrons in the target material) and nuclear energy loss (transfer to the nuclei). In both cases, the interaction is basically of a Coulomb type, although electron screening must be taken into account for the nuclear case. The energy dependence is different for the two processes. For light ions, electronic loss peaks at ion energies of the order of 0.1–1.0 MeV; nuclear loss peaks at lower energies, of the order of 0.1–10 keV (Figure 11.2). IBA techniques, and RBS in particular, involve ion energies of the order of MeV, where electronic loss largely dominates (Figure 11.2). Therefore, in the following, we will focus on electronic stopping power as the mechanism providing depth-resolving capabilities. Note that nuclear loss is nevertheless an important process: not only does it contribute to the total energy loss for lower energy and heavier ions, it is essentially the process that originates the particles detected by RBS (ions scattered from a target nucleus), ERDA (recoiled target atoms) and NRA (particles originating from a close encounter between incident and target nuclei). High-energy, light ions incident on a solid (e.g., MeV ${}^4\mathrm{He}^+$ in silicon) lose kinetic energy mostly by colliding with the target electrons, i.e., via excitation and ionization of the atoms. Although it is a discrete process, in the

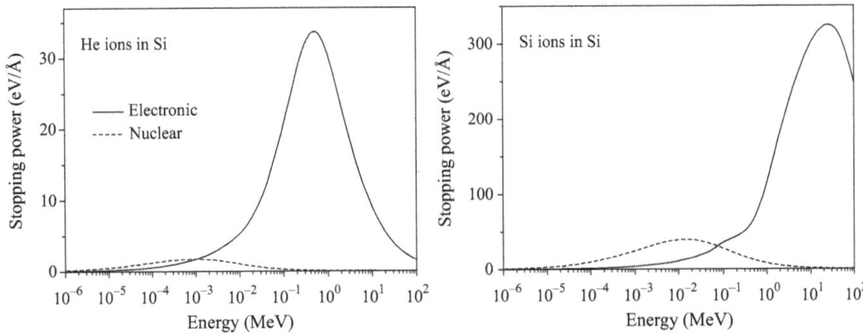

Figure 11.2 Stopping power (electronic and nuclear) for He (left) and Si (right) ions traversing a Si target, calculated using SRIM2008 [4]

context of IBA techniques, we can assume that it occurs continuously and defines an average energy loss known as stopping power: $S(E) = -dE/dx$, where x is the distance traversed by the ions in the solid and the minus sign makes $S(E)$ positive (since dE/dx is negative). The total $S(E)$ is a sum of electronic (S_e) and nuclear (S_n) contributions, with S_n being negligible for high-energy light ions (Figure 11.2). There are essentially two distinct energy regimes of electronic stopping power, below and above the peak. Below, known as the Lindhard regime, $S_e(E)$ increases with increasing energy, as the electron stripping of the incident ion becomes more efficient, i.e., as its effective (positive) charge and therefore the strength of the Coulomb interaction with the target electrons increase. Above the peak, known as the Bethe–Bloch regime, $S_e(E)$ decreases with increasing energy, as all the electrons have been stripped, and the time of interaction (and therefore the momentum transfer) between a fast ion and an atomic electron decreases with increasing ion velocity. In the Bethe–Bloch regime, considering direct collisions of the ion with the electrons in the solid, as well as distant resonant energy transfer, it can be shown that $S_e(E)$ of an elemental target is given by [1]:

$$S_e(E) = \frac{2\pi Z_1^2 e^4}{E} N Z_2 \frac{M_1}{m} \ln\left(4 \frac{m}{M_1} \frac{E}{I}\right) \tag{11.5}$$

where m is the electron mass and I is the average excitation energy of the element composing the target, which is approximately given by $I = 10Z_2$ (in eV). Since the stopping power is a characteristic of the ion as well as of the target material, it is often expressed as *stopping cross section* ε, normalized to the atomic density N (number of atoms per unit volume) of an elemental target $\varepsilon = S/N$. Typical units of S and ε are eV/Å and eV cm^2, respectively.

11.1.1.3 Elemental analysis

Understanding the underlying mechanisms, such as those introduced earlier, allows us to extract quantitative compositional information from experimental IBA spectra. Figure 11.3 is a calculated RBS spectrum illustrating the effect and

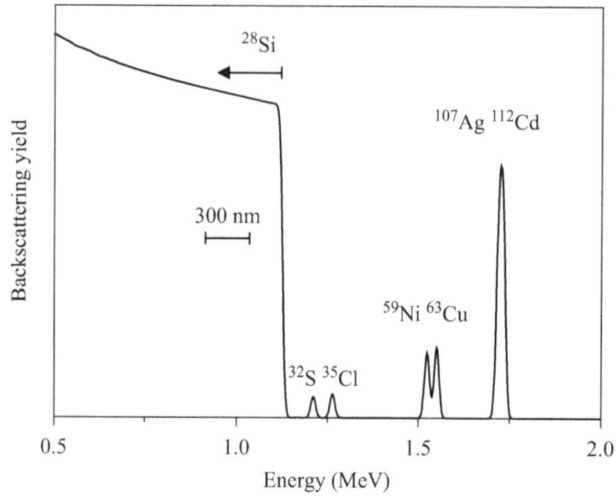

Figure 11.3 Simulated RBS spectrum for ^{32}S, ^{35}Cl, ^{59}Ni, ^{63}Cu, ^{107}Ag and ^{112}Cd (10^{15} atoms per cm^2 for each isotope) deposited on the surface of a ^{28}Si substrate. The simulation was performed using DataFurnace (NDF) [5], for a 2-MeV $^4He^+$ beam, 5° incidence, backscattering angle $\theta = 170°$ and 15 keV energy resolution. The bar within the ^{28}Si signal indicates the relation between energy depth scales: a variation in energy of 125 keV corresponds to approximately 300 nm, and an energy resolution of 15 keV corresponds to depth resolution better than 40 nm (under these specific conditions)

dependencies of the parameters described earlier (kinematic factor, cross section and stopping power). Since the kinematic factor (11.1) increases with M_2, the He ions scattered from the heavier elements in the target reach the detector with a higher energy E_1. However, since dE_1/dM_2 decreases with M_2 (proportional to $1/M_2^2$), the ability to distinguish neighboring elements (i.e., similar M_2) decreases with increasing M_2. In other words, the mass resolution deteriorates for heavier elements. For example, while the peaks corresponding to ^{32}S and ^{35}Cl, differing in M_2 by three unified atomic mass units (u), can be easily distinguished, some overlap is observed for ^{59}Ni and ^{63}Cu (4 u apart), and for ^{107}Ag and ^{112}Cd (5 u apart), the peaks are completely indistinguishable. On the other hand, the sensitivity, i.e., the signal intensity for the same amount of target atoms and impinging ions, is significantly better for heavier elements, since the cross section (11.3) increases with Z_2^2. Finally, knowing the stopping power in a given target allows converting an energy scale into a depth scale (Figure 11.3). Compared to RBS, ERDA typically provides a much better depth resolution, due to the higher stopping power ((11.5) and Figure 11.2) associated with heavier ions (incident or recoiled). For RBS, since the energy of the detected particle carries information on both M_2 and depth, it is not always possible to disentangle the two without additional knowledge on the sample composition. In particular, light elements (e.g., ion implanted or composing a thin film) in or on a thick and heavy substrate are particularly challenging due to the overlap of their weak signal (low cross section) on a strong substrate signal. This limitation is less of an issue (or not at all) for ERDA, PIXE or NRA, although depth-resolved information is never accessible with PIXE, since it is based on the detection of X-rays. Photons are associated with an energy-dependent absorption length, but not an incremental energy loss that can be converted to a path length. Further considerations regarding the use of IBA for elemental analysis are beyond the scope of this chapter, which focuses on IBA channeling techniques to study the atomic structure of defects. In principle, all of the IBA techniques introduced earlier can be used (and may offer problem-specific advantages) in channeling geometry. Currently, however, RBS channeling is by far the most commonly used.

11.1.2 Channeling techniques

11.1.2.1 Introduction to ion channeling

The interaction of the impinging ions with the substrate drastically changes when investigating single crystals. Indeed, when aligning the ion beam with a major symmetry direction of the crystal—either a crystal axis or a plane—the ions are "gently steered" through the crystal lattice (a phenomenon referred to as *ion channeling*) and the probability for close encounter nuclear interactions decreases significantly. Consequently, the measured yield drops accordingly.

The channeling of ions in a crystal was first noticed in the late 1950s when determining the depth profile of implanted impurities [6]. In an attempt to explain the unexpectedly deep profiles, computer simulations were used to investigate the role of the alignment of the beam with respect to the crystal direction, thus revealing the first clear proof of channeling [7]. Very soon after, three groups independently provided unambiguous experimental evidence of ion channeling [8–10], followed by a full theoretical treatise a year after [11]. Essentially, an ion that channels through a crystal undergoes a very large number of consecutive small-angle collisions, none of them deviating the ion sufficiently for it to become dechanneled (Figure 11.4).

An important step in Lindhard's theoretical approach was to implement a continuum potential (first suggested by Lehmann and Leibfried [12] and Nelson and Thompson [9]), which means that the potential of a row (or plane) of atoms is "smeared out" evenly along the row (resembling a charged rod), rather than considering a large number of collisions with individual atoms. From his theory, Lindhard could deduce two important quantities: the *critical angle* ψ_1 and the *minimum yield* $\chi_{\min}$. The former sets an upper limit to the misalignment between the ion beam and the crystal axis/plane that can be tolerated in order to still have channeling. On the other hand, the minimum yield (the lowest yield obtained when the ion beam is optimally aligned) is a measure for the fraction of ions that remains channeled, i.e., that do not undergo a close-encounter interaction with the atomic nuclei within the crystal. Any deviation from a perfect string of atoms, hence from a perfect lattice structure, will be reflected in the critical angle and minimum yield.

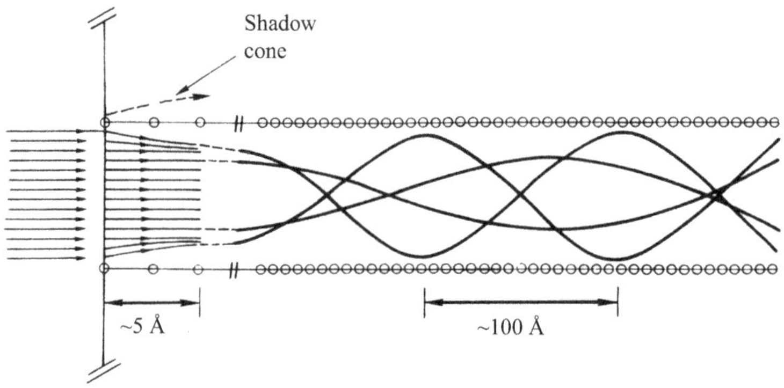

Figure 11.4 Trajectory of ions impinging along a crystal axis, illustrating a large number of small-angle collisions occurring. Note the different length scale in the horizontal and vertical direction. © 2007 Springer Nature. Reprinted, with permission, from [1]

Such deviations obviously include not only defects of all kinds but also thermal vibrations and elastic strain.

Whatever ion beam technique used, the consequence of the significantly lower close-encounter probability for channeling ions is that a lower yield is detected. Examples of this are a lower backscattering yield in RBS experiments, a lower X-ray production yield during PIXE or a lower particle yield when performing NRA. Moreover, the channeling spectra contain identical information as is obtainable when using the ion beam technique in non-channeled (so-called random) mode, e.g., depth resolution, mass resolution, element or isotope sensitivity and (absolute) quantification, as discussed in Section 11.1.1.

In the mid-1960s, when channeling was discovered and understood, RBS came to full maturity as well. Ever since, the two techniques showed to be a perfect marriage. However, as will be shown in Section 11.1.2.3 (and as already predicted by Lindhard in 1965 [11]), the rule of reversibility states that an equivalent channeling process can occur for particles emitted from an atom within a crystal, as is the case for ions from an external beam bombarded onto the crystal. The inverse approach is the basis of so-called EC (Section 11.1.2.3), as already demonstrated in 1965 [13]. In the remainder of this tutorial, we will therefore focus on RBS channeling (RBS/C) and EC.

An RBS/C experiment consists of measuring a number of consecutive RBS energy spectra in the vicinity of axial alignment [Figure 11.5(a)]. As the ion beam gets gradually better aligned with the channel, the backscattering yield starts decreasing, reaching its minimum when "perfect" alignment is achieved. Subsequently, the backscattering yield—averaged over a certain depth region (i.e., a certain energy range) in the near-surface region—is plotted as a function of the angle between the ion beam and the channel [Figure 11.5(b)].

Such a plot, typically referred to as *angular scan*, shows how the yield first drops when approaching ideal alignment and subsequently increases again when

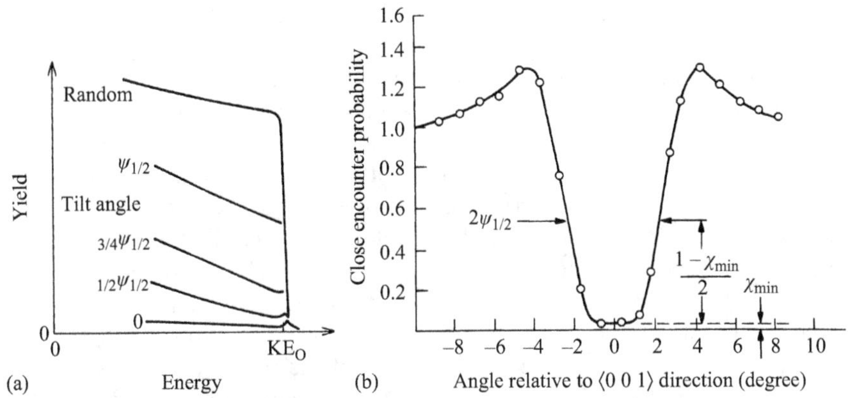

Figure 11.5 (a) RBS energy spectra at several angles of incidence with respect to a crystal axis. The tilt angle is normalized to the critical angle. © 1978 Elsevier. Reprinted, with permission, from [14]
(b) Integrated yield in the near-surface region plotted as a function of tilt angle with respect to the crystal axis. © 1982 Elsevier. Reprinted, with permission, from [15]

misaligning along the other side of the channel. The two basic channeling quantities, the minimum yield and the critical angle, can be directly deduced from the angular scan as the measured yield for an optimal channeling geometry normalized to the yield for random beam incidence, and the half-width at half-minimum of the angular scan, respectively [Figure 11.5(b)]. For high-quality crystals, experimental values typically match reasonably well with the prediction by Lindhard's theory [16]:

$$\psi_1 = \sqrt{\frac{2Z_1Z_2e^2}{dE}} \tag{11.6}$$

$$\chi_{min} \approx \pi a^2 dN \tag{11.7}$$

where Z_1 and Z_2 are the atomic number of the particle and substrate atoms, respectively, e is the elementary charge, E is the ion energy, a is the Thomas–Fermi screening length, N is the atomic density and d is the interatomic spacing along the direction of the ion beam. Although the Lindhard equations describe the dependence on the various parameters very well, the absolute values of the critical angle and the minimum yield are typically overestimated by approximately 20% and underestimated by a factor of 3, respectively. Although these ad hoc "correction factors" are often used, computer codes have been used as early as 1971 to simulate the channeling of ions through a crystal by following a large number of binary collisions, thereby also considering thermal vibrations of the atoms in the crystal [17]. The outcome of this pioneering work by John Barrett is a proportionality factor to the Lindhard expressions ((11.6) and (11.7)), using the two-dimensional (2D) vibration amplitude and the Thomas–Fermi interatomic screening length as input parameters. We refer the interested reader to the literature for further details [18].

In order to obtain a complete spatial view of the ion channeling around the channel under scrutiny, one can repeat the angular scan for various tilt angles in a direction perpendicular to the angular scan. Doing so, a 2D *image scan* is constructed, revealing not only the crystal axis but also the crystal planes constituting the axis. Whereas such time-consuming experiments may be overkill in the vast majority of studies, they have been of valuable importance for specific investigations, such as materials containing large concentrations of ordered vacancies [19], or for the lattice location of impurities or displaced host atoms [20] in a crystal. On the other hand, one has to be aware that the extended exposure to MeV He ions required for an image scan may induce defects in the crystal. As will be discussed in Section 11.1.2.3, for channeling techniques using particle emission, the concept of an image scan is implemented by the use of position-sensitive detectors.

As mentioned, channeling can occur for most of the current IBA techniques, e.g., also for PIXE. This technique is in many ways complementary to RBS, in a sense that it is element sensitive (mass resolution is not such a big issue for PIXE) but lacks depth sensitivity. Although the optimal experimental conditions will slightly vary between RBS and PIXE, performing both simultaneously with the same ion beam is feasible, as illustrated in Figure 11.6 for a ZnO sample doped with Mn, Fe and Co. Whereas the low concentrations of transition metal dopants cannot be distinguished by RBS [Figure 11.6(a)], these elements are clearly discerned by PIXE [Figure 11.6(b)], thus allowing to study their channeling signal.

The entire discussion so far considers a "perfect" lattice. However, any deviation from a perfect crystal structure will enhance the probability for close-encounter nuclear interactions—picture it as atoms *sticking out* of their crystal row. In turn, such an increase is typically revealed via a decrease of the critical angle and/or an increase of the minimum yield, i.e., the angular scan becomes narrower and less deep. As will be illustrated in the second part of this chapter, a comparison

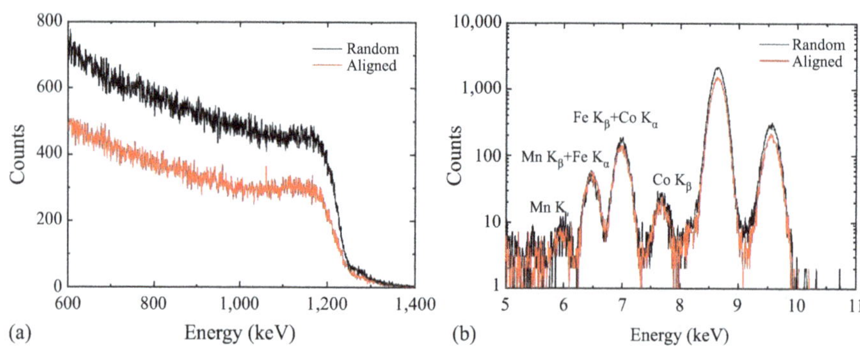

Figure 11.6 Random and channeled spectra of a ZnO sample doped with Mn, Fe and Co, measured simultaneously with RBS (a) and PIXE (b)

of the experimental data to theoretical models or computer simulations allows deducing quantitative information on the (defect) structure of the crystal.

Besides capturing the full angular scan, additional valuable information is often also contained in the RBS spectra with random and channeled beam incidence. Indeed, these spectra reveal the depth profile of the lattice imperfections, as opposed to angular scans, which are based on depth-integrated data. Specifically, the first atom of each row (i.e., the surface plane of the crystal) is always fully exposed to the ion beam, whereas the following atoms are to a large extent in the shadow of the first atom. As a consequence of the enhanced scattering probability with atoms in the surface plane, a surface peak is normally observed in the channeling geometry. This peak can be further enhanced by significant contributions from the second, third, etc. atoms in the row, e.g., as a result of surface reconstruction. In a very similar way, defects at an interface (which can be considered as a buried surface), such as misfit dislocations, result in an enhanced scattering probability as well. This is illustrated in Figure 11.7, which shows the random and aligned spectra of a $CoSi_2$ thin film on a Si(1 1 1) substrate. Both the Si (channel 550–710) and the Co (channel 760–860) signals show an enhanced surface peak, at channel 700 and 850, respectively, as well as an interface peak at channel 620 and 775, respectively.

In conclusion, any deviation from a perfect atomic stacking in the crystal will be spotted in channeling experiments, allowing access to damage, defects, precise lattice location of impurities in the crystal, static displacement of lattice atoms or

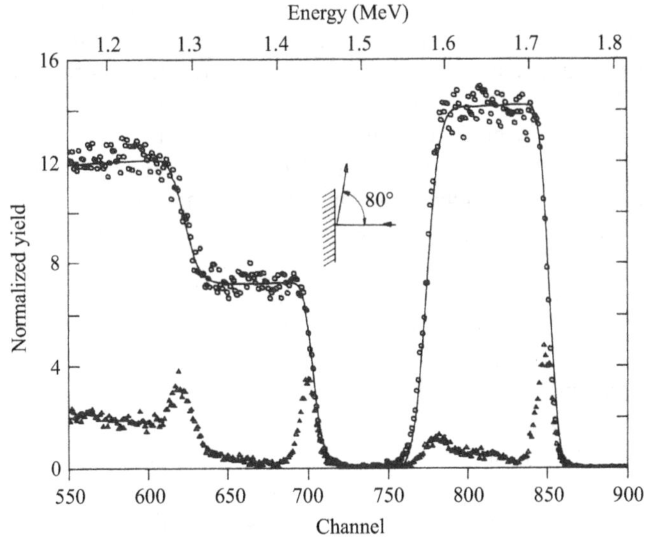

Figure 11.7 Random and aligned backscattering spectra of a $CoSi_2$ thin film on a Si(1 1 1) substrate, revealing the excellent crystal quality as well as the presence of a surface and interface peak. © 1991 Elsevier. Reprinted, with permission, from [21]

impurities, lattice vibrations, etc. In other words, basically any structural imperfection will be detected—be it that only the *structural* impact will be noticed, in contrast to many other techniques treated throughout this book (e.g., DLTS and PAS), which are sensitive to the electrical properties of the material, vacancies, etc. Moreover, one has to be aware that, due to the nature of the technique, channeling requires single crystalline samples. When investigating defects in semiconductors, this condition is nearly always fulfilled. Therefore, ion channeling has reached a prominent position in this research field, to provide complementary information (at the subatomic scale) to the other techniques described in this book. In the following sections, we will focus on how channeling provides qualitative and quantitative information on damage and defect concentrations (Section 11.1.2.2) and on the lattice site of impurities in semiconductors (Section 11.1.2.3).

11.1.2.2 Characterization of defects and damage

Semiconductors can contain a large variety of structural defects, ranging from intrinsic defects during the growth or processing of the wafers, implantation-induced defects resulting from doping processes, to surface or interface defects in (multi-)layered samples, quantum dots or quantum wells, e.g., as a result of thin-film reaction or thermal treatment. Studying defect concentrations and depth profiles has been a workhorse application for channeling, ever since the early days. Using ion implantation as an example, we will show how the defect concentration as a function of depth can be determined from the channeling spectra. Still, the approach we use can be applied to any kind of damage, whatever the mechanism for inducing the defects was, and whatever the type of defects is—as long as atoms have been displaced. Even vacancies or impurities can play a major role in this respect, since their presence can result in a displacement of the atoms in the vicinity.

One of the first key applications of channeling, still routinely used nowadays, is the generation and recovery of implantation-induced defects in semiconductors. The abundant nuclear collisions occurring when an implanted dopant slows down in the lattice result in a large number of displaced host atoms—part of which recovers quasi-instantaneously. Figure 11.8 shows the random RBS spectrum of a Ge substrate implanted with 160 keV Ar ions, along with the channeling spectra for a variety of implantation fluences (i.e., various Ar concentrations) [22].

First, a peak is observed in the energy region between 1.00 and 1.35 MeV, which is the implanted near-surface region of the sample. The area of this peak directly scales with the number of displaced Ge atoms and increases as the implanted fluence increases, until finally reaching the random level, indicating amorphization of the crystal (all host atoms have been displaced). Moreover, it is clear that the defects extend deeper and deeper into the substrate (i.e., lower backscattering energy) as the Ar concentration increases. Finally, the increasing yield in the unimplanted substrate region (0.4–0.9 MeV) deserves special attention. As no defects have been induced in this part of the sample, no change in the defect concentration is expected. However, the defects present in the implanted region will dechannel the impinging ions, as a result of which a slightly diverted ion beam

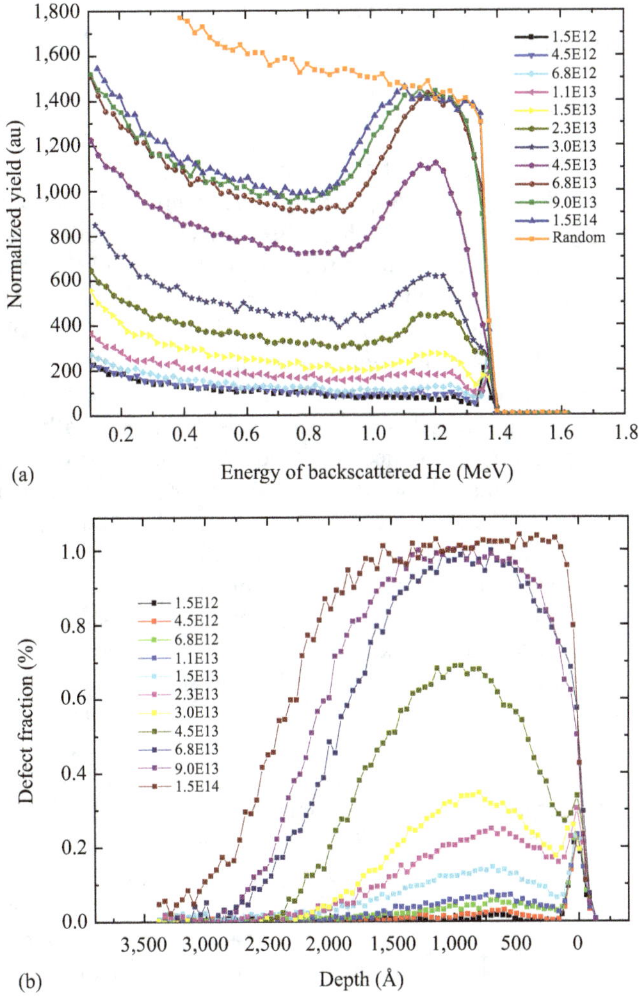

Figure 11.8 (a) Random and aligned RBS spectra of a Ge substrate implanted with 160 keV Ar ions, for various Ar fluences. (b) Defect depth profile deduced from the channeling spectra in (a). © 2009 IOP Publishing. Reprinted, with permission, from [22]

enters the deeper regions of the sample. This beam divergence, which results in an enhanced scattering probability, increases with increasing defect concentration in the damaged layer. Therefore, the increasing backscattering yield in the *virgin* part of the substrate is not a token of defects being present but of dechanneling of the ion beam. To obtain the real defect concentration, this dechanneling contribution must be subtracted. The latter can be relatively easily done by applying a two-beam model, whereby the channeled beam $(1 - f_R(x))$ provides the defect density and the random beam ($f_R(x)$) the dechanneled fraction, respectively [23]:

$$\chi(x) = f_R(x) + (1 - f_R(x))f\frac{n_D(x)}{N} \tag{11.8}$$

where x is the depth, f is the defect scattering factor, N is the atomic density and $n_D(x)$ is the defect depth profile. Applying this procedure to the spectra shown in Figure 11.8(a) yields the defect depth profiles $n_D(x)$ of Figure 11.8(b), which clearly reveal the defect concentration (decreasing to a zero-level in the virgin, deeper region of the sample), the saturation level, as well as the depth range. Several computer codes, e.g., DICADA [24], use this two-beam model to deduce the direct scattering depth profile from the experimental spectra.

It is important to point out that a quantitative assessment of the defect profile can be readily done (for each element in the substrate individually) without the need of complex modeling or the need of standards for comparison. On the other hand, it is by far less straightforward to unravel the *nature* of the defects. One possibility is to repeat the measurements using several ion energies, since the energy dependence of the scattering cross section differs from one type of defect to the other [15]. For example, with increasing beam energy, the dechanneling probability decreases for point defects and amorphous clusters, increases for dislocations and remains constant for stacking faults. However, it should be emphasized that channeling is by far less sensitive than techniques such as transmission electron microscopy (TEM), for which a relatively high defect concentration is required (e.g., ~1% in case of point defects), and that the obtained information is averaged over the entire beam spot (of order 1 mm^2) and averaged over all defects present in the sample. In this respect, channeling is complementary to many other techniques, which either exhibit problems *quantifying* defects or obtaining their *depth profile* or *saturate* for defect concentrations approaching the per cent level.

From the examples that will be described in Section 11.2.1, it is clear that disentangling different types of defects from one spectrum can be a complex and challenging task. To obtain the ultimate information from the measurements, detailed simulations are often required, as well as complementary measurements, e.g., X-ray diffraction [25]. In particular, the simple two-beam approximation discussed earlier [23] is often insufficient. Indeed, rather than merely separating the dechanneling contribution from direct scattering (from *randomly* displaced atoms), more dedicated codes, such as McChasy [26], allow isolating contributions from extended defects (dislocations, stacking faults, etc.) as well. Ignoring the latter may result in a significant overestimation of the defect concentrations in the deeper parts of the sample (Figure 11.9).

In order to be discerned with ion channeling, it is not required that the atoms are fully removed from their equilibrium lattice site. Indeed, as soon as atoms are displaced by a distance as small as ~0.1 Å, they can have an impact on the minimum yield and (particularly) on the critical angle. Such displacements can be a result of doping with atoms that are larger or smaller than the host atoms (see Section 11.1.2.3), of the presence of vacancies in the vicinity of the atom, or

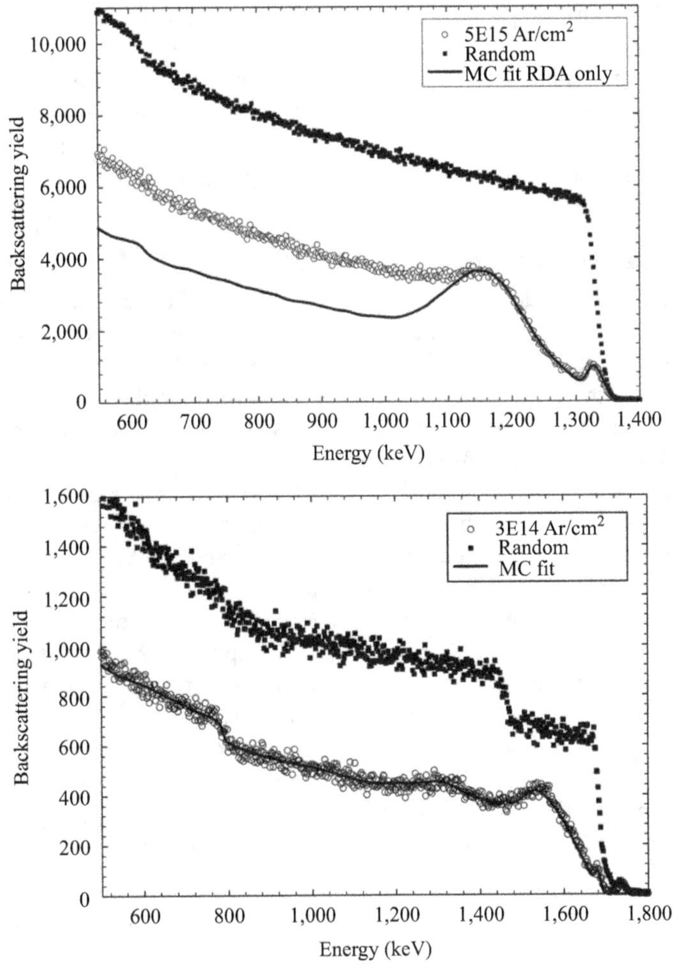

Figure 11.9 Random and aligned spectra for implanted ZnO (top) and $SrTiO_3$ (down) samples, along with two simulations, based on (top) the two-beam model and (down) a McChasy simulation including extended defects. Assessment of extended defects is indispensable for making a good estimate of the defect concentration in the deeper parts of the sample. © 2014 Elsevier. Reprinted, with permission, from [26]

anything alike. These types of displacements are *static*, i.e., they do not vary with time, and can be either randomly directed or in a preferential direction. From a detailed analysis of measurements along noncoplanar directions, complemented with computer simulations, both the magnitude and direction of the displacement can be determined.

As opposed to these *static* displacements, thermal vibrations (phonons) are *dynamic* in nature. Still, they have a similar impact on the channeling behavior.

Since the flight time of the ion in the crystal is one to two orders of magnitude shorter than typical phonon periods, the ion "encounters" a static lattice, with the atoms being randomly displaced with an average magnitude determined by the thermal vibration amplitude. Therefore, temperature-dependent ion channeling measurements allow investigating the phonon behavior for every constituting element individually, as a function of depth (as long as its signal is clearly distinguished in the RBS spectra).

11.1.2.3 Lattice location

Besides characterizing defects and damage, the ability of the crystal lattice to steer the motion of charged particles and the corresponding inhomogeneous particle flux densities at different lattice sites in single crystals can also be used to investigate the lattice location of impurities. Depending on whether the source of charged particles is located outside or inside the crystal, one can distinguish between external beam and particle emission methods.

When a single crystal is bombarded by a homogeneous flux of MeV ions from the outside, the particle flux inside the crystal depends in a very characteristic way on the incidence angle and initial energy of the ions and can be prepared in such a way that it exhibits large differences at distinct lattice sites (Figure 11.10). If one monitors suited interaction rates of charged particles and impurity atoms as a function of incidence angle from major crystallographic directions, information on lattice sites and root mean square displacement of impurity atoms is obtained. In the case of foreign atoms which are heavier than the crystal atoms, one may apply RBS as probing interaction, since the energies of ions backscattered from the impurities are in this case above those backscattered from crystal atoms. For light element impurities, on the other hand, this is generally not possible because the few ions backscattered from light impurities cannot be separated from ions which have

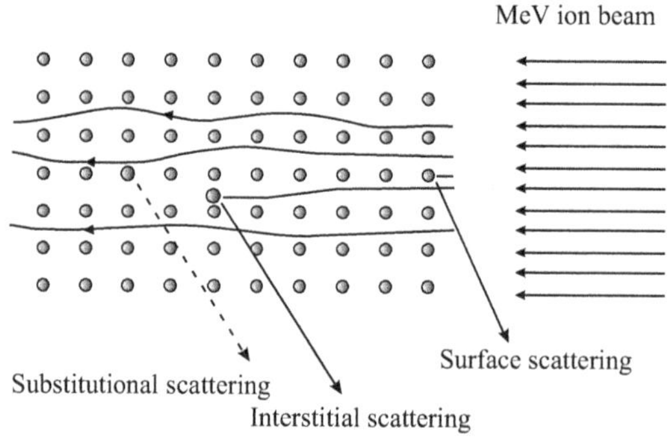

Figure 11.10 Principle of lattice location by means of ion beam channeling. A homogeneous flux of MeV particles is incident parallel to a major crystallographic direction of a single crystal. Inside the single crystal, the channeling effect enhances the particle flux density in the interstitial region, while it is reduced along the rows of atoms

undergone collisions with heavier lattice atoms deeper inside the sample. An alternative is to use inner shell X-ray-generating reactions for light impurity detection, i.e., the PIXE technique, which, however, has typically an order of magnitude lower efficiency than RBS and, furthermore, is not depth sensitive. Some specific light elements like B, N or H can also be probed by means of nuclear reactions, e.g., $^{11}B(p, \alpha)^{8}Be$, $^{15}N(p, \alpha)^{12}C$, $D(^{3}He, p)^{4}He$ [3]. In this case, parameters like detection sensitivity, depth resolution, lattice location accuracy, etc. depend very strongly on the energy-dependent cross section of the nuclear reaction available, the energy loss of projectile and ejectile, and on the depth profile of the impurity.

Several approaches of varying sophistication are possible in order to quantitatively assess the location of impurity atoms from an ion beam channeling experiment. Let us assume that certain impurity atoms are occupying the following two sites in a single crystal: (i) ideal substitutional sites, i.e., in case of 100% of the impurities on that site, the minimum yield χ^{imp} for an ion incidence along a crystallographic direction would be equal to the one from the host atoms χ^{host}, cf. (11.7), and (ii) so-called random sites, which are supposed to be randomly distributed in the unit cell so that the normalized yield for all angles of the incoming beam is equal to one. In this case, the fraction f_S^{imp} of substitutional impurity atoms found in an experiment can be estimated from the following simple relation, see, e.g., [15,27]:

$$f_S^{\mathrm{imp}} = \frac{1 - \chi^{\mathrm{imp}}}{1 - \chi^{\mathrm{host}}} \tag{11.9}$$

Many lattice location studies can be found in the literature where the RBS or PIXE yields for lattice and impurity atoms were only measured for random and aligned particle incidence and the quantitative evaluation using χ^{host} and χ^{imp} relied on the simple two-site assumption mentioned earlier. However, this approach can lead to entirely wrong results if the impurities do in fact *not occupy* a mixture of ideal substitutional and random sites. It is therefore already a much better practice to measure one-dimensional (1D) angular scans of the backscatter yield [15,18] (as shown in Figure 11.5(b)), since in this case, e.g., a narrowing of the impurity angular scan compared to the host atom dip will indicate that the foreign atoms do not occupy ideal substitutional sites, hence (11.9) would lead to inaccurate results. Equation (11.9) is also not applicable to the case of interstitial impurity atoms, where a comparably simple estimate for the impurity yield is not possible since (i) there is no reference signal from interstitial lattice atoms, and (ii) due to the so-called flux peaking effect [28], the flux density inside the channels (and hence the normalized yield from interstitial foreign atoms) may reach values that exceed unity several times. In any case, for a more accurate analysis of the specific lattice sites and the number of impurity atoms which occupy them, it is indispensable to compare the experimental angular scan data to theoretical channeling yields calculated for a variety of lattice sites in order to see which sites fit best [15,18]. All state-of-the-art ion beam lattice location studies and the examples discussed in Section 11.2.2 follow this approach.

While a considerable number of computer codes are available to simulate the ion beam channeling effect, the ones which are most commonly used in lattice location studies are FLUX [29], CASSIS [30,31] and McChasy [32]. They all use the Monte Carlo approach, i.e., the individual tracks of a large number of particles are followed inside the single crystal, which allows calculating the probability of close encounters with both substitutional lattice atoms and impurities on specific lattice sites. The choices of interaction potentials between the moving ion and the lattice atoms are usually exponentially screened Coulomb potentials, most commonly the Molière or the Ziegler–Biersack–Littmark potential, see, e.g., [17,18,29]. While the individual ion tracks are calculated by classical Newtonian mechanics, they are subject to random variables, the probability distributions of which are known. The major random variables are the point of impact of the ion on the surface of the crystallographic unit cell, an initial small spread of the incidence angle due to imperfect collimation of the ion beam, the exact positions of the lattice atoms which are subjected to temperature-dependent Gaussian probability distributions around the ideal lattice sites and the angular deflections caused by collisions with target electrons, which are coupled to the electronic energy loss. Since such theoretical modeling of the motion of ions in crystals has reached a high degree of accuracy, often very precise lattice location information may be derived from a comparison of experimental results and theoretical predictions.

The approach for the quantitative analysis of ion beam lattice location experiments by means of fitting experimental data to simulated angular yields for impurities on different lattice sites is in principle the same as in the case of EC experiments, which will be discussed later. The major differences are the way how the theoretical angular yields are obtained, and the fact that in ion beam studies, usually only 1D angular scans are measured, while the particle emission methods use complete 2D emission patterns.

In the particle emission lattice location methods, charged particles are emitted directly from the sites where probe atoms are located inside the crystal and experience channeling or blocking on their way out of the crystal, as first reported by Domeij in 1965 [13]. This effect is called EC and blocking and manifests itself outside the single crystal via a highly anisotropic emission of particles in the vicinity of crystallographic axes and planes. While emission of charged particles can be induced by means of nuclear reactions from host crystal or impurity atoms, also including RBS, most experiments are performed by incorporating radioactive probe atoms which decay by the emission of charged particles, an approach which is unique with respect to the small number of probe atoms required and hence its unrivalled efficiency [33–40]. The method can be used for all types of radioactive decays that are accompanied by the emission of charged particles, i.e., α, β^-, β^+, and also conversion electrons (CE). However, α active isotopes are mostly found for elements heavier than Pb, with the exception of some rare earth (RE) (e.g., ^{149}Tb, ^{150}Dy or ^{153}Er) and very light isotopes (e.g., ^{8}Li, ^{8}B, ^{20}Na). For applications of the EC technique in semiconductors, as α emitters so far only ^{8}Li has been relevant, which was reviewed in detail [38] and shall not be addressed here. While β^- and β^+ or electron capture decays exist for most elements of the periodic system, β^+ emitting isotopes that are suitable for EC are in fact quite rare since it is much more common to find electron capture decays. However, since electron capture decays are not associated with the emission of β particles, this type of decay is not applicable to the EC method unless it involves the emission of CEs. Conversion electron emission often follows radioactive decays that populate excited levels in the daughter nucleus, where it is an alternative process to gamma particle emission. The vast majority of EC experiments have been performed with electron-emitting isotopes, either in the form of β^- or CE emitters and in the following we will focus on this approach.

In comparison to conventional lattice location techniques by means of ion beam channeling, e.g., RBS/C, the main benefits of EC are roughly four orders of magnitude higher efficiency and the ability to study also elements lighter than the host atoms. These facts allow performing detailed lattice location studies with very good statistical accuracy at low fluences of implanted probe atoms, usually as a function of implantation or annealing temperature of the very same sample, which is not feasible by other methods. While typical areal densities of impurities in ion beam channeling experiments are $\sim 10^{14}$ cm^{-2} for RBS/C and $\sim 10^{15}$ cm^{-2} for PIXE, EC lattice location has been reported already in the 10^{10} cm^{-2} range [41].

Figure 11.11 illustrates the principle of electron EC. Electrons which are emitted from probe atoms on substitutional lattice sites are always guided by the crystal potentials, resulting in emission maxima along all axial and planar directions. The situation is more complicated for probe atoms which are located interstitially since they produce a mixture of channeling and blocking effects.

Three key ingredients played a crucial role in establishing EC as a technique that nowadays accounts for a significant part of lattice location studies found in the literature:

1. Facilities where beams of radioactive ions can be produced and implanted into materials.
2. Position-sensitive detectors suitable for the specific particles (electrons or alphas) emitted in nuclear decay.
3. Simulation codes that allow to reliably predict the angular-dependent emission patterns from probe atoms on different lattice sites.

As was already mentioned, EC requires detecting the angular-dependent emission yield around major crystallographic axes. The simplest approach for that and which was common till the 1990s is by means of tilting the crystal with respect to a well-collimated particle detector which is placed at a certain distance from the sample in order to achieve sufficient angular resolution, a similar approach to the one illustrated in Figure 11.5 [33,34]. More efficient, however, is the use of position-sensitive detectors [38,39,42–44], which allows measuring a whole range of emission angles simultaneously, as is indicated in Figure 11.12.

In order to introduce the radioisotopes into the single crystalline samples, almost exclusively ion implantation has been used. While some experiments with diffused isotopes have been reported, this approach is not very suitable because it does not result in depth profiles that are sufficiently well defined for an accurate

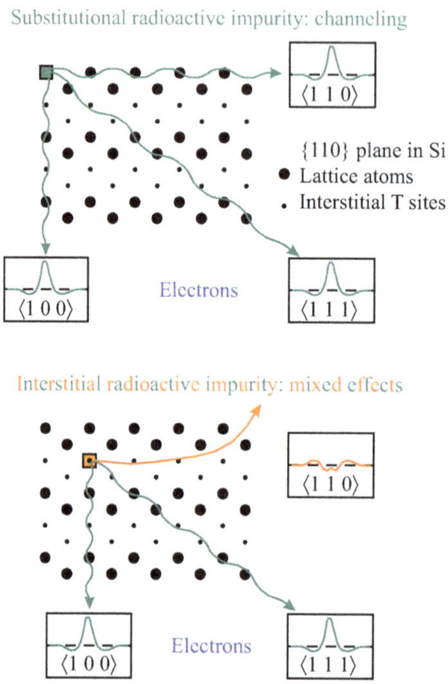

Figure 11.11 *Principles of the emission channeling effect of electrons emitted from radioactive probe atoms. Shown are the {1 1 0} plane of the Si lattice and the schematic angular-dependent emission yields of electrons along ⟨1 0 0⟩, ⟨1 1 0⟩ and ⟨1 1 1⟩ axes for substitutional and tetrahedral interstitial emitter atoms*

quantitative analysis (cf. discussion of dechanneling below). For the first EC studies reported in the literature, long-lived radioisotopes, either locally produced or commercially acquired, were introduced into the samples at the home laboratory by means of ion implanters. However, in order to make the method more versatile, access to a wider range of radioactive isotopes, including also short-lived ones, is needed (Figure 11.13), and since 1985 for that purpose, the ISOLDE online isotope separator facility at CERN [45] is used. At ISOLDE, currently more than 800 different radioisotopes of 60–70 chemical elements can be produced by nuclear reactions and are available to the user as 30–60 keV ion beams, which is a very suitable energy range for implantation of EC probes. Due to the interaction of channeling electrons with electrons of the sample and thermally displaced lattice atoms along the atomic rows, an exponential depopulation of channeling states occurs, with "dechanneling lengths" (i.e., the mean free path of channeling electrons) in most materials typically of the order of ~100–2,000 Å [33,36,37]. Implantation energies of 30–60 keV, which lead to implanted depths around 100–1,000 Å, are hence sufficiently low to avoid excessive dechanneling; on the

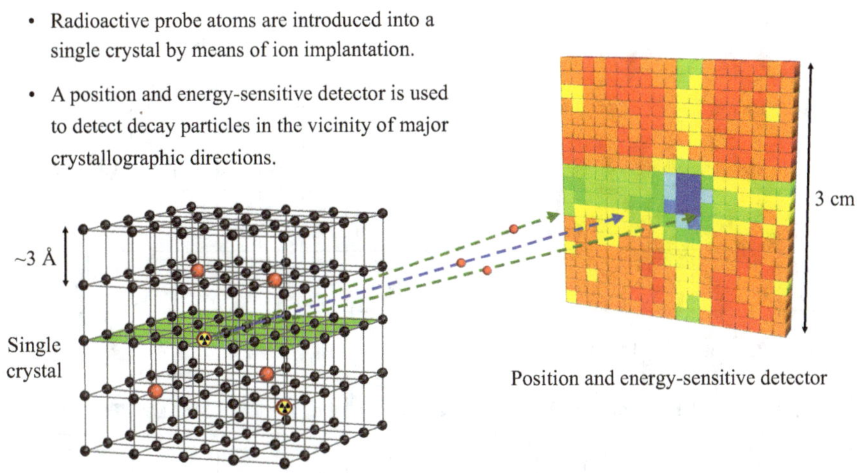

Figure 11.12 Schematics illustrating the principle of an emission channeling experiment from implanted radioactive isotopes using a position-sensitive detector. The distance from the detector to the sample is ~30 cm so that an angular range of roughly ±3° around a specific crystallographic direction is covered

other hand, they are high enough to place the probes well below potentially disordered surface layers.

The half-lives of the radioactive isotopes are crucial with respect to what kind of experiments can be conducted. Long-lived isotopes (typically above 12 h of half-life) are most conveniently measured in off-line setups, i.e., the sample is implanted in a simple implantation chamber at room temperature and subsequently mounted in an external EC setup, where thermal annealing and measurement take place. As a rule of thumb, the time scale for a typical offline experiment is about two half-lives, which means that experiments can take from several days to months. Shorter lived isotopes must be measured in a setup which is directly attached to the beam line of the isotope separator. In this case, if the half-life is in the range of several hours, the sample can be implanted at room temperature and one or several annealing steps followed by subsequent channeling measurements performed "in situ." Since the typical duration of annealing is 10 min at elevated temperature plus the time it takes to heat and cool the sample, for isotopes with very short half-lives, below ~15 min, the only efficient way to perform EC experiments is in "online" modus [44], i.e., the angular emission patterns must be measured simultaneously with the implantation as a function of the implantation temperature.

Due to the fact that the movement of electrons inside a crystal is subject to diffraction, their trajectories are not well described by the laws of classical mechanics but require a quantum mechanical treatment. Moreover, typical energies of β particles from nuclear decay reach up to several MeV, hence demanding a relativistic description. Both aspects are combined using the so-called many-beam

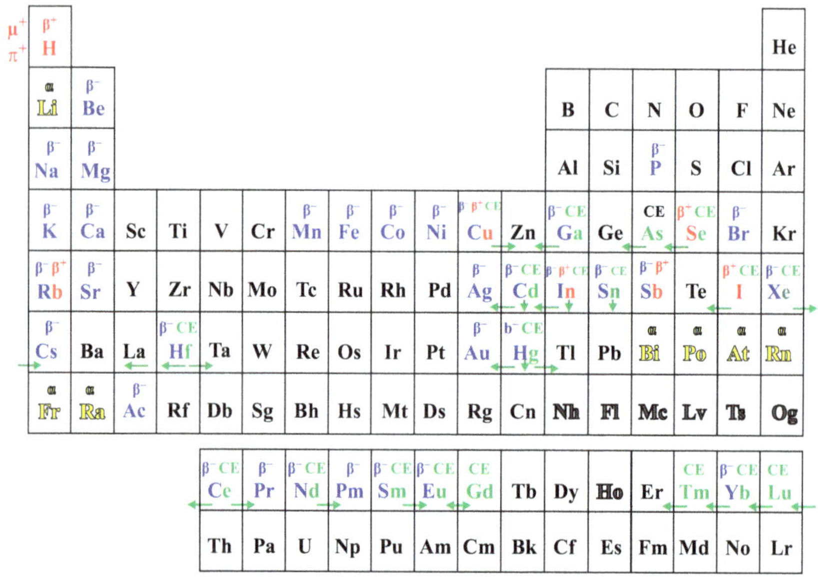

Figure 11.13 Periodic table showing the elements for which radioactive isotopes were successfully used in emission channeling experiments. The decay type used (β^-, β^+, CE or α) is indicated. Elements marked with CE acted as mother isotopes for conversion electron emitters, with the arrows indicating the decay to the CE active daughter isotope. Elements with an arrow pointing downwards were studied by means of CE emission from isomeric states (^{111m}Cd, ^{112m}In, ^{114m}In, ^{117m}Sn, ^{119m}Sn, ^{197m}Hg). Note that H does of course not have a positron-emitting isotope but has been included in the list due to the experiments using the decay of μ^+ or π^+, which can be regarded as equivalent to light isotopes of H

diffraction theory of electron channeling in single crystals [33,36,37,39]. In these calculations, the crystal potential is approximated by a superposition of atomic Doyle–Turner potentials. Since for a description of the EC effect only electrons moving under relatively small angles with respect to the channeling axis are relevant and these move at relativistic velocities, the potential along the channeling direction can be averaged (continuum potential) and the electron wave function splits into a longitudinal component along the channeling direction, which obeys a relativistic Klein–Gordon equation, and a transverse component perpendicular to the axis, which is given by the solution of a 2D Schrödinger equation in a periodic potential. The β particle flux density inside the crystal can then be calculated after obtaining the transverse electron wave function using standard Fourier decomposition and Bloch wave techniques. The many-beam approach has proven to accurately describe EC effects in a multitude of cases [33,36,37,39]. Still, the approach is challenging since in order to treat the continuous electron energy spectra resulting from β decay and which stretch from zero up to several MeV, simulations have to be performed at a multitude of electron energies and the results averaged according to the β^- energy distributions.

Quantitative analysis of electron EC patterns [39,43] (and, as was already mentioned above, also angular scans in ion beam channeling) is performed by fitting experimental angular yields in the vicinity of major crystallographic directions by theoretical patterns calculated for emitter atoms on different lattice sites; nowadays, it is not uncommon to consider 200–300 different sites. We define $N(\theta, \phi)$ as the number of events (or counts) measured at angles θ, ϕ, from a crystallographic direction, N_R as the number of events measured under the same conditions in case of an amorphous sample, f_1, f_2 as the fractions of emitter atoms on two lattice sites 1, 2, with $\chi_1(\theta, \phi)$, $\chi_2(\theta, \phi)$ the theoretically calculated normalized angular yields for these sites. Since it is the anisotropy $\chi_{\text{exp}}(\theta, \phi) - 1$, i.e., the deviation of the normalized yield from isotropy, which is proportional to the fraction of emitter atoms on a certain lattice site, we have the following

$$\frac{N(\theta,\phi)}{N_R} - 1 = \chi_{\text{exp}}(\theta,\phi) - 1 = f_1[\chi_1(\theta,\phi) - 1] + f_2[\chi_2(\theta,\phi) - 1] \quad (11.10)$$

In this context, $f_R = 1 - f_1 - f_2$ can be identified with the so-called random fraction f_R, i.e., the fraction of emitter atoms that do not cause any anisotropy in the patterns, e.g., because they are located in amorphous regions of the sample or on very low-symmetry lattice sites. However, any flat background in the pattern from other sources, e.g., from backscattered electrons or gamma background in the detector, will also be forced into the random fraction which then has to be appropriately corrected.

Note that any quantitative analysis of channeling patterns or angular scans requires at least an approximate knowledge of the number of counts N_R measured under total absence of any channeling effects, as would be the case for an amorphous sample under the same conditions. In ion beam channeling, N_R is usually approximately determined by measuring the RBS yield under conditions that are well away from a channeling direction; with N_R kept fixed, only the f_i are obtained by means of chi-square minimization routines. In EC, N_R is treated as a fit parameter that is the same for all angles in a 2D pattern. In addition, the relative transverse and azimuthal orientation x_0, y_0, ϕ_0 of the experimental pattern resulting from the exact position of the detector towards the sample are optimized as well. Typically, only the two lattice sites which cause the most prominent channeling effects are accessible by this method, only in some special cases, meaningful three-site analyses were possible.

11.2 Examples

In the following section, we will demonstrate the power of IBA when studying defects in semiconductors, in particular to investigate defects induced in the semiconductor structure by, e.g., ion implantation (for doping purposes), the recovery of these defects and activation of the dopants, and the precise lattice location of the implanted dopants. We will mainly focus on GaN and ZnO as host materials. Since the goal is to illustrate the ion beam approaches with a few specific examples, this chapter does not aim at providing a full review of the defect formation mechanisms as such.

11.2.1 Damage in semiconductors

11.2.1.1 Damage in GaN

Stimulated by the exciting developments in epitaxial growth of thin GaN layers, leading to a manifold of optical devices, particular attention was devoted to ion implantation into this material around the turn of the century. Indeed, in contrast to the well-known case of Si, doping of nitrides was still confronted with many open questions, such as finding a good p-type dopant, inserting RE metals that are optically active, etc. Many research groups addressed this topic, reporting on implantation with a wide range of ions, energies and experimental conditions.

One of the first striking findings is that *dynamic annealing* in GaN is very efficient, i.e., a major fraction of the induced defects recovers quasi-instantaneously after bombardment [46–48]. The Frenkel pairs generated in the collision cascade during implantation are mobile in GaN—even at liquid-nitrogen temperature (LN_2)—and largely annihilate [49]. As a result, chemical effects can play an important role in the damage buildup and eventual amorphization, as the implanted fluence increases.

A second specific observation when implanting GaN concerns the defect profile. Figure 11.14 shows the depth profile of displaced Ga atoms as a function of depth, as derived from channeling analysis (using the procedure described above,

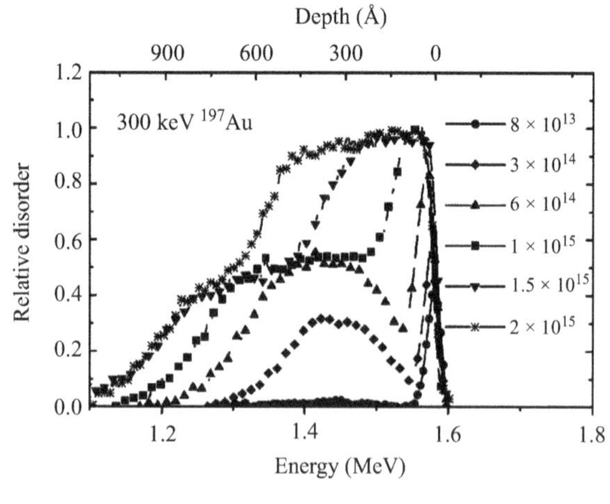

Figure 11.14 Defect depth profile of GaN implanted at RT with 300 keV Au ions to various fluences. © 2001 by the American Physical Society. Reprinted, with permission, from [46]

see Figure 11.8), after implantation of Au at 300 keV. As shown in the figure, implantation to a fluence of 8×10^{13} at./cm^2 hardly results in any defects near the end-of-range (i.e., around 300 Å depth) as a result of the dynamic annealing. In contrast, a significant fraction of defects accumulates at the surface. Upon increasing the fluence, both the surface and bulk defect fraction increase. However, for room temperature implantation, the defect concentration in the bulk peak saturates at a level below amorphization, the value of which is ion dependent. In contrast, the surface defects evolve toward amorphization, since defects are efficiently trapped at the surface [50]. Subsequently, the amorphization proceeds layer by layer, starting from the surface and gradually proceeding deeper into the sample. Based on the assumption that the amorphous/crystalline interface acts as a perfect sink for mobile point defects, Titov *et al.* have modeled the saturation of the bulk-defect concentration and the amorphous layer thickness [51].

The saturation behavior of the bulk-defect fraction strongly depends on the implanted species (Figure 11.15). As expected, a higher fluence is needed to reach saturation for lighter ions. However, in contrast to what is common belief, the saturation level of the bulk peak is higher for light ions, even reaching to amorphization. This finding is explained by the chemical effects of the implanted species, which efficiently trap the implantation-induced migrating point defects [46]. It should be pointed out that secondary phase formation, precipitation due to the very high implanted fluence and N_2 bubble formation, can play a role as well [46,52].

GaN dissociation is a general concern, not only due to ion irradiation but also during annealing, either in vacuum or in N_2 at ambient pressure. It leads to loss of nitrogen and the formation of Ga droplets at the surface. Furthermore, GaN erosion can also occur as a result of selective sputtering of amorphous GaN from the surface [53]. RBS in combination with channeling is ideally suited to determine the (near-surface) composition of the nitride film, its overall thickness and the effect on the crystalline quality.

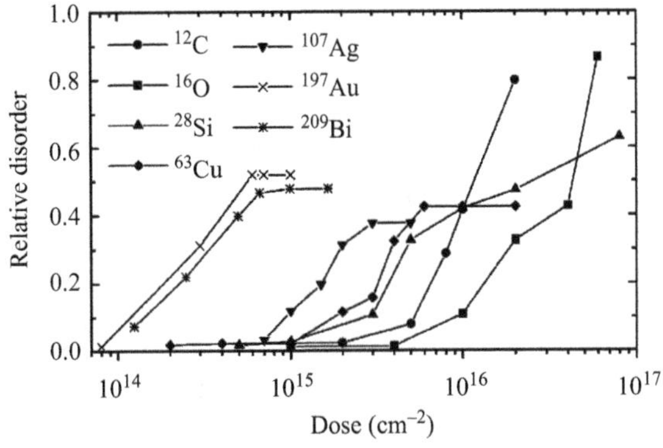

Figure 11.15 Maximum defect concentration in the bulk peak as a function of fluence, after RT implantation. © 2001 by the American Physical Society. Reprinted, with permission, from [46]

The above findings clearly illustrate the importance of the surface in processes such as amorphization and nitride deterioration. This conclusion is further underlined by the fact that a very-thin AlN cap—even as thin as 10 nm—deposited onto the GaN film prevents the surface amorphization during implantation as well as the dissociation during high-temperature annealing, even up to 1,300°C [54]. The backscattering spectra in channeling geometry in Figure 11.16 clearly indicate that, upon implanting with 2×10^{15} Eu/cm^2, the surface amorphization is completely suppressed when an AlN cap is deposited.

The proximity of the surface has a strong impact on the relative disorder in the bulk peak. Using ion channeling with enhanced depth resolution, Charnvanichborikarn *et al.* demonstrated that the retained damage in the bulk region increases as the defect profile is further away from the surface, in other words, when annihilation at the surface becomes less probable [55]. Conversely, point defects generated closer to the surface experience efficient recombination.

Throughout the many investigations of implantation-induced damage in GaN, a lot of attention was devoted to the defect accumulation as a function of implanted fluence. In order to compare experiments with different ions, performed under different experimental conditions, the fluence is typically expressed as a number of *displacements per atom* (dpa), a direct measure for the number of ballistic displacements that occurred in the collision cascade. Overall, three main stages have been found: after an initial increase (dominated by nuclear energy deposition and pronounced recombination effects), the defect concentration quasi-stagnates, to finally result in the nucleation of amorphous zones at very high fluence (Figure 11.17); see, e.g., [47,57]. By measuring the relative concentration of uncorrelated displaced atoms, Elke Wendler has mathematically modeled the damage buildup, which eventually results in amorphization [56]. To this end, she used data for various ion species, implantation temperatures, ion fluences, ion

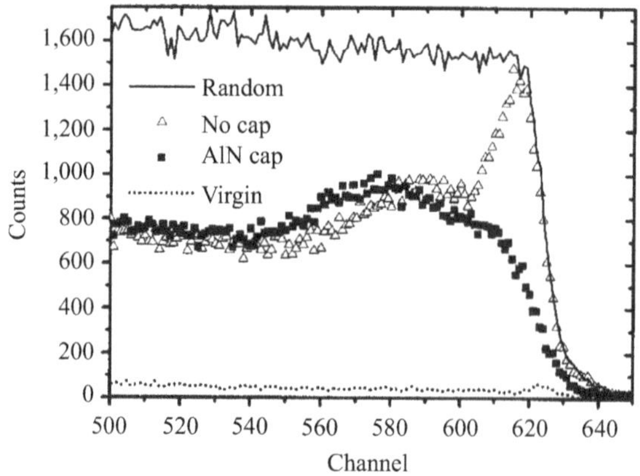

Figure 11.16 Random and aligned spectra for Eu-implanted GaN samples, with and without a 10 nm AlN cap prior to the implantation.

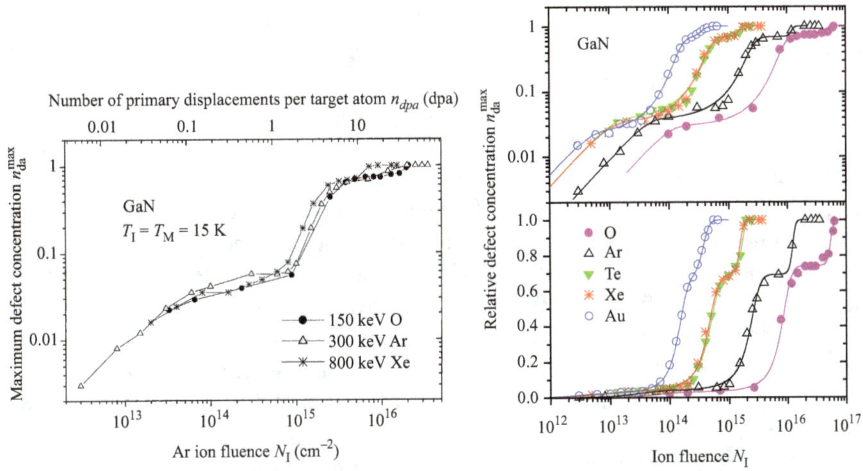

Figure 11.17 (Left panel) Relative defect concentration (in the maximum of the depth profile) as a function of Ar fluence after low-temperature implantation. The top axis, expressed in dpa, allows comparing directly to implantation with other species, at other energies. © 2003 Elsevier. Reprinted, with permission, from [47] (Right panel) Relative defect concentration (in the maximum of the depth profile) as a function of fluence, after low-temperature implantation (both logarithmic and linear scale). The lines are fits to the model mentioned in the text. © 2009 Elsevier. Reprinted, with permission, from [56]

fluxes, ion energies and a range of semiconductors. The model is based on a balance between defect production and defect recombination and correctly describes a three-step amorphization process. It is interesting to point out that this model predicts a critical temperature, above which amorphization cannot occur. However, among the exceptions to this behavior is GaN.

Obviously, when relying on ion channeling to determine the defect concentration, one should be aware that the dechanneling rate depends on the nature of the defect (see tutorial Section 11.1.2.2). More specifically, the dechanneling probability for correlated displacements (e.g., dislocations and stacking faults) is likely to be higher than for random displacements (e.g., point defects). Therefore, the defect concentration in the deeper layers can be somewhat overestimated.

In many of the studies, raising the implantation temperature, hence modifying the defect recombination rate, is used as a means to minimize damage accumulation. However, since the temperature window is limited (due to the stability of the nitride layer) and because dynamic annealing is already very efficient at low temperature, the impact of an elevated implantation temperature in GaN is rather modest. An alternative way to tune the defect concentration is to vary the implantation geometry. In a similar way as described in Section 11.1.2.1, ion channeling can be used during implantation, by aligning the implantation beam to the normal crystal axis, typically GaN ⟨0 0 0 1⟩. Although a different formalism should be used compared to (11.6) [11], the critical angle for heavy ions with an energy of several tens to hundreds of keV is very large, of order 4°–6°. Consequently, the alignment is easily achieved without the use of a high-precision goniometer. As described earlier, the channeling effect minimizes the probability for close encounter collisions with the target atoms. Rather than resulting in less backscattering (as is the case for RBS), the lower close encounter probability leads to a lower defect generation probability during channeled implantation [48]. Indeed, gradually varying the implantation angle from a random incidence, i.e., approximately 7°–10° away from the ⟨0 0 0 1⟩ *c*-axis, toward an aligned geometry, results in a reduction of the accumulated defect density by about an order of magnitude (Figure 11.18). In particular, the effect remains prominent after high-temperature annealing and results in a drastically enhanced activation of the (optical) dopants [58]. It should be added that the most drastic reduction of defect density occurs in the bulk peak, whereas it applies to a lesser extent to the surface damage peak.

For channeled implantation, the same three regimes as a function of implanted fluence are found as discussed earlier for random implantation, with the most drastic defect reduction occurring in regime 2 [59]. However, once sufficient damage starts accumulating, the channeling effect reduces, in turn leading to a higher defect production—finally resulting in amorphization (regime 3).

Finally, the reduced defect fraction in channeled implantation results in a drastic reduction of electrically active defects in Eu-implanted GaN, as measured by DLTS [60]. In particular, the Eu_2 electron trap at E_c −0.36 eV, which is not

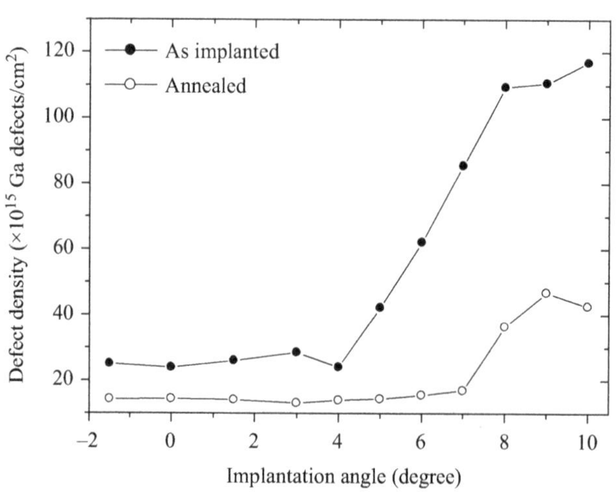

Figure 11.18 Defect density as a function of the angle between the implanted ion beam and the GaN c-axis, for Er implantation into GaN at 80 keV, to a fluence of 2.5 × 10^{14} *at./cm*^{2}*, both after implantation and after annealing at 950°C in* N_2 *flow. © 2003 Elsevier. Reprinted, with permission, from [58]*

observed after electron irradiation or implantation of light ions, vanishes in the channeling geometry.

11.2.1.2 Damage in ternary nitrides

Given the relatively small lattice mismatch of the three wurtzite nitride lattices, alloying GaN with either AlN or InN is achievable and allows tuning the band gap. As expected, $In_xGa_{1-x}N$ becomes less damage resistant as the In concentration increases. Due to the inherent mass resolution of RBS, the defect evolution can be studied for the In atoms and Ga atoms separately [61,62]. Despite the different displacement energy for In versus Ga, no difference is found in the damage evolution for In and Ga atoms. On the other hand, for InGaN, the defect concentration continuously increases as a function of fluence (i.e., regime 2 is not observed) and amorphization occurs at a lower fluence compared to GaN [62].

Besides the mass resolution of RBS, its depth resolution can be exploited as well as illustrated for InGaN/GaN multilayer stacks [63]. In a study of the stability against ion beam mixing, the authors did find a different behavior between the In and Ga sublattice, as well as the occurrence of the second regime with increasing fluence. Analysis by a similar approach as used by Wendler (see above), complemented by a detailed TEM study, indicates the presence of extended defects in the InGaN/GaN multilayers (Figure 11.19).

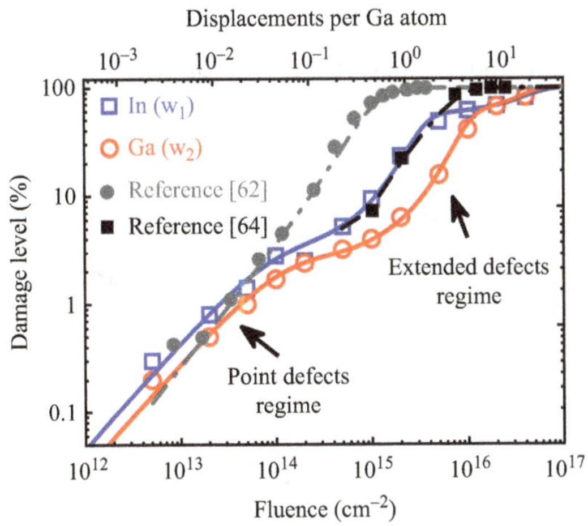

Figure 11.19 Relative defect concentration as a function of fluence, for InGaN/GaN multilayer stacks, implanted with 35 keV N^+ at room temperature. For comparison, data for Au implantation at low temperature [62] and at room temperature [64] in InGaN single layers are shown as well, clearly illustrating the different regime 2 behavior. © 2015 IOP Publishing. Reprinted, with permission, from [63]

Finally, a similar study in $Al_xGa_{1-x}N$ reveals that the defect accumulation is more complex, i.e., four regimes are observed [65]. Even with the very high fluences used in their study, amorphization is not reached. Furthermore, as opposed to common belief, the defect fraction increases with increasing Al fraction, rather than decreasing.

11.2.1.3 Damage in ZnO

In a similar way as reviewed above for GaN, implantation-induced defects, as a result of doping, have been under intense scrutiny for ZnO as well. Although the observed mechanisms show, to a certain extent, some overlap with those in GaN, the situation is much more complicated for ZnO. Amongst the most striking differences are (i) the pronounced role played by extended defects, (ii) the—yet—stronger dynamic annealing, (iii) the presence of a "middle" defect peak for some species, in between the surface and bulk peak and (iv) four regimes in the fluence dependence of the damage accumulation [56,66,67].

Several studies indicated the importance of the implanted ion on the damage stabilization. Using implantation of elemental versus compound species (e.g., B^+ versus BF_2^+) with the same energy per atomic mass unit, Azarov *et al.* separated the purely ballistic effect from the chemical effect [67]. They found that the amount of disorder is rather determined by the number of B atoms in the ZnO sample than by the ballistic damage they induce (expressed in *dpa*). Interaction of B with defects plays a pivotal role [67], in a similar way as also shown for Si [66].

A striking observation is the "inverse temperature effect" found for high fluences (regimes 3 and 4 in Figure 11.20): dynamic annealing is more efficient at low temperature than at room temperature. This surprising observation is explained

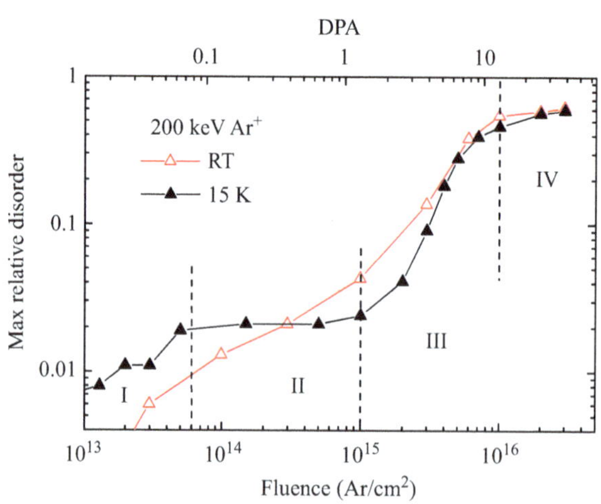

Figure 11.20 Maximum defect fraction in the bulk peak as a function of fluence, for Ar-implanted ZnO. © 2014 AIP Publishing LLC. Reprinted, with permission, from [67]

by the insufficient mobility of defects at low temperature, which prevents them from reaching a region where they can stabilize—consequently they rather annihilate [67].

To study the specific role of point defects versus extended defects, Azarov *et al.* used a combination of deep Ag implantation and shallow B implantation [68]. Their channeling analysis evidenced that the Ag implantation generates extended defects, presumably as a result of Ag agglomeration. Furthermore, using a Li-based monitoring technique, they showed that Zn_i released from B-induced defects is efficiently trapped by the Ag-extended defects. Therefore, extended defects are believed to play an important role in the damage saturation at high fluence. Subsequently, this trapping mechanism was used to study defect formation during implantation of nitrogen, an important p-type dopant in ZnO. In order to separate the N-related defects from the preexisting extended defects generated by the Ag implantation, the depth sensitivity of ion channeling along with dedicated data analysis (using the DICADA code [24]) played an important role. Figure 11.21(b)

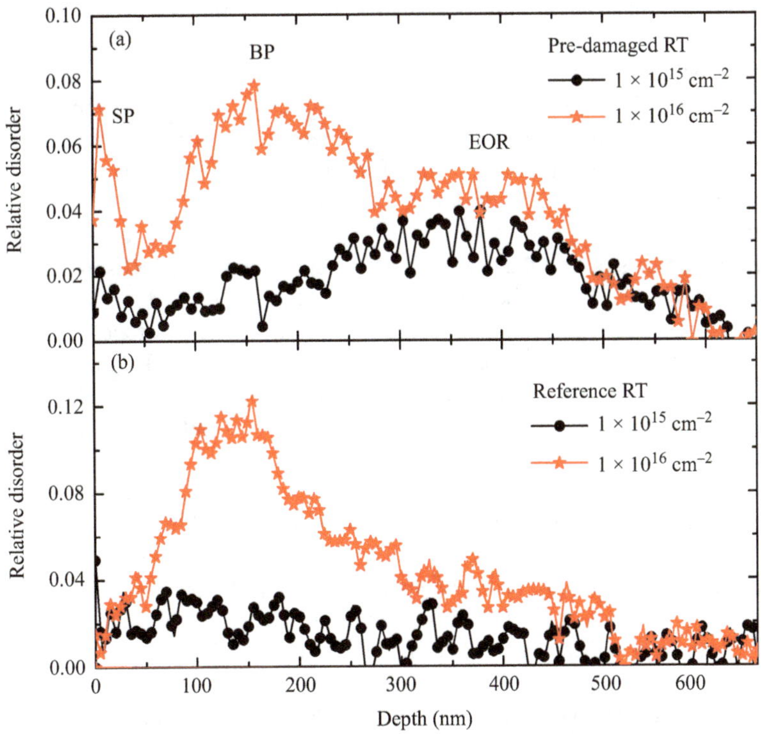

Figure 11.21 Depth profile of the defect fraction after room temperature N implantation into ZnO: (a) pre-damaged sample, containing a buried layer of extended defects, and (b) virgin sample. © 2017 AIP Publishing LLC. Reprinted, with permission, from [69]

shows the defect depth profile of N implanted at 80 keV to two fluences. In agreement with SRIM simulations, the damage peaks are located close to the projected range. However, for the pre-damaged sample [Figure 11.21(a)], the damage starts accumulating much deeper, i.e., mobile point defects resulting from the N implantation are trapped at the Ag-induced extended defects [69]. At higher fluence (1×10^{16} at./cm^2), a multimodal defect profile develops, with defects both at the sample surface, the projected range and at the extended defects. In a broader context, this approach can be used in a defect engineering scheme to swipe defects away from the active zone of a device.

11.2.2 Lattice location of dopants

11.2.2.1 Optical dopants in GaN

RE-doped semiconductors have been investigated during the last decades on account of their optical properties, where sharp luminescence lines in the infrared or visible due to intra-4f shell atomic transitions of the REs can be observed. The quenching of RE luminescence at room temperature decreases with an increasing band gap of the material; therefore, many recent studies have focused on RE-doped wide band gap semiconductors, in particular GaN. Since the RE luminescent properties depend on its atomic environment and hence on its lattice site and interaction with other defects, there has been ongoing interest in characterizing the RE lattice sites in GaN by means of ion beam techniques. GaN can be doped with REs either during growth or by means of ion implantation, and depending on the RE concentration and method of doping, lattice location is accessible by different techniques.

The lattice sites of low-fluence ($1\text{–}2 \times 10^{13}$ cm^{-2}) implanted REs in GaN can be well studied with the EC method discussed in Section 11.1.2.3. To date, EC experiments have been performed in GaN using the radioactive RE isotopes ^{141}Ce, ^{143}Pr, ^{147}Nd, ^{149}Pm, ^{149}Gd, ^{153}Sm, ^{155}Eu, ^{167}Tm, ^{169}Yb and ^{170}Lu, cf. [70,71] and references therein. As a general feature (Figure 11.22), the EC experiments revealed that following RT implantation, large fractions of REs occupy substitutional Ga sites; however, they are subject to rms displacements from the ideal Ga sites in the range $u_1 \sim 0.14\text{–}0.23$ Å, i.e., significantly larger than the RT thermal vibrations of Ga atoms, for which values around 0.057–0.084 Å can be found in the literature. The displacements of RE probes were interpreted as being the consequence of implantation defects (cf. Section 11.2.1) in the immediate neighborhood that causes a shift of the RE from the ideal Ga sites. A clear major effect of annealing the RE-implanted samples up to 900°C was to reduce these rms displacements (Figure 11.22).

The lattice location of higher fluence ($5 \times 10^{14}\text{–}1.5 \times 10^{15}$ at./cm^2) implanted REs in GaN is accessible by studies using the RBS/C technique. As an example, Figure 11.23 summarizes RBS/C $\langle 0\,0\,0\,1 \rangle$ and $\langle 1\,0\,\bar{1}\,1 \rangle$ angular scans for Ce, Pr, Eu, Dy, Er, Tm and Lu-implanted GaN. It was found that the majority of implanted RE atoms occupy substitutional Ga sites; however, the substitutional fraction was anticorrelated with the implanted fluence. At the used annealing temperature of 1,000°C, no differences of the rms displacements of RE atoms with those assumed

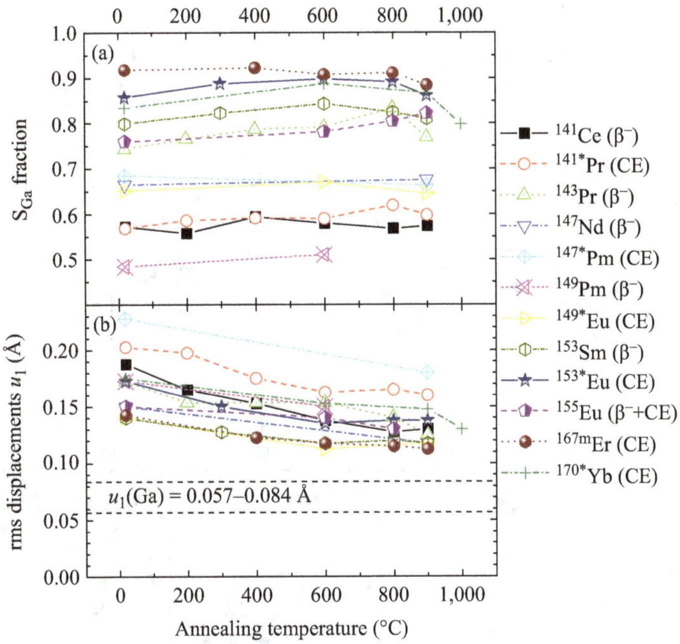

Figure 11.22 Comparison of emission channeling results using radioactive RE probes; β^- indicates β-emitting isotopes, while "CE" means that conversion electrons emitted by excited states of the mentioned daughter nuclei were used. Part (a) shows the fraction of RE atoms on substitutional Ga lattice sites as a function of annealing temperature, while (b) is the rms displacement from the perfect Ga site. The region between the dotted lines indicates the spread in the RT thermal vibration amplitude of Ga atoms found in the literature. © 2010 Springer Nature. Reprinted, with permission, from [71]

for Ga atoms in the FLUX simulations (0.070 Å) could be established by RBS/C, except in the case of Eu, where the angular width of the ⟨1 0 $\bar{1}$ 1⟩ Eu scan was reduced in comparison to the Ga scan and could be fitted by assuming a Eu rms displacement of 0.14 Å when projected perpendicular to the ⟨1 0 $\bar{1}$ 1⟩ axis. Since the ⟨0 0 0 1⟩ angular scan showed approximately the same width as the Ga scan, a possible interpretation is that Eu is shifted by around 0.20 Å along the *c*-axis.

The lattice location of Eu was also investigated in GaN samples that have been doped during growth with ~0.1 atomic per cent of Eu at various growth temperatures (Figure 11.24) [72]. For a growth temperature of 1,000°C, a perfect match of the Ga and Eu angular scans was observed, corresponding to 100% incorporation of Eu on ideal substitutional Ga sites. On the other hand, for a sample grown at 900°C, the width of the Eu angular scans was reduced in comparison to Ga, a clear proof of nonideal lattice sites. Since the effect was stronger for the ⟨1 0 $\bar{1}$ 1⟩ scan than for

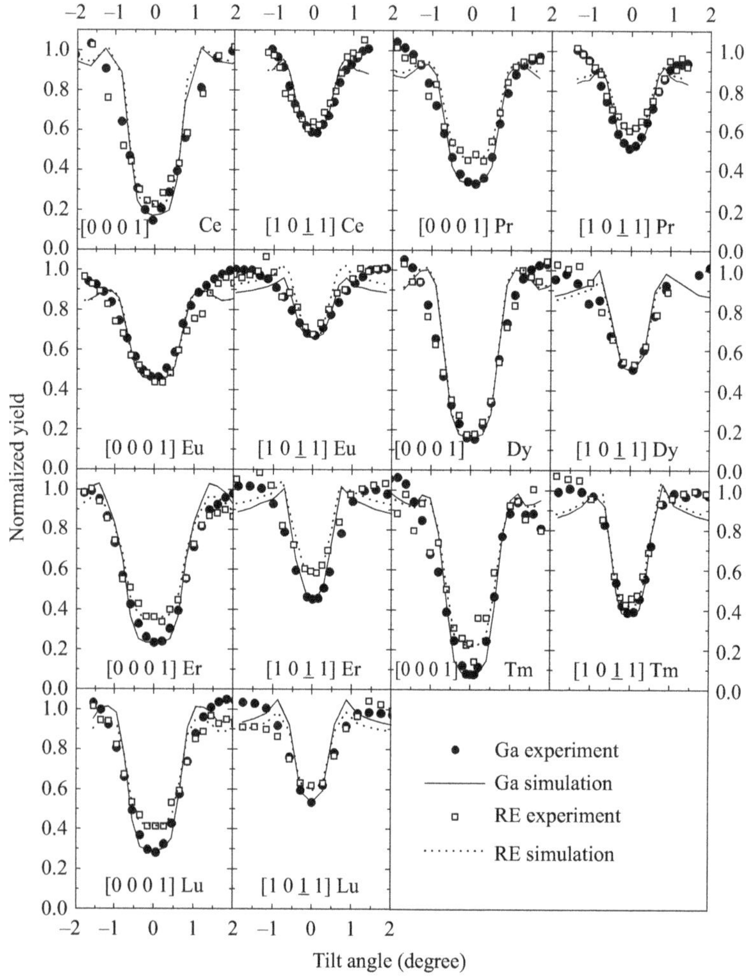

Figure 11.23 *$\langle 0\,0\,0\,1\rangle$ and $\langle 1\,0\,\bar{1}\,1\rangle$ RBS/C scans for Ga and various REs in GaN along with the best fits of FLUX Monte Carlo simulations for substitutional Ga sites. REs were implanted with 160 keV at RT at fluences ranging from 5×10^{14}–1.5×10^{15} at./cm^2 followed by annealing at 1,000°C. The RE rms displacement u_1 used in the simulations was set equal to a representative value of the thermal vibration amplitude of Ga ($u_1 = 0.070$ Å), except for Eu $\langle 1\,0\,\bar{1}\,1\rangle$ where an rms value of 0.14 Å was used. © 2003 Elsevier. Reprinted, with permission, from [70]*

the $\langle 0\ 0\ 0\ 1\rangle$ direction, a possible explanation was a 0.24-Å displacement of Eu along the *c*-axis; however, data were also compatible with an isotropic displacement of rms value 0.17 Å. The results of this study showed that the incorporation of Eu on nonideal Ga sites may also occur for other doping techniques than

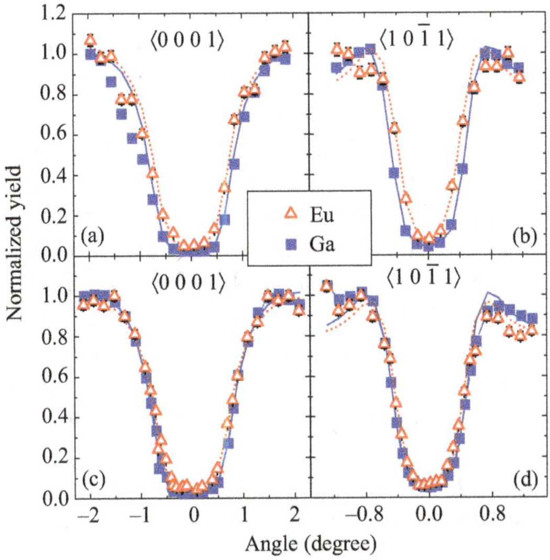

Figure 11.24 *The symbols show experimental RBS/C Ga and Eu angular scans for GaN doped with Eu during growth at two different temperatures: Parts (a) and (b) grown at 900°C, (c) and (d) at 1,000°C. While (a) and (c) are scans across the ⟨0 0 0 1⟩ surface direction, (b) and (d) show the off-surface direction ⟨1 0 1̄ 1⟩. The solid and dotted lines are the best fits of FLUX simulation results for Ga and Eu scans, respectively. © 2010 AIP Publishing LLC. Reprinted, with permission, from [72]*

ion implantation, although for samples doped during growth, the nonideal Eu lattice sites were attributed to the formation of Eu clusters rather than implantation defects.

11.2.2.2 Electrical dopants in wide-gap semiconductors: Mg in GaN and AlN and Na in ZnO

Mg-doped p-type GaN is nowadays a core component of many optoelectronic devices, e.g., light-emitting diodes for solid-state white lighting or blue lasers. Yet, there are some basic properties related to p-type doping of GaN that are still poorly understood and also limit the performance of devices. One such major problem is an inherent doping limit: once the Mg concentration in GaN surpasses ~10^{19}–10^{20} cm^{-3}, further introduction of Mg does not lead to an increase in the hole concentration. Recently, a theory has been put forward [73] that explains this behavior by suggesting an amphoteric nature of Mg: once the doping limit has been reached, additional Mg atoms are not incorporated on substitutional Ga sites anymore (where they act as acceptors) but on interstitial sites (where they form compensating double donors). The alkali metal Na in the II–VI semiconductor ZnO is the analogue case to the earth alkali Mg in GaN. ZnO suffers from even more severe p-type doping problems than GaN, and Na has since long been discussed as an acceptor candidate. However, also in the case of ZnO:Na, theoretical predictions suggest Na interstitial donors as "the leading killer defects which prevent p-type doping" [74]. Until recently, no experimental data on the lattice sites of Mg in GaN and AlN or Na in ZnO existed. Ion beam lattice location by means of RBS/C is not applicable in these cases since Mg and Na are lighter than the cation host atoms Ga, Al and Zn, and using PIXE or NRA fails at the low concentrations of the dopants.

The EC technique with short-lived isotopes has provided unique opportunities of studying the lattice locations of Mg in GaN [41] and AlN [75] and of Na in ZnO [76], using the isotopes ^{27}Mg ($t_{1/2}$ = 9.5 min) and ^{24}Na (14.96 h) produced at CERN's ISOLDE facility, giving direct evidence that besides substitutional cation sites, these light impurities can also occupy interstitial sites. Moreover, the exact locations of the interstitial sites could be determined with an accuracy better than 0.15 Å.

Figure 11.25 shows the major substitutional and interstitial lattice sites in the wurtzite structure. As natural candidates for interstitial impurity positions, one may

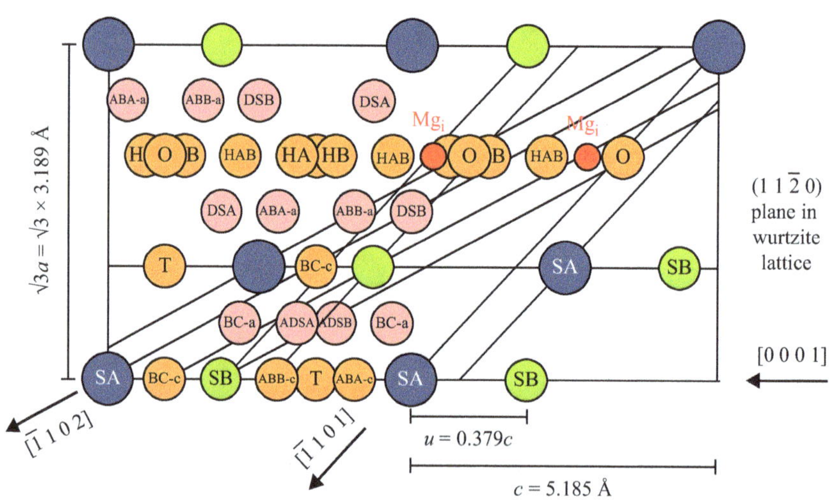

Figure 11.25 *(1 1 $\bar{2}$ 0) plane in the wurtzite lattice, showing the cation and anion atom positions (i.e., substitutional sites SA and SB) and the major interstitial sites. Note that the HA and HB sites are quite close (+0.32 and −0.32 Å) to the octahedral interstitial sites O so that the corresponding circles overlap in the picture. The position of the Mg interstitial sites Mg$_i$ in GaN, which was determined by EC experiments to be +0.60 Å from O sites, is indicated by the red circles. Along the ⟨$\bar{1}$ 1 0 2⟩ and ⟨$\bar{1}$ 1 0 1⟩ directions, the double rows of cation and anion atoms are indicated. The lattice constants a and c, as well as the bond length parameter u given, are those of GaN*

consider sites in the wide open interstitial region parallel to the c-axis, and also the tetrahedral T sites, which are located along the c-axis atomic rows. In order to fully distinguish those sites, EC measurements need to be performed along several crystallographic directions that also include other directions than the ⟨0 0 0 1⟩. The theoretical β^- emission patterns that are expected in the case of ^{27}Mg on substitutional Ga, substitutional N and tetrahedral interstitial T sites are displayed in Figure 11.26. As will be shown in the following, the most likely sites for Mg interstitials are sites that are displaced by +0.60 Å along the c-axis from ideal octahedral O sites toward the HAB sites, i.e., they are located in between the HA and HAB sites and indicated as Mg_i in Figure 11.25; their theoretical channeling patterns are also shown in Figure 11.26.

In order to determine the exact lattice sites of Mg in GaN and AlN and of Na in ZnO, experimental β^- emission yields in the vicinity of ⟨0 0 0 1⟩, ⟨$\bar{1}$ 1 0 2⟩, ⟨$\bar{1}$ 1 0 1⟩ and ⟨$\bar{2}$ 1 1 3⟩ directions were fitted by theoretical patterns allowing a linear combination of contributions from two lattice sites: (i) substitutional cation sites SA and (ii) interstitial sites. While allowing for T positions did not result in an improvement of fit quality, the chi-square of fit lowered considerably once interstitial sites parallel to the c-axis were introduced. Plotting the resulting chi-square of the fit against the position of the interstitial site, which was varied parallel to the c-axis in steps of 0.03–0.05 Å, allowed identifying the most likely interstitial sites (Figure 11.27). The positions for which best overlap of χ^2 minima was found were for Mg in GaN $+0.60 \pm 0.14$ Å from O sites, i.e., in between HA and HAB sites, and for Na in ZnO $+0.23 \pm 0.04$ Å from O sites, i.e., in between O and HA sites. The interstitial site of Mg in AlN was also located +0.23 Å from O toward HA sites [75]. Experimental channeling patterns are compared to best-fit theoretical ones for these lattice site combinations in Figure 11.28. The fact that the majority of ^{24}Na occupies the interstitial sites in ZnO following RT implantation is responsible for the absence of channeling effects along axial ⟨0 0 0 1⟩ and ⟨$\bar{2}$ 1 1 3⟩ and planar (0 1 $\bar{1}$ 0) directions, all of which still show pronounced channeling for ^{27}Mg in GaN since the majority of Mg occupies substitutional Ga sites. Note that for ^{24}Na in ZnO, the EC experiments were capable of determining the position of the Na interstitial site more precisely since in this case, the majority of RT-implanted Na (~70%) occupied the interstitial positions. In contrast, for Mg in GaN, while it was possible to initially obtain fractions up to ~30% of interstitial ^{27}Mg in p-type samples, this situation was not stable under prolonged irradiation since the created implantation damage was quite effective in destroying the p-type character of the samples (Figure 11.29). As a consequence, detailed lattice location experiments measuring several crystallographic directions of the same sample could only be performed under sample conditions that were already subject to radiation damage, resulting in ^{27}Mg interstitial fractions that are characteristically the same as for undoped GaN, i.e., only 9%–13% of interstitial ^{27}Mg for RT implantation.

In Figure 11.29, the fraction of interstitial Mg_i found in four different doping types of GaN (undoped, Si-doped n-type, Mg-doped-activated p-type by annealing

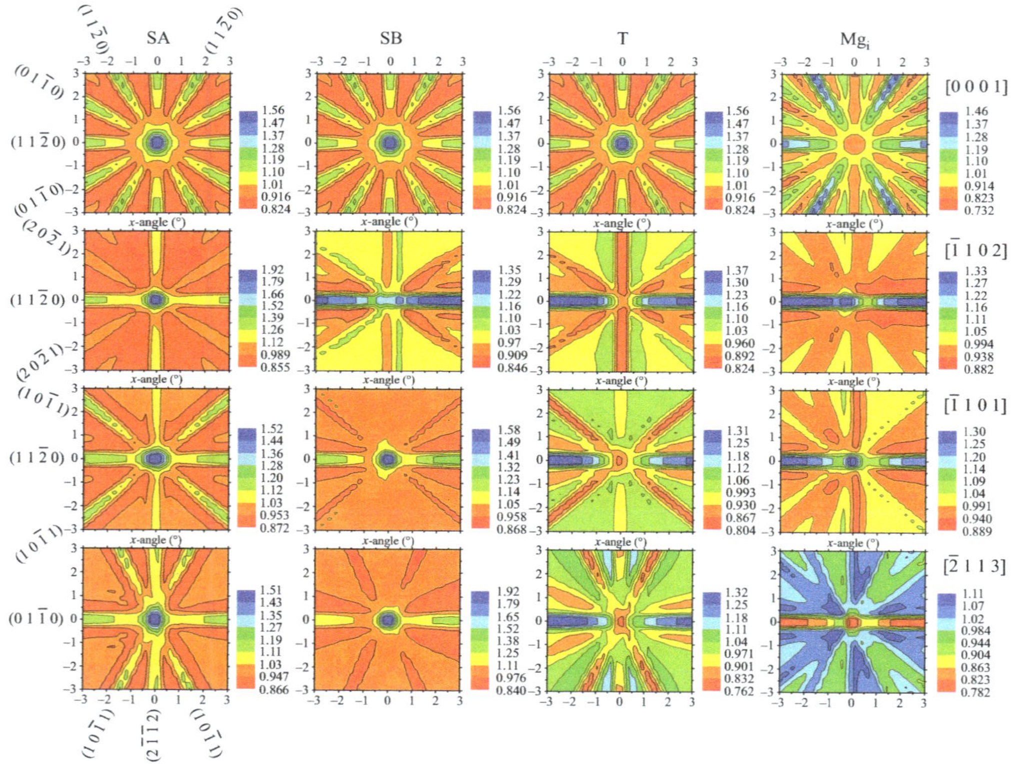

Figure 11.26 *Theoretical β^- emission patterns from ^{27}Mg on substitutional Ga (SA), substitutional N (SB), tetrahedral interstitial T and the Mg$_i$ interstitial sites shown in Figure 11.25 around the $\langle 0\,0\,0\,1\rangle$, $\langle \bar{1}\,1\,0\,2\rangle$, $\langle \bar{1}\,1\,0\,1\rangle$ and $\langle \bar{2}\,1\,1\,3\rangle$ axes of GaN. Note that the SA, SB and T sites are aligned with the* c-axis *atomic rows, so their $\langle 0\,0\,0\,1\rangle$ patterns are identical. On the other hand, patterns for the Mg$_i$ sites can be understood by the fact that these sites are located in the wide open interstitial regions of the lattice, where they are not aligned with $\langle 0\,0\,0\,1\rangle$, $\langle \bar{2}\,1\,1\,3\rangle$ and $(0\,1\,\bar{1}\,0)$, but fully aligned with $(1\,1\,\bar{2}\,0)$ planes. This causes for Mg$_i$ sites blocking along $\langle 0\,0\,0\,1\rangle$, $\langle \bar{2}\,1\,1\,3\rangle$ and $(0\,1\,\bar{1}\,0)$, while $(1\,1\,\bar{2}\,0)$ planes remain fully channeling*

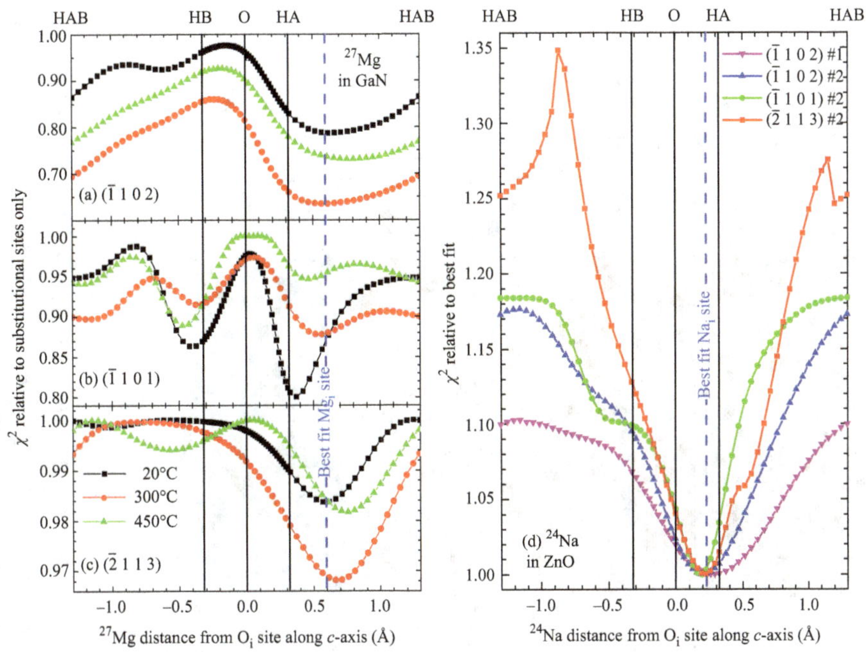

Figure 11.27 *(a–c) Relative χ^2 of the fits to the experimental $\langle\bar{1}\ 1\ 0\ 2\rangle$, $\langle\bar{1}\ 1\ 0\ 1\rangle$ and $\langle\bar{2}\ 1\ 1\ 3\rangle$ patterns of p-GaN:Mg at implantation temperatures of 20°C, 300°C and 450°C as a function of displacement of the ^{27}Mg atoms from the ideal interstitial O sites parallel to the c-axis. Each data point corresponds to a two-site fit where the first site was kept fixed at S_{Ga}. The reduced χ^2 was normalized to the one considering only Mg on S_{Ga} sites. The adopted average value of the Mg_i displacement of $+0.60 \pm 0.14$ Å is shown by the dashed blue line. (d) Relative chi-square of fit χ^2 as a function of displacement of the major ^{24}Na fraction in ZnO from the ideal octahedral interstitial O site along the c-axis, as derived from the fits of the off-c-axis patterns of two samples in the RT as-implanted state. The chi-square of the fit for all three directions was minimized for a position that is shifted by $+0.23 \pm 0.04$ Å from ideal O sites toward the HA position, as shown by the dashed blue line. Note that for Na in ZnO, χ^2 could not be normalized to the one considering only Na on S_{Zn} sites since the majority of Na occupies interstitial sites and hence no meaningful fit that only considers S_{Zn} sites is possible*

and Mg-as-grown) is displayed as a function of implantation temperature. Four major conclusions can be drawn:

1. The interstitial fraction of Mg is generally highest in both Mg-doped samples, lowest in n-type and intermediate in undoped GaN.

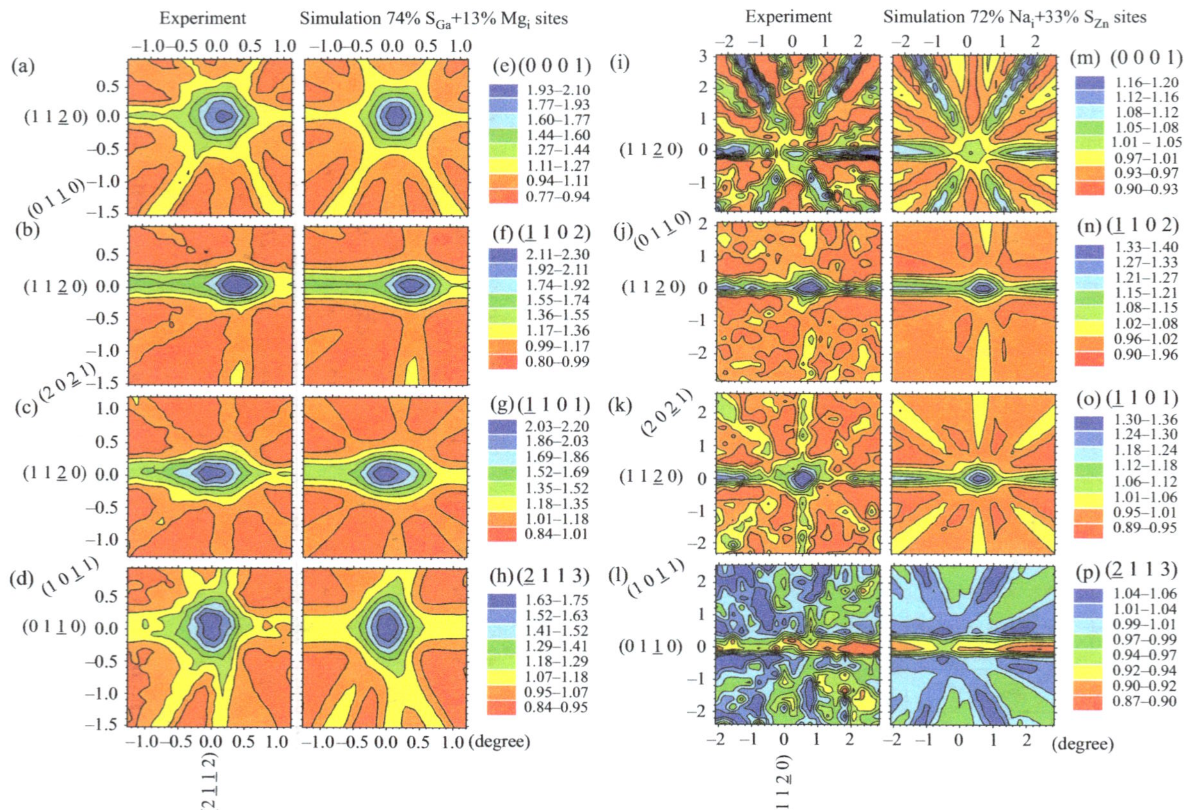

Figure 11.28 *Parts (a)–(d) are experimental β^- emission patterns from ^{27}Mg in p-GaN:Mg in the vicinity of $\langle 0\ 0\ 0\ 1\rangle$, $\langle \bar{1}\ 1\ 0\ 2\rangle$, $\langle \bar{1}\ 1\ 0\ 1\rangle$ and $\langle \bar{2}\ 1\ 1\ 3\rangle$ axes obtained at 20°C; (e)–(h) are best fits of simulated patterns, corresponding to 74% on S_{Ga} and 13% on Mg_i sites. The equivalent patterns for ^{24}Na in ZnO are shown in panels (i)–(p), corresponding to 72% of ^{24}Na on interstitial Na_i and 33% on S_{Zn} sites. Note that ^{27}Mg patterns were measured with a high angular resolution with the detector at a larger distance from the sample, hence their angular range is only half of that of ^{24}Na*

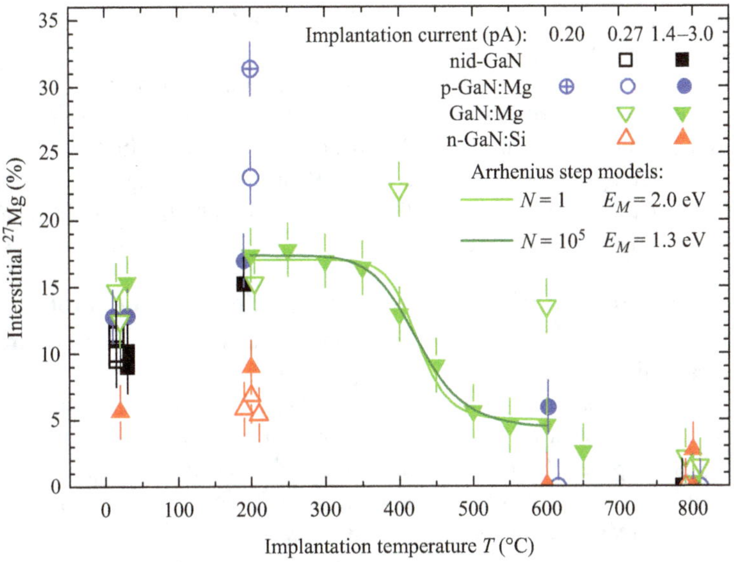

Figure 11.29 Interstitial fractions of ^{27}Mg in four different types of GaN as a function of the implantation temperature. Open symbols were implantations at low beam currents of 0.20–0.27 pA, while filled symbols were measured at 1.4–3.0 pA. The solid lines show the calculated fractions using two Arrhenius models for the site change of interstitial Mg_i to substitutional Ga sites: N is the assumed number of jumps required for Mg_i before encountering a Ga vacancy and E_M is the activation energy for migration of Mg_i. © 2017 by the American Physical Society. Reprinted, with permission, from [41]

2. An increase in implantation current and the resulting accumulation of damage lowers the interstitial Mg fractions in p-GaN but slightly increases them in n-GaN, thus in both cases, changes are directed towards the situation in undoped GaN.
3. Increasing the implantation temperature above RT first results in an increase of the interstitial fraction, which is a consequence of the fact that at elevated temperatures, less Ga vacancies are created, which reduces the substitutional incorporation of Mg.
4. Further increase of the implantation temperature above 350°C lowers the Mg interstitial fraction, which almost completely disappears at 800°C.

This latter effect is attributed to a site change of Mg_i to substitutional Ga sites which is the result of interstitial Mg becoming mobile so that it can diffuse and react with Ga vacancies. By using Arrhenius models for this site change, the activation energy E_M for the migration of interstitial Mg can be estimated to be 1.3–2.0 eV in GaN. Similar site changes were also observed for ^{27}Mg in AlN [75] and ^{24}Na in ZnO [76]; however, the temperature ranges and associated migration energies differ considerably: for ^{27}Mg in GaN at 400°C (E_M = 1.3–2.0 eV), ^{27}Mg in AlN at 350°C (1.1–1.7 eV) and ^{24}Na in ZnO at 100°C (0.8–1.3 eV).

Summarizing, the EC results fully confirmed the amphoteric nature of Mg in GaN and AlN, and of Na in ZnO by identifying the coexistence of cation substitutional and interstitial sites. The location of the interstitial sites could be determined with high accuracy. As a common feature, site changes of interstitial to substitutional cation sites were identified in all cases and attributed to the onset of interstitial migration of Mg and Na.

11.2.2.3 Magnetic dopants in dilute magnetic semiconductors

In a similar approach to electrical or optical doping, magnetic atoms can be introduced into nonmagnetic semiconductors to make them ferromagnetic. In these materials, known as *dilute magnetic semiconductors* (DMS), magnetic dopants randomly substitute a few per cent of the atoms in the host matrix of the nonmagnetic semiconductor. A *magnetic dopant* is, in this context, an element which is not only different from the host's constituents but which also carries a nonvanishing magnetic moment. These magnetic moments may originate, for example, in unfilled 3d or 4f shells of transition metals or REs, respectively. The usefulness of a DMS material system depends on two conditions: (i) by some long-range mechanism, the localized magnetic moments order ferromagnetically and the ferromagnetic state is attained up to typical operation temperatures (room temperature and above); (ii) the charge carriers in the semiconductor are spin-polarized, in order to allow, for example, the control of magnetization by an electric field or the use of these semiconductors as efficient spin injectors. Condition (ii) indirectly poses a third condition on the distribution and structure of the magnetic impurities in the host semiconductor. Efficient spin polarization of the carriers requires that these must be involved in the order mechanism, which is the case for carrier-mediated ferromagnetism. This excludes materials where the ferromagnetic order and the semiconductor bands involved in the conduction are decoupled, which is the case, for example, for materials composed of small ferromagnetic precipitates embedded in a semiconductor matrix. Phase segregation is a critical problem in the context of DMS materials, even more than in other typical doping contexts, since typical concentrations of magnetic dopants are several orders of magnitude higher than those of electric and optical dopants (Sections 11.2.2.1 and 11.2.2.2). Therefore, these effects are introduced in more detail in this section. A generic structural phase diagram of a DMS system is shown in Figure 11.30, consisting of a nonmagnetic compound A_yB_z (e.g., GaAs) doped with a magnetic element X (e.g., Mn). The main coordinates in such a diagram are processing temperature (of growth, implantation or annealing) and the nominal concentration (x) of the transition metal dopant. In the single-phase region, the magnetic atoms substitute one of the host constituents (e.g., element A), maintaining the host's crystal structure, i.e., forming a $A_{y-x}X_xB_z$ alloy (e.g., $Ga_{1-x}Mn_xAs$). The dopants may be randomly distributed (region 1) or, for sufficiently high concentration or temperature, aggregate into impurity-rich regions (region 2), e.g., regions with a local Mn concentration higher than x. For sufficiently high processing temperature or concentration, the magnetic atoms may segregate into small precipitates of $A_yX_xB_z$ secondary phases

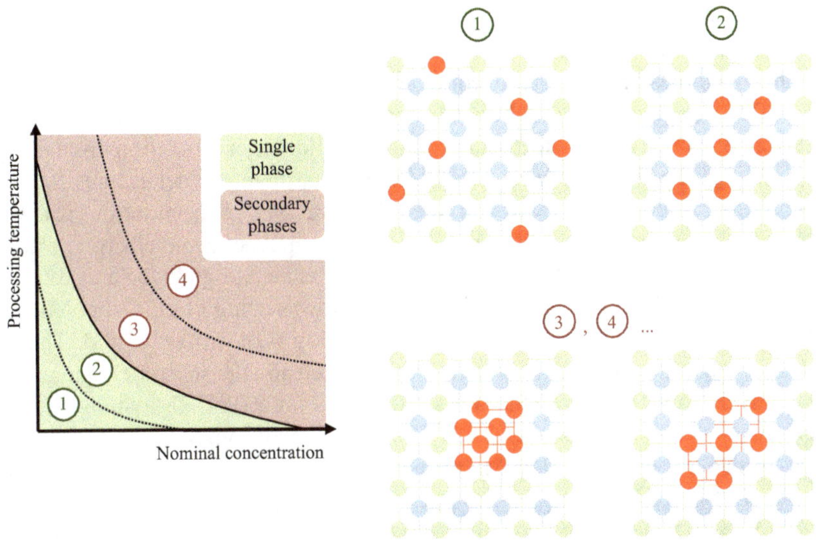

Figure 11.30 Generic structural phase diagram of a dilute magnetic semiconductor. © 2017 IOP Publishing. Reprinted, with permission, from [77]

(e.g., Mn or MnAs) with a structure that is incommensurate with that of the host (regions 3 and 4). This distinction between randomly substituted DMS materials (commonly referred to as *intrinsic* DMS) from those where the ferromagnetism originates from secondary phases embedded in the semiconductor matrix is of fundamental importance. When synthesizing DMS materials, it may happen that the impurities are not randomly incorporated but instead segregate into impurity-rich regions or even in secondary phases (Figure 11.30). These non-DMS regions or phases may themselves be ferromagnetic, and this ferromagnetism is commonly referred to as *not intrinsic*. In this section, we give an overview of channeling studies carried out on the two main types of DMS systems: (i) GaAs and InAs doped with Mn and (ii) wide-gap semiconductors such as ZnO and GaN doped with 3d transition metals. The first are well-established *intrinsic* DMS, although the maximum achieved Curie temperature (T_C) remains well below room temperature (i.e., partially not fulfilling condition (i)). The second are now widely accepted to exhibit ferromagnetic-like properties at room temperature only when phase segregation occurs [77,78], i.e., not fulfilling condition (ii). Channeling techniques played important roles leading to the current understanding of both types of DMS, as discussed in the following.

Mn-doped GaAs and InAs have become model systems for exploring the physics of carrier-mediated ferromagnetism in semiconductors and the associated spintronic phenomena [78–80]. In particular, (Ga,Mn)As (Mn-doped GaAs) is a perfect example of how the magnetic behavior is strongly dependent on local structure. (Ga,Mn)As thin films are typically grown by low-temperature molecular beam epitaxy. In high T_C films, with several per cent Mn, while the majority of the

Mn atoms substitute for Ga (Mn_s), a minority fraction occupies interstitial sites (Mn_i) [81–83]. Mn_s provides both the localized magnetic moment and the itinerant hole that mediates the magnetic coupling, whereas Mn_i has a 2-fold compensating effect: magnetically, as Mn_i–Mn_s pairs and triplets (Figure 11.31) couple antiferromagnetically, and electrically, since double donor Mn_i compensates Mn_s acceptors [82]. For a given Mn_s concentration, Mn_i therefore determines the hole concentration, the Fermi level and the effective Mn_s concentration (of non-compensated Mn_s moments), all of which define the magnetic and electric behavior of (Ga,Mn)As, including the T_C [78,79,82,83]. Ion channeling played a central role in this rich research history of (Ga,Mn)As, since it was using PIXE/C that Yu *et al.* established the presence and role of interstitial Mn in (Ga,Mn)As [81,83]. The fact that Mn is lighter than Ga and As (composing the film and the substrate), and present in a relatively low concentration, raises difficulties for RBS/C in a similar way to the examples shown in Section 11.2.2.2. The weak signal from Mn overlaps with the signals from Ga and As and thus Mn cannot be independently probed. PIXE, on the other hand, is based on detecting characteristic X-rays [cf. Figure 11.6(b)]. The Mn K_α line (5.9 keV), even with a low intensity due to the small amount of Mn (few per cent within a thin film), is well separated from the L (<2 keV) and K (>9 keV) lines of Ga and As, considering a typical resolution better than 200 eV (below 10 keV) for conventional X-ray detectors (e.g., Si drift detectors or high-purity Ge detectors). By measuring and comparing the minimum yield for Mn (χ_{Mn}) with PIXE/C and for GaAs (χ_{GaAs}) with RBS/C, Yu *et al.* estimated that approximately 17% of the Mn atoms in a $Ga_{0.80}Mn_{0.20}As$ thin film-occupied tetrahedral interstitial (T) sites in the as-grown state. This estimate was based on measurements for various crystal axes ($\langle 1\ 0\ 0\rangle$, $\langle 1\ 1\ 1\rangle$ and $\langle 1\ 1\ 0\rangle$), considering that the T interstitial is shadowed by the host Ga and As atom along $\langle 1\ 0\ 0\rangle$ and $\langle 1\ 1\ 1\rangle$, but exposed along the $\langle 1\ 1\ 0\rangle$ channel, and assuming a flux peaking effect of 1.5 for the $\langle 1\ 1\ 0\rangle$ direction [81]. More recently, similar experiments were carried out by a different group, using RBS/C instead of PIXE/C [85]. Despite the overlap between the Mn and GaAs signals, it was possible to determine the Mn contribution to the minimum yield by fitting the channeled spectrum in the vicinity of the Mn peak in the (Ga,Mn)As, assuming that the GaAs contribution follows a third-order polynomial. Assuming no flux peaking effect, the authors estimated an interstitial fraction of 30%. The discrepancy between the two results—17% in [81] versus 30% in [85]—for the same composition ($Ga_{0.80}Mn_{0.20}As$) can originate from different growth conditions, including different film thicknesses (110 versus 20 nm). Nevertheless, it is likely that the assumptions regarding flux peaking effects (1.5 versus 1) play a role as well, since flux peaking is only expected to have a significant effect above 100 nm [85]. Ultimately, the highest accuracy can only be reached by analyzing the experimental data with the support of channeling simulations. Moreover, measuring the full angular dependence in the vicinity of the channeling axis (i.e., a 2D pattern), instead of the conventional 1D scans, and numerically fitting such patterns to simulations (i.e., instead of just comparing minimum yield values), provides additional control over possible sources of error. This approach (2D patterns and fitting to simulations) is uncommon in the context of ion channeling (RBS/C and PIXE/C) but has become standard for EC (see

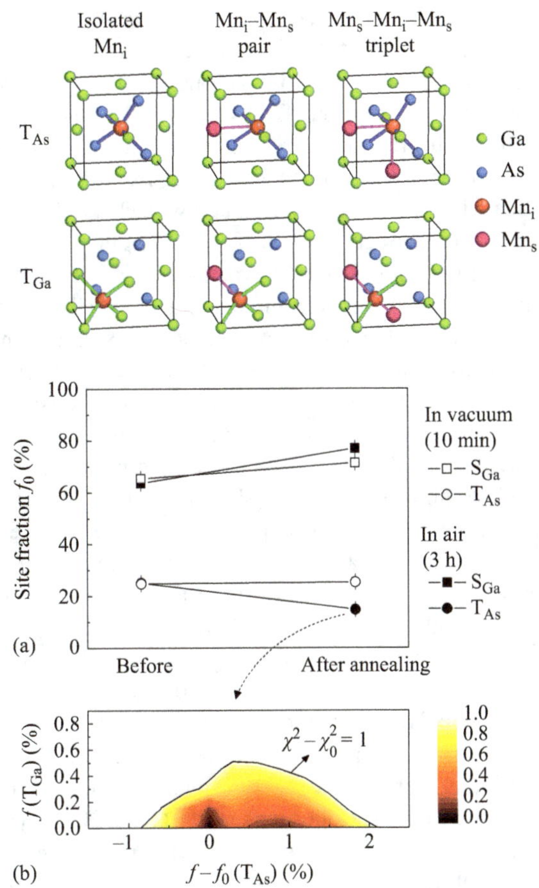

Figure 11.31 *(Top) Representation of the relevant Mn_i sites and coordination in GaAs: T_{As} (top row) and T_{Ga} (bottom row) sites, for isolated Mn_i, Mn_i–Mn_s pairs and Mn_s–Mn_i–Mn_s triplets. (Bottom) (a) Fractions of ^{56}Mn atoms on S_{Ga} and T_{As} sites in (Ga,Mn)As obtained from emission channeling measurements before and after thermal annealing at 200°C. (b) Error associated with the fit: $\chi^2 - \chi_0^2$, with χ_0^2 being the value for the best fit and χ^2 the value obtained as the fractions f on T_{As} and T_{Ga} sites varied in the vicinity of the best fit values f_0 (15% for T_{As} and 0% for T_{Ga}), leaving all other fit parameters free. The black line contour defines the uncertainty from the fit, given by $\chi^2 - \chi_0^2 = 1$, corresponding to $f(T_{As}) = (15 \pm 2)\%$ and $f(T_{Ga}) = (0.0 \pm 0.5)\%$.*

Section 11.1.2.3). In particular, in the context of (Ga,Mn)As, this approach allowed to unambiguously identify which of the T sites are occupied by interstitial Mn: the tetrahedral interstitial sites coordinated by As atoms (Figure 11.31) [84]. These results are also consistent with previous EC experiments on Mn-implanted GaAs in the ultra-dilute (<0.05% Mn) regime [86,87]. This approach (2D patterns and fitting to simulations) is so powerful that it even allowed to exclude the occupation of the other T site (coordinated by Ga atoms, T_{Ga}) to below 0.5% of all Mn (Figure 11.31), contradicting previous reports of comparable T_{As} and T_{Ga} occupancy based on X-ray scattering techniques [84]. An additional advantage of EC, and particularly in this context, is the short measuring time required per pattern (few minutes, currently limited by the maximum count rate of the detector). This measuring speed makes EC an ideal technique for dynamic measurements, e.g., as a function of annealing temperature. In the context of (Ga,Mn)As, this has recently allowed to study the site-specific thermal stability of substitutional and interstitial Mn [88],[1] closing an important gap in the understanding of self-compensation mechanisms in (Ga,Mn)As. It was relatively well established that, for Mn concentrations in the few per cent regime, interstitial Mn out-diffuses and is passivated at the surface during thermal annealing in the 160°C–200°C range, resulting in a significant increase in T_C [89–91]. The highest T_C (188 K) was achieved for a Mn concentration of 12% after an activating thermal annealing in air at 160°C [90,91]. Annealing above 160°C (up to 220°C) still increases T_C compared to the as-grown state, but the improvement is lower than for 160°C [89], which can be interpreted as due to the onset of segregation of substitutional Mn. However, these temperatures are far below the well-established regime of secondary phase formation (>400°C). The earlier PIXE/C results had indicated that part of the substitutional Mn is converted to a non-substitutional (random) component upon annealing at 282°C [81]. The recent EC studies on (Ga,Mn)As with 5% Mn [88] showed that, indeed, although Mn_i diffuses at slightly lower temperatures than Mn_s, segregation of substitutional Mn does occur already at around 200°C, thus explaining why the optimal annealing temperature (in the few per cent Mn regime) is so narrow. These results finally established that the complex dependence of T_C on annealing temperature in the vicinity of 200°C is a consequence of the competition between diffusion of interstitial Mn and segregation of substitutional Mn.

Motivated to some extent by the inability to increase T_C of (Ga,Mn)As to room temperature, other types of DMS systems have been investigated. Since the first report of high-temperature ferromagnetism in co-doped TiO_2 [92] and the prediction by Dietl *et al.* that highly p-type Mn-doped ZnO and GaN could reach high T_C [93], room temperature ferromagnetism has been reported in a rapidly growing

[1]Note that the site fractions determined using emission channeling correspond to the implanted radioactive ^{56}Mn probes, which are not necessarily (and most likely are not) the same as those of the stable ^{55}Mn atoms incorporated during MBE growth. Nevertheless, the migration and diffusion behavior does not depend on the isotope nor on the incorporation kinetics (as long as the effect of implantation-induced defects can be neglected). Thus, although the absolute site fractions for ^{56}Mn and ^{55}Mn may differ, the thermally induced changes are accurately probed by ^{56}Mn.

number of wide-gap DMS materials [77,78]. However, thanks to the continuous development of growth and doping techniques and of advanced characterization techniques, it has become established that for the most representative of these systems (e.g., GaN doped with Mn and ZnO doped with Fe and Co), *intrinsic* ferromagnetism only emerges at low temperature (below 20 K), if at all [77,78]. The often-reported room temperature ferromagnetism is *not intrinsic*, i.e., it originates from secondary phases (Figure 11.30) and other artifacts [77,78]. Although less prominent than with (Ga,Mn)As, channeling techniques also played an important role in the context of wide-gap DMS. In addition to routine applications of RBS/C, e.g., to characterize ion damage in systems based on ion implantation of the transition metals dopants [94–96], some notable results have been reported on the lattice location of the transition metal dopants [94,97–100].

For example, based on electron EC measurements (Figure 11.32), it was shown that for Fe-implanted ZnO, even for relatively high Fe concentration (fluence), up to 90% of the Fe atoms occupy Zn sites in well-ordered ZnO, with the remainder in disordered (*random*) sites [94]. In particular, it was observed that, as-implanted, the Fe atoms exhibit a small displacement from the ideal substitutional site (observed as an increase in root-mean-square displacement of up to 0.04 Å), indicating the formation of defect complexes between Fe atoms and neighboring point defects. Upon annealing at 700°C, the Fe atoms exhibit a root-mean-square displacement around 0.08 Å, similar to the Zn atoms in the ZnO lattice, indicating that the defect complexes were annealed. Moreover, in this regime of *ideal* Zn substitution (upon thermal annealing), for sufficiently high concentrations (fluence), magnetometry measurements showed signs of Fe segregation into superparamagnetic inclusions. Such a regime of Fe segregation while maintaining *ideal* Zn substitution corresponds to region 2 in Figure 11.30. Although channeling techniques are not suited to probe incommensurate secondary phases (regions 3 and 4), it is a powerful technique to combine with others when dilute and segregated phases may coexist. For example, in Fe-implanted $SrTiO_3$, dilute Fe (region 1) was shown to coexist with segregated Fe clusters (region 3) [101]. The dilute component was characterized using EC as Fe substituting for Ti, whereas the clusters were characterized as γ-Fe nanoparticles of a few nm diameter, using synchrotron X-ray diffraction and TEM. This detailed and unambiguous characterization of both dilute and segregated components allowed then for additional characterization and quantification using Mössbauer spectroscopy and extended X-ray absorption fine structure (EXAFS), which would have been ambiguous without the input from EC and the other techniques. In general, the combination of channeling and EXAFS techniques is particularly attractive, as it provides a complete characterization of the local structure: channeling probes the exact position of the dopant atom with respect to the host lattice (e.g., substitutional versus interstitial), while EXAFS probes the configuration of the nearest neighbors (e.g., local relaxation effects or the formation of dopant-vacancy complexes).

Another notable example of the application of EC to wide-gap DMS is the observation of coexistence of cation and anion substitution by implanted transition metals in ZnO and GaN: Mn in both Ga and N sites in GaN [98] and Mn, Co and Ni

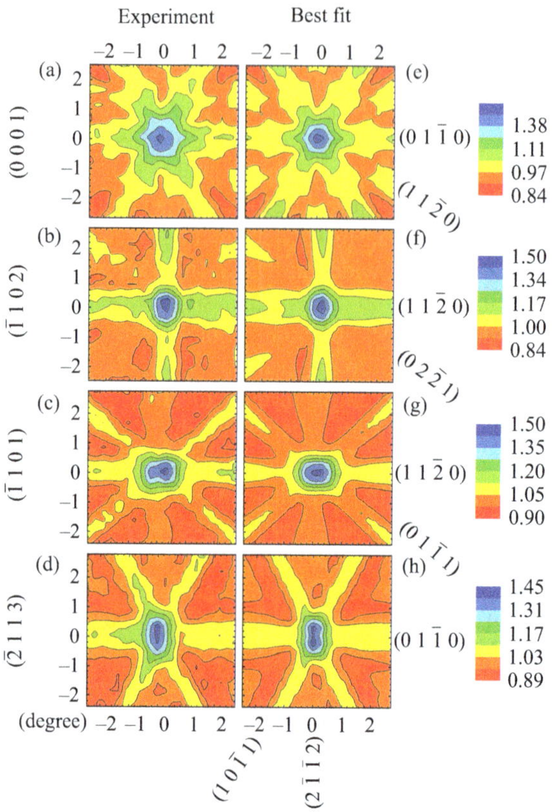

Figure 11.32 Emission channeling measurements and best fit for ^{59}Fe-implanted ZnO: (a)–(d) experimental emission patterns along the $\langle 0\ 0\ 0\ 1\rangle$, $\langle \bar{1}\ 1\ 0\ 2\rangle$, $\langle \bar{1}\ 1\ 0\ 1\rangle$ and $\langle \bar{2}\ 1\ 1\ 3\rangle$ directions; (e)–(h) corresponding best fits of theoretical patterns, obtained by varying the fraction of Fe on substitutional Zn sites (best fit for a fraction of 87(5)%) and varying its root-mean-square displacement u_1 (best fit for $u_1 = 0.08(1)$ Å). © 2013 IOP Publishing. Reprinted, with permission, from [94]

in both Zn and O sites in ZnO [97,99,100]. The observed anion substitution (N sites in GaN and O sites in ZnO), although a minority fraction (below 30%), goes against the expectation that 3d transition metals occupy only cation-substitutional sites, based on the fact that, chemically, 3d transition metals are more closely related to the post-transition metals Ga and Zn than to N and O. Figure 11.33 (top) illustrates how these two components can be distinguished when fitting EC data, for the case of ZnO implanted with ^{61}Co and ^{56}Mn. Each data point corresponds to the reduced χ^2 obtained for the best fit using two given sites, with the corresponding two fractions as free parameters. The site pairs are composed of Zn-substitutional (S_{Zn}) plus each of the simulated sites along the *c*-axis: the O-substitutional (S_O) and the tetrahedral (T) sites, the bond centered and antibonding sites along the *c*-axis, and

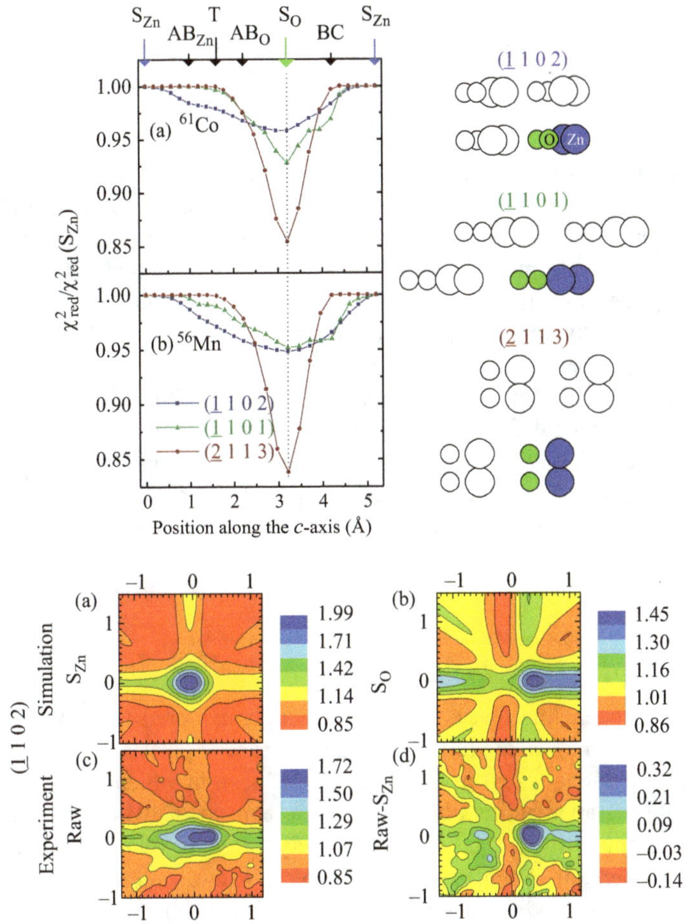

Figure 11.33 (Top) Reduced χ^2 of the fits to emission channeling data for ZnO implanted with ^{61}Co and ^{56}Mn, in the vicinity of the $\langle \bar{1}\ 1\ 0\ 2 \rangle$, $\langle \bar{1}\ 1\ 0\ 1 \rangle$ and $\langle \bar{2}\ 1\ 1\ 3 \rangle$ crystal axes. The projection of the ZnO lattice onto the plane perpendicular to each direction is also shown. (Bottom) Comparison of the experimental and theoretical $\langle \bar{1}\ 1\ 0\ 2 \rangle$ patterns for ^{56}Mn-implanted ZnO: calculated patterns for 100% ^{56}Mn probes on S_{Zn} (a) and S_O (b) sites; (c) normalized experimental pattern; (d) pattern resulting from subtracting the fitted S_{Zn} component from the experimental pattern. © 2011 by the American Physical Society. Reprinted, with permission, from [97]

various intermediate positions. The reduced χ^2 (y-axis) of these two-site fits is normalized to that of the one-site (S_{Zn}) fit. The x-axis corresponds to the position (along the c-axis) of the non-S_{Zn} site used in each fit. The projection of the ZnO lattice, onto the plane perpendicular to each direction, shows that the separation between Zn and O atom rows (sublattices) is maximized for $\langle\bar{2}\ 1\ 1\ 3\rangle$, which is consistent with the observed higher sensitivity of the χ^2 for that direction. This highly unexpected result illustrates how powerful channeling techniques can be when 2D patterns are measured and numerically fitted to accurate simulations. In simpler cases, such as interstitial Mn in GaAs, minority sites may be inferred from the conventional 1D scans, even without comparison to simulations (e.g., if the minority interstitial site is not shadowed by a substitutional row along certain axes). However, even in such simpler cases, the amount (and reliability) of information that can be extracted from 2D measurements and simulations is significantly larger. Moreover, in the case of mixed cation and anion substitution in ZnO and GaN, it would be virtually impossible to convincingly identify and quantify the dual occupancy simply based on 1D scans (i.e., from minimum yield and critical angle values). With the 2D approach, the minority anion substitution is not only established by the numerical fit, it is even evident from the visual inspection of the patterns (Figure 11.33, bottom): comparing patterns (a), (b) and (d) provides a clear visual identification of the S_O occupancy, i.e., pattern (d) reproduces the features of the S_O pattern (b) and is completely distinct from the S_{Zn} pattern (a).

11.3 Outlook

11.3.1 Technical developments

One aspect of channeling that is being increasingly explored is analysis based on 2D measurements, i.e., on the full angular dependence (e.g., as a function of two tilt angles with orthogonal axes) in contrast to a 1D scan in which only one angle is varied. While the approach is ubiquitous in the context of EC (Section 11.1.2.3), 2D measurements have so far been relatively rare for other IBA techniques. Notable examples of recent applications of 2D measurements include RBS [20,102,103] and its medium-energy counterpart, MEIS (medium energy ion scattering) [104–110]. 2D RBS patterns can be recorded in the form of "image scans" using conventional channeling setups [19,20] or with position-sensitive detectors in blocking geometry [102,103]. Using 2D patterns has the two big advantages that besides axial effects also planar directions are included and that errors in properly aligning the sample (which are common in case of measuring 1D scans only, where often the azimuthal orientation is not well enough characterized) are completely eliminated. Analyzing 2D patterns hence offers considerably increased reliability and statistical significance when comparing experimental and simulated results. Applications of 2D patterns in MEIS-channeling are also emerging, in particular for strain depth-profiling with nm and even sub-nm resolution [107–109]. Computer simulation of channeling behavior has also been a subject of recent developments, with the use of molecular dynamics (MD) calculations [111,112]. So far, calculation and simulation of channeling effects have relied on analytical theories and atomistic computer simulations of relatively simple crystal structures (perfect crystals or crystals with defects that preserve a large degree of periodicity). MD simulations are now emerging as a powerful alternative, in particular, for cases of more arbitrary atomic structure, such as extended defects and highly disordered structures [112].

Another IBA aspect that has been through major development is high-resolution PIXE, either using wavelength dispersive X-ray spectrometers (WDS) [113,114] or transition-edge sensors (TES) [115,116]. Compared to conventional energy dispersive spectrometers (EDSs), i.e., Si and Ge detectors with energy resolution in the 100–200-eV range for few keV X-rays, WDS and TES detectors can reach resolutions of few eV, or even sub-eV for WDS [113]. While these developments have mostly targeted the improvement in detection limits for compositional analysis and even for chemical state identification, they also create new opportunities for PIXE channeling. Channeling techniques such as PIXE channeling, in which the channeled particles are impinging ions (the beam), require that measurements are carried out as a function of sample orientation. A typical 2D channeling pattern in the vicinity of a major crystal axis is composed of at least $20 \times 20 = 400$ individually measured spectra. This need for hundreds of individually measured spectra poses even stronger constraints (on detection limits or measuring time) for channeling measurements compared to conventional elemental analysis. Since lower energy lines are significantly more intense (e.g., L compared to K lines), and resolutions of the order of few eV allow to distinguish low-energy (few keV) X-ray lines from different elements (which would otherwise overlap with EDS detection), the use of WDS and TES detectors would greatly increase the range of materials that can be studied with PIXE channeling, including low concentration dopants in semiconductors.

11.3.2 Emerging applications

While IBA channeling methods will remain playing a relevant role in defect characterization of more conventional semiconductors, various related classes of materials and phenomena are emerging which can particularly benefit from the capabilities of these techniques. In the following, we discuss two topical examples: defects in diamond for quantum technologies and magnetic doping of topological insulators for spintronics and topological quantum computation.

Point defects in diamond are being intensively investigated for their applications in processing and communication of quantum information, as well as for metrology. So far, the negatively charged nitrogen-vacancy center (NV^-) has been the most studied defect [117]. Thanks to its efficient optical spin polarization and spin-state dependent fluorescence, it is being exploited, for example, in the context of high-sensitivity magnetometers [118]. More recently, owing to their superior optical properties, the group IV-vacancy centers have emerged as the leading type of point defects for quantum computing and networking applications. Most of the research has focused on the negatively charged centers: SiV^- [119,120], GeV^- [121], SnV^- [122], PbV^- [123]. Recent work has also addressed the neutrally charged SiV^0 defect, which is more closely related to the NV^- center in terms of electronic structure and behavior [124,125]. Although largely unexploited, IBA channeling techniques are well suited for the characterization of these defects. For example, it is generally accepted that, whereas the N atom in the NV^- center occupies a substitutional C site, the group IV atoms in group IV-vacancy centers occupy bond-centered sites, in a split-vacancy configuration. However, experimentally, these structural configurations have only been indirectly established. A detailed, direct characterization of the structure of these defects is especially important for the field, since the superior properties of the group IV-vacancy centers are to a large extent a consequence of the inversion symmetry of the split-vacancy configuration. EC is particularly well suited to characterize and quantify the various possible defect configurations (lattice sites), as well as the changes induced upon thermal annealing (activation of the quantum defects). In particular, the doping with group IV elements is generally performed using ion implantation, i.e., compatible with the current implementation of EC at ISOLDE-CERN. Moreover, the fluence regime is typically of the order of 10^{12} cm^{-2} or lower, making EC better suited than RBS or PIXE channeling, and suitable EC probes could be ^{31}Si (β^-, $t_{1/2} = 157$ min); ^{73}Ge from ^{73}As (CE, 80 day); ^{121}Sn (β^-, 27 h); ^{209}Pb (β^-, 3.2 h).

Phenomena emerging from nontrivial band-structure topology in solids have become a topical subject in condensed matter physics. The traditional classification of materials as insulators (with a large band gap), semiconductors (small band gap) or metals (gapless) has been a useful concept for nearly a century, explaining a wide range of phenomena based on the existence and size of a band gap. In recent years, however, the topological character of the band structure in both gapped and gapless systems has been shown to encompass an even wider and unexplored landscape for intriguing new physics and novel electronic functionalities. Among a wealth of new physics, several classes of materials and phenomena have emerged which are essentially newly discovered states of quantum matter. Topological insulators, such as Bi_2Se_3, Bi_2Te_3, Sb_2Te_3 and their alloys, are a notable example of such materials, exhibiting topologically protected 2D surface states with Dirac cones as energy momentum dispersion [126]. One of the most striking discoveries associated with these materials is the realization of the quantum anomalous Hall (QAH) effect, induced by doping with transition metals (e.g., Cr or V) [127,128]. These quantum anomalous Hall insulators (QAHI) possess a chiral edge state associated with dissipationless, spin-polarized current on the edge of a thin QAHI film. In addition, QAH insulators are a key building block in device structures envisioned for topological quantum computation [129]. As for any doping case, and for magnetic doping in particular, the effect of the dopants on the electronic structure and functionality of the doped material strongly depends on the lattice site(s) occupied by the dopant atoms, and if the doping results in single- or secondary-phase formation (see Section 11.2.2.3 for the comparable case of DMS). Topological insulators such as Bi_2Se_3 have a complex crystal structure with a large hexagonal unit cell consisting of three quintuple layers (QLs) separated by van der Waals (vdW) gaps. Therefore, numerous nonequivalent lattice sites exist for the incorporation of the transition metal dopants. These include substitutional incorporation in cation or anion sites as well as interstitial incorporation within the QLs or in the vdW gap. There is, therefore, tremendous unexplored potential for IBA channeling techniques in this context, and these materials have been shown to be robust against beam damage during IBA [130]. RBS/C can in principle be used, as long as the QAHI film is sufficiently thin and the substrate is composed of light elements (lighter than the dopant) so that the energy of the ions that are backscattered from the transition metal dopants does not overlap with that from the other elements in the film and substrate. More generally, PIXE channeling is a suitable alternative.

References

[1] Alford T.L., Feldman L.C., Mayer J.W. Fundamentals of nanoscale film analysis. New York: Springer Science & Business Media; 2007.

[2] Assmann W., Huber H., Steinhausen C., Dobler M., Glückler H., Weidinger A. 'Elastic recoil detection analysis with heavy ions'. *Nuclear Instruments and Methods in Physics Research Section B: Beam Interactions with Materials and Atoms*. 1994;**89**(1–4):131–9.

[3] Vizkelethy G. 'Nuclear reaction analysis: Particle-particle reactions' in Tesmer J.R., Nastasi M. (eds.) *Handbook of modern ion beam materials analysis*. Pittsburgh: Materials Research Society; 1995. pp. 139–65.

[4] Ziegler J.F., Biersack, J.P., Ziegler M.D. *SRIM, the stopping and range of ions in matter*. Morrisville: Lulu Press; 2009.

[5] Barradas N.P., Marriott P.K., Jeynes C., Webb R.P. 'The RBS data furnace: Simulated annealing'. *Nuclear Instruments and Methods in Physics Research Section B: Beam Interactions with Materials and Atoms*. 1998;**136**:1157–62.

[6] Davies J.A., Sims G.A. 'The range of Na^{24} ions of kiloelectronvolt energies in aluminum'. *Canadian Journal of Chemistry*. 1961;**39**:601–10.

[7] Robinson M.T., Oen O.S. 'The channeling of energetic atoms in crystal lattices'. *Applied Physics Letters*. 1963;**2**(2):30–2.

[8] Piercy G.R., McCargo M., Brown F., Davies J.A. 'Experimental evidence for the increase of heavy ion ranges by channeling in crystalline structure'. *Physical Review Letters*. 1963;**10**(9):399–400.

[9] Nelson R.S., Thompson M.W. 'The penetration of energetic ions through the open channels in a crystal lattice'. *Philosophical Magazine*. 1963;**8**(94): 1677–90.

[10] Lutz H., Sizmann R. 'Super ranges of fast ion in copper single crystals'. *Physics Letters*. 1963;**5**(2):113–14.

[11] Lindhard J. 'Influence of crystal lattice on motion of energetic charged particles'. *Matematisk-fysiske Meddelelser det Kongelike Danske Videnskabernes Selskab*. 1965;**34**(11):1–63.

[12] Lehmann C., Leibfried G. 'Long-Range Channeling Effects in Irradiated Crystals'. *Journal of Applied Physics*. 1963;**34**:2821–36.

[13] Domeij B., Björkqvist K. 'Anisotropic emission of α-particles from a monocrystalline source'. *Physics Letters*. 1965;**14**(2):127–8.

[14] Chu W.K., Mayer J.W., Nicolet M.A. Backscattering spectroscopy. New York: Academic Press; 1978.

[15] Feldman L.C., Mayer J.W., Picraux S.T. *Materials analysis by ion channeling – Submicron crystallography*. New York: Academic Press; 1982.

[16] Lindhard J. 'Motion of swift charged particles as influenced by strings of atoms in crystals'. *Physics Letters*. 1964;**12**(2):126–8.

[17] Barrett J.H. 'Monte Carlo channeling calculations'. *Physical Review B*. 1971;**3**(5):1527–47.

[18] Swanson M. 'Channeling' in Tesmer J.R., Nastasi M. (eds.) *Handbook of modern ion beam materials analysis*. Pittsburgh: Materials Research Society; 1995. pp. 231–300.

[19] Wu M.F., Vantomme A., De Wachter J., *et al*. 'Comprehensive Rutherford backscattering and channeling study of ion-beam-synthesized $ErSi_{1.7}$ layers'. *Journal of Applied Physics*. 1996;**79**(9):6920–5.

[20] Wendler E., Becker G., Rensberg J., Schmidt E., Wolf S., Wesch W. 'Direction-dependent RBS channelling studies in ion implanted $LiNbO_3$'. *Nuclear Instruments and Methods in Physics Research Section B: Beam Interactions with Materials and Atoms*. 2016;**379**:195–9.

[21] Wu M.F., Vantomme A., Langouche G., Vanderstraeten H., Bruynseraede Y. 'Strain and orientation in epitaxial $CoSi_2$(111) layers formed by ion implantation'. *Nuclear Instruments and Methods in Physics Research B*. 1991;**54**(4) 444–52.

[22] Decoster S., Vantomme A. 'Implantation-induced damage in Ge: Strain and disorder profiles during defect accumulation and recovery'. *Journal of Physics D: Applied Physics*. 2009;**42**:165404.

[23] Bøgh E. 'Rutherford scattering of protons in the surface layers of a tungsten single crystal'. *Physical Review Letters*. 1967;**19**(2):61–3.

[24] Gärtner K. 'Modified master equation approach of axial dechanneling in perfect compound crystals'. *Nuclear Instruments and Methods in Physics Research B*. 2005;**227**:522–30.

[25] Turos A., Jozwik P., Wojcik M., Gaca J., Ratajczak R., Stonert A. 'Mechanism of damage buildup in ion bombarded ZnO'. *Acta Materialia*. 2017;**134**:249–56.

[26] Turos A., Jozwik P., Nowicki L., Sathish N. 'Ion channeling study of defects in compound crystals using Monte Carlo simulations'. *Nuclear Instruments and Methods in Physics Research B*. 2014;**332**:50–5.

[27] Vantomme A. '50 years of ion channeling in materials science'. *Nuclear Instruments and Methods in Physics Research Section B: Beam Interactions with Materials and Atoms*. 2016;**371**:12–26.

[28] Andersen J.U., Andreasen O., Davies J.A., Uggerhøj E. 'The use of channeling-effect techniques to locate interstitial foreign atoms in silicon'. *Radiation Effects*. 1971;**7**(1–2):25–34.

[29] Smulders P.J.M., Boerma D.O. 'Computer simulation of channeling in single crystals'. *Nuclear Instruments and Methods in Physics Research B*. 1987;**29**(3): 471–89.

[30] Kling A. 'CASSIS – A new Monte-Carlo computer program for channelling simulation of RBS, NRA and PIXE'. *Nuclear Instruments and Methods in Physics Research B*. 1995;**102**(1–4):141–4.

[31] Kling A. 'Refurbishment of the CASSIS code for channeling simulations'. *Nuclear Instruments and Methods in Physics Research B*. 2012;**273**:88–90.

[32] Nowicki L., Turos A., Ratajczak R., Stonert A., Garrido F. 'Modern analysis of ion channeling data by Monte Carlo simulations'. *Nuclear Instruments and Methods in Physics Research B*. 2005;**240**:277–82.

[33] Hofsäss H., Lindner G. 'Emission channeling and blocking'. *Physics Reports*. 1991;**201**(3):121–83.

[34] Winter S., Hofsäss H., Jahn S.G., Lindner G., Wahl U., Recknagel E. 'Emission channeling' in Langouche G. (ed.) *Hyperfine interaction of defects in semiconductors*. Amsterdam: Elsevier; 1992. pp. 157–85.

[35] Hofsäss H., Winter S., Jahn S.G., Wahl U., Recknagel E. 'Emission channeling studies in semiconductors'. *Nuclear Instruments and Methods in Physics Research B*. 1992;**63**(1–2):83–90.

[36] Hofsäss H., Wahl U., Jahn S.G. 'Impurity lattice location and recovery of structural defects in semiconductors studied by emission channeling'. *Hyperfine Interactions*. 1994;**84**(1):27–41.

[37] Hofsäss H. 'Emission channeling'. *Hyperfine Interactions*. 1996;**97–98**(1): 247–83.

[38] Wahl U., ISOLDE Collaboration. 'Emission channeling studies of Li in semiconductors'. *Physics Reports*. 1997;**280**:145–285.

[39] Wahl U. 'Advances in electron emission channeling measurements in semiconductors'. *Hyperfine Interactions*. 2000;**129**(1–4):349–70.

[40] Wahl U., Correia J.G., Rita E., *et al.*, ISOLDE Collaboration. 'Recent emission channeling studies in wide band gap semiconductors'. *Hyperfine Interactions*. 2005;**158**(1–4):363–72.

[41] Wahl U., Amorim L.M., Augustyns V., *et al.* 'Lattice location of Mg in GaN: A fresh look at doping limitations'. *Physical Review Letters*. 2017;**118**:095501.

[42] Wahl U., Correia J.G., Cardoso S., Marques J.G., Vantomme A., Langouche G., ISOLDE Collaboration. 'Electron emission channeling with position-sensitive detectors'. *Nuclear Instruments and Methods in Physics Research B*. 1998;**136–138**:744–50.

[43] Wahl U., Correia J.G., Czermak A., *et al.* ISOLDE Collaboration. 'Position-sensitive Si pad detectors for electron emission channeling experiments'. *Nuclear Instruments and Methods in Physics Research A*. 2004;**524**:245–56.

[44] Silva M.R., Wahl U., Correia J.G., Amorim L.M., Pereira L.M.C. 'A versatile apparatus for on-line emission channeling experiments'. *Review of Scientific Instruments*. 2013;**84**:073506.

[45] Catherall R., Andreazza W., Breitenfeldt M., *et al.* 'The ISOLDE facility'. *Journal of Physics G: Nuclear and Particle Physics*. 2017;**44**:094002.

[46] Kucheyev S.O., Williams J.S., Jagadish C., Zou J., Li G., Titov A.I. 'Effect of ion species on the accumulation of ion-beam damage in GaN'. *Physical Review B*. 2001;**64**:035202.

[47] Wendler E., Kamarou A., Alves E., Gärtner K., Wesch W. 'Three-step amorphisation process in ion-implanted GaN at 15 K'. *Nuclear Instruments and Methods in Physics Research B*. 2003;**206**:1028–32.

[48] Vantomme A., Hogg S.M., Wu M.F., *et al.* 'Suppression of rare-earth implantation-induced damage in GaN'. *Nuclear Instruments and Methods in Physics Research B*. 2001;**175–177**:148–53.
[49] Kucheyev S.O., Williams J.S., Jagadish C., Zou J., Li G. 'Damage buildup in GaN under ion bombardment'. *Physical Review B*. 2000;**62**(11):7510–22.
[50] Kucheyev S.O., Williams J.S., Jagadish C., Li G., Pearton S.J. 'Strong surface disorder and loss of N produced by ion bombardment of GaN'. *Applied Physics Letters*. 2000;**76**(26)3899–901.
[51] Titov A.I., Karaseov P.A., Kataev A.Y., Azarov A.Y., Kucheyev S.O. 'Model for radiation damage buildup in GaN'. *Nuclear Instruments and Methods in Physics Research B*. 2012;**277**:80–3.
[52] Ishimaru M., Zhang Y., Weber W.J. 'Ion-beam-induced chemical disorder in GaN'. *Journal of Applied Physics*. 2009;**106**:053513.
[53] Lorenz K., Wahl U., Alves E., *et al.* 'High temperature annealing of rare earth implanted GaN films: Structural and optical properties'. *Optical Materials*. 2006:**28**:750–8.
[54] Lorenz K., Wahl U., Alves E., *et al.* 'High-temperature annealing and optical activation of Eu-implanted GaN'. *Applied Physics Letters*. 2004;**85**(14): 2712–14.
[55] Charnvanichborikarn S., Myers M.T., Shao L., Kucheyev S.O. 'Interface-mediated suppression of radiation damage in GaN'. *Scripta Materialia*. 2012;**67**(2):205–8.
[56] Wendler E. 'Mechanisms of damage formation in semiconductors'. *Nuclear Instruments and Methods in Physics Research B*. 2009;**267**:2680–9.
[57] Lorenz K., Barradas N.P., Alves E., *et al.* 'Structural and optical characterization of Eu-implanted GaN'. *Journal of Physics D: Applied Physics*. 2009;**42**:165103.
[58] Pipeleers B., Hogg S.M., Vantomme A. 'Influence of the implantation angle on the generation of defects for Er implanted GaN'. *Nuclear Instruments and Methods in Physics Research B*. 2003;**206**:95–8.
[59] Pipeleers B., Hogg S.M., Vantomme A. 'Defect accumulation during channeled erbium implantation into GaN'. *Journal of Applied Physics*. 2005;**98**:123504.
[60] Mamor M., Matias V., Vantomme A., Colder A., Marie P., Ruterana P. 'Defects induced in GaN by europium implantation'. *Applied Physics Letters*. 2004;**85**(12):2244–6.
[61] Alves E., Wahl U., Correia M.R., Pereira S., De Vries B., Vantomme A. 'Annealing behavior and lattice site location of Er implanted InGaN'. *Nuclear Instruments and Methods in Physics Research B*. 2003;**206**:1042–6.
[62] Wendler E., Wesch W., Alves E., Kamarou A. 'Comparative study of radiation damage in GaN and InGaN by 400 keV Au implantation'. *Nuclear Instruments and Methods in Physics Research B*. 2004;**218**:36–41.
[63] Redondo-Cubero A., Lorenz K., Wendler E., *et al.* 'Analysis of the stability of InGaN/GaN multiquantum wells against ion beam intermixing'. *Nanotechnology*. 2015;**26**:425703.

[64] Kucheyev S.O., Williams J.S., Zou J., Jagadish C. 'Dynamic annealing in III-nitrides under ion bombardment'. *Journal of Applied Physics*. 2004;**95**(6): 3048–54.

[65] Faye D.N., Wendler E., Felizardo M., *et al.* 'Mechanisms of implantation damage formation in $Al_xGa_{1-x}N$ compounds'. *Journal of Physical Chemistry C*. 2016;**120**(13):7277–83.

[66] Kucheyev S.O., Williams J.S., Jagadish C., *et al.* 'Ion-beam-produced structural defects in ZnO'. *Physical Review B*. 2003;**67**:094115.

[67] Azarov A.Y., Wendler E., Kuznetsov A.Y., Svensson B.G. 'Crucial role of implanted atoms on dynamic defect annealing in ZnO'. *Applied Physics Letters*. 2014;**104**:052101.

[68] Azarov A., Rauwel P., Hallén A., Monakhov E., Svensson B.G. 'Extended defects in ZnO: Efficient sinks for point defects'. *Applied Physics Letters*. 2017;**110**:022103.

[69] Azarov A., Wendler E., Lorenz K., Monakhov E., Svensson B.G. 'Effect of buried extended defects on the radiation tolerance of ZnO'. *Applied Physics Letters*. 2017:**110**:172103.

[70] Wahl U., Alves E., Lorenz K., *et al.* 'Lattice location and optical activation of rare earth implanted GaN'. *Materials Science and Engineering B*. 2003;**105**:132–40.

[71] Vantomme A., Wahl U., De Vries B. 'Lattice location of rare earth impurities in III-nitrides'. *Topics in Applied Physics*. 2010;**124**:55–98.

[72] Lorenz K., Alves E., Roqan I.S., *et al.* 'Lattice site location of optical centers in GaN:Eu light emitting diode material grown by organometallic vapor phase epitaxy'. *Applied Physics Letters*. 2010;**97**:111911.

[73] Miceli G., Pasquarello A. 'Self-compensation due to point defects in Mg-doped GaN'. *Physical Review B*. 2016;**93**:165207.

[74] Park C.H., Zhang S.B., Wei S.H. 'Origin of p-type doping difficulty in ZnO: The impurity perspective'. *Physical Review B*. 2002;**66**:073202.

[75] Amorim L.M., Wahl U., Pereira L.M.C., *et al.* 'Precise lattice location of substitutional and interstitial Mg in AlN'. *Applied Physics Letters*. 2013;**103**(26): 262102.

[76] Wahl U., Correia J.G., Amorim L., Decoster S., da Silva M.R., Pereira L.M.C. 'Lattice sites of Na dopants in ZnO'. *Semiconductor Science and Technology*. 2016;**31**:095005.

[77] Pereira L.M.C. 'Experimentally evaluating the origin of dilute magnetism in nanomaterials'. *Journal of Physics D: Applied Physics*. 2017;**50**(39): 393002.

[78] Dietl T. 'A ten-year perspective on dilute magnetic semiconductors and oxides'. *Nature Materials*. 2010;**9**(12):965.

[79] Dietl T, Ohno H. 'Dilute ferromagnetic semiconductors: Physics and spintronic structures'. *Reviews of Modern Physics*. 2014;**86**(1):187.

[80] Jungwirth T., Wunderlich J., Novák V., *et al.* 'Spin-dependent phenomena and device concepts explored in (Ga,Mn)As'. *Reviews of Modern Physics*. 2014;**86**(3):855.

[81] Yu K.M., Walukiewicz W., Wojtowicz T., *et al.* 'Effect of the location of Mn sites in ferromagnetic $Ga_{1-x}Mn_xAs$ on its Curie temperature'. *Physical Review B.* 2002;**65**(20):201303.

[82] Jungwirth T., Wang K.Y., Mašek J., *et al.* 'Prospects for high temperature ferromagnetism in (Ga,Mn)As semiconductors'. *Physical Review B.* 2005; **72**(16):165204.

[83] Dobrowolska M., Tivakornsasithorn K., Liu X., *et al.* 'Controlling the Curie temperature in (Ga,Mn)As through location of the Fermi level within the impurity band'. *Nature Materials.* 2012;**11**(5):444.

[84] Lima T.A.L., Wahl U., Augustyns V., *et al.* 'Identification of the interstitial Mn site in ferromagnetic (Ga,Mn)As'. *Applied Physics Letters.* 2015;**106**(1): 012406.

[85] Benzeggouta D., Khazen K., Vickridge I., *et al.* 'Quantitative determination of the Mn site distribution in ultrathin $Ga_{0.80}Mn_{0.20}As$ layers with high critical temperatures: A Rutherford backscattering channeling investigation'. *Physical Review B.* 2014;**89**(11):115323.

[86] Pereira L.M.C., Wahl U., Decoster S., *et al.* 'Direct identification of interstitial Mn in heavily p-type doped GaAs and evidence of its high thermal stability'. *Applied Physics Letters.* 2011;**98**(20):201905.

[87] Pereira L.M., Wahl U., Decoster S., *et al.* 'Stability and diffusion of interstitial and substitutional Mn in GaAs of different doping types'. *Physical Review B.* 2012;**86**(12):125206.

[88] Lima T.A.L., Wahl U., Costa A., *et al.* 'Thermal stability of interstitial and substitutional Mn in ferromagnetic (Ga,Mn)As' *submitted* 2019.

[89] Wang M., Campion R.P., Rushforth A.W., Edmonds K.W., Foxon C.T., Gallagher B.L. 'Achieving high Curie temperature in (Ga,Mn)As'. *Applied Physics Letters.* 2008;**93**(13):132103.

[90] Wang M., Edmonds K.W., Gallagher B.L., *et al.* 'High Curie temperatures at low compensation in the ferromagnetic semiconductor (Ga, Mn)As'. *Physical Review B.* 2013;**87**(12):121301.

[91] Němec P., Novak V., Tesařová N., *et al.* 'The essential role of carefully optimized synthesis for elucidating intrinsic material properties of (Ga,Mn) As'. *Nature Communications.* 2013;**4**:1422.

[92] Matsumoto Y., Murakami M., Shono T., *et al.* 'Room-temperature ferromagnetism in transparent transition metal-doped titanium dioxide'. *Science.* 2001;**291**(5505):854–6.

[93] Dietl T., Ohno H., Matsukura F., Cibert J., Ferrand E.D. 'Zener model description of ferromagnetism in zinc-blende magnetic semiconductors'. *Science.* 2000;**287**(5455):1019–22.

[94] Pereira L.M.C., Wahl U., Correia J.G., *et al.* 'Paramagnetism and antiferromagnetic interactions in single-phase Fe-implanted ZnO'. *Journal of Physics: Condensed Matter.* 2013;**25**(41):416001.

[95] Pereira L.M.C., Araújo J.P., Wahl U., *et al.* 'Searching for room temperature ferromagnetism in transition metal implanted ZnO and GaN'. *Journal of Applied Physics.* 2013;**113**(2):023903.

[96] Pereira L.M.C., Som T., Demeulemeester J., Van Bael M.J., Temst K., Vantomme A. 'Paramagnetism and antiferromagnetic interactions in Cr-doped GaN'. *Journal of Physics: Condensed Matter*. 2011;**23**(34):346004.

[97] Pereira L.M.C., Wahl U., Decoster S., *et al.* 'Mixed Zn and O substitution of Co and Mn in ZnO'. *Physical Review B*. 2011;**84**(12):125204.

[98] Pereira L.M.C., Wahl U., Correia J.G., *et al.* 'Evidence of N substitution by Mn in GaN'. *Physical Review B*. 2012;**86**(19):195202.

[99] Pereira L.M.C., Wahl U., Correia J.G., *et al.* 'Minority anion substitution by Ni in ZnO'. *Applied Physics Letters*. 2013;**103**(9):091905.

[100] Pereira L.M.C., Wahl U., Correia J.G., *et al.* 'Emission channeling studies on transition-metal doped GaN and ZnO: Cation versus anion substitution'. *Nuclear Instruments and Methods in Physics Research Section B: Beam Interactions with Materials and Atoms*. 2014;**332**:143–7.

[101] Augustyns V., van Stiphout K., Joly V., *et al.* 'Evidence of tetragonal distortion as the origin of the ferromagnetic ground state in γ-Fe nanoparticles'. *Physical Review B*. 2017;**96**(17):174410.

[102] Miranda P.A., Wahl U., Catarino N., da Silva M.R., Alves E. 'Performance of resistive-charge position-sensitive detectors for RBS/Channeling applications'. *Nuclear Instruments and Methods in Physics Research Section A: Accelerators, Spectrometers, Detectors and Associated Equipment*. 2014;**760**:98–106.

[103] Redondo-Cubero A., David-Bosne E., Wahl U., *et al.* 'Strain detection in crystalline heterostructures using bidimensional blocking patterns of channeled particles'. *Journal of Physics D: Applied Physics*. 2018;**51**(11): 115304.

[104] Shimoda S., Kobayashi T. 'Development of three-dimensional medium-energy ion scattering spectroscopy'. *Nuclear Instruments and Methods in Physics Research Section B: Beam Interactions with Materials and Atoms*. 2004;**219–220**:573–7.

[105] Shimoda S., Kobayashi T. 'Structure of an Er silicide film on a Si(1 1 1) substrate using three-dimensional medium-energy ion scattering'. *Journal of Applied Physics*. 2004;**96**:3550–2.

[106] Kobayashi T. 'Development of three-dimensional medium-energy ion scattering using a large solid angle detector'. *Nuclear Instruments and Methods in Physics Research Section B: Beam Interactions with Materials and Atoms*. 2006;**249**:266–9.

[107] Jalabert D., Pelloux-Gervais D., Béché A., *et al.* 'Depth strain profile with sub-nm resolution in a thin silicon film using medium energy ion scattering'. *Physica Status Solidi (a)*. 2012;**209**(2):262–5.

[108] Jalabert D. 'Real space structural analysis using 3D MEIS spectra from a toroidal electrostatic analyzer with 2D detector'. *Nuclear Instruments and Methods in Physics Research Section B: Beam Interactions with Materials and Atoms*. 2012;**270**:19–22.

[109] Avila T.S., Fichtner P.F., Hentz A., Grande P.L. 'On the use of MEIS cartography for the determination of $Si_{1-x}Ge_x$ thin-film strain'. *Thin Solid Films*. 2016;**611**:101–6.

[110] Trombini H., Grande P.L., Hentz A., Vos M., Winkelmann A. 'A comparison of the analysis of non-centrosymmetric materials based on ion and electron beams'. *Nuclear Instruments and Methods in Physics Research Section B: Beam Interactions with Materials and Atoms*. 2018;**431**:31–7.

[111] Nordlund K., Djurabekova F., Hobler G. 'Large fraction of crystal directions leads to ion channeling'. *Physical Review B*. 2016;**94**(21):214109.

[112] Zhang S., Nordlund K., Djurabekova F., Zhang Y., Velisa G., Wang T.S. 'Simulation of Rutherford backscattering spectrometry from arbitrary atom structures'. *Physical Review E*. 2016;**94**(4):043319.

[113] Kavčič M., Budnar M., Mühleisen A., *et al.* 'Design and performance of a versatile curved-crystal spectrometer for high-resolution spectroscopy in the tender X-ray range'. *Review of Scientific Instruments*. 2012;**83**(3): 033113.

[114] Kavčič M., Petric M., Gasser F., *et al.* 'New parallel beam wavelength dispersive X-ray emission spectrometer at Ljubljana microprobe'. *Nuclear Instruments and Methods in Physics Research Section B: Beam Interactions with Materials and Atoms*. 2017;**404**:81–6.

[115] Palosaari M.R., Kinnunen K.M., Julin J., *et al.* 'Transition-edge sensors for particle induced X-ray emission measurements'. *Journal of Low Temperature Physics*. 2014;**176**(3–4):285–90.

[116] Palosaari M.R., Käyhkö M., Kinnunen K.M., *et al.* 'Broadband ultrahigh-resolution spectroscopy of particle-induced X rays: Extending the limits of nondestructive analysis'. *Physical Review Applied*. 2016;**6**(2):024002.

[117] Doherty M.W., Manson N.B., Delaney P., Jelezko F., Wrachtrup J., Hollenberg L.C. 'The nitrogen-vacancy colour centre in diamond'. *Physics Reports*. 2013;**528**(1):1–45.

[118] Rondin L., Tetienne J.P., Hingant T., Roch J.F., Maletinsky P., Jacques V. 'Magnetometry with nitrogen-vacancy defects in diamond'. *Reports on Progress in Physics*. 2014;**77**(5):056503.

[119] Sipahigil A., Evans R.E., Sukachev D.D., *et al.* 'An integrated diamond nanophotonics platform for quantum optical networks'. *Science*. 2016; **354**(6314):847–50.

[120] Sukachev D.D., Sipahigil A., Nguyen C.T., *et al.* 'Silicon-vacancy spin qubit in diamond: A quantum memory exceeding 10 ms with single-shot state readout'. *Physical Review Letters*. 2017;**119**(22):223602.

[121] Bhaskar M.K., Sukachev D.D., Sipahigil A., *et al.* 'Quantum nonlinear optics with a germanium-vacancy color center in a nanoscale diamond waveguide'. *Physical Review Letters*. 2017;**118**(22):223603.

[122] Iwasaki T., Miyamoto Y., Taniguchi T., *et al.* 'Tin-vacancy quantum emitters in diamond'. *Physical Review Letters*. 2017;**119**(25):253601.

[123] Trusheim M.E., Wan N.H., Chen K.C., *et al.* 'Lead-Related Quantum Emitters in Diamond'. *Physical Review B*. 2019;99(7):075430

[124] Green B.L., Mottishaw S., Breeze B.G., *et al.* 'Neutral silicon-vacancy center in diamond: Spin polarization and lifetimes'. *Physical Review Letters*. 2017;**119**(9):096402.

[125] Rose B.C., Huang D., Zhang Z.H., *et al.* 'Observation of an environmentally insensitive solid-state spin defect in diamond'. *Science.* 2018; **361**(6397):60–3.

[126] Hasan M.Z., Kane C.L. 'Colloquium: Topological insulators'. *Reviews of Modern Physics.* 2010;**82**(4):3045.

[127] Chang C.Z., Zhang J., Feng X., *et al.* 'Experimental observation of the quantum anomalous Hall effect in a magnetic topological insulator'. *Science.* 2013;**340**(6129):167–70.

[128] Chang C.Z., Zhao W., Kim D.Y., *et al.* 'High-precision realization of robust quantum anomalous Hall state in a hard ferromagnetic topological insulator'. *Nature Materials.* 2015;**14**(5):473.

[129] He Q.L., Pan L., Stern A.L., *et al.* 'Chiral Majorana fermion modes in a quantum anomalous Hall insulator–superconductor structure'. *Science.* 2017;**357**(6348):294–9.

[130] Alarcon-Diez V., Eddrief M., Vickridge I. 'Rutherford Backscattering Spectrometry analysis of iron-containing Bi_2Se_3 topological insulator thin films'. *Nuclear Instruments and Methods in Physics Research Section B: Beam Interactions with Materials and Atoms.* 2016;**371**:224–9.

Index

ABC method 58
Abragam function 217
absorption coefficient 98
acceptor–hydrogen pairs 126–7
acceptor-like defects in wide bandgap semiconductors 277
  cation and oxygen vacancies in metal oxides 279–83
  cation vacancies and acceptor impurities in GaN and ZnO 277–9
AC field-driven resonance (RF-μSR) 228–9
activation energy 11, 22, 24, 33, 55, 250, 461
  measurement of 481–6
activators 46
AlSb, resonant interaction in 124–5
amorphous Si (a-Si) 440, 462, 464
amorphous state, recrystallization of 465
angular scan 508–11
anharmonic coupling 110, 118
anharmonic potential 103
anomalous muonium 234
anticrossing 110
anti-Stokes process 111
apodization function 114
Arrhenius behavior 35–6, 245
Arrhenius plot 5, 10–12, 32, 183, 483–4, 486–7
atomic collisions 452, 454, 502–4
atomic displacement 372–3, 482
atomic Doyle–Turner potentials 522
atomic force microscopy (AFM) 346–8
  for dislocation density quantification 383–4
atom probe tomography (APT) 417
  generalities and history 419
  one-dimensional defects 440–1
  point defects 442
    in situ photoluminescence 442–4
    three-dimensional field ion microscopy 442
  principles and performances 419–24
  versus secondary ion mass spectrometry (SIMS) 424–7
  three-dimensional defects 427
    impurity clusters 431–3
    nanocrystals (NCs) and quantum dots (QDs) 427–30
  two-dimensional defects
    intergranular segregation of dopants in silicon 438–9
    segregation of dopants to gate interfaces in MOSFET transistors 439–40
    stacking faults (SFs), segregation to 433–5
    stacking faults (SFs) in heterostructures 435–8
atoms, determining number of 104–5
avoided level-crossing resonance (ALCR) 229, 232

backscattered electrons (BSEs) 356–7
band filling for shallow defects 316–17
basal configurations 192
beam sputtering 372–3, 403

Bethe–Bloch regime 505
Bethe–Salpeter equation (BSE) approach 308
binary collision approximation (BCA) 459
Bloch equations 164
Bois–Chantre fits 26
Bois–Vincent–Chantre model 6
bond-centred configuration 126
bond-centred hydrogen 107, 114–15
Born–Oppenheimer approximation 299
boron–oxygen complex 20
Bradley–Harper theory 470
Bragg angle 357, 361–2
Bragg scattering 371
bright-field imaging 377, 379
Brillouin zone 98, 317
Burgers circuit 388–9, 440
Burgers vector 360, 383, 385, 388–90

capacitance spectroscopic techniques, variations of 13
  junction DLTS 14–15
  Laplace-DLTS (LDLTS) 13–14
  minority carrier transient spectroscopy (MCTS) 16
  optical DLTS (ODLTS) 15–16
capture coefficients 50
capture rates 4–5, 13, 17
carbon in GaAs 121–2
carbon monoxide (CO) 103
carrier concentrations 32, 247
cathodoluminescence (CL) 364–6
CdTe, impurities in 125
C-face silicon carbide, electron channelling contrast imaging (ECCI) examination 402
channeling maps 454
channeling techniques 507
  defects and damage, characterization of 512–16
  ion channeling 507–12
  lattice location 516–23
charge-coupled devices (CCDs) 112, 362
charge transition levels 273, 289–90, 295, 310–12, 323–5, 330
chemical vapour transport (CVT) growth 128
Cliff–Lorimer analysis 381
coincidence Doppler broadening spectroscopy 270–1
complementary metal-oxide-semiconductor (CMOS) 362
complete spatial randomness (CSR) 385
conduction band minimum (CBM) 290
conductive AFM (C-AFM) 348, 350
configuration-coordinate (CC) model 46, 63–7, 295–6
contact mode 347–9
continuous wave (CW) EPR 165
conversion electrons (CE) 518–19
copper-related green band 84–5
correlation energy 301
Cottrell atmospheres 418, 440
Coulomb interaction 68–9, 154, 266, 299–300, 307, 309–10, 331, 503, 505
critical angle 507, 509–10, 514
crystal field resonance (CFR) 326–7
crystalline semiconductors, positrons in 264
  kinetic trapping model and positron lifetime spectroscopy 267–8
  positron thermalization and diffusion 265–6
  positron trapping 266
  radioactive sources, positrons obtained from 264
  slow monoenergetic positron beams for thin-layer studies 264–5
Czochralski-grown silicon 116

"damage overlap" fraction 462
dangling-bond hybrid (DBH) 326
dark-field imaging 379

deep-level optical spectroscopy (DLOS) 2, 16–19
deep-level transient spectroscopy (DLTS) 2, 7, 79, 199, 461
  measuring capture cross section 12–13
  profiling 12
defect levels 289–90
defect relaxation time constants, measurement of 479–80
density functional theory (DFT) 271, 298, 301–3
  achieving higher accuracy with 303
    fitted chemical potentials 313–14
    generalized Koopman's theorem 305–7
    Green's function approaches 308–11
    hybrid functionals 311–13
    LDA+U and PBE+U approaches 307–8
    observations 303–5
depth profile 12, 514
diamagnetic muonium 235, 241, 245
diatomic model 102, 105, 107
DICADA code 514, 531
diffraction contrast 371
diffraction contrast imaging of defects, techniques for 377–9
diffuse X-ray scattering (DXS) studies 459
diffusion Monte Carlo (DMC) method 303
dilute magnetic semiconductors (DMS), magnetic dopants in 542–50
dislocation density quantification
  atomic force microscopy (AFM) methods for 383–4
  scanning electron microscope (SEM) methods for 384–6
dislocation density quantitation
  transmission electron microscope (TEM) methods for 386
    core identification by HR-STEM 388–9
    $g \cdot b$ analysis 387–8
displacement energy 276, 372
displacements per atom (dpa) 526
divacancies 21
donor–acceptor-pair (DAP) transitions 47–8, 68
donor and acceptor levels, measuring 245–7
donor–hydrogen complexes 126–7
dopants, lattice location of 532
  GaN 532–5
  magnetic dopants in dilute magnetic semiconductors 542–50
  Mg in GaN and AlN and Na in ZnO 535–42
Doppler broadening techniques 267, 280
  coincidence Doppler broadening spectroscopy 270–1
  defect identification, supporting theory for 271–3
  Doppler broadening spectroscopy 269–70
Doppler shift 263, 269
dose-rate effect (DRE) 473
dumbbell defect in GaN 183–6
DX centres 123
dynamic annealing (DA) 458, 473, 530
  in GaN 524–5

e-A transitions 47–8, 57–8
elastic recoil-detection analysis (ERDA) 501–2, 504
electrical characterisation techniques in AFM 348–51
electrical dopants in wide-gap semiconductors 535–42

electrically active defects 1
  capacitance spectroscopic techniques, variations of 13
    junction DLTS 14–15
    Laplace-DLTS (LDLTS) 13–14
    minority carrier transient spectroscopy (MCTS) 16
    optical DLTS (ODLTS) 15–16
  characteristics of 3–6
  deep-level optical spectroscopy 16–19
  deep-level transient spectroscopy (DLTS) 7
    measuring capture cross section 12–13
    profiling 12
  extending junction spectroscopic techniques beyond traditional methodology 36–8
  hydrogen-related defect level 29–33
  iron-related defect level 28–9
  irradiation-induced point defects in crystalline silicon 20
  mono-vacancy 20–5
  oxygen vacancy-related defect level 33–6
  vacancy-related defects 21–5
electrically detected magnetic resonance (EDMR) 172–3
electron backscatter diffraction (EBSD) 357, 361–4
electron backscatter patterns (EBSPs) 357, 362
electron beam-induced current (EBIC) 366–8
electron-capture coefficient 58
electron channelling contrast imaging (ECCI) 354, 357–61, 402
electron channelling patterns (ECPs) 359
electron–electron Coulomb interaction 300
electron emission channeling (EC) 508, 547
electron energy loss spectroscopy (EELS) 372
electron guns 375
electron–hole pairs 52, 364
electronic structure theory 289
electron microscopy 462
electron nuclear double resonance (ENDOR) 170–1, 200
electron paramagnetic resonance (EPR) 19, 151–3, 186, 199–200, 321, 461
electron–phonon coupling 65
electron–positron correlation energy 271
electron probe microanalyser (EPMA) 369
electron spin resonance spectroscopies 151
  electrically detected magnetic resonance (EDMR) 172–3
  electron nuclear double resonance (ENDOR) 170–1
  electron paramagnetic resonance (EPR) 152–3
  experimental setup 165
    choice of microwave frequency 168–70
  hyperfine interactions 158–9
  optically detected magnetic resonance (ODMR) 172–3
  pulsed spectroscopies 171–2
  resonance 159
    selection rules 160–1
  spin Hamiltonian formalism 153–6
    zero-field splitting 156–8
  transition probability: relaxation phenomena 161–5
elemental analysis 505–7
emission rate 5
ENDOR (electron nuclear double resonance) 159
energy- and wavelength-dispersive X-ray spectroscopy (EDX/WDX) 368–9
energy dispersive spectrometers (EDSs) 551
energy loss 504–5

energy spectrum 205
equilibrium Fermi energy 296
Eshelby twist 389
Euler–Lagrange equations 300, 302
Everhart–Thornley detector 355
exchange–correlation energy 302
exchange energy 300
extended X-ray absorption fine structure (EXAFS) 547

facetted ELOG (FACELOG) 391
Faraday cup systems 458
^{59}Fe-implanted ZnO
  emission channeling measurements and best fit for 548
Fermi–Dirac function 4, 310
Fermi energy 292, 296
Fermi level 5, 7, 24, 48, 54–5, 132, 273, 401
Fermi resonance 110, 120, 124
ferromagnetic resonance (FMR) 152
ferromagnetism 543, 546–7
field emission gun (FEG) 375
field-enhanced emission 11
field factor 420
field ion microscopy (FIM) 419
finite size effect 290, 297–8, 314
  band filling for shallow defects 316–17
  charged defects 315–16
first nearest neighbour (1NN) distances 433
first principles methods for defects 289
  defect clustering and aggregation (example) 329–33
  defect levels 289–92
  defect thermochemistry
    configuration coordinate diagrams and optical excitation 295–6
    defect equilibria and carrier concentrations 296–7
    defect formation enthalpy 292–3
    defect formation enthalpy diagrams 294–5
    phase stabilities and chemical potential space 293–4
  density functional theory (DFT), achieving higher accuracy with 303
    fitted chemical potentials 313–14
    generalized Koopman's theorem 305–7
    Green's function approaches 308–11
    hybrid functionals 311–13
    LDA+U and PBE+U 307–8
    observations 303–5
  finite size effects 314
    band filling for shallow defects 316–17
    charged defects 315–16
  Mg and other possible acceptors in GaN (example) 317–21
  oxygen vacancies and other defects in metal oxides (example) 321
  sources of uncertainty in defect formation energy calculations 297–8
  total energies using first principles electronic structure calculations
    density functional theory (DFT) 301–3
    Hartree–Fock (HF) 299–301
    quantum Monte Carlo 303
    Schrödinger equation 298–9
  transition metal impurities (example) 326–9
fluencimetry 458
flux peaking effect 517
Fourier analysis 225
Fourier transform infrared spectroscopy (FTIR) 19, 113–14, 309
Franck–Condon shift 11, 65, 321
Frenkel pair 31, 273, 372, 486, 524
Freysoldt, Neugebauer, and Van de Walle (FNV) correction 316
full width at half maximum (FWHM) 65

GaAs 1
  impurities in 121
    carbon 121–2
    oxygen 124
    silicon 123
GaN
  damage in 524–9
  dynamic annealing in 524
  Mg-doped p-type GaN 535
  MgH complex in 127–8
  optical dopants in 532–5
GaN, radiative defects in 71
  defect-related PL bands in undoped GaN 71–4
  photoluminescence in Mg-doped GaN 81–3
  yellow luminescence 74
    shape and fine structure of the YL1 band 74–9
    thermal quenching of the YL1 band 79–80
    YL1 band in GaN heavily doped with Si 80–1
Gaussian curve 67, 87
Gaussian distribution 213–14, 434
Gaussian shape 45, 65, 77, 212
generalized gradient approximations (GGAs) 272, 302
geometrically necessary dislocations (GNDs) 363
Gibbs free energy 5
glow curve 2
green luminescence (GL1) band 63
Green's function 298, 305, 308–11
g tensor 154, 156, 158–9, 175
GW method 298, 308–11

Hall-effect measurements 58, 188
Hartree energy 300, 302
Hartree–Fock (HF) equations 298–301
h-D-type transitions 48
Hedin's equations 309–10
Hedin's formulation 309
Heitler–London type interaction 329
Hermite polynomials 108–9
hexagonal boron nitride (h-BN) 402–6
Heyd-Scuseria-Ernzerhof (HSE06) 312
high electron mobility transistors (HEMTs) 382
high-field LF spectrometer (HIFI) 231
(high frequency) capacitance 18
high-resolution transmission electron microscopy (HR-TEM) imaging 371
homogeneous amorphization 463
Huang–Rhys factor 65–6
Hubbard-*U* 307–8
hydride vapor phase epitaxy (HVPE) 71, 74
hydrogen 199
  molecules 136
  -related defect level 29–33
  in silicon 125–6
  in strontium titanate 134–6
hydrogen donors in oxide semiconductors 130
  In_2O_3 132
  SnO_2 132
  TiO_2 132–3
  ZnO 130–2
hydrostatic pressure 115–16, 132
hyperfine interactions 153, 158–9

IMAGO 418
impurity-hydrogen pairs 126–7
impurity–ligand interaction 330
In_2O_3 132
infrared (IR)
  absorption 105–6
  activity 103–4
  pump–probe 114–15
  vibrational spectroscopy 200
in situ photoluminescence 442–4
interferogram 113–14
intergranular segregation of dopants in silicon 438–9
interstitial defects 21

intrinsic DMS 543
intrinsic ferromagnetism 547
inverse Laplace transform 13
inverse temperature effect 530
ion beam analysis (IBA) 501, 550–1
  atomic collisions 502–4
  elemental analysis 505–7
  emerging applications 551–3
  energy loss 504–5
ion-beam modification of semiconductors 451
  damage production and annealing 458
    defect migration and interaction 461–2
    high-fluence damage overlap and amorphization 462–4
    low-fluence primary damage 459–60
    recrystallization and dopant activation, defect evolution during 465–6
    recrystallization of amorphous state 465
  ion implantation experiments 457–8
  ripple formation on semiconductors 466
    experimental background 467–8
    theoretical models 469–73
  theory of ion stopping and energy deposition 452–6
  time-resolved experiments to probe defect evolution
    comparison of radiation defect dynamics in different semiconductors 486–8
    dynamic annealing (DA) 473
    pulsed-ion-beam method 478
    traditional dose-rate effect experiments 473–8
ion channeling 507–12
iron-related defect level 28–9
irradiation-induced point defects in crystalline silicon 20
ISOLDE 520

Jahn–Teller distortion 291
Janak's theorem 306
junction DLTS 14–15
junction spectroscopic techniques, extending
  beyond traditional methodology 36–8
junction spectroscopic techniques for characterizing deep-level defects 7
  deep-level optical spectroscopy 16–19
  deep-level transient spectroscopy (DLTS) 7
    measuring capture cross section 12–13
    profiling 12
  variations of capacitance spectroscopic techniques 13
    junction DLTS 14–15
    Laplace-DLTS (LDLTS) 13–14
    minority carrier transient spectroscopy (MCTS) 16
    optical DLTS (ODLTS) 15–16, 35

Kelvin probe force microscopy (KPFM) 350
Keren function 217–18
Kikuchi bands 361–2, 378
Kikuchi lines 359
Kinchin–Pease equation 459
kinematic factor 502
kinetic trapping model and positron lifetime spectroscopy 267–8
Klein–Gordon equation 522
Kohn–Sham approach 301
Kohn–Sham DFT Hamiltonian 309
Kohn–Sham equations 302
Kohn–Sham Hamiltonian 305
Kohn–Sham orbitals 311

Koopmans' condition 328
Koopman's theorem 305–7, 312, 318
Kossel cones 361–2
Kubo–Toyabe function 214, 216–17, 227, 245

Lande tensor 154, 156
Laplace-DLTS (LDLTS) 13–14
Larmor frequency 193, 217, 223, 243
laser-assisted APT 425
laser pulses 421
lattice location 516–23
  of dopants 532
    GaN 532–5
    magnetic dopants in dilute magnetic semiconductors 542–50
    Mg in GaN and AlN and Na in ZnO 535–42
Lehigh group 132
level-crossing resonance (LCR) 229
light and elevated temperature-induced degradation 20
light-emitting diodes (LEDs) 366, 382
light-induced degradation (LID) 20
Lindhard's theoretical approach 507
linear chain 98–102
Linear Combination of Molecular Orbitals (LCAO) theory 179
linewidth and temperature dependence 106–8
lithium-related orange band 85
local density approximation (LDA) 302
local density of states (LDOS) 351
localons 124
local vibrational modes (LVMs) 97–8, 100–2, 126, 136–8
lock-in amplifier 168
lock-in weighting function 10
longitudinal field μSR 226–7
longitudinal mode 122
longitudinal relaxation time 163
long-lived isotopes 521
Lorentzian-damped oscillation 214
Lorentzian line 168
low-energy muons (LEM/LE-μSR) 231
luminescence spectra 461

Madelung energy of point charge 316
magnesium in GaN 81–3, 317–21
magnesium oxide 321–2
magnetic dopants, in dilute magnetic semiconductors 542–50
magnetic resonance methods 151
  electrical and optical activities (examples) 186
    origin of the residual conductivity of Ga_2O_3 188–90
    P-type doping of GaN 186–8
    SiC defects as quantum bits 190–2
  electron spin resonance (ESR) spectroscopies 151
    electrically detected magnetic resonance (EDMR) 172–3
    electron nuclear double resonance (ENDOR) 170–1
    electron paramagnetic resonance (EPR) 152–3
    experimental setup 165–70
    hyperfine interactions 158–9
    microwave frequency, choice of 168–70
    optically detected magnetic resonance (ODMR) 172–3
    pulsed spectroscopies 171–2
    resonance 159–61
    spin Hamiltonian formalism 153–8
    transition probability 161–5
  N dumbbell in GaN (illustrative examples) 183–6
  SiC/oxide interface defects (illustrative examples) 173
    energy level position 180–2
    passivation 182
magnification 421
Magnéli phases 332

Makov–Payne correction 316
many-beam diffraction theory of electron channeling 521–2
materials properties, imaging of the impact of dislocations on
  combining ECCI and CL to investigate the influence of defects on light emission 394–6
  interaction of dislocations with doping 390–4
MEIS (medium energy ion scattering) 550
  2D patterns in 550
memory effect 71
metal–insulator transition 328
metal-organic chemical vapor deposition (MOCVD) 71, 74, 82, 121
metal-organic vapour-phase epitaxy (MOVPE) 383
metal oxides, oxygen vacancies and other defects in 321
  N_O in ZnO 324–5
  V_O in MgO 321–2
  V_O in ZnO 322–4
metal-oxide-semiconductor field effect transistors (MOSFET) transistors 418
  segregation of dopants to gate interfaces in 439–40
MgH complex in GaN 127–8
Mg in GaN and AlN 535–52
Michelson interferometer 113
microscopy techniques 345
  2D materials, imaging defects in 400
    addressing the electrical properties of defects 400–2
    electron channelling contrast imaging (ECCI) examination 402
    hexagonal boron nitride (h-BN) 402–6
  dislocation densities and Burgers vectors in GaN (example) 382–3
    AFM methods for dislocation density quantification 383–4
    intercomparison of the various methods 389–90
    SEM methods for dislocation density quantification 384–6
    TEM methods for dislocation density quantitation 386–9
  materials properties, imaging of the impact of dislocations on
    combining ECCI and CL to investigate the influence of defects on light emission 394–6
    interaction of dislocations with doping 390–4
  multi-microscopy example 396
    practical aspects 396–9
    results 399–400
  scanning electron microscopy (SEM)
    backscattered electrons (BSEs) 356–7
    cathodoluminescence (CL) 364–6
    electron backscatter diffraction 361–4
    electron beam-induced current (EBIC) 366–8
    electron channelling contrast imaging 357–61
    energy- and wavelength-dispersive X-ray spectroscopy (EDX/WDX) 368–9
    resolution 354
    secondary electrons (SEs) 355
    signals used to generate images in the SEM 355
  scanning probe microscopy
    atomic force microscopy (AFM) 346–8
    electrical characterisation techniques in AFM 348–51
    scanning tunnelling microscopy (STM) 351–2

transmission electron microscopy
  beam damage 372–3
  compositional analysis near defects, techniques for 379–81
  diffraction contrast imaging of defects, techniques for 377–9
  electron-matter interaction in a TEM 369–72
  high-resolution imaging of defects, techniques for 381–2
  sample preparation 373–5
  transmission electron microscopy equipment 375–7
microwave frequency, choice of 168–70
microwave resonator 165
minimum yield 507, 509
minority carrier transient spectroscopy (MCTS) 16
Mn-doped GaAs and InAs 543
modified random element isodisplacement (MREI) model 102
molecular-beam epitaxy (MBE) 74
molecular dynamics (MD) calculations 550
Molière potential 518
mono-vacancy 20–5
Monte Carlo method 303, 385
Morse potential 103, 129
MOS (metal/oxide/semiconductor) structure 173
Mössbauer spectroscopy 547
multi-microscopy example 396–400
muonium centres
  identifying 234
  processes and dynamics involving 237
    muonium in motion 241–4
    neutral and charged muonium, transitions between 238–41
    two-state bidirectional transition 244–5
muonium states 233–4
muon spin research (μSR) 200–1
  contributions to the field 247
    group II–IV–V_2 chalcopyrites 252–3
    group II–VI 252
    group III–V 250–1
    group-IV 249–50
    oxides 253–4
  general principles 203–10
  illumination and muon spectroscopy of excited states 231–2
  longitudinal field μSR 226–7
  low-energy muons (LEM/LE-μSR) 231
  polarisation functions 210–23
  resonance-based μSR techniques 228
    AC field-driven resonance (RF-μSR) 228–9
    avoided level-crossing resonance (ALC or LCR) 229
    quadrupolar level-crossing resonance (QLCR) 229–31
  transverse field μSR 223–6
  zero-field μSR 227–8
muons' role in semiconductors 232
  carrier concentrations, effects of 247
  measuring donor and acceptor levels 245–7
  muonium centres, identifying 234
  muonium states 233–4
  processes and dynamics involving muonium centres 237
    muonium in motion 241–4
    neutral and charged muonium, transitions between 238–41
    two-state bidirectional transition 244–5

nanocrystals (NCs) 427–30
narrow bandgap semiconductors, temperature-dependent measurements in 273

positron traps in low temperature irradiated GaSb 274–7
vacancy annealing in Ge 273–4
near-band-edge (NBE) emission 46
NH complex in ZnO 128–30
nitrogen-related red band 85–6
non-contact imaging 348
nonradiative recombination 48
nuclear magnetic resonance (NMR) 151, 200
nuclear reaction analysis (NRA) 501–2, 504

O atom 120
optical absorption 461
optical DLTS (ODLTS) 15–16
optically detected magnetic resonance (ODMR) 172–3
orange luminescence (OL) band 85
oxide-based materials 253–4
oxygen 124
interstitial 116
germanium 118–19
silicon 116–18
silicon under pressure 119–20
oxygen vacancies and other defects in metal oxides 321
N_O in ZnO 324–5
V_O in MgO 321–2
V_O in ZnO 322–4
oxygen vacancy 321
-related defect level 33–6

Pässler model 6
Pauli exclusion 300–1
Pauli principle 152
Pb defects 173, 180
PeakForce TUNA 348
"PeakForce" tapping mode 348
Perdew-Burke-Ernzerhof (PBE) 302
Perdew-Wang (PW91) 302
persistent photoconductivity (PPC) 135
phase stabilities and chemical potential space 293–4
photoluminescence (PL) 45, 429
AlN, PL bands in 86–8
configuration-coordinate model 63–7
diagonal transitions between localized states 67
donor–acceptor pairs (DAPs) 68
potential fluctuations, effect of 68
phenomenological models 48–68
radiative defects in GaN 71
defect-related PL bands in undoped GaN 71–4
photoluminescence in Mg-doped GaN 81–3
yellow luminescence 74
rate equations model 49
abrupt and tunable quenching of PL 55–6
determination of the absolute internal quantum efficiency of PL 58–61
determination of the concentration of radiative defects 61–3
PL lifetime 56–8
temperature dependence of PL intensity 52–5
types of transitions related to 46–8
zinc oxide, PL bands in 84
copper-related green band 84–5
lithium-related orange band 85
nitrogen-related red band 85–6
zinc vacancy, PL bands related to 86
photon flux 6, 15–16, 18, 52
point defect 7, 20, 289, 442, 526, 551
poisoners 46
polarisation functions 210–23
Poole–Frenkel effect 11–12, 28
position-sensitive detector (PSD) 419, 421–2
positron annihilation spectroscopy (PAS) 19, 183, 263

acceptor-like defects in wide bandgap semiconductors 277
cation and oxygen vacancies in metal oxides 279–83
cation vacancies and acceptor impurities in GaN and ZnO 277–9
crystalline semiconductors, positrons in 264
kinetic trapping model and positron lifetime spectroscopy 267–8
positron thermalization and diffusion 265–6
positron trapping 266
radioactive sources, positrons obtained from 264
slow monoenergetic positron beams for thin-layer studies 264–5
Doppler broadening techniques 267
coincidence Doppler broadening spectroscopy 270–1
defect identification, supporting theory for 271–3
Doppler broadening spectroscopy 269–70
temperature-dependent measurements in narrow bandgap semiconductors 273
positron traps in low temperature irradiated GaSb 274–7
vacancy annealing in Ge 273–4
proton (or particle)-induced X-ray emission (PIXE) 501–2, 504, 510, 551
proton-implanted GaN 133
p-type doping of GaN 186–8, 535
p-type semiconductor 55, 57, 70
pulsed-ion-beam method 478
activation energies and cascade density effects, measurements of 481–6
defect diffusion lengths, measurement of 481
defect relaxation time constants, measurement of 479–80
pulsed spectroscopies 171–2
pump–probe experiments 114

Q-Band spectrometers 166–7, 171
QP energies 309
quadrupolar level-crossing resonance (QLCR) 229–32
quantum anomalous Hall insulators (QAHI) 552
quantum-based density matrix approach 218
quantum dots (QDs) 427–30
quantum Monte Carlo (QMC) 298, 303
quantum wells (QWs) 421
quasiparticle energy 290
quintuple layers (QLs) 552

radioactive sources, positrons obtained from 264
radio frequency (RF) electromagnetic field 170
Raman scattering 97, 111–12
Raman spectroscopy 112, 136, 138, 462
random fraction 523
rapid thermal annealing (RTA) 465
rate equations model 49
abrupt and tunable quenching of PL 55–6
determination of absolute internal quantum efficiency of PL 58–61
determination of concentration of radiative defects 61–3
PL lifetime 56–8
temperature dependence of PL intensity 52–5
recrystallization and dopant activation, defect evolution during 465–6
recrystallization of amorphous state 465
RE-doped semiconductors 532

resonance 159
  selection rules 160–1
resonance-based μSR techniques 228
  AC field-driven resonance (RF-μSR) 228–9
  avoided level-crossing resonance (ALC or LCR) 229
  quadrupolar level-crossing resonance (QLCR) 229–31
reststrahlen band 123
Ripley's *K*-function 385
ripple formation on semiconductors 466
  experimental background 467–8
  theoretical models 469–73
rotating reference frame (RRF) transformation 225
Rutherford backscattering/channeling (RBS/C) 462
Rutherford backscattering spectrometry (RBS) 501–4, 508, 510, 529
Rutherford scattering 379
Rydberg states 266

scanning electron microscopy (SEM) 37, 349–50
  atomic force microscopy (AFM) 346–8
  backscattered electrons (BSEs) 356–7
  cathodoluminescence (CL) 364–6
  for dislocation density quantification 384–6
  electrical characterisation techniques in AFM 348–51
  electrical properties of defects in 2D transition metal dichalcogenides by 400–2
  electron backscatter diffraction 361–4
  electron beam-induced current (EBIC) 366–8
  electron channelling contrast imaging 357–61
  energy- and wavelength-dispersive X-ray spectroscopy (EDX/WDX) 368–9
  resolution 354
  scanning tunnelling microscopy (STM) 351–2
  secondary electrons (SEs) 355
  signals used to generate images in 355
scanning spreading resistance microscopy (SSRM) 348, 350
scanning transmission electron microscopy (STEM) 429
  STEM with a high-angle annular dark-field detector (STEM-HAADF) 371
scanning tunneling microscopy (STM) 193, 346, 351–2
scanning tunnelling spectroscopy (STS) 352
Schottky barrier 2, 12
Schottky contact 7, 15
Schottky emitter 354–5
Schrödinger equation 289, 291, 299, 522
Schön–Klasens mechanism 54
screened Coulomb interaction 310
secondary electrons (SEs) 355
secondary ion mass spectrometry (SIMS) 29, 70, 417
  atom probe tomography (APT) versus 424–7
Seitz–Mott mechanism 54
self-interaction 21, 290, 300, 302
self-regulating response (SRR) 326
semiconductors, damage in 524
  GaN 524–9
  ternary nitrides 529–30
  ZnO 530–2
Shockley–Reed–Hall approach 48
signal-to-noise ratio 114, 165
silicon 1, 19, 123
  hydrogen in 125–6
silicon divacancy–hydrogen complex 133

single particle Green's function 309–10
Si–O–Si defect 119
Slater determinant 300
SnO_2 132
spin Hamiltonian formalism 153–6, 160
  zero-field splitting 156–8
spin–lattice relaxation 163, 243
spin–orbit coupling 154
spin–spin relaxation time 164
stacking faults (SFs)
  in heterostructures 435–8
  segregation to 433–5
steady-state photo-capacitance (SSPC) 17
stochastic Monte Carlo approaches 303
Stokes process 111–12
Stranski–Krastanov growth mode 427
stress, applied 115–16
stretch mode 110, 122
"strongly constrained and appropriately normed" (SCAN) 302
strontium titanate, hydrogen in 134–6
super-hyperfine 158
surface ripples 467
"surface" channels 204

tapping mode 347
technical developments 550–1
temperature-dependent ion channeling measurements 516
temperature-dependent measurements in narrow bandgap semiconductors 273
  positron traps in low temperature irradiated GaSb 274–7
  vacancy annealing in Ge 273–4
ternary nitrides, damage in 529–30
thermal quenching of PL 53
thermal vibrations (phonons) 515
thermionic gun 375
thermochemistry, defect
  configuration coordinate diagrams and optical excitation 295–6
  defect equilibria and carrier concentrations 296–7
  defect formation enthalpy 292–3
  defect formation enthalpy diagrams 294–5
  phase stabilities and chemical potential space 293–4
threading dislocations (TDs) 358–9, 365, 394
3d-derived in-gap orbitals 326
three-dimensional field ion microscopy 442
Tikhonov regularization method 13
time differential (TD) 207
time integral (TI) 207
time-resolved experiments to probe defect evolution
  comparison of radiation defect dynamics in different semiconductors 486–8
  dynamic annealing (DA) 473
  pulsed-ion-beam method 478
    activation energies and cascade density effects, measurements of 481–6
    defect diffusion lengths, measurement of 481
    defect relaxation time constants, measurement of 479–80
  traditional dose-rate effect experiments 473–8
TiO_2 132–3
topological insulators 552
total energies using first principles electronic structure calculations
  density functional theory (DFT) 301–3
  Hartree–Fock (HF) 299–301
  many-particle Schrödinger equation 298–9
  quantum Monte Carlo 303
total energy 290, 299–300, 305

traditional dose-rate effect
experiments 473–8
transient bleaching 114
transient capacitance spectroscopy 19
transient-enhanced diffusion 466
transition-edge sensors (TES) 551
transition metal impurities 326–9
transition probability 161–5
transmission electron microscopy
(TEM) 459, 514
beam damage 372–3
compositional analysis near defects,
techniques for
energy-dispersive X-ray
spectroscopy 380–1
high-angle annular dark-field
STEM 379–80
diffraction contrast imaging of
defects, techniques for 377–9
for dislocation density quantitation
386
core identification by HR-STEM
388–9
$g \cdot b$ analysis 387–8
electron-matter interaction in
369–72
equipment 375–7
high-resolution imaging of defects,
techniques for
high-resolution STEM 381–2
HR-TEM 381
sample preparation 373–5
transverse field muon spin rotation
(TF-mSR) 223–6
transverse mode 122
transverse relaxation time 164
Tungsten 264
tunnelling AFM (TUNA) 348
2D image scan 510
2D materials, imaging defects in 400
dislocation and point defect
identification 402–6
electrical properties of defects in 2D
transition metal
dichalcogenides 400–2
electron channelling contrast
imaging (ECCI) examination
402

ultrahigh vacuum (UHV) 182
ultraviolet luminescence (UVL) 71, 81
uncertainty, sources of
in defect formation energy
calculations 297–8
uniaxial stress 115
units 97–8

vacancy-hydrogen complexes 133–4
vacancy–oxygen pair (VO) 21
vacancy-related defects 21–5
valence band maximum (VBM) 290
variational Monte Carlo (VMC)
framework 303
vibrational spectroscopy 97
examples 116
hydrogen donors in oxide
semiconductors 130–3
hydrogen in silicon 125–6
hydrogen in strontium titanate
134–6
hydrogen molecules 136
impurities in CdTe 125
impurities in GaAs 121–4
impurity-hydrogen pairs: trends
126–7
interstitial oxygen 116–20
from LVM to phonon 136–8
MgH complex in GaN 127–8
NH complex in ZnO 128–30
resonant interaction in AlSb
124–5
vacancy-hydrogen complexes
133–4
experiment 112
applied stress 115–16
Fourier transform infrared
spectroscopy 113–14
infrared (IR) pump–probe
114–15
Raman spectroscopy 112

theory
anharmonic coupling 110
anharmonic potential 103
defect vibrational modes 102–3
infrared (IR) activity 103–4
infrared absorption 105–6
linear chain 98–102
linewidth and temperature dependence 106–8
number of atoms 104–5
Raman scattering 111
units 97–8
wave functions and symmetry 108–10

wag modes 110
wave functions and symmetry 108–10
wavelength dispersive X-ray spectrometers (WDS) 551
weak beam dark-field (WBDF) imaging 377, 379
wetting layers 427
wide-bandgap semiconductors 45, 47, 55
acceptor-like defects in 277
cation and oxygen vacancies in metal oxides 279–83
cation vacancies and acceptor impurities in GaN and ZnO 277–9
wide-gap semiconductors, electrical dopants in 535–42
Wien filter 223

X-Band spectrometers 166–7, 171
X-ray absorption fine structure (XAFS) spectra 37
x-ray diffraction (XRD) 462

yellow luminescence 74
shape and fine structure of the YL1 band 74–9
thermal quenching of the YL1 band 79–80
YL1 band in GaN heavily doped with Si 80–1

Zeeman interaction 151–3, 158, 235
Zeeman levels 161, 165, 192, 229
Zeeman resonance 159
Zeeman splitting 190, 219
zero-field splitting 156–8
zero-field μSR 227–8
zero-phonon line (ZPL) energy 65
Ziegler–Biersack–Littmark potential 518
zincblende crystal structure 363
Zincblende III–V systems 250
zinc oxide 20, 130–2
damage in 530–2
electrically active defects in 25
hydrogen-related defect level 29–33
iron-related defect level 28–9
oxygen vacancy-related defect level 33–6
Na in 535–42
NH complex in 128–30
PL bands in 84
copper-related green band 84–5
lithium-related orange band 85
nitrogen-related red band 85–6
zinc vacancy, PL bands related to 86

www.ingramcontent.com/pod-product-compliance
Lightning Source LLC
LaVergne TN
LVHW011538200926
848484LV00001B/283
* 9 7 8 1 7 8 5 6 1 6 5 5 6 *